Metal-Catalysed Reactions of Hydrocarbons

FUNDAMENTAL AND APPLIED CATALYSIS

Series Editors: M. V. Twigg
Johnson Matthey
Catalytic Systems Division
Royston, Hertfordshire, United Kingdom

M. S. Spencer
Department of Chemistry
Cardiff University
Cardiff, United Kingdom

CATALYST CHARACTERIZATION: Physical Techniques for Solid Materials
Edited by Boris Imelik and Jacques C. Vedrine

CATALYTIC AMMONIA SYNTHESIS: Fundamentals and Practice
Edited by J. R. Jennings

CHEMICAL KINETICS AND CATALYSIS
R. A. van Santen and J. W. Niemantsverdriet

DYNAMIC PROCESSES ON SOLID SURFACES
Edited by Kenzi Tamaru

ELEMENTARY PHYSICOCHEMICAL PROCESSES ON SOLID
SURFACES
V. P. Zhdanov

HANDBOOK OF INDUSTRIAL CATALYSTS
Lawrie Lloyd

METAL-CATALYSED REACTIONS OF HYDROCARBONS
Geoffrey C. Bond

METAL–OXYGEN CLUSTERS: The Surface and Catalytic Properties of
Heteropoly Oxometalates
John B. Moffat

SELECTIVE OXIDATION BY HETEROGENEOUS CATALYSIS
Gabriele Centi, Fabrizio Cavani, and Ferrucio Trifirò

SURFACE CHEMISTRY AND CATALYSIS
Edited by Albert F. Carley, Philip R. Davies, Graham J. Hutchings,
and Michael S. Spencer

A Continuation Order Plan is available for this series. A continuation order will bring delivery of each new volume immediately upon publication. Volumes are billed only upon actual shipment. For further information please contact the publisher.

Metal-Catalysed Reactions of Hydrocarbons

Geoffrey C. Bond

Emeritus Professor
Brunel University
Uxbridge, United Kingdom

With 172 illustrations

Geoffrey C. Bond
59 Nightingale Road
Rickmansworth, WD3 7BU
United Kingdom

Library of Congress Cataloging-in-Publication Data

Bond, G.C. (Geoffrey Colin)
    Metal-catalysed reactions of hydrocarbond/Geoffrey C. Bond.
        p.   cm. — (Fundamental and applied catalysis)
    Includes bibliographical references and index.
    ISBN 0-387-24141-8 (acid-free paper)
        1. Hydrocarbons.   2. Catalysis.   3. Metals—Surfaces.   4. Reaction mechanisms
    (Chemistry)   I. Title.   II. Series.

    QD305.H5B59 2005
    547′.01—dc22

                                                                            2004065818

ISBN-10: 0-387-24141-8   e-ISBN: 0-387-26111-7   Printed on acid-free paper.
ISBN-13: 987-0387-24141-8

Printed in the United States of America

9   8   7   6   5   4   3   2   1

springeronline.com

ACKNOWLEDGMENTS

No work such as this can be contemplated without the promise of advice and assistance from one's friends and colleagues, and I must first express my very deep sense of gratitude to Dr Martyn Twigg, who more than anyone else has been responsible for this book coming to completion. I am most grateful for his unfailing support and help in a variety of ways. I am also indebted to a number of my friends who have read and commented (sometimes extensively) on drafts of all fourteen chapters: they are Dr Eric Short, Professor Vladimir Ponec, Dr Adrian Taylor, Professor Norman Sheppard, Professor Zoltan Paál and Professor Peter Wells (who read no fewer than six of the chapters). Their advice has saved me from making a complete ass of myself on more than one occasion. As to the remaining errors, I must excuse myself in the words of Dr Samuel Johnson, who when accused by a lady of mis-defining a word in his dictionary gave as his reason: *Ignorance, Madam; pure ignorance.*

One of the most pleasing aspects of my task has been the speed with which colleagues world-wide, some of whom I have never met, have responded promptly and fully to my queries about their work; Dr Andrzej Borodziński and Professor Francisco Zaera deserve particular thanks for their extensive advice on respectively Chapters 9 and 4. Dr Eric Short has been especially helpful in teaching me some of the tricks that have made the use of my pc easier, and Mrs Wendy Smith has skillfully typed some of the more complex tables.

Finally, I could not have completed this work without the patient and loving support of my wife Mary.

PROLOGUE

There must be a beginning of any good matter ...

SCOPE AND PURPOSE OF THE WORK

It is important at the start to have a clear conception of what this book is about: I don't want to raise false hopes or expectations. The science of heterogeneous catalysis is now so extensive that one person can only hope to write about a small part of it. I have tried to select a part of the field with which I am familiar, and which while significant in size is reasonably self-contained. Metal-catalysed reactions of hydrocarbons have been, and still are, central to my scientific work; they have provided a lifetime's interest. Age cannot wither nor custom stale their infinite variety.

Experience now extending over more than half a century enables me to see how the subject has developed, and how much more sophisticated is the language we now use to pose the same questions as those we asked when I started research in 1948. I can also remember papers that are becoming lost in the mists of time, and I shall refer to some of them, as they still have value. Age does not automatically disqualify scientific work; the earliest paper I cite is dated 1858.

It is a complex field in which to work, and there are pitfalls for the unwary, into some of which I have fallen with the best. I shall therefore want to pass some value-judgements on published work, but in a general rather than a specific way. While there is little in the literature that is actually wrong, although some is, much is unsatisfactory, for reasons I shall try to explain later. I have always tried to adopt, and to foster in my students, a healthy scepticism of the written word, so that error may be recognised when met. Such error and confusion as there is arises partly from the complexity of the systems being studied, and the great number of variables, some uncontrolled and some even unrecognised,[1] that determine catalytic performance. Thus while in principle (as I have said before[2]) all observations are valid within the context in which they are made, the degree of their validity is circumscribed by the care taken to define and describe that context. In this respect, heterogeneous catalysis differs from some other branches of physical chemistry, where fewer variables imply better reproducibility, and therefore more firmly grounded theory.

Nevertheless it will be helpful to try to identify what constitutes the solid, permanent core of the subject, and to do this we need to think separately about observations and how to interpret them. Interpretation is fluid, and liable to be changed and improved as our knowledge and understanding of the relevant theory grows. Another source of confusion in the literature is the attempt to assign only a single cause to what is seen, whereas it is more likely that a number of factors contribute. A prize example of this was the debate, now largely forgotten, as to whether a metal's ability in catalysis was located in geometric or in electronic character, whereas in fact they are opposite sides of the same coin. It was akin to asking whether one's right leg is more important than one's left. Similar misconceived thinking still appears in other areas of catalysis. So in our discussion we must avoid the temptation to over-simplify; as Einstein said, *We must make things as simple as possible – but not simpler.*

THE CATALYSED REACTIONS OF HYDROCARBONS

This book is concerned with the reactions of hydrocarbons on metal catalysts under reducing conditions; many will involve hydrogen as co-reactant. This limitation spells the exclusion of such interesting subjects as the reactions of syngas, the selective hydrogenation of α,β-unsaturated aldehydes, enantioselective hydrogenation, and reactions of molecules analogous to hydrocarbons but containing a hetero-atom. For a recent survey of these areas, the reader is referred to another source of information[3]. There will be nothing about selective or non-selective oxidation of hydrocarbons, nor about the reforming of alkanes with steam or carbon dioxide. That still leaves us plenty to talk about; hydrogenation, hydrogenolysis, skeletal and positional isomerisation, and exchange reactions will keep us busy. Reactions of hydrocarbons by themselves, being of lesser importance, will receive only brief attention.

Most of the work to be presented will have used supported metal catalysts, and a major theme is how their structure and composition determine the way in which reactions of hydrocarbons proceed. Relevant work on single crystals and polycrystalline materials will be covered, because of the impressive power of the physical techniques that are applicable to them. There are however important differences as well as similarities between the macroscopic and microscopic forms of metals.

This may be an appropriate time to review the metal-catalysed reactions of hydrocarbons. The importance of several major industrial processes which depend on these reactions – petroleum reforming, fat hardening, removal of polyunsaturated molecules from alkene-rich gas streams – has generated a great body of applied and fundamental research, the intensity of which is declining as new challenges appear. This does not of course mean that we have a perfect understanding of hydrocarbon reactions: this is not possible, but the decline in the publication rate provides a window of opportunity to review past achievements and the present status of the field.

I shall as far as possible use IUPAC-approved names, because although the writ of IUPAC does not yet apply universally I am sure that one day it will. Trivial names such as isoprene will however be used after proper definition; I shall try to steer a middle course between political correctness and readability.

You must be warned of one other restriction; this book will not teach you to *do* anything. There will be little about apparatus or experimental methods, or how to process raw results; only when the method used bears strongly on the significance of the results obtained, or where doubt or uncertainty creeps in, may procedures be scrutinised.

Some prior knowledge has to be assumed. Elementary concepts concerning chemisorption and the kinetics of catalysed reactions will not be described; only where the literature reveals ignorance and misunderstanding of basic concepts will discussion of them be included. Total linearity of presentation is impossible, but in the main I have tried to follow a logical progression from start to finish.

UNDERSTANDING THE CAUSES OF THINGS

I mentioned the strong feeling I have that there is much in the literature on catalysis that is unsatisfactory: let me try to explain what I mean. I should first attempt a general statement of what seems to me to be the objectives of research in this field.

> *The motivation for fundamental research in heterogeneous catalysis is to develop the understanding of surface chemistry to the point where the physicochemical characteristics of active centres for the reactions of interest can be identified, to learn how they can be modified or manipulated to improve the desired behaviour of the catalyst, and to recognise and control those aspects of the catalyst's structure that limit its overall performance.*

If this statement is accepted, there is no need for a clear distinction to be made between pure and applied work: the contrast lies only in the strategy adopted to reach the desired goals. In applied work, the required answer is often obtained by empirical experimentation, now sometimes aided by combinatorial techniques; in pure research, systematic studies may equally well lead to technically useful advances, even where this was not the primary objective.

In the past, the work of academic scientists has concentrated on trying to understand known phenomena, although there has been a progressive change of emphasis, dictated directly or indirectly by funding agencies, towards the discovery of new effects or better catalyst formulations. I have no wish to debate whether or not this is a welcome move, so I will simply state my own view, which is that it is the task of academic scientists to uncover scientific concepts and principles, to rationalise and to unify, and generally to ensure that an adequate infrastructure of methodologies (the so-called 'enabling technologies') is available to support and sustain applied work. Industrial scientists must build on and use this corpus of knowledge so as to achieve the practical ends. The cost of scaling-up and developing promising processes is such that academic institutions can rarely afford to undertake it; this sometimes means that useful ideas are stillborn because the credibility gap between laboratory and factory cannot be bridged.

The objective of the true academic scientist is therefore to *understand,* and the motivation is usually a strictly personal thing, sometimes amounting to a religious fervour. It is no consolation to such a person that someone else understands, or thinks he understands: and although some scientists believe they are granted uniquely clear and divinely guided insights, many of us are continually plagued by doubts and uncertainties. In this respect the searches for religious and scientific truths resemble one another. With heterogeneous catalysis, perhaps more than with any other branch of physical chemistry, absolute certainty is hard to attain, and the sudden flash of inspiration that brings order out of chaos is rare. It says much for the subject that the last person to have heterogeneous catalysis mentioned in his citation for a Nobel Prize was F.W. Ostwald in 1909.

For many of us, what we require is expressed as a *reaction mechanism* or as a statement of how physicochemical factors determine activity and/or product selectivity. What constitutes a reaction mechanism will be discussed later on. *What is however so unsatisfactory about some of what one reads in the literature is that either no mechanistic analysis is attempted at all, or that the conclusions drawn often rest on a very insubstantial base of experimental observation;* magnificent edifices of theoretical interpretation are sometimes supported by the flimsiest foundation of fact, and ignore either deliberately or accidentally much information from elsewhere that is germane to the argument. I particularly dislike those papers that devote an inordinate amount of space to the physical characterisation of catalysts and only a little to their catalytic properties. Obtaining information in excess of that required to answer the questions posed is a waste of time and effort: it is *a work of supererogation.*[4] Full characterisation should be reserved for catalysts that have interesting and worthwhile catalytic behaviour, and adequate time should be devoted to this.

This book is not intended as an encyclopaedia, but I will try to cite as much detail and as many examples as are needed to make the points I wish to make. Three themes will pervade it.

(1) The dependence of the chemical identity and physical state of the metal on its catalytic behaviour; integration of this behaviour for a given metal over a series of reactions constitutes its catalytic profile.

(2) The effect of the structure of a hydrocarbon on its reactivity and the types of product it can give; this is predicated on the forms of adsorbed species it can give rise to.

(3) The observations on which these themes are based will wherever possible be expressed in quantitative form, and not merely as qualitative statements.

Lord Kelvin said we know nothing about a scientific phenomenon until we can put numbers to it. However, with due respect to his memory, numbers are the raw material for understanding, and not the comprehension itself. We must chase the origin and significance of the numbers as far into the depths of theoretical chemistry as we can go without drowning. We shall want to see how far theoretical chemistry has been helpful to catalysis by metals. For most chemists there are however strict limits to the profundity of chemical theory that they can understand and usefully deploy, and it is chemists I wish to address. If however you wish to become better acquainted with the theoretical infrastructure of the subject, please read the first four chapters of a recently published book;[3] for these my co-author can claim full credit.

The foregoing objectives do not require reference to all those studies that simply show how the *rate* varies with some variable under a single set of experimental conditions, where the variable may for example be the addition of an inactive element or one of lesser activity, the particle size or dispersion, the addition of promoters, or an aspect of the preparation method. Such limited measurements rarely provide useful information concerning the mechanism, and many of the results and the derived conclusions have recently been reviewed elsewhere.[3] We look rather to the determination of kinetics and product distributions to show how the variable affects the reaction mechanism.

To explore the catalytic chemistry of metal surfaces, and in particular of small metal particles, we shall have to seek the help of adjacent areas of science. These will include the study under UHV conditions of chemisorbed hydrocarbons, concerning which much is now known; homogeneous catalysis by metal complexes, and catalysis by complexes adsorbed on surfaces (to a more limited extent); organometallic chemistry in general; and of course theoretical chemistry.

CONCERNING THE USEFULNESS OF MODELS AND MECHANISMS

The training of chemists inculcates a desire to interpret the phenomena of chemistry through the properties of individual atoms and molecules. To this end they have devised a variety of ways of symbolising and visualising their composition, size and shape. The purely symbolic method of identifying elements, while successfully distinguishing a hundred or so by means of at most two letters, requires subscripts and superscripts to define atomic mass, nuclear charge and oxidation state, but there is no means of showing size or chemical character.

Structural formulae of various degrees of sophistication may be used to show how atoms are linked in a molecule, what the bond angles and lengths are, and ultimately how orbitals are employed in bonding, but depictions of adsorbed species and surfaces processes of a very elementary kind are still often used, and all too frequently there is no diagram or sketch at all to show what is in the writer's mind. This is a pity, because most chemists have pictorial minds, and a simple sketch can speak volumes. A flexible and informative symbolism for surfaces states and events is urgently needed, because our ability to think innovatively and imaginatively is limited by the techniques we have to express our thoughts. Words are very imperfect vehicles for ideas and emotions. Perception of the third dimension is helped by molecular graphics, but such displays are impermanent until printed, when the extra dimension is lost. Often there is no alternative to the use of some kind of atomic model to convey the structures of surface phases.

Our belief that we can meaningfully describe the transformations of molecules by a few squiggles on a sheet of paper is a major act of faith. Acts of faith have their place in science as in religion, and our ability to create a conceptual model or hypothesis is however no more than a set of statements, either formal or informal, that increases the probability of successfully predicting an event or the outcome from a given situation. Karl Popper asserted that no hypothesis can ever be proved correct; it only remains plausible as long as no evidence is found to contradict it. A few scientific ideas have graduated from speculation through theory to the status of immutable and universal law; the Periodic Classification of the Elements and General Theory of Relativity are two such, but unfortunately there is as yet little in catalysis of which we can say 'It will always be thus'.

1. D. Rumsfeld: *There are things we do not know we do not know* (2003).
2. *Catalysis by Metals* (Preface), Academic Press: London (1962).
3. V. Ponec and G. C. Bond, *Catalysis by Metals and Alloys*, Elsevier: Amsterdam (1996).
4. See Article XIV of the Articles of Religion in the 1662 English Prayer Book.

CONTENTS

CHAPTER 3. CHEMISORPTION AND REACTIONS
            OF HYDROGEN

CHAPTER 4. THE CHEMISORPTION OF HYDROCARBONS

CHAPTER 5. INTRODUCTION TO THE CATALYSIS OF HYDROCARBON REACTIONS

CHAPTER 6. EXCHANGE OF ALKANES WITH DEUTERIUM

CHAPTER 7. HYDROGENATION OF ALKENES AND RELATED PROCESSES

CHAPTER 8. HYDROGENATION OF ALKADIENES AND POLY-ENES

CHAPTER 9. HYDROGENATION OF ALKYNES

CHAPTER 10. HYDROGENATION OF THE AROMATIC RING

CHAPTER 11. HYDROGENATION OF SMALL ALICYCLIC RINGS

CHAPTER 12. DEHYDROGENATION OF ALKANES

CHAPTER 13. REACTIONS OF THE LOWER ALKANES WITH HYDROGEN

CHAPTER 14. REACTIONS OF HIGHER ALKANES WITH HYDROGEN

1

METALS AND ALLOYS

PREFACE

This book in some ways resembles a detective story, but the criminal that we seek is the answer to the question: which solid-state properties of metals determine their behaviour as catalysts for the reactions of hydrocarbons? The search will lead us from the bulk metallic state through the small supported metal particles whose greater area makes them more fit to catalyse in a useful way; and from the reactions of hydrogen and hydrocarbon molecules with both sorts of metal to their catalytic interactions. The chain of cause and effect may not be straightforward.

In the metals of the Transition Series, where our attention will be focused, the strength of interatomic bonding and all the parameters which reflect it vary greatly: only six nuclear charges and their compensating electrons separate tungsten from mercury. The clear physicochemical differences that separate the metals of the first Transition Series from those in the second and third Series will be reflected in their chemisorptive and catalytic properties, as will the subtler differences between the second and third Series, for the understanding of which we are indebted to Albert Einstein and Paul Dirac. Gold has always been seen as the ultimate in nobility and iron as most liable to corrode; indeed this contrast was invoked by Geoffrey Chaucer's village priest, who in describing the high qualities needed in one of his calling, asked rhetorically *If gold rust, what shall iron do?*

Metal surfaces are the place where chemical changes start and even on large pieces of many metals the surface atoms are not quiescent but in a state of permanent agitation; this has quite a lot to do with their reactivity. They are, as Flann O'Brien remarked, *livelier than twenty leprechauns dancing a jig on a tombstone.*

While there are only some seventy-five metals, there are an infinite number of binary alloys, and it is little wonder that some are better catalysts than the pure metals that comprise them. In telling the story of catalysis by alloys we shall see how suspects were wrongly identified, and how the real truth was discovered: but

1

first we have to know something about the structure of metals and the metallic surface.

1.1. THE METALLIC STATE

1.1.1. Characteristic Properties

Of the first one hundred elements in the Periodic Table (Figure 1.1), about seventy-five are metals: in bulk form, most of them exhibit the characteristic physical properties of the metallic state, namely, strength, hardness, ductility, malleability and lustre, as well as high electrical and thermal conductivity. They owe their chemical and physical properties to their having one or more easily removed valence electrons: they are therefore electropositive, and most of their inorganic chemistry is associated with their simple or complex cations.[1,2] Metallic character in certain Groups of the Periodic Table increases visibly with increasing atomic number: while all the *d*-block elements in Groups 3 to 13 are obviously metals,

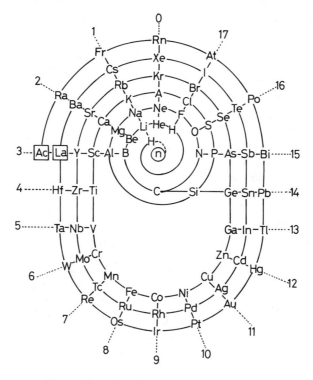

Figure 1.1. Periodic Classification of the elements.

of the Groups containing elements of the short Series, i.e. the *sp*-elements, this is only true of Group 1 (Figure 1.1). In Groups 2, 3, 14 and 15, the early elements are either clearly non-metallic or are semi-metals (e.g. beryllium, boron). The transition from non-metallic to semi-metallic to wholly metallic behaviour is most evident in Group 14, in which silicon and germanium are semi-metals, tin is ambivalent (the grey allotrope, α-Sn, is a semi-metal, while the much denser white form (β-Sn) is metallic), and lead is of course a metal. In Group 15, arsenic and antimony are semi-metals, but bismuth is a metal; in Group 16, tellurium and polonium are semi-metals.[3]

It is however not always easy to decide what substances show metallic behaviour.[4-6] One criterion for distinguishing semi-metals from true metals under normal conditions is that the co-ordination number of the former is never greater than eight, while for metals it is usually twelve (or more, if for the body-centred cubic structure one counts next-nearest neighbours as well). Other criteria have been proposed. Which category an element falls into also depends upon the conditions employed; thus for example some metals lose their metallic character above their critical temperature (e.g. mercury) or when in solution (e.g. sodium in liquid ammonia). Interatomic separation is then large and valence orbitals cannot overlap, so electrical conduction is impossible. On the other hand, the application of pressure causes some substances that are normally insulators or semiconductors to behave like metals; thus for example α-Sn changes into β-Sn, in accordance with Le Chatelier's Principle.[2] Similar changes also occur with other semi-metals (e.g. silicon and germanium), and even hydrogen under extreme pressure shows metallic character. Electrical conduction takes place when metal atoms are close enough together for extensive overlap of valence orbitals to occur. All metals when sufficiently subdivided fail to show the typical characteristics of the bulk state; the question of the minimum number of atoms in a particle for metallic character to be shown will be considered in Chapter 2.

The physical and structural attributes of the metals vary very widely: tungsten for example melts only at about 3680 K, while mercury is a liquid at room temperature (m.p. 234 K), this change being produced by increasing the nuclear charge only by six. The way in which the outermost electrons are employed in bonding ultimately determines all aspects of the metallic state. This is a question which is poorly treated if at all in text books of inorganic chemistry,[1,2,7] so some further description of the relevant facts and theories will be necessary. This information bears closely on the chemisorptive and catalytic properties of metal surfaces, which are our principal concern. When a metal surface is created by splitting a crystal, bonds linking atoms are broken, and in the first instant the dangling bonds or free valencies thus formed have some of the character of the unbroken bonds. We may therefore expect to see some parallelism between the behaviour of metals as shown by the chemical properties of their surfaces and the manner in which their valence electrons are used in bonding.

The metals of interest and use in catalysis are confined to a very small area of the Periodic Table, so that most of our attention will be given to the nine metals in Groups 8 to 10,[8] with only occasional mention of neighbouring elements in Groups 7 and 11, and of the earlier metals of the Transition Series (Figure 1.1). A principal object of our enquiry will be to understand why catalysis is thus restricted and our gaze will therefore be limited largely to the trends in metallic properties that occur in and immediately after the three Transition Series.

The properties of metals that disclose how their valence electrons are used may be put into four general classes: (i) mechanical, (ii) geometric, (iii) energetic, and (iv) electronic. The *mechanical* class (hardness, strength, ductility and malleability) may be quickly dismissed, because in polycrystalline materials these are mainly controlled by interactions at grain boundaries, and are influenced both by adventitious impurities that lodge there, and by deliberate additions that result in grain stabilisation, with consequent improvement in strength and hardness. With single crystals, they are described by the plastic and elastic moduli, which in turn are governed by the ease of formation and mobility of defects within the bulk under conditions of stress. They bear some relation to the strength of metallic bonding, but are of lesser interest than other properties. Metals having the body-centred cubic structure are less ductile than those that have close-packed structures (see below), because they lack the planes of hexagonal symmetry that slide easily past each other.

Bulk *geometric* parameters are those that describe the arrangement of the positive nuclei in space, and the distances separating them: the former is conveyed by the crystal structure and co-ordination number, and the latter by the metallic radius. Most metals crystallise in either the face-centred cubic (fcc) or the close-packed hexagonal (cph) or the body-centred cubic (bcc) structure; the first two are alternative forms of closest packing (Figure 1.1). Four other structures are known: rhombohedral (distorted fcc: mercury, bismuth), body-centred tetragonal (A4, e.g. grey tin[9]), face-centred tetragonal (indium, manganese), and orthorhombic (distorted cph: gallium, uranium). Many metals exhibit allotropy, i.e. they exhibit different structures in different regimes of temperature but in catalysis our only concern is with the form stable below about 770 K; of the metals of catalytic interest, only cobalt suffers a phase transition below this temperature (from cph to fcc at 690 K). Within the Transition Series there is a strikingly regular periodic variation in crystal structure; most metals in Groups 3 and 4 are cph (aluminium is fcc), those in Groups 5 and 6 are bcc, those in Groups 7 and 8 are again mainly cph (excepting iron, which is bcc, and manganese), while those in Groups 9 to 11 (except cobalt) are all fcc at ordinary temperatures.[3,10] Explanation of this regularity will be a prime requirement for theories of the metallic state (Section 1.12).

As the nuclear charge increases on moving across each Transition Series, the number of valence electrons forming covalent bonds at first rises rapidly, then

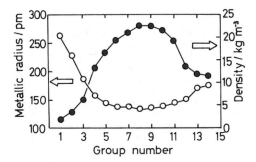

Figure 1.2. Periodic variation of metallic radius and density in the Third Transition Series.

remains almost constant in Groups 5 to 10, and afterwards starts to fall. This effect is clearly shown in metallic radius and density, values for which for the third Transition Series are shown in Figure 1.2. To relate these two parameters precisely, it is necessary to correct for changes in atomic mass. The plot of atomic density (i.e. density/atomic mass) versus the reciprocal of the cube of the radius (Figure 1.3) shows two good straight lines, one for the close-packed metals and another of slightly lower slope for the more open bcc metals. Figure 1.4 shows the periodic variation of the reciprocal cube of the radius for all three Transition Series: in the First Series iron, cobalt and nickel have almost the same bond lengths, while in the later Series the minimum bond length is shown at ruthenium and osmium. The similarity between the bond lengths in the second and third Series is only partly a consequence of the Lanthanide Contraction (see below).

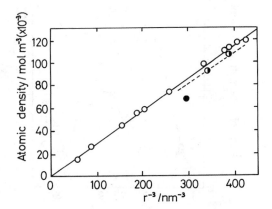

Figure 1.3. Dependence of atomic density on the reciprocal of the cube of the radius for metals of the Third Transition Series; open points, close-packed structures; half-filled points, bcc structure; filled point, Hg.

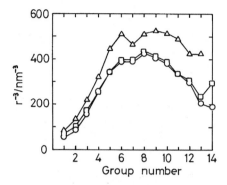

Figure 1.4. Periodic variation of the reciprocal of the cube of the radius for metals of Groups 1 to 14; triangles, First Series; squares, Second Series; circles, Third Series.

While geometric parameters reflect indirectly the strength of bonds between atoms, a more direct approach is provided by *energetic* parameters relating to phase change, i.e. melting and vaporisation or sublimation. Accurate values for melting temperature are available for most metals, their boiling points being in some cases less certain,[11] but the sublimation energy is the most useful quantity, this being the energy needed to secure complete atomisation of a given mass of metal. Division by the bulk co-ordination number gives the average bond strength. Figure 1.5 shows the periodic variation of sublimation heat for metals in Groups 1 to 14: there

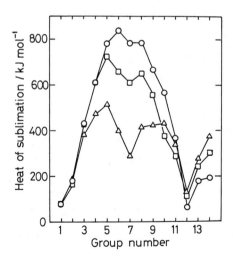

Figure 1.5. Periodic variation of the heats of sublimation of metals of Groups 1 to 14 (see Figure 1.4 for meaning of points).

are curious differences between this figure and figure 1.4, in that maximum bond strength now appears at Group 5 or 6 instead of at Group 8 or thereabouts. The effect may originate in the variable contributions that the energy changes accompanying electronic reorganisation make to the energetics of sublimation. There is clearly no uniquely reliable way of measuring the strength of bonds between metal atoms.

We may pause at this point to review in a qualitative manner those factors that influence the strength of intermetallic bonds. Clearly the number of valence electrons is of prime importance, but there are three other effects, not all equally obvious or apparent, that have to be noted. *The Lanthanide Contraction* has already been mentioned. This arises from the shape of the f-orbitals, which become filled between lanthanum and hafnium; they do not afford efficient shielding either of themselves or of other outer electrons from the nuclear charge, and hence they are all drawn towards the nucleus. This contraction, which is also shown to a minor extent as d-electron shells are filled, makes atomic sizes in the second and third Transition Series almost the same in corresponding Groups (Figures 1.3 and 1.4). Bond *strengths* however differ quite considerably, especially after Group 5 (Figure 1.5).

A second important effect is the stability of the half-filled d-shell. This is responsible for the unusual structure and chemistry of manganese, and for its low sublimation enthalpy (Figure 1.5) and melting temperature. The effect is also present, but less marked, in the second Transition Series, and is barely observable in the third; it is somehow anticipated by chromium and molybdenum in Group 5 (Figure 1.5), which have lower sublimation enthalpies than might otherwise have been expected.

The third factor is the most subtle and least well appreciated. In consequence of the Special Theory of Relativity, the mass m of a moving object increases with its speed v:

$$m = m_o/\sqrt{(1 - (v/c)^2)} \tag{1.1}$$

where m_0 is its rest mass and c the speed of light. For atoms with atomic number greater than about 50, the $1s$ electrons are sufficiently influenced by the nuclear mass that their speed becomes a substantial fraction of that of light (for mercury, $Z = 80$, $v/c = 0.58$) and their mass increases correspondingly.[12-18] The size of the orbital contracts (by 23% in the case of mercury); and outer s shells also shrink in consequence, since orthogonality must be preserved. Electrons in p-orbitals are also affected, but d and f electrons less so, because their probability of being found near the nucleus is low. However, their effective potentials are *more* efficiently screened because of the relative contraction of the s and p shells; they therefore increase in energy and expand radially. The Schrödinger equation is non-relativistic, and in effect assumes the speed of light to be infinite, and for heavier atoms the relativistic Dirac equation should be used.[19] Although

Dirac himself dismissed the idea that relativity could impinge on chemistry, he was in fact mistaken, and certainly all the third Transition Series and later elements are subject to its influence. The resulting orbital contraction is additional to the non-relativistic Lanthanide Contraction, and is partly responsible for the close correspondence in sizes between the second and third Series, already noted. It also accounts for the difference in colour and in chemistry between silver and gold, for the unusual structure and weak bonding in mercury, and for many other facets of the chemistry of the heavier elements traditionally associated with the stability of the $6s^2$ electron pair.[1,12–18,20–22] Gold is the most electronegative metal, and forms salt-like compounds with very electropositive elements (e.g. Cs^+Au^-)

So far we have considered only those properties of metals that are attributable to the strength of the bonds between the atoms: however, towards the end of each Transition Series there are more valence electrons than can be accommodated in bonding orbitals, and those in excess are in effect localised on individual atoms. These also contribute importantly to the *electronic* properties of metals. All metals are good conductors of electricity, but some are better than others. There is little regularity in the variation of atomic conductance (i.e. specific conductance/atomic volume) across the Periodic Table; values are high in Groups 1 and 2, and exceptionally so in Group 11, but are very low for manganese and mercury, due no doubt to their unusual structures.[10] Thermal conductance closely parallels electrical conductance, in line with the Wiedemann-Franz Law, which states that for all metals their ratio is a constant at a fixed temperature; its value is proportional to absolute temperature (the Wiedemann-Franz-Lorentz Law[23]).

Metals also show a range of magnetic properties.[24] The magnetic susceptibility κ measures the ease of magnetisation:

$$I = \kappa H \tag{1.2}$$

where I is the intensity of magnetisation and H the field strength. Paramagnetic substances have positive values of κ, and diamagnetic materials negative values. All metals of the Transition Series show weak, temperature-independent paramagnetism, except for the ferromagnetic iron, cobalt and nickel, which can be permanently magnetised below the Curie temperature and show the normal paramagnetism above it. The saturation moment of magnetisation (or the atomic magnetic moment) when expressed in Bohr magnetons gives the average number of unpaired electrons at zero Kelvin; this is another fixed point for explanation by theories of the metallic state. Manganese, technetium and palladium all have very high magnetic susceptibilities.

An interesting and potentially very useful property of the metallic state is *superconductivity*: the conductivity of a number of metals and alloys increases dramatically at very low temperatures, as the electrons pass through the rigid lattice of nuclei almost without obstruction. The phenomena is however of little relevance to catalysis.

1.1.2. Theories of the Metallic State[10,25–30]

It is no mere accident that the human race is designated as *Homo sapiens*, because it is the desire to know, and to understand the causes of things, that distinguishes us from other living creatures. The value of a good theory or explanation lies in its ability to correlate a wide range of observed phenomena by a simple model which is derived from the behaviour of the basic constituents of matter with the fewest possible assumptions. Theoretical descriptions of the metallic state rest on a knowledge of how the valence electrons behave; the correspondence between expectation and observation tests the precision of this knowledge and the methodology used to apply it.

A firm foundation for the theory of metals only became possible with the advent of the Quantum Theory and the application of the Schrödinger equation to electron waves: in particular the realisation, embodied in the Pauli Exclusion Principle, namely, that within a given system no more than two electrons can exist in the same energy state, was of fundamental importance. In a free atom in the ground state, electrons occupy definite energy *levels* corresponding to orbitals designated s, p, d, or f, according to the relevant quantum numbers. When however a number of atoms of the order of 10^{20} come together to form a metal crystal, their valence electrons cannot all continue to be in precise levels because by the Pauli Exclusion Principle no two electrons can have exactly the same values of all four quantum numbers; each is therefore compelled to take a microscopically different energy, but the energy difference between adjacent levels is however only about 10^{-40} J, and to all intents and purposes we may think of them occupying an *energy band*.[23,31,32] The width of the band depends on the interatomic distance, as shown in Figure 1.6, and the number of levels within the band is determined by the number of atoms in the assembly. The inner electron levels still behave as such, because the interatomic spacing is too great for them to interact (Figure 1.6); for them each atom is an isolated system, which is why sharp K_α emission lines are obtained when transitions occur between K and L shells, and why the frequency of such lines is not affected by the state of chemical combination of the element. Bands may overlap to form hybrid bands as shown in Figure 1.6.

We now need to know how the probability of finding an electron of specified energy varies across the permitted band. In the first and simplest version of the *Electron Band Theory*, electrons were assumed to move in a field of uniform positive potential (i.e. ion cores were neglected), and mutual electrostatic repulsion was ignored. Application of the Schrödinger equation and Fermi-Dirac statistics leads to the conclusion that a collection of N electrons at the absolute zero occupies the $N/2$ lowest levels, those at the maximum being said to be at the *Fermi surface E_F*.

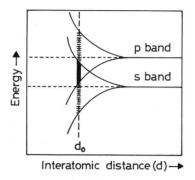

Figure 1.6. Overlap of s and p bands as a function of interatomic distance: d_0 shows the normal separation of atoms in the solid.

The number of energy states in a minuscule interval dE is termed the *electron level density* or *density of states n(E)* and this is proportional to $E^{1/2}$ (Figure 1.7). This highly simplified theory worked quite well for metals having only s and p electrons (sodium, magnesium, aluminium, tin), and provided the first reasonable interpretation of their electronic specific heats: it also led to a precise expression for the Wiedemann-Franz ratio.[23,33]

Extension of the Band Theory to the metals of the Transition Series required the introduction of ion cores into the argument. While for sodium the ion core is about 10% of the atomic volume, for Transition metals it is a much larger fraction, and cannot be ignored. Although in the case of an alkali metal the nature of the ion core is unambiguously defined, with Transition metals the core will not always have an inert gas configuration, and its structure has to be *assumed* before the potential field of the crystal can be defined. Moreover the location of the nuclei has to be precisely defined. It is a major weakness of the Band Theory that it does not address the directional nature of bonding between metal atoms

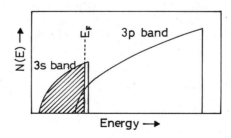

Figure 1.7. Electron level density diagram for magnesium based on simple band theory; the overlap of the $3s$ and $3p$ bands allows electrical conduction.

and so the lattice has to be regarded as a datum in the further development of the Theory.

A number of procedures have been devised for obtaining wave functions for the valence electrons.[10] It is unnecessary to describe these in detail, as they are thoroughly expounded in texts on solid state chemistry.[31,32] The first was the *Cellular Method*, due to Wigner and Seitz,[34] in which the solid was notionally divided into cells, each containing one ion core. The *Augmented Plane Wave Method*[35] used a muffin tin model of the crystal potential, in which the unit cell is divided into two regions by spheres drawn about each ion core. The potential inside the spheres is spherically symmetrical, and resembles that for the isolated atom, whereas outside the potential is constant, so that an electron here would behave as a plane wave. The *KKR Method* (Korringa,[36] Kohn and Rostoker[37]) supposes that electrons as plane waves undergo diffraction as they encounter ion cores, in a way which permits the wave to be reconstructed so that it can proceed through the lattice. The wave functions derived from the latter two methods are virtually equivalent.

The theoretician is now in a position to calculate a density of states curve for any element, by selecting a method for formulating the wave function and applying it to the appropriate crystal potential. These choices are not always straightforward: it has been said that it is far easier to give descriptions than advice on how to choose them. Once the choices are made, however, solution of the wave equations leads to an energy band diagram, and hence by integration to a density of states curve.

It is unnecessary to provide details of the results of such calculations, or of their comparison with experimental determinations by for example soft X-ray spectroscopy:[23,31,32] band structures for Transition Metals can adopt quite complex forms,[10] so we must content ourselves with a few qualitative observations. For the metals of catalytic interest, the nd-electron band is narrow but has a high density of states (Figure 1.8), because these electrons are to some degree localised about each ion core, whereas the $(n + 1)s$ band is broad with a much lower density of states because s-electrons extend further and interact more. On progressing from iron through to copper, the d-band occupancy increases quickly, and the level density at the Fermi surface falls. The extent of vacancy of the d-band is provided by the saturation moment of magnetisation; thus for example the electronic structure of metallic nickel is (Ar core) $3d^{9.4}4s^{0.6}$, and is said to have 0.6 'holes in the d-band'. There have been many attempts to correlate the outstanding chemisorptive and catalytic properties of the Groups 8-10 metals with the presence of an incomplete d-band or unfilled d-orbitals. According to the Band Theory, electrical conduction requires excitation to energy levels above the Fermi surface, so that substances that have only completely filled bands will be insulators. A metal such as magnesium for example is a good conductor because it possesses a partly filled hybrid sp band. By the same token, it is easier to carry a full bottle of mercury than a half-full one, because it doesn't slop about so much.

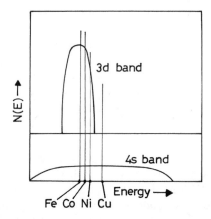

Figure 1.8. Schematic band structures for metals at the end of the First Transition Series according to the Rigid Band Model.

It needs to be stressed that models of the metallic state on which the Band Theory is based suppose an infinite three-dimensional array of ion cores, so that the band structure cannot be expected to persist unchanged to the surface. Moreover the ion cores must be precisely located before theoretical analysis starts, and we shall shortly see that interatomic distances and vibrational amplitudes in the surface differ somewhat from those in the interior. These factors certainly complicate the useful application of Band Theory to the properties of surfaces.

While the Band Theory is based on the concept of a free electron gas obeying the appropriate statistical mechanical rules, the Valence Bond Theory, due to Pauling,[9,10,33,38,39] takes the view that the behaviour of metals is adequately described by essentially covalent bonds between neighbouring atoms. It distinguishes between those electrons which take part in cohesive binding, and those which are non-bonding and responsible for example for magnetic properties. Pauling's model first recognises that d-electrons can participate in bonds between atoms; it then supposes that nd-electrons can be promoted into $(n + 1)s$ and $(n + 1)p$ orbitals, with the formation of hybrid $d^x sp^y$-orbitals. From potassium to vanadium the number of bonding electrons increases from one to five, accounting for the increase in cohesive strength described above. Since the covalent bonds require electrons to be paired, these elements are neither ferromagnetic nor strongly paramagnetic despite the d-shell being incomplete.

Of the five d-orbitals, it is assumed that only 2.56 are capable of bonding, the remaining 2.44 being localised *atomic d-orbitals*, which are non-bonding, and capable of receiving electrons with parallel spins as long as is permitted by Hund's Rule. With chromium the sixth electron is divided as shown in Table 1.1. Now the *dsp*-hybrid orbital should in theory accommodate 6.56 electrons

TABLE 1.1. Electronic Structures of Some First Row Transition Metals According to Valence Bond Theory

Metal	Total electrons	Electrons in hybrid dsp-orbital	Electrons in atomic d-orbital Spin+	Spin−	$\sum$	Saturation moment Calc.	Obs.
Cr	6	5.78	0.22	0	0.22	0.22	—
Mn	7	5.78	1.22	0	1.22	1.22	—
Fe	8	5.78	2.22	0	2.22	2.22	2.2
Co	9	5.78	2.44	0.78	3.22	1.66	1.61
Ni	10	5.78	2.44	1.78	4.22	0.66	0.61

(i.e.$1 + 3 + 2.56$), but in fact it is necessary to assume that the maximum number is 5.78, the remaining 0.78 orbitals being *metallic orbitals*; these are said to be needed to effect the unrestricted synchronous resonance of the bonding orbitals. Although it may look as if the numbers are pulled like rabbits out of a hat, they are in fact selected to account for the saturation moments of magnetisation for iron, cobalt and nickel, as given by the number of *unpaired* electrons in the atomic d-orbital (Table 1.1). Their non-integral nature represents a time-average of an atom in one of two states.

Finally it is possible to calculate the fractional d-character of the covalent bonds for the Transition Series metals; these numbers were formerly much used by chemists to explain trends in catalytic activity, but are now little used. It is recognised that, while the model gives a qualitatively realistic picture of how the valence electrons are employed, it is an interpretation rather than an explanation, and its quantitative conclusions are unreliable. The role of the metallic orbitals is particularly mysterious: they are reminiscent of the Beaver who

> ... *Paced on the deck,*
> *Or sat making lace in the bow;*
> *Who had often (the Captain said) saved them from wreck,*
> *But none of the sailors knew how.*

A detailed critique of Pauling's theory has been given in reference 10.

What is lacking in the theoretical analyses dealt with so far is any attempt to rationalise the regularity of changes in structure of the elements as one passes through the Transition Series (Section 1.1.1). It appears that this may be determined by the fraction of unpaired d-electrons in the hybrid dsp-bonding orbitals;[33,40] this is thought to increase to a maximum in Group 7, and then to decrease. Metals in Groups 2, 9 and 10, where fraction is about 0.5 are fcc; in Groups 3, 4, 7 and 8, the fraction is about 0.7 and the structures are usually cph; and in Groups 5 and 6, the fraction is about 0.9 and structures are bcc. It is not however clear *how* the composition of the hybrid orbitals determines their direction in space and hence the crystal structure. The idea of the importance of bonding d-electrons in deciding structure has been further developed by Brewer[41] and Engel[42]; the application of the concept to alloys and intermetallic compounds will be considered below.

One further theoretical approach deserves to be rescued from oblivion. In 1972, Johnson developed an *Interstitial-Electron Theory* for metals and alloys;[43] it emphasises the spatial location of electrons, but also incorporates quantum mechanical aspects of bonding, such as electron correlation and spin. The interstices between the ion cores are the location of valence or itinerant electrons,[11] and are thus 'binding regions', and the Hellmann-Feynman theorem provides a rigorous basis for analysing forces between electrons and ion cores in these regions.[44] Electrons occupy interstices so that they provide maximum screening of positive ion cores, and suffer minimum electron-electron repulsions. In close-packed structures, there are only three interstices per ion core, and some vacant interstices are needed to account for metallic properties such as conductance. Thus before the number of valence electrons rises to six, some must be localised as d-states on the ion cores, while the rest remain itinerant. These latter act as ligands and determine the degeneracy of the localised electrons, and hence the magnetic properties.

The Interstitial-Electron Theory has been applied to the structure of metals, alloys and interstitial compounds, to their magnetic and superconducting properties, as well as to a range of surface phenomena.[45] This work has seemingly not come to the attention of the wider scientific community perhaps because it was published only in Japanese journals. It merits wider recognition and a critical evaluation.

The reader may be confused by the number of different theoretical models that have been advanced to explain the properties of metals. Each type of approach has concentrated on a limited aspect. Electron Band Theory looks at the collective properties of electrons, especially their energy; the prediction of structure is not a prime target, and the location of electrons in the energy dimension is thought to be more important than finding where they are in real space. It is possible to gain the impression that theoreticians with a leaning to physics regard the existence of atoms as a complication if not a positive nuisance. The more chemically-oriented theories are less worried about electron energies, and cannot yield density of states curves, but they provide a generally satisfying qualitative picture of the behaviour of metals, which if not derived from fundamental theory is nevertheless useful to the practising chemist.

1.2. THE METALLIC SURFACE[10,33,46]

1.2.1. Methods of Preparation

Very many different forms of metals are used as catalysts: the size of the assembly of metal atoms varies from the single crystal, which may contain an appreciable fraction of a mole of the metal, to the tiniest particle containing only

5 to 10 atoms. For practical catalysis it is usually desirable for the metal to be in a such highly divided form, exposing a large surface are to the reactants; however, the smaller the particle the more unstable it becomes, and special measures need to be adopted to prevent loss of area by aggregation or sintering. The best way of doing this is to form the particles on a *support*, but the subject of supported metal catalysts is of such importance and size that a large part of the next chapter is devoted to them. There are however other means of making and using quite small metal particles, without the assistance of a support; these are briefly described in this section, but their characterisation and properties will be considered in Section 2.2, alongside supported metal particles.

For fundamental studies there is much to be said for using the metal in a massive form;[47,48] the disadvantage is the very limited surface areas that are obtained. Historically, polycrystalline wires, foils and granules were used,[10,33] and indeed these forms still find application in major processes, such as ammonia oxidation and oxidative dehydrogenation of methanol, which are not within the scope of this work. A major advance in the formation of clean metal surfaces for catalysis research was the introduction of evaporated metal films[49−51] (more properly called *condensed* metal films). First used in the 1930's, Otto Beeck and his associates subsequently developed them,[52−55] and they were quickly adopted by other scientists. By conductive heating of a wire of the catalytic metal, or of a fragment of the metal attached to an inert wire, in an evacuated vessel, atoms of the metal evaporated and then condensed on the walls of the vessel, forming first islands and later a continuous film. A major strength of the technique was the ability to apply a range of techniques to the study of chemisorption on the film; these included calorimetry, electrical conductance, work function measurement and changes in magnetisation.[51] We shall refer below to important results obtained on hydrocarbon reactions using metal films, although they are however no longer much employed.

The more recently favoured form for fundamental research is the single crystal, made by slowly cooling the molten metal. By judicious cutting, an area of about 1cm^2 of a well-defined crystal surface is exposed, and when placed within a UHV chamber it can be heated and cleaned by ion bombardment.[56] A particular danger with some metals is the slow emergence at the surface of dissolved impurities, particularly sulphur; this is a problem that has been recognised since the early 1970s.[57]

Two other forms of massive metal deserve a mention. Extremely fine metal tips have been used for Field-Emission Micrpscopy (FEM) and Field-Ion Microscopy (FIM);[58] by the latter technique, atomic resolution of the various planes near the tip can be obtained,[59] and the process of surface migration closely can be studied.

Considerable interest has been shown in the recent past in amorphous or glassy metals,[10,60,61] made by extremely rapid cooling of the molten metal; the product lacks long-range order, and it was believed that their study would reveal the importance of crystallinity in catalysis. However, pure metals are difficult to make in the amorphous state, because of the ease with which they recrystallise.

The tendency is much less with binary alloys and intermetallic compounds, but catalytic activity is generally low before 'activation', which roughens or otherwise disturbs the surface. Interest in their use seems to be declining.

It would be logical at this stage to consider the techniques[9,47,58,62-64] that can be used to characterise the metal forms and their surfaces listed above. The problems we face may be classified as follows. (i) With any metal form, it is desirable to establish the surface cleanliness: this is best done by techniques such as X-ray photoelectron spectroscopy (XPS) and the associated Auger spectroscopy (AES), or most sensitively by secondary-ion mass spectrometry (SIMS) or ion-scattering spectroscopy (ISS). These methods[9,10,62-64] necessitate placing the material in vacuo, where one hopes it remains stable and unaffected by the radiation used; they are not often applied to the unsupported forms such as blacks or powders.[65] (ii) With the more dispersed forms, it is useful to know the size, size distribution and shape of the particles; many of the techniques that are appropriate here are also applied to the study of supported metal catalysts, and will therefore be treated in Chapter 2. (iii) The structure of metal surfaces at the atomic level can only really be examined using single crystals; the predominant method is low-energy electron diffraction (LEED), which can give surface structures, at least for those areas where the atoms experience long-range order.[10,62-64] Other methods capable of providing atomic resolution include scanning-tunnelling microscopy (STM) and atomic force microscopy (AFM)[10,64,66], use of which is becoming more popular.

1.2.2. Structure of Metallic Surfaces[30,67]

Certain things are easy to define, and we have already met a few; other things are more easily recognised than defined. Someone once remarked: *I cannot define an elephant, but I'm sure if I saw one I should recognise it.* It is much the same with surfaces. It is simple to say that the surface of a solid is the interface between the bulk and the surrounding fluid phase or vacuum; it is also straightforward, if somewhat more complicated, to assign thermodynamic properties to the 'surface phase'. It is however when one starts to examine a metal surface at atomic resolution that the problems start.

A plane occupied by atoms or ions within a crystal, or at its surface, is defined by its Miller index, which consists of integers that are the reciprocals of the intersections of that plane with the system of axes appropriate to the crystal symmetry.[63,68,69] The procedure was not in fact devised by Miller, but by Whewell (1825) and Grossman (1829), and only popularised by Miller in his textbook on crystallography (1829)[68]: it served to characterise visually observable crystal planes at surfaces long before their atomic structure was known. Consider the three low-Miller-index planes of an fcc metal (Figure 1.9). In the (111) plane, the atoms are close-packed and have a co-ordination number (CN) of nine; while these atoms

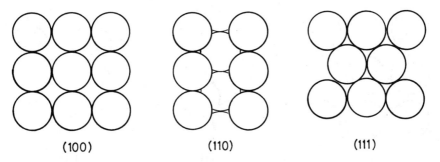

Figure 1.9. Arrangements of atoms in low-index planes for the fcc structure.

are undoubtedly in the surface, some of the properties of atoms in parallel planes beneath the surface plane are not quite the same as those truly in the bulk, but the effect of the interface dies away, usually quite rapidly, as one moves towards the interior. The problem of defining the surface is seen even more clearly with the (100) and (110) planes (Figure 1.9). With the former, atoms in the top plane have CN of 8, and the atoms of the next layer form the bottom of the octahedral holes in the surface and peep through the gap. Yet more obviously, with the (110) surface the atoms actually forming the plane (CN7) are separated by rows of atoms in the next plane down (CN10) which are readily accessible from above. This plane can in fact be represented as a highly stepped (111) surface in which both types of atom participate. Certainly any atom that does not have the full quota of 12 nearest neighbours has to be regarded as part of the surface; the lower its CN the greater is its contribution to it.

Most attention is usually paid to the surfaces of metals of fcc structure because this group contains the best catalysts. The problem of identifying surfaces is greater with the cph structure, where the $(10\bar{1}0)$ and $(11\bar{2}0)$ planes have second layer atoms that are almost totally exposed (Figure 1.10): the (0001) and $(30\bar{3}4)$ planes[70] are

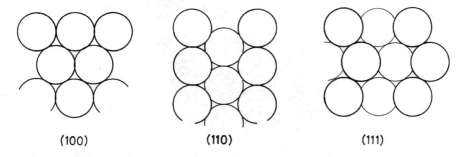

Figure 1.10. Arrangements of atoms in low-index planes for the cph structure.

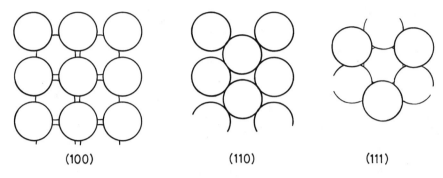

(100)   (110)   (111)

Figure 1.11. Arrangements of atoms in the low-index planes of the bcc structure.

however respectively the same as the (111) and (100) planes of the fcc structure. The (100) and (211) planes of the bcc structure also contain second-layer atoms that are substantially exposed (Figure 1.11). Thus, except for the close-packed planes of the fcc and cph structures having hexagonal or cubic symmetry, all other surfaces contain atoms of different CN.

Ordered arrays of atoms of low CN can be produced by cutting a single crystal at a slight angle to a low-index plane;[63,71] this will produce (at least in theory) a series of single atom steps separated by plateaux the width of which depends on the angle selected (Figure 1.12). In a further elaboration of this concept, cutting at a slight angle to *two* low-index planes produces a surface that is both stepped and kinked (Figure 1.13); atoms of unusually low CN are then exposed. The structure of such surfaces may be defined by the Miller index of the plane formed by the atoms at the steps or kinks, or more simply by the indices of the plateau and at the step, together with the number of atoms between the steps (e.g. 5(111) × (100)). This procedure, conceived and exploited by G.A. Somorjai,[63] has helped to reveal

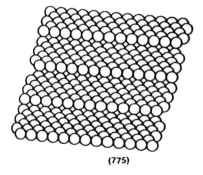

(775)

Figure 1.12. Representation of stepped crystal surface (fcc (755)).[63]

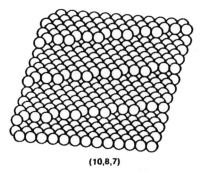

(10,8,7)

Figure 1.13. Representation of a kinked crystal surface fcc (10,8,7).[63]

the role of low CN atoms in chemisorption and catalysis, and in a way it models the characteristics of small metal particles. However, unless the steps are quite close together, the contribution of the atoms in the plane defined by the Miller indices will be swamped by that of the low-index plateaux. Incidentally, scanning-tunnelling microscopy (STM) has shown that even surfaces giving a seemingly perfect LEED pattern for a low-index plane may nevertheless have quite a high density of steps and other defects.[62] Kink sites lack symmetry when step lengths and faces on either side are unequal; their mirror images are therefore not superimposable, and they possess the quality of chirality.[72,73] Representations of many normal and stepped surfaces are to be found in Masel's book.[30]

Surface atoms, being defined as having a CN less than the bulk value, are said to be co-ordinatively unsaturated, and, lacking neighbours above them, they experience a net inward force: this effect is equivalent to the more readily sensed surface tension of liquid surfaces, and may be thought of as due either to multiple bonding between atoms in the surface layer, using the surface free valencies, or to a wish to maximise interatomic bonding. It is expressed quantitatively as the *surface tension* γ, which is the energy needed to create an extra unit of surface area:[11,63,70,74] its units are therefore $J \, m^{-2}$. It is a periodic function of atomic number (Figure 1.14), following closely the pattern set by sublimation heat (Figure 1.5). For a single-component system at constant temperature and pressure,

$$\gamma = G^s \tag{1.3}$$

where G^s is the specific surface free energy. Conventional thermodynamic formulae can be applied to give the enthalpy, entropy and heat capacity of the surface layer.[63]

Even when surface atoms have found their stable places, they will oscillate about their mean positions with a frequency which increases with temperature: thus the signals given by techniques such as LEED, EXAFS and Mössbauer

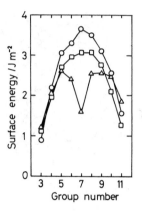

Figure 1.14. Periodic variation of the surface tension (specific surface work) at 0 K for metals of Groups 1 to 14 (see Figure 1.4 for meaning of points).

spectroscopy weaken as temperature rises, as fewer and fewer atoms are to be found at their exact lattice sites.[24,58,63] Surface atoms experience a greater vibrational amplitude than those in the bulk, since they have no neighbours above them to restrain them. Atoms at step and kinks, having fewest neighbours, vibrate most freely, and rising temperature affects surface atoms more than bulk atoms; in this way surface phenomena can sometimes be distinguished from things happening in the bulk. It also follows that the surface is a weaker scatterer of radiation than the bulk.

These concepts may be quantified as follows. A quantum of lattice vibration is termed a *phonon*, and the mean deviation of an atom from its lattice position is the mean-square displacement $\langle u^2 \rangle$. Phonons are detected by vibrational spectroscopy by absorption peaks below $500 \, cm^{-1}$. According to the Debye model, atoms vibrate as harmonic oscillators with a distribution of frequencies, the highest of which is ω_D: then the *Debye temperature* θ_D is defined as

$$k\theta_D = h\omega_D/2\pi \tag{1.4}$$

A high θ_D betokens a rigid lattice, and vice versa: it will be lower for surface atoms than for bulk atoms by a factor of 1/3 to 2/3. It can be measured for bulk atoms by XRD, EXAFS and by the scattering of neutrons or high-energy electrons, and for surface atoms by varying the energy of electrons in LEED to obtain by extrapolation the scattering characteristic of zero energy. By the Lindemann criterion, melting bgins when $\langle u^2 \rangle$ exceeds a quarter of the interatomic distance: surface melting therefore precedes melting of the bulk.

We may expect surface energy to be less for planes having the greater density of atoms, such as the fcc(111), because their creation by splitting a large crystal requires less work to be done than with more open planes such as the fcc (100). One important consequence of the skin effect is that bond lengths normal to the surface are usually shortened between the first and second layers, typically by a few percent, while there is expansion again between the second and third, and contraction again between the third and fourth layers. Contraction is greater the lower the co-ordination number of the surface atoms, i.e. the greater the surface energy, so that atoms at kinks are most noticeably affected, as are less densely packed planes in general.

The desire of a system to minimise its total energy determines that the surface energy should be as small as possible, so that where the potential benefit of changing from a high to a lower surface energy is sufficiently great the surface may reconstruct to achieve this saving. This may mean that the surface layer has to adopt a structure which is out of register with that of the underlying crystal; thus for example the (100) surfaces of iridium, platinum and gold are covered by a layer of (111) geometry, and the (110) surfaces can reconstruct to form more extended areas of (111) plane by 'losing' alternate rows.[63,75–79] These types of reconstruction occur with clean surfaces *only* in the case of the Third Transition Series fcc metals, because they have higher heats of sublimation than the earlier metals (Figure 1.5), and therefore stand to gain more by forming closely-packed surfaces: this is a further manifestation of the relativistic effect.[13,21] Ease of reconstruction depends on the density of packing the surface metal atoms; surfaces containing atoms of relatively low CN are most mobile, and reconstruct more easily than close-packed planes. With the surfaces of elements other than iridium, platinum and gold, reconstruction occurs only in the presence of a chemisorbed layer of atoms (e.g. carbon, oxygen, sulphur), which is able to mobilise the surface metal atoms by weakening their bonds to the atoms beneath. With molecular species that are strongly chemisorbed (e.g. ethylidyne, ethyne, see Chapter 4), a small group of metal atoms involved in the bonding may be drawn outwards; this process is sometimes described as *extractive chemisorption*, and the consequences of this disturbance may be felt by atoms even more remote.

The relative surface energies of different crystal planes determine the equilibrium shape of a metal crystal, and this can be predicted from knowledge of the energy terms by use of the Wulff construction.[10] In this procedure the surface energies are drawn as vectors normal to the planes described by the indices, and these planes set at the tops of the vectors define the crystal shape.

The activation energies for surface self-diffusion are also much lower than sublimation energies, since here again only a limited number of bonds need to be broken to allow movement. Diffusion coefficients are not surprisingly much greater, and activation energies are lower, the more closely-packed the atoms in the plane: thus for example values of the latter for rhodium atom migration vary from 84 kJ mol^{-1} on the (100) plane, through 58 kJ mol^{-1} on the (110), to as little as 15 kJ mol^{-1} on the (111) surface.[63]

The work that has to be done to remove an electron from a metal into vacuum with zero kinetic energy at zero K is termed the *work function*,[24,80] and this is the same as the ionisation potential, but is larger than that of the free atom because of the space charge or surface dipole that exists at the surface, due to the asymmetry of electron density. Work function is greatest at planes having a high concentration of atoms (i.e. generally low-index planes), and decreases with step density at stepped surfaces.[63] Variation of work function with crystal plane underlies the technique of Field-Emission Microscopy (FEM).

This short survey suggests that, in the absence of a chemisorbed layer, atoms at the surface of massive metal are in a state of considerable agitation that belies the static impression that structural images convey.[77,78] We shall see in the following chapters how this conclusion is affected by particle size and by adsorbed entities.

1.2.3. Theoretical Descriptions of the Metal Surface[29]

Our concern now is with the form of theoretical analysis appropriate to the ex-tended metal surface; corresponding approaches suitable for small metal particles will receive attention in Chapter 2, although some of the concepts may apply there also. The literature is somewhat coy in dealing with this problem. We know that the necessary conditions for the Band Model cannot obtain at the surface, and indeed electrons emerging at near-grazing angles in X-ray photoelectron spectroscopy (XPS), and those stimulated by lower energy UV radiation (UPS), indicate bands narrower than those of the bulk, as well as additional features due to 'surface states', i.e. to electrons localised on surface atoms.[63,81] It has to be accepted that the elec-tronic structure of surface atoms, i.e. the extent to which their various energy levels or the bands in which they participate are occupied, will differ from that of atoms below the surface, and will be unique to each co-ordination number.[82] Thus for example atoms at the tops of steps may have a lower density than those at the foot, and this may be the source of greater or even excessive activity in chemical process, although the evidence on this point is equivocal. Structure can be ex-pressed as a *local density of states* (LDOS)[83], which can now be calculated using the tight-binding approximation, and particular interest attaches to the LDOS at the Fermi surface (E_F-LDOS) because it appears[84] that electron density outside the surface (defined as the plane through the nuclei) comprises progressively more of these energetic electrons as distance increases. The E_F-LDOS may well therefore determine surface reactivity in chemisorption and catalysis (see also Section 2.55).

The chemist's mind may have some difficulty in grasping the concept of a band structure and LDOS for a single atom (e.g. at a kink site), but it may be helpful to remember that band width is a variable feast, increasing with the number of adjacent atoms of like type. There can therefore be all widths from the full width for bulk atoms to zero widths for isolated atoms, but this kind of information does not however seem to have been much applied to understanding and interpreting catalytic phenomena.

An alternative procedure is to attempt a molecular orbital description of surface atoms. It starts with the simple-minded view of an unrelaxed surface, such as would exist immediately after cleavage of a metal crystal.[85] It was noticed some years ago[86,87] that the arrangement of atoms in a metal of fcc structure, i.e. 12 near neighbours and 6 next-nearest neighbours, was precisely matched by the disposition of, respectively, t_{2g} and e_g orbital lobes of the d-electrons employed in octahedral complexes such as $Pt^{II}Cl_6^{2-}$. Bands may then be formed by a symmetry-adapted linear combination of these atomic orbitals.[24,88,89] It is a straightforward matter to map out how these orbital lobes project from any crystal plane[85,90] (for an example, see Figure 1.15). The procedure has been criticised as overly simplistic,[91] as indeed it is: it does not employ hybrid dsp-orbitals, as it should, although whatever the actual hybridisation it must generate orbitals sterically similar to those of the d-electrons. Furthermore, whatever the composition of the hybrid, the emerging orbitals must, according to this picture, be congruent with those that determine the bulk structure. Although such descriptions are unlikely to apply to the clean equilibrated (i.e. relaxed) surface, it is possible that a molecular orbital picture along the above lines may be appropriate when chemisorbed species have counteracted the relaxation. With all its imperfections, however, it has received some

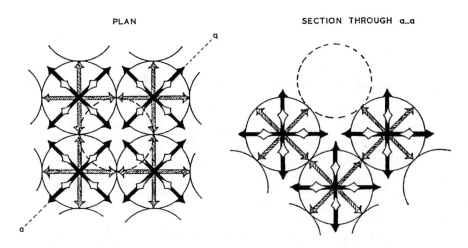

PLAN   SECTION THROUGH a...a

Figure 1.15. The emergence of orbitals from the (100) face of an fcc metal. Filled arrows, e_g orbitals in the plane of the paper: hatched arrows, t_{2g} orbitals in the plane of the paper: open arrows, t_{2g} orbitals emerging at 45° to the plane of the paper. The dashed circle shows the position of an atom in the next layer above the surface layer. In both the plan and the section an e_g orbital emerges normal to the plane of the paper from each atom.

support,[59,88,89,92] especially from R.L. Augustine and his associates[92-94] who have extended it to the various types of surface atom having a low CN, and have further adapted the Angular Overlay Model to predict the energies of their s-,p- and d-electrons. A more rigorous and theoretical respectable quantum mechanical treatment of emergent orbitals has been developed, and applied to hydrogen chemisorption (Chapter 3).

The original work[85] and these further developments relate only to the fcc structure, which accounts for the most catalytically interesting and useful metals, but the concept can be extended to the bcc and cph structures using the orbital assignments given many years ago by Trost[95] and others.[40] In these cases however it is essential to use hybrid orbitals having d, s and p components.

Returning for a moment to the description of bonding inside the crystal,[24] those d-orbitals whose interactions are responsible for bonding nearest neighbours (viz. the t_{2g} family) will form a band which is broader than that formed by the e_g family, since interactions between next-nearest neighbours are less strong. Extending this concept to surface atoms, we see on the (100) surface for example that the absence of atoms above the plane means that the overlap of d_{xz} and d_{xy} orbitals has decreased and their band is narrowed, while the d_{yz} orbitals in the surface plane are unaffected, and their band remains broader. Similar but smaller effects will occur with the e_g and s-orbitals. The modification of electronic structure of atoms at steps and kinks is then easily rationalised,[96] and the story will be resumed in Chapter 2, where other concepts developed in the context of small metal particles will be considered.

1.3. ALLOYS[96]

1.3.1. The Formation of Alloys[11,23,31,41,50,97]

This section is not primarily concerned with the mechanics of making alloys, but rather with the physical chemistry that determines whether they are formed or not. The term 'alloy' has been used indiscriminately in the literature,[10] but we shall restrict its meaning to a material containing two or more elements in the zero-valent state that are mixed at the atomic level. What happens when two metals are brought together depends on the thermodynamic functions that describe their interaction, and on temperature; the former depends on their relative sizes and electronic structures.

We consider first bimetallic *substitutional alloys*, where atoms of either kind can occupy the same lattice site. Now an ideal solution is one for which the enthalpy of mixing is zero, and the process of alloy formation is purely entropy-driven;

$$\Delta S_{\text{mix}} = R[x \ln x + (1-x) \ln (1-x)] \tag{1.5}$$

where x is the mole fraction of one component. Deviations from ideal behaviour are expressed by excess functions. Near-ideal solid solutions are formed between metals having the same crystal structure, their atoms being of almost the same size and having similar electronic structures; this usually means they have to be in the same or adjacent Groups of the Periodic Table. Such pairs have complete mutual solubility and are described as monophasic. There are many examples of this class, for example, alloys formed between silver and gold or between palladium and silver. For solutions formed endothermically (i.e. ΔH_{mix} is positive), there is a limit to the solubility of each in the other even if both have the same structure, and there is a range of composition where the alloy comprises various amounts of two phases of constant composition: such alloys are said to be biphasic. The nickel-copper system is of this type[10,49,98] and it has been very thoroughly studied. The free energy of mixing at 473 K has been calculated from the excess functions (Figure 1.16), and the two minima define the limits of the two-phase region in which there is a physical mixture of alloy phases containing respectively 3 and 85% copper. Thus a mixture containing equal numbers of nickel and copper atoms will have 40% of the copper-rich alloy and 60% of the nickel rich alloy. If temperature is increased, the entropy term $-T\Delta S_{mix}$ becomes more important, ΔG_{mix} becomes more negative, and the two-phase region contracts, until the critical solution temperature is reached, above which the components are miscible in all proportions. The platinum-gold system is also of this type.

For alloys that are formed endothermically or only slightly exothermically, there are many indications of a mutual perturbation of the electronic structure. As attractive interaction increases and the process becomes more exothermic,[99] random arrangements are replaced at certain compositions by ordered superlattices

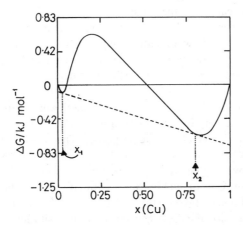

Figure 1.16. Free energy of formation of nickel-copper alloys at 473 K as a function of composition.

(or superstructures) such as are formed in the platinum-copper and gold-copper systems (e.g. Cu_3Pt, Cu_3Au). Other examples occur in the platinum-tin system (Pt_3Sn, Pt_2Sn) and in the aluminium alloys used for making Raney metals.[100] Even stronger interaction leads to the very exothermic formation of *intermetallic compounds*[101,102] between quite unlike elements (e.g. $HfIr_3$, Ce_2Ni, $CeRh_3$, $ZrPd_3$). Compositions used for forming *amorphous alloys* by melt spinning are often of this type.[61] The terminus of this progression is of course the formation of recognisable compounds such as oxides and sulfides.

Interstitial alloys[101] are formed between metals and non-metallic or semi-metallic elements such as boron, phosphorus, carbon and nitrogen: the latter occupy holes in the metal structure, which may however have to expand or re-arrange to accommodate them.[2,7] Carbides and nitrides of metals of the first Transition Series form spontaneously during catalytic reactions where the reactants contain these atoms, and are themselves catalytically active.[101,103] They do not exist as stable compounds of the noble metals of Groups 8 to 10.

Many of the physical properties of monophasic alloys are intermediate between those of the pure components. Lattice parameters, readily determined by X-ray diffraction and constituting a sensitive means of checking alloy composition, often show only slight deviations from Vegard's Law, which states that they should be linear functions of composition. The silver-gold system is unusual in that the lattice parameter passes through a maximum, but the two metals are of almost the same size. Certain binary alloys exhibit a number of intermediate phases differing in structure as well as composition; according to the Hume-Rothery Rules,[23,31] the structure depends on the ratio of valence electrons to atoms, and such alloys (occurring for example in the Cu-Zn (brass) system) are termed *electron compounds*. The systematic alteration of structure that is independent of the atomic mass of the metal is reminiscent of that found on passing through the Transition Series metals, but the explanation cannot be the same.

Electronic properties such as electrical conductance, magnetic behaviour and band structure typically show dramatic changes with alloy composition, especially where the electronic structures of the pure components differ greatly, as happens for example when the *d*-shell is filled. Alloys of this type (Ni-Cu, Pd-Ag, Pd-Au) were the subject of intensive research in the period 1945-1970, as it was believed that the presence of an incompletely-filled *d*-shell was an important feature in determining catalytic activity, and that filling would occur at some composition that could be deduced from electronic properties. The experimental results and the theoretical models that form our present state of understanding of the behaviour of electrons in alloys will be considered in the following section.

One quite new way of making alloys suitable for fundamental study is to condense atoms of one metal onto the surface of a single crystal of the other: there results a 'two-dimensional alloy'. The surface composition is easily changed, the problem of surface segregation (see following text) is avoided, and bimetallic systems having little or no mutual bulk solubility (e.g. Ru-Ag) are available for study. Extensive information is now available on how metals interact when brought together in this way,[10] and on the way in which atoms of the second metal agglomerate as their coverage is increased.[10] One surface which has received much attention is the isotropic Ru(0001) (Figure 1.10); the coinage metals have all been deposited on it to give bimetallic catalysts simulating the dispersed forms which have also proved of great catalytic interest (see Chapter 2). However the film of the deposited metal may suffer strain through mismatch of atomic sizes, with consequential effects on its physical properties.[104]

1.3.2. Electronic Properties of Alloys and Theoretical Models[105]

The paramagnetism or ferromagnetism shown by the metals of Group 10 is progressively lost on alloying with a metal of Group 11. Nickel, which has a saturation magnetic moment of 0.606 Bohr magnetons, was thought to have about 0.54 d-band holes per atom, these two numbers being in fair agreement, and on addition of copper the magnetic moment falls linearly to a minimum value at 60% copper. Parallel changes were observed[23,106] with the low-temperature electronic specific heat coefficient and with the Curie temperature (at which ferromagnetic elements become paramagnetic), and analogous changes were seen with other Group 10-Group 11 alloys.[10,33] It was therefore perfectly logical to suppose that the s-electrons of the Group 11 metal atoms entered and filled the d-band of the Group 10 metal: this simple and satisfying picture, first advanced by Mott and Jones,[106] came to be called the Rigid Band Model, because it was assumed that the band shape of the Group 10 metal was unaltered by alloying. The valence electrons were thought to share a common band system, and the metals to lose their identity, except in regard of their nuclear charge. Unfortunately, as Oscar Wilde said, *The truth is rarely pure and never simple*: this model proved to be incorrect. *Errare humanum est.*

Early work on electron band structure by soft X-ray spectroscopy was concentrated on pure metals,[23] and it was not until the advent of photoelectron spectroscopies that alloys started to be examined. It soon became clear that small additions of nickel to copper resulted in the appearance of electrons having energies close to the Fermi value; there was no common d-band, but each component exhibited its own band structure[10,58] (Figure 1.17). Many other kinds of physical measurement confirmed this, and corresponding behaviour was observed with the palladium-silver system (Figure 1.18). It became necessary to find a new and better theory.

The heart of the problem is this: a copper atom in a nickel matrix does not wish to lose its $4s$ electron totally, nor does a nickel atom wish to accept, it as this would be tantamount to forming an ionic bond Ni^-Cu^+. While electrostatic bonding can contribute to the stability of some intermetallic compounds, as in the

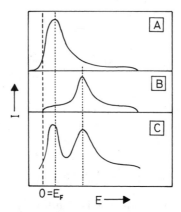

Figure 1.17. Band structures of (A) nickel, (B) copper and (C) a nickel-copper alloy as revealed by the intensity of photoemission current as a function of energy.

Cs^+Au^- type of compound, which we have already met, it is not a suitable basis for explaining the formation of alloys. An early attempt to solve at least part of the problem involved looking at a nickel atom in a copper matrix. The $3d$ electrons of the nickel atom occupied highly localised energy levels around it,[107] but they were broadened by resonant interaction with the $4s$ electrons of copper.[108] The width of the d-band should increase with nickel concentration as the d-electrons

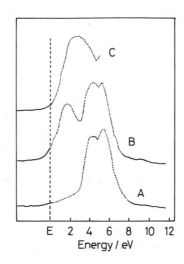

Figure 1.18. Valence band spectra for (A) silver, (B) palladium + silver and (C) palladium supported on silica.

start to interact; this is what is observed experimentally. The d-electrons thus occupy a *virtual bound state*, this being the name applied to the model, which is associated with the names of Friedel and Anderson.[109] The d-band holes (about 0.6 of them on average), remain associated with the nickel atoms,[110] and alloys are *paramagnetic* up to the highest copper contents. XANES measurements (Section 2.42) on nickel-copper powders using L_{II} and L_{III} absorption edges have confirmed that the number of d-states on nickel atoms decreases only slowly with increasing copper content, and at Ni_5Cu_{95} is still 95% of that in pure nickel. However, even at $Ni_{15}Cu_{85}$, nickel atoms start to form clusters, which show *superparamagentism*: these start to overlap at about $Ni_{40}Cu_{60}$ and *ferromagnetism* appears. In going from this composition to pure nickel the overlap increases, giving a linear dependence of ferromagnetic saturation moment on nickel content.[106] It was this that misled early workers into thinking that copper's electrons filled the d-band holes in nickel. Short-range ordering is slight in this system, but increases in importance as Periodic Group separation of the elements forming the alloy becomes greater.

Proper quantum mechanical approaches to understanding the properties of alloys encounter a real difficulty at the start, because of the inhomogeneous potential field through which the electron waves must move. The random arrangement of ion cores of different nuclear and electronic charge density causes the electron waves to suffer multiple scattering, which causes electrical conductance to increase: the essence of the theoretical problem lies in how to tackle this. One quite successful method[10,111] is to suppose that the electron wave moves through a uniform or coherent potential, and is affected only by a single selected scatterer. The effective medium is chosen self-consistently, and the Coherent Potential Approximation results from the self-consistent solution to a multiple-scattering version of the Schrödinger equation within a single site approximation. Density of states curves and other physical properties are well described by this theory.[10]

We now have to think how the chemical interaction of the components of the alloy when at the surface affects their ability in chemisorption. Before we can look at this, however, we must address the problem of surface segregation.

1.3.3. The Composition of Alloy Surfaces[74,99]

In general the ratio of the two components of a bimetallic alloy is not the same in the surface layer (or layers, see preceding section) as in the bulk. The reason for this is that at equilibrium the configuration is adopted which minimises the total energy, and, since this includes surface energy, that component having the lower surface energy will tend to concentrate in the surface. If it does not, it is because diffusion normal to the surface is slow, and the system is not at equilibrium. To a first approximation, therefore, we may use values of surface tension (specific surface work) or heat of sublimation to predict which partner will segregate at the surface.

The formalism used for liquid solutions can also be applied to solids.[74] For an ideal solution, the enthalpy of mixing is zero, and by application of the Gibbs equation we can deduce that the ratio of the mol fractions of the two components in the surface x_1^s/x_2^s is given by

$$(x_1^s/x_2^s) = (x_1^b/x_2^b) \exp[(\gamma_2 - \gamma_1)a/RT] \tag{1.6}$$

where superscript b stands for bulk and a is the molar area of component 2. This equation permits the surface composition to be estimated if the surface tensions of the two components are known. We may note that the extent of segregation will decrease with rising temperature. The next refinement is to allow for non-ideality by introducing the regular solution parameter Ω, which is in the nature of an excess heat and may be defined as

$$\Omega = \Delta H_{\text{mix}}/x_1 x_2 \tag{1.7}$$

When Ω is positive, bonds between unlike atoms are preferred and greater enrichment occurs at low bulk concentrations, but if it is negative bonds between like atoms are preferred and greater enrichment occurs at high bulk concentrations. The somewhat complex equations[10] which describe surface enrichment for real solutions, i.e. when Ω is not zero, can also predict concentration differences in second, third and fourth layers.

For example, when Ω is negative, weak enrichment of one component in the surface is accompanied by an increased amount of the other in the second layer. We have already considered the difficulty of deciding which atoms are actually in the surface layer. With ideal solid solutions, it is only the concentration in the first layer that differs from that of the bulk.

There are several other factors that deserve mention before this subject is left. First, the surface tensions of different crystal planes may vary quite considerably; the value is greater for less densely packed planes (e.g. for fcc(100) it is less than for fcc(111), Figure 1.9) because of the smaller number of bonds that need to be broken to create new surface. Preferential segregation at these planes therefore minimises the system's energy, and by extension of this principle atoms of unusually low CN, such as occur at steps and kinks, are particularly favoured sites for segregated atoms. The reader wishing to explore further the question of surface enrichment in binary alloys should consult the classic paper by Williams and Nason.[112]

Amongst the main experimental techniques that have been deployed to determine the composition and structure of binary alloy surfaces, Auger electron spectroscopy (AES), XPS, LEED and ISS feature most prominently. There is now a very large literature describing the results obtained, which, after some early inconsistencies had been resolved, are now generally in line with theoretical predictions. One further factor has however emerged as being important. With the ordered superlattices Cu_3Au, $CuAu$ and $CuAu_3$, gold segregation has been detected,[113] although this is not what the sublimation energies would lead one to predict. The additional factor is the relative sizes of the atoms, those of gold being squeezed out of the bulk because they are the larger; the strain is thereby relieved.[105] This also explains the unexpected sense of segregation (based on Gibbs theory) in the cases of platinum alloyed with iron, cobalt, nickel and rhodium. The simple application of ideal or regular solution theory cannot be expected to work perfectly when the sizes of the atoms are much different.

For the purposes of chemisorption and catalysis, we need to know the composition and structure of the surface. The average composition is now easily measured, and can be checked against relevant theory, but for structure we have to find the arrangement of the two kinds of atoms at the atomic level. Assuming an average composition and the equality of all interaction energies in the surface, it is a straightforward use of binomial theorem to calculate the population of groups of two, three etc. of atoms of one kind.[114] Unfortunately this simple procedure does not apply where there are atoms of different CN or where there is a tendency to cluster formation.[115] Other computational and experimental methods are required, and these will feature in the next chapter in the context of small metal particles. We shall also discover in due course how the presence of a chemisorbed layer can alter the surface composition of an alloy.

REFERENCES

1. J.E. Huheey, E.A. Keitner and R.L. Keitner, *Inorganic Chemistry*, Harper Collins: New York (1993).
2. N.N. Greenwood and A. Earnshaw, *Chemistry of the Elements*, 2nd. edn., Butterworth-Heinemann: Oxford (1997).
3. R.T. Sanderson, *Chemical Periodicity*, Reinhold: New York (1960).
4. P.P. Edwards and M.J. Sienko, *Internat. Rev. Phys. Chem.* **3** (1983) 83; *Acc. Chem. Res.* **15** (1982) 87; *J. Chem. Educ.* **60** (1983) 691.
5. P.P. Edwards, R.L. Johnston and C.N.R. Rao, in: *Metal Clusters in Chemistry*, Vol. 3 (P. Braunstein, L.A. Oro and P.R. Raithby, eds.), Wiley-VCH: Weinheim (1998), p.1454,
6. P.P. Edwards and M.J. Sienko, *Chem. Brit.* 39 (1983, Jan.).
7. F.A. Cotton and G. Wilkinson, *Advanced Inorganic Chemistry*, 4th. edn., Wiley: Chichester (1980).
8. J.W. Arblaster, *Platinum Metals Rev.* **40** (1996) 62.
9. L. Pauling, *Nature of the Chemical Bond*, Cornell Univ. Press: Ithaca, N.Y. (1960).
10. V. Ponec and G.C. Bond, *Catalysis by Metals and Alloys*, Elsevier: Amsterdam (1995).
11. F.R. DeBoer, R. Boom, W.C.M. Mattens, A.R. Miedema and A.K. Niessen, *Cohesion in Metals*, North Holland: New York (1988).
12. G.C. Bond and E.L. Short, *Chemistry and Industry*, 12 (2002, June 3).
13. G.C. Bond, *J. Molec. Catal. A: Chem.* **156** (2000) 1; *Platinum Metals Rev.* **44** (2000) 146.
14. P. Pyykkö, *Chem. Rev.* **88** (1988) 563.
15. P. Schwerdtfeger, *Angew. Chem.* **108** (1996) 2973; *Heteroatom Chem.* **13** (2002) 578.

16. K.S. Pitzer, *Acc. Chem. Res.* **12** (1979) 271.

17. P. Pyykkö and J. P. Desclaux, *Acc. Chem. Res.* **12** (1979) 276.

18. N. Kaltsoyannis, *J. Chem. Soc. Dalton Trans.* (1997) 1.

19. F. Wilczek in: *It Must Be Beautiful: Great Equations of Modern Science*, (G. Farmelo, ed.), Granta: London (2002).

20. K. Balasubramanian, *Relativistic Effects in Chemistry*, Wiley: New York (1997).

21. G.C. Bond and D.T. Thompson, *Catal. Rev.-Sci. Eng.* **41** (1999) 319.

22. H. Schmidbaur, *Gold Bull.* **23** (1990) 11.

23. W. Hume-Rothery, *Electronic Theory for Students of Metallurgy*, Institute of Metals: London (1947).

24. J.W. Niemansverdriet, *Spectroscopy in Catalysis,* VCH: Weinheim (1993).

25. R. Hoffman, *Rev. Mod. Phys.* **60** (1988) 601.

26. L. Salem, *J. Phys. Chem.* **89** (1985) 5576.

27. R. Hoffman, *Solids and Surfaces: A Chemist's View of Bonding in Extended Structures*, VCH: Weinheim (1988).

28. W. Romanovski, *Highly Dispersed Metals*, Ellis Horwood: Chichester (1987) (see Appendix to Ch. 4, pp.106–118, by H. Chojnacki).

29. A. Clark, *The Chemisorption Bond: Basic Concepts*, Academic Press: New York (1974).

30. R.I. Masel, *Principles of Adsorption and Reaction on Solid Surfaces*, Wiley: New York (1996).

31. P.A. Cox, *The Electronic Structure and Chemistry of Solids*, Oxford U. P.: Oxford (1987).

32. L. Smart and E. Moore, *Solid State Chemistry*, Chapman and Hall: London (1992).

33. G.C. Bond, *Catalysis by Metals*, Academic Press: London (1962).

34. E. Wigner and F. Seitz, *Phys. Rev.* **43** (1933) 804.

35. J.C. Slater, *Phys. Rev.* **45** (1934) 794; **92** (1953) 603.

36. J. Korringa, *Physica* **13** (1947) 392; *J. Phys. Chem. Solids* **7** (1958) 252.

37. W. Kohn and N. Rostoker, *Phys. Rev.* **94** (1954) 1111.

38. L. Pauling, *J. Chem. Soc.* (1948) 1461.

39. L. Pauling, *Proc. Roy. Soc. A* **196** (1949) 343.

40. S.L. Altmann, C.A. Coulson and W. Hume-Rothery, *Proc. Roy. Soc. A* **240** (1957) 145.

41. L. Brewer, *Electronic Structure and Alloy Chemistry of the Transition Elements*, Wiley Interscience: New York (1963).

42. N. Engels, *Amer. Soc. Metals Trans.* **57** (1964) 610; *Acta Met.* **15** (1967) 557.

43. O. Johnson, *Bull. Chem. Soc. Japan* **45** (1972) 1599, 1607; **46** (1973) 1919, 1923, 1929, 1935.

44. T. Berlin, *J. Chem. Phys.* **19** (1951) 208.

45. O. Johnson, *J. Res. Inst. Catal. Hokkaido Univ.* **20** (1972) 95, 109, 125; **21** (1973) 1.

46. G.C. Bond, *Heterogeneous Catalysis: Principles and Applications*, 2nd. edn., Oxford U.P.: Oxford (1987).

47. *Surface Analysis: The Principal Techniques*, (J.C. Vickerman, ed.), Wiley-VCH: Chichester (1997).

48. J. Hudson, *Surface Science: an Introduction*, Wiley: Chichester (1998).

49. *Chemisorption and Reactions on Metallic Films* (J.R. Anderson, ed.), vols.1 and 2, Academic Press: London (1971).

50. D.R. Rossington in: *Chemisorption and Reactions on Metal Films*, Vol. 2 (J.R. Anderson, ed.), Academic Press: London, (1971), p. 211.

51. T. Wissman (ed.), *Thin Metal Films and Gas Chemisorption*, Elsevier: Amsterdam (1987).

52. O. Beeck, A.E. Smith and A. Wheeler, *Proc. Roy. Soc. A* **177** (1941) 62.

53. O. Beeck and A.W. Ritchie, *Discuss. Faraday Soc.* **8** (1950) 159.

54. O. Beeck, *Adv. Catal.* **2** (1950) 151.

55. A.W. Adamson and A.P. Gast, *The Physical Chemistry of Surfaces*, 6th. Edn., Wiley-VCH: Chichester (1997).

56. H.E. Farnsworth, *Adv. Catal.* **15** (1964) 31.

57. R.C. Pitkethly, in: *Chemisorption and Catalysis*, (P. Hepple, ed.), Inst. Petroleum: London, (1971), p. 98.

58. J.B. Hudson, *Surface Science: an Introduction*, Butterworth-Heinemann: New York (1992).

59. Z. Knor and E.W. Müller, *Surf. Sci.* **10** (1968) 21.

60. A. Molnár, G.V. Smith and M. Bartók, *Adv. Catal.* **36** (1989) 32.

61. A. Baiker in: *Handbook of Heterogeneous Catalysis* Vol. 2 (C. Erte, H. Knözinger, & J. Weitkamp, eds.), VCH: Weinheim (1997), p. 803.

62. K. Christmann, *Introduction to Surface Physical Chemistry*, Steinkopff: Darmstad (1991).

63. G.A. Somorjai, *Introduction to Surface Chemistry and Catalysis*, Wiley: New York (1994).

64. J.M. Thomas and W.J. Thomas, *Principles and Practice of Heterogeneous Catalysis*, VCH: Weinheim (1997).

65. Z. Paál, R. Schlögl and G. Ertl, *Catal. Lett.* **12** (1992) 331; *J. Chem. Soc. Faraday Trans.* **88** (1992) 1179.

66. King Lun Yeung and E. E. Wolf, *J. Catal.* **143** (1993) 409.

67. G.A. Somorjai, *Catal. Rev.* **7** (1973) 87.

68. A.W. Adamson, *Textbook of Physical Chemistry*, Academic Press: New York/London (1973).

69. R.A. Alberty, *Physical Chemistry*, 6th edn., Wiley: New York (1983).

70. G.C. Bond and G. Webb, *Platinum Metals Rev.* **6** (1962) 12.

71. H.-C. Jeong and E.D. Williams, *Surf. Sci. Rep.* **34** (1999) 171.

72. C.F. McFadden, P.S. Cremer and A.J. Gellman, *Langmuir* **12** (1996) 21.

73. A.J. Gellman, J.D. Horváth and M.T. Buelow, *J. Molec. Catal. A: Chem.* **167** (2000) 3.

74. S.K. Overbury, P.D. Bertrant and G.A. Somorjai, *Chem. Rev.* **75** (1975) 547.

75. S. Titmuss, A. Wander and D.A . King, *Chem. Rev.* **96** (1996) 1291.

76. D.A. King and D.P. Woodruff (eds.), *The Chemical Physics of Solid Surfaces*, Vol. 7, Elsevier: Amsterdam (1994).

77. G.A. Somorjai and G. Rupprechter, *J. Chem. Educ.* **75** (1998) 171.

78. G.A. Somorjai, *Ann. Rev. Phys. Chem.* **45** (1994) 721.

79. G.A. Somorjai, *Langmuir* **7** (1991) 3176.

80. J.C. Rivière, in: *Solid State Surface Science,* Vol. 1, (M. Green, ed.), Dekker: New York (1969).

81. J.R. Smith, in: *Interactions in Metal Surfaces*, Topics in Applied Physics, Vol. 4, (R. Gomer, ed.), Springer,: Berlin, (1975), p. 1.

82. A.R. Cholac and V.M. Tapilin, *J. Molec. Catal. A: Chem.* **181** (2000) 181.

83. J.J. van der Klink, *Adv. Catal.* **44** (1999) 1.

84. Y.Y. Tong, A.J. Renouprez, G.A. Martin and J.J. van der Klink, *Proc. 11th. Internat. Congr. Catal.* (J.W. Hightower, W.N. Delgass, E. Iglesia and A.T. Bell, eds.), Elsevier: Amsterdam **B** (1996) 911.

85. G.C. Bond, *Discuss. Faraday Soc.* **41** (1966) 200.

86. S. Carrà and R. Ugo, *Inorg. Chim. Acta Rev.* **1** (1967) 49.

87. J.B. Goodenough, *Magnetism and the Chemical Bond*, Interscience: New York (1963).

88. N.H. March, *Chemical Bonds Outside Metal Surfaces*, Plenum: New York (1986).

89. W.F. Banholzer, Y.O. Park, K.M. Mak and R.I. Masel, *Surf. Sci.* **128** (1983) 176; **133** (1983) 623.

90. G.C. Bond in: *Proc. 4th Internat. Congr. Catal.*, Editions Technip: Paris **2** (1968) 266.

91. E. Yagasaki and R.I. Masel, in: *Specialist Periodical Reports: Catalysis* **11** (1994) 165.

92. R.L. Augustine, *Heterogeneous Catalysis for the Synthetic Chemist*, Dekker: New York (1996), Chs. 3 and 4.

93. R.L. Augustine and K.M. Lahanas, in: *Catalysis of Organic Reactions* (J.R. Kosak and T.A. Johnson, eds.), Dekker: New York (1994), p. 279.

94. R.L. Augustine, K.M. Lahanas and F. Cole, in: *Proc. 10th. Internat. Congr. Catal.*, (L. Guczi, F. Solymosi and P. Tétényi, eds.), Akadémiai Kiadó: Budapest **C** (1993) 1567.

95. W.R. Frost, *Canad. J. Chem.* **37** (1959) 460.

96. V. Ponec, *Adv. Catal.* **32** (1983) 149.

97. E.G. Allison and G.C. Bond, *Catal. Rev.* **7** (1977) 37.

98. W.M.H. Sachtler and G.J.H. Dorgelo, *J. Catal.* **4** (1965) 654, 665.

99. M. Masai, K. Honda, A. Kubota, S. Ohnaka, Y. Nishikawa, K. Nakahara, K. Kishi and S. Ikeda, *J. Catal.* **50** (1977) 419.

100. L. Brewer, *J. Phys. Chem.* **94** (199) 1196.

101. S.T. Oyama and G.L. Haller, in: *Specialist Periodical Reports: Catalysis* Vol. 5 (G. C. Bond and G. Webb, eds.), *Roy. Soc. Chem.* (1982) p. 333.

102. *Intermetallic Compounds*, Vol.1, (J.H. Westbrook and R.L. Fleischer, eds.), Wiley-VCH: Chichester (1994).

103. A. Baiker and M. Maciejewski, *J. Chem. Soc. Faraday Trans. I* **80** (1984) 2331.

104. V. Ponec, *J. Molec. Catal. A: Chem.* **133** (1998) 221.

105. L.A. Rudnitskii, *Russ. J. Phys. Chem.* **53** (1979) 1727.

106. N.F. Mott and H. Jones, *The Theories and Properties of Metals and Alloys*, Oxford U.P.: London (1936).

107. S. Hunter, G.K. Wertheim, R.L. Cohen and J.H. Wernicle, *Phys. Rev. Lett.* **28** (1972) 488.

108. G.A. Martin, *Catal. Rev.–Sci. Eng.* **30** (1998) 519.

109. P.W. Anderson, *Phys. Rev.* **124** (1961) 41.

110. G. Meitzner, D.A. Fischer and J.H. Sinfelt, *Catal. Lett.* **15** (1992) 219.

111. P. Soven, *Phys. Rev.* **156** (1967) 809; **178** (1969) 1136.

112. F.L. Williams and D. Nason, *Surf. Sci.* **33** (1974) 254.

113. J.M. McDavid and S. C. Fain Jr., *Surf. Sci.* **52** (1975) 161.

114. D.A. Dowden, in: *Proc. 5th. Internat. Congr. Catal.*, (A. Farkas, ed.), Academic Press: New York. **1** (1973) 621.

115. N.T. Anderson, F. Topsøe, I. Alstrup and J.R. Rostrop-Nielsen, *J. Catal.* **104** (1987) 454.

2

SMALL METAL PARTICLES AND SUPPORTED METAL CATALYSTS

PREFACE

Studies of chemisorption and catalysis on metal surfaces fall into two categories: (i) those made with massive or macroscopic metals, either monocrystalline or polycrystalline, and (ii) those using very small metal particles containing from a few tens to a few thousands of atoms. The first category affords access to a much wider range of investigational techniques, many of which are unfortunately only applicable to massive metals because of the necessity of working under ultrahigh vacuum (UHV) conditions: such surfaces are however of limited practical utility. A few techniques are only usable when the extended surface area provided by highly dispersed metals is available; thus it is difficult to obtain an adsorption isotherm on a single crystal, although heats of adsorption can now be measured calorimetrically. Some procedures (particularly spectroscopic ones) are of course applied to both types of surface. Small metal particles are extremely useful, but are difficult to characterise, and because they are inherently unstable it is usually necessary to affix them to a *support*, typically a high surface area oxide, in order to prevent their aggregation. The desired adherence of metal support does however lead to a number of ambiguities in the interpretation of what is observed. The relevance of information obtained with massive metals to the behaviour of small metal particles has been hotly debated, but techniques such as X-ray absorption spectroscopy and nuclear magnetic resonance will however provide detailed descriptions of even the smallest particle.

2.1. INTRODUCTION

2.1.1. Microscopic Metals[1,2]

Chemisorption and catalysis are surface phenomena, and to optimise the catalytic activity of a given mass of a metal it is necessary to increase its surface area to the greatest possible extent, by forming it into very small particles. Terms used to describe what has been achieved by doing this are (i) the *dispersion* (or more properly the degree of dispersion), (ii) the *dispersity* (favoured by European scientists) and (iii) the *fraction exposed* (a term which has not found widespread acceptance). All three terms express the ratio of surface atoms to total atoms in the particle, but there are of course other ways of expressing the degree of subdivision. As this is increased, the surface area of unit mass will increase, as will the numbers of particles per unit mass, while the number of atoms per particle decreases. These statements apply to the somewhat artificial situation when all particles have the same size or shape. Approximate relations between all these quantities are easily calculated by assuming the particles are all spheres of uniform size[3,4] or are uniform cubes exposing five faces. Some results for palladium and platinum are shown in Figures 2.1 and 2.2. We may note that while the area per unit mass for a given

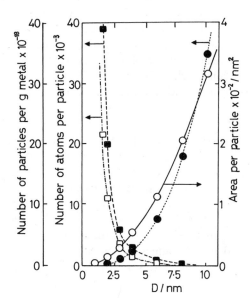

Figure 2.1. Calculations based on the Uniform Sphere Model for palladium and platinum: dependence of (i) number of particles per g of metal, (ii) number of atoms per particle, and (iii) surface area per particle, on particle size.[3,4] Note: (ii) and (iii) are the same for both metals (atoms of each are about the same size).

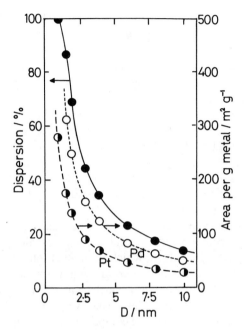

Figure 2.2. Further calculations based on the Uniform Sphere Model: dependence of (i) surface area per g metal and (ii) degree of dispersion, on particle size.[3,4]

particle size increases as atomic mass decreases, the dispersions shown by a sphere and by a five sided cube are about the same for bodies of the same diameter or length of side, although the cube contains almost twice as many atoms. Indeed it hardly matters whether the pyramid, the hemisphere, the octahedron or any other shape is considered; the dependence of dispersion on size is much the same.[5-9] A useful benchmark is that a spherical particle of platinum having 230 atoms will be 2 nm in diameter and will have a dispersion of about 60%.

The conclusions of these rough calculations should not however be pressed too far, for several reasons: (i) they pay no regard to the way in which atoms are packed in the solid; (ii) for small particles in particular, the uncertainty of knowing which atoms to count as being on the surface affects the estimation of dispersion, (unless all the faces are close packed); and (iii) for the same reason the estimation of surface area becomes difficult for very small particles. Nevertheless they have some qualitative value in directing our thinking towards the size of particles likely to be found useful in catalysis.

Such model calculations are in addition divorced from the real world for other reasons. In practice the members of a small collection of small particles are most unlikely all to have exactly the same size, i.e. the set will not be *monodisperse* nor will their shape be a single geometrical form.[5] Practical methods for determining mean size, size distribution and shape will be treated in Section 2.4.

Microscopic bimetallic particles have been the subject of much interest and attention because of their superiority over pure metals in a number of catalysed hydrocarbon reactions. The term 'alloy' is not applied to them, because it implies a very intimate mixing of the components, in the manner discussed in Chapter 1; this is not always the case with very small supported bimetallic particles. It has been necessary to try to establish the surface concentration of each component by physical and theoretical methods. Pairs of metals forming a continuous range of solid solutions, (e.g. Pd-Ag and Pd-Au) are readily formed into small bimetallic particles, although their surfaces are expected[10] and found[11] to be enriched with the Group 11 metal, having the lower surface energy and sublimation enthalpy. Those pairs showing a miscibility gap in some range of temperature (e.g. Ni-Cu, Pt-Au) are more problematic and the Ni-Cu system in particular had to be carefully examined before the truth emerged. Useful bimetallic particles can however also be fabricated with pairs of metals for which the mutual solubility is very low, (e.g. Ru-Cu, Os-Cu), and here again the Group 11 metal is found at the surface as if it was chemisorbed on the surface of the Group 8 metal. Efforts to analyse the surface composition of small bimetallic particles are partly vitiated by the observation that, due to the flexibility of the structure, the component interacting most strongly with a chemisorbed species can draw it to the surface against the thermodynamic force that would apply in a vacuum. Methods of preparing small bimetallic particles are considered in Sections 2.2 and 2.3.2; theoretical studies are mentioned in Section 2.5.5.

2.1.2. Instability of Small Metal Particles

The self-evident characteristic of small metal particles, which determines most of their properties, is that many of the atoms are on the surface and are therefore atypical. Their total free energy is greater than that of the same amount of macroscopic metal because of free energy (see Sections 1.2.2 and 2.5.3) is additional to the other forms. Another way of visualising the greater energy of small particles is to remember how much energy would have to be used to break a large lump of metal into tiny particles. Very many strong chemical bonds would have to be broken, and the energy used to do this remains in them. The total free energy would therefore decrease if particles were to grow again, and the proportion of surface atoms (i.e. the dispersion) becomes smaller. This indeed happens in an exothermic process, which releases the energy used in the subdivision: this is known as *sintering*. It is a process that is important in powder metallurgy, where metal powders are compacted by pressure and heated to form the desired strong shape. The ease with which this occurs depends on the atmosphere in which it is conducted. A layer of chemisorbed oxygen atoms both stabilises small particles by lowering the surface energy, and inhibits the migration of surface metal atoms that is needed to form links between particles at the start of the growth process. Removal of the oxide layer by a reducing gas facilitates sintering. Sometimes it is desirable to prevent the further growth of metal crystallites; this can be done by coating them with a thin layer of an oxide such as zirconia, which even more effectively than oxygen cuts down the surface movement of metal atoms. This effect is called *grain stabilisation*. It is apparent that the stronger the bond between the metal atoms, the greater will be the surface energy of small particles, and the greater their tendency to sinter. Thus ease of sintering increases with the enthalpy of sublimation of the metal.

The above remarks are couched in very general terms, and apply to some extent to most forms of small metal particle. The available forms are: (i) powders, a term which embraces a set of particles of any size, as long as it flows freely (Section 2.2), (ii) aerosols, (iii) colloidal dispersions (Section 2.2), and (iv) supported metals (Section 2.3), including particles formed by condensation of metal atoms onto a flat surface: this leads ultimately to a condensed metal film.

2.2. PREPARATION OF UNSUPPORTED METAL PARTICLES

We may return to consider briefly the methods available for making small metal particles in an unsupported form. Because of handling problems and difficulty in restraining their movement, they are chiefly used in liquid media: even so, their separation by filtration may be difficult or indeed impossible. *Colloidal dispersions* containing metal particles between 2 and 20 nm in size may be made by careful reduction of a dilute aqueous solution of an appropriate salt[12–16] (see also Further Reading section at the end of the chapter). Colloidal gold has been very much studied;[17] stable platinum sols can be made by citrate ion reduction, but not all metals are capable of forming stable dispersions, the base metals of Groups 8 to 10 being particularly intractable.[18] Sols of platinum, palladium, rhodium and ruthenium can be made by reducing solutions of their salts with silanes such as $(EtO)_3SiH$.[19] Stabilisation is helped by the introduction of protective agents,[16] but their presence is likely to inhibit catalytic properties. Recent advances in colloid chemistry include the use of non-aqueous solvents[19] and reverse micelles[20,21] (i.e. a microemulsion in which very small droplets of aqueous solution are dispersed in a hydrophobic solvent); sonochemical reduction has also been used.[22] Gold and gold-silver bimetallic colloidal particles form ordered arrays by self-organisation on removal of the dispersant, especially when stabilised by organic thiols.[23–28]

Less well dispersed metal powders are made by reducing solutions of salts with reducing agents such as formaldehyde (methanal), formic (methanoic) acid,[29] hydrazine and the tetrahydridoborate (BH_4^-) ion,[30–32] although in this last case some boron may be incorporated in the product.[13] The term *metal black* is often applied to powders made in this way:[33,34] they are used for basic research on gas-phase reactions but they sinter easily. Metal powders are also made by reduction of oxides (or sulfides[35]) by hydrogen or by thermal decomposition of metal formates. There are two routes to well dispersed metal that are particularly useful for three phase reactions. *Raney* or *skeletal metals*[36-39] are made by forming an alloy of the active metal with aluminium, and then treating it with strong base so as to dissolve most of the aluminium, leaving the desired metal as a sponge like material.[40] Nickel has been much used in this form, but many other metals can be treated similarly.[41] Finally, *Adams oxides* of the noble metals are made by treating a metal salt in molten sodium nitrate; they are reduced *in situ* under ambient conditions by hydrogen, thus avoiding the need to form the metal outside the reactor, with the attendant risks of sintering and poisoning. All the above procedures can be equally applied to mixtures of two or more metal compounds, leading to bimetallic or multimetallic products in which the elements are intimately mixed.[20]

Metallic *aerosols* can be made by passing a massive amount of electricity through a thin wire by discharging a bank of capacitors, whereupon the wire explodes.[42] The technique, to which the name *deflagration* has been applied, was pioneered (as were so many things) by Michael Faraday,[43] who converted a gold wire into powder by using the current produced by a Leiden battery.

2.3. SUPPORTED METAL CATALYSTS[1,44-46]

2.3.1. Scope

The need to employ very small particles to secure high dispersion, with their irritating habit of sintering, poses a major problem in the design of a technically usable metal catalyst. Fortunately it has been resolved by the simple expedient of affixing the particles to a thermally stable material usually known as a *support*, but sometimes a *carrier*. The following statement has been offered[45] as a definition and statement of the scope of the resulting substance.

Supported metal catalysts comprise 0.1–20% by weight of a metal of Groups 8-11 dispersed over the surface of a support, which is typically a high-surface-area oxide. They are widely used on an industrial scale and in research laboratories. These materials are effective as catalysts because the metallic phase is present as extremely small particles, having a degree of dispersion of 10 to 100%. They are firmly anchored to the surface and are widely separated from each other, and hence do not readily coalesce or sinter.

Like all attempts at generalisations, the above statements, while reasonably accurate, provide only the barest outline of the wealth of information available in the patent and the open literature on these materials. Their study, and descriptions of the science (or art) of their preparation, has engendered an enormous literature, as numerous books and review articles attest. We shall have to content ourselves now with a quite short discussion of their most relevant features to the matter in hand.

The earliest material to resemble a supported metal catalyst was made by Döbereiner, who mixed platinum black with clay in order to dilute its catalytic action. This remains a significant objective, since for most purposes it would be quite impractical to employ undiluted metal. The use of supported metals facilitates handling and minimises metal loss, a particularly important consideration with the noble metals: by appropriate choice of the physical form of the support, they can be used in various types of catalytic reactor, such as fixed-bed or fluidised-bed configurations. The support has often been regarded as catalytically inert, but in addition to those cases such as bifunctional catalysts, where the acidic support has long been known to play a vital role, there is a growing number of examples of participation by the support in catalytic processes. The support surface also facilitates the incorporation of modifiers, such as promoters or selective poisons.

Much attention has been given in recent years to the general question of how the composition and structure of a supported metal affects its ability as a catalyst, in terms of activity, stability and product specificity. The need to address this question has stimulated very fundamental studies of the processes happening in preparation, and the size, shape and location of within the support body of the metal particles,[47] to which matters the most refined (and costly) physical techniques have been applied. If progress is measured by the precision of the questions that we can now ask, we are indeed making rapid progress.

2.3.2. Methods of Preparation

This has been the subject of a number of recent surveys[48–57] (see also Further Reading section), and this fact combined with the need to be brief means that the following account will have to be in the nature of a framework that readers must fill in according to their needs, using the cited references, of which there is a generous number (see the 'Further Reading' sections at the end of the chapter).

For catalyst supports that are to be used in industrial processes, the principal considerations[58] are: (i) chemical stability, (ii) mechanical strength and stability, (iii) surface area and porosity, (iv) cost, (v) physical form and (vi) coöperation (if any) with the active phase. These apply equally to reactions in the gas phase, the liquid phase and to three phase systems, but some of them are of small importance in fundamental research (e.g., i, iii, and vi). Nevertheless because of the general desire (for obvious reasons) to perform basic work that has a detectable relevance to industrial problems, those supports that feature most commonly in large-scale operation also appear most often in academic laboratories.

For many purposes it is very desirable to use a *support*[59] having a high surface area, which we may take to mean greater than 100 m^2g^{-1}. This ensures that the metal particles can be well separated from each other, to minimise sintering, but it implies the presence of *microporosity*, i.e. of pores less than 2 nm in diameter. Areas greater than about 50 m^2g^{-1} cannot generally be obtained by lowering the size of non-porous particles. *Mesoporosity* (pore diameter 2–50 nm) is due to cracks between the primary particles, and *macroporosity* to voids between larger aggregates (pore size greater than 50 nm). Surface area and pore structure are assessed by the physisorption of nitrogen or other inert gas,[60–62] or by mercury porosimetry. The methodology is very well documented, so it is only necessary to consider the results.

Only two simple oxides have the ability to become microporous: alumina and silica. *Alumina* exists in a number of different phases, derivable by calcination from different oxyhydroxides, the nature of which depends upon the procedure for precipitation form solution of an aluminium salt.[63] The most commonly used forms of alumina are the γ (area $\sim$100 m^2 g^{-1}) and the low-area α (1–10 m^2 g^{-1}), formed by high-temperature calcination. The chemistry of *silica*[64] is simpler: amorphous silica can be made in a number of ways, with areas in the range 500–600 m^2g^{-1}. Several low-area forms occur naturally, or may be obtained by high-temperature calcination (quartz, cristobalite).

The introduction of about 13% alumina into the silica lattice generates new acidic centres, which have Lewis character in the absence of water but Brønsted character when water is around. The high surface area of silica is maintained, and amorphous materials of this type have been much used as catalysts in their own right, where carbocationic species are involved, or as supports for bifunctional petroleum reforming catalysts.

This leads us directly to *zeolites*, which are crystalline aluminosilicates: some occur naturally and many others have been made artificially, showing a large and bewildering variety of crystal habits. All have microporous channels permeating the crystals, with cavities of various sizes at regular intervals, where are found the cations, which balance the negative charge on the lattice, caused by the partial substitution of silicon by aluminium. Protonic (i.e. Brønsted) acidity also occurs when alkali metal cations are replaced by protons, so in summary we have materials having *molecular sieve* properties, with a controllable level of acidity or alkalinity.[65] They have found numerous applications (e.g., as drying agents and water softeners), and are catalysts in their own right as well as being supports for metals.[66–71]

Other substances exhibit the characterisations of zeolites. *Silicalite* is the zeolitic form of silica, and an hexagonal mesoporous silica (H_I-SiO_2) has been prepared and characterised.[65] Aluminium phosphate gives rise to a family of ALPO zeolites, but they lack useful acidity. Much interest has been shown in the incorporation of small amounts of other elements into aluminosilicate and ALPO structures in order either to increase acidity or to generate new catalytic capabilities (e.g. in oxidation). We may expect further significant developments in the design of microporous structures having utility as catalyst supports: the creation of wide-pore zeolites such as MCM-41[72] indicates the possibilities.[73] The modification of the properties of small metal particles by interaction with the zeolite framework and its ions will be discussed in Section 2.6.

This brief survey concludes with mention of other classes, support of particular relevance to the catalysis of hydrocarbon reactions. Basic supports such as magnesia, ferroelectrics such as $CaTiO_3$ and $BaTiO_3$, and hydrotalcites[74,75] have interesting and useful properties. *Carbon*[76−79] is available in a number of physical forms: amorphous carbon (activated charcoal) has a very high surface area (up to ca.1000 m^2g^{-1}) and is particularly suitable for use in liquid media. The preparation of carbon as 'nanofibres'[80,81] and as 'nanotubes'[82,83] has also been reported. Graphite is less useful because of its very low area, but C_{60} buckminsterfullerene has been used in laboratory work.[83−86] Organic polymers containing functional groups (silk, Nylon and other polyamides) have also been used in three-phase systems. Oxides other than silica and alumina, and their combinations, do not form microporous structures, although low levels of various elements have been introduced into zeolitic structures. *Titania* has been made as very small particles with an area of $\sim$180 m^2g^{-1};[87] it has been the focus of much basic research in recent years, for reasons that will appear later. It exists in two main polymorphic forms, viz., anatase and rutile, the former changing into the latter at high temperatures; the third form (brookite) is only rarely used. It is important to know how much of each form is present: pigmentary anatase ($\sim$10 m^2g^{-1}) contains additives (e.g. K^+ and PO_4^{3-}) and trace impurities (e.g. Cl^- or SO_4^{2-}) that may have disastrous consequences for a supported metal catalyst made from it. The commonly used Degussa P-25 ($\sim$50 m^2g^{-1}) is purer, but contains both anatase and rutile as separate particles in the ratio of about 3:1. Quite generally it is helpful to characterise a support as fully as possible, both chemically and physically, before using it to make a catalyst, if the process of preparation and the structure of the product is to be understood.[88]

The physical form of the support has to be chosen with a view to the type of reactor in which its use is intended. Silica and alumina are available as coarse granules or fine powders, and may be formed into various shapes with the aid of a binder (stearic acid, graphite): they can then be used in fixed bed reactors. For fluidised beds, or for use in liquid media, fine powders are required.[89] Ceramic monoliths having structures resembling a honeycomb are used where (as in vehicle exhaust treatment) very high space velocities have to be used, but they are made of a non-porous material (α-alumina, mullite) and have to have a thin wash-coat of high area alumina applied, so that the metal can be firmly affixed.

Methods of preparation fall into two classes: (i) the support and precursor to the metal are formed at the same time, or (ii) the metal precursor is applied to the pre-formed support. The first, less used category embraces *co-precipitation*, which works well, for example, for Ni/Al_2O_3, where $Ni(OH)_2$ and $Al(OH)_3$ can be coprecipitated from a mixed solution, but it is somewhat limited in scope. Similar precursors can be made by the *sol-gel* method[90−96] and the active metal ion protected against premature reduction by complexation. Mixed aqueous solutions can also be sprayed into a hot zone (873-1273 K) to form a useful precursor. The second, more widely used method has a number of manifestations, choice of which is conditioned by the kind of support, and the desired size and location of the metal particles at the end.[97] With *impregnation*, an aqueous or organic[98] solution containing the metal salt or complex[99−101] is drawn by osmosis into the pores of the support, which acts like a sponge. With *ion exchange*,[102] protons of acidic hydroxyl groups are exchanged for cationic precursors of the metal. In *deposition-precipitation*, a metal precursor such as a hydroxide is precipitated onto the support, which nucleates its formation. In each case, the solvent (which is usually but not necessarily water) is removed by heating and/or evacuation, after washing when appropriate. The dried material may be calcined to convert the precursor to its oxide[103] or it may be reduced directly with hydrogen, to form the metal. All of these operations, so briefly described, are capable of infinite variation and refinement, of which the following are simply examples. (i) The location of the metal within the pore system is controllable by the rate of drying or use of competitive adsorbates[47] in the case of impregnation. (ii) With ion exchange, a better dispersion of the precursor ion through the support particle is achieved if a competitive cation such as NH_4^+ is also present. (iii) Smaller metal particles are formed by ion exchange than by impregnation but the loading is limited by the number of surface hydroxyl groups. (iv) Direct reduction of the adsorbed precursor sometimes gives smaller metal particles than reduction after calcination, but if the precursor is a metal chloride it may leave chloride ions on the support,[104−106] whereas calcination will eliminate them. (v) The use of microwave radiation for heating supported catalyst precursors has produced interesting and unexpected results.[107] Metal particles so formed differ form those made conventionally with respect to selectivities shown in certain hydrocarbon reactions (see Chapter 14). Commonly used precursor salts are the chlorides of the noble metals, (H_2PtCl_6,[97] $RhCl_3$, $PdCl_2$, $HAuCl_4$) and nitrates of the base metals.

Much use has also been made of the zero-valent metal complexes as metal precursors, where it is thought advisable to avoid chloride or other possible harmful ions: the carbonyls of ruthenium, rhodium, osmium, and iridium (of which there are many) have been used either as vapour if that is possible, or more often as solutions in organic solvents.[68,100,108,109] Acetylacetonate (acac) complexes and π-allylic complexes have also been used. The term *chemical vapour deposition* (CVD) is used when the vapour of a volatile complex reacts with a support surface, and deposits something that may be decomposed to a metal; the method has been reviewed.[110] These complexes often react with partial decomposition with hydroxylated support surfaces, and the conversion to metal is completed by heating. The processes occurring have been scrutinised using spectroscopic techniques, and with care the original architecture of the complex can be preserved; thus for example the Ir_4 cluster has been studied in detail.[111] Where particle growth occurs, it is frequently limited so that very high dispersions can be obtained: the use of $Ru(acac)_3$ with alumina gives metal particles containing only 12–15 atoms.[112]

Although gold is still only of limited interest as a catalyst for hydrocarbon reactions,[113,114] it exhibits outstanding activity for the oxidation of carbon monoxide and similar processes, and for this purpose it has to be made as 2 to 4 nm particles supported on an oxide of the first row Transition Metals; this is done by co-precipitation or deposition-precipitation.[115]

Unless the metal is introduced as such, e.g. as a colloid or by metal-atom-vapour deposition[116,117] (see later), the final and critical step is inevitably a *reduction*, performed either *ex situ* or *in situ* (or both). Molecular hydrogen is most often used, although carbon monoxide has a thermodynamic advantage, which is useful for less easily reducible species because the carbon dioxide produced is less effective than water in reversing the process. Reduction of a base metal oxide can be effected by hydrogen atoms spilling over (see Section 3.34) from reduced noble metal particles.[118] More exotic reductants (e.g. Cr^{II} ions,[119] oxirane[120] and hydrogen atoms[121]) have been tried. Particle size[122] and surface morphology[123] are affected by reduction conditions (i.e. temperature ramping rate, time at maximum temperature, hydrogen pressure, flow-rate and purity etc). One very simple but useful technique to monitor the process is temperature-programmed reduction[124,125] (TPR; see Further Reading section), which can help to determine the temperature at which precursors become reduced, the number of different steps or species involved, and the stoichiometry of the process. Exposure of metal particles on ceramic oxides to hydrogen at high temperatures (>773 K) creates a strongly held form of hydrogen that is inimical to catalytic activity.[126,127] It can also lead to the partial formation of intermetallic compounds such as $PtAl_x$[128] and $PtSi_x$, and with reducible oxides to the Strong Metal-Support Interaction (see Sections 2.6 and 3.35). Reduction is probably the most important (and capricious) step in making a supported metal catalyst.

It is now possible to delve deeper into the chemical processes occurring during all stages of catalyst preparation by employing the whole panoply of techniques now available.[129–139] However it cannot be stressed too strongly that very great care is needed to control each step in the preparation if reproducible results are to be obtained. This warning seems to apply particularly to preparations made on a small scale ($\sim$5 g). Lack of such control means that it is very difficult to compare results from different laboratories, and the tendency for each group to make its own catalysts has materially hindered progress. In contrast the availability of a number of *standard supported metal catalysts* in the USA, in Europe (e.g. EUROPT-1, 6% Pt/SiO$_2$), in Japan, and in Russia, has proved beneficial.[10,140,141]

The term *model* catalyst[142-144] (see Further Reading section) is applied to a material prepared in UHV conditions by condensing metal atoms formed by evaporating from a heated source onto the flat surface of an oxide either as a single crystal (e.g. MgO) or as a cleavage plane of crystal (e.g. mica, quartz, diamond) or as a thin oxide layer formed on the surface of amother metal.[145] This simulates a supported metal catalyst in a way that facilitates examination by physical methods, especially electron microscopy (see Section 2.42), and thus allows direct study of matters such as effect of gas atmosphere on particle shape, and the mechanism and kinetics of particle growth by sintering, and dependence of catalytic rates on particle size. In a similar way, small particles of oxide can be created on the surface of a metal single crystal; this is a useful way of exploring the possible role of sites at the metal-support interface.

We turn now to the problem of preparing supported *bimetallic catalysts*[146-147] (see Further Reading section). Catalysts having two metallic components have found great use in industrial processing, especially in petroleum reforming, and this has lead to extensive academic research especially on PtRe/A1$_2$O$_3^{103,148,149}$ and PtSn/A1$_2$O$_3^{150,151}$ catalysts. Information on bulk alloys is of marginal relevance except when large particles exist on the support, when their structure can be determined by X-ray diffraction. In small bimetallic particles, the normal constraints on solubility no longer apply, and very useful bimetallic catalysts can be made using metals that have little tendency to interact in the bulk.[152] A further system that has attracted much attention is *ruthenium-copper*,[72,85,152,153] where copper segregates to the surface of the particle, preferentially occupying sites of low co-ordination number: the effect of ensemble size of the active metal can then be studied by changing the composition.

Two types of method are commonly used to make supported bimetallic catalysts: (i) impregnation, usually by simultaneous introduction of both precursor ions, although sequential addition with an intervening calcination is sometimes used; (ii) impregnation with solutions of organo-bimetallic complexes,[155] especially carbonyls in an organic solvent.[146] Choice of support is important in achieving conjunction of the compounds; silica is preferred to alumina because metal ions and atoms are more freely mobile on it,[156] and calcination before reduction is generally to be avoided, because, if separate oxide particles are formed and these make separate metal particles on reduction, a high temperature treatment will be needed to homogenise the system. Techniques for answering the vexed question[157-159] of whether your procedure has or has not succeeded in bringing the two metals into the desired degree of intimacy will be described later. (Section 2.4.2)

Esoteric methods of forming supported metals or their analogues include the use of aerogels and xerogels,[160] photolithography[161] and electron-beam lithography,[162] and a 'solvated metal-atom dispersion' method.[163]

Two final words of caution are necessary. (i) It is *essential* to analyse the finished catalyst chemically, because not all of the active components will have ended up where you would like – on the support. (ii) It is desirable to measure the total surface area and porosity of the finished catalyst, because they may have been changed by the preparation. This is particularly necessary when silica is the support, because it readily undergoes hydrothermal sintering during drying, calcination and reduction, leading to the sealing off of internal pores.

2.4. MEASUREMENT OF THE SIZE AND SHAPE OF SMALL METAL PARTICLES[1,2,6,46,73,164–169]

2.4.1. Introduction: Sites, Models, and Size Distributions

Understanding the fundamental connections between the chemical composition and physical structure of a catalyst and its ability to perform chemical reactions is the central problem of catalysts: it has exercised the minds of scientists over a long period of years. In the field of our immediate concern, exploration of the links between them requires amongst other things the fullest possible knowledge of the sizes and shapes of the metal particles responsible. Each of the many available physical techniques addresses a somewhat different aspect of the problem. Sometimes good luck or skill ensures that several different methods give convergent answers; at other times, conflicting answers are obtained, but this can be just as informative if the reason for it is discovered.

Selection of the techniques for investigation is conditioned by the information required. There is a general need to be able to express the catalytic rate under defined experimental conditions in as meaningful a way as possible. We may start with the rate per unit mass of catalyst; then, with knowledge of the analysis, per unit mass of metal. An estimate of the dispersion allows the specific or areal rate to be stated, i.e. the rate per unit area of metal. The dispersion may have been derived from some technique giving the average particle size or the size distribution: assuming we know how the atoms are packed on the surface, we can guess the total number of exposed atoms, a figure which is also obtainable with some further assumptions from a chemical titration (Section 2.43). We can then quote a rate per exposed atom, which is termed the *turnover frequency*. The number of active sites may however be less than the number of surface atoms, if each site is as ensemble of more than one atom: if each site comprises N_B atoms, (the *Balandin number*), then the *Taylor fraction*, which is the fraction of sites in a given number of surface atoms, equals N_B^{-1} (see also Section 5.4). Estimates of metal area by physical techniques may exceed those by chemical titration if the surface is partly obscured by some strongly held but unreactive species. In such cases the lower values are the more reliable.

A great deal of enjoyment, and some enlightenment, is to be had by constructing models of small metal particles with plastic spheres (e.g. table tennis balls[170]) or by simulating them with computer based molecular graphics or even simply by performing calculations.[171–173] These exercises are usually conducted using the close-packed fcc structure,[5,9,168,174] although the cph and bcc structures have also been studied. Using perfect geometric forms, such as cubes, octahedra, tetrahedra or cubo-octahedra (Figure 2.3A), the surfaces of which contain atoms having only a few different co-ordination numbers (CN), it is at once clear that the fraction of surface atoms having low CN decreases quickly, while that of atoms having high CN rises, with increasing particle size.[5–7,9,175] These changes occur within what Poltorak called[171] the *mitohedrical region*, although the term has not caught on. For complete regular fcc octahedra having m atom along each side, the fractions of atoms having CN four (apex), seven (edge) and nine (plane) are shown in Figure 2.3B. These model studies also allow attention to be given to multi-atom sites, which may have importance as active sites in catalytic reactions; the B_5 site (of which there are two kinds) occurs on the incomplete surfaces of fcc particles[174] (Figure 2.3C), and has been implicated in certain unusual chemisorptions and reactions.[8] All possible low CN fcc surface atoms have been identified, codified and illustrated.[175–178]

Work of this kind is easily (and cheaply) performed, but it has serious limitations: (1) as noted above (Section 2.1.1), real catalysts will have a distribution of particle sizes, and not contain simply particles of all the same size; (2) real particles will only rarely have a complete outer shell, because this requires a very specific number of atoms. Almost always the surfaces will be rough, and contain atoms of various CN. While this situation can be modelled to a limited extent, the use of random procedures to add atoms to a perfect shape shows that the number of atoms of a given CN changes irrationally and irreproducibly. Only a time-average over all possible configurations of surface atoms has any significance, but the *average* CN of $s$urface atoms (defined as all whose CN is less than the bulk value), for arrangements that maximise contacts, changes more smoothly with total number of atoms. In the case of octahedral and tetrahedral, exposing only (111) facets, for example, this will tend to a value of nine for large sizes, and to a mean over all atoms of 12 (Figure 2.4). Considerations based on co-ordination numbers also avoid the problem, mentioned above (Section 2.11), of deciding what atoms are on the surface and what are not. For this reason a *'free–valence'* dispersion D_{fv} defined for particles having fcc structure as

$$D_{fv} = \Sigma(12 - CN)/12\, N_T \qquad (2.1)$$

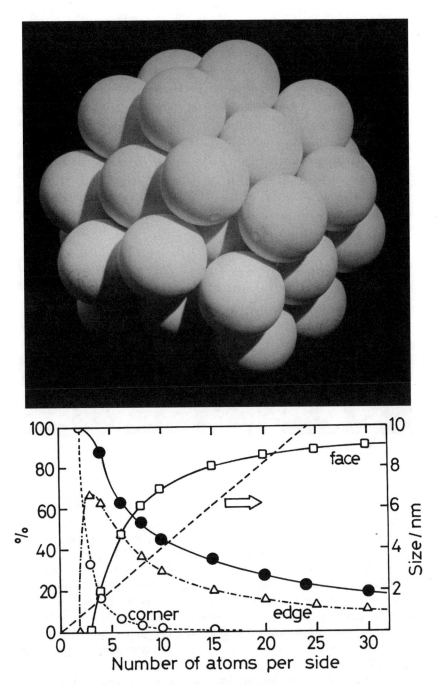

Figure 2.3. (A) Model of a small cubo-octahedron; (B) fractions of surface atoms on perfect fcc octahedra having CN four (corner, circle), seven (edge, triangle) and nine (face, square), together with the corresponding dispersions (N_S/N_T, filled circle) and sizes (—);[5] (C) model of large cubo-octahedron having an incomplete outer layer.

Figure 2.3. (Cont.)

where N_T is the total number of atoms (Figure 2.5), would perhaps be more significant than that based on an arbitrary definition of 'surface atom'. (3) A further reason caution in using surface models is that small particles will be in a state of constant agitation, with surface atoms in particular in rapid motion (Sections 1.2.2 and 2.5). Models create an unwarranted impression of rigidity and indeed the fluid nature of surfaces of small crystals calls into question the whole concept of geometric factors in catalysis. It also appears that quite different structures can have similar stabilities, and oscillation between, for example, cubic and pentagonal forms (e.g. cubo-octahedra and icosahedra) can happen easily (see also Section 2.5): indeed the order of stability may depend upon the strength of the metal support interaction (Section 2.6).

The purpose of studying the physical parameters of supported metal particles is not only to refer the measured rate to unit area of active surface, but also to see whether the rate so expressed (or TOF, see Section 5.2.3) is itself a function

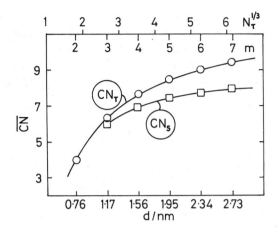

Figure 2.4. Mean CN of surface (CN_s) and all atoms (CN_T) as a function of size as given by (i) $N_T^{1/3}$ where N_T = total number of atoms, and (ii) number of atoms per side m, for perfect octahedral particles.

of particle size; the resulting enquiries as to whether there is, in any particular case, a *particle-size effect* have occupied the minds of numerous researchers,[179] and the conclusions reached will require attention in all chapters from Chapter 6 onwards. The inference of 'particle size' from a physical measurement is not straightforward. To start with, theoretical treatments assume that the inevitable size distribution is *mononodal* (that is, it has a single maximum), but this is not always the case. A mononodal distribution will be *skewed* and not symmetrical, because there is no upper limit to size, so the *mean* size is taken as the central tendency.[180] There are several ways of defining the mean, depending on the type of measurement

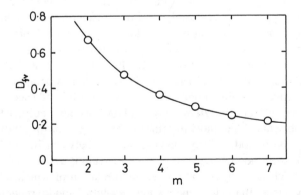

Figure 2.5. 'Free-valence' dispersion as a function of size ($N_T^{1/3}$) for perfect octahedral particles.

made;[164,165] gas chemisorption (Section 2.4.3), for example, measures surface area, and this provides a *volume-area* (or *volume-surface*[181]) *mean diameter* d_{vs}. Initially this may be taken as the reciprocal of the dispersion D, but the proportionality constant contains the average area of each surface atom, which as we have seen is hard to define. Theoretical considerations based on fractal analysis have led to the relation

$$\ln d_{rel} = \ln k_c - (3 - D_c)^{-1} \ln D \tag{2.2}$$

where d_{rel} is d_{vs}/d_{at} (d_{at} = atomic diameter) and D_c is the 'chemisorption dimension'. A plot of $\ln d_{rel}$ vs. $\ln D$ for various approximately spherical crystal forms is satisfactorily linear with $D_c = 2$ up to $d_{rel} = 0.2$, but for higher dispersions (up to 0.92) the size of the atom relative to that of the particle becomes significant, and the slope of the plot becomes more negative ($D_c = 2.19$). Above 0.92, the concept of size loses much of its meaning, but nevertheless the logarithmic plot can be used as a universal curve for obtaining an estimate of mean size from a measurement of dispersion, irrespective of the exact crystal form. The study by Borodziński and Bonarowska[181] is strongly recommended for a penetrating discussion of this subject; other arguments also lead to the expectation that the logarithmic plot should not be linear over the entire range.[182] The task of divining the number of *multiatomic* sites that may constitute active centres is even more demanding,[182,183] and for very small particles is almost impossible.

2.4.2. Physical Methods for Characterising Small Metal Particles[164,168,169,184]

Half a century ago there was simple no way of knowing the size of metal particles in a supported metal catalyst. X-ray diffraction had been used in the 1940s to obtain information on particles larger than about 5 nm, and the BET method could be used on metal powders. Questions concerning size effects were therefore rarely considered. Two developments in the 1950s and 1960s changed that: *Transmission Electron Microscopy* (TEM) began to be applied,[185] and the use of the hydrogen chemisorption titration was developed[186] (Section 3.2).

Transmission Electron Microscopy[187] (TEM; see also Further Reading section) is probably the single most informative technique for the study of catalytic materials. In this method, electrons having energies between 20 and 1000 keV are fired at a thin specimen mounted in ultrahigh vacuum; transmitted electrons are converted into visual images, the proper interpretation of which calls for much skill and experience. In the low-energy, low-resolution mode, *scanning transmission EM* (STEM) simply reveals gross morphology, e.g., the shapes and sizes of support particles. At higher energy and better resolution, small metal particles become visible, and, at the highest electron energy, *high-resolution TEM* (HRTEM) affords rewarding detail of the structure of metal particles,[188,189] with resolution of lattice planes and detection of surface contamination (Section 2.5). Work of this nature requires the instrument to be firmly located in a vibration free situation, and is now routinely carried out. Various methods are available for making thin layers of supported metal catalysts.

The technique, which is capable of much refinement, can supply other information: examination of the diffraction pattern identifies the phase, space group and unit cell dimension;[190,191] scattered X-rays observed in energy dispersive spectroscopy (EDS) provide information on chemical composition, this enabling concentration gradients of metal within support particles to be seen; and the electroenergy loss spectrum reveals the electronic structure and atomic environment of the material. The most common use is however just to obtain a particle size distribution; the form of the distribution is found if a sufficient number of particles (typically 1000) is measured. This tedious procedure can be automated, and the results may reveal a binodal or even trinodal distribution, which other techniques might miss. It is important to view images taken at different points in the sample in order to get a representative answer; lack of uniformity in the density of particles at different places is often seen, and may indicate a failure in the preparative method. Observation of metal particles on the surface and within the cavities of zeolites helps to explain catalytic behaviour. Somewhat large particles can be formed inside the zeolite particle by disruption of the lattice.[67]

When X-ray quanta hit a crystalline material, at certain angles of incidence θ those that are reflected reinforce one another, while at other angles (due to their wave character) they interfere and cancel: this creates the phenomenon of *X-ray diffraction*,[73,168,192,193] described by the equation

$$n\lambda = 2d \sin \theta \qquad (2.3)$$

where λ is the wavelength of the X-rays, d the distance between adjacent layers of atoms and n is the order of the reflection. This leads readily to identification of the phase causing the diffraction. If however the size of the particle responsible is less than about 100 nm, the diffraction line is broadened, and for particles between about 5 and 50 nm this effect can be used to give a volume-averaged particle diameter d with the help of the Scherrer equation:

$$d = 0.9\lambda/\beta\cos \theta \qquad (2.4)$$

where β is the peak width at half-height in radians. This approximate relation has to be corrected for instrumental broadening and the non-monochromaticity of the X-ray beam. Analysis of the diffraction peak profile can yield a particle size distribution.

X-ray diffraction (see Further Reading section) may now be performed in controlled gas atmospheres and at elevated temperatures (up to 1270 K) permitting the identification of phases after chemisorption or during catalysis: in this way, for example, the conversion of palladium to the β-hydride phase during hydrogenation reactions has been followed. X-rays may be scattered by the surface as well as reflected, and their analysis is the basis of *Small Angle X-ray Scattering*[7,194,195] (SAXS), which can also give size estimates, provided any pores of the support are the first eliminated by filling or compression. The use of *anomalous SAXS*[196–199] (ASAXS) has also been described. Interaction of X-rays with amorphous materials leads to scattering at wide angles, from which structural information is extracted by recording intensity as a function of angle: this is *Wide Angle X-ray Scattering* (WAXS), and use of an equation due to Debye (i.e., Debye function analysis[200]) gives a radial distribution function (RDF). This method has been applied to metal particles in zeolites, revealing changes in their structure with conditions of treatment. Debye function analysis has also led to determination of the structure of platinum particles in the standard Pt/SiO_2 catalyst EUROPT-1.[201,202] *Anomalous XRD*[190] is informative, but not yet widely practised; diffuse anomalous X-ray scattering[203] and wide-angle anomalous scattering[199] are claimed to be superior to EXAFS (see below) for studying small metal particles.

The majority of X-rays impinging on a solid surface result in the formation of a photoelectron, and a plot of the dependence of absorption coefficient against X-ray photon energy shows an extended fine structure above each absorption edge, K-, L-, M-, etc. The K- shell edge is due to the onset of ejection of $1s$ electrons, and the three L edges to the start of ejection of $2s$ (L_I) and $2p_{1/2}$ (L_{II}) and $2p_{3/2}$ (L_{III}). The K edges of most of the $3d$ and $4d$ metals are accessible as are the L_{III} of the $5d$ metals. In a solid the emerging photoelectron is scattered by interference with adjacent atoms or ions, and constructive interference between the outgoing and back–scattered waves creates a diffraction effect, which shows itself as the fine structure above the absorption edge. The technique is therefore known as *Extended X-ray Absorption Fine Structure* (EXAFS; see Further Reading section). The method provides information on the atoms or ions surrounding the source of the electron, but extraction of quantitative information from the fine structure is a complex process, and the reader wishing to do this will need to consult specialised texts; numerous summaries are also available[63,73,165,193,204,207] (see also Further Reading section).

The outgoing wave is spherical and is scattered by all species adjacent to the absorbing atom; unlike X-ray diffraction, long-range order is not a prerequisite, so amorphous samples can be studied. The method does not directly identify the nature of the neighbours, which can only be inferred from the interatomic distances.

TABLE 2.1. References to EXAFS Studies of the Metal-Support Interface

| Metal | Support | | | | |
	Al_2O_3	SiO_2	TiO_2	MgO	Zeolites*
Pt	307	288, 431	—	431	419, 308, 420, 316, 288, 366, 393
Rh					
Ru	439, 112	—	—	—	
Ir	—	—	—	433	
Pd	—	288	—	—	

* Neutral and acidic

The number of neighbours of a given type (i.e. the CN of the absorber) contributes to the intensity of the scattered wave at a particular energy, and this enables the structure, and for the small enough particles also the size, to be determined. The method has been applied to many pure supported metals (Table 2.1), as well as to bimetallic and promoted or modified catalysts(see Further Reading section). Its limitation lies in its sensing all atoms of a given kind in the sample, and it only shows a degree of surface sensitivity when the particle size is so small that surface atoms predominate. Pitfalls encountered in extracting particle size estimates from EXAFS results have been discussed.[208] Interpretation is greatly helped comparison with compounds of known structure, and the combination of EXAFS with XRD is particularly powerful.[72] The important information arising from the analysis of peaks in the Fourier transform of EXAFS spectra at very short distances[209] (AXAFS) is mentioned in a later section.

Further significant developments of the EXAFS technique are envisaged.[210] Energy-dispersive X-ray absorption fine structure (DXAFS) allows collection of a spectrum in much less than 1 s, and its extension to the μs or even the ns time scale is not impossible. Polarisation-dependent total-reflection fluorescence X-ray absorption fine structure (PTRF-XAFS) permits high spatial resolution and has been applied to the copper trimer Cu_3 on a Ti(110) surface. The use of 'time- and space-resolved XAFS observation for studies of dynamic aspects of the local structure at catalyst surfaces under working conditions' is forseen[210]. Note that the acronym XAFS is used, rather than EXAFS.

Ejected electrons of low kinetic energy (<30 eV) interact with the valence electrons of other atoms and often show complex scattering paths. This leads to an additional peak near the absorption edge; it is known as a *white line*, and the *X-ray Absorption Near-Edge Structure* (NEXAFS or XANES) is caused by transitions of (for example) $2p$ electrons to unoccupied $5d$ levels.[65,211] The white line intensity can therefore sense the occupancy of d-levels in certain transition metals (particu- larly platinum; see Further Reading section), so no line appears with the Group 11 metals, which have filled d-shells. Changes in its intensity have sometimes been connected with the presence of chemisorbed hydrogen atoms,[212–214] and sometimes not.[215] Its possible extension to recognising the *positions* of chemisorbed hydrogen atoms will be noted in the next chapter.

A further manifestation of the power in X-rays in the analysis of solids is *X-ray Photoelectron Spectroscopy*[73,193,216,217] (XPS; see Further Reading section). Originally used purely to identify elements present in a sample, as its first name (Electron Spectroscopy for Chemical Analysis, ESCA) indicates, it identifies the binding energies of various electron levels for each element after correction for sample charging. For many elements the technique is almost surface specific, since the escape depth of electrons for many transitions is only a few nm, and the ability to etch the sample by ion bombardment enables depth profiling to be performed. The binding energy is somewhat sensitive to the oxidation state of the element, and peak shape analysis may reveal the presence of more than one state. It also appears to be sensitive to particle size when this is small enough, but the interpretation of the observed effect, which is small, has been disputed,[218–220] and it has not become a routine method for particle size estimation. Auger transitions are readily observed particularly for the lighter elements, and can give helpful information.

Mössbauer spectroscopy[193,221] is based on the observation that nuclei held rigidly in a lattice can undergo recoil-free emission and absorption of X-radiation; the separation of nuclear energy levels can be measured with great accuracy, and it is possible to detect weak interactions between a nucleus and its electronic environment. This may reveal the chemical state of the atom or ion, but only a few nuclei are susceptible to the effect, most work having been done with iron (^{57}Fe) and tin (^{119}Sn) and a little with ruthenium (^{99}Ru).[222,223]

Atomic nuclei having an odd number of protons and neutrons have a non-zero nuclear spin I, and thus a magnetic moment μ which equals $\gamma \hbar I$, where γ is the gyromagnetic ratio, which is a quantity specific to each nucleus. When such a nucleus is placed in an external magnetic field B_0 of strength typically 2–14 T, the Zeeman interaction leads to quantised orientation of the nuclear magnetic moments, and the nucleus adopts $(2I + 1)$ magnetic eigenstates having energies E_m given by:

$$E_m = -m \, \gamma \, \hbar B_o \tag{2.5}$$

where n takes values of I, $I-1 \ldots -1$. Transitions between adjacent states ($\Delta m = \pm 1$) can be effected by electromagnetic radiation having the Larmor frequency V_0, where

$$V_0 = \gamma B_0 / 2\pi \tag{2.6}$$

These are in the radio frequency range of 1–600 MHz. The usefulness of this *nuclear magnetic resonance* (NMR; see Further Reading section), which is central to determining the structures of organic molecules including chemisorbed species (see Chapter 4), hinges on the phenomenon of the chemical shift σ, which is due to the perturbation of the Larmor frequency by the influence on the specified nucleus of the diamagnetic field about it. For a variety of reasons the number of metals of interest in catalysis and having nuclei suitable for study by NMR is small: they include hydrogen, carbon, silicon and aluminium, the last two being much used in work on zeolites, but of the metals only platinum (^{195}Pt) is really relevant to our interests. Little has been done with other possible nuclei (^{103}Rh, ^{109}Ag, ^{61}Ni, ^{63}Cu), but effective use has been made of proton NMR in the study of chemisorbed hydrogen (see Chapter 3). The theory and scope of the method has been reviewed on a number of occasions (see Further Reading section).

The spectral lines in the NMR of solids are very broad, due to static anisotropic interactions to which nuclei are subjected, whereas molecular motion in liquids averages these out, and narrow lines result. The same effect is produced in solids by rotating the sample at high speed at an angle θ to the direction of B_o of 54° 44′ since the term $(3 \cos^2\theta - 1)$, which appears in equations for dipolar interaction, chemical shift anisotropy and quadrupolar interactions of first order, is zero for this angle. This value of θ is termed the 'magic angle' and the technique is therefore named *Magic Angle Spinning NMR* (MASNMR).

In the case of metals, an additional effect on the Larmor frequency arises because of the polarisation of conduction electrons by the magnetic field of the nucleus; this creates the *Knight shift*, which can be much greater than the chemical shift.[224] The spin-lattice relaxation rate T_1^{-1} and the associated Korringa constant ($T_1 T$, where T is the absolute temperature), also provide useful information. Marked changes in NMR line shape of platinum in Pt/Al$_2$O$_3$ catalysts with dispersion have been observed[225–228](see also Further Reading setion). The use of spin-echo double resonance[169,229] (SEDOR) gives a high degree of surface specificity, and the combination of ^{195}Pt on the surface of a particle with a nucleus such as ^{13}C in a chemisorbed species allows determination of its structure (See Chapter 4). Chemical shifts on ^{129}Xe NMR caused by interaction with ^{1}H nuclei has led to estimation of the size of metal particles in zeolite cavities,[230] and even on conventional supports.[231,232]

The intensity of magnetisation M of a substance is proportional to the magnetic field strength H:[7,164,168]

$$M = \chi H \tag{2.7}$$

where χ is the magnetic susceptibility. If this is positive, the material is paramagnetic; if negative, it is diamagnetic. Many metals have weak temperature-independent paramagnetism, but iron, cobalt and nickel (and their alloys) are ferromagnetic, that is, they can be permanently magnetised below the Curie temperature. Small particles (<20 nm) of a ferromagnetic material consist of single magnetic domains with a magnetic moment μ proportional to the volume. In the absence of a magnetic field, these moments are randomly oriented, but they align themselves in the direction of an applied field, the largest particles responding most quickly. The saturation moment M_s is reached when all particles are so oriented: the rate of increase of M/M_s with H is thus characteristic of a specific particle size, and the experimentally measured curve can be manipulated to give a particle-size distribution.[168] Experimental procedures for examining the magnetic properties of small metal particles have been described.[164,168]

Cyclic voltammetry applied to Pt/graphite allows an estimate of the relative amounts of the exposed low-index planes of the metal particles through comparison with results found with single-crystal surfaces;[233] the method is of course limited to conducting supports. *Time-differential perturbed angular correlation*[234] (TDPAC) is a nuclear spectroscopy that permits characterisation of materials on an atomic scale through hyperfine interactions due to interactions between the nuclear electrical quadruole moment of a suitable radioactive probe isotope and the electric field gradients originating in its neighbourhood: characteristic parameters obtained through detecting emitted γ-rays originating in the nuclear cascade of the isotope are (i) the nuclear quadrupole frequency interaction and (ii) asymmetry parameters. The method has been used to examine the structure of the PtIn/Nb_2O_5 system, using the ^{111}In isotope.

A number of other techniques have been used in the study of metal catalysts, but some of them only rarely by reason of their difficulty, cost or other limitation.[193] Scanning-tunnelling and atomic-force microscoies (STM/AFM), which permit the visual representation of small particles are however quite widely available.[235–240] Other methods meriting note include Rutherford back-scattering (RBS),[220,236] inelastic neutron scattering (INS),[77] electron holography,[241,242] electron paramagnetic resonance (EPR or ESR),[243,244] secondary-ion mass-spectroscopy (SIMS),[245] and plasma atomic emission spectroscopy.[246]

The advantages of using two or more methods on the same material are almost self-evident,[72,153] as each method fails on occasion or has constraints that are not immediately apparent. When different methods have been compared there has often been a pleasing consistency between the results; major discrepancies sometime find a logical explanation.

2.4.3. Measurement of Dispersion by Selective Gas-Chemisorption[7,164,169,180,247]

Unlike the physical methods of characterisation outlined above, the theory underlying this procedure is very straightforward and the equipment needed is relatively cheap,[247–250] but the practice is surrounded by pitfalls for the unwary and the interpretation is fraught with difficulties. Starting with a clean surface, one measures the number of molecules needed to form a monolayer just on the metal, and from this number, which is taken to be equal or proportional to the number of surface metal atoms, and a knowledge of the total amount of metal present from chemical analysis, a figure for the dispersion is at once obtained; and hence also, knowing the size of the metal atom, the surface area and the mean size (if the particle shape is assumed). This seemingly simple procedure is however capable of much elaboration.[251,252] This section offers a short review of the technique; a more detailed coverage of methods involving hydrogen will be found in Section 3.3.

The first problem is to clean the surface of the metal particles. They will usually start with a layer of chemisorbed oxygen on them, and this cannot be removed merely by heating and evacuation, because the metal-oxygen bond is too strong. The best course is to treat with hydrogen, which reduces the oxygen to water and leaves chemisorbed hydrogen atoms, which are more easily persuaded off. High temperatures are not usually needed, and indeed are to be avoided with reducible supports such as titania. Evacuation of gases from the pores of a microporous solid can however be tedious, and getting a good dynamic vacuum is no guarantee that it will hold when pumping ceases. Other contaminants (carbon, sulfur) less easily removed even than oxygen may also be present, and should be eliminated: chloride ion remaining from a preparation using a chloride salt may hold fast to metal and especially to support and can only be eliminated by prior washing or steam treatment *in situ*.

Selection of the adsorbate molecule requires care. Hydrogen[247] and carbon monoxide are by far the most often used, but both have some disadvantages. Hydrogen taken up is not always used only to form atoms on the surface metal atoms; it may 'spill over' onto or into the support, or it may break metal-oxygen bonds at the metal-support interface (Section 3.3), or it may form weakly-held species above the monolayer point, or in the case of palladium may dissolve into the metal (Section 3.1). Most of these difficulties can be avoided by suitable choice of experimental conditions, but it is by no means unusual to find that the number of atoms taken up exceeds the *total* number of metal atoms present (i.e. $H/M_{tot} > 1$).[253–256] Use of carbon monoxide avoids most of these difficulties, but it may chemisorb either in a linear or bridged form, and may even on some metals dissociate entirely; parallel use of IR spectroscopy will reveal what has happened. Other molecules are rarely used: oxygen atoms are prone to dissolve into metals except at low temperatures,[257,258] although nitrous oxide is successfully used as a source of oxygen atoms to chemisorb on copper and ruthenium, its use for other metals has not been explored.

There are various ways in which the monolayer volume can be measured. In the static method, successive small doses of the adsorbate are admitted, and the number of adsorbed molecules after equilibrium has been reached are deduced either gravimetrically (possible with carbon monoxide, difficult with hydrogen) or volumetrically from the residual pressure or by some other technique such as NMR or XANES.[211] The procedure is repeated until no further uptake occurs, when the monolayer capacity will be known. If chemisorption on the metal is strong and specific to the metal, the adsorption isotherm will show an initial rapid increase of uptake with pressure, followed by a linear region of low slope. The monolayer capacity is often reported (when anything at all is said) as the intercept at zero pressure obtained by extrapolation of the linear part. Alternatively, and perhaps better, the results are plotted according to the one of the linearised forms of the Langmuir equation from which the monolayer volume can be obtained.

Metal dispersion is also measurable in a dynamic system, in which the surface is initially cleansed after reduction by flowing pure helium or other inert gas. Injection of small doses of adsorbate is started and the non-retained part measured (e.g., by a thermal conductivity or other detector); this is continued until no more adsorbate is retained, and the aggregate of the adsorbed gas is found. This may be taken as a measure of monolayer capacity, but it is a function of the time gap between injections: increasing the time allows more weakly adsorbed species to desorb, and by varying the interval information is obtained on their amount. Although this procedure can be used with hydrogen, it is difficult to exclude oxygen entirely, and spuriously high retention may result from the reaction to form water. A safer method is to inject carbon monoxide into a stream of hydrogen, which does not compete with its chemisorption, and maintains a fully reduced surface.

It is claimed that measurement of the saturation amounts of hydrogen, oxygen and carbon monoxide can quantify the relative contributions of the three low index planes exposed by palladium particles: this however assumes that all adsorbing atoms belong to one or other of these planes.[141,259,260]

2.5. PROPERTIES OF SMALL METAL PARTICLES[2,6,164]

2.5.1. Variation of Physical Properties with Size: Introduction

Since chemisorption and catalysis are surface phenomena, their importance and usefulness increase with dispersion; great interest has therefore be shown in the physical and chemical properties of small metal particles (see Further Reading section). Since the limit of dispersion is the single atom, there must come a point at which, as dispersion is increased, metallic character is lost (Section 2.5.4). It may of course disappear slowly rather than suddenly, like the Cheshire cat, and it is necessary to explore several parameters that might characterise the metallic state, since not all may very with size in the same way. For example metallic behaviour may be manifested by Pauli-like paramagnetic susceptibility and NMR Knight shift even when the separation of energy levels is greater than kT, and there is no periodic-in-space Bloch wave.[224] Although single metal atoms (and ions) can undoubtedly act as catalytic centres, workers in the field of catalysis have often wondered how important metallic character (however defined) is to catalytic activity, and whether there comes a point at which further efforts to improve dispersion become self-defeating. They have therefore made attempts to relate activity (best expressed per unit area or per surface atom, Section 2.4.1) to particle size.

This work, conducted on a variety of catalytic systems, has revealed clear trends, which have enabled systems to be broadly classified as either size-dependent or size-independent. Quantification of these trends has however proved difficult for two reasons: (1) it is almost impossible to produce monodisperse metal particles on a support, so the best that can be done is to relate activity to *average* size; (2) those means that are deployed to change the size (e.g. altering metal concentration, thermal treatment) may introduce other factors that will affect activity (e.g. concentration of impurities, surface structure); and (3) the reaction may not proceed on all or indeed any of the metal surface, because specific sites at the metal-support interface of the support itself, by 'spillover catalysis', may be the site of catalytic activity.[56,261] Notwithstanding the improbability of small particles remaining rigid under catalytic conditions (Section 2.4.1) there have been many proposals to attribute activity to specific types of atom (e.g. edge, corner) or of site (e.g. the B_5 site). Because of the ease with which structural models can be created and analysed, the *geometric factor in catalysis* has featured more prominently in the literature than the *energetic factor*, which is harder to evaluate.

There is one further important general point to make. It is almost impossible to investigate the size dependence of any physical or chemical parameter in the absence of support, although certain properties of small vapour-phase 'clusters' are susceptible to study.[262] If metal particles are formed by metal evaporation or ion implantation onto an oxide support, we may suppose that the effect of the support on the metal particle is minimal. Particles may grow epitaxially on single-crystal oxide surfaces, but this method is as near as we can come to looking at intrinsic size effects. Certainly, as we shall see shortly, when metal particles are created by chemical means on high area supports their properties and behaviour may be considerably, indeed sometimes profoundly, affected by interaction with the support. Although such interactions may be of small importance with supports that are neutral (e.g. carbon) or are of low acidity (e.g. silica), the old idea that the support was simply an inert vehicle for the metal has long since been rejected as having universal validity. Understanding the extent and nature of the metal-support interaction is essential for rationalising the catalysis of hydrocarbon reactions by metals. The essence of the problem is this: supported metal particles may be expected to show intrinsic size effects, but superimposed on them are support interaction effects which themselves may be size-dependent, i.e. small particles will be more liable than large ones to experience the consequences of propinquity to the support.

Although it is widely accepted that it is difficult to prepare metal particles in a narrow size range, and while several techniques are known for determining size distribution, it is often *assumed* that sizes are distributed about a single mean value.

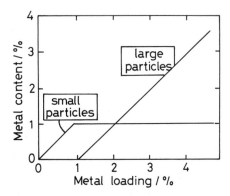

Figure 2.6. Amounts of small and large particle forms as a function of total metal loading.

There is nevertheless evidence[261,263] from various sources that certain modes of preparation of some catalyst systems may afford a *binodal* distribution of sizes, i.e. both large particles easily sensed by TEM and other techniques, and very small particles or even single atoms that may well escape detection by conventional means, but which may none the less make a major contribution to chemisorption and catalysis. What you see is not necessarily all you have. The proportions of each typically vary with metal loading as in Figure 2.6. With Ir/TiO$_2$,[264] temperature-programmed reduction has shown two distinct reduction events, the first of which corresponded to small particles formed from precursor ions chemically bonded to a limited number of sites on the support, and which reached a limit at about 1% metal, while the second, related to large particles formed from non-adsorbed ions, increased progressively thereafter. This complication which would not have been revealed by measurement of hydrogen chemisorption or other methods that integrate over the whole sample, may occur quite often, as each type of particle may have its own characteristic set of properties the meaning of size effects may become ambiguous.

All the manifestations of the size-dependent physical properties of very small metal particles arise from the self-evident fact that a substantial fraction of the atoms are on the surface, and being there they differ from atoms inside simply because they have fewer neighbours and more unused valencies. This difference was quantified by defining a free-valence dispersion (Section 2.4.1), which depends upon the number of atoms in the particle in a similar way to that predicted by the equation

$$G(N) = C_0 + C_{-1}/N^{1/3} \tag{2.8}$$

where $G(N)$ is the value of some physical property shown by a particle of N atoms, C_0 and C_{-1} being constants appropriate to the system. The decrease in the mean number of bonds formed by surface atoms to their neighbours below as N diminishes also leads to greater surface mobility and flexibility of surface structures.

2.5.2. Structure

The principal points of interest here are (i) crystal structure and (ii) interatomic distances. Important considerations for the former are the mechanisms of nucleation and growth (i.e. whether these occur in the vapour phase or on the surface, the atmosphere (if any) in which particles are formed and examined, and the energy of the radiation used for their study.

Particles formed by condensation of metal atoms onto a crystalline support have been observed to have pentagonal symmetry (see Further Reading section), i.e. to exist as either icosahedra or decahedra,[189,265–269] under the conditions indicated in Figure 2.7.[270] Theoretical calculations have supported the conclusion that for small particles these may be stabler than any of the forms expected for metals that normally enjoy cubic or hexagonal forms (e.g. octahedron, tetrahedron, cubooctahedron, square pyramid etc): such confirmation is however a *vatecinium post eventum*. Burton[269] has shown how a particle of five-fold symmetry might begin, although the process ignores the basic rules of stereochemistry: it could develop zones of 'normal' structure, exposing faces of 'normal' type, joined to each other so as to constitute a multiply-twinned crystallite. Observations in the electron microscope showed that small particles are flexible and can change their structure under the heating influence of the electron beam;[270] a kind of Uncertainty Principle

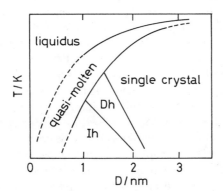

Figure 2.7. Phase diagram indicating regions of existence of liquid, quasimolten, multiply-twinned icosahedra (Ih) and dodecahedra (Dh); D = particle diameter.

is clearly at work, where the act of observation distorts what is being examined. The harder you look, the less you see. The stabilities of the various forms are therefore probably not greatly different; exposure to different gas atmospheres can also affect particle morphology, because the adsorbate may change the relative surface free energies of crystal planes, and the heat released by chemisorption can trigger change from a metastable to a stable structure.[6,271] The equilibrium shape of a free-floating single crystal is expressed by the Wulff construction;[6] it defines the proportions of different crystallographic planes that minimise the total energy. It is not clear however how such a crystal can be formed and studied in practice.

Anomalous structures (e.g. bcc gold, fcc lithium) have sometimes been found; these seem to occur most frequently with metals of low sublimation enthalpy, and less often with palladium and platinum.[6] Their formation may be linked to an epitaxial effect of the support on which they are formed and grow. Clearly developed crystal planes were only shown by particles larger than about 2 nm;[70] Mössbauer spectroscopy showed a platinum particle with 309 atoms to have the normal fcc structure,[222] but palladium-platinum particles suffered electron-beam-induced change from fcc to cph.[272]

In parallel with the usually observed contraction in interatomic spacing at the surface of single crystals (Section 1.2.2), it is commonly found that interatomic distances are less in small metal particles than in bulk metals.[6,232,273–275] Very accurate estimation of the lattice perimeter of aluminium particles on magnesia using the moiré fringes of individual particles showed that contraction increases as size decreases, and that the effect (which is usual only a few percent at most) is greatest at the surface; particles larger than about 20 nm gave the bulk value. Atomic separation in platinum particles deposited on $Al_2O_3/NiAl(100)$ began to decrease at 3 nm size, and at 1 nm was only 90% of the bulk value.[276] EXAFS measurements on supported ruthenium and rhodium particles also revealed static disorder, as well as decreased bond length, which was increased to values greater than the bulk value after chemisorption of hydrogen.[275] The Debye temperature of Rh/TiO_2 rose from 241 K to 341 K on passing into the SMSI state (see Section 3.3.5). The relevance of shorter bond length to catalysis may however only be slight, because it is usually observed under UHV and is apparently negated by chemisorption.[6] Indeed the driving force for this is the surface energy residing in atoms having less than the bulk CN, and when this is utilised by chemisorption the surface atoms relax towards their bulk positions, or may (as noted above) move even further outwards if bonding to the adsorbate is strong enough.

Platinum 'nanowires' have been formed in the channels of NaY and FSM-16 (wide pore) zeolites,[277] and 'nanosheets' between the layers of graphite.[278]

There seems to be comparatively little information available on the *structures* of bimetallic particles,[147] and although certain systems of particular industrial relevance (e.g. Pt-Re and Pt-Sn), and others having great scientific interest (e.g. Ni-Cu, Ru-Cu), have been intensively studied, the emphasis has been on *chemical* structure, i.e., the phases present and the surface composition as a function of total composition.[279] These considerations were briefly treated in Section 2.1.1, and theoretical methods applied to surface composition are mentioned in Section 2.5.5. Segregation of the component of lower surface energy is expected to occur, although it does not always with NiCu/SiO$_2$;[280] the minimum concentration of that component needed to secure complete coverage increases as the particle size falls. Colloidal ruthenium-platinum particles were examined by TEM and EXAFS, and were shown to retain the fcc structure at all compositions.[281]

2.5.3. Energetic Properties

Notwithstanding the shorter bond lengths usually found between surface atoms and those beneath them, implying, as they do, stronger bonds and a 'skin' effect analogous to the surface tension of liquids, such atoms experience larger vibration amplitudes and frequencies, and enhanced mobility relative to bulk atoms[269,270,282] (Section 1.2.2). Melting temperature decreases with particle size, and for gold it may approach ambient temperature for the smallest particles.[113,218,283,284] The activation energy for surface mobility is much less than the sublimation energy (typically $\sim$15%), so that even for metals of quite high melting temperature the surface layer or zone may be semi-fluid (or quasi-molten) at high temperatures. The regions in which such a phase and other metastable phase can exist are shown schematically as a phase diagram in Figure 2.7. If the surface of a particle has passed through a semi-fluid stage (for example, during reduction), it may appear to be amorphous when examined at room temperature. These observations form an additional cogent reason for mistrusting arguments about particle size effects that are based solely on geometric considerations.

The excess free energy associated with small platinum particles can be measured by the potential at which they are oxidised to Pt^{4+} in the presence of chloride ion.[285] An auxiliary redox system (Fe^{3+}/Fe^{2+}) had to be used, and the platinum particles then acted as a microelectrode, taking the reversible potential of the ferrous-ferric equilibrium, which was calculated by the Nernst equation. Use of different concentrations of ferric ion allowed the potential at which platinum atoms were oxidised to be determined, and from the equation

$$\Delta G = -nF\Delta E \qquad (2.9)$$

where $n = 4$ and $F = $ the Faraday, the excess free energy ΔG compared to the bulk metal was determined. For alumina-supported particles of mean size 21 nm this was found to be -54 kJ mol^{-1} and for 3.7 nm particles having a broad size distribution -108 kJ mol^{-1}, i.e. small particles were more easily oxidised than large ones.

2.5.4. Electronic Properties[209,286]

Since the role of the geometric factor in the behaviour of small metal particles may be obscured by the effects described above, we may expect the electronic factor to assume a larger importance. Qualitatively it is apparent that lowering particle size must mean progressive loss of metallic character as gauged by commonality of the valence electrons, that is to say, by their bandwidth; ultimately the band structure relapses to an energy level structure when the particle contains only a few atoms (Section 2.5.1). This argument is the reverse of that used to explain why and how bands appear when particle size in increased.[3] To a first approximation the spacing δ between energy levels is given by

$$\delta = E_F/n \tag{2.10}$$

where E_F is the Fermi level energy[224] and n the number of atoms in the particle. When separation between adjacent levels is greater than the thermal energy kT, they begin to act individually, and metallic character starts to disappear. For a value of E_F of 10 eV (a typical value), the critical size at room temperature is about 400 atoms: but note that at the borderline raising the temperature should result in an *increase* in metal like behaviour. By some criteria full bulk metallic character may start to be seen in particles which have no more than 150–200 atoms;[6] at this size about 60–70 % of the atoms are still on the 'surface', but only about 20–25 % of valences are not used in bonding atoms to each other (Figure 2.5).

Changes in band structure with particle size may be followed by *ultraviolet photoelectron spectroscopy* (UPS). They can be summarised by saying that (i) the density of states at the Fermi energy E_F increases, (ii) the band becomes less sharp and develops features at lower energy, and (iii) the centre of 'gravity' of the band moves to lower energies, as particle size increases. These are the expected consequences of the development of the collective behaviour of electrons, and are supported by measurements of valence-electron binding energies by *X-ray photoelectron spectroscopy* (XPS). These decrease with growing particle size,[276,287] but the effects are so small (usually 1–2 eV), and it was initially unclear whether they were caused by initial- or final-state effects. The former seems now to be the generally accepted explanation.[218]

The energy quantum used in NMR is much smaller than that needed for electron spectroscopy,[288,289] but the response is less sensitive, and a large sample is needed; also the interpretation is even less straightforward. Supported metal catalysts are very suitable for study and the ^{195}Pt nucleus has been extensively examined (Section 2.4.2). This dependence of NMR amplitude on field/frequency shows separate, if not well resolved, peaks at 1.138 and 1.10 G kHz^{-1}, corresponding respectively to Knight shifts of –3.34 and zero percent and thus to bulk and surface atoms. The plots have been imaginatively deconvoluted, but no use has been made of the individual peak areas. The results have been manipulated to yield estimates of the local density of states (LDOS) at the Fermi surface[224] (see the next section); this is much less for surface atoms than for those below, as would be expected from the UPS and XPS results quoted above. It is also stated to be independent of the type of support (SiO_2, TiO_2, Al_2O_3) and method of preparation (impregnation, ion exchange, colloidal process), and to depend solely on dispersion.

There have been a number of studies on the particle size dependence of ferromagnetic behaviour, but fewer on paramagnetic. Small palladium particles have however been shown to have lower paramagnetic susceptibility than large particles.[168]

Heats of adsorption of hydrogen on, and of dissolution in, supported palladium particles were size-independent down to 3 nm, but thereafter decreased significantly;[290] those of oxygen on rhodium, palladium and platinum fell with increasing size.[291]

2.5.5. Theoretical Methods

Small metal particles have been a happy hunting ground for theoreticians. By performing calculations by a variety of theoretical procedures (LCAO-MO Xα scattered wave, density functional theory with various approximations), it has been possible to explore two main points: (i) the relative stabilities of structures of five-fold symmetry and those having the normal bulk form, and (ii) the emergence of electron energy bands with increasing particle size. The conclusions of this body of work have been summarised, and it is not surprising to find that experimental observations find a good measure of support from them. Indeed the cynic might say that a theory could be produced to account for any experimental result. Nevertheless it is a welcome check on the validity of computational procedures that they are able to confirm so closely what is in fact seen.

Particular attention has also been given to deriving the local density of states at the Fermi level (E_F-LDOS).[224,288] Calculations performed using the tight binding approximation confirm the result noted above, that this quantity decreases as all particles become smaller, i.e. with decreasing mean CN. These calculations have been pursued to their logical limit, by assigning values of E_F-LDOS to atoms of each individual CN at the surface of a complete cubo-octahedron: they also decrease with CN. The chemically minded reader, if worried by the use of Band Theory language to describe the electronic structure of small particles, can change LDOS to a statement of the energies and occupancies of the valence orbitals.

Although as we shall see there is much experimental work that bears on the particle size dependence of catalytic behaviour, its interpretation is speculative, and there is little that informs *directly* on how surface atoms differ from those inside. Perhaps the most immediate evidence comes from examining the vibration frequency in the carbon monoxide molecule decreases with particle size on Ir/Al_2O_3[292] and Pt/SiO_2[293] the which may be due to increasing free-atom like character of surface atoms, leading to enhanced electron density and back donation into $\pi*$ antibonding orbitals.[294] Interpretation of the spectra of chemisorbed carbon monoxide can also lead to estimates of the relative areas of different crystal planes[295] and of the occurrence of atoms differing in their co-ordination number.[141]

The contribution that a surface atom makes to surface energy increases as its CN decreases. In a bimetallic particle, the atoms of the metal of lower surface energy (due to its lower heat of sublimation) will therefore be expected to congregate at low CN sites. Monte Carlo calculations[10,296,297] based on pair-wise bond interactions leading to surface configurations that minimise the total energy have been carried out for a number of Groups 8–10 and Group 11 pairs (Pt-Cu, Rh-Ag etc.).[10,298] They confirm that for cubo-octahedra it is edges and vertices that are first occupied by Group 11 atoms, followed by (100) planes; (111) planes are filled last. These observations may vitiate conclusions on the sizes of active centres deduced by altering the composition of an alloy, if it is assumed that the inactive atoms are *randomly* distributed over the surface.[280] Considerations of size and of specific chemical bond formation[299] may also determine the location of the inactive element.

The molecular orbital approach to describing free valencies at metal surfaces (Section 1.23) has been extended to treat small particles.[175,178,300,301]

2.5.6. Conclusions

The purpose of the last sections was to create an overview of the changes which occur in the physical properties of metal particles as their sizes alter, and the ways in which these may be explained, in order to provide a basis for understanding particle size effects in chemisorption and catalysis. The complications that may arise from the having a binodal distribution of particles size have however always to be kept in mind. The main conclusions have been summarised by Burch[6] in the following way.

(1) Proportions of atoms having unusually low CN will change relatively little as size is increased beyond about 2 nm (ca. 60 % dispersion). (2) Bulk electronic properties are not likely to be shown by particles having less than about 150 atoms (1.7 nm, $\sim$70 % dispersion) (3) Typical *surface* electronic properties are however probably shown by particles with only 25–30 atoms (90 % dispersion). (4) Unusual crystallographic structures are rare. (5) Since the heat released by chemisorption is enough to convert one structure into another, metastable structures are unlikely to survive under reaction conditions, and surface reconstruction may occur frequently. (6) A small metal particle may comprise a solid core and a semi-fluid surface layer.

To this catalogue may be added the worry (7) that the size distribution may be binodal.

It is usual but not necessarily sensible to regard geometric and electronic structures as quite separate things, whereas in reality they are closely connected. The form adopted by a metallic or bimetallic particle will depend on the bonds formed between the component atoms, that is to say, on its electronic structure; and, because of the mobility of surface atoms, the properties of very small particles (dispersion >50%) will be dominated by electronic factors.

2.6. METAL-SUPPORT INTERACTIONS

2.6.1. Causes and Mechanisms

The point has already been made that it is almost impossible to study metal particles unless they are supported in the same way. The extent of interaction between the metal and support may be classified as either *weak* or *moderate* or *strong*. With metal particles formed by condensation of atoms from the vapour or by chemical means onto supports that are neutral in character (e.g. graphite), the interaction will probably be weak and the particles will show normal intrinsic effects of size. Complications such as they are will be limited to pseudomorphism caused by epitaxial growth on crystalline supports.[273] Larger effects occur with particles formed by chemical means on supports such as alumina, where the forces responsible for anchoring them to the support have often been discussed, and have now been characterised (see later). Even larger consequences for the metal arise if it finds itself in contact with or near to protons (e.g. in zeolites) or basic cations (e.g. in zeolite LTL or hydrotalcites), but the largest effects of all happen when the support is to some extent reduced by hydrogen: a phenomenon then occurs that has been given the name *Strong Metal-Support Interaction* (SMSI).[91,199,243,244,302–305] This interaction is a consequence of the migration of entities from the support to the metal, with consequent blocking of the active surface. This happens with a number of reducible oxides, and is a result of hydrogen atoms moving from the metal to the support, i.e. of *hydrogen spillover*, which will be discussed in Chapter 3. Consideration of the striking consequences of the SMSI will also be deferred to that place.

The strength of the metal-support interaction is however a continuously varying commodity, depending in each system on the pre-treatment conditions, especially temperature. Since a metal-support interaction is most likely to be visible at small particle sizes and to diminish as size increases, it is natural to try to explore the effect by systematically altering the size: this of course means assuming the absence of intrinsic effects. There is however another problem: the methods commonly used to produce different particle sizes (alteration of metal loading, calcination, change of reduction temperature etc), may lead to other consequences

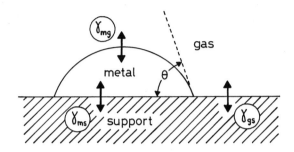

Figure 2.8. Contact angle and interfacial energies for a structureless metal particle on a support.

affecting the number or type of catalytically active sites, e.g. various concentrations of anions (e.g. Cl⁻ which is strongly retained by alumina and titania), formation of toxic species from impurities in the support (e.g. SO_4^{2-} which generates H_2S during reduction) or, even with supports as hard to reduce as alumina, silica or magnesia, the partial obscuration of the metal through one of the several causes (see later). There are ways (some simple) of avoiding or at least recognising all these potential problems, although they are not always adopted. Intrusion of such phenomena can create a false impression of particle size and metal support effects. The intended change is therefore not always that which is produced, or not only that. *The good that I would, I do not. . . .*

For a structureless particle or liquid drop on a support, the contact angle θ that it makes at equilibrium depends upon the surface energies at the three interfaces (see Figure 2.8) according to the equation[306]

$$\cos \theta = (\gamma_{sg} - \gamma_{ms})/\gamma_{mg} \tag{2.11}$$

The corresponding forms adopted by crystalline particles as the interaction across the metal/support interface increases are also shown. In some systems, especially those subject to SMSI, particles are occasionally seen to form raft-like structures[307] indicative of a very strong interaction, but for others where the interaction is less strong they are usually described as hemispheres, half-cubo-octahedra, or square pyramids that may be truncated.[6,164] In the presence of an adsorbing gas, the particle shape will depend on the adsorption energy through the γ_{mg} term, and the differential energies at the various faces will determine the extents of their exposure; the effect on the γ_{sg} term will usually be negligible.

The structure at the metal-support interface is the critical factor in controlling particle shape and stability.[295] A number of careful studies using EXAFS have illuminated it, and the changes that occur when reduction temperature is increased. Systems subjected to particular study are listed in Table 2.1. In addition to

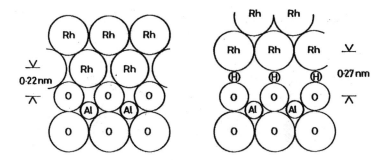

Figure 2.9. Changes in the metal-oxygen ($M \ldots O^{2-}$) distances observed by EXAFS with Rh/Al$_2$O$_3$: the longer distance occurs when H atoms are present at the interface (right-hand part), and the shorter distance when they are removed by evacuation or oxidation (left-hand part). The sizes of the various components are only approximately to scale.

metal-metal distances, metal-oxygen distances have also been determined; these decrease at high reduction temperature or (in the case of alumina[306,307]) with outgassing, due it is thought to the irretrievable loss of hydrogen emanating from the surface hydroxyls (Figure 2.9).[308] The closer contact between metal (or interfacial metal ions) and oxygen, coupled with the loss of protons, may constitute the 'chemical glue', which has long been suspected to account for the stability of supported metal particles. The same changes are often seen with very small palladium, platinum or ruthenium particles in acidic or basic zeolites,[47,68−70,157,191,309,310] such particles frequently appearing to have an electron-deficient character.[68,69,311]

Several explanations have been suggested for these effects. (1) There may develop a degree of chemical bonding, covalent or more likely ionic, between the bottom of metal atoms and the oxide ions of the support; this might give these atoms some positive charge and make the whole particle electron-deficient (Figure 2.10A):

$$S-OH \ldots .. M^0 \xrightarrow{\frac{1}{2}H_2} S-O-M^0 \longleftrightarrow S-O^-M^+ \qquad (2.A)$$

The existence of metal cations close to small metal particles in Ru/Al$_2$O$_3$ has been noted,[112] but they are not seen in Rh/Al$_2$O$_3$.[306] They are however commonly seen with palladium catalysts, and electron-deficient palladium and Pd^{x+} cations are assigned unusually high catalytic activity in hydrocarbon reactions[218] (Figure 2.10B). (2) Such species (and similar platinum ones[212,213,215,312,313]) could arise in acidic zeolites or on other acidic supports in the following way:

$$S-OH \ldots .. M^0 \rightarrow S-O^- \ldots .. H-M^+ \qquad (2.B)$$

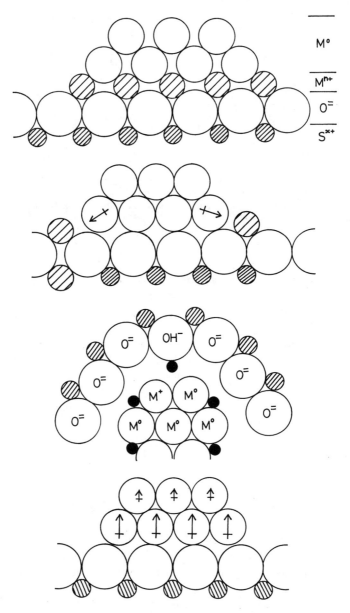

Figure 2.10. Modes of metal-support interaction: (A) by formation of a layer of cations at the interface: (B) by the influence of unreduced cations at the periphery; (C) by proton transfer within a zeolite cavity; (D) by Coulombic attraction between metal and oxide ions.

No desorption of hydrogen is needed for this (Figure 2.10C). (3) More recently the study of zeolite LTL and of silica with very small pallium or platinum particles, and various levels of alkalinity in the form of K^+, has suggested yet another explanation.

This arises from analysis of peaks in the Fourier transform of EXAFS spectra at distances less than 0.15 nm, due to the scattering of the emergent photoelectron against the atomic potential of the absorber atom itself: the effect is named *Atomic XAFS* (AXAFS).[65] It is noted that the observed changes in XPS binding energies, XANES shape resonance (see Chapter 3) and AXAFS with basicity cannot be due to a simple polarisation or redistribution of charge caused by a neighbouring cation and that, although charge transfer from oxide ions of the support would increase electron density on the metal, this is unlikely to occur, especially when the support is an insulator. The proposed model relies on a change in the energies of the metal valence electrons, with the metal's ionisation potential decreasing with increasing alkalinity. The primary interaction is Coulombic interaction between metal and oxide ions, affecting the interatomic potential but avoiding the need for actual electron transfer (Figure 2.10D).[314] Two questions remain. (i) Is this the basis of a *general* explanation of metal support interactions? Does it help to explain the different activities reported for platinum supported on alumina, silica and other ceramic and semi-conducting supports through a variable elecron density on the oxygen? (ii) How exactly do the suggested changes in electron energy levels affect catalytic activity? Increasing alkalinity seriously depresses activity of ruthenium[315] and platinum[65,316] for hydrogenloysis and skeletal isomerisation; this does not however agree with the high activity of electron-deficient palladium,[218] suggesting that Pd^{x+} ions have an opposite effect to that caused by K^+ ions.

To confuse the situation further, differences in the vibration frequencies of M–C bonds in chemisorbed carbon monoxide and M–H bonds in chemisorbed hydrogen, between those found with small platinum particles in NaX zeolite[317] and in magnesium aluminium hydrotalcite,[74] and those for larger particles or bulk metal, have been interpreted to imply that the metal is *negatively* charged. It is however difficult to see how a particle can be both negatively charged and electron deficient at one and at the same time.

Attempts to eliminate residual metal cations by very high temperature reduction[318] are usually unsuccessful because of the harmful effects caused to the already reduced metal[307] (Section 2.32). It has often been found that this treatment produces a loss of catalytic activity larger than can be explained by sintering,[91,305,319,320] and that rates decrease faster than hydrogen chemisorption capacity, thus giving apparently larger values of TOF. These effects have been found with Pt/SiO_2 (EUROPT–1),[305] Pt/Al_2O_3,[128,321,322] Pt/MgO[323] and Pd/SiO_2,[218,324] as well as other systems, and there is good evidence for at least some formation of intermetallic compounds with silicon or aluminium, or other elements:[325] partial or substantial coverage of the metal by the support is another possibility. Reduction of support cations must be preceded by migration of hydrogen atoms from the metal by *hydrogen spillover* (Section 3.34). Where support reduction is easier we find the SMSI (see Section 3.3.5), so consideration of titania and similar supports is deferred until then. It has also been thought that metals can retain some very strongly held hydrogen after high temperature treatment, and that this acts as a poison (Section 2.3.2); the effects of forming palladium hydrides on catalytic activities is also well appreciated. These matters are considered again in Section 3.3.4 and elsewhere.

2.6.2. Particle Size Effects and Metal-Support Interactions: Summary

- It is difficult if not impossible to separate these two effects completely because they are so closely interlocked.
- Apparent size effects and interactions can be caused by poisons originating in the support or metal precursor, or can be a consequence of thermal treatment leading for example to the formation of intermetallic compounds through reduction of cations of the support.
- Bifunctional and spillover catalysis and reaction at interfacial sites may occur to an extent depending upon the components of the catalyst.
- Particle size, shape and relative areas of different crystal planes can be determined by the way in which metal precursor reacts with the support surface and by epitaxial effects of the support on the metal particle during reduction.
- The number of metal particles formed, and hence their mean size, may be conditioned by the concentration of defect sites on the support surface that act as nucleating points.
- Small metal particles are more liable to feel support effects than large ones.
- Particle size effects can be caused by (i) consequences of structural epitaxy; (ii) variation of electronic constitution due to occurrences at the metal-support interface; (iii) alteration in the extent to which electron energy levels overlap and form bands, so that the particle acquires metallic character; (iv) changes in the population of atoms of specified coordination number, or of ensembles of atoms having specified characteristics; (v) alteration in the mobility of surface atoms associated with the variation of melting temperature with size; (vi) a difference in the number or proportion of atoms having special character due their being at the periphery of the particle, and hence able to collaborate with the support in achieving the reaction.
- Models for metal-support interactions occurring at moderate temperatures include (i) M^{n+} cations adjacent to the metal particle (Figure 2.9A), (ii) unreduced M^{n+} cations by proton acquisition in acidic zeolites (Figure 2.9B); and (iii) formation of M^{n+} cations by proton acquisition in acidic zeolites (Figure 2.9C): these all give the metal an electron-deficient

character. (iv) Support acidity/basicity may determine the strength of Coulombic interaction across the oxide ion-metal atom interface, causing changes in electronic structure within the metal atoms without actual change transfer (Figure 2.9D).

2.7. PROMOTERS AND SELECTIVE POISONS

We now encounter a semantic problem of considerable size. It has been recognised for a very long time that the activity of metal catalysts can be helped by the presence of quite small amounts of substances that of themselves have no or little activity. This concept first achieved prominence in the development of iron catalysts for ammonia catalysts, and of iron and cobalt catalysts for Fischer-Tropsch synthesis, and the term *promoter* was applied to these substances. They were of two kinds: (i) structural promoters such as alumina, which acted as grain stabilisers and prevented metal particle sintering and (ii) electronic promoters such as potassium that entered the metallic phase and actually enhanced its activity. In these cases the metal is the major component, so that the catalyst is a *promoted metal* rather than a supported metal.

The problem of terminology arises when the idea is extended to supported metal catalysts, where a number of tricks have evolved to preserve, protect and defend the activity of the metal, and to direct the reaction into more desirable paths by improving selectivity to the wanted product and eliminating or minimising formation of side products and of species that remain on the surface as catalyst poisons. The term 'promoter' can be sensibly used to describe some of these auxiliary agents, but the same or similar effects are produced by other substances to which this term by tradition or convention is *not* used. Promoters comprise two types: (1) those chiefly associated with the support, e.g. stabilising it for use at very high temperatures (occupying holes in the γ-Al_2O_3 structure) or neutralising acidic centres that might initiate carbocationic polymerisation of alkenes; and (2) those that mainly interact with metal particles or perhaps the metal-support interface. These need a fuller analysis, which is attempted in Table 2.2, and which for the sake of completeness mentions reactions that are not strictly in our terms of reference. Here we can see two effects at work; one, the less common, where the modifying species[245,286,315,326−321] may exert an electronic influence on neighbouring metal atoms, either by tending to donate electrons to the metal or to withdraw them: the consequences are not unlike those shown by altering the acidity or alkalinity of the support. The second, the more common, is where the modifier by occupying places on the surface of the metal eliminates the larger groups of atoms which may be the active centres for dissociative chemisorption or decomposition of a hydrocarbon reactant. There seems to be a general rule applying to hydrocarbon reactions that *the reaction most likely to occur is that by which the surface free energy is most greatly*

TABLE 2.2. Selectivity Promoters for Metal Catalysts

Category	Examples	Effects	References
Electron-rich metals	Cu, Ag, Au, Hg, Sn, Pb, Ge	(a) Improved selectivity in the hydrogenation of multiple bonds (b) Suppression of hydrogenolysis in alkane transformations (c) Suppression of carbon deposition (d) Alteration of hydrogenolysis selectivity	286,327,328,417
Other electron-rich species	Cl^-, S^{2-}, N bases, CO	(a) Suppression of hydrogenolysis in petroleum reforming (b) Improved selectivity in the hydrogenation of multiple bonds (Lindlar Catalyst)	
Oxides of electropositive elements	Li^+, Na^+, K^+, Cs^+, Ca^{2+}, Mg^{2+}, La^{3+}	(a) Suppression of carbon deposition (b) Control of chain growth in Fischer-Tropsch synthesis	
Mid-Transition Series elements, and oxo-compounds	Re, MoO_x, VO_x, TiO_x, CeO_x, GeO_x	(a) Improved yields of oxygenates in Fischer-Tropsch synthesis (b) Improved alkene/alkane ratio in Fischer-Tropsch synthesis	

reduced, either because the adsorbed intermediates engage the largest number of metal atoms or because they prefer to react with atoms of low CN: this has thermodynamic logic and explains the tendency of hydrocarbons to dehydrogenate and form strongly held multiply bonded species (see Chapter 12). The corollary is that reactions that are content to proceed on small active centres, even on single atoms or on low index planes, can only do so when larger or more reactive sites are removed by the addition of a partial monolayer of inert atoms or ions. Thus for example Ru/TiO$_2$ becomes quite selective for the skeletal isomerisation of n-butane at 633 K when partially poisoned by sulfur. The widespread interest in and use of bimetallic systems where mutual solubility is low (e.g. Ru-Cu, Pt-Re, Ru-Sn) is largely explained by these effects; species that moderate chain growth in Fischer-Tropsch synthesis, or direct the reaction towards oxygenated products by limiting the dissociation of carbon monoxide, operate in the same way.[256] Unfortunately, since such selectivity promoters deactivate part of the surface, some loss of activity has to be accepted as the necessary price to pay for a cleaner reaction: they therefore fall into the category of *selective poisons*.[332–334] Only rarely (as for example with alkaloid-mediated enantioselective hydrogenation) does the modifier actually lead to a faster rate.

The definition of promotion suggested above excludes cases where the support itself provides an essential service (e.g., in bifunctional and spillover catalysis) or where the promotion occurs unintentionally. Carbon deposition can however sometimes result in a selectivity improvement and of course a species may act at the same time as both an electronic and a selectivity promoter. For these reasons it is not always easy to classify what a particular promoter does, or how it works.

2.8. SINTERING AND REDISPERSION[145]

Sintering is the process whereby at high temperature small metal particles grow into big ones, with consequent loss of metal area and activity (see Further Reading section). It has been widely studied, both experimentally and theoretically, because of its great nuisance value particularly in reactions that need somewhat high temperatures. Two mechanisms have been considered: (i) whole particle migration and (ii) movement of single atom species governed by Ostwald ripening considerations. The thermodynamic driving force for particle growth has already been discussed (Section 2.1.2). Both processes have been observed, but the former is thought to be more important for practical catalysts having rough surfaces. Mobility of intermediate species involved in catalyst preparation (e.g. Pt/(NH$_3$)$_2$O) can be responsible for particle growth during drying and calcination. Sintering rate depends very much on the gas atmosphere; it is much faster under oxidising than under reducing conditions.

Once sintering has taken place, it is no easy task to reverse the process, which is thermodynamically up-hill, unless the metal-support interaction is stronger than that between the metal atoms, in which case sintering ought not to have occurred. It is usually therefore necessary to use a chemical driving force to disaggregate the metal, that is, to convert it via oxidation and reduction back to highly dispersed metal. Most of the work in the open literature concerns the platinum/alumina system because of its relevance to petroleum reforming (Chapter 14), and it mostly appeared in the 1970s and 1980s. In this system, platinum-containing species that are mobile over the surface of the support are formed by treating the used catalyst with oxygen and a chlorine-containing compund (Cl_2, HCl or a chlorinated hydrocarbon). The exact composition of this mobile species does not seem to have been established, and is usually represented just as PtO_xCl_y. A procedure of this type also appears to be effective in reconstituting bimetalluc particles (PtIr, PtRe etc), but much of the relevant information is hidden in the patent literature, and does not contribute much to the scientific understanding of the process; neither does the open literature inform us much about means of redispersing other types of catalyst.

REFERENCES

1. G.C. Bond, *Chem. Soc. Rev.* **20** (1991) 441.
2. Y. Takasu and A.M. Bradshaw, in: *Specialist Periodical Reports: Chem. Phys. Solids Surf.*, Vol. 7 (J.M. Thomas and M.W. Roberts, eds.), The Chemical Society: London, (1978), p. 59.
3. G.C. Bond, *Surf. Sci.* **156** (1985) 966.
4. G.C. Bond in: *Electronic Structure and Reactivity of Metal Surfaces*, (E.G. Derouane and A.A. Lucas, eds.) Plenum: New York (1976), p. 523.
5. R. van Hardeveld and F. Hartog, *Adv. Catal.* **22** (1972) 75.
6. R. Burch in: *Specialist Periodical Reports: Catalysis*, Vol.7 (G.C. Bond and G. Webb, eds.), Roy. Soc. Chem. (1985), p. 149.
7. G. Bergeret and P. Gallezot in: *Handbook of Heterogeneous Catalysis*, Vol.2 (G. Ertl, H. Knözinger and J. Weitkamp, eds.), Wiley-VCH, Weinheim: (1997), p. 439.
8. R. van Hardeveld and A. van Montfoort, *Surf. Sci.* **17** (1969) 90.
9. R. van Hardeveld and F. Hartog, *Surf. Sci.* **15** (1969) 189.
10. M. Muhler, Z. Paál and R. Schögl, *Appl. Surf. Sci.* 47 (1991) 281.
11. B. Heinrichs, F. Noville, J.P. Shoebrechts and J.-P. Picard, *J. Catal.* **192** (2000) 108.
12. L.N. Lewis, *Chem. Rev.* **93** (1993) 2693.
13. V. Ponec and G.C. Bond, *Catalysis by Metals and Alloys,* Elsevier: Amsterdam (1996), Ch. 7.
14. M. El-Sayed. *Chem. Mater.* **8** (1996) 1161; *Science* **272** (1996) 1824.
15. G. Schmidt, V. Maihack, F. Lantermann and S. Peschel, *J. Chem. Soc. Dalton Trans.* (1996) 589.
16. BASF, A.G., *Eur. Appl.,* 920, 912A (1999).
17. J. Turkevich, P.C. Stevenson and J. Hillier, *Discuss. Faraday Soc.* **11** (1951) 55.
18. E. Matijevíc, *Faraday Discuss. Chem. Soc.* **92** (1991) 229.
19. L.N. Lewis and N. Lewis, *Chem. Mater.* **1** (1989) 106; *J. Amer. Chem. Soc.* **108** (1986) 743, 7228.
20. M. Boutonnet, J. Kizling, P. Stenius and G. Maire, *Colloids Surf.* **5** (1982) 209.

21. H. Bönnemann and W. Brijoux, in *Advanced Catalysts and Nanostructured Materials,* Academic Press: New York (1996), p. 165.
22. Y. Mizuboshi, R. Oshima, Y. Maeda and Y. Nagata, *Langmuir* **15** (1999) 2733.
23. M. Brast, D. Bethell, D.J. Schiffrin and C.J. Kiely, *Adv. Mater.* **7** (1995) 795.
24. D. Bethell, M. Brast, D.J. Schiffrin and C.J. Kiely, *J. Electroanalyt. Chem.* **409** (1996) 137.
25. J. Fink, C.J. Kiely, D. Bethell and D.J. Schiffrin, *Chem. Mater.* **10** (1998) 922.
26. C.J. Kiely, J. Fink, M. Brast, D. Bethell and D.J. Schiffrin, *Nature* **396** (1998) 494.
27. M. Aindow, S.N. Williams, R.E. Palmer, J. Fink, and C.J. Kiely, *Phil. Mag. Lett.* **79** (1999) 569.
28. C.J. Kiely, J. Fink, J.-G. Zheng, M. Brast, D. Bethell and D.J. Schiffrin, *Adv. Mater.* **12** (2000) 640.
29. G. Rienäcker and S. Engels, *Monatsber.* **4** (1962) 716.
30. W. Palczewska, M. Cretti-Bujnowska, J. Pielaszek, J. Sobczak and J. Stachurski, in: *Proc. 9th Internat. Congr. Catal.* (M.J. Phillips and M. Ternan, eds.), Chem. Inst. Canada: Ottawa (1988), 1410.
31. M. Arai, K. Usui, M. Shirai and Y. Nishiyama, in: *Preparation of Catalysts VI*, (G. Poncelet et al., eds.) Elsevier: Amsterdam (1995), p. 923.
32. H.C. Brown and C.A. Brown, *Tetrahedron, Suppl. 8, Pt. I, 149; J. Org. Chem.* **31** (1966) 3989.
33. Z. Paál, R. Schlögl and G. Ertl, *J. Chem. Soc. Faraday Trans.* **88** (1992) 1179.
34. J.A. Don, A.P. Pijpers and J.J.F. Scholten, *J. Catal.* **80** (1983) 296.
35. A.J.S. Chowdhury, A.K. Cheetham and J.A. Cairns, *J. Catal.* **95** (1985) 353.
36. J. Freel, W.J.M. Pieters and R.B. Anderson, *J. Catal.* **14** (1969) 247; **16** (1970) 281.
37. B.V. Aller, *J. Appl. Chem.* **8** (1958) 163, 492.
38. J.H.P. Tyman, *Chem. and Ind.* (1964) 404.
39. R. Sassoulas and Y. Trambouze, *Bull. Soc. Chim. France* (1964) 985.
40. M. Raney, *Ind. Eng. Chem.* **32** (1940) 1199.
41. R. Adams and V. Voorhees, *J. Amer. Chem. Soc.* 44 (1922) 1683; J.A. Stanfield and P.E. Robbins, *Proc. 2nd.Intenat. Congr. Catal.* Editions Technip: Paris **2** (1960), 2570.
42. G. Webb and S.J. Thomson, *J. Chem. Soc. Chem. Comm.* **20** (1965) 473.
43. M. Faraday, *Phil. Trans. Roy. Soc.* **147** (1858) 145.
44. A.D. O Cinneide and J.K.A. Clarke, *Catal. Rev.* **7** (1973) 213.
45. G.C. Bond, *Acc. Chem. Res.* **26** (1993) 490.
46. K. Foger, in: *Catalysis – Science and Technology,* Vol. 6 (J.R. Anderson and M. Boudart, eds.) Springer-Verlag: Berlin, (1984), p. 227.
47. P. Papageorgiou, D.M. Price, A. Gavriilidis and A. Varma, *J. Catal.* **158** (1996) 439.
48. J.A. Schwarz, C. Contescu and A. Contescu, *Chem. Rev.* **95** (1995) 477.
49. F. Pinna, *Catal. Today* **41** (1998) 129.
50. R. Psaro and S. Recchia, *Catal. Today* **41** (1998) 139.
51. *Catal. Today* **42** (1–2) (1998) 1–158 (M.C. Kung, R.D. Gonzalez, E.I. Vio and L.T. Thompson, eds.)
52. A.D. O Cinneide and J.K.A. Clark, *Catal. Rev.* **7** (1972–3) 213.
53. *Preparation of Solid Catalysts* (G. Ertl, H. Knözinger and J. Weitkamp, eds.), Wiley-VCH: Weinheim (1999).
54. A.B. Stiles, *Catalyst Manufacture: Laboratory and Commercial Preparations*, Dekker: New York (1983).
55. *Progress in Catalyst Preparation,* (R.E. Resasco and R. Miranda, eds.), *Catal. Today* **15** (1992) 339.
56. *Active Centres in Catalysis*, (R. Burch, ed.), Catal. Today **10** (1991) 233.
57. M. Campanati, G. Fornasari and A. Vaccari, *Catal. Today,* 77 (2003), 299.
58. A.T. Bell, in: *Catalyst Design: Progress and Perspectives*, (L.L. Hegedus, ed.), Wiley: Chichester (1987), Ch. 4, p. 103.

59. G. Leofanti, M. Padovan, G. Tozzola and B. Venturelli, *Catal. Today* **41** (1998) 207.
60. F. Rouquerol, J. Rouquerol and K. Sing, *Adsorption by Powders and Porous Solids,* Academic Press: London (1998).
61. The SCI/IUPAC/NPL Project on Surface Area Standards, *J. Appl. Chem. Biotechnol.* **24** (1974) 199.
62. IUPAC Commission on Colloid and Surface Chem., *Manual on Catalyst Characterisation; Pure & Applied Chem.,* **63** (1991) 1227; *Pure & Applied Chem.* **67** (1995) 1257.
63. G.C. Bond, *Catalysis by Metals,* Academic Press: London 1962.
64. R.K. Iler, *Chemistry of Silica,* Wiley: New York (1979).
65. B.L. Mojet, J.T. Miller, D.E. Ramaker and D.C. Koningsberger, *J. Catal.* **186** (1999) 373.
66. W.M.H. Sachtler and Z.C. Zhang, *Adv. Catal.* **39** (1993), 129.
67. L. Surdelli, G. Martra, R. Buro, C. Dossi and S. Coluccia, *Topics in Catal.* **8** (1999) 237.
68. R.A. Dalla Betta and M. Boudart, *Proc. 5th Internat. Congr. Catal.,* (J.W. Hightower, ed.), North Holland: Amsterdam **2** (1973) 1329.
69. K. Foger and J.R. Anderson, *J. Catal.* **54** (1978) 318.
70. A. Tonschiedt, P.L. Ryder, N.I. Jaeger and G. Schulz-Egloff, *Surf. Sci.* **281** (1993) 51.
71. J. de Graaf, A.J. van Dillen, K.P. de Jong and D.C. Koningsberger, *J. Catal.* **203** (2001) 307.
72. G. Sankar and J.M. Thomas, *Topics in Catal.* **8** (1999) 1.
73. J.M. Thomas and W.J. Thomas, *Principles and Practice of Heterogeneous Catalysis,* VCH: Weinheim,(1997).
74. V.B. Kazansky, V. Yu. Borovlov, A.I. Serykh and F. Figuéras, *Catal. Lett.* **49** (1997) 35.
75. F.-B. Li and G.-Q. Yuan, *Catal. Lett.* **89** (2003) 115.
76. L.R. Radovic and F. Rodríguez-Reinoso, in: *Chemistry and Physics of Carbon,* Vol. 25 (P.A. Thrower, ed.), Marcel Dekker Inc.: New York (1996)
77. P. Albers, R. Burmeister, K. Seibold, G. Prescher, S.F. Parker and D.K. Ross, *J. Catal.* **181** (1999) 145.
78. V.A. Semikotenov, V.I. Zaikovski and G.V. Plaksin, in *Abstracts, 9th Internat. Symp. on Relations between Homogeneous and Heterogeneous Catalysis,* Roy. Soc. Chem.: London (1998), p. 46.
79. W. Airey, S.I. Ajiboye, P.A. Barnes, D.R. Brown, S.C.J. Buckley, E.A. Dawson, K.F. Gadd and G. Midgley, *Catal. Today* **7** (990) 179.
80. J.-M. Nhut, R. Vieira, L. Pesant, J.-P. Tessonnier, N. Keller, G. Ehret, C. Pham-Huu and M.J. Ledoux, *Catal. Today* **76** (2002) 11.
81. J.H. Bitter, M.K. van der Lee, A.G.T. Slotboom, A.J. van Dillen and K.P. de Jong, *Catal. Lett.* **89** (2003) 139.
82. M.L. Toebes, J.H. Bitter, A.J. van Dillen and K.P. de Jong, *Catal. Today* **76** (2002) 33.
83. K.P. de Jong and G.W. Geus, *Catal. Rev.-Sci. Eng.* **77** (2000) 359.
84. T. Braun, M. Wohlers, G. Nowitzke, G. Wortmann, Y. Uchida, N. Pfänder and R. Schlögl, *Catal. Lett.* **43** (1997) 167; T. Braun, M. Wohlers, T. Belz and R. Schlögl, *Catal. Lett.* **43** (1997) 175.
85. B. Coq, V. Brotous, J.M. Planeix, L.C. de Ménorval and R. Dutartre, *J. Catal.* **176** (1998) 358.
86. H. Nagashima, A. Nakaoka, S. Tajima, Y. Saito and K. Itah, *Chem. Lett.* (1992) 1361.
87. R. Kozlowski, R.F. Pettifer and J.M. Thomas, *J. Phys. Chem.* **87** (1983) 5176.
88. J.A. Schwarz, C. Contescu and J. Jagiello, in: *Specialist Periodical Reports: Catalysis,* vol. 11, (J.J. Spivey and S.K. Agarwal, eds.), *Roy. Soc. Chem.* (1994), p. 127.
89. F. Kapteijn, J.H. Heiszwolf, T.A. Nijhuis and J.A. Moulijn, *Cat-Tech.* 5, Aug. (1999), 18.
90. R.D. Gonzalez, L. Lopez and R. Gomez, *Catal. Today* **35** (1997) 293.
91. T. Ueckert, R. Lamber, N.I. Jaeger and U. Schubert, *Appl. Catal. A: Gen.* **155** (1997) 75.
92. Ihl Hyun Cho, Seung Bin Park, Sung June Cho and Ryong Ryoo, *J. Catal.* **171** (1998) 295.
93. S.L. Anderson, A.K. Datye, T.A. Wark and M.J. Hampden-Smith, *Catal. Lett.* **8** (1991) 345.
94. K. Balakrishnan and R.D. Gonzalez, *J. Catal.* **144** (1993) 395.
95. T. Lopez, R. Gomez, O. Novaro, A. Ramirez-Sotis, E. Sanchez-Mora, S. Castillo, E. Poulain and J.M. Martinez-Mayadan, *J. Catal.* **14** (1993) 114.

96. C.K. Lambert and R.D. Gonzalez, *Appl. Catal. A: Gen.* **172** (1998) 233.
97. J.R. Regalbuto, A. Navada, S. Shadid, M.L. Bricker and Q. Chen, *J. Catal.* **184** (1999) 335.
98. Sui-wen Ho and Yu-Shu Su, *J. Catal.* **168** (1997) 51.
99. F. Boccuzzi, A. Chiorino, M. Gargano and N. Ravazio, *J. Catal.* **165** (1997) 129.
100. K. Akasura, W.-J. Chua, M. Shirai, K. Tomishige and Y. Iwasawa, *J. Phys. Chem. B.* **101** (1997) 5549.
101. P. Reyes, M. Oportus, G. Pecchi, R. Rety and B. Moraweck, *Catal. Lett.* **37** (1996) 193.
102. A. Gouget, M. Aouine, F.J. Cadete Santos Aires, A. De Hallmann, D. Schweich and J.P Candy, *J. Catal.* **209** (2002) 135.
103. R.W. Joyner and E.S. Shpiro, *Catal. Lett.* **9** (1991) 239.
104. G.C. Bond, R.R. Rajaram and R. Burch, *Appl. Catal.* **27** (1986) 379.
105. A.E. Newkirk and D.W. McKee, *J. Catal.* **11** (1968) 370.
106. G. Marquez de Cruz, G. Djega-Mariadassou and G. Bugli, *Appl. Catal.* **17** (1985) 205.
107. S. Ringler, Ph.D. thesis, Univ. L. Pasteur, Strasbourg, 1998.
108. M. Ichikawa, *Plat. Met. Rev.* **44** (2000) 3.
109. D. Alexeev and B.C. Gates, *J. Catal.* **176** (1998) 310.
110. P. Serp, P. Kalck and R. Feuer, *Chem. Rev.* **102** (2002) 3085.
111. B.C. Gates, *Chem. Rev.* **95** (1995) 511.
112. B. Coq, E. Crabb, M. Warawdekar, G.C. Bond, J.C. Slaa, S. Galvagno, L. Mercadante, J. Garcia Ruiz and M.C. Sanchez Sierra, *J. Molec. Catal.* **99** (1994) 1.
113. G.C. Bond and D.T. Thompson, *Catal. Rev.–Sci. Eng.* **41** (1999) 319.
114. J. Jia, K. Haraki, J.N. Kondo, K. Domen and K. Tamaru, *J. Phys. Chem. B* **104** (2000) 11153.
115. M. Okumura, T. Akita and M. Haruta, *Catal. Today* **74** (2002) 265.
116. A. Berthet, A.L. Thomann, F.J. Cadete Santos Aires, M. Brun, C. Deranlot, J.L. Bertolini, J.P. Rozenbaum, P. Brault and P. Andreazza, *J. Catal.* **190** (2000) 49.
117. P. Serp, R. Feurer, R. Morancho and P. Kalck, *J. Molec. Catal. A: Chem.* **101** (1995) L107.
118. I.C. Brownlie, J.R. Fryer and G. Webb, *J. Catal.* **14** (1969) 263.
119. G.C. Bond and G. Hierl, *J. Catal.* **61** (1980) 348.
120. G. Henrici-Olivé and S. Olivé, *Chimia* **27** (1967) 87.
121. M. Che, M. Richard and D. Olivier, *J. Chem. Soc. Faraday Trans. I* **76** (1980) 1526.
122. M. Gueaia, M. Breysse and R. Frety, *J. Molec. Catal.* **25** (1984) 119.
123. Chor Wong and R.W. McCabe, *J. Catal.* **107** (1987) 535.
124. A. Jones and B.D. McNicol, *Temperature-Programmed Reduction for Solid Materials Characterization,* Dekker: New York (1986).
125. P. Malet and A. Caballero, *J. Chem. Soc. Faraday Trans.* **84** (1988) 2369.
126. D.E. Damiani and A.J. Rouco, *J. Catal.* **100** (1986) 512.
127. P.G. Menon and G.F. Froment, *Appl. Catal.* **1** (1981) 31; *J. Catal.* **59** (1979) 138.
128. R.-Y. Tang, R.-A. Wu and L.-W. Liu, *Appl. Catal.* **10** (1984) 163.
129. A.I.M. van den Broek, J. van Grondelle and R.A. van Santen, *J. Catal.* **167** (1997) 417.
130. B. Shelimov, J.-F. Lambert, M. Che and B. Didillon, *J. Catal.* **183** (1999) 462.
131. T. Shido, M. Lok and R. Prins, *Topics in Catal.* **8** (1999) 223.
132. A. Muñoz-Páez and D.C. Koningsberger, *J. Phys. Chem.* **99** (1995) 4193.
133. D. Bazin, I. Kovács, L. Guzci, P. Parent, C. Laffon, F. De Groot, O. Ducreux and J. Lynch, *J. Catal.* **189** (2000) 456.
134. S.C. Chan, S.C. Fung and J.H. Sinfelt, *J. Catal.* **113** (1988) 164.
135. B.M. Shelimov, J.-F. Lambert, M. Che and B. Didillon, *J. Molec. Catal. A. Chem.* **158** (2000) 91.
136. Y. Udagawa, S. Tanabe, A. Veno and K. Toji, *J. Amer. Chem. Soc.* **106** (1984) 612.
137. A.K. Datye, *J. Catal.* **216** (2003) 144.
138. J.-P. d'Espinose de la Caillerie, M. Kermerac and O. Clause, *J. Am. Chem. Soc.* **117** (1995) 11471.
139. S. Boujday, J. Lehman, J.-F. Lambert and M. Che,*Catal. Lett.* **88** (2003) 23.

140. G.C. Bond, in: *Handbook of Heterogeneous Catalysis*, Vol. 3 (G. Ertl. H. Knözinger and J. Weitkamp, eds.), VCH: Weinheim, p. 1489.
141. M.J. Kappers and J.H. van der Maas, *Catal. Lett.* **10** (1991) 365.
142. D.W. Goodman, *Chem. Rev.* **95** (1995) 523.
143. P.L.J. Gunter, J.W. Niemantsverdriet, F.H. Ribeiro and G.A. Somorjai, *Catal. Rev.–Sci. Eng.* **39** (1997) 77.
144. C.T. Campbell, *Adv. Catal.* **36** (1989) 2.
145. B.K. Min, A.K. Santra and D.W. Goodman, *Catal. Today* **85** (2003) 113.
146. L. Guczi and A. Sárkány, in: *Specialist Periodical Reports: Catalysis,* Vol. 11, (J.J. Spivey and S.K. Agarwal, eds.), *Roy. Soc. Chem.* (1994), p. 318.
147. J.H. Sinfelt, *Acc. Chem. Res.* **20** (1987) 134.
148. K.E. Coulter and A.G. Sault, *J. Catal.* **154** (1995) 56.
149. C. Betizeau, C. Bolivar, H. Charcosset, R. Frety, G. Leclerq, R. Maurel and L. Tounayan, in: *Preparation of Catalysts* (B. Delmon, P.A. Jacobs, G. Poncelet, eds.), Elsevier: Amsterdam (1976), p. 425.
150. C. Kappenstein, M. Saouabe, M. Guérin, P. Marecot, I. Uszkurat and Z. Paál, *Catal. Lett.* **31** (1995) 9.
151. C. Audo, J.F. Lambert, M. Che and B. Didillon, *Catal. Today* **65** (2001) 157.
152. A. Giroir-Fendler, D. Richard and P. Gallezot, *Faraday Discuss. Chem. Soc.* **92** (1991) 69.
153. J.H. Sinfelt, *Internat. Rev. Phys. Chem.* **7** (1988) 281.
154. G.C. Bond and Yide Xu, *J. Molec. Catal.* **25** (1984) 141.
155. B.D. Chandler, A.B. Schnabel and L.H. Pignolet, *J. Catal.* **193** (2000) 186.
156. G.M. Nuñez and A.J. Rouco, *J. Catal.* **111** (1988) 41.
157. M. Fernández-García, J.A. Anderson and G.L. Haller, *J. Phys. Chem.***100** (1996) 16247.
158. L. Guczi, R. Sundararajan, Zs. Koppáry, Z. Zsoldos, Z. Schay, F. Mizakami and S. Niwa, *J. Catal.* **167** (1997) 482.
159. O.S. Alexeev, G.W. Graham, M. Shelef and B.C. Gates, *J. Catal.* **190** (2000) 157.
160. B. Heinrichs, F. Noville and J.-P. Pirard, *J. Catal.* **170** (1997) 366.
161. I. Zuburtikudis and H. Saltsburg, *Science* **258** (1992) 1337.
162. G.A. Somorjai, *Catal. Lett.* **76** (2001) 1.
163. G. Cárdenas, R. Oliva, P. Reyes and B.L. Rivas, *J. Molec. Catal. A: Chem.* **191** (2003) 87.
164. M. Che and C.O. Bennett, *Adv. Catal.* **36** (1989) 55.
165. *Handbook of Heterogeneous Catalysis*, Vol. 2 (G. Ertl, H. Knözinger and J. Weitkamp, eds.), Wiley-VCH: Weinheim (1997).
166. . *Faraday Discuss. Chem. Soc.* **92** (1991).
167. *Handbook of Heterogeneous Catalysis*, Vol. 3 (G. Ertl, H. Knözinger and J. Weitkamp, eds.), VCH: Weinheim (1997).
168. W. Romanowski, *Highly Dispersed Metals,* Ellis Horwood: Chichester (1987).
169. R.L. Burwell Jr., *Langmuir* **2** (1986) 2.
170. O.L. Pérez, D. Romen and M.J. Yacamán, *Appl. Surf. Sci.* **13** (1982) 402; *J. Catal.* **79** (1983) 240.
171. O.M. Poltorak and V.S. Boronin, *Russ. J. Phys. Chem.* **40** (1966) 1436; **39** (1965) 781, 1329.
172. O.-L. Perez and M.J. Yacaman, *J. Catal.* **79** (1983) 240.
173. J.M. Montejano-Carrizales and J.L. Moran-Lopez, *Nanostr. Mater.* **1** (1992) 397.
174. E.G. Schlosser, *Ber. Bunsenges. Phys. Chem.* **73** (1969) 358.
175. R.L. Augustine, *Heterogeneous Catalysis for the Synthetic Chemist*, Dekker, New York: (1996), Chs. 3 and 4.
176. R.L. Augustine and J.F. Van Pepper, *Anal. New York Acad. Sci.* **172** (1970) 244.
177. S. Siegel, J. Outlaw Jr. and N. Garti, *J. Catal.* **52** (1978) 102.
178. R.L. Augustine and K.M. Lahanas, in: *Catalysis of Organic Reactions,* (J.R. Kosak and T.A. Johnson, eds.), Dekker: New York (1994), p. 279.

179. J.R. Anderson, *Sci. Prog.,* Oxford, **69** (1985) 461.
180. J.R. Anderson, *Structure of Metallic Catalysts,* Academic Press: New York (1975).
181. A. Borodziński and M. Bonarowska, *Langmuir* **13** (1997) 5613.
182. M. Che and C.O. Bennett, *J. Catal.* **120** (1989) 293.
183. G.C. Bond, *J. Catal.* **136** (1992) 631.
184. A.Y. Stakheev and L.M. Kustov, *Appl. Catal. A: Gen.* **188** (1999) 3.
185. G.C. Bond and J. Turkevich, *Trans. Faraday Soc.* **52** (1956) 1235.
186. J.E. Benson and M. Boudart, *J. Catal.* **4** (1965) 704.
187. S. Bernal, J.J. Calvins, M.A. Cauqui, J.M. Gatica, C. López Cartes, J.A. Pérez Omil and J.M. Pintado, *Catal. Today,* **77** (2003) 385.
188. A. Howie, *Faraday Discuss. Chem. Soc.* **92** (1991) 1.
189. P.-A. Buffat, M. Flüeli, R. Spycher, P. Stadelmann and J.-P. Borel, *Faraday Discuss.Chem. Soc.* **92** (1991) 173.
190. D.J.C. Yates and E.B. Prestridge, *J. Catal.* **106** (1987) 549.
191. P. Gallezot, A.I. Bienenstock and M. Boudart, *Nouv. J. Chim.* **2** (1978) 263.
192. G. Bergeret, in: *Handbook of Heterogeneous Catalysis,* Vol. 2 (G. Ertl, H. Knozinger and J. Weitkamp, eds.), Wiley-VCH: Weinheim, (1997), p. 464.
193. J.W. Niemantsverdriet, *Spectroscopy in Catalysis,* VCH: Weinheim (1993).
194. H. Brumberger, F. DeLaglio, J. Goodisman, M.G. Phillips, J.A. Schwarz and P. Sen, *J. Catal.* **92** (1985) 199.
195. J. Goodisman, H. Brumberger and R. Cupelo, *J. Appl. Crystallogr.* **14** (1981) 305.
196. A. Benedetti, L. Bertoldo, P. Cantan, G. Goerigk, F. Pinna, P. Riello and S. Polizzi, *Catal. Today* **49** (1999) 485.
197. A. Benedetti, S. Polizzi, R. Riello, F. Pinna and G. Goerigk, *J. Catal.* **171** (1997) 345.
198. F.B. Rasmussen, A.M. Moelenbroek, B.S. Clausen and R. Feidenhans, *J. Catal.* **190** (2000) 205.
199. P. Georgopoulos and J.B. Cohen, *J. Catal.* **92** (1985) 211.
200. D.A.H. Cunningham, W. Vogel, R.M. Torres Sanchez, K. Tanaka and M. Harata, *J. Catal.* **183** (1999) 24.
201. V. Gutzmann and W. Vogel, *J. Phys. Chem.* **94** (1990) 4991.
202. G.C. Bond and Z. Paál, *Appl. Catal. A Gen.* **86** (1992) 1.
203. K.S. Liang, S.S. Laderman and J.H. Sinfelt, *J. Phys. Chem.* **86** (1987) 2352.
204. J.H. Sinfelt, in: *Catalysis-Science and Technology,* (J.R. Anderson and M. Boudart, eds.), Springer-Verlag, Berlin: (1981), p. 257.
205. G. Vlaic, D. Andreatta and P.E. Colavita, *Catal. Today* **41** (1998) 261.
206. R.W. Joyner and P. Meehan, *Vacuum* **23** (1983) 691.
207. M. Vaarkamp and D.C. Koningsberger in: *Handbook of Heterogeneous Catalysis,* Vol. 2 (G. Ertl, H. Knözinger and J. Weitkamp, eds.), Wiley-VCH: Weinheim, (1997), p. 475.
208. B.S. Clausen, *Catal. Today* **39** (1998) 293.
209. M.G. Mason, *Phys. Rev. B: Condens. Matter* **27** (1983) 748.
210. Y. Iwasawa, *J. Catal.* **216** (2003) 165.
211. K. Akasura, T. Kubota, N. Ichikuni and Y. Iwasawa, *Proc. 11ᵗʰ Internat. Congr. Catal.* (J.W. Hightower, W.N. Delgass, E. Iglesia and A.T. Bell, eds.), Elsevier: Amsterdam **B** (1996) 911.
212. N. Ichikuni and Y. Iwasawa, *Catl. Lett.* **20** (1993) 87.
213. R.H. Lewis, *J. Catal.* **69** (1981) 511.
214. D.C. Koningsberger, M.K. Oudenhuijzen, J. de Graaf, J.A. van Bokhoven and D.E. Ramaker, *J. Catal.* **216** (2003) 178.
215. R.A. Dalla Betta, M. Boudart, P. Gallezot and R.S. Weber, *J. Catal.* **69** (1981) 514.
216. A.M. Venezia, *Catal. Today,* **77** (2003) 359.
217. S.D. Jackson, J. Willis, G.D. McLellan, G. Webb, M.B.T. Keegan, R.B. Moyes, S. Simpson, P.B. Wells and R. Whyman, *J. Catal.* **139** (1993) 191.

218. Z. Karpiński, *Adv. Catal.* **37** (1990) 45.
219. B. Bellamy, A. Masson, V. Degouveia, M.C. Desjoquères, D. Spanjaard and G. Tréglia, *J. Phys.: Condens. Matter* **1** (1989) 5875.
220. A. Masson, B. Bellamy, Y. Hadj Romdhane, M. Che, H. Roulet and G. Dufour, *Surf. Sci.* **173** (1986) 479.
221. J.W. Niemantsverdriet and J. Butz, in: *Handbook of |Heterogeneous Catalysis,* Vol. 2 (G. Ertl. H. Knözinger and |J. Weitkamp, eds.), Wiley-VCH: Weinheim, (1997), p. 512.
222. F.M. Mulder, R.C. Thiell, L.J. de Jongh and P.C.M. Gubbens, *Nanostruct. Mat.* **7** (1996) 269.
223. S. Bernal, J.J. Calvino, M.A. Cauqui, J.M. Gatica, L. Larese, J.A. Pérez Omil and J.M. Pintado, *Catal. Today* **50** (1999) 175.
224. J.J. van der Klink, *Adv. Catal.* **44** (1999) 1.
225. J.P. Bucher, J. Buttet, J.J. van der Klink, M. Graetzel, E. Newson and T.B. Truong, *J. Molec. Catal.* **43** (1987) 213.
226. C.P. Slichter, *Surf. Sci.* **106** (1981) 382.
227. H.T. Stokes, C.D. Makowka, P.K. Wang, S.L. Rudaz and C.P. Slichter, *J. Molec. Catal.* **20** (1983) 321.
228. C.D. Makowka and C.P. Slichter, *Phys. Rev. B: Condens. Matter* **31** (1985) 5663.
229. K. Sakaie, C.P. Slichter and J.H. Sinfelt, *J. Magnet. Res. A* **119** (1996) 235.
230. L.C. de Menorval, J.P. Fraissard and T. Ito, *J. Chem. Soc. Faraday Trans. I* **78** (1982) 403.
231. J. Fraissard, *Catal. Today* **51** (1999) 481.
232. Sung June Cho, Wha-Seung Aha, Suk Bong Hong and Ryong Ryoo, *J. Phys. Chem.* **100** (1996) 4996.
233. G.A. Attard, D.J. Jenkins, O.A. Hazzai, P.B. Wells, J.E. Gillies, K.G. Griffin and P. Johnson in: *Catalysis in Application* (S.D. Jackson, J.S.J. Hargreaves and D. Lennon, eds.), Roy. Soc. Chem.: London (2003), p. 70.
234. F.B. Passos, I.S. Lopes, P.R.J. Silva and H. Saitovvitch, *Catal. Today* **78** (2003) 411.
235. J. Winterlin, *Adv. Catal.* **45** (2000) 131.
236. B. Bellamy, S. Meckken and A. Masson, *Z. Phys. D – Atoms, Mols, and Clusters* **26** (1993) 561.
237. C. Xu, X. Lai and D.W. Goodman, *Faraday Disc. Chem. Soc.* **105** (1996) 247.
238. A. Berkó and F. Solymosi, *J. Catal.* **183** (1999) 91.
239. A. Berkó, G. Klivényi and F. Solymosi, *J. Catal.* **182** (1999) 511.
240. King Lun Yeung and E.E. Wolf, *J. Catal.* **143** (1993) 409.
241. A.K. Datye, D.S. Kalakkad, E. Völke and L.F. Allard, in: *Electron Holography,* (A. Tonomura, L.F. Allard, G. Pozzi, D.C. Joy and Y.A. Ono, eds.), Elsevier: Amsterdam (1995), p. 199.
242. S. Schimpf, M. Lucas, C. Mohr, U. Rodenerck, A. Brückner, J. Radnik, H. Hofmeister and P. Claus, *Catal. Today* **72**, 63 (2002).
243. J. Saaz, M.J. Rojo, P. Malet, G. Munuera, M.T. Blasco, J.C. Conesa and J. Soria, *J. Phys. Chem.* **89** (1985) 5427.
244. M.T. Blasco, J.C. Conesa, J. Soria, A.R. González-Elipe, G. Munuera, J.M. Rojo and J. Sanz, *J. Phys. Chem.* **92** (1988) 4685.
245. N. Takahashi, T. Mori, A. Farata, S. Komai, A. Miyamoto, T. Hattori and Y. Murakami, *J. Catal.* **110** (1988) 410.
246. S. Tschudin, T. Shido, R. Prins and A. Wokaun, *J. Catal.* **181** (1999) 113.
247. C.H. Bartholomew in: *Specialist Periodical Reports: Catalysis,* Vol. 11, (J.J. Spivey and S.K. Agarwal, eds.), *Roy. Soc. Chem.* (1994), p. 93.
248. R.A. Dalla Betta, *J. Catal.* **34** (1974) 57.
249. R. Gomez, S. Guentes, F.J. Fernandez del Valle, A. Campero and J.M. Ferreira, *J. Catal.* **38** (1975) 47.
250. G.C. Bond and Lou Hui, *J. Catal.* **147** (1994) 346.
251. Dong Jin Suh, Tae-Jin Park and Song-Ki Ihm, *J. Catal.* **149** (1994) 438.

252. L. Maffussi, P. Iengo, M. Di Serio and E. Santacesaria, *J. Catal.* **172** (1997) 485.
253. A. Frennet and P.B. Wells, *Appl. Catal.* **18** (1985) 243.
254. B.J. Kip, F.B.M. Duivenvoorden, D.C. Koningsberger and R. Prins, *J. Catal.* **106** (1986) 26.
255. B.J. Kip, F.B.M. Duivenvoorden, D.C. Koningsberger and R. Prins, *J. Am. Chem. Soc.* **108** (1986) 5633.
256. J.A. Norcross, C.P. Slichter and J.H. Sinfelt, *Catal. Today* **53** (1999) 343.
257. Y.Y. Tong, A.J. Renouprez, G.A. Martin and J.J. van der Klink, *Proc. 11th. Internat. Congr. Catal.* (J.W. Hightower, W.N. Delgass, E. Iglesia and A.T. Bell, eds.), Elsevier: Amsterdam **B** (1996) 901.
258. S.K. Masthan, K.V.R. Chary and P. Kanta Rao, *J. Catal.* **124** (1990) 289.
259. A. Corma, M.A. Martn and J. Pérez, *J. Chem. Soc. Chem. Comm.* (1983) 1512.
260. A. Corma, M.A. Martín and J. Pérez-Pariente, *Surf. Sci.* **136** (1984) L31.
261. G.C. Bond and R. Burch in: *Specialist Periodical Reports: Catalysis*, Vol. 6, (G.C. Bond and G. Webb, eds.), *Roy. Soc. Chem.* (1983), p. 27.
262. T.H. Lee and K.M. Ervin, *J. Phys. Chem.* **98** (1994) 10,023.
263. I.W. Bassi, F.W. Lytle and G. Parravano, *J. Catal.* **42** (1976) 139.
264. Le Van Tiep, M. Bureau-Tardy, G. Bugli, G. Djega-Mariadassou, M. Che and G.C. Bond, *J. Catal.* **99** (1986) 449.
265. Liqiu Yang and A.E. DePristo, *J. Catal.* **149** (1994) 223.
266. A.S. McLeod and L.F. Gladden, *J. Catal.* **173** (1997) 43.
267. W. Romanowski, *Surf. Sci.* **18** (1969) 373.
268. Y. Fukano and C.M. Wayman, *J. Appl. Phys.* **40** (1969) 1656.
269. J.J. Burton, *Catal. Rev.-Sci. Eng.* **9** (1974) 209.
270. L.D. Marks and N. Doraiswamy, in: *The Chemical Physics of Solid Surfaces* (D.A. King and D.P. Woodruff, eds.), Vol. 7, Elsevier: Amsterdam (1994), Ch. 11, p. 443.
271. A. Rodriguez, C. Amiens, B. Chaudret, M.-J. Casanove, P. Lecante and J.S. Bradley, *Chem. Mater.* **8** (1996) 1978.
272. J. Lynch, *J. Microsc. Spectrosc. Electron.* **8** (1983) 479.
273. C.R. Henry, *Surf. Sci. Rep.* **31** (1998) 231.
274. H.-J. Freund, *Faraday Discuss. Chem. Soc.* **114** (1994) 1.
275. H. Kuroda, T. Yokoyama, K. Asakura and Y. Iwasawa, *Faraday Discuss. Chem. Soc.* **92** (1991) 189.
276. H.-J. Freund, M. Bäumer and H. Kuhlenbeck, *Adv. Catal.* **45** (2000) 334.
277. A. Fakuoka, N. Highasimoto, Y. Sakamoto, M. Sasaki, N. Sugimoto, S. Inagaki, Y. Fakushima and M. Ichikawa, *Catal. Today* **66** (2001) 23.
278. M. Shirai, K. Igeta and M. Arai, *Chem. Comm.* (2000) 623.
279. J.K. Strohl and T.S. King, *J. Catal.* **116** (1989) 540.
280. G.A. Martin, *Catal. Rev.-Sci. Eng.* **30** (1988) 519.
281. D. Richard, J.W. Couves and J.M. Thomas, *Faraday Discuss. Chem. Soc.* **92** (1991) 109.
282. G.A. Somorjai, *Introduction to Surface Chemistry and Catalysis,* Wiley: New York (1994).
283. G. Schmid and B. Corain, *Eur. J. Inorg. Chem.* (2003) 3081.
284. K. Dick, T. Dhanasekaran, Z.-Y. Zhang and D. Meisel, *J. Am. Chem. Soc.* **124** (2002) 2312.
285. A. Dollidah, P. Marécot, S. Szabo and J. Barbier, *Appl. Catal. A: Gen.* **225** (2002) 21.
286. Li Yongtue, Wu Shuguang and Qin Xin, *React. Kinet. Catal. Lett.* **38** (1989) 63.
287. T. Huizinga, H.F.J. Van't Blick, J.C. Vis and R. Prins, *Surf. Sci.* **135** (1983) 580.
288. Y.Y. Tong, A.J. Renouprez, G.A. Martin and J.J. van der Klink, *Proc. 11th Internat. Congr. Catal.* (J.W. Hightower, W.N. Delgass, E. Iglesia and A.T. Bell, eds.), Elsevier: Amsterdam **B** (1996) 901.
289. J.P. Bucher, J. Buttet, J.J. van der Klink and M. Graetzel, *Surf. Sci.* **214** (1989) 347.
290. Pen Chou and M.A. Vannice, *J. Catal.* **104** (1987) 1.

291. Z.-A. Quan and J.M. Miller, *Appl. Chem. A: Gen.* **209** (2001) 1.

292. F.J.C.M. Toolenaar, A.G.T.M. Bastein and V. Ponec, *J. Catal.* **82** (1983) 35.

293. R.K. Brandt, M.R. Hughes, L.P. Bourget, K. Truszowska and R.G. Greenler, *Surf. Sci.* **286** (1993) 15.

294. V. Ponec, in: Metal-Support and Metal-Additive Effects in Catalysis (B. Imelik, C. Naccache, G. Coudurier, H. Praliaud, P. Meriaudeau, P. Gallezot, G.A. Martin and J.C. Védrine, eds.), Studies in Surface Science and Catalysis, Elsevier: Amsterdam, **11** (1982) 63.

295. L. Marchese, M.R. Bocutti, S. Coluccia, S. Lavagnino, A. Zecchina, L. Bonneviot and M. Che, Structure and Reactivity of Surfaces (C. Morterra, A. Zecchina and G. Costa, eds.), Studies in Surface Science and Catalysis', Elsevier: Amsterdam, **48** (1989) 653.

296. T. Baird, in: *Specialist Periodical Reports: Catalysis*, Vol. 5, (G.C. Bond and G. Webb, eds.), *Roy. Soc. Chem.* (1982), p. 172.

297. Ling Zhu, K.S. Liang, B. Zhang, J.S. Bradley and A.E. DePristo, *J. Catal.* **167** (1997) 412.

298. Liqiu Yang and A.E. DePristo, *J. Catal,* **148** (1994) 575.

299. H.R. Aduriz, P. Boduariuk, B. Coq and F. Figueras, *J. Catal.* **119** (1989) 97.

300. R.L. Augustine, K.M. Lahanas and F. Cole, in *Proc. 10th Internat. Congr. Catal.* (L. Guczi, F. Solymosi and P. Tétényi, eds.), Akadémiai Kiadó: Budapest **B** (1993) 1567.

301. G.C. Bond, in: *Proc.4th Internat. Congr. Catal.*, Editions Technip: Paris **2** (1968) 266.

302. G.L. Haller and D.E. Resasco, *Adv. Catal.* **36** (1989) 173.

303. J. van de Loosdrecht, A.M. van der Kraan, A.J. van Dillen and J.W. Geus, *J. Catal.* **170** (1997) 217.

304. N.S. de Resende, J.-G. Eon and M. Schmal, *J. Catal.* **183** (1999) 6.

305. G.A. Martin, R. Dutartre and J.A. Dalmon, *React. Kinet. Catal. Lett.* **16** (1981) 329.

306. G.C. Bond in: Handbook of Heterogeneous Catalysis, Vol. 2 (G. Ertl, H. Knözinger, and J. Weitkamp, eds.), VCH: Weinham (1997) p. 752.

307. H. Vaarkamp, J.T. Miller, F.S. Modica and D.C. Koningsberger, *J. Catal.*163 (1996) 294.

308. J.T. Miller, B.L. Meyers, F.S. Modica, G.S. Lane, M. Vaarkamp and D.C. Koningsberger, *J. Catal.* **143** (1993) 395.

309. Y.Y. Tong, J.J. van der Klink, G. Clugnet, A.J. Renouprez, D. Lamb and P.A. Buffat, *Surf. Sci.* **292** (1993) 276.

310. S.B. Sharma, T.E. Laska, P. Balaraman, T.W. Root and J.A. Dumesic, *J. Catal.* **150** (1994) 225.

311. T.J. McCarthy, C.M.P. Marquez, H. Treviño and W.M.H. Sachtler, *Catal. Lett.* **43** (1997) 11.

312. A. Sárkány, G. Steffler and J.W. Hightower, *Appl. Catal. A: Gen.* **127** (1997) 77.

313. Z.C. Zhang and B.C. Beard, *Appl. Catal. A: Gen.* **188** (1999) 229.

314. D.E. Ramaker, J. de Graaf, J.A.R. van Veen and D.C. Koningsberger, *J. Catal.* **203** (2001) 7.

315. G.C. Bond and A.D. Hooper, *React. Kinet. Catal. Lett.* **68** (1999) 5.

316. D.E. Ramaker, B.L. Mojet, M.T.G. Oostenbrink, J.T. Miller and D.C. Koningsberger, *Phys. Chem. Chem. Phys.* **1** (1999) 2293; J.T. Miller. B.L. Mojet, D.E. Ramaker and D.C. Koningsberger, *Catal. Today* **62** (2000) 101.

317. V.B. Kazansky, V.Yu. Borovkov, N. Sokolova, N.I. Jaeger and G. Schulz-Egloff, *Catal. Lett.* **23** (1994) 263.

318. T.L.M. Maesen, M.J.P. Botman, T.M. Slaghek, L.-Q. She, J.Y. Zhang and V. Ponec, *Appl. Catal.* **25** (1986) 35.

319. D.J. Smith, D. White, T. Baird and J.R. Fryer, *J. Catal.* **81** (1983) 107.

320. D. White, T. Baird, J.R. Fryer, L.A. Freeman, D.J. Smith and M. Day, *J. Catal.* **81** (1983) 107, 119.

321. F.M. Dautzenberg and H.B.M. Wolters, *J. Catal.* **51** (1978) 26.

322. K. Kuninori, Y. Ikeda, M. Soma and T. Uchijima, *J. Catal.* **79** (1983) 185.

323. *Strong Metal–Support Interactions* (R.T.K. Baker, S.J. Tauster and J.A. Dumesic, eds.), American Chemical Society: Washington DC (1986), ACS Symposium Series 298.

324. R. Lamber, N.I. Jaeger and G. Schulz-Ekloff, *J. Catal.* **123** (1990) 285.
325. E. Ruckenstein and H.Y. Wang, *J. Catal.* **190** (2000) 32.
326. T. Fukunaga and V. Ponec, *Appl. Catal. A: Gen.* **154** (1997) 207.
327. L.F. Liotta, A.M. Venezia, A. Martorano, A. Rossi and G. Deganello, *J. Catal.* **171** (1997) 169.
328. G. Deganello, D. Duca, A. Martorano, G. Fagherazzi and A. Benedetti, *J. Catal.* **171** (1994) 127.
329. B. Tesche, T. Bentel and H. Knözinger, *J. Catal.* **149** (1994) 100.
330. M.C. Sanchez Sierra, J. Garca Ruiz, M.G. Proietti and J. Blasco, *J. Molec. Catal. A: Chem.* **96** (1995) 65; *Physica B* **208–209** (1995) 705.
331. J. El Fallah, S. Boujana, H. Dexpert, A. Kiennemann, J. Majerus, O. Touret, F. Villain ans F. Le Normand, *J. Phys. Chem.* **98** (1994) 5522.
332. M.J. Sterba and V. Haensel, *Ind. Eng. Chem. Prod. Res. Dev.* **15** (1976) 2.
333. J.-R. Chang, S.-L. Chang and T.-B. Lin, *J. Catal.* **169** (1997) 338.
334. R. Maurel, G. Leclercq and J. Barbier, *J. Catal.* **37** (1975) 324.
335. *Metal-Support Interactions in Catalysis, Sintering and Redispersion,* (S.A. Stevenson, J.A. Dumesic, R.T.K. Baker and E. Ruckenstein, eds.), Van Nostrand Reinhold: New York (1987).
336. H. Lieske and J. Völter, *J. Phys. Chem.* **89** (1985) 1841.
337. M. Kishida, K. Umakoshi, J.-I. Ishiyama, H. Nagata and K. Wakabagashi, *Catal. Today* **29** (1996) 355.
338. R.W. Devenish, T. Goulding, B.T. Heaton and R. Whyman, *J. Chem. Soc. Dalton Trans.* (1996) 673.
339. B.M.I. van der Zande, M.R. Böhmer, L.G.J. Fokkink and C. Schoenenberger, *J. Phys. Chem. B* **101** (1997) 852.
340. R.M. Wilenzilck, D.C. Russell, R.H. Morriss and S.W. Marshall, *J. Chem. Phys.* **47** (1967) 533.
341. A. Beck, A. Horváth, A. Szűcs, Z. Schay, H.E. Horváth, Z. Zsoldos, I. Dékány and L. Guczi, *Catal. Lett.* **65** (2000) 33.
342. J.T. Richardson and M.V. Twigg, *Appl. Catal. A: Gen.* **167** (1998) 57.
343. C. Bigey, L. Hilaire and G. Maire, *J. Catal.* **184** (1999) 406.
344. C. Ciciarello, A. Benedetti, F. Pinna, G. Strakul, W. Juszczyk and H. Brumberger, *Phys. Chem. Chem. Phys.* **1** (1999) 367.
345. K. Hadjiivanov, M. Mihaylov, D. Klissurski, P. Stefanov, N. Abadjieva, E. Vassileva and L. Mintchev, *J. Catal.* **185** (1999) 314.
346. G.S. Attard, J.C. Clyde and C.G. Göltner, *Nature* **378** (1995) 366.
347. R.J. Best and W.W. Russell, *J. Am. Chem. Soc.* **76** (1954) 838.
348. Chin-Pei Hwang and Chuin-Tih Yeh, *J. Molec. Catal. A: Chem.* **112** (1996) 295.
349. M.J. Jacaman, S. Fuentes and J.M. Dominguez, *Surf. Sci.* **106** (1981) 472.
350. S. Fuentes, A. Váquez, J.G. Pérez and M.J. Yacaman, *J. Catal.* **99** (1986) 492.
351. C.-P. Hwang and C.-T. Yeh, *J. Catal.* **182** (1999) 48.
352. S. Engels, J. Hangt and H. Lausch, *Z. Phys. Chem. (Leipzig)* **269** (1988) 463.
353. T.E. Huizinga, J. van Grondelle and R. Prins, *Appl. Catal.* **10** (1984) 199.
354. B. Baranowski and S. Majchrzak, *Roczn. Chem.* **42** (1968) 1137.
355. S.D. Robertson, B.D. McNicol, J.H. de Baas, S.C. Kloet and J.W. Jenkins, *J. Catal.* **37** (1975) 424.
356. R. Berte, A. Bossi, F. Garbassi and G. Petrini, in: *Proc. 7th Internat. Conf. Thermal Anal.* Vol. 2, (1982), p. 1224.
357. P.G.J. Koopman, A.P.G. Kieboom and H. van Bekkum, *J. Catal.* **69** (1981) 172.
358. A. Masson, in: *NATO ASI Series,* (J. Davenas and P.M. Rabette, eds.), Martinus Nijhoff Publishers, (1986), p. 295; *Structure and Reactivity of Surfaces, SSSC* **48** (1989) 665.
359. C.T. Campbell, in: *Handbook of Heterogeneous Catalysis* Vol. 2 (G. Ertl, H. Knozinger and J. Weitkamp, eds.), Wiley-VCH: Weinheim, (1997), p. 814.
360. G.A. Somorjai, *Appl. Surf. Sci.* **121/122** (1997) 1.

361. H. Poppa, *Vacuum* **34** (1984), 1081.
362. J.W. Niemartsverdriet, A.F.P. Engelen, A.M. de Jong, W. Wieldraaijer and G.J. Kramer, *Appl: Surf. Sci.* **144–145** (1999) 366.
363. D.R. Rainer and D.W. Goodman, *J. Molec. Catal. A: Chem.* **131** (1998) 259.
364. C.T. Campbell, *Surf. Sci. Rep.* **27** (1997) 1.
365. G. Rupprechter, G. Seeker, H. Goller and K. Hayek, *J. Catal.* **186** (1999) 201.
366. M. Klimenkov, S. Nepijko, H. Kuhlenbeck, M. Bäumer, R. Schlögl and H.- J. Freund, *Surf. Sci.* **391** (1997) 27.
367. D.W. Goodman, *J. Catal.* **216** (2003) 213.
368. J.K.A. Clarke, *Chem. Rev.* **75** (1975) 291.
369. R.L. Moss in: *Specialist Periodical Reports: Catalysis,* Vol. 4, (C. Kemball and D.A. Dowden, eds.), *Roy. Soc. Chem.* (1981) p. 31.
370. J.H. Larsen and I. Chorkendorf, *Surf. Sci. Rep.* **35** (1999 163.
371. J.H. Sinfelt and J.A. Cusumano, *Adv. Mater. Catal.* (1977) 1 (see *Chem. Abs.* **89**, 31353j).
372. R.D. Gonzalez, *Appl. Surf. Sci.* **19** (1984) 181.
373. J.H. Sinfelt, *Acc. Chem. Res.* **20** (1987) 134.
374. E.G. Allison and G.C. Bond, *Catal. Rev.* **7** (1973) 233.
375. J. Gretz, M.A. Volpe, A.M. Sica, C.E. Gigola and R. Touroude, *J. Catal.* **167** (1997) 314.
376. Ling Zhu and A.E. DePristo, *J. Catal.* **167** (1997) 400.
377. A. Guerrero-Ruiz, B. Bachiller-Baeza, P. Ferreira-Apericio and I. Rodríguez-Ramos, *J. Catal.* **171** (1997) 374.
378. Z. Huang, J.R. Fryer, C. Park, D. Stirling and G. Webb, *J. Catal.* **175** (1998) 226.
379. A.J. Renouprez, J.F. Trillat, B. Moraweck, J. Massardier and G. Bergeret, *J. Catal.* **179** (1998) 390.
380. R. Prestvik, K. Moljord, K. Grande and A. Holmen, *J. Catal.* **174** (1998) 119.
381. M.J. Yacamán, *Appl. Catal.* **13** (1984) 1; M.J. Yacamán, G. Díaz and G. Gómez, *Catal. Today* **23** (1995) 161.
382. J.M. Thomas, *Faraday Discuss. Chem. Soc.* **105** (1996) 1.
383. P. Hawkes in: *Microscopy and Analysis,* Jan. (1998), p. 9.
384. P.L. Gai, *Topics in Catal.* **8** (1999) 97.
385. A.K. Datye, in: *Handbook of Heterogeneous Catalysis*, Vol. 2 (G.Ertl. H.Knozinger and J. Weitkamp, eds.), Wiley-VCH: Weinheim, (1997), p. 493.
386. P.L. Gai and E.D. Boyes, *Electron Microscopy in Heterogeneous Catalysis*, Inst. Phys. Publ.: Bristol (2003).
387. A.K. Datye, D.S. Kalakkad, M.H. Yao and J. Smith, *J. Catal.* **155** (1995) 148.
388. Z. Huang, J.F. Fryer, C. Park, D. Stirling and G. Webb, *J. Catal.* **148** (1994) 478.
389. S. Giorgia, C.R. Heary, C. Chapon and C. Roucau, *J. Catal.* **148** (1994) 534.
390. A.S. Ramachandran, S.L. Anderson and A.K. Datye, *Ultramicroscopy* **51** (1993) 282.
391. J.H.A. Martens, R. Prins, H. Zandbergen and D.C. Koningsberger, *J. Phys. Chem.* **92** (1988) 1903.
392. S. Fuentes, A. Váquez, R. Silva, J.G. Pérez-Ramírez and M.J. Yacaman, *J. Catal.* **111** (1988) 353.
393. F.W.H. Kampers, C.W.R. Engelen, J.H.C. van Hooff and D.C. Koningsberger, *J. Phys. Chem.* **94** (1990) 8574.
394. E.B. Prestridge, G.H. Via and J.H. Sinfelt, *J. Catal.* **50** (1977) 115.
395. M. Fernández-Garcia, F.K. Chong, J.A. Anderson, C.H. Rochester and G.L. Haller, *J. Catal.* **182** (1999) 199.
396. D. White, T. Baird, J.R. Fryer and D.J. Smith, *Inst. Phys. Conf. Ser. No. 61,* EMAG: Cambridge (1981), Ch. 8, p. 403.
397. N.R. Avery and J.V. Sanders, *J. Catal.* **18** (1970) 132.
398. M. Pan, J.M. Cowley and I.Y. Chan, *J. Appl. Crystallogr.* **20** (1987) 300.

399. M.J. Yacamán and A. Gómez, *Appl. Surf. Sci.* **19** (1984) 348.
400. J.J.F. Scholten and A. van Montfoort, *J. Catal.* **1** (1962) 85.
401. W.M. Targos and H. Abrevaya in: *Microbeam Analysis - 1986,* (A.D. Romig Jr. and W.F. Chambers, eds.), San Francisco Press: San Francisco (1986), p. 605.
402. A.D. Logan, E.J. Braunschweig, A.K. Datye and D.J. Smith, *Langmuir* **4** (1988) 827.
403. R. Lamber and G. Schulz-Ekloff, *J. Catal.* **146** (1994) 601.
404. R. Lamber and G. Schulz-Ekloff, *Surf. Sci.* **258** (1991) 107.
405. R. Lamber and N.I. Jaeger, *Surf. Sci.* **289** (1993) 247.
406. M. Vaarkamp, J.V. Grondelle, J.T. Miller, D.J. Sajkowski, F.S. Modica, G.S. Lane, B.C. Gates and D.C. Koningsberger, *Catal. Lett.* **6** (1990) 369.
407. P. Canton, F. Menegazzo, S. Polizzi, F. Pinna, N. Pernicone, P. Tietto and G. Fagherazzi, *Catal. Lett.* **88** (2003) 141.
408. S.D. Jackson, J. Willis, G.J. Kelly, G.D. McLellan, G. Webb, S. Mather, R.B. Moyes, S. Simpson, P.B. Wells and R. Whyman, *Phys. Chem. Chem. Phys.* **1** (1999) 2573.
409. T. Mallát, S. Szabó and J. Petró, *Appl. Catal.* **29** (1987) 117.
410. J. Pielaszek in: *X-Ray and Neutron Structure Analysis,* (J. Hasek, ed.), Plenum (1989), 209.
411. W. Palczewska, A. Jablonski, Z. Kaszkur, G. Zuba and J. Wernisch, *J. Molec. Catal.* **25** (1984) 307.
412. B.N. Das, A.D. Sarney-Loomis, F. Wald and G.A. Wolff, *Materials. Res. Bull.* **3** (1968) 649, 705.
413. R.A. van Nordstrand, A.J. Lincoln and A. Carnevale, *Anal. Chem.* **36** (1964) 819.
414. T.A. Dorling and R.L. Moss, *J. Catal.* **5** (1966) 111; **7** (1967) 378.
415. V. Ponec, *Mat. Sci. Eng.* **42** (1980) 135.
416. F. Bozon-Verduraz, A. Omar, J. Escard and B. Pontvianne, *J. Catal.* **53** (1978) 126.
417. A.M. Venezia, A. Rossi, D. Duca, A. Martorana and G. Deganello, *Appl. Catal. A: Gen.* **125** (1995) 113.
418. *X-Ray Absorption Studies in Catalysis,* (G.W. Coulston and R.W. Joyner, eds.), *Catal. Today* **39** (1994) 261.
419. M. Vaarkamp, B.L. Mojet, M.J. Kappers, J.T. Miller and D.C. Koningsberger, *J. Phys. Chem.* **99** (1995) 16067.
420. M.Vaarkamp, F.S. Modica, J.T. Miller and D.C. Koningsberger, *J. Catal.* **144** (1993) 611.
421. A. Borgna, F. Le Normand, T. Garetto, C.R. Apesteguia and B. Moraweck, *Catal. Lett.* **13** (1992) 175.
422. F.B.M. Duivenvoorden, B.J. Kip, D.C. Koningsberger and R. Prins, *J. Phys. Colloq.* C8 **1** (1986) 227.
423. Sun Hee Choi and Jae Sung Lee, *J. Catal.* **167** (1997) 364.
424. P.V. Menacherry, M. Fernández-García and G.L. Haller, *J. Catal.* **166** (1997) 75.
425. L.M.P. van Gruijthuijsen, G.J. Howsman, W.N. Delgass, D.C. Koningsberger, R.A. van Santen and J.W. Niemantsverdriet, *J. Catal.* **170** (1997) 331.
426. Ihl Hyung Cho, Seung Bin Park, Sung June Cho and Ryong Ryoo, *J. Catal.* **173** (1998) 295.
427. A. Munoz-Paez and D.C. Koningsberger, *J. Phys. Chem.* **99** (1995) 4193.
428. D.C. Koningsberger and B.C. Gates, *Catal Lett.* **14** (1992) 271.
429. P. Prins, J.H.A. Martens and D.C. Koningsberger, *Structure and Reactivity of Surfaces* **48** (1989) 759.
430. D.C. Koningsberger, J.H.A. Martens, R. Prins, D.R. Short and D.E. Sayers, *J. Phys. Chem.* **90** (186) 3047.
431. S.K. Purnell, K.M. Sanchez, R. Patrini, J.-R. Chang and B.C. Gates, *J. Phys. Chem.* **98** (1994) 1205.
432. M. Vaarkamp, B.L. Mojet, M.J. Kappers, J.T. Miller and D.C. Koningsberger, *J. Phys. Chem.* **99** (1995) 16067.
433. F.B.M. Van Zon, S.D. Maloney, B.C. Gates and D.C. Koningsberger, *J. Am. Chem. Soc.* **115** (1993) 10317.
434. J.R. Shapley, W.S. Uchiyama and R.A. Scott, *J. Phys. Chem.* **94** (1990) 1190.

435. R.W. Joyner, *J. Chem. Soc. Faraday Trans. I* **76** (1980) 357.
436. A.N. Mansour, J.W. Cook Jr., D.E. Sayers, R.J. Emrich and J.R. Katzer, *J. Catal.* **89** (1984) 462.
437. K.I. Zamaraev and D.I. Kochubey, *Kinet. Katal.* **27** (1986) 1034; K.I. Zamarev, *Topics in Catal.* **3** (1996) 1.
438. M.C. Sanchez Sierra, J. Garcia Ruiz, M.G. Proietti and J. Blasco, *J. Molec. Catal. A: Chem.* **108** (1996) 95.
439. G. Vlaic, J.C.J. Bart, W. Cavigialo, A. Furesi, V. Ragaini, M.G. Cattania Sabbadini and E. Burattini, *J. Catal.* **107** (1987) 263.
440. S.J. Cho, W.-S. Ahn, S.B. Hong and R. Ryoo, *J. Phys. Chem.* **100** (1996) 4996.
441. F. Engelke, R. Vincent, T.S. King and M. Pruski, *J. Chem. Phys.* **102** (1994) 7262.
442. D.P. VanderWiel, M. Pruski and T.S. King, *J. Catal.* **188** (1999) 186.
443. P.P. Edwards and M.J. Sienks, *Internat. Rev. Phys. Chem.* **3** (1983) 83; *Acc. Chem. Res.* **15** (1982) 87; *J. Chem. Educ.* **60** (1983) 691.
444. P.P. Ewards, R.L. Johnston and C.N.R. Rao, in: *Metal Clusters in Chemistry,* (P. Braunstein, L.A. Oro and P.R. Raithby, eds.), Wiley-VCH: Weinheim, Vol. 3, p. 1454.
445. V. Ponec, *Adv. Catal.* **32** (1983) 149.
446. R.C. Baetzold and J.F. Hamilton, *Prog. Solid State Chem.* **15** (1983) 1.
447. M. Boudart, in: *Precious Metals Science and Technology,* (L.S. Benner, T. Suzuki, K. Meguror and S. Tanaka, eds.), IPMI Press: Allentown PA (1991), Ch. 6.
448. N. Wada, *Jap. J. Appl. Phys.* 7 (1968), 1287.
449. *Metal-Support and Metal-Additive Effects in Catalysis,* (B. Imelik, C. Naccache, G. Coudurier, H. Praliaud, P. Meriaudeau, P. Gallezot, G.A. Martin and J.C. Védrine, eds.), Studies in Surface Science and Catalysis, Elsevier: Amsterdam, **11** (1982).
450. G.C. Bond in: *Metal-Support and Metal-Additive Effects in Catalysis,* (B. Imelik, C. Naccache, G. Coudurier, H. Praliaud, P. Meriaudeau, P. Gallezot, G.A. Martin and J.C. Védrine, eds.), Studies in Surface Science and Catalysis, Elsevier: Amsterdam, **11** (1982) 1.
451. R.W. McCabe, R.K. Usmen, K. Ober and H.S. Gandhi, *J. Catal.* **151** (1995) 385.
452. H. Lieske, G. Lietz, H. Spindler and J. Völter, *J. Catal.* **81** (1983) 8.
453. G. Lietz, H. Lietz, H. Spindler, W. Haake and J. Völter *J. Catal,* **81** (1983) 17.
454. G.H. Bartholomew and W.L. Sorensen, *J. Catal.* **81** (1983) 131.
455. J.A. Anderson, *Catal. Lett.* **13** (1992) 363.
456. S.C. Fung, *Catal. Today* **53** (1999) 325.
457. G.B. McVicker, R.L. Garten and R.T.K. Baker, *J. Catal.* **54** (1978) 129.
458. J.L. Contreras and G.A. Fuentes, *Proc. 11th Internat. Congr. Catal.* (J.W. Hightower, W.N. Delgass, E. Iglesia and A.T. Bell, eds.), Elsevier: Amsterdam (1996) 95.
459. K. Wong, S. Johansson and B. Kasemo, *Faraday Disc. Chem Soc.* **105** (1996) 237.
460. G. Rupprechter, K. Hayek and H. Hofmeister, *J. Catal.* **173** (1998) 409.
461. S. Bernal, J.J. Calvino, M.A. Cauqui, J.M. Gatica, C. Larese, J.A. Pérez Omil and J.M. Pintado, *Catal. Today* **50** (1999) 175.
462. F.B. Passos, D.A.G. Aranda and M. Schmal, *Catal. Today* **57** (2000) 283.
463. H. Müller, C. Opitz and L. Skala, *J. Molec. Catal.* **54** (1989) 389.
464. Sun Hee Choi and Jae Sung Lee, *J. Catal.* **193** (2000) 176.
465. D. Buzin, C. Mottet and G. Tréglia, *Appl. Catal. A: Gen.* **200** (2000) 47.
466. E.A. Sales, J. Jove, M. de J. Mendes and F. Bozon-Verduraz, *J. Catal.* **195** (2000) 88, 96.
467. M. Bonorowska, J. Pielaszek, W. Juszczyk and Z. Karpiński, *J. Catal.* **195** (2000) 304.
468. S.U. Troitski, M.A. Serebriakova, M.A. Fedotov, S.V. Ignashin, A.L. Chuvilin, E.M. Moroz, B.N. Novogorodov, V.A. Likholobov, B. Blanc and P. Gallezot, *J. Molec. Catal. A. Chem.* **158** (2000) 461.
469. W. Tu and H. Liu, *J. Mater. Chem.* **10** (2000) 2207.
470. J. Batista, A. Pintar, D, Mandrino, M. Jenko and V. Martin, *Appl. Catal. A: Gen.* **15** (1992) 219.

471. B.D. Chandler, L.I. Rubinstein and L.H. Pignolet, *J. Molec. Catal. A: Chem.* **133** (1998) 267.
472. B.D. Chandler, A.B. Schabel, C.F. Blanford and L.H. Pignolet, *J. Catal.* **187** (1999) 367.
473. B.D. Chandler, A.B. Schabel and L.H. Pignolet, *J. Catal.* **193** (2000) 186.
474. B.D. Chandler, A.B. Schabel and L.H. Pignolet, *J. Phys. Chem. B.* **105** (2001) 149.
475. T. Fujimoto, S. Terauchi, H. Umehara, L. Kojima and W. Henderson, *Chem. Mater.* **13** (2001) 1057.
476. M.J. Williams, P.P. Edwards and D.P. Tunstall, *Faraday Discuss. Chem. Soc.* **92** (1991) 229.
477. C. Mihut, C. Descorme, D. Duprez and M.D. Amiridis, *J. Catal.* **212** (2002) 125.
478. G.B. Sergeev, *Russ. Chem. Rev.* **70** (2001) 809.
479. U. Heiz, A. Sauchez, S. Abbet and W.-D. Scheider, *J. Amer. Chem. Soc.* **121** (1999) 3214.

FURTHER READING

Colloids 19, 41, 185, 271, 337–341, 468, 469, 475

Methods of catalyst preparation 31, 47–55, 57, 70, 90, 91, 97–99, 101, 122, 129–131, 161, 164, 342–347

Temperature-programmed reduction 104, 123, 156, 158, 217, 333, 336, 343, 348–357, 458

'Model' catalysts 100, 142–144, 145, 168, 219, 220, 236, 273, 274, 276, 358–367, 459, 460

Bimetallic catalysts 103, 146, 157, 158, 204, 298, 368–380, 465–467, 470–474, 477

Transmission electron microscopy 11, 34, 70, 77, 102, 112, 123, 137, 158, 187, 217, 246, 248, 296, 306, 319, 320, 329, 349, 350, 378, 380–407

X-ray diffraction and related techniques 158, 208, 333, 343, 344, 400, 408–414

Photoelectron spectroscopy 10, 30, 33, 34, 65, 77, 112, 158, 216, 244, 246, 287, 328, 343, 351, 375, 408, 411, 415–417

Extended X-ray absorption fine structure (EXAFS) 42, 65, 92, 100, 102, 112, 130, 131, 133, 136, 208, 210, 217, 232, 246, 263, 307, 308, 316, 330, 333, 343, 395, 391, 406, 4118–439, 464, 476

XANES 65, 133, 157, 316, 395, 434, 436, 438, 440, 464

Nuclear magnetic resonance using[195]**Pt** 130, 211, 224, 225, 227–229, 243,244, 256, 257, 288, 309, 310, 438–442

Properties of small metal particles 6, 113, 144, 164, 179, 184, 269, 270, 273, 443–446, 463

Particles of pentagonal symmetry 6, 184, 265–269, 273, 447, 448

Metal-support interactions 6, 7, 58, 65, 66, 164, 168, 184, 261, 294, 295, 302, 306, 316, 323, 335, 369, 429, 449–451, 461, 462

Sintering 145, 313, 319, 320, 321, 335, 336, 344, 395, 452–457

CHEMISORPTION AND REACTIONS OF HYDROGEN

PREFACE

The hydrogen molecule is dissociatively chemisorbed with great facility by a number of metals within the Transition Series: elsewhere the process is more highly activated, but chemisorbed atoms once formed are stable. The contrast between hydrogen and the alkanes is very striking: the σ H—H bond can sometimes be broken at temperatures close to 20 K, the activation energy being minimal, whereas with methane each σ C—H bond is effectively shielded by the other hydrogen atoms, so that much higher temperatures are needed for its chemisorption by dissociation.

The strengths of hydrogen-metal bonds at the surfaces of metals of Groups 8 to 10 are of similar order, but they tend to increase on moving towards the centre of each Transition Series. These trends are shown more distinctly by other molecules such as oxygen and nitrogen, where a clear parallelism with analogous bulk compounds is shown: but within the Transition Series stoichiometric hydrides are formed only by metals of Groups 3, 4 and 5. Because of commercial interest and experimental convenience, certain metals have commanded disproportionate attention: tungsten, platinum and nickel in the massive form and platinum also in the microscopic form. For these same reasons, other metals have escaped attention almost entirely (e.g. manganese and osmium).

All the catalysed reactions to be considered in the following chapters involve hydrogen (or deuterium) as reactant or product, and a knowledge of its chemisorption, and of reaction between its isotopic and spin variants, will prove helpful in understanding its reaction with hydrocarbons. For this reason most emphasis will be placed on the interaction of hydrogen with metals of Groups 8 to 10.

While the utility of hydrogen in catalysed reactions is due in part to its small size and easy migration, these features add greatly to the difficulty of describing the process of adsorption and desorption and the structure of the adsorbed layer under any specified conditions, especially when the adsorbent is a supported metal.

3.1. THE INTERACTION OF HYDROGEN WITH METALS

As almost all the catalysed reactions to be considered in the following chapters involve hydrogen (or deuterium) as reactant or product, it is first necessary to explore how this molecule interacts with metals in general, and those metals of most interest to catalysis in particular.[1-17] The hydrogen-metal system has been extremely thoroughly studied, and a whole book has been devoted to the relevant surface chemistry and catalysis. The interactions that occur are also of great importance in metallurgical processing, but that is beyond the scope of this work.

Hydrogen interacts with metals in three principal ways: (i) by dissociative chemisorption at the surface; (ii) by physical adsorption as molecules at very low temperatures; and (iii) by dissolution[18] or occlusion.[19,20] As we shall see, to these three extreme forms have been added numerous intermediate states of various lifetimes and stabilities, some of which may have importance in catalysis. There is for example clear evidence for a molecular state formed at about 100 K on stepped surfaces saturated with atomic hydrogen (on Ni(510),[21,22] Pd(510)[21] and (210)[23]): this is distinct from a molecular *precursor* state such as that seen with deuterium on Ni(111) at 100 K. The role of such states will be discussed further below (Sections 3.2.2 and 3.3.3). The small size and electronic simplicity of the hydrogen atom formed by dissociation enable it to bond to metal surfaces in different ways, and simple-minded notions about its forming only a single covalent bond to another atom have to be abandoned.

The earliest studies of the chemisorption of hydrogen by metals were made using powders, foils and wires, which were polycrystalline, exposing various crystal planes, and by today's standards far from clean.[24] Nevertheless this early work started to delineate some of the main features of the interactions. During the late 1940s, Otto Beeck and his associates at the Shell Development Company developed the use of metal films[25] (Section 1.2.1), and these were extensively used in the following decades for studies of chemisorption and catalysis.[26-28] Although still polycrystalline, it was possible to persuade them to expose a specific crystal plane preferentially,[29] and bimetallic (alloy) films were also made. Such films had quite clean surfaces and yielded results that have been useful for correlating strength of adsorption with catalytic activity.[30] The next important development was field-emission microscopy (FEM),[31] followed by field-ion microscopy (FIM)[26,32-34] (Section 1.2.1), techniques which revealed the preference of adsorbates for different crystal planes at the tip of very fine needles, albeit under stressful conditions.[31,35] The most recent advance has been the use of single crystal metal surfaces,[36-38] exposing chiefly a single crystal plane or a predetermined variety of configuration on stepped or kinked surfaces[37,39,40] (Section 1.21). Their surfaces may be thoroughly cleaned by ion bombardment and identified by LEED, although initial cleanliness is not always maintained (e.g. dissolved impurities such as sulfur may emerge slowly) and STM reveals defects and imperfections that LEED misses. Nevertheless their employment has caused a revolution in our understanding of the dynamics of formation and structure of chemisorbed layers. This information, some aspects of which will be summarised in the next section, is of limited relevance to the behaviour of small metal particles, although pertinent to those catalysed reactions that can be observed on single crystals: the degree of relevance is increased by using stepped and kinked surfaces.

The way in which hydrogen interacts with supported metal catalysts is necessarily more complex than is the case with single crystals. Small particles expose different crystal planes and many edge and corner atoms, and are probably more defective than massive forms (Section 2.5.2). Moreover, although hydrogen only rarely interacts directly with the materials commonly used as supports, it can nevertheless migrate with surprising ease from metal to support, producing a number of interesting and occasionally useful consequences (Section 3.34). Reaction at the metal-support interface has also been detected. The frequent use of hydrogen chemisorption and of hydrogen-oxygen titration to estimate metal area and particle size[41] is justified by the low cost and simple operation of these techniques, but great experimental and theoretical care is needed before all ambiguities and uncertainties are removed, and complete confidence in the answer is obtained.

The physical adsorption of molecules at surfaces is due to dispersion or Van der Waals forces of the type that hold them together in the liquid state, so that this form is not in general stable much above the boiling point of the liquid. In the case of hydrogen, its physical adsorption at ambient temperature and above can safely be ignored, although its existence in principle provides a means of bringing the molecule close to the surface without dissociation, thus easing the breaking of the bond (Section 3.2.1).

Hydrogen dissolves into a number of the Transition Series metals with the formation of metallic hydrides;[6,11,14,42-44] with platinum however, the process is endothermic and occurs only to a small extent and is therefore unimportant. Apart from palladium, only in the case of iridium has its retention or occlusion within the grain structure been implicated in catalytic behaviour, being thought responsible for the low selectivity which this metal usually gives in the hydrogenation of alkynes and alkadienes[19,20] (Chapters 8 and 9). Palladium is however unique in its capacity to absorb hydrogen;[6,14,43,45-47] on raising the pressure at room temperature, the α-phase is first formed, and it co-exists with the β-phase in the H/Pd range 0.05 to 0.58, after which only the β-phase exists[48] (Figure 3.1). The range of

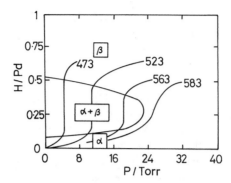

Figure 3.1. Phase diagram for the palladium-hydrogen system (note the resemblance to the CO_2 phase diagram).

co-existence of the two phases decreases with increasing temperature. The β-phase is a semiconductor, but mechanical strength is maintained to quite high H/Pd. A major application of this property is to produce hydrogen of high purity,[49] by diffusion through a palladium sheet or membrane: it is preferable to use an alloy with silver ($Pd_{77}Ag_{23}$) to avoid mechanical failure through embrittlement due to passage in and out of the β-phase, the formation of which causes a volume expansion of 10%. Much significant catalytic work has been done with palladium alloys of the Group 11 metals,[49,50] and the solubility of hydrogen and deuterium in them has been investigated.[47,51–53] The concentration of hydrogen atoms in palladium at saturation corresponds to about Pd_4H_3 and approaches that in liquid hydrogen. It now appears however that such a concentration is insufficient to sustain nuclear fusion at ordinary temperatures when hydrogen is replaced by deuterium. Solubility decreases as particle size becomes smaller,[15,44,54,55] and reduction of small palladium oxide particles by hydrogen gives metallic palladium, while larger particles lead to the hydride phase. There is circumstantial evidence to show that hydrogen atoms can dissolve into and diffuse quite easily through gold;[56–58] they also dissolve into nickel[59] and nickel-copper alloys[60] under electrochemical persuasion. The extraordinary properties of the palladium-hydrogen system have been thoroughly studied, and are described in an extensive literature.[15,46,48]

Earlier elements in the Transition Series (Groups 1 to 5) and intermetallic compounds containing a Rare Earth element (e.g. $LaNi_5$) form stoichiometric hydride phases which have possible application in hydrogen storage, but they are loath to release their hydrogen to acceptor molecules and even in the presence of hydrogen gas are undistinguished as catalysts.[15] Many of the pre- and post-Transition Series elements form molecular hydrides,[43] some of which (notably elements in Groups 3, 4 and 14) are polymeric.

3.2. CHEMISORPTION OF HYDROGEN ON UNSUPPORTED METALS AND ALLOYS[61,62]

3.2.1. Introduction

The nucleus of the hydrogen atom (^{1}H) is a single proton; addition of one neutron gives a deuteron (D or ^{2}H, natural abundance 0.0156%), and of two neutrons gives a tritium atom (T or ^{3}H): the D nucleus is stable, but the T nucleus is a weak β-emitter (half-life 12.35 years). The diatomic molecules HD, H_2 and D_2 differ significantly in their mass and other physical characteristics, making analysis of their mixtures quite straightforward, and the mass distinction in particular has allowed deuterium used as a stable isotopic tracer for more than half a century; tritium atoms are easily located by standard radiochemical methods. Isotopically pure samples are hard to obtain, and in admixture species such as hydrogen deuteride (HD) or hydrogen tritide (HT) will prevail. The equilibrium constant for the system

$$H_2 + D_2 \rightleftharpoons 2HD \tag{3.A}$$

is about 3.2 at ambient temperature, due to the zero-point energy effect. Nuclear spin isomers also exist, in which the spins are either parallel or anti-parallel: parallel spins give the ortho-forms, which at equilibrium constitute 75% of hydrogen and tritium, and 66.7% of deuterium, while the anti-parallel spins give the para-forms. The stability of the hydrogen molecule (dissociation energy 436 kJ mol^{-1}) is such that atoms only exist at equilibrium with it to very small extents (0.08% at 2000 K). Dissociation is achieved more easily by mercury-sensitised photolysis and especially by metallic and other catalysts. Interconversion of spin isomers is not diagnostic for dissociation and recombination because this can be induced by a strong magnetic field; equilibration of molecular hydrogen and deuterium, forming hydrogen deuteride of course demands bond breaking and re-formation. Basic studies of chemisorption are usually made with hydrogen, although subtle but significant differences are sometimes seen when isomers are compared.[63,64]

Most of the results to be reviewed in this section will have been obtained on single crystals cut to exposed a defined place which may be flat, stepped or kinked and on surface alloys: where necessary (because of the lack of information on the relevant single crystal), or helpful for purposes of comparison, we shall refer to results obtained on polycrystalline forms (films, powders etc). This attention to single crystals is justified by the reliability and fundamental nature of the measurements, which involve a large number of sophisticated surface-sensitive techniques or procedures for monitoring the solid state as it responds to chemisorption at its surface. Many of the methods referred to in the preceding Chapter for characterising clean surfaces also provide information on hydrogen chemisorption, some of the more important including LEED and related methods, ^{1}H and ^{129}Xe

NMR, EXAFS/XANES,[65,66] and STM[67] (see Section 2.4.2). Procedures specific to chemisorbed states[3,17,29,30,68–70] include measurement of changes in electrical or magnetic character, vibrational spectroscopies, (HREELS, DRIFTS/RAIRS etc), calorimetry[71] and thermal desorption.[72,73] This short list is far from being comprehensive, and the reader is reminded that the purpose of this work is not to instruct in the use of these techniques, but rather to present and evaluate the results they generate. The principles concerned are only mentioned where understanding of the results necessitates it. This is somewhat in the spirit of Jerome K. Jerome, who in his Preface to 'Three Men in a Boat (to say nothing of the dog)' advised his readers not to use it as a manual for a River Thames holiday; they would, he said, be wiser to stay ay home.

It is not easy to decide how to order a summary of the vast amount of information available. The theme which we seek to illustrate is how the processes of chemisorption and desorption and the properties of the adsorbed state depend upon the chemical composition of the surface and its atomic structure. The rest of this section (3.2) is sub-divided into the chemisorption process (3.2.2), structural aspects of the chemisorbed state (3.2.3) and energetic aspects including desorption (3.2.4). Such subdivision is highly artificial as there are close interrelations between all subsets; however, each technique is chiefly directed towards obtaining information of a specific type, and integration of the various results is largely an intellectual or modelling exercise.

A convenient way of entering the detail of the subject is through the one dimensional potential energy diagram due originally to Sir John Lennard-Jones:[74] it is one dimensional in the sense that the H—M distance is the only variable considered, whereas the H—H distance will also increase as dissociation occurs and a fuller representation of the process is possible by two-dimension diagrams containing both distances as variables.[3,75] The basic one-dimensional picture[3,6] (Figure 3.2) depicts a molecule approaching the surface and falling into a shallow energy trough, the depth of which gives the heat of physical adsorption (typically $10–20 \text{ kJ mol}^{-1}$); the minimum is quite far from the surface, at a distance equal to the sums of the covalent and Van der Waals radii of the metal and hydrogen atoms. This process is not activated and this state would not be significantly occupied at ordinary temperatures.

The second curve is for two separate atoms coming towards the surface: it therefore starts at the right well above the energy zero, as 436 kJ mol^{-1} have to be found to atomise the molecule in free space. The depth of the trough gives the heat of chemisorption and the minimum is quite close to the surface, the distance being roughly the sum of the covalent radii. In Figure 3.2 the curves are drawn so that the transition state occurs very close to this potential energy zero, which means that the process of chemisorption shown here has only a very small activation energy. It does however have a finite value, because the chemisorption of deuterium is slower than that of hydrogen at 87 K, the rate being sensitive to the zero-point energy

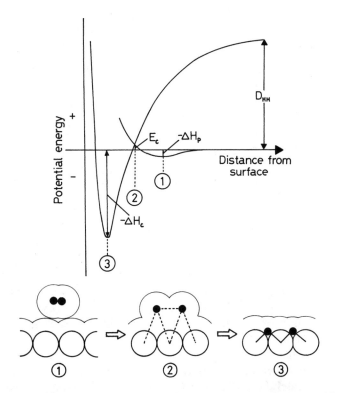

Figure 3.2. Lennard-Jones potential energy diagram for the interaction of hydrogen with the surface of a metal of Groups 8–10 (see text for description). The lower part of the diagram shows possible configuration at three points in the chemisorption process.

difference. Small activation energies are the general rule for metals of Groups 8 to 10 (E_c < 10 kJ mol^{-1}), but for the coinage metals in Group 11 they are large and quite easily measurable[76] (E_c ~20–60 kJ mol^{-1}). We can see that the state of physical adsorption provides a pathway for a molecule to approach the surface and to attain a position where interaction of its orbitals with those of the metal can proceed and lead to dissociation, without a disconcertingly large input of energy; therein lies much of the secret of catalysis. Some other forms of energy (thermal, radiative etc) can also induce dissociation, but it is the great strength of the H—H bond that determines that in the absence of a catalyst it can be broken only at a very high temperature. The presence of a catalyst is therefore *essential* for reactions involving hydrogen to proceed under mild conditions. Figure 3.3 shows the intervention of a significant activation energy in the case of Group 11 metals.

The potential energy diagram can be extended to show features of the process of dissolution of hydrogen atoms into the bulk.[3] There is evidence in some systems

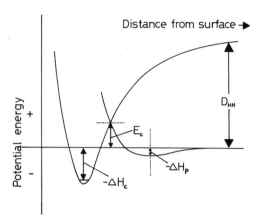

Figure 3.3. Potential energy diagram for the interaction of hydrogen with the surface of a metal of Group 11.

that sites immediately below the surface atoms can accommodate them in a stable way,[77,78] whence (especially with palladium and its alloys) they migrate into the interior. Figure 3.4 shows the heat content changes and activation energies involved.

These diagrams are helpful in understanding how chemisorption of hydrogen occurs, but they have their limitations: they imply for example that all hydrogen atoms are chemisorbed in the same way and with the same strength, but this as we shall see is rarely true. The initial collision may, for example, give atoms at places where they are 'uncomfortable' and from which they may diffuse to more energetically favourable positions. This initial condition constitutes an *atomic precursor state*.[4]

3.2.2. The Process of Chemisorption[26,32,33,68,75,78]

The fundamental measure of the reactivity of a metal surface towards hydrogen is the initial sticking probability S_0 defined simply as the fraction of molecules colliding with the surface that remain chemisorbed, extrapolated to zero coverage. As the *exposure* (defined as pressure × time) increases, so does the coverage θ, and the sticking probability S must fall; if a molecular beam gas source is used, the increase of the background pressure (measured by an ion gauge or a mass spectrometer) after a particular exposure, ratioed to the pressure rise when no further adsorption occurs, gives the value of S at that point. The commonly used unit of exposure is the Langmuir, which is 10^{-6} Torr s; one Langmuir (1 L) will give a full coverage of a typical metal surface if S is unity. Coverage is calculated from the number of molecules that have chemisorbed and the estimated number of surface atoms.

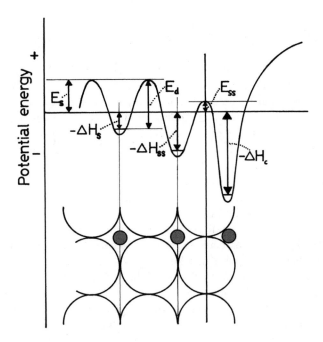

Figure 3.4. Potential diagram showing the process of dissolution of hydrogen atoms into a metal: ss = subsurface site; s = dissolution; d = diffusion.

It is worth recalling from Section 1.2.2 that a surface is defined by the crystallographic plane in which the outermost atoms reside, although with certain atomically rough or 'open' surfaces, such as the fcc(110) or the bcc(111), atoms from a second or even a third underlying layer are also substantially exposed. It is usual to express coverage in terms only of the atoms in the strict surface plane and the result is that with such surfaces, values of the coverage θ may rise well above unity since the partly exposed atoms also contribute to the bonding capability of the surface. Absolute coverages of deuterium on Pt(111) have been measured[79] by nuclear microanalysis using the reaction $D(^3He,p)^4He$.

The sticking probability S of the hydrogen molecule being the simplest measure of metal surface reactivity, its dependence upon crystal face and coverage may provide ideas for understanding other reactions. In the absence of a precursor state and on an energetically homogeneous surface S will depend upon θ as

$$S = S_0(1 - \theta)^n \tag{3.1}$$

where n is the number of atoms making up the site, but where a precursor state exists S will initially fall less quickly, but more rapidly as the coverage approaches completion. The initial value S_0 is determined by the temperature of the surface, by the translational and internal energies of the molecule as it hits it, and by its chemical nature and structure. Much information has been obtained on the effect of these variables in the case of the Group 11 metals[64,80] and their alloys,[81] where the existence of a significant activation energy E_{ads} (see Figure 3.3) and a low value of S_0 makes experimentation more straightforward than with metals of Groups 8 to 10. Surface temperature matters when some intermediate adsorbed state exists for sufficiently long to attain thermal equilibrium, but not if dissociation is a direct, activated process. Excitation of both translational and vibrational energies of the colliding molecule assists dissociation, but rotational energy seems to have little effect. The different behaviours shown by hydrogen and deuterium help to resolve details of the process of dissociation.[63,64]

In the case of the Groups 8 to 10 metals, a major factor is the surface structure.[62] With the atomically compact surfaces such as the fcc(100) and (111) values of S_0 are usually small (<0.1), but with the more open fcc(110) and cph(1010) the incoming molecule experiences a softer and deeper potential well (and a greater variety of emergent orbitals) and S_o values are close to unity.[1,17,75] Their roughness may also help to dispose of the heat of adsorption, which is probably used to excite lattice phonons (i.e., vibrational modes of the solid).[1,75] The literature states that defects, and intentional irregularities such as steps and kinks, encourage sticking,[82] but in the case of platinum this has been disputed;[83] preadsorbed oxygen on Pt(110) lowers the sticking coefficient of hydrogen,[84] and the effect of potassium on Ni(111)[85] and on Pt(111) is similar.[86] There is clearly much yet to be learned about this seemingly simple process.[75]

There is a useful application of the Principle of Microscopic Reversibility (or Detailed Balancing) in the study of surface processes. This is a principle that requires that, when carried out under identical conditions, the reverse of any process should proceed by *exactly* the same route as the forward process: thus whatever energy input is needed for the chemisorption of a hydrogen molecule will be recovered and released when the two atoms recombine and desorb. Measurement of the relaxation of the vibrational and translational energy of the desorbing molecule therefore provides information on the needs in dissociation , and values of S can also be derived.[80]

As we shall see in what follows, in very many cases there are distinct 'phases' formed as the coverage by hydrogen atoms increases; these show characteristic values of heat of adsorption, sticking probability S and other physical properties. The fine structures of S versus θ plots that are observed when the process of chemisorption is not energetically uniform will be considered in the following section.

3.2.3. The Chemisorbed State: Geometric Aspects[1,3,17,75]

We have seen that the kind of metal used and the atomic structure of its surface both affect the ease with which the hydrogen molecule becomes dissociatively

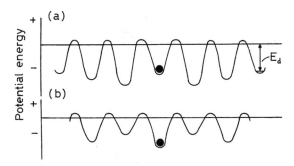

Figure 3.5. Potential profiles through the surface illustrating (a) attractive and (b) repulsive interactions at sites adjacent to a hydrogen atom.

chemisorbed on it. We now turn our attention to the nature of the adsorbed phase, concentrating mainly on the metals of Groups 8 to 10 and thinking first of 'geometric aspects', particularly the location and disposition of the adsorbed atoms.

In order to try to summarise and make a little sense of the very extensive literature, it is necessary to start with a few basic considerations. (1) The presence of one chemisorbed hydrogen atom may influence the tendency for adjacent sites to accept another: this influence may be either positive or negative, that is to say the two atoms may either attract or repel each other (see Figure 3.5). Thus on certain surfaces the atoms are observed by LEED to adopt an expanded array (Figure 3.6), signifying a repulsive interaction extending over as many as five or six atomic diameters (>3 nm); this must take place *through* the metal rather than directly between the atoms. On the fcc(110) surface, which has two-fold symmetry, the interaction may be anisotropic, being repulsive across the rows and attractive along them (see Figure 3.6c). Easy surface diffusion is needed to allow hydrogen atoms to move from their initial positions where they occupy a precursor state (Figure 3.7) to positions of greatest stability, but this happens at all but the lowest temperatures because the activation energy for diffusion is only about one tenth that for desorption[75] (typically 7–10 kJ mol $^{-1}$).

(2) We saw in Chapter 2 that the (110) faces of the fcc 5*d* metals (Ir and Pt) undergo spontaneous reconstruction in the clean state to give a 'missing row' (MR) structure: this also happens with the same surfaces of nickel, rhodium[87] and palladium when hydrogen atoms are present in high enough coverage and when the temperature is sufficiently elevated. These *adsorbate-induced reconstructions* are seen with many other atoms, (O, S, alkali metals) and serve to remind us of the dynamic character of a metal surface. Passage of palladium single crystal and films through the hydride phase results in the appearance of (111) facets.[88] Reaction models based on static surfaces are surely over-simplified, and similar structural modifications may be expected with small particles when hydrogen is

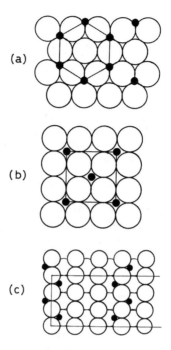

Figure 3.6. Maps of expanded arrays of adsorbed hydrogen atoms at low coverages caused by repulsive (a and b) or repulsive + attractive interactions (c).
(a) Ni(111)c(2 × 2)2H at <270 K: $\theta = 0.67$ (note both types of tetrahedral hole are used).
(b) Pd(110)c(2 × 2) at <260 K: $\theta = 0.5$.
(c) Ni(110)c(2 × 2) at <200 K.

present. The effects produced by sulfur and alkali metal atoms may have relevance to the phenomena of poisoning and of promotion (Section 2.7). Although neither copper nor gold retains a significant coverage by hydrogen atoms at 300 K,[76] in their presence their (110) surfaces are both affected; hydrogen also enhances the diffusion of platinum atoms in channels on the reconstructed (110) surface.[89]

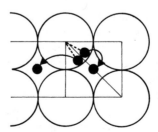

Figure 3.7. Calculated initial position of adsorption of a hydrogen molecule on an fcc(110) surface, and movement of atoms on dissociation to octahedral sites.[355]

These changes must result in a lowering of the surface energy, but it is not clear whether this stabilisation is caused by the presence of the chemisorbed atoms, or whether they simply facilitate it by weakening metal-metal bonds in the surface.[1] Spontaneous restructuring in the case of iridium and platinum but not with the 3d and 4d metals was attributed (Section 1.22) to the greater energy advantage that resulted, i.e. to their higher sublimation heats: this would suggest that the former explanation is the more probable. On the more densely packed fcc(100) and (111) surfaces, where chemisorption produces no changes parallel to the surface, it partly counteracts the 'skin' effect (Section 1.2.2), and M—M bond lengths normal to the surface relax towards their bulk values; however with the (5 × 20) hexagonal reconstruction of the Pt(100) surface hydrogen chemisorption restores the original square lattice.[90] Some examples of the structures formed by hydrogen atoms on fcc metals at low coverage are shown in Figure 3.6. These reveal an important characteristic of the chemisorbed phase, namely, that the stablest position of the atom is where it makes most contacts with the metal atoms, i.e. where the 1s electron overlaps most with the emergent orbital lobes or where its wave function integrates most completely with the unfilled surface electron band. These positions are therefore four-fold sites on (100) surfaces and three-fold sites on (111) and (110) surfaces (Figure 3.6). The small size of the hydrogen atom relative to that of a metal atom means that the electron interaction is extensive, and indeed on the (100) surface the atom almost disappears into the surface structure[91] (Figure 3.8).

As exposure is increased, other structures appear in which the hydrogen atoms are ever more densely packed (Figure 3.9). At low temperature on the unreconstructed Ni(110) surface, four other structures follow that shown in Figure 3.6c, their coverages being 0.5, 0.67, 1.0 and 1.5; this last is shown in Figure 3.9b. At exposures giving coverages between these values, two phases will co-exist. On unreconstructed Rh(110) there is a similar sequence of phases, but the final one, formed only above 120 K under an atmosphere pressure of hydrogen, corresponds to a coverage of 2.0. The location of hydrogen atoms is also obtained by *high-resolution electron-energy-loss spectroscopy* (HREELS),[5,17,92] which detects M—H vibrations (subject to certain impact selection rules as well as the more familiar dipole selection rules).

There is much evidence to suggest that the M—H bond is essentially covalent in character, but that the hydrogen atom is slightly the more electronegative,[1,93] carrying a charge of about $0.1e^-$: charge is therefore withdrawn *from* the metal, and the work function is *increased*, provided the centre of charge density is above the image plane of the metal. This is not so with Pt(110) and (111) and Ir(110), where the low location of the atom causes the sense of the work function change $\Delta\varphi$ to be reversed. Work function changes are complex quantities, since the perturbation of the electronic structure of the metal by relaxation or restructuring alters its chemical potential, and hence its contribution to the work function: firm conclusions about the polarity of the M—H bond are therefore not easy to derive, although changes in the slope of $\Delta\varphi$ versus coverage reflect the occurrence of different phases,[1]

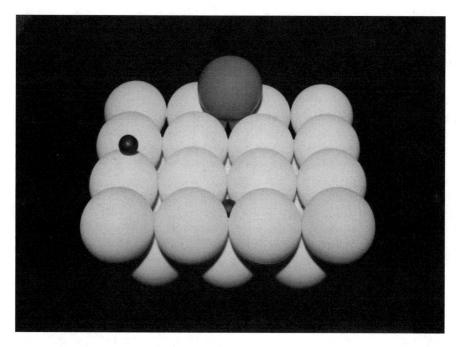

Figure 3.8. Model showing hydrogen atoms chemisorbed on the (100) face of an fcc metal (the large sphere shows the location of a metal atom in the next layer).

or the existence of step sites (e.g. Pt(997), which are occupied first and with an *increase* in $\Delta\varphi$.[17] However, after proper calibration, measurement of $\Delta\varphi$ provides a handy means of estimating surface coverage. The covalent nature of M—H bonds is confirmed by numerous estimates of their length by LEED structural studies; values equate closely to the sum of their covalent radii if the hydrogen atom radius is taken as 40–60 pm.

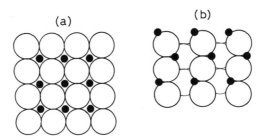

Figure 3.9. Arrangements of hydrogen atoms at high coverage: (a) Pd(100)(2 × 2) at $\theta = 1$; (b) Ni(110)(2 × 1)p2mg.

Ultraviolet photoelectron spectroscopy (UPS), in either the angle-integrated or angle-resolved (ARUPS) mode, reveals changes in the metal's electronic band structure caused by the chemisorption of hydrogen atoms.[1] New resonance states at 5–10 eV below the Fermi level are regularly seen, and these 1s derived states extend over more than 4 eV, confirming their free-electron-like form. Close to the Fermi surface there is a suppression of *d*-band states, showing their participation in the bonding and in the case of nickel there may be magnetic consequences as well. Although the picture of a delocalised M—H bond is clearly consistent with the atoms occupying a three or four-fold site, it needs a modicum of mental gymnastics to square this with the conclusion that the bond is 'essentially covalent'. Perhaps the safest conclusion to draw from the proliferation of results obtained by all these different techniques is that expressed in the old adage: *what you see depends on where you look.*

Thus far we have been concerned almost exclusively with low index surfaces of single crystals of pure metals, for good reason. In the older work on metal powders[24] and films,[26,27,29] their polycrystalline nature forbade the drawing of conclusions about surface geometry, and their doubtful cleanliness was a continual worry. Only the work on FEM and FIM produced results of comparable importance, although these techniques are now rarely used.[26,34] There have however been a number of studies of stepped surfaces. The behaviour of chemisorbed hydrogen on Ni(997), which is a stepped version of Ni(111), is dominated by the plane areas, but the order-disorder transition from the (2 × 2)2H phase to the randomly located lattice gas was raised by about 40 K by the presence of the steps, which stabilise the ordered structure.

There is much less information available on geometric aspects of hydrogen chemisorption on bimetallic systems.[94,95] Addition of an 'inert' metal (of Group 11) to a metal of Groups 8 to 10 not surprisingly lowers the sticking probability, and nickel-aluminium and platinum-tin intermetallic phases seem to be totally inactive.[96–98] Most of the work that has been done has had the purpose of characterising the surface composition, and the principal conclusion reached was that the 'inert' element acted more or less as a site-blocking agent without very much affecting the reactivity of the active atoms, i.e., no ligand effect was detected. Thus for example while the p(2 × 2)$Sn_{25}Pt_{75}$/Pt(111) surface will not chemisorb *molecular* hydrogen, presumably because of an activation energy constraint, the Pt—H bond formed when atomic hydrogen was used had about the same strength as when on Pt(111).[94] Similarly on p(2 × 2)NiCu(111), the binding energy of hydrogen atoms residing selectively on nickel sites was not affected by the presence of the copper. A more recent molecular beam study has compared sticking probabilities and desorption characteristics of deuterium on the p(2 × 2)SnPt(111) and ($\sqrt{3}$ × $\sqrt{3}$)R30°SnPt(111) surfaces (see Figure 4.5), and concluded that the absence of a triangle of platinum atoms in the latter led to an increased barrier for dissociation and a decreased sticking probability. Formation of the bimetallic surface ruthenium-copper phases by evaporation of the latter onto the former appeared to show that three or four ruthenium atoms were needed to bind one hydrogen atom, but hydrogen atoms migrated from ruthenium to copper sites.

3.2.4. The Chemisorbed State: Energetic Aspects

While the geometric aspects of hydrogen chemisorption reviewed above reveal fascinating details of the surface chemistry of metals, their relevance to the phenomenon of catalysis is not easily perceived. When reactions occur, it is the strengths of chemisorptive bonds that matter, and so although (as we shall see) there are close connections between geometric and energetic factors, it is the latter that can provide numbers that will assist our understanding of catalytic processes.

Once again the detail needs to be prefaced by some general considerations. The process of chemisorption is in essence a chemical reaction: unfortunately, like many chemical reactions, it is not a simple process, as it does not always lead to a well-defined product. The occurrence of the different structures of the hydrogen atom adlayer exemplifies this. Nevertheless the fact that chemisorption takes place means that the Gibbs free energy of the system must decrease, and that because of the loss of translational entropy there has to be a decrease in the system's heat content: chemisorption is thus of necessity *exothermic*. All manner of thermodynamic parameters can therefore be ascribed to the process and to the resulting state.[4,17,75]

If we represent the chemisorption of hydrogen as

$$H_2 + 2M \rightarrow 2H - M \tag{3.A}$$

we can define the equilibrium constant, which is here called the *adsorption coefficient b* as

$$b = \theta_H / P_H^{1/2} \tag{3.2}$$

and from its temperature dependence the adsorption free energy, heat and entropy may be derived using the Van't Hoff isochore and the Clausius-Clapeyron equation. The variation of θ_H with pressure at constant temperature constitutes an *adsorption isotherm*; this is an experimental observation and may be modelled by an *adsorption equation*, the simplest of which is associated with the name of Langmuir and takes the form

$$\theta_H = b^{1/2} P_H^{1/2} / \left(1 + b^{1/2} P_H^{1/2}\right) \tag{3.3}$$

We must note however that this equation rests on simplifying assumptions, namely, that each 'site' can acquire only one atom, and that all 'sites' are energetically

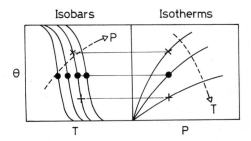

Figure 3.10. Dependence of surface coverage by hydrogen atoms on (a) temperature at various fixed pressures (isobars), and (b) on pressure at various fixed temperatures (isotherms). Isosteric heats of adsorption at (say) half-coverage are derived from the pressures giving this coverage at each temperature.

equivalent. Despite these limitations, which ought to make the equation unusable, it often gives a fair description of what happens, and more complex equations that avoid the assumptions inherent in the Langmuir equation are rarely used. Because chemisorption has to be exothermic, by Le Chatelier's Principle the value of b will have to decrease with increasing temperature and the variation of surface coverage with temperature at variable constant pressure is illustrated in Figure 3.10: the curves are isobars and the corresponding plots of coverage against pressure (i.e. the isotherms) are shown in the same Figure. Application of the Clausius-Clapeyron equation in the form

$$\left(\partial \ln P_{\mathrm{H}}^{1/2} / \partial T \right)_{\theta} = -q_i / RT^2 \tag{3.4}$$

leads to q_i, the *isosteric* heat of adsorption, i.e. the heat released at constant θ, and hence by selecting different values of θ the dependence of heat on coverage may be derived. As noted above, work function measurements are often used to estimate coverage. It is also possible to measure heats of adsorption calorimetrically; techniques for doing this with films were pioneered by Beeck[25,99−101] and by Trapnell,[26] and more recently King has succeeded to do this with single crystals (see Chapter 4). Heat released over a range of coverage yields an *integral* heat; if successive increments of the gas admitted are small enough, one obtains a *differential* heat, which is termed a *true* differential heat $-\Delta H_{\mathrm{ads}}$ if no work is done in the adsorption. Then

$$q_i = \Delta H_{\mathrm{ads}} + RT \tag{3.5}$$

since RT is the maximum work done in adsorbing one mol of ideal gas at temperature T. Heats of adsorption may also be obtained from the careful analysis of temperature-programmed desorption (TPD) results (see below).

Heats of adsorption depend on the nature of the metal, the surface geometry,[82] the coverage and the perfection of the surface. As concerns the effect of coverage, four aspects are to be noted: (i) in some cases, the heat is constant over a considerable range of coverage[3] (e.g. up to $\theta_H \sim 0.5$ for Rh (110) and to $\theta_H \sim 1.0$ for Pd(100)); (ii) sometimes there is an initial sharp fall (e.g. Ni(100)) or rise (e.g. Ni(110)) before a constant value is attained; (iii) at higher coverages the heat decreases, but sometimes this only occurs as θ_H approaches unity (e.g. Pd(100)) and for other surfaces there is a progressive fall from the beginning (e.g. Pt(111)); and (iv) occasionally there is fine structure, i.e. plateaux and intervening decreases, that betoken the existence of different structures. On the whole however the heat is remarkably insensitive to the surface concentration of hydrogen atoms, showing that the magnitudes of the attractive and repulsive interactions are relatively small. An attractive effect at low coverage causes the heat to rise on Ni(110) (Figure 3.6c), while an initial fall is probably due to some step sites or defects: the decreases at high coverage are almost certainly a consequence of repulsive interactions.

Values of heats of adsorption at the flat regions or the initial value where there is no plateau are given in Table 3.1. This shows that for nickel, palladium and iron there is no strong dependence on surface geometry and indeed mean values are all between ca. 95 and 100 kJ mol^{-1}. Analysis of the TPD spectra for deuterium on

TABLE 3.1. Initial Heats of Hydrogen Chemisorption on Single Crystal Faces,[3,62,102] Films[26] and Powders

Metal	fcc (100)	(110)	(111)	Film	Powder
Ni	95	90	95	121–134	85[105]
Cu	—	—	39[a]	—	—
Rh	—	92	78	109	—
Pd	99	103	88	113	—
Ag	—	—	~15	—	—
Ir	—	77	—	—	—
Pt	79[b]	75[b]	71[b]	89[104]	90

	bcc (100)	(110)	(111)		
Fe	100	109	88	134–151	—

	cph (1010)		(0001)		
Co	80		67	—	
Ru	80		80, 120	—	65[165] 90[188,327]

[a]Cu(311)
[b]Approximate average values

Ni(111) by the isostere method gave slightly lower values ($\sim$90 kJ mol^{-1}).[102] With cobalt, rhodium, platinum (which is also structure insensitive) and iridium, values are somewhat lower (75$-$90 kJ mol^{-1}) and for copper and manganese lower still. The heat of adsorption ($-\Delta H_{ads}$) is converted into an M$-$H bond strength D_{MH} by the relation

$$D_{MH} = {}^{1}\!/_{2}(\Delta H_{ads} - D_{HH}) \qquad (3.6)$$

where D_{HH} is the bond strength of the hydrogen molecule (436 kJ mol^{-1}): in Groups 8 to 10, D_{MH} values lies between 250 and 275 kJ mol^{-1} and are typical of covalent bonds.[103] This equation actually holds only where relaxation or reconstruction of the surfaces makes no contribution to the energy released: loss of strain will be exothermic and will appear in the heat term.

It is of interest to compare the results formed with single crystals with those for films obtained chiefly by Beeck[99] and by Trapnell[26] and others.[104] Although similar in form, values of the heat at zero coverage (Table 3.1) are considerably greater, due perhaps to a higher concentration of defects, edges and steps, although the heat liberated with a stepped platinum surface is *less* than that for the corresponding plane. Values for powders are often below those for single crystals (Table 3.1), and this may be due to lack of cleanliness: electronegative impurities cause the heat for hydrogen to decrease (e.g. O on Ni (100)). Little work appears to have been done with bimetallic surfaces, perhaps for fear of chemisorption-induced restructuring; values for heats of adsorption on powders of the interstitial alloys NiB and NiP are less than those for pure nickel powder.[105] Measurements with supported metals and theoretical calculations are considered later.

Extensive work on the nature of chemisorbed hydrogen has been performed using temperature programmed desorption (TPD)[1,3,17,26,41,75,106,107]. What is done is as follows: after a very small exposure (e.g. 0.01L) the specimen is heated and the TPD spectrum recorded, using, for example, a mass spectrometer. The procedure is repeated on the cleaned surface with successively larger exposures, until finally the coverage becomes as close to saturation as possible. What is typically seen is that the small doses desorb only at relatively high temperatures; then as the dose is increased, a second peak appears at a somewhat lower temperature and ultimately a third (sometime sharp) peak may appear at much lower temperature. Results for the Ni(110) surface showing this behaviour are presented in Figure 3.11; here the three events are coded respectively β_2, β_1 and α. 'Open' surfaces such as the fcc(110) exhibit more peaks than tightly packed surfaces because of the greater variety of phases they can accommodate (see Section 3.23).

The object of the game is then to relate what is seen to a model of what is on the surface. This is quite a complicated exercise and the technique has spawned a large literature on its interpretation; only a brief summary is possible here. There are two issues: (i) to relate the areas under each peak, obtained by deconvolution, in a

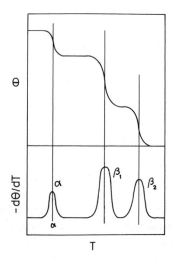

Figure 3.11. Temperature-programmed desorption (TPD) of hydrogen from Ni(110) showing integral and differential plots.

qualitative way to the various surface phases shown for example by LEED: and (ii) in a quantitative way to infer from the values of T_{max} what are the activation energies for desorption (E_d), since if there is no activation energy for chemisorption these are equal in size but of opposite sign to the heats of adsorption. Reliable values are however only obtained by rigorous application of the isostere method,[102] rather than rough and ready methods based on T_{max}.

A note on terminology is in order: the term 'adsorption' means 'chemisorption' unless prefixed by the word 'physical'; the term 'physisorption' is not much used by people working on catalysis. Purists have pointed out that the word 'desorption' is etymologically unsound: the prefix meaning 'away from' is *dis-*. Thus we have associate and dissociate: so the opposite of adsorption should be dissorption; but it is probably too late to do much about it. The solid is referred to as the *adsorbent* and the gaseous adsorbing species as the *adsorptive;* this species when adsorbed is referred to as the *adsorbate.*

Returning to the interpretation of TPD spectra, the early exposures place the hydrogen atoms in the stablest lattice sites (for example, see Figure 3.7) and this continues until the first state (β_2) is fully occupied. Further exposure begins to fill the β_1 state, and when this is complete the least stable α state is started. TPD reverses this process (see Figure 3.12); desorption at low temperature destroys the α state and leaves atoms forming the β_1 state: then further desorption leaves only the β_2 state. The various peaks in TPD spectra do not signify *a priori* heterogeneity of the surface. Hydrogen atoms formed at low coverage will if mobile migrate to occupy sites at which they are most stable, and, as coverage increases, interactions

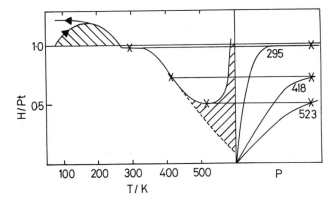

Figure 3.12. Isobar and isotherms for chemisorption of hydrogen on supported platinum cataly-sts.[122,126,135,136] The left-hand part shows the variation in the amount adsorbed (expressed as H/Pt$_s$) at a fixed pressure, and the right-hand part the corresponding variations with pressure at fixed temperature (see text for full description).

within the adsorbed layer may cause the heat of adsorption to vary; in some cases, at higher coverages, different sites to those occupied at low coverage will be favoured by the *whole* adlayer (Figure 3.10). During *desorption*, as the coverage falls, reverse site-switching may occur, and the interactions will decrease; both effects lead to structure in the TPD spectrum even from a fcc(111) single-crystal surface. TPD does not necessarily resolve every state identified by LEED, as in some cases the adsorption energies are not sufficiently different: the techniques can however with suitable calibration provide estimates of the number of atoms in each resolvable state. Inherent heterogeneity as with stepped surfaces will create extra desorption peaks.

Kinetic analysis of desorption is fairly straightforward where only a single state of uniform energy (i.e., obeying the Langmuir equation) exists. The rate is given by

$$-d\theta_H/dt = k\theta_H{}^x \tag{3.7}$$

where x is the order of reaction (usually $x = 2$ for desorption of H atoms, but sometimes 1) and by analogy with the Arrhenius equation

$$k = A_x \exp\left(-E_{des}/RT\right) \tag{3.8}$$

where A_x is a pre-exponential factor for reaction of order x. This leads to the *Wigner-Polanyi equation*

$$-(d\theta_H/dt) = A_x \exp(-E_{des}/RT)\theta_H^x \qquad (3.9)$$

where σ_o; this is the basis for all further analysis of TPD results. For second-order processes, peaks are sometimes symmetrical; the value of T_{max} depends upon the heating rate β, and the theoretical treatment[17] leads finally to the equation

$$\ln\left(\theta_{0,H} T^2_{max}\right) = (E_{des}/RT_{max}) + \ln(A_2 R/\beta E_{des}) \qquad (3.10)$$

where $\theta_{0,H}$ is the initial concentration of adsorbed hydrogen atoms. E_{des} is then easily extracted from this equation. More complex procedures allow evaluation of E_{des} in those cases where it is a function of coverage (T_{max} then depends on β) and where there is a multiplicity of states.

There have been a number of studies of the TPD of chemisorbed hydrogen from alloy single crystal surfaces,[98] e.g. Ni-Cu[108] and Pt-Au. The principal conclusion seems to be that the inclusion of the Group 11 metals alters the population of the various states but not very greatly their character (i.e. not their T_{max} values), although small changes in T_{max} may suggest interference with attractive interactions between hydrogen atoms.[109]

As with every technique, there are pitfalls for the unwary. It is a destructive technique, and processes other than simple desorption may occur during the heating (and cooling) stages. Surface reconstruction is thermally activated and may take place on a similar time-scale to the heating; its degree of completion at the top temperature may be a function of exposure, and the structure may not revert to its pristine state on cooling. False effects due to the sample holder and other instrumental causes are well understood and easily corrected. *Caveat operator!*

3.3. CHEMISORPTION OF HYDROGEN ON SUPPORTED METALS[41,110–114]

3.3.1. Introduction: Determination of Metal Dispersion

The chemisorption of hydrogen onto supported metals is inevitably more complex than onto unsupported macroscopic forms for three reasons: (i) small supported metal particles differ even from their unsupported counterparts because of the very presence of the support to which they are attached and with which they may interact;[115,116] (ii) they will have a higher proportion of low co-ordination number atoms, and for the smallest particles[117,118] this may cause the loss of typical metallic properties (see Section 2.5); and (iii) some hydrogen may become adsorbed onto the support by a process other than direct chemisorption. Thus although what has been learnt by studying single crystals, films, wires and powders is extremely useful, its relevance to the kind of particles found in a typical supported metal catalyst needs close examination: such particles are indeed *sui generis*.

The importance of the manner of interaction of hydrogen with supported metals is two-fold: (1) it provides means for estimating the number of surface metal atoms, and hence with certain assumptions the exposed metal area and the mean particle size; and (2) it informs us of possible states in which hydrogen may exist on the catalyst during catalytic reactions in which it is a partner. The first consideration is treated in this section, and the second in the following section.

Anyone intending to use hydrogen chemisorption to estimate metal dispersion or any derived quantity is faced with a daunting number of decisions that must be taken before experimentation can start. There are three basic procedures to choose from. (A) Volumetric determination of the amount of hydrogen chemisorbed in a static system (the hydrogen atom is too light for gravimetric determination to work well). (B) A titration method in which chemisorbed oxygen reacts with hydrogen or vice versa. (C) Measurement of hydrogen chemisorption by some means in a dynamic system. Each method has advantages and drawbacks, and although each is simple to operate they all require very great care in the choice of conditions and procedures if significant results are to be obtained.

Whatever method is used, it is necessary to prepare the catalyst surface thoughtfully. The conditions for the initial reduction of the precursor will depend on the nature of the metal and salt or complex used in the preparation, and whether or not it has been calcined: in any event it is essential to ensure complete reduction and removal of any species (e.g. Cl^-) that might interfere with the adsorption. It has to be remembered that particle size is critically dependent upon the preparation procedure and also upon the *in situ* treatment preceding adsorption, especially if reduction has not been performed during preparation. By far the larger part of published work concerns platinum, for which detailed standard protocols are available:[119,120] hydrogen reduction of Pt/SiO_2 made from H_2PtCl_6 should be for 10 h at 773 K (but not higher), with subsequent outgassing at the same temperature for 4 h to remove adsorbed hydrogen,[121] but high dispersion of Pt/SiO_2 made from the ammine is secured by reduction at 573 K using hydrogen at only 1 Torr pressure.[122]

Early work on the volumetric method by Russian[122,123] and American[124–126] scientists in the 1950s laid the foundations on which all later work has been based. It is hard to realise that in the late 1940s and early 1950s, work on supported metal catalysts was performed in total ignorance of the metal dispersion, although catalysts were occasionally made using colloidal platinum, the particle size of which was known.[127] The central difficulty with the volumetric method however arises from the several forms that adsorption can take, and the consequent problem of identifying and isolating that form which tells about the metal dispersion. The literature[110,112,128,129] makes frequent reference to 'strong' and 'weak' forms, with the implication that only the former relates to chemisorption of the molecule as atoms upon the metal. Thus the 'strong' form resists evacuation, while the 'weak' is removed; in temperature-programmed desorption (TPD) or in measuring isobars (i.e. the amount adsorbed at constant pressure as a function of temperature), the 'weak' form is detected at low temperature and the 'strong' is retained at temperatures above which the 'weak' is lost. Unfortunately, and this is the great dilemma with the method, the two forms can rarely be clearly distinguished, their ratio depending upon operating conditions such as adsorption temperature, pressure and time. Moreover TPD often, indeed usually, reveals a plethora of overlapping peaks,[7,8,41,110,118,130−134] the deconvolution of which sometimes appears to be more of an art form than a scientific exercise. To obtain reproducible and meaningful results it is therefore necessary to adopt very strict operational protocols, defined with an understanding of the idiosyncrasies of the particular system under study; and although attempts are often made to assign states and locations to these various forms, their appearance depends upon operational factors and their identity remains elusive. Some may be purely adventitious as with Pt/Al_2O_3, where a peak at 723 K has been attributed[133] to the decomposition of ammonia, which was a ligand in the precursor salt used in the preparation.

It is helpful first to determine adsorption isobars at several pressures and adsorption isotherms at several temperatures in order to decide what conditions to use:[122,126] Figure 3.12 shows the forms that the results may take. At subambient temperatures, hydrogen is weakly adsorbed on highly polar supports such as silica and alumina; this is removed by the time room temperatures is reached, but by then the 'strong' form may be starting to desorb as well; indeed a 'substantial part' of the hydrogen adsorbed on Pt/SiO_2 is reversible at this temperature.[135] With certain supports, there is an additional uptake seen especially at high temperatures: this is due to *hydrogen spillover* onto the support, a subject to be discussed in Section 3.3.4. By good fortune the amount of hydrogen adsorbed at equilibrium in the region of ambient temperature normally provides a good compromise: there will usually be little physically adsorbed gas remaining and little spillover, and between 273 and 323 K the isobar is comparatively flat,[136] so that close control of temperature is unnecessary, and the common assumption that H/Pt_s is about unity under these conditions is quite well justified. Extensive measurements of isobars and isotherms on EUROPT-1 have been reported.[136]

Some however prefer to work at low temperature (e.g. 77 K) to ensure that a complete monolayer remains on the metal and that spillover is absent. It is then necessary to estimate the amount of physical adsorption that has to be subtracted from the total, either by measuring the adsorption isotherm for the support alone, or, more reliably the isotherm on the catalyst, following the first adsorption and a short evacuation. This method is referred to as the *back titration method* (Figure 3.13), and even at ambient temperature the second isotherm often reveals some 'weak adsorption', the nature of which may vary from one system to another, and which needs to be taken from the total in order to isolate the amount of the 'strong form'. This often leads to an isotherm that is flatter in the high-pressure

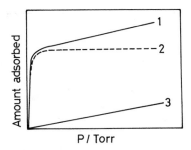

Figure 3.13. Measurement of hydrogen chemisorption isotherm by the back-titration method: (1) first isotherm, (2) second isotherm measured after short evacuation, (3) the difference.

region, making extrapolation to zero pressure easier: this procedure is frequently used to obtain a 'monolayer volume' from which metal area etc. can be derived. Yet further decisions have to be made. Although at room temperature and below some 95% of each dose is taken up very quickly, the rest may be taken up only quite slowly so that true equilibrium uptakes are only achieved after perhaps 1 h. The determination of a full isotherm with 20 or so points can therefore be a task for two days. Some of the slow uptake may be caused by diffusion to the least accessible metal, but very slow and prolonged uptake probably signifies a poisoned surface.

Over what range of pressure should the isotherm be measured? Careful work over an extended pressure range shows that there may be no sharp 'knee' corresponding to the completion of a monolayer on the metal, but rather continuing uptake which appears to approach some saturation limit without ever quite getting there. This implies that the zero-pressure monolayer volume obtained by extrapolation depends upon the pressure range over which the isotherm has been measured (see Figure 3.14).

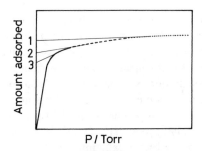

Figure 3.14. Variation of the uptake extrapolated to zero pressure with the pressure range over which the measurements are made.

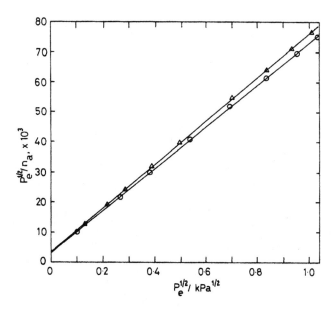

Figure 3.15. Hydrogen chemisorption on EUROPT-1 reduced at 758 K plotted according to the Langmuir equation[137] (see text; the upper line is for results obtained with oxidation before the reduction).

One way of circumventing this problem is to fit the results to the appropriate linearised form of the Langmuir equation for dissociative adsorption:[120,137]

$$P_e^{1/2}/n = (b^{1/2}n_m)^{-1} + \left(P_e^{1/2}/n_m\right) \tag{3.11}$$

where n is the molar amount of hydrogen chemisorbed at the equilibrium pressure P_e, n_m the monolayer amount and b the adsorption coefficient. Excellent straight line plots are obtained (see Figure 3.15), the inverse slope of which gives n_m; this procedure utilises all the data points, the only assumption being the applicability of the equation.

The volumetric method has very often been used with platinum catalysts for which quite satisfactory results are generally obtained: it is usual to assume that the monolayer volume or amount, obtained as just described or by extrapolation corresponds to an $H:M_s$ (hydrogen atom to metal surface atom) ratio of 1:1. Some justification for this assumption is to be found, at least for particles of moderate size, in the adsorption stoichiometry shown by films and single crystals, but for very small particles and at high pressures the H/M_s ratio can exceed unity quite substantially: this is especially so with rhodium[41,121,128] and iridium[41,138] (see below). Care is however needed with palladium[41,45,115,139−142] because of the risk of forming the hydride; however, monolayer coverage is obtained at pressures below which dissolution starts. The base metals iron, cobalt and nickel have been intensively studied,[8,110,143] but copper presents a special problem because it does not absorb hydrogen very strongly,[64,144,145] and adsorption of oxygen atoms either from molecules or from nitrous oxide has to be resorted to. Similar difficulties are met with silver and gold. Particular problems also attend hydrogen chemisorption on ruthenium (see below) and on any of the Group 8–10 metals, supported on reactive oxides of the Transition Series metals (see Section 3.3.3). There are however numerous studies of hydrogen chemisorption on metal particles in/on zeolites.[117,118,132,146–151]

The development of reliable procedures for volumetric determination of metal dispersion was somewhat hampered by the necessity to use homemade apparatus, although commercial equipment is now available (at a price). One important factor which varied from one apparatus to another, but was often ignored, was the pumping speed. This governs the rate at which desorbed molecules can be removed from the neighbourhood of the sample, and is critically dependant on the diameter of the tubing used. Assessment of the importance of this and many other factors has been greatly helped by the availability of *standard catalysts*,[152] by the use of which both equipment and operators could be calibrated. A range of such catalysts was prepared in Japan, and there were more limited programmes in the United States and Russia. In Europe a 6.3% Pt/SiO_2 (EUROPT-1), a typical petroleum reforming catalyst (0.3% Pt/Al_2O_3: EUROPT-3) and a 20% Ni/SiO_2 were made and widely distributed. The EUROPT-1 in particular has been extremely thoroughly examined (see Further Reading section); it has a mean particles size of 1.8 nm and a hydrogen monolayer volume by the Langmuir method of $\sim$190 μmol g_{cat}^{-1}.

The volumetric method may be carried out in several different ways. As noted above, the usual procedure of admitting a series of small doses and allowing each to reach equilibrium is a slow and tiresome process (although it can now be automated), and errors in measurement accumulate as the work progresses. This latter difficulty is avoided by introducing a single massive dose, sufficient to saturate both 'strong' and 'weak' sites, and then either monitoring the isotherm in the desorption mode (where equilibrium is attained more quickly[8,153]) or changing the size of the single dose so that the amount of the 'weak' form is altered and can therefore be obtained by extrapolation.[128,139] With EUROPT-1 this procedure gives a monolayer volume of 188 μmol g_{cat}^{-1} in good agreement with that found using mulitple injections. Error accumulation is also avoided by "pressure programming", where the pressure over the sample is continuously increased by heating a large reservoir attached to the sample vessel.[154] Comparison of the pressure-temperature curve with and without a catalyst being present permits derivation of the adsorption isotherm.

Having defined what seems to be a suitable method for estimating the monolayer capacity, that is, the number of hydrogen atoms strongly adsorbed per unit mass of metal, it still remains to be shown that this equates to the number of surface metal atoms; although this has often been *assumed* in the past, the variable co-ordination number of atoms at or close to the surface renders this a somewhat unreliable guess for calculating dispersion and mean particle size. Some further enlightenment concerning the nature of the 'weak' form is also desirable.

Progress was made when the results for hydrogen chemisorption were compared with those for the chemisorption of other gases (e.g. O_2, CO) and especially with those for physical methods such as TEM and EXAFS, which are independent of chemical factors. Although there are some instances in the literature where good agreement has been reported between chemical and physical methods,[41,153] there are also cases where the agreement has been poor and these turn out to be the more informative. The experience with EUROPT-1 is a case in point. The hydrogen monolayer quantity of $\sim$190 μmol g_{cat}^{-1} noted above[137] corresponds to an H/Pt_{tot} ratio of about 1.2, while TEM results show quite clearly that the mean size is about 1.8 nm which equates to a dispersion of only 60%. Reference to the TPD measurements showed that the largest of the three peaks, together with an assumed 1/1 H/Pt stoichiometry, led to a dispersion of 65%: the highest temperature peak was assigned to spillover, and the remaining peak to hydrogen taken up at the metal/support interface by breaking M—O— bonds. This accounted for the excess hydrogen chemisorbed. It cannot therefore be safely assumed that all strongly held hydrogen resides on the exposed metal surface.

Although the suggestion of some hydrogen being located at the interface lacked independent confirmation at the time, it has received unequivocal support from the EXAFS work of Koningsberger's group.[118,151,153,155–158] The subject of metal-support interaction was treated in considerable detail in Section 2.6, so it only remains to emphasise the role that hydrogen plays. In the case of Pt/γ-Al_2O_3,[151,158] reduction at 573 K created small three dimensional metal particles at a distance of 0.27 nm from the oxide ions of the support: this quite large Pt—O bond length suggested that hydrogen atoms had inserted themselves at the interface, creating a Pt—H . . . HO situation. Reduction at 723 K formed raft-like particles exposing mainly ⟨100⟩ facets and having a smaller tendency to chemisorb hydrogen, the Pt—O distance was now only 0.22 nm, and XANES measurements showed that there were 95% more holes in the d-band. It was inferred that the hydrogen at the interface had been lost and that particles were held by a coulombic attraction between the metal and the oxide ions. These results, together with those on silica and various zeolites, reinforce the need to appreciate the structural and electronic changes brought about by reduction and evacuation, and the effects that these may have on adsorption stoichiometry and catalytic behaviour. Recognition of these effects might go far towards the resolving many of the anomalies and contradictions that the literature reveals.

It remains to offer a further comment on the nature of 'weakly' chemisorbed hydrogen. We saw (Section 3.2.2) when discussing single metal crystals that adsorption was often detected at coverages greater than unity, when expressed in terms of the number of metal atoms in the exact surface plane. Above 'complete' coverage, heats of adsorption fell,[159] and hydrogen atoms occupied sites where

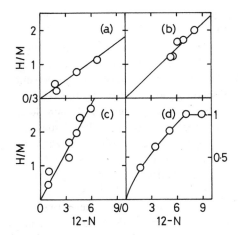

Figure 3.16. Hydrogen chemisorption on supported metals: plots of H/M_{tot} versus $(12 - N)$ where N is the first–shell coordination number determined by EXAFS; (a) Pt/Al_2O_3; (b) Rh/Al_2O_3; (c) Ir/SiO_2 and Ir/Al_2O_3; (d) curve for estimating dispersion D from $(12-N)$ (see text).

they made fewer contacts with metal atoms, i.e. atop or two-fold sites. Such excess adsorption was especially prevalent with rhodium. This may also happen with supported metal particles and in addition there may be two or even more hydrogen atoms attached to edge or corner atoms. Careful measurements[41,111,146,151,158] by EXAFS and hydrogen chemisorption on supported platinum, rhodium and iridium catalysts, differing in metal loading and reduction temperature have shown that even for dispersions less than 100% the H/M_{tot} ratio can rise to quite high values (Figure 3.16). For the largest particles showing a calculated 100% dispersion, corresponding to a first shell co-ordination number of 5, H/M_{tot} values are for platinum, 1.25; for rhodium, 1.7; and for iridium, 3.3. Quite clearly the propensity for different metals to chemisorb hydrogen differs enormously and use of the same stoichiometry for all would give quite incorrect results. It is therefore unacceptable to use the H/M_{tot} ratio as a measure of size or dispersion, without knowing the proper proportionality factor to use.

In this connection we may note that naked gas-phase Transition Metal clusters react with hydrogen, at rates that vary markedly with size, to form adducts MH_n where the maximum value of n is three for palladium, five for nickel and platinum and eight for iridium.[160] Platinum also forms the $(PtH_4)^{2-}$ anion, which is planar and stable below 195 K, but at higher temperature the four hydrogens oscillate around the six octahedral positions; it is converted to the unstable $(PtH_6)^{2-}$ anion at high hydrogen pressures and temperature.[161]

A great deal of work has been carried out on unsupported[162] and supported ruthenium catalysts,[163–169] as well as on single crystals;[1,170,171] this interest arises from this metal's high activity in methanation of carbon monoxide and in alkane hydrogenolysis (see Chapters 13 and 14) and in forming high molecular weight hydrocarbon in Fischer-Tropsch synthesis. In this last reaction, the presence of alkaline species (especially K^+) can direct the reaction towards forming oxygenated products (e.g. C_2H_5OH). What is of interest in the present context is that it is frequently difficult to obtain good isotherms by the volumetric method, because of the very long times needed for the uptake of each dose to finish. If enough time is not allowed, the isotherm above the 'knee' may have a high slope, and the monolayer amount obtained by extrapolation to zero pressure suggests much larger particles than TEM shows. The problem is largely, perhaps entirely, due to poisons arising from the preparation (or from the atmosphere); these, especially chloride ion,[168,172-174] oxygen[168] and sulfur[175] species are firmly retained by the metal and inhibit hydrogen chemisorption. Their slow removal determines the rate at which hydrogen is taken up.[176] Potassium ion also blocks hydrogen uptake, interfering particularly with the 'weak' state formed at high pressure.[163] The dispersion of Ru/Al_2O_3 has also been determined using low-temperature oxygen chemisorption[177] and dynamic pulse reaction with nitrous oxide.[178]

Far less use has been made of hydrogen chemisorption in the study of supported bimetallic catalysts, for which there are several possible reasons. (i) The process of chemisorption may lead to a restructuring of the metal particle, in which the component is drawn to the surface: this certainly happens with carbon monoxide and may occur with hydrogen, especially with small particles. (ii) Where one component's atoms are inactive towards hydrogen (e.g. a Group 11 metal), hydrogen atoms may migrate to them from the active metal (Section 3.23) or there may be sites where interaction occurs with atoms of both types. (iii) Even if it is assumed, as is often done, that hydrogen chemisorption counts the number of exposed atoms, this cannot safely lead to an estimate of dispersion or particle size, since there is no means of knowing the number of inactive surface atoms. Physical methods such as TEM are preferred where the size needs to be known, and XPS or AES can give information on surface composition.[179,180] Despite these uncertainties there have been a number of studies of hydrogen chemisorption on bimetallic catalysts,[49,181-183] especially those of ruthenium with either copper[184,185] or silver.[163,186,187] In general, provided other factors do not obtrude, the amount of hydrogen adsorbed falls progressively as the surface concentration of the inert element rises: there is little evidence for electronic interaction[188] or of migration to the inactive partner.

Hydrogen chemisorption on supported metals can also be studied in a dynamic mode.[130,189] The sample is first reduced and then cleansed by passing a stream of very pure inert gas at high temperature. Successive small doses of hydrogen are then admitted at regular intervals at the desired temperature; the first few are wholly retained, some later ones only partially, and finally when the surface is fully covered the doses pass through unchanged. The amount of hydrogen retained under these circumstances is easily calculated. This amount may however comprise both 'strong' (H_s) and 'weak' forms (H_w) forms; the two can be resolved by altering the time between doses, since the longer the interval the more of the 'weak' form will be lost between doses. Thus from a plot of amount retained versus time interval, ($H_s + H_w$) is obtained by extrapolation to zero time while H_s alone is found at long time ($\sim 1/2$ h). It is necessary to use very pure carrier gas to avoid error due to traces of oxygen.

There are two variants of this method. If the carrier gas is suddenly replaced by hydrogen, the manner in which it is eluted (i.e., the change in its concentration with time at the outlet) allows estimation of the amount adsorbed at whatever hydrogen pressure is used: this is the *frontal elution method*.[135,190] In the *chromatographic method*,[191,192] a long column is packed with catalyst and pre-treated as before: if small doses of hydrogen are admitted, it is found that after the surface is saturated each further dose passes quickly, because it does not interact with the surface. Then at higher temperatures (~ 500 K for Pt/SiO_2) the retention time increases because the surface is now partly bare, and the dosed molecules adsorb and desorb as they pass. Finally retention time falls again as the adsorption equilibrium moves more and more towards the gas side. These results can be manipulated to give a value for the heat of adsorption.[191]

The final procedure requiring attention is the *titration method*.[41,103,110] If a catalyst sample is pre-treated with oxygen in a standard way, so that the O/M_s ratio is known or can be safely assumed, then admission of hydrogen leads quickly to the formation of water, which is adsorbed on the support, and to a layer of chemisorbed hydrogen atoms, e.g.

$$Pt_s{-}O + \tfrac{3}{2} H_2 \rightarrow Pt_s{-}H + H_2O \qquad\qquad (3.B{:}HT)$$

The amount of hydrogen used is therefore thrice the amount that would be used just for chemisorption, and this can be measured more accurately. The cycle is completed by titrating the chemisorbed hydrogen with oxygen:

$$2Pt_s{-}H + \tfrac{3}{2} O_2 \rightarrow 2\, Pt_s{-}O + H_2O \qquad\qquad (3.C{:}OT)$$

If then the usual volumetric uptake for hydrogen (HC) is also measured, the relative amounts of HC: OC: HT should be as $1{:}1{:}3$. These ratios are in fact observed,[7,41,110] but they depend of course on the assumed $Pt_s{-}H$ and $Pt_s{-}O$ stoichiometry. We have seen how hard it is to obtain a reliable value for the hydrogen monolayer amount, and with oxygen it is even harder because the stoichiometry is size-dependent and partial oxidation of the metal can occur. Very careful work with platinum catalysts[41] has shown that the HC:OC:HT ratios should be taken as $1.1{:}0.71{:}2.42$. This method, which has been widely used,[135,162,190,193−196] has the advantage that it is not necessary to start with a totally clean surface, so that catalysts that are sensitive to high temperature reduction and evacuation can be examined. Also it appears that only the hydrogen in the exposed metal reacts quickly with oxygen; that trapped at the interface or elsewhere is not sensed. Care must be taken to ensure that the support is able to adsorb the water formed.[190] Titrations may be performed either volumetrically or dynamically (using frontal elution).[135]

Variants on this procedure are possible. Carbon monoxide has been used in place of hydrogen,[197] but although this eliminates some problems, it often creates others.[198] Chemisorbed hydrogen has been titrated with alkenes:[53,199,200] these reactions may be followed using frontal elution and gas chromatography to detect products, or pulse-wise.[199] Chemisorbed hydrogen also reacts almost instantly with excess deuterium:[24,201]

$$Pt_s{-}H + D_2 \rightarrow Pt_s{-}D + HD \qquad (3.D)$$

and this leads to a simple but effective means for estimating the adsorbed amount;[201] spillover hydrogen exchanges more slowly (see below). This method, developed many years ago, has not enjoyed the popularity it deserves.

Our conclusion must be that although hydrogen chemisorption is a powerful tool for assessing the degree of metal dispersion in supported metal catalysts, the results obtained by any single method may not be reliable unless they are cross-checked by comparing them with purely physical methods. A profound understanding of the changes that can occur during reduction and outgassing is also needed if the results are to be related to valid models of the catalyst structure.

3.3.2. Characterisation of Chemisorbed Hydrogen

Much has already been learned about the nature of chemisorbed hydrogen on supported metals from the foregoing discussion of its quantitative use to determine metal area and particle size, but in the main the techniques used have been relatively simple and straightforward, although requiring good experimental skills. The chief exception is the use of EXAFS to monitor the structural changes occurring at the metal-support interface when hydrogen is introduced or removed; these studies have added importantly to our understanding of the metal-support interaction. It remains to see what confirmation or extension of ideas presented above are derived from the use of other sophisticated techniques such as vibrational spectroscopies and nuclear magnetic resonance spectroscopy, and to enquire what information is there is concerning the strengths of M—H bonds on supported metals.

It might have been logical to start a disquisition on the interaction of hydrogen with supported metals with a paragraph or two on the *rates* of chemisorption, as was the case with unsupported metals (Section 3.22), but the quantitative measurement of rates of adsorption (and of desorption) on highly porous materials presents formidable difficulties, and the general opinion seems to be that the reward does not justify the effort needed. It is true that some years ago it was discovered that the rate of chemisorption on many supported metal catalysts was readily measurable, the rate decreasing with the progress of the adsorption and being described by the Elovich equation:[24,26,202,203]

$$+d\theta/dt = a \exp(-\alpha\theta/RT) \qquad (3.12)$$

where a and α are constants, the intergrated form being

$$\alpha\theta = RT \ln((t + t_o)/t_o) \qquad (3.13)$$

where t_o is $RT/a\alpha$. Theses studies are particularly associated with the names of H. Austin Taylor[204,205] (a brother of H.S. Taylor) and Manfred Low,[206,207] and a model consistent with the Elovich equation was developed by Nathaniel Thon.[204] It is however now thought doubtful to what physical process the rate limiting step should be assigned, and this approach has therefore been largely abandoned. While there is much qualitative and anecdotal evidence for the importance of foreign substances in determining rate (some of it mentioned earlier in this chapter), there is only the occasional mention in the recent literature on the qualitative effect of inactive metals on rates (e.g. of Ag on Ru/SiO$_2$). The use of temperature-programmed desorption has however generated values for heats of adsorption (see following text).

Proton (^{1}H) nuclear magnetic resonance (NMR) has been extensively applied to characterising chemisorbed hydrogen,[208−211] because of the high NMR sensitivity of this element. With Pt/SiO$_2$ (homemade, $\bar{d} = 10$ nm) the NMR spectrum showed[129,208] an α peak at 0 ppm due to the support hydroxyl groups, and a β peak Knight-shifted to higher frequencies, the size of which correlated with hydrogen coverage. At pressures greater than 50 Torr, gaseous hydrogen (oNH$_2$) made a growing contribution, but there was no evidence of hydrogen spillover. The results in the low-pressure region were interpreted in terms of strongly- and weakly-bonded atoms, the former being 3-coordinate and the latter, arising later, occupying atop positions; chemical shifts (resonance frequencies) were respectively about −48 and +37 ppm. A further (γ) peak was also detected at −20 ppm and ascribed to hydrogen atoms at the metal-support interface, the presence of which has also been implied by EXAFS studies (see above). EUROPT-1 gave essentially similar behaviour,[212] with two adsorbed states showing rapid exchange at all temperatures; the smaller particle size ($d_{av} = 1.8$ nm) caused a change in chemical shifts, which were now respectively −31 and +45 ppm. The greater change for the strongly held species was thought to be due to its being more deeply embedded in the surface, and therefore more influenced by the electronic structure of the surface metal atoms, than were the weaker species. The apparent independence of chemical shift upon coverage in the low-pressure region has been ascribed[209] to an inhomogeneous distribution of hydrogen within the sample, but this explanation has been strongly criticised.[213] The NMR spectrum of ^{129}Xe is affected by whether it is physically adsorbed on bare metal or on hydrogen-covered metal; this effect has been used to measure the hydrogen monolayer amount, and hence particle size, particularly for zeolite supports.[209]

Use has also been made of the NMR behaviour of the ^{195}Pt nucleus in following the consequences of hydrogen chemisorption on platinum catalysts.[2,214–216]

There have been a number of proton or deuteron NMR studies involving ruthenium catalysts, especially Ru/SiO$_2$, and Ru/TiO$_2$(see Further Reading section). With the former, in the pressure range 10^{-4} to 600 Torr, three hydrogen species designated α_I (I = immobile), α_M (M = mobile) and β appeared progressively as pressure was increased. With silica as support, the technique has been used to observe the effect of chloride ion on the metal; it affects the ^{1}H resonance frequency and the spin-lattice relaxation time, weakening the Ru$-$H bond by suppressing the amount of strongly held hydrogen through site blocking and local electronic effects. Sulfur from hydrogen sulfide has the opposite effect,[175] perhaps by interacting selectively with low CN atoms: sulfur coverages of about 0.5 inhibit hydrogen adsorption entirely. Studies have also been made with bimetallic ruthenium-silver[187,188] and –copper[217] catalysts, and with potassium-promoted materials.[187,188]

The ^{2}H NMR spectrum for deuterium on Pd/Al$_2$O$_3$ showed[181] a chemical shift moving from -78 to -20 ppm as coverage was raised to 0.6, with a further movement to -12 ppm at saturation. Strong and weak forms, rapidly exchanging, were invoked to explain the results. With Rh/Al$_2$O$_3$ the chemical shift rose from -180 to -100 with increasing coverage, the difference enabling the surface composition of a bimetallic Rh-Pd/Al$_2$O$_3$ catalyst to be estimated. Hydrogen chemisorptions on Pd/NaY zeolite[209] and on the Cu/MgO[208] and Cu/Al$_2$O$_3$[144] have also been reported.

^{1}H NMR is clearly an informative and sensitive technique for observing the forms of hydrogen on and around supported metals; although in the main it has only served to confirm existing concepts, it deserves further development, as the interpretation of what is seen is not always obvious, and extension to other systems is desirable.

There are three vibrational spectroscopies for studying chemisorption of hydrogen on supported metals: (1) infrared spectroscopy, (2) Raman spectroscopy[5,110,206] and (3) Incoherent Inelastic Neutron Scattering (IINS)[9,110]. For the first of these, the equipment is relatively cheap and readily available, but unfortunately the small dynamic dipole moment and the small polarisability of metal-hydrogen bonds renders recording of their vibrational spectrum difficult, and in the case of infrared spectroscopy it is necessary to use the transmission mode. Nevertheless, in spite of such difficulty, spectra have been published for a number of supported metals, and these show bands in a high-frequency region (1850–2100 cm^{-1}) associated with the weaker form on atop positions, and others at lower frequencies (700-950 cm^{-1}) due to the more strongly-held form occupying multi-atom sites.[10] In a given system, the band frequencies do not change with increasing coverage. The great advantage of infrared spectroscopy is that it may be used with high hydrogen pressures as the molecule itself is not infrared active.

IINS is less widely used because it requires a source of thermal neutrons, but its advantage lies in hydrogen's very high incoherent inelastic cross-section; it has however been mainly applied to metal powders[9,218] (Raney Ni,[219] Pt and Pd blacks), but the adsorption of hydrogen on Ru/C[220] and Pt/C[141] catalysts has been studied. Adsorption sites were determined by comparison with known structures (e.g. hydridocarbonyls[218]). HREELS cannot be used with supported metals.

Some attention was given in Section 2.4.2 to the use of magnetic characteristics for determining metal particle size, the technique being limited to paramagnetic metals. The chemisorption of hydrogen induces a loss of saturation magnetisation with ferromagnetic nickel due to the coupling of unpaired spins on surface atoms with the hydrogen's $1s$ electron. In times past these effects were intensively investigated,[11] using supported nickel catalysts and films, and more limited work was done with iron and cobalt catalysts because of the difficulty of obtaining them in a completely reduced state. The technique has however commanded little interest in recent years, perhaps, because of the limited range of metals having catalytic interest to which it can be applied. (See however references 11 and 182).

The strength of the metal-hydrogen bond D_{MH} as derived from the heat of adsorption by the equation

$$-\Delta H_a = 2D_{HM} - D_{HH} \qquad (3.14)$$

is an important characterising parameter, and the latter may be obtained with supported metals using either (i) TPD (somewhat unsatisfactorily) or (ii) (better) calorimetry,[221,222] a technique not yet applied to single crystals with hydrogen, or (iii) temperature-dependence of isotherms. There is an extensive literature on the results obtained, particularly by calorimetry, and this has been reviewed:[15,25,61,71] values obtained, even for a single type of catalyst are disconcertingly variable, often for unknown causes. Some very general statements can however be made.

Method (iii) yields isosteric heats from the temperature dependence of the pressure needed to produce a constant coverage (provided this can be measured): it can certainly reveal the presence of strong and weak forms, adsorption on the support, and, in the case of palladium, dissolved hydrogen also. Indeed it is stated that this information can all be extracted by analysis of the adsorption isochore.[223] Calorimetry provides either an integral value, i.e., the average value to saturation or whatever coverage is reached, or a differential value, obtained by sensing the heat released as small doses are admitted: the latter procedure permits extrapolation to zero coverage and may also reveal steps and plateaux indicative of different states of adsorption.[224] Care is however necessary to ensure uniform distribution of the gas through the catalyst sample and this may not be achieved at low temperature.[224]

Integral values are normally less than initial differential values to an extent that depends on the coverage dependence in the range examined.

The most striking and surprising observation is that heats of adsorption of comparable type vary so little either with the type of metal, the nature of the support (if any) and the degree of dispersion. Much of the literature[225,226] concerns palladium and platinum, for which integral values have been summarised[71] as, respectively, 63 ($\pm$ 4) and 56 ($\pm$10) kJ mol^{-1}: the latter range is validated by values of 59 kJ mol^{-1} for platinum powder and 56 kJ mol^{-1} for Pt/Al$_2$O$_3$.[227] For 20% Ni/SiO$_2$, the adsorption heat for the strong (irreversible) form was 78 kJ mol^{-1}, and the value falling to 29 kJ mol^{-1}, characteristic of the weak (reversible) form, at high pressure. The H$_w$/H$_s$ ratio decreased with rising temperature and increasing particle size, and was greater for Ni/Al$_2$O$_3$ than Ni/SiO$_2$.[159] The initial heat on NiCu/SiO$_2$ catalysts (105 kJ mol^{-1}) started to fall when the copper content rose above 15%, due it was thought to the hydrogen atom being forced to occupy energetically less favourable sites such as bridge sites, or three-fold sites where one of the atoms was copper.[228]

The differential heat for Pt/SiO$_2$ (EUROPT-1) was[229] 110 kJ mol^{-1}, independent of coverage below the isotherm knee; inclusion of gold,[221] or of tin[230,231] and/or potassium,[232] into Pt/SiO$_2$ did not alter the initial value[233] ($\sim$110 kJ mol^{-1}), but the rate of decrease with increasing coverage was faster the higher the concentration of the additive, due mainly to blockage of surface sites rather than electronic effects. There was however much less electronic interaction in the case of PtAu/SiO$_2$ than with PtSn/SiO$_2$. Integral values for Pt/SiO$_2$ showed no systematic dependence on particle size,[234] and were similar to those quoted above for powder and Pt/Al$_2$O$_3$. For platinum in various zeolites the heat of adsorption varied somewhat with structure and with cation[71] (Pt/NaY, 95 kJ mol^{-1}; Pt/KL, 84 kJ mol^{-1}; Pt/BaL, 105 kJ mol^{-1}).

With Pd/SiO$_2$ the heat fell from 84 kJ mol^{-1} initially to 12 kJ mol^{-1} for the weak form, rising again to 33 kJ mol^{-1} at higher pressures as the hydride phase was formed.[223] Very similar heats were recorded[222] for palladium powder and for palladium on various supports ($\sim$63 kJ mol^{-1}), but this value rose sharply to $\sim$100 kJ mol^{-1} when the particle size fell below 3 nm; at this point the heat of formation of the hydride phase ($\sim$36 kJ mol^{-1}) also increased. For Pd/Al$_2$O$_3$, it was higher at 97% dispersion than at 26% dispersion[140] (125 compared to 110 kJ mol^{-1}), in qualitative agreement with the previous study. The activation energy for hydrogen desorption from Pt/SiO$_2$ increased from 34 to 50 kJ mol^{-1} as the particle size increased from 1.6 to 4.3 nm.[235]

Some of the clearest quantitative evidence for the different strengths of adsorption of the various forms of hydrogen has been shown by Ru/SiO$_2$.[188,236] The heat of adsorption for the strong α_I form was 80–90 kJ mol^{-1}, for the weaker α_M form 40–43 kJ mol^{-1} and for the weakest β form at high pressure about 10 kJ mol^{-1}. Very high ratios (>5) of H/Ru$_s$ were noted at high pressure (>500 Torr), and the state of adsorbed hydrogen was then described graphically as constituting a cloud or fog.[236] A lower value (57 kJ mol^{-1}) has been given[165] for Ru/TiO$_2$, and much lower values (26–29 kJ mol^{-1}) for Cu/MgO.

Thus comparable values for the differential heats shown by supported metals of Groups 8 to 10 are remarkably uniform in the range 100 ± 25 kJ mol^{-1}(for SiO$_2$ as support at low coverage). This range is almost identical that reported many years ago for silica-supported nickel, ruthenium, rhodium, iridium, platinum and copper (100–117 kJ mol^{-1}). *Plus ça change, plus c'est la même chose.* Values for unsupported powders also fall in this area. Many of the values reported in Table 3.1 for single crystals and films are also in the same region, although films tend to show higher values and single crystals (especially iridium and platinum) lower values, perhaps reflecting the greater perfection of their surfaces. The strength of the metal-hydrogen bond formed at low coverage is quite strikingly insensitive to the nature of the metal, and it is perhaps only at higher coverages that larger differences occur: it is of course here that the forms likely to be active in catalysis occur, but where also it is more difficult to obtain reliable experimental results. The most useful information comes when the isotherm is measured as well as the heat released, and it is here that the isosteric method scores. There are unfortunately few cases where energetic measurements are accompanied by those of other techniques (e.g. spectroscopic, NMR etc, see also reference 236).

We noted in Section 2.6 that alteration of the acidity/basicity of the support produced changes, recognised by atomic XAFS (AXAFS), in the population of the valence orbitals of the surface atoms, through a charge rearrangement between the $6s$ orbital and the adjacent oxide ions; the effect of this on surface reactivity was not however developed. It is well known that XANES is sensitive to the presence of hydrogen atoms on the surface, and a recent development claims to analyse the change that they produce in the spectrum to give information not only on their number, but also on the types of site they occupy. The allocation is also sensitive to the nature of the support, and has been used to attempt a better understanding of the mechanism of alkane hydrogenolysis. This interesting thought will be considered again in Chapter 14.

Electrochemical methods have also been applied to characterising adsorbed hydrogen on supported metals.[12,237–239]

3.3.3. Theoretical Approaches[24,240–246]

The earliest theoretical exercises attempted to calculate the initial heat of adsorption of hydrogen and the corresponding bond energies D_{MH} in order to compare them with observed values, using an adaptation of the Pauling equation:

$$D_{MH} = {}^1\!/_2(D_{MM} - D_{HH}) + (x_M - x_H)^2 \qquad (3.15)$$

where D_{MM} is the energy of the metal-metal bond assumed to be broken, taken as one-sixth of the heat of sublimation, and x_M and x_H are the electronegativities of the metal and hydrogen atoms respectively in appropriate units. Several methods for estimating the electronegativity difference were tried, and more by luck than judgement, answers approximating to the observed values (sometimes surprisingly closely) were obtained. In an extension of this approach, Tanaka and Tamaru observed[247] a close parallelism between the heats of sublimation of metals and the heats of formation of oxides (usually in the highest oxidation state), and regarding the chemisorbed state as a surface compound then showed that adsorption heats for hydrogen (and other molecules) were also linear functions of the M—M bond strength. These effects suggested that the H—M bond was qualitatively similar to the M—M bond. Other empirical methods have been described.[248]

There have been many publications describing theoretical analyses of the potential energy surface for hydrogen chemisorption, with particular attention to the size and location of the energy barrier, and the importance of precursor states. Density functional theory (DFT) has been widely applied,[240,244,249,250] and shows clearly the distinction between the metals of Groups 10 and 11; however the conclusion[244] that chemisorption on gold is impossible may require modification as it does happen to some extent under some circumstances.[251] Other recent applications have been to the interaction of hydrogen with palladium[67,252] and nickel[67,253] and their alloys,[67] and to fcc(100) planes in general,[254] where the place at which the molecule is presumed to sit first has been identified (see Figure 3.8). M—H bond energies for all Transition Series metals have been calculated with various degrees of sophistication,[244,250] and show at least the same trends as the measured values. A statistical mechanical calculation of the kinetic parameters for the forward and reverse reactions of hydrogen on the metals of Group 10 has been described.[255]

The prior question, which has only recently received attention,[78] is why the hydrogen molecule desires to chemisorb at all, since the atom's valencies are already mutually satisfied. Its truly remarkable propensity for chemisorbing on metals having unfilled d-orbitals and its reluctance to do so on metals having either no d-electrons (Mg,Al) or filled d-shells (Groups 11, 12 etc) imply a controlling role for vacant d-orbitals. Two alternative models have been suggested to explain this. The first, which originated almost half a century ago, took note of the extensive evidence then available for a form of chemisorption of hydrogen on d-metals intermediate between the physically adsorbed and dissociated states; this was referred to as *Type C chemisorption*.[24] It was then supposed that the potential energy curves for both classes of metal were essentially similar, since hydrogen *atoms* chemisorb quite strongly on sp-metals, but that the intervention of an additional curve for the Type C state overcame the barrier imposed by the intersection of the curves for physical adsorption and chemisorption well above the zero of potential energy, thus effectively removing the activation energy that is shown in Figure 3.3. The absence of Type C chemisorption on sp-metals allowed this barrier to remain, thus accounting for the sizable activation energies they display. The nature of the Type C state was considered, and it was thought it might be molecular in form, and be due to some weak electronic interaction between the undissociated molecule and the metal's surface orbitals. Although the molecular states observed on stepped surfaces saturated with hydrogen atoms appear to be *sui generis*,[21,22] a molecular precursor state has been identified on Ni(111),[39] and the evidence for the Type C state cannot be ignored (although the term itself is no longer used).

The difficulty with this model is that the filled $1s_g$ binding level of the hydrogen molecule lies well below that of the typical Fermi level for d-metals, so that transfer of an electron from molecule to metal cannot happen. The vacant antibonding $1s_u$ lies above the Fermi level, so that transfer from metal to molecule, which has in any event to precede dissociation, is however feasible. Electronic charge extending into space beyond the cores of the surface atoms is rich in electrons having energies close to the Fermi level, so that the difficulty cannot be evaded by arguments based on local densities of states (LDOS; see Section 2.54). In one of the models proposed by Harris,[256,257] as a hydrogen molecule approaches the surface, the two sets of levels move to facilitate electron transfer in both directions, leading to dissociative chemisorption. The process of chemisorption can therefore be regarded as 'push' and 'pull', and thus shows some resemblance to the Chatt model for alkene coordination and chemisorption (see Section 4.5). Clearly if there are no d-band vacancies, the 'pull' element cannot operate; thus with sp metals it is necessary to create them by thermal excitation, or to find them at defects or atoms of unusual CN, and this is the origin of the activation energy. With d-metals it was suggested that transfer of the metal's sp-electrons into vacant d-levels lowered the Pauli repulsion between them and the approaching $1s_g$ orbital of the molecule, thus allowing the potential energy curve for physical adsorption to get closer to the surface, with the consequent almost complete removal of the activation energy.[257] The curve for physical adsorption (Figure 3.2) then becomes indistinguishable from that which would represent Type C chemisorption. It may however be that the mere existence of d-band vacancies is sufficient to provide the 'push' element necessary for completion of the process. This model does not include a molecular precursor state, but does not necessarily rule it out. The extensive evidence for its existence, and for the weak states observed at high coverage on supported metals (Section 3.3.1), suggest that the last word has not yet been written on this formally simple process.

Theoretical work on the adsorption and desorption of hydrogen is perhaps of greatest interests to theoreticians who are honing their skills to devise ever more complex and possibly accurate procedures to describe the behaviour of physical systems. To experimentalists in heterogeneous catalysis however their work can seem arcane and remote from the realities of the natural world, and only rarely does it tackle questions about small metal particles, such as their interaction with the support and the dependence of shape on the reduction protocol. Theoretical insights into matters such as these would be really worthwhile.

3.3.4. Hydrogen Spillover

It has been known for some years that hydrogen in some form or other can under favourable circumstances migrate from the metal particle on which it is chemisorbed onto the support, where it may initiate one or more of several possible consequences; the process is known as *spillover*[258] (in French épandage). This subject has been deeply researched (see Further Reading section), and a number of substantial reviews have appeared; indeed, so wide and various are the implications of this and cognate processes that there have been a series of international conferences devoted to it, the first in 1983.[259–262] It is not easy to summarise this large body of work in a page or so, especially since the phenomenon is hard to study, and so many different and conflicting explanations have been offered; but the attempt must be made, because of its probable importance to the proper understanding of metal-catalysed reactions. A large number of factors can affect the rate and extent of the process: these include (i) the type and composition of the metallic phase, (ii) its dispersion, (iii) the kind of support, (iv) impurities thereon, (v) temperature, (vi) hydrogen pressure and coverage of the metal by hydrogen, and (vii) water content. Once again, almost every study employs only one technique and a unique material, making comparisons difficult, and the mechanism is also likely to vary from system to system, so that the distillation of any universal truths from the available literature is unlikely to happen.

We may construct the taxonomy of hydrogen spillover in the following way.

(A) Spillover to chemically inert supports, including the ceramic oxides (Al_2O_3, SiO_2, MgO etc), acidic oxides (including zeolites), and carbon (activated charcoal, graphite etc).

(B) Spillover leading to reduction of impurities in or on the support (e.g. Fe^{3+} in Al_2O_3, SO_4^{2-} in anything).

(C) Spillover from a metal (supported or unsupported) to a reducible oxide (which may itself be supported).

(D) Spillover to a support that contains cations that are partially reducible, the former leading very often to the 'Strong Metal-Support Interaction' to be discussed in the next section.

(E) Spillover at high temperature, leading to the partial reduction of ceramic oxides (see the following section).

(F) Spillover leading to a solid state reaction in which hydrogen atoms are retained in some form in the support or in the acceptor phase; the product is generically called a *hydrogen bronze*.[263–265] This process is distinct from those in B, C and D, where the hydrogen appears as water.

The various configurations in which spillover can be observed are illustrated in Figure 3.17.

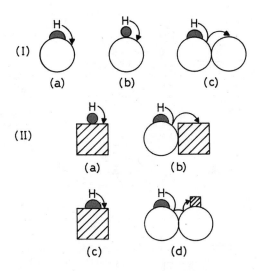

Figure 3.17. Configurations for the observations of hydrogen spillover.
(I) (a) Supported metal; (b) physical mixture of metal + support; (c) physical mixture of supported metal + support.
(II) (a) Physical mixture of metal + acceptor; (b) physical mixture of supported metal + acceptor; (c) metal supported on acceptor; (d) physical mixture of supported metal + supported acceptor.

For each class, one or more of the following questions needs to be asked, and if possible answered.

(1) Under what conditions does the process occur?
(2) To what extent and over what distance?
(3) At what rate? With what Arrhenius parameters?
(4) How is the process to be detected?
(5) By what mechanism does it occur?
(6) Is it reversible?
(7) What are the consequences especially for catalysis?

This framework should be helpful in the following survey of the five classes of behaviour.

When the support contains cations that are not easily reducible (i.e. those of Al, Si, Mg, Zr), hydrogen spillover occurs above 573 K without observable chemical reaction (Class A). However if it contains ferric ions as impurities, as is often the case with alumina, reduction to ferrous ion is detectable by EPR; and if it contains sulfate ion, as may be the case with titania, reduction of the precursor with hydrogen automatically generates hydrogen sulfide which poisons the metal (Class B). If deuterium is used in place of hydrogen, support hydroxyls

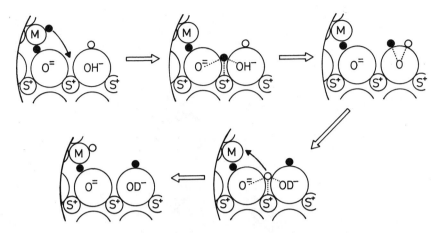

Figure 3.18. A possible mechanism for the exchange of protons on the support by deuterium atoms on a metal particle: the small open circles are H atoms and the small filled circles are D atoms.

near metal particles are exchanged[266,267] (Figure 3.18), and the process which can be followed spectroscopically, provides estimates of the diffusion coefficient D. Unfortunately, measured values of D vary over a very large range (10^{-7} to 10^{-20} m^2 s^{-1}), as do the associated activation energies,[14,262,268] even for Class A supports, showing the sensitivity of the process to the precise material used, and certainly also the method of measurement. An intermediate value (7×10^{-10} m^{2}s^{-1}) has been determined by ^{1}H NMR for the Pt/WO$_3$ system.[269] One simple manifestation of spillover is the H/M$_{tot}$ attained at high pressures; even allowing for the values above unity shown by very small particles (Section 3.3.1), the cause of which is unlikely to be spillover, ratios greater than three have often be observed and can only be the result of spillover. The amount of spillover hydrogen can also be detected by alkene titration, in which the slow diffusion of hydrogen back to the metal is monitored as it is subsequently picked up by the alkene, and appears as alkane.[270] Reverse spillover is also responsible for the catalytic effect of platinum on the decomposition of germanium hydride;[271] desorption of hydrogen atoms left on the surface of the germanium film is accelerated by their diffusion to and recombination at adjacent platinum.[272] Spillover is also recognised as a slow uptake of hydrogen, following fast chemisorption on the metal, in a volumetric apparatus.[273] High temperature peaks in TPD spectra have been attributed[131,132] to spillover hydrogen, which has also been identified by ^{1}H NMR spectroscopy, electrical conductivity and calorimetry.[268] The occurrence of spillover is a potential threat to the use of hydrogen chemisorption to estimate metal dispersion, but for measurements at room temperature it is rarely very significant.[274]

There was a time when much interest was shown in the possible catalytic activity of spillover hydrogen.[260,268,275,276] Questions of the following kind were posed. (i) Can this form of hydrogen react with acceptors (e.g. alkenes) arriving through the gas phase, or adsorbed alongside? (ii) Does the spillover process generate active centres at which pairs of reactants (e.g. alkenes and hydrogen) can chemisorb and react? The answer to both questions is a qualified 'yes' but the concentration of spiltover species on ceramic oxides is at best very low (<1% coverage), and the contribution of this 'spillover catalysis' to the whole is usually very small. That is not to say that the possibility of its occurrence should be ignored, as in certain reactions (not involving hydrocarbons) it assumes a dominant role.[277] As to the second question, elegant experimentation[268,275−279] has shown that removal of the catalyst from pure support with which it was admixed allows, after appropriate treatment, a continuing mild catalytic activity.

In a recent and perhaps more significant development, the classical mechanism for the bifunctional cracking of alkanes on acidic zeolites has had to be revised to provide a role for the involvement of spiltover hydrogen at the acidic centres of a Pt/erionite catalyst.[280]

With carbon as support, very high H/M_{tot} ratios have been reported,[13,268] and much higher concentrations of hydrogen on the support are suspected. Activated charcoal of course contains a variety of reactive surface groupings ($>C{=}O$, $-CHO$ etc) with which hydrogen could react, and spillover to clean graphite has been studied by thermal analysis,[281] where attachment to 'dangling bonds' at the edges of the graphite sheets is possible.

The chemical nature of the migrating species has been much discussed,[13,262,268,275,276,282,283] without definitive conclusions emerging. The presence of hydroxyl groups on ceramic supports is clearly beneficial, and their partial replacement by chlorine, or neutralisation by potassium ion, is detrimental. Migrants evidently have reducing properties, so they are most probably atoms (H) or hydride ions (H^-). Protons (H^+) have also been considered,[283] and could move across an hydroxylated support surface by a kind of Grotthuss conduction; it is conceivable that hydrogen atoms could move in a similar way,[262] and this is the simplest and most probable explanation for spillover at low and moderate temperatures.[268,284] However the use of ^{18}O and deuterium as tracers show that on Rh/γ-Al_2O_3 free of chloride at 473 K the rates of oxygen and hydrogen migration are the same, so that spillover by motion of hydroxyl groups is a further possibility.[262]

The reducing power of splitover hydrogen is very clearly shown in classes C, D and E. In class C, there are many examples[276,285−295] of the catalysis by a noble metal of Group 10 of the hydrogen reduction of an oxide not easily reduced, in one of the configurations shown in Figure 3.18; Group 11 metals have low activity.[268] Examination of a number of oxides showed that the catalytic effect was usually greater when the cation had an accessible oxidation state one less than the initial value[296] (e.g., Co^{III}, Fe^{III}, Mo^{VI}, W^{VI} etc). This suggested that the active species was a one electron reductant, and there were indications that unusual oxides (e.g. of Mo^V and Cr^V) could be synthesised in this way, being stable at near ambient temperature but unstable at higher temperature.[292,296] Gaseous hydrogen atoms are effective and useful reductants of metal oxides.[297]

If the activating metal is placed on or very close to the reducible oxide, careful reduction may lead to mobile atoms which attack the activating metal particles: thus reduction of Pd/ZnO leads to PdZn alloy particles. Reduction may cease at an intermediate stage, or be limited to the neighbourhood of the metal particle, or to the surface as with Pd/CeO$_2$.[298] Interesting consequences can follow; these are recognised by the somewhat misleading title of 'Strong Metal-Support Interaction' (SMSI), which is the subject of the next section.

Perhaps the most fascinating exemplar of hydrogen spillover, and certainly the most visually convincing is the process whereby it reacts with certain oxides to be incorporated in their structures without the elimination of water. The products, which are known collectively as *hydrogen bronzes*, are formed especially by the oxides of tungsten, molybdenum, vanadium and rhenium (see Further Reading section). The highly coloured products (W blue; Mo violet) contain ions of lower oxidation state (e.g. W^V, Mo^V, V^{IV}) formed for example as

$$W^{VI}O^{2-} \rightarrow W^VOH^- \longleftrightarrow W^VO^{2-}H^+ \tag{3.E}$$

but the electron is itinerant and is not fixed to any specific cation. The products are therefore analogous to the better known alkali metal bronzes (e.g. Na_xWO_3), which in fact resemble bronze in colour; the name has stuck, although the colours are different. It was in fact with Pt/WO$_3$ that the phenomenon of hydrogen spillover was first encountered;[299] the extents of hydrogen uptake make it an easy system to study, and it has been used to elucidate the spillover mechanism, and show the importance of water and other polar molecules in enabling the process to occur.[300−302] Quite enormous amounts of hydrogen can be accommodated in these lattices (for H_xXO_y, X = W, $x < 0.6$; X = Mo, $x = 0.3-2.0$, X = V, $x = 1.5-1.9$): when $x \simeq 2$, the concentration of hydrogen approaches that in the liquid state. These materials have many interesting solid-state properties;[270,303] the tungsten and molybdenum bronzes are for example metallic conductors, and all are able to release their hydrogen to acceptors such as alkenes[270,303,304] (Section 7.2.7). The small size and electronic simplicity of the hydrogen atom permits it to undertake numerous missions denied to larger and more complex species: these enrich its chemistry but frustrate the aims of the scientific investigator intent on finding definitive answers to simple questions. Hydrogen spillover is a prime example of the mobility of the atom (or ion or whatever), and this irritating habit is elegantly expressed in a sentence to be found in the Proceedings of the 1993 International

Conference:[260] *We have such a view that the hydrogen sneaked out from palladium to carbon.*

3.3.5. The "Strong Metal-Support Interaction"

Before the late 1970's there was comparatively little research on supported metal catalysts, except on methods for estimating particle size. There then occurred an apocalyptic event in the form of two papers[305,306] by scientists of the Exxon Research and Engineering Laboratory (Linden, New Jersey), following an earlier patent.[307] This work attracted enormous attention rather because of its unexpected and inexplicable nature than its practical importance. It had the effect however of drawing attention to interesting but difficult questions concerning supported metals simply as materials, and it unleashed a flood of publications in which all manner of investigational methods were applied, and theoretical explanations suggested. The flood has now subsided and it is possible to regard it in perspective. The mass of publications on the subject is however so great that only a few original works can be cited, and many of the references given are to books and reviews of which there are plenty. References 277, 308 and 309 are particularly rich in information, as are the reports of conference proceedings[259-262] (see also Further Reading section).

The basic observations were simple enough. It was observed that when a metal of Groups 8–10 supported on certain oxides of Transition Metals was heated in hydrogen above about 473 K, it progressively lost its ability to chemisorb hydrogen at ambient temperature.[273,277] The original work was performed with iridium,[306] but was later extended to the other noble metals,[304] and by others to the base metals of these Groups: on titania, on which most work has been done, all six noble metals lost almost all their chemisorption capacity for hydrogen (somewhat less so with osmium and ruthenium than the others, but their initial dispersions were lower[309,310]). Ability to chemisorb carbon monoxide was also diminished. The supports showing this effect included the oxides of tantalum (Ta_2O_5), niobium (Nb_2O_5), vanadium (V_2O_3[311]), titanium (TiO_2), and manganese (MnO); others (SiO_2, Al_2O_3, ZrO_2, HfO_2, Sc_2O_3, MgO) did not show it to any degree not explained by a little sintering.[306] The possibility that the major effect was due to catastrophic sintering was eliminated by TEM observations, which showed that it did not occur, and by the fact that it was readily reversed by an oxidative treatment. Metal particles that had suffered this loss of chemisorptive virility were said to have experienced a *'Strong Metal-Support Interaction'* or SMSI; the term has stuck although it may not describe accurately what has taken place. One consequence of the obstruction to hydrogen chemisorption was that its *dissolution* into palladium was inhibited.[15]

Not surprisingly, it was soon found that metals in this state had also lost much if not all their activity for reactions performed under strictly reducing conditions;[273,277] the only class of reaction which seemed to benefit was that involving reduction of carbon monoxide[277] (methanation, Fischer-Tropsch synthesis[312,313]) and the water-gas shift,[314,315] and much of the subsequent literature was focused on this area. It seems strange with hindsight that a phenomenon so lacking in practical utility should have commanded such attention, but enquiring minds needed to understand the cause, and the hope (ever springing eternal) was that there might be some great benefit still to be discovered. We should briefly review some of the other pertinent observations, and try to assess the several explanations that have been advanced. Our main interest will lie, not in the many simple observations of how entry into the SMSI state decreases catalytic *activity*, but rather in the changes produced in the *selectivities* shown of hydrocarbon reactions, since this can illuminate the dependence of reaction path on the structure and composition of the active centre.

Inspection of the above lists of oxides that furnish the SMSI and of those that do not at once reveals that the former have one or more readily accessible lower oxidation states, the cations of which (Ti^{3+}, Nb^{4+}, Ta^{4+}, V^{2+} etc) can be formed by spiltover hydrogen. The titanous ion has frequently been detected[273,316,317] by EPR spectroscopy in catalysts in the SMSI condition, and its presence has been confirmed or deduced by many other techniques (e.g. electrical conductivity[318] and EXAFS[319,320]). By electron diffraction it has been shown that the Ti_4O_7 phase is formed around platinum particles,[273] and model studies indicate that entry into the SMSI state is easier when the monoxide (TiO) or the sequioxide (Ti_2O_3) is used as support. There is less information about the other reduced cations. The necessity for having lower valent cations is reinforced by the reversal of the effect under oxidising conditions,[321] and even with some metals during reactions that generate water (e.g. CO methanation, Fischer-Tropsch synthesis), but the ease of reversal varies considerably from one metal to the another (Ru > Rh > Pt). A further early observation, often confirmed, was that passage into the SMSI state led to a change in metal particle morphology, from three-dimensional to thin rafts;[273,310] this may result from a change in the interfacial energy caused by reduction of the support.

The initial reduction of a support cation may be represented as

$$Pt^0 \; Ti^{4+} \; O^{2-} \; \rightarrow \; Pt^0 \; Ti^{3+} \; OH^- \; \longleftrightarrow \; Pt^{-1} \; Ti^{4+} \; OH^- \qquad (3.F)$$

and the thought that in this way the metal particles might become electron rich, and thus more gold-like, was the first attempted explanation, inevitably backed by theoretical analysis.[320,322,323] However, XPS measurements[324−326] have shown little change in binding energy, and it is now thought that any slight tendency for the metal to acquire an electron would immediately create a Schottky barrier to the transfer of further charge.[273] In fact the effects of SMSI are uncannily like those produced when the active metal is admixed with a Group 11 metal,[273] where the arguments against electron transfer are very strong. There is much more evidence, some of a very direct kind (e.g. TEM[327−329]) for the migration of species such as TiO or TiO^+ from the support onto the metal, leading ultimately to the burial of the metal particles within the support.[324] This would be aided by the creation of anion vacancies by loss of water, although in the SMSI state only 0.008 oxygens per titanium ion have been lost.[330] Many of the observations are well explained by a simple "site blocking" model, although it is not always thought quite adequate.[212] A variant on this is the idea of TiO species dissolving in the metal.[313,332] There are several claims that intermetallic phases (e.g. Pt_3Ti) are formed,[277] and there is the additional possibility noted in the last Section of strongly held hydrogen being retained within the metal. Magnetic studies on Ni/TiO_2 have shown that the SMSI is a purely surface matter, and although the explanation may not be the same for all systems and all circumstances the 'site blocking' picture remains the most attractive, together with a possible rehybridisation of the surface metal orbitals of the kind advanced to explain the characteristics of metal particles in zeolites, and which is indicated by changes in the shapes of L_{II} and L_{III} absorption edges.[212,308]

There have been a number of studies in which small amounts of reducible oxides have been added as 'promoters' to metals supported on 'inert' oxides (SiO_2, Al_2O_3), e.g. Rh/Nb_2O_5-SiO_2,[262,333,334] Ru/TiO_2-SiO_2,[335,336] Ni/TiO_2-Al_2O_3.[260] In this way the extent of the SMSI effect can be controlled by varying the amount of 'promoter', and indeed SMSI phenomena are produced in this way. There have been many studies of model systems, mainly formed by evaporating the oxide or its parent metal onto the active metal;[335,337] nickel has also been deposited on the rutile (100) surface to generate an SMSI.[338] This paper reminds us that little attention has been paid to the relevance of the titania phase (anatase, rutile) to the onset of the SMSI. Frequent use is made of Degussa P-25, which contains both phases, but a systematic study of this matter has not been undertaken.

One of the consequences of hydrogen spillover briefly noted in the last section (see also Further Reading section) is the partial reduction at high temperatures (773–1273 K) of ceramic oxides (Al_2O_3, SiO_2, MgO, BeO, CaO, La_2O_3, Y_2O_3, ZrO_2), leading to effects resembling the SMSI, namely, loss of inclination to chemisorb hydrogen and *a fortiori* diminution of catalytic activity. It is not always clear what has happened; with alumina and silica there is definite evidence for the formation of stoichiometric aluminides and silicides,[273] the driving force being the very exothermic nature of the process: the energy barrier for this is high, so high temperatures are needed. With magnesia and possibly also alumina, the effect may be caused by reduction of sulfate impurity, giving a partial coating of the metal by sulfur.[339] In other cases (including titania), there may be intermetallic phases formed, or simply surface decoration. There is also a considerable literature to show that treatment with hydrogen at about 770 K can create some strongly retained hydrogen on the catalyst, and that this also has a toxic effect, by entering sub-surface sites.[110,340]

This brief survey must suffice to introduce a pervasive phenomenon which must often affect the structure and composition of small metal particles, and hence their catalytic behaviour: this will be a recurring item in later chapters. *Nunc est bibendum.*

TABLE 3.2. Catalysed Reactions of Hydrogen and its Analogues

Designation	Reaction
1 Dissociation	$H_2 + 2_* \rightarrow 2H_* \rightarrow 2H\cdot + 2_*$
2 Atom recombination	$2H\cdot + 2_* \rightarrow 2H_* \rightarrow H_2 + 2_*$
3 Isotopic equilibration	$H_2 + D_2 + 4_* \rightarrow 2H_* + 2D_* \rightarrow 2HD + 4_*$
4 Spin-isomer equilibration	$p\text{-}H_2 + 2_* \rightarrow 2H_* \rightarrow o\text{-}H_2 + 2_*$
5 Isotopic exchange	$H_* + D_2 \rightarrow D_* + HD$
6 Hydrogen ion discharge	$H^+ + _* + e^- \rightarrow H_* \rightarrow {}^1\!/_2\,H_2 + _*$

· Gaseous H atom
* Adsorption site

3.4. REACTIONS OF HYDROGEN

The concern of this section is with incestuous reactions of hydrogen, that is to say, with reactions between the various distinguishable forms of the molecule rather than of the molecule with the catalyst or anything else. They will be treated only briefly (see Further Reading section), because they do not bear importantly on reactions of hydrocarbons although they may sometimes occur in parallel with them.

The reactions that hydrogen can undergo are listed in Table 3.2. The dissociation of molecules on metal surfaces has already been discussed at length; at temperatures above about 1270 K, hydrogen atoms desorb from the surface of metal wires and can be trapped (e.g. by MoO_3[24,341]). Rates are proportional to $P_H^{0.5}$ and activation energies are about 200 kJ mol^{-1}. The reverse process of atom recombination has been a popular reaction to study, as it was thought to sense directly the bonding propensity of the surface responsible, and the relative efficiencies of a number of metals,[24] and alloys (Pd-Ag, Pd-Au,[342] Cu-Ni) have been determined. High recombination efficiency correlates with low heat of adsorption. A related process is the discharge of hydrogen ions in solution,[343] where acquisition of an electron from the metal gives adsorbed atoms, which then recombine. Gaseous hydrogen atoms reacted with chemisorbed deuterium atoms on Pt(111) at 110 K to give gaseous hydrogen deuteride.[344]

The processes of isotopic and spin isomer equilibration have also been much studied in the past. The latter reaction can be followed simply by changes in thermal conductivity, and was attractive for this reason,[345] but it does not necessarily require the molecule to be dissociated, as nuclear spins can reverse in the presence of strong paramagnetic centres. The hydrogen-deuterium reaction (homomolecular exchange) is however diagnostic for dissociation and recombination of atoms.[346] The mechanism of these deceptively simple reactions is however not at all straightforward. The early work has been fully reviewed,[24,49,342,337] and is unnecessary to rehearse this at length; two mechanisms have been considered, the first of which is that shown in Table 3.2 and is associated with the names of Bonhoeffer and

Farkas, while the second is a chain mechanism:

$$
\begin{array}{ccccccc}
\text{D--D} & & \text{D} & & \text{D--H} & & \\
& \text{H} & \text{D} & \text{H} & \text{D} & & \\
* & * & * & * & * & * &
\end{array}
\tag{3.G}
$$

proposed by Eley and Rideal. It has proved extraordinarily difficult to distinguish between these mechanisms on kinetic grounds, but it seems likely that the former applies on platinum below 110 K and the latter in the range 110–200 K.[348] There is however also a contribution from a paramagnetic mechanism due to unpaired electrons at the surface, because the conversions of para-hydrogen and of ortho-deuterium are faster than the isotopic equilibration.[349-351] Analogous findings have been reported sfor nickel.

A particular objective of the early studies was to establish a connection between electronic structure and catalytic activity,[345] and they have been extensively reviewed.[24,49] In the nickel-copper system,[352,353] the activation energy was independent of composition, but the pre-exponential factor decreased with increasing copper concentration, in line with the AES-determined surface concentration. With palladium-silver, the activation energy *increased* with silver content in the range 48–100% from exceedingly low values ($\sim$1.7 kJ mol^{-1}) to about 25 kJ mol^{-1}, in line with the heat of adsorption derived from the kinetics.[354] An excellent compensation plot was obtained, except for one point (50% Ag) where the film may have been poisoned. These well-intentioned but ultimately misguided studies failed because the results were interpreted by a simplistic model of the electronic structures of alloys, which as we have seen (Section 1.3.2) was incorrect. A re-interpretation based on a knowledge of surface composition and structure is still awaited. Para-hydrogen conversion is hardly ever used now, but the hydrogen-deuterium reaction is still used as a rapid means of showing that reversible dissociation of the molecules occurs.[355,356] "Model" studies with Pd/C catalysts suggested[357] that the reaction was particle-size dependent with this system, the rate being maximal at a mean size of 1.3 nm; the activation energy was also size dependent, falling from 50 kJ mol^{-1} at 1.1 nm to 20 kJ mol^{-1} at 1.9 nm, in parallel with the decrease in the $3d_{5/2}$ binding energy. This reaction is particularly useful with the Group 11 metals,[253,358-360] the reaction on Au/MgO and Au/SiO$_2$ being accelerated by adsorbed oxygen atoms.[359] Similar promotion occurs with sulfur on Pt(111).[360,361] It is a classic case of a structure-insensitive reaction over a wide range of particle size.

Isotopic exchange (see Table 3.2) in which an atom one isotope is replaced by another is a useful way of assessing the available pool of atoms, either to determine the dispersion of metal particles or the extent of hydrogen spillover (e.g. into hydrogen bronzes) or of dissolution into palladium.[362] The rate of hydrogen evolution from the discharge of solvated protons is also determined by the strength of the H–M bond. Extensive results for rates of para-hydrogen conversion, hydrogen atom recombination and hydrogen evolution show a pleasing parallelism,[363,364] and especially highlight the low activity of manganese, on which hydrogen is only weakly adsorbed. They also emphasise the superior activity of iron, cobalt and nickel in the first Transition Series.

REFERENCES

1. K. Christmann, *Molec. Phys. B* **66** (1989) 1.
2. *Hydrogen Effects in Catalysis,* (Z. Paál and P.G. Menon, eds.), Marcel Dekker: New York (1988).
3. K.R. Christmann, in: *Hydrogen Effects in Catalysis* (Z. Paál and P.G. Menon, eds.), Marcel Dekker: New York (1988), p. 3.
4. A.W. Adamson, *Physical Chemistry of Surfaces,* Wiley: New York (1990).
5. C.M. Mate, B.E. Bent and G.A. Somorjai, in: *Hydrogen Effects in Catalysis* (Z. Paál and P.G. Menon, eds.), Marcel Dekker: New York (1988), p. 57.
6. J.W. Geus, in: *Hydrogen Effects in Catalysis* (Z. Paál and P.G. Menon, eds.), Marcel Dekker: New York (1988), p. 85.
7. P.G. Menon, in: *Hydrogen Effects in Catalysis* (Z. Paál and P.G. Menon, eds.), Marcel Dekker: New York (1988), p. 117.
8. C.H. Bartholomew, in: *Hydrogen Effects in Catalysis* (Z. Paál and P.G. Menon, eds.), Marcel Dekker: New York (1988), p. 139.
9. T.J. Udovic and A.D. Kelley, in: *Hydrogen Effects in Catalysis* (Z. Paál and P.G. Menon, eds.), Marcel Dekker: New York (1988), p. 167.
10. T. Szilágyi, in: *Hydrogen Effects in Catalysis* (Z. Paál and P.G. Menon, eds.), Marcel Dekker: New York (1988), p. 183.
11. J.W. Geus, in: *Hydrogen Effects in Catalysis* (Z. Paál and P.G. Menon, eds.), Marcel Dekker: New York (1988), p. 195.
12. J. Petró, T. Mallát, E. Polyánszky and T. Máthé, in: *Hydrogen Effects in Catalysis* (Z. Paál and P.G. Menon, eds.), Marcel Dekker: New York (1988), p. 225.
13. W.C. Conner Jr., in: *Hydrogen Effects in Catalysis* (Z. Paál and P.G. Menon, eds.), Marcel Dekker: New York (1988), p. 311.
14. R. Burch, in: *Hydrogen Effects in Catalysis* (Z. Paál and P.G. Menon, eds.), Marcel Dekker: New York (1988), p. 347.
15. W. Palczewska, in: *Hydrogen Effects in Catalysis* (Z. Paál and P.G. Menon, eds.), Marcel Dekker: New York (1988), p. 373.
16. J. Halpern, *Adv. Catal.* **11** (1959) 301.
17. K. Christmann, *Introduction to Surface Physical Chemistry,* Steinkopff: Darmstad (1991).
18. Y. Ebisuzaki and M. O'Keefe, *Prog. Solid State Chem.,* **4** (1967) 187.
19. P.B. Wells, *J. Catal.* **52** (1978) 498.
20. A.G. Burden, J. Grant, J. Martos, R.B. Moyes and P.B. Wells, *Discuss. Faraday Soc.* **72** (1981) 95.
21. C. Nyberg, K. Svensson, A.-S. Mårtensson and S. Anderson, *J. Elec. Spec. Rel. Phen.* **64** (1993) 51; A.-S. Mårtensson, C. Nyberg and A. Anderson, *Phys. Rev. Lett.* **57** (1986) 2045.
22. P.K. Schmidt, K. Christmann, G. Kreese, J. Hafner, M. Lischka and A. Gross, *Phys. Rev. Lett.* **87** (2001) 96.
23. J.N. Russell Jr., S.M. Gates and J.T. Yates Jr., *J. Chem. Phys.* **85** (1986) 6792.
24. G.C. Bond, *Catalysis by Metals,* Academic Press: London (1962).
25. O. Beeck, *Adv. Catal.* **2** (1950) 151.
26. D.O. Hayward and B.M.W. Trapnell, *Chemisorption,* 2nd ed., Butterworths: London (1964).
27. *Thin Metal Films and Gas Chemisorption,* (P. Wissman, ed.), Elsevier: Amsterdam (1987).
28. *Chemisorption and Reactions on Metallic Films,* (J.R. Anderson, ed.). Vols. 1 and 2, Academic Press: London (1971).

29. D.R. Rossington in: *Chemisorption and Reactions on Metallic Films,* Vol. 2 (J.R. Anderson, ed.), Academic Press: London (1971), p. 211.
30. G.C. Bond, *Heterogeneous Catalysis: Principles and Applications,* 2nd edn. Oxford University Press: Oxford 1987.
31. B.E. Nieuwenhuys, *Surf. Sci.* **59** (1976) 430.
32. F.C. Tompkins, *Chemisorption of Gases on Metals,* Academic Press: London (1978).
33. M.W. Roberts and C.S. McKee, *Chemistry of the Metal-Gas Interface,* Clarendon Press: Oxford (1978).
34. R.H.P. Gasser, *An Introduction to Chemisorption and Catalysis by Metals,* Clarendon Press: Oxford (1985).
35. Z. Knor and E.W. Müller, *Surf. Sci.* **10** (1968) 21.
36. K. Christmann, *Bull. Soc. Chem. Belg.* **88** (1979) 519.
37. R. Lang, R.W. Joyner and G.A. Somorjai, *Surf. Sci.* **39** (1972) 440.
38. K. Christmann, E. Ertl and T. Pignet, *Surf. Sci.* **54** (1976) 365.
39. A.T. Hanbicki, S.B. Darling, D.J. Gaspar and S.J. Sibener, *J. Phys. Chem.* **111** (1999) 9053.
40. K. Christmann and G. Ertl, *Surf. Sci.* **60** (1976) 365.
41. C.H. Bartholomew, in: *Specialist Periodical Reports: Catalysis,* Vol. 11 (J.J. Spivey and S.K. Agarwal, eds.), *Roy. Soc. Chem.* (1994), p. 93.
42. *Hydrogen in Metal Systems II,* (F.A. Lewis and A. Aladjem, eds.), Scitech Publ. Uetikon-Zurich (2000).
43. N.N. Greenwood and A. Earnshaw, *Chemistry of the Elements,* 2nd edn. Butterworth-Heinemann: Oxford (1997).
44. Sh. Shaikhutdinov, M. Heemeier, M. Bäumer, T Lear, D. Lennon, R.J. Oldham, S.D. Jackson and H.-J. Freund, *J. Catal.* **200** (2001) 330.
45. Z. Karpiński, *Adv. Catal.* **37** (1990) 45.
46. W. Palczewska, *Adv. Catal.* **24** (1975) 245.
47. M. Shamsuddin and O.J. Kleppa, *J. Chem. Phys.* **71** (1979) 5154.
48. A.G. Knapton, *Platinum Metals Rev.* **21** (1977) 44.
49. G.C. Bond and E.G. Allison, *Catal. Rev.* **7** (1972) 233.
50. W. Palczewska, M. Cretti-Bujnowska, J. Pielaszek, J. Sobczak and J. Stachurski, in: *Proc. 9th Internat. Congr. Catal.* (M.J. Phillips and M. Ternan, eds.), Chem. Inst. Canada: Ottawa (1988) 1410.
51. F.A. Lewis and W.H. Schurter, *Naturwiss.* **8** (1960) 177.
52. A. Maeland and T.B. Flanagan, *J. Phys. Chem.* **69** (1965) 3575.
53. K. Allard, A. Maeland, J.W. Simons and T.B. Flanagan, *J. Phys. Chem.* **72** (1968) 136.
54. M. Boudart and H.S. Hwang, *J. Catal.* **39** (1975) 44.
55. D.H. Everett and P.A. Sermon, *Zeit. Phys. Chem. (Wiesbaden)* **114** (1979) 109; R.K. Nandi, G. Georgopoulos, J.B. Cohen, J.B. Butt and R.L. Burwell Jr., *J. Catal.* **77** (1982) 421.
56. F. Chao and M. Costa, in: *Hydrogen in Metals,* (Proc. 2nd Internat. Congr., Paris, (1977)), Pergamon: Oxford, (1978), Paper 58A.
57. B.J. Wood and H. Wise, *J. Catal.* **5** (1966) 135.
58. G.C. Bond and T. Mallát, unpublished work.
59. A. Janko and J. Pielaszek, *Bull. Acad. Pol. Sci. Ser. Sci. Chim.* **15** (1967) 569.
60. B. Baranowski and S. Majchrzak, *Roczn. Chem.* **42** (1968) 1137.
61. Z. Knor, in: *Catalysis: Science and Technology,* Vol. 3 (J.R. Anderson and M. Boudart, eds.), Springer-Verlag: Berlin (1982), p. 231.
62. K. Christmann, *Surf. Sci. Rep.* **9** (1988) 1.
63. D.O. Hayward and A.O. Taylor, *Chem. Phys. Lett.* **146** (1988) 221.
64. M.J. Murphy and A. Hodgson, *J. Chem. Phys.* **108** (1998) 4199.
65. K. Asakura, T. Kubota, N. Ichikuni and Y. Iwasawa, *Proc. 11th. Internat. Congr. Catal.* (J.W. Hightower, W.N. Delgass, E. Iglesia and A.T. Bell, eds.), Elsevier: Amsterdam, Vol. **B** (1996), p. 911.
66. M.M. Otten, M.J. Clayton and H.H. Lamb, *J. Catal.* **149** (1994) 211.

67. S. Romanowski, W.M. Bartczak and R. Wesotkowski, *Langmuir* **15** (1999) 5773.
68. G. Wedler, *Chemisorption: an Experimental Approach,* (Trans. D.F. Klemperer), Butterworths: London (1976).
69. G.A. Somorjai, *Introduction to Surface Chemistry and Catalysis,* Wiley: New York (1994).
70. J.M. Thomas and W.J. Thomas, *Principles and Practice of Heterogeneous Catalysis,* VCH: Weinheim (1997).
71. N. Cardona-Martinez and J.A. Dumesic, *Adv. Catal.* **38** (1992) 149.
72. J.B. Miller, H.R. Siddiqi, S.M. Gates, J.N. Russell Jr., J.T. Yates Jr., J.C. Tully and M.J. Cardillo, *J. Chem. Phys.* **87** (1987) 6725.
73. L.P. Ford, H.L. Nigg, P. Blowers and R.I. Masel, *J. Catal.* **178** (1998) 163.
74. J.E. Lennard-Jones, *Trans. Faraday Soc.* **28** (1932) 333.
75. H.-J. Freund, in: *Handbook of Heterogeneous Catalysis,* Vol. 3 (G. Ertl, H. Knözinger and J. Weitkamp, eds.), VCH: Weinheim, (1997), p. 911.
76. J.M. Campbell and C.T. Campbell,*Surf. Sci.* **259** (1991) 1.
77. G. Lee and E.W. Plummer, *Surf. Sci.* **498** (2002) 229.
78. R.I. Masel, *Principles of Adsorption and Reactions on Solids,* Wiley: New York (1996).
79. P.R. Norton, J.A. Davies and T.E. Jackman, *Surf. Sci.* **121** (1982) 103.
80. A. Hodgson, *Prog. Surf. Sci.* **63** (2000) 1.
81. B.E. Hayden and A. Hodgson, *J. Phys. Cond. Mat.* **11** (1999) 8397.
82. K. Christmann and G. Ertl, *Surf. Sci.* **66** (1976) 365.
83. L.P. Ford, H.L. Nigg, P. Blowers and R.I. Masel, *J. Catal.* **179** (1998) 163.
84. R.W. McCabe and L.D. Schmidt, *Surf. Sci.* **60** (1976) 85.
85. C. Resch, V. Zhukov, A. Lugstein, H.F. Berger, A. Winkler and K.D. Rendulic, *Chem. Phys.* **177** (1993) 421.
86. J.K. Brown, A.C. Luntz and P.A. Schultz, *J. Chem. Phys.* **95** (1991) 3767.
87. G.A. Somorjai, *Langmuir* **7** (1991) 3176.
88. A. Janko, W. Palczewska and I. Szymerska, *J. Catal.* **61** (1980) 264.
89. S. Horch, H.T. Lorensen, S. Helveg, E. Lægsgaard, I. Stensgard, K.W. Jacobsen, J.K. Nørskov and F. Besenbacher, *Nature* **398** (1999) 134.
90. P.R. Norton, J.A. Davies, D.P. Jackson and N. Matsunami, *Surf. Sci.* **85** (1979) 269.
91. A.K. Datye, D.S. Kalakkad, E. Völke and L.F. Allard, in: *Electron Holography,* (A. Tonomura, L.F. Allard, G. Pozzi, D.C. Joy and Y.A. Ono, eds.), Elsevier: Amsterdam, (1993), p. 199.
92. M. Nishijima, S. Masuda, M. Jo and M. Onchi, *J. Electron Spectrosc. Relat. Phenom.* **29** (1983) 273.
93. R.V. Culver and F.C. Tompkins, *Adv. Catal.* **11** (1959) 68.
94. *Strong Metal-Support Interactions* (R.T.K. Baker, S.J. Tauster and J.A. Dumesic, eds.) ACS Symposium Series 298, Am. Chem. Soc.: Washington, D.C. (1986).
95. C.T. Campbell, *Ann. Rev. Phys. Chem.* **41** (1990) 775.
96. P. Samson, A. Nesbitt, B.E. Koel and A. Hodgson, *J. Chem. Phys.* **109** (1998) 3255.
97. V. Ponec and G.C. Bond, *Catalysis by Metals and Alloys,* Elsevier; Amsterdam (1995).
98. B.E. Nieuwenhuys, in: *Physics of Covered Surfaces*, Landolt-Börnstein, Group 42 (subvolume A), (H.P. Engel, ed.), Springer-Verlag: Berlin (2001), p. 362.
99. O. Beeck, *Discuss. Faraday Soc.* **8** (1950) 118.
100. O. Beeck, A.E. Smith and A. Wheeler, *Proc. Roy. Soc.* **A177** (1941) 62.
101. O. Beeck and A.W. Ritchie, *Discuss. Faraday Soc.* **8** (1950) 159.
102. J.B. Miller, H.R. Siddiqui, S.M. Gates, J.N. Russell Jr., J.T. Yates Jr., J.C. Tully and M.J. Cardillo, *J. Chem. Phys.* **87** (1987) 6725; A.O. Taylor and D.O. Hayward, *Surf. Sci.* **346** (1996) 222.
103. J.E. Benson and M. Boudart, *J. Catal.* **4** (1965) 704.
104. S. Černy, M. Smutek and F. Buzek, *J. Catal.* **38** (1975) 245.
105. J. Shen, B.E. Spiewak and J.A. Dumesic, *Langmuir* **13** (1999) 2735.
106. L.D. Schmidt, *Catal. Rev.– Sci. Eng.* **9** (1974) 115.

107. J.L. Falconer and J.A. Schwarz, *Catal. Rev.-Sci. Eng.* **25** (1983) 141.
108. V. Ponec and G.C. Bond, *Catalysis by Metals and Alloys,* Elsevier: Amsterdam, 1996, p. 404ff.
109. B. Tardy and S.J. Teichner, *J. Chim. Phys. Physicochim. Biol.* **67** (1970) 1968.
110. Z. Paál and P.G. Menon, *Catal. Rev.– Sci. Eng.* **25** (1983) 229.
111. J.W. Niemantsverdriet, *Spectroscopy in Catalysis,* VCH: Weinheim (1993).
112. M. Che and C.O. Bennett, *Adv. Catal.* **36** 1989) 55.
113. G.C.A. Schuit and L.L. van Reijen, *Adv. Catal.* **10** (1958) 243.
114. G. Bergeret and P. Gallezot in: *Handbook of Heterogeneous Catalysis*, Vol.2 (G. Ertl, H. Knözinger and J. Weitkamp, eds.), Wiley-VCH: Weinheim (1997), p. 439.
115. R.N. Reifsnyder and H.H. Lamb, *Catal. Lett.* **40** (1996) 155.
116. B. Coq, E. Crabb, M. Warawdekar, G.C. Bond, J.C. Slaa, S. Galvagno, L. Mercadante, J. Garcia Ruiz and M.C. Sanchez Sierra, *J. Molec. Catal.* **99** (1994) 1.
117. B.L. Mojet, J.T. Miller, D.E. Ramaker and D.C. Koningsberger, *J. Catal.* **186** (1999) 373.
118. M. Vaarkamp, F.S. Modica, J.T. Miller and D.C. Koningsberger, *J. Catal.* **144** (1993) 611.
119. *Determination of the Specific Surface Areas of powders,* BS4359:Part 4 (1995). (Recommendations for Methods of Determination of Metal Surface Area using Gas Adsorption Techniques)
120. ASTM, D3908-80.
121. P.C. Flynn and S.E. Wanke, *Canad. J. Chem. Eng.* **53** (1975) 636.
122. O.M. Poltorak and V.S. Boronin, *Russ. J. Phys. Chem.* **39** (1965) 781, 1329; **40** (1966) 1436.
123. G.K. Boreskov and A.P. Kharnaukov, *Zhur. Fiz. Khim.* **26** (1952) 1814.
124. L. Spenadel and M. Boudart, *J. Phys. Chem.* **64** (1960) 204.
125. S.F. Adler and J.J. Keavney, *J. Phys. Chem.* **64** (1960) 208.
126. H.L. Gruber, *J. Phys. Chem.* **66** (1962) 42.
127. G.C. Bond and J. Turkevich, *Trans. Faraday Soc.* **52** (1956) 1235.
128. R. Giannantonio, V. Ragaini and P. Magni, *J. Catal.* **146** (1994) 103.
129. M.A. Chesters, K.J. Packer, D. Lennon and H.E. Viner, *J. Chem. Soc. Faraday Trans.* **91** (1995) 2191.
130. H. Ehwald and U. Leibnitz, *Catal. Lett.* **38** (1996) 149.
131. M. Arai, M. Fukushima and Y. Nishiyama, *Appl. Surf. Sci,* **99** (1996) 145.
132. J.T. Miller, B.L. Meyers, F.S. Modica, G.S. Lane, M. Vaarkamp and D.C. Koningsberger, *J. Catal.* **143** (1993) 395.
133. J.T. Miller, B.L. Meyers, M.K. Barr, F.S. Modica and D.C. Koningsberger, *J. Catal.* **159** (1996) 41.
134. F. Lai, D.-W. Kim, O.S. Alexeev, G.W. Graham, M. Shelef and B.C. Gates, *Phys. Chem. Chem. Phys.* **2** (2000) 1997.
135. F.P. Netzer and H.L. Gruber, *Z. Phys. Chem. NF* **96** (1975) 25; **85** (1973) 159.
136. A.R. Berzins, M.S.W. Lau Vong, P.A. Sermon and A.T. Wurie, *Ads. Sci. Technol.* **1** (1984) 51.
137. G.C. Bond and Lou Hui, *J. Catal.* **147** (1994) 346.
138. G.B. McVicker, R.T.K. Baker, R.L. Garten and E.L. Kugler, *J. Catal.* **65** (1980) 207.
139. V. Ragaini, R. Giannantonio, P. Magni, L. Lucarelli and G. Leofanti, *J. Catal.* **146** (1994) 116.
140. A. Guerrero, M. Reading, Y. Grillet, J. Rouquerol, J.P. Boitiaux and J. Cosyns, *Z. Phys. D* **12** (1989) 583.
141. P. Albers, E. Auer, K. Ruth and S.F. Parker, *J. Catal.* **196** (2000) 174.
142. P.A. Sermon, *J. Catal.* **24** (1972) 467.
143. C.H. Bartholomew, *Catal. Lett.* **7** (1990) 27.
144. J.B.C. Cobb, A. Bennett, G.C. Chinchen, L. Davies, B.T. Heaton and J.A. Iggo, *J. Catal.* **164** (1996) 268.
145. J. Tabatabaei, B.H. Sakakini, M.J. Watson and K.C. Waugh, *Catal. Lett.* **59** (1999) 157.
146. P. Mériaudeau, A. Thangaraj, C. Naccache and S. Narayanan, *J. Catal.* **148** (1994) 617.
147. D.C. Koningsberger, J. de Graaf, B.L. Mojet and D.E. Ramaker, *Appl. Catal. A: Gen.* **191** (2000) 205.

148. J.T. Miller, B.L. Mojet, D.E. Ramaker and D.C. Koningsberger, *Catal. Today* **62** (2000) 101.

149. D.E. Ramaker, B.L. Mojet, M.T.G. Oostenbrink, J.T. Miller and D.C. Koningsberger, *Phys. Chem. Chem. Phys.* 1 (1999) 2293.

150. M. Vaarkamp, *Catal. Lett.* **6** (1990) 369.

151. M. Vaarkamp, B.L. Mojet, M.J. Kappers, J.T. Miller and D.C. Koningsberger, *J. Phys. Chem.* **99** (1995) 16067.

152. G.C. Bond in: *Handbook of Heterogeneous Catalysis,* Vol. 3 (G. Ertl, H. Knözinger and J. Weitkamp, eds.), Wiley-VCH: Weinheim (1997), p. 1489.

153. B.J. Kip, F.B.M. Duivenvoorden, D.C. Koningsberger and R. Prins, *J. Catal.* **106** (1986) 26; *J. Am. Chem. Soc.* **108** (1986) 5633; A. Renouprez, C. Hoang-Van and P.A. Compagnon, *J. Catal.* **34** (1974) 411.

154. D.I. Hall, V.A. Self and P.A. Sermon, *J. Chem. Soc. Faraday Trans. I* **83** (1987) 2693.

155. G.C. Bond in: *Handbook of Heterogeneous Catalysis,* Vol. 2 (G. Ertl, H. Knözinger and J. Weitkamp, eds.), VCH: Weinheim, (1997), p. 752.

156. A.Y. Stakheev and L.M. Kustov, *Appl. Catal. A: Gen.* **188** (1999) 3.

157. O. Alexeev, Du-Woan Kim, G.W. Graham, M. Shelef and B.C. Gates, *J. Catal.* **185** (1999) 170.

158. M. Vaarkamp, J.T. Miller, F.S. Modica and D.C. Koningsberger, *J. Catal.* **163** (1996) 294.

159. J.T. Richardson and T.S. Gale, *J. Catal.* **102** (1986) 419.

160. A. Kaldor and D.M. Cox, *J. Chem. Soc. Faraday Trans. I* **86** (1990) 2459.

161. W. Bronger and G. Auffermann, *Angew. Chem. Internat. Edn.* **33** (1994) 1112.

162. H. Kubicka, *React. Kinet. Catal. Lett.* **5** (1976) 223.

163. D.O. Uner, M. Pruski, B.C. Gerstein and T.S. King, *J. Catal.* **146** (1994) 530.

164. R.K. Thampi, L. Lucarelli and J. Kiwi, *Langmuir* **7** (1991) 2642.

165. V.P. Londhe and N.M. Gupta, *J. Catal.* **169** (1997) 415.

166. P. Jonsen, *Coll. and Surf.* **36** (1989) 127.

167. F. Engelke, R. Vincent, T.S. King and M. Pruski, *J. Phys. Chem.* **102** (1994) 7262.

168. J.A. Don, A.P. Pijpers and J.J.F. Scholten, *J. Catal.* **80** (1983) 296.

169. R.A. Dalla Betta, *J. Catal.* **34** (1974) 57.

170. K. Christmann, G. Lauth and E. Schwarz, *Vacuum* **41** (1990) 293; *J. Chem. Phys.* **91** (1989) 3729.

171. H. Shi, P. Geng and K. Jacobi, *Surf. Sci.* **315** (1994) 1.

172. G.C. Bond, R.R. Rajaram and R. Burch, *Appl. Catal.* **27** (1986) 379.

173. T. Narita, H. Miura, M. Ohira, H. Hondu, K. Sugiyama, T. Matsuda and R.D. Gonzalez, *Appl. Catal.* **32** (1987) 185.

174. Xi Wu, B.C. Gerstein and T.S. King, *J. Catal.* **135** (1992) 68.

175. S. Bhatia, B.C. Gerstein and T.S. King, *J. Catal.* **134** (1992) 572.

176. T.E. Hoost and J.G. Goodwin Jr., *J. Catal.* **130** (1991) 283.

177. H. Berndt and V. Müller, *Appl. Chem. A: Gen.* **180** (1999) 63.

178. S.K. Masthan, K.V.R. Chary and P. Kanta Rao, *J. Catal.* **124** (1990) 289.

179. L. Guczi and A. Sárkány in: *Specialist Periodical Reports: Catalysis*, Vol. 11 (J.J. Spivey and S.K. Agarwal, eds.), *Roy. Soc. Chem.* (1994), p. 318.

180. V. Ponec, *Mat. Sci. Eng.* **42** (1980) 135.

181. Tsong-huei Chang, Chen Rjeng Cheng and Chuin-tih Yeh, *J. Chem. Soc., Faraday Trans.* **90** (1994) 1157.

182. J.A. Dalmon, G.A. Martin and B. Imelik, *Jap. J. Appl. Phys.* Suppl. 2, Pt. 2, (1974) 261.

183. J.R. Anderson, K. Foger and R.J. Breakspere, *J. Catal.* **57** (1979) 458.

184. J.H. Sinfelt, Y.L. Lam, J.A. Cusumano and A.E. Barnett, *J. Catal.* **42** (1976) 227.

185. A.J. Hong, A.J. Rouco, D.E. Resasco and G.L. Haller, *J. Phys. Chem.* **91** (1987) 2665.

186. N. Savargaonkar, R.L. Narayan, M. Pruski, D.O. Uner and T.S. King, *J. Catal.* **178** (1998) 26.

187. D.P. VanderWiel, M. Pruski and T.S. King, *J. Catal.* **188** (1999) 186.

188. R.L. Narayan, N. Savaragaonkar, M. Pruski and T.S. King, in: *Proc. 11ᵗʰ Internat. Congr. Catal.*, (J.W. Hightower, W.N. Delgass, E. Iglesia and A.T. Bell, eds.), Elsevier: Amsterdam **B** (1996) 921.

189. L. Carballo, C. Serrano, E.E. Wolf and J.J. Carberry, *J. Catal.* **52** (1978) 507.

190. R. Bacaud, G. Blanchard, H. Charcosset and L. Tournayan, *React. Kinet. Catal. Lett.* **12** (1979) 357.

191. S. Gönenç and Z.I. Önsan, *Doğa TUJ Chem.* **12** (1988) 249.

192. A. Gervasini and C. Flego, *Appl. Catal.* **72** (1991) 153.

193. Dong Jin Suh, Tae-Jin Park and Son-Ki Ihm, *J. Catal.* **149** (1994) 486.

194. G. Prelazzi, M. Cerboni and G. Leofanti, *J. Catal.* **181** (1999) 73.

195. J.P. Bucher, J. Buttet, J.J. van der Klink, M. Graetzel, E. Newson and T.B. Truong, *J. Molec. Catal.* **43** (1987) 213.

196. J. Prasad, K.R. Murthy and P.G. Menon, *J. Catal.* **52** (1978) 515.

197. F. Notheisz, A.G. Zsigmond and M. Bartok, *J. Chromatogr.* **241** (1982) 101.

198. P. Wentrcek, K. Kimoto and H. Wise, *J. Catal.* **33** (1973) 279.

199. E. Choren, J. Hernandez, A. Arteaga, G. Arteaga, H. Lugo, M. Arrúez, A. Parra and J. Sanchez, *Catal. Lett.* **1** (1988) 283.

200. G.C. Bond and P.A. Sermon, *J. Chem. Soc. Faraday Trans. I*, **72** (1976) 745; R.L. Augustine, K.P. Kelly and R.W. Warner, *J. Chem. Soc. Faraday Trans. I* **79** (1983) 2639.

201. J. Barbier and J. Rivière, *Bull. Soc. Chim. Fr.* **7-8** (1978) I-409.

202. C. Aharoni and F.C. Tompkins, *Adv. Catal.* **21** (1970) 1.

203. I.S. McLintock, *J. Catal.* **16** (1970) 126; *Nature* **216** (1967) 1204.

204. H.A. Taylor and N. Thon, *J. Am. Chem. Soc.* **74** (1952) 4169.

205. M.J.D. Low and H.A. Taylor, *Canad. J. Chem.* **37** (1959) 544.

206. M.J.D. Low, *Chem. Rev.* **60** (1960) 267.

207. M.J.D. Low, *Canad. J. Chem.* **38** (1960) 588.

208. J.J. van der Klink, *Adv. Catal.* **44** (1999) 2.

209. J. Fraissard, *Catal. Today* **51** (1999) 481.

210. C.P. Slichter, *Ann. Rev. Phys. Chem.* **37** (1986) 25.

211. F. Engelke, R.Vincent, T.S. King and M. Pruski, *J. Chem. Phys.* **102** (1994) 7562.

212. M.A. Chesters, K.J. Packer, H.E. Viner, M.A.P. Wright and D. Lennon, *J. Chem. Soc. Faraday Trans.* **92** (1996) 4709.

213. M.A. Chesters, K.J. Packer, H.E. Viner, M.A.P. Wright and D. Lennon, *J. Chem. Soc., Faraday Trans.* **91** (1995) 2203.

214. Y.Y. Tong, A.J. Renouprez, G.A. Martin and J.J. van der Klink, *Proc. 11ᵗʰ. Internat. Congr. Catal.* (J.W. Hightower, W.N. Delgass, E. Iglesia and A.T. Bell, eds.), Elsevier: Amsterdam **B** (1996) 901.

215. Y.Y. Tong and J.J. van der Klink, *J. Phys. Chem.* **98** (1994) 11011.

216. J.P. Bucher, J.J. van der Klink, and M. Graetzel, *J. Phys. Chem.* **94** (1990) 1209.

217. X. Wu, B.C. Gerstein and T.S. King, *J. Catal.* **121** (1990) 271.

218. D. Graham, J. Howard and T.C. Waddington, *J. Chem. Soc. Faraday Trans. I* **79** (1983) 1281.

219. H. Jobic and A. Renouprez, *J. Chem. Soc. Faraday Trans. I* **80** (1984) 1991.

220. P.C.H. Mitchell, S.F. Parker, J. Tomkinson and D. Thompsett, *J. Chem. Soc. Faraday Trans.* **94** (1998) 1489.

221. J.-Y. Chen, J.M. Hill, R.M. Watwe, S.G. Podkolzin and J.A. Dumesic, *Catal. Lett.* **60** (1991) 1.

222. Pen Chou and M.A. Vannice, *J. Catal.* **104** (1987) 1.

223. S. Decker and A. Frennet, *Catal. Lett.* **46** (1997) 145.

224. B.E. Spiewak, R.D. Cortright and J.A. Dumesic, *J. Catal.* **176** (1998) 405

225. C.T. Campbell, *Adv. Catal.* **36** (1989) 2.

226. N. Cardona-Martinez and J.A. Dumesic, *Adv. Catal,* **38** (1992) 149.

227. F.B. Passos, S. Schmal and M.A. Vannice, *J. Catal.* **160** (1996) 106.

228. G.A. Martin, *Catal. Rev. – Sci. Eng.* **30** (1988) 519.

229. P.N. Aukett in: *Structure and Reactivity of Surfaces*, SSSC **48** (1989) 11.
230. R.D. Cortright and J.A. Dumesic, *J. Catal.* **148** (1994) 771.
231. M.A. Natal-Santiago, S.G. Podkolzin, R.D. Cortright and J.A. Dumesic, *Catal. Lett.* **45** (1997) 155.
232. R.D. Cortright, E. Bergene, P. Levin, M. Natal-Santiago and J.A. Dumesic, *Proc. 11th. Internat. Congr. Catal.*, (J.W. Hightower, W.N. Delgass, E. Iglesia and A.T. Bell, eds.), Elsevier: Amsterdam **B** (1996) 1185.
233. R.D. Cortright, R.M. Watwe, B.E. Spiewak and J.A. Dumesic, *Catal. Today* **53** (1999) 395.
234. B. Sen and M.A. Vannice, *J. Catal.* **130** (1991) 9.
235. Y. Takasu, M. Teramoto and Y. Matsuda, *J. Chem. Soc. Chem. Comm.* (1983) 1329.
236. S. Bhatia, F. Engelke, M. Pruski, B.G. Gerstein and T.S. King, *J. Catal.* **147** (1994) 455.
237. J.W. Jenkins, *Platinum Metals Rev.* **28** (1984) 98.
238. T. Mallát, É. Polyánszky and J. Petró, *J. Catal.* **44** (1976) 345.
239. J. Barbier, E. Lamy-Pitara and P. Marécot,*Bull. Chem. Soc. Belg.* **105** (1996) 99.
240. G.J. Hutchings, G.W. Watson and D.J. Willock, *Chem. and Ind.* (1997) 603.
241. G.P. Brivio and M.I. Trioni, *Rev. Mod. Phys.* **71** (1999) 231.
242. R. Hoffmann, *Rev. Mod. Phys.* **60** (1988) 601.
243. L. Salem, *J. Phys. Chem.* **89** (1985) 5576.
244. B. Hammer and J.K. Nørskov, *Adv. Catal.* **45** (2000) 71.
245. M. Neurock and R.A. van Santen, *J. Phys. Chem. B* **104** (2000) 11127.
246. A. Clark, *The Chemisorption Bond: Basic Concepts*, Academic Press: New York (1974).
247. K.-I. Tanaka and K. Tamara, *J. Catal.* **2** (1963) 366.
248. E. Miyazaki and I. Yasumori, *Surf. Sci.* **55** (1976) 747.
249. G.W. Watson, R.P.K. Wells, D.J. Willock and G.J. Hutchings, *J. Chem. Soc. Chem. Comm.* (2000) 705.
250. G. Papoian, J.K. Nørskov and R. Hoffmann, *J. Am. Chem. Soc.* **122** (2000) 4129.
251. G.C. Bond and D.T. Thompson, *Catal. Rev. – Sci. Eng.* **41** (1999) 319.
252. J.W. Andzelm, A.E. Alvarado-Swaisgood, F.U. Axe, M.W. Doyle, G. Fitzgerald, C.M. Freeman, A.M. Gorman, J.-R. Hill, C.M. Kölmel, S.M. Levine, P.W. Saxe, K. Stark, L. Subramanian, M.A. van Daelen, E. Wimmer and J.M. Newsam, *Catal. Today* **50** (1999) 451.
253. G. Kresse and J. Hafner, *Surf. Sci.* **459** (2000) 287.
254. P.J. Feibelman and J. Harris, *Nature* **372** (199) 135.
255. H. Sellers and J. Gislason, *Surf. Sci.* **426** (1999) 147.
256. J. Harris, S. Anderson, P. Nordlander and C. Holmberg, *J. Vac. Sci. Technol. A* **5** (1993) 1.
257. J. Harris, *Faraday Discuss. Chem. Soc.* **96** (1993) 1.
258. M. Boudart, M.A. Vannice and J.E. Benson, *Z. Physikal. Chem. NF* **64** (1969) 171.
259. *Spillover Phenomena in Solid State Chemistry and Catalysis*, (P. Grange, ed.) *Appl. Catal. A: Gen.* **202** (2000).
260. *Spillover of Adsorbed Species* (G.M. Pajorik, S.J. Teichner and J.E. Germain, eds.), Studies in Surface Science and Catalysis, Elsevier: Amsterdam, Vol. **17** (1983).
261. *New Aspects of Spillover Effect in Catalysis* (T. Inui, K. Fujimoto, T. Uchijima and M. Masai, eds.), Studies in Surface Science and Catalysis, Elsevier: Amsterdam **77** (1993).
262. *Proc. 2nd Internat. Conf. Spillover, Leipzig 1989*, (K.-H. Steinberg, ed.), Karl-Marx-Universität: Leipzig (1989).
263. J.J. Fripiat in: *Surface Properties and Catalysis by Non-Metals*, (J.P. Bonnelle, B. Delmon and E.G. Derouane, eds.), Reidel: Dordrecht (1983), p. 477.
264. G.C. Bond in: *Spillover of Adsorbed Species* **17** (1983) 1.
265. D.B. Hibbert and C.R. Churchill, *J. Chem. Soc. Faraday Trans. I* **80** (1984) 1977.
266. A.S. McLeod in:*Catalysis in Application* (S.D. Jackson, J.S.J. Hargreaves and D. Lennon, eds.), Roy. Soc. Chem.: London (2003), p. 86.
267. J.T. Kiss, I. Pálinkó and A. Molnár, *J. Mol. Struct.* **293** (1993) 273.

268. G.M. Pajonk, in: *Handbook of Heterogeneous Catalysis,* Vol. 3 (G. Ertl, H. Knözinger and J. Weitkamp, eds.), Wiley-VCH: Weinheim, (1997), p. 1064.
269. M.A. Vannice, M. Boudart and J.J. Fripiat, *J. Catal.* **17** (1970) 359.
270. G.C. Bond and P.A. Sermon, *J. Chem. Soc. Faraday Trans. I* **72** (1976) 730.
271. K. Tamaru, *Dynamic Heterogeneous Catalysis,* Academic Press: London (1978).
272. J.C. Kuriacose, *Indian J. Chem.* **5** (1967) 646.
273. *Metal-Suport Interactions in Catalysis, Sintering and Redispersion* (S.A. Stevenson, J.A. Dumesic, R.T.K. Baker and E. Ruckenstein, eds.), Van Nostrand Reinhold: New York (1987).
274. G.C. Bond and P.A. Sermon, *Catal. Rev.* **8** (1973) 211.
275. W.C. Conner Jr. and J.L. Falconer,*Chem. Rev.* **95** (1995) 759.
276. D.A. Dowden, in: *Specialist Periodical Reports: Catalysis*, Vol. 3 (C. Kemball and D.A. Dowden, eds.), *Roy. Soc. Chem.* (1980), p. 136.
277. G.C. Bond and R. Burch, in *Specialist Periodical Reports: Catalysis*, Vol. 6 (1983), p. 27.
278. D. Bianchi, G.E.E. Gardes, G.M. Pajonk and S.J. Teichner, *J. Catal.* **38** (1975) 135.
279. M.S.W. Lau and P.A. Sermon, *J. Chem. Soc. Chem. Comm.* (1978) 891.
280. F. Roessner, U. Roland and T. Braunschweig, *J. Chem. Soc. Faraday Tran. I* **91** (1995) 1539.
281. G.C. Bond and T. Mallát, *J. Chem. Soc. Faraday Trans. I* **77** (1981) 1743.
282. F. Roessner and U. Roland, *J. Molec. Catal. A: Chem.* **112** (1996) 401.
283. U. Roland, T. Braunschweig and F. Roessner, *J. Molec. Catal. A: Chem.* **127** (1997) 61.
284. M. Stoica, M. Caldărăru, F. Rusu and N.I. Ionescu, *Appl. Catal. A: Gen.* **183** (1999) 287.
285. N.I. Il'chenko, *Russ. Chem. Rev.* **41** (1972) 47.
286. H. Charcosset and B. Delmon, *Ind. Chim. Belg.* **38** (1973) 481.
287. R.P. Viswanath, B. Viswanathan and M.V.C. Sastri, *React. Kinet. Catal. Lett.* **2** (1975) 51.
288. H. Charcosset, R. Frety, A. Soldat and Y. Trambouze, *J. Catal.* **22** (1971) 204.
289. C. Blejeau, P. Boutry and R. Montarnal, *C.R. Acad. Sci., Paris, Ser. C*, **270** (1970) 257.
290. N.I. Il'chenko, *Kinet. Katal.* **8** (1967) 215.
291. J. Masson, B. Delmon and J. Nechtschein, *C.R. Acad. Sci., Paris, Ser. C.* **266** (1968) 428.
292. G.C. Bond and J.B.P. Tripathi, *Proc. Climax 1st Internat. Congr. Chem. Uses Molybdenum,* (P.C.H. Mitchell, ed.), Climax: London, (1974), p. 17; *J. Less-Common Metals* **36** (1974) 31.
293. R. Frety, H. Charcosset and Y. Trambouze, *Ind. Chim. Belg.* **38** (1973) 501.
294. C. Bolivar, H. Charcosset, R. Frety, M. Primet, L. Tournayan, C. Betizeau, G. Leclercq and R. Maurel, *J. Catal.* **39** (1975) 249.
295. E.J. Nowak, *J. Phys. Chem.* **73** (1969) 3790.
296. G.C. Bond and J.B.P. Tripathi, *J. Chem. Soc. Faraday Trans. I* **72** (1976) 233.
297. M. Che, B. Conosa and A.R. Gonzalez-Elipe, *J. Chem. Soc. Faraday Trans.* **78** (1982) 1043.
298. S. Bernal, A. Pintar, G.A. Cifredo, J.M. Rodriguez-Izquierdo, V. Perrichon and A. Laachir, *J. Catal.* **137** (1992) 1.
299. S. Khoobiar, *J. Phys. Chem.* **69** (1964) 149.
300. J.E. Benson and M. Boudart, *J. Catal.* **5** (1966) 307.
301. H.W. Kohn and M. Boudart, *Science* **145** (1964) 149.
302. R.B. Levy and M. Boudart, *J. Catal.* **32** (1974) 304.
303. J.P. Marcq, G. Poncelet and J.J. Fripiat, *J. Catal.* **87** (1984) 339.
304. R. Benali, C. Hoang-Van and P. Vergon, *Bull. Soc. Chem. France* (1985) 417.
305. S.J. Tauster, S.C. Fung and R.L. Garten, *J. Am. Chem. Soc.* **100** (1978) 170.
306. S.J. Tauster and S.C. Fung, *J. Catal.* **55** (1978) 29.
307. S.J. Tauster, L.L. Murrell and S.C. Fung, *USP* 1,576, 848 (1976).
308. H. Knözinger and E. Taglauer in:*Handbook of Heterogeneous Catalysis,* Vol. 2 (G. Ertl, H. Knözinger and J. Weitkamp, eds.), VCH: Weinheim, (1997), p. 216.
309. G.L. Haller and D.E. Resasco, *Adv. Catal.* **36** (1989) 173.
310. S.J. Tauster, S.C. Fung, R.T.K. Baker and J.A. Horsley, *Science* **211** (1981) 1121.
311. You-Jyh Lin, D.E. Resasco and G.L. Haller, *J. Chem. Soc. Faraday Trans. I* **83** (1987) 2091.

312. M.A. Vannice, *J. Catal.* **74** (1982) 199.
313. C.H. Bartholomew, R.B. Pannell, J.L. Butler and D.G. Mustard, *Ind. Eng. Chem. Prod. Res. Dev.* **20** (1981) 296.
314. Z. Wei, Y. Chen and Y. Chen, *Ziran Zazhi* **6** (1983) 476 (*Chem. Abs.* **25** (1983) 99: 19733k).
315. R. Ruppert, J.-P. Sauvage, J.-M. Lehn and R. Ziessel, *Nouv. J. Chim.* **6** (1982) 235.
316. J.C. Conesa and J. Soria, *J. Phys. Chem.* **86** (1982) 1392.
317. L. Bonneviot and G.L. Haller, *J. Catal.* **113** (1988) 96.
318. J.-M. Hermann and P. Pichat, *J. Catal.* **78** (1982) 425.
319. D.C. Koningsbeger, J.H.A. Martens, R. Prins, D.R. Short and D.E. Sayers, *J. Phys. Chem.* **90** 1986 3047.
320. J.H.A. Martens, R. Prins, H. Zandbergen and D.C. Koningsberger, *J. Phys.Chem.* **92** (1988) 1903.
321. A.J. Ramirez-Cuesta, R.A. Bennett, P. Stone, P.C.H. Mitchell and M. Bowker, *J. Molec. Catal. A: Chem.* **167** (2001) 171.
322. R. Galicia, *Rev. Inst. Mex. Petr.* **14** (1982) 26.
323. B.-H. Chen and J.M. White, *J. Phys. Chem.* **86** (1982) 3534.
324. A.R. Gonzalez-Elipe, P. Malet, J.P. Espinos, A. Caballero and G. Munuera in: *Structure and Reactivity of Surfaces,* SSSC Vol. 48 (1989), p. 427.
325. T. Huizinga, H.F.J. van't Blik and R. Prins, *Surf. Sci.* **135** (1983) 580.
326. V. Andera, *Appl. Surf. Sci.* **51** (1991) 1.
327. T. Komaya, A.T. Bell, Zara Weng-Sieh, R. Gronsky, F. Engelke, T.S. King and M. Pruski, *J. Catal.* **149** (1994) 142.
328. R.T.K. Baker in: (B. Imelik, C. Naccache, G. Coudurier, H. Praliaud, P. Meriaudeau, P. Gallezot, G.A. Martin and J.C. Védrine, eds.), Studies in Surface Science and Catalysis, Elsevier: Amsterdam, **11** (1982) 37.
329. C.C.A. Riley, P. Jonsen, P. Meehan, J.C. Frost, K.J. Packer and J.P.S. Badyal, *Catal. Today* **9** (1991) 121.
330. K. Kunimori, S. Matsui and T. Uchijima, *J. Catal.* **85** (1984) 253.
331. S. Tang, G.-X. Xiong and H.-L. Wang, *J. Catal.* **111** (1988) 136.
332. C.S. Ko and R.J. Gorte, *J. Catal.* **90** (1984) 59.
333. S.I. Ito, T. Fujimori, K. Nagashima, K. Yuzaki and K. Kunimori, *Catal. Today* **57** (2000) 247.
334. K. Kunimori, Y. Doi, K. Ito and T. Uchijima, *J. Chem. Soc. Chem. Comm.* (1986) 965.
335. J.P.S. Badyal, R.M. Lambert, K. Harrison, C.C.A. Riley and J.C. Frost, *J. Catal.* **129** (1991) 429.
336. G.C. Bond, R.R. Rajaram and R. Burch, *Proc. 9th. Internat. Congr. Catal.*, (M.J. Phillips and M. Ternan, eds.), Chem. Inst. Canada: Ottawa **3** (1988) 1130.
337. T. Suzuki and R. Sonda, *Surf. Sci.* **448** (2000) 33.
338. C.-C. Kao, S.-C. Tsai and Y.-W. Chung, *J. Catal.* **73** (1982) 136.
339. K. Kunimori, Y. Ikeda, M. Soma and T. Uchijima, *J. Catal.* **79** (1983) 185.
340. P.G. Menon and G.F. Froment, *Appl. Catal.* **1** (1981) 31.
341. D. Brennan, *Adv. Catal.* **15** (1964) 1.
342. D.D. Eley, *J. Res. Inst. Catal., Hokkaido Univ.* **16** (1968) 101.
343. P. Stonehart and P.N. Ross, *Catal. Rev. -Sci. Eng.* **12** (1975) 1.
344. H. Busse, M.R. Voss, D. Jerdev, B.E. Koel and M.T. Paffett, *Surf. Sci.* **490** (2001) 133.
345. G. Rienäcker, J. Völter and S. Engels, *Zeit. f. Chem.* **9** (1970) 321.
346. H.E. Farnsworth and R.F. Woodcock, *Adv. Catal.* **9** (1957) 123.
347. E.K. Rideal, *J. Res. Inst. Catal., Hokkaido Univ.* **16** (1968) 45.
348. R.J. Breakspeare, D.D. Eley and P.R. Norton, *J. Catal.* **27** (1972) 215.
349. R.P.H. Gasser, K. Roberts and A.J. Stevens, *Surf. Sci.* **20** (1970) 123.
350. V.I. Stavchenko, G.K. Boreskov and V.V. Gorodetskii, *Kinet. Katal.* **1** (1971) 766.
351. D.D. Eley and P.R. Norton, *Discuss. Faraday Soc.* **41** (1966) 135.
352. J.J. Byrne, P.F. Carr and J.K.A. Clarke, *J. Catal.,* **20**, (1971), 412.
353. Y. Takasu and T. Yamashina, *J. Catal.* **28** (1973) 174.

354. D.R. Rossington in: *Chemisorption and Reactions on Metallic Films,* Vol. 2 (J.R. Anderson, ed.) Academic Press, London, (1971), p. 211.
355. I.E. Wachs and R.J. Madix, *Surf. Sci.* **58** (1976) 430.
356. Z. Schay and P. Tétényi, *J. Chem. Soc. Faraday Trans. I* **75** (1979) 1001.
357. Y. Takasu, T. Akimaru, K. Kasahara, Y. Matsuda, H. Miura and T. Toyoshima, *J. Am. Chem. Soc.* **104** (1982) 5249.
358. M.A. Chesters, K.J. Packer, H.E. Viner and M.A.P. Wright, *J. Phys. Chem. B* **101** (1997) 9995.
359. G.K. Boreskov, V.I. Savchenko and V.V. Gorodetskii, *Dokl. Akad. Nauk* **189** (1969) S. Naito and M. Tanimoto, *J. Chem. Soc. Chem. Comm.* (1988) 832.
360. J. Oudar in: *Metal-Support and Metal-Additive Effects in Catalysis,* SSSC, (B. Imelik, C. Naccache, G. Coudurier, H. Praliaud, P. Meriaudeau, P. Gallezot, G.A. Martin and J.C. Védrine, eds.), Studies in Surface Science and Catalysis, Elsevier: Amsterdam, **11** (1982), 255.
361. C.M. Pradier, Y. Berthier and J. Oudar, *Surf. Sci.* **130** (1983) 229.
362. D.A. Outka and G.W. Foltz, *J. Catal.* **30** (1991) 268.
363. Y. Takasu and Y. Matsuda, *Electrochim. Acta* **21** (1976) 133.
364. G.C. Bond, A.D. Kuhn, J. Lindley and C.J. Mortimer, *J. Electroanal. Chem.* **34** (1972) 1.
365. G.C. Bond and Z. Paál, *Appl. Catal. A: Gen.* **86** (1992) 1.
366. A. Frennet and P.B. Wells , *Appl. Catal.* **18** (1985) 243.
367. S. Bernal, G. Blasco, A.J. Franco, J.M. Gatica, C. Larese and M. Pozo, *SECAT,* 99, (G. Blasco and C. Mira, eds.), p. 31.
368. A. Frennet and P.B. Wells, *Appl. Catal.* **18** (1985) 243.
369. G.A. Martin, R. Dutartre and J.A. Dalmon, *React. Kinet. Catal. Lett.* **16** (1981) 329.
370. R.W. Joyner, *J. Chem. Soc. Faraday Trans. I* **76** (1980) 357.
371. Y.-N. Yang, J.-C. Pan, N. Zheng, X.-Q. Lin and J.-Y. Zhang, *Appl. Catal.* **61** (1990) 75.
372. M. Muhler, Z. Paál and R. Schlögl, *Appl. Surf. Sci.* **47** (1991) 281.
373. S. Bhatia, F. Engelke, M. Pruski and T.S. King, *Catal. Today* **21** (1994) 129.
374. M. Boudart, M.A. Vannice and J.E. Benson, *Z. Phys. Chem.* (NF) **64** (1969), 17.
375. W.C. Conner Jr., G.M. Pajonk and S.J. Teichner, *Adv. Catal.* **34** (1986) 1.
376. E. Bittner and B. Bockrath, *J. Catal.* **170** (1997) 325.
377. G. Fröhlich and W.M.H. Sachtler, *J. Chem. Soc. Faraday Trans.* **94** (1998) 1339.
378. D. Bianchi, M. Lacroix, G.M. Pajonk and S.J. Teichner, *J. Catal.* **68** (1981) 411.
379. M. Lacroix, G. Pajonk and S.J. Teichner, *Proc. 7thInternat. Congr. Catal.* (T. Seiyama and K. Tanabe, eds.), Elsevier: Amsterdam (1981) 279.
380. M. Larsson, M. Hultén, E.A. Blekkan and B. Andersson, *J. Catal.* **164** (1996) 44.
381. P.G. Dickens, R.H. Jarman, R.C.T. Slade and C.J. Wright, *J. Phys. Chem.* **77** (1982) 575.
382. P.G. Dickens, J.J. Birthill and C.J. Wright, *J. Solid State Chem.* **28** (1979) 185.
383. R. Benali, C. Hoang-Van and P. Vergnon, *Bull. Soc. Chim. France* (1985) 417.
384. S. Horiuchi, N. Kimizuka and A. Yamamoto, *Nature* **279** (1979) 226.
385. H. Kubicka, *Rhenium, Technetium and the Platinum Metals as Catalysts for Reactions involving Hydrogen,* Inst. Low Temp. and Str. Res., PAN: Wrocław (1978).
386. G.C. Bond, *Catalysis by Metals,* Academic Press: London, 1962, Ch 8.
387. J.R. Anderson and B.G. Baker in: *Chemisorption and Reactions on Metallic Films,* Vol. 2 (J.R. Anderson, ed.) Academic Press: London, (1971), p. 64.
388. B.M.W. Trapnell in: *Catalysis,* (P.H. Emmett, ed.), Reinhold, New York, **5** (1957) 1.
389. B.H. Davis in: *Handbook of Heterogeneous Catalysis,* Vol. 1 (G. Ertl, H. Knözinger and J. Weitkamp, eds.), VCH: Weinheim, (1997), p. 13.
390. S. Bernal, J.J. Calvino, M.A. Cauqui, J.M. Gatica, C. Laresse, J.A. Pérez Omil and J.M. Pintado, *Catal. Today* **50** (1999) 175.
391. C. Niklasson and B. Andersson, *Ind. Eng. Chem. Res.* **27** (1988) 1370.
392. T. Engel and G. Ertl in: *Chemical Physics of Solid Surfaces and Heterogeneous Catalysis* (D.A. King and D.P. Woodruff, eds.), Elsevies: Amsterdam (1982), vol. 4, p. 195.

393. S. Bernal, J.J. Calvino, M.A. Cauqui, J.M. Gatica, C. López Carles, J.A. Pérez Omil and J.M. Pintado, *Catal. Today* **77** (2003) 385.

394. G.C. Bond, *Chem. Soc. Rev.* **20** (1991) 441.

395. T.E. Huizinga, J. van Grondelle and R. Prins, *Appl. Catal.* **10** (1984) 199.

396. E.I. Ko, J.M. Hupp, F.H. Rogan and N.J. Wagner, *J. Catal.* **84** (1983) 85.

397. G. Sankar, S. Vasudevan and C.N.R. Rao, *J. Phys. Chem.* **92** (1988) 1878.

398. Y.-W. Chung and Y.-B. Zhao, *ACS Symp. Ser.* **298** (1986) 54; *J. Catal.* **106** (1987) 369.

399. B.A. Tatarchuk and J.A. Dumesic, *J. Catal.* **70** (1981) 335.

400. S.W. Ho, J.M. Cruz, M. Houalla and D.M. Hercules, *J. Catal.* **135** (1992) 173.

401. G. Marcelin and J.E. Lester, *J. Catal.* **93** (1985) 270.

402. G.C. Bond in: *Metal-Support and Metal-Additive Effects in Catalysis* (B. Imelik, C. Naccache, G. Coudurier, H. Praliaud, P. Meriaudeau, P. Gallezot, G.A. Martin and J.C. Védrine, eds.), Studies in Surface Science and Catalysis, Elsevier: Amsterdam, **11** (1982) 1.

403. K. Kunimori, Y. Ikeda, M. Soma and T. Uchijima, *J. Catal.* **79** (1983) 185.

404. A.J. Renouprez, T.M. Tejero and J.P. Candy, in: *Proc. 8th Internat. Congr. Catal.,* Verlag Chemie: Weinheim **3** (1985), 47.

405. R.U. Tang, R.G. Wu and L.W. Lin, *Appl. Catal.* **10** (1984) 163.

406. Z.-C. Hu, A. Maeda, K. Kunimori and T. Uchijima, *Chem. Lett.* (1986) 2079.

407. F.M. Dautzenberg and H.B.M. Wolters, *J. Catal.* **51** (1978) 26.

408. J.W. Sprys and Z. Mencik, *J. Catal.* **40** (1975) 290.

409. H. Praliaud and G.A. Martin, *J. Catal.* **72** (1981) 394.

410. J. van de Loosdrecht, A.M. Van der Kraan, A.J. van Dillen and J.W. Geus, *J. Catal.* **170** (1997) 217.

411. R. Lamber, N.I. Jaeger and G. Schulz-Egloff, *J. Catal.* **123** (1990) 285.

412. T. Ueckert, R. Lamber, N.I. Jaeger and V. Schubert, *Appl. Catal. A: Gen.* **155** (1997) 75.

413. J. Ademiec, S. Beszterda and R. Fiedorov, *React. Kinet. Catal. Lett.* **31** (1986) 371.

414. O.H. Ellestad and C. Naccache in: *Perspectives in Catalysis,* (R. Larsson, ed.) C.W.K. Gleerup: Lund, 1981, p. 95.

415. E. Ruckenstein and H.Y. Wang, *J. Catal.* **190** (2000) 32.

416. L. Tournayan, A. Auroux, H. Charcosset and R. Szymanski, *Ads. Sci. Technol.* **2** (1985) 55.

417. R.J. Mikovsky, M. Boudart and H.S. Taylor, *J. Am. Chem. Soc.* **76** (1954) 3814.

418. D.R. Jennison, O. Dulub, W. Hebenstreit and V. Diebold, *Surf. Sci.* **492** (2001) L677.

419. J.W.E. Coenen, *Appl. Catal.* **54** (1989) 59, 61; **75** (1991) 193.

420. X.-C. Guo and R.J. Madix, *J. Catal.* **153** (1995) 336.

FURTHER READING

Standard catalysts: EUROPT-1 128, 136, 137, 190, 195, 213, 365–372

Standard catalysts: EURONI-1 419

^{1}H and ^{2}H NMR studies of supported ruthenium catalysts 166, 167, 208, 217, 236, 327-329, 373

Hydrogen spillover 13, 259, 262, 266, 268, 274–276, 374–380, 402, 418, 421

Hydrogen bronzes Tungsten: 258, 270, 299-302; molybdenum: 270, 292, 381–383; vanadium: 303; rhenium: 384.

The "Strong Metal-Support Interaction" 14, 155, 156, 273, 278, 308–310, 325, 393–401

Effects of high-temperature hydrogen treatment Alumina: 339, 369, 403 –408; silica: 53, 367, 369, 404, 409–412; others: 413–417

Reactions of hydrogen 76, 82, 113, 354, 385–392, 417, 420

4

THE CHEMISORPTION
OF HYDROCARBONS

PREFACE

We have considered at some length the ways in which hydrogen can interact with metals and supported metal catalysts, and with itself when isotopic analogues are used, and because our main theme is the reactions that hydrogen undergoes with hydrocarbons it is logical that we should next examine how they interact with metal surfaces. The great variety of types of hydrocarbon (alkanes, alkenes, alkynes, aromatics etc.) make this potentially a much larger subject than that of the last Chapter, and a further complication is that the forms adopted by a specific hydrocarbon depend on the nature of the metal, the crystal plane, the particle size, and particularly the temperature. These factors have acted as a challenge to experimentalists and theoreticians, and there is a very extensive literature, on which it will be necessary to try to impose some order; selection and compression will also be called for.

Our aim will be to try to identify those adsorbed species most likely to be intermediates in the reactions we shall meet later, and to recognise species derived from the original molecule that may inhibit the desired reaction, or indeed be required for reaction under more forcing conditions. In some cases the structure of the relevant intermediate looks like that of the product, and, just as with hydrogen, the essential part of the process is accomplished in the chemisorption. The very precise and detailed information available on the structures of certain adsorbed species and their effects on the arrangement of surface metal atoms will excite our admiration.

4.1. INTRODUCTION

4.1.1. Types of Alkane[1]

The principal structural types of alkane are (i) linear, (ii) branched, (iii) cyclic and (iv) polycyclic, and two or more of these may appear within a single molecule. Almost all the work done on their chemisorption has been with the lower members of the linear series, i.e. methane and ethane.

4.1.2. Types of Unsaturated Hydrocarbon

There are three basic types of carbon-carbon unsaturation, namely, (i) the double bond as in alkenes, (ii) the triple bond as in alkynes, and (iii) the aromatic bond as in benzene. If this were the end of the story, chemistry would be simpler but less interesting. The properties and reactivity of the double bond in particular are sensitive to its environment; when it is exocyclic, i.e. one of the sp^2 carbons forms part of an alicyclic ring, its stretching frequency changes by about 130 cm^{-1} as the size of the ring is increased from three to six, while when endocyclic it hardly changes at all unless there are other alkyl substituents present (see Figure 4.1). These various differences are no doubt explicable in terms of electron delocalisation (which used to be called hyperconjugation). The carbon-carbon frequency of the triple bond increases markedly with alkyne substitution, but the vibrations in the aromatic ring are less dependent on substituents.

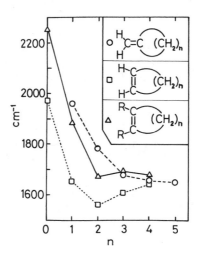

Figure 4.1. Dependence of C=C vibration frequency in cyclic alkenes on ring size for the structures shown.

The double bond may also be conjugated with another, and in the series propadiene (allene), 1,3-butadiene, 1,4-pentadiene etc. the extent of interaction decreases as the number of single bonds between the points of unsaturation increases. So, as we shall see later, in catalysed reactions propadiene and 1,3-butadiene often resemble the corresponding alkynes. There is a conjugation between double and triple bonds in ethenylethyne, and between these bonds and the aromatic ring in, for example, styrene (phenylethene) and phenylethyne.

The cyclopropane ring represents an interesting intermediate between alkanes and alkenes. Theoreticians have had great fun with this molecule, for which no simple (i.e. non-wave mechanical) picture is satisfactory. There is certainly some electron density in the middle of the ring, and it can be hydrogenated to propane quite easily, although less easily than propene. Alkyl- and alkenylcyclopropanes react in interesting ways (Chapters 7 and 11).

4.1.3. The Literature

In view of the almost endless ways in which carbon-carbon unsaturation can manifest itself, it is perhaps fortunate that students of chemisorption have confined themselves in the main to looking at only a few molecules and only a few metals. Nevertheless the interactions of these few molecules especially with single crystal surfaces have proved an irresistible attraction,[2,3] and a quite enormous literature developed, particularly in the late 1970s and throughout the 1980s. Interest has diminished since, although the lesser frequency of publication is compensated by the occasional use of more profound and informative experimental techniques. Nevertheless the fact that review articles[4,5] published in 1995 and 1996 contain respectively 899 and 480 references underscores that magnitude of the task of summarising what is known in the space of a few pages.

It is worth noting the outstanding features of the relevant literature. Of the alkenes, at least 90% of papers deal only with ethene, and less than 10% with higher alkenes; the great simplicity of the ethene molecule, with only one kind of carbon-carbon and carbon-hydrogen bond, is at once its usefulness and limitation, because like so many first members of homologous series it is quite atypical. Similarly there are hardly any papers dealing with the higher alkynes, or for that matter with dienes. Of the metals, by far the most work has been done with platinum, presumably because of its importance in petroleum reforming: there is no obvious scientific explanation. Nickel, palladium and rhodium also feature substantially, but other metals of Groups 8 to10 have not had corresponding attention. In consequence there is a very good understanding of the chemisorption and reactions of a few hydrocarbon molecules with platinum, but with other metals of interest our knowledge is more sketchy. Nevertheless we are not short of material to discuss.

4.2. THE CHEMISORPTION OF HYDROCARBONS: AN OVERVIEW[6,7]

The purpose of this section is to try to construct a map to guide us through the jungle that is the literature on the chemisorption of hydrocarbons, to evolve - if indeed it is possible - some general principles, to delineate the main features, and to outline the strategy to be adopted in later sections.

The study of the interactions of hydrocarbons with metal surfaces proceeds in the following way. At very low temperatures, molecules are *physically adsorbed* without significant effects on their structure: this state has very little relevance to catalysis, and may be passed over quickly. Unsaturated molecules are *chemisorbed* on metals of Groups 8 to 11 in many different ways: at low temperatures and higher pressures, forms are seen in which there is minimum disruption of the multiple bond, the bonding being due to overlap of π orbitals with orbitals projecting from the metal. Next there are forms in which one or more of the multiple bonds is undone, and σ-bonds between carbon and metal are formed. Both of these types are *non-dissociative* and are very probably those that are reactive in catalytic reactions such as hydrogenation.

The major difficulty in this field is the ease with which carbon-hydrogen bonds are broken, even at quite low temperatures, but progressively as temperatures are increased. Much is known about the routes followed in these thermal decompositions[4,5,8-10] (Section 4.6); the species formed may be toxic to some catalytic reactions, but they may also be relevant intermediates in reactions such as hydrogenolysis that require higher temperatures. The ultimate fate of all adsorbed hydrocarbons is some type of carbon on the surface, and hydrogen in the gas phase (Section 4.8). With base metals, carbon atoms can easily dissolve to form a carbide phase; ethene seems likely to react at high temperature only with *large* palladium particles,[11] and with Pd_6 clusters it did not react at all at 323 K.[12] As we shall see (for example in Chapter 9), small amounts of dissolved carbon have significant effects on catalytic reactions.

The motivation for these changes is of course thermodynamic in origin, and parallels those found in the gas phase, but in the presence of a metallic surface there are the additional factors of (i) the stability of the gaseous hydrogen molecules compared to adsorbed atoms at high temperature, and (ii) the lowering of the surface free energy by the formation of carbon-metal bonds. One guiding general principle may be therefore stated as follows. *Provided sufficient thermal energy is available to overcome any activation barrier, those changes will occur to an adsorbed hydrocarbon species that lead to the formation of the greatest number of chemical bonds to the metal, and hence to the greatest reduction of surface free energy.* This principle can be illustrated by a system of potential energy curves (Figure 4.2), and we shall encounter many examples of it in due course. Reactions within the chemisorbed layer may lead to gaseous products

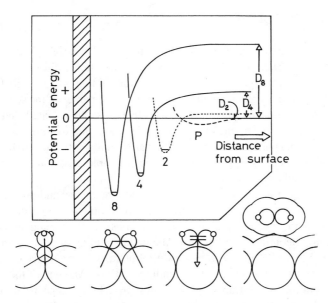

Figure 4.2. A simplified schematic picture of potential energy curves describing the adsorption of ethene in a typical metal surface as being physically adsorbed **P**, π-adsorbed(**2**), σ-diadsorbed (**4**) or in the form of ethylidyne ($\equiv$C—CH$_3$, **8**). The numbers are those used in Table 4.2. The exact locations of the curves with respect to each other and to the zero of potential energy will differ from one system to another, and further intermediate forms may sometimes appear (see text).

other than hydrogen: as has long been known,[13,14] ethane may be a product of the chemisorption of ethene, but such reactions are not catalytic, because a dehydrogenated species is left behind blocking the surface. Reactions of a chemisorbed hydrocarbon by and with itself are for this reason treated in this chapter (Section 4.6).

The next section will deal briefly with experimental techniques: many of these have been introduced already, but the use of vibrational spectroscopy and of sum-frequency generation call for some further description. Section 4.4.1 describes the principal types of adsorbed hydrocarbon structure that have been found with alkenes and alkynes (aromatic hydrocarbons and cyclic C$_6$ species will be considered in Chapters 10 and 12 respectively); Section 4.4.2 discusses the conditions under which the several chemisorbed forms of alkenes make their appearance. In Section 4.5 we look at detailed structural studies of a few adsorbed molecules, and Section 4.6 deals somewhat briefly with interconversions and decompositions of adsorbed alkenes, and structures of species formed. Finally there are sections on theoretical approaches (4.7), on the chemisorption of alkanes (4.8), and carbonaceous deposits that are the ultimate product of the decomposition process (4.9).

4.3. THE TECHNIQUES[15]

We have already met a number of the techniques that have been used to study the structure and reactivity of chemisorbed hydrocarbons. Table 4.1 lists most of the important ones together with a few references that will help the reader to access the wider literature, and to draw attention to the names of the scientists who have made major contributions to their use. It will be appreciated that many of them are only applicable to one class of material, i.e. either to single crystals or to finely divided metal particles: techniques that are equally applicable to both classes (e.g. vibrational spectroscopy) are particularly valuable in enabling comparisons between the two classes to be drawn.

Recalling that this is not a handbook to teach the theory and operation of the techniques used, it is necessary only to note and comment on certain constraints that apply to them: such constraints cover matters such as cost and availability of the equipment, theoretical limitations to the understanding of what is observed or what can be observed. Those factors are well illustrated by reference to the field of vibrational spectroscopy (see Further Reading section at the end of the chapter).

For a vibration to be infrared-active, it must produce a change in the molecule's electric dipole moment: the intensity of the absorption band is proportional to the square of the size of the change that occurs between the extreme points of the motion. However at a flat metal surface, the very high polarisability of the conduction electrons leads to an image of opposite sign which offsets the effect of the dipole moment change, so only those fully symmetrical vibrations that have a component of the dipole moment normal to the surface will give an absorption band. The RAIRS method (Table 4.1) used for work with single crystals is therefore subject to this *metal surface selection rule*.[16-19] If however the metal particle is sufficiently small (<2 nm), the rule is relaxed, and with organometallic complexes it does not apply at all. Whereas in infrared spectroscopy the molecule absorbs a quantum of radiation, in the Raman effect the photon is scattered but loses energy which is transferred to the molecule; for a vibration to be Raman active, the event has to produce a change in the polarisability of the molecule. Although widely used to study oxides of catalytic interest, Raman spectroscopy has contributed much less to the study of hydrocarbons on metals.[9]

HREELS (Table 4.1) has the advantage of detecting all types of vibration: this is because there are two excitation mechanisms, viz. dipole scattering and impact scattering. The former is subject to the same selection rules as RAIRS and gives strong features on-specular, but the latter excites all vibrational modes. There are however supplementary selection rules that apply to impact scattering in the on-specular direction.[17] As noted earlier, this technique is not applicable to supported metal catalysts.

TABLE 4.1. Methods Applicable to the Study of Chemisorbed Hydrocarbons

Acronym	Name	To single crystals etc.?	To small particles?
—	(Micro) Calorimetry	Yes[23,24,310–314]	Yes[25,133,136,149,208]
FTIR(S)	Fourier-transform infrared (spectroscopy)	Yes[16,17]	Yes[9,17,209]
RAIRS/IRAS	Reflection - absorption infrared spectroscopy	Yes[17,207,135]	No
DRIFTS	Diffuse-reflectance infrared Fourier-transform spectroscopy	No	Yes[17,209]
(HR)EELS/ VEELS	(High resolution) or vibrational electron energy-loss spectroscopy	Yes[17,74,210,211,316–320]	No
SFG	Sum-frequency generation	Yes[9,10,212–218]	No
LEED	Low-energy electron diffraction	Yes[17,32,50,211,219,321]	No
NEXAFS (XANES)	Near-edge X-ray absorption fine structure	Yes[5,17,190,220–222]	No
AES	Auger electron spectroscopy	Yes[33,190,223–227]	?
PED	Photoelectron diffraction	Yes[5,17,21,22,128,147]	No
UPS/XPS	Ultraviolet (X-ray) photoelectron spectroscopy	Yes[32–34,38,40,322,323]	Possible
ARUPS	Angle-resolved ultraviolet photoelectron spectroscopy	Yes[5,41,232,324,325]	No
INS	Inelastic neutron scattering	No	Yes[17,233,234]
—	Work function change	Yes[134,191,211,224,235–237]	No
—	High-field magnetic methods	No	Yes[238–240]
TPD/LID/ CID	Temperature-programmed (laser-induced, collision-induced) desorption	Yes[17,241,242]	No
(M)MB	(Modulated) molecular beam	Yes[243,244]	No
STM	Scanning-tunnelling microscopy	Yes[17,130,213,245–247,327,328]	No
—	Stable and radioactive isotopes	Yes[329,330]	Yes[248–250]
SIMS	Secondary-ion mass-spectroscopy	Yes[251]	Yes[142]
—	Ellipsometry	Yes[192]	No
SERS	Surface-enhanced Raman spectroscopy	Yes[331]	Yes[43,331]
NMR (SEDOR)	Nuclear magnetic resonance (spin-echo double resonance)	Yes[17,28,143,154,228–231,353]	No
—	Chromatography	No	Yes[274]

Column 3 also covers foils and films; column 4, metal blacks and powders as well as supported metals. Only a few indicative references are given; for further sources of information, see for example 17, 25, 108, 254.

A relatively new method for studying chemisorbed species is *sum-frequency generation (SFG)* (see Table 4.1 for references). This is a second-order non-linear process, requiring both a fixed visible and a tuneable laser: the selection rules determine that a vibrational mode must result in changes both to dipole moment and to polarisability for the effect to occur, and this limits it to a medium which lacks inversion symmetry, i.e. to the surface and not the gas phase. This, coupled with the fact that excitation is by photons, not electrons, leads to the inestimable benefit of being usable in the presence of a high gas pressure, and therefore enables *in situ* examination of the surface under reaction conditions.

It may be of interest to compare and contrast the usefulness of SFG and IR spectroscopy in this field. (i) IRS is experimentally much easier and the equipment needed is much cheaper: anybody can do it. (ii) It covers a wide range of frequencies whereas SFG is limited by the frequency range of the tuneable laser (originally to the C–H stretch region). (iii) Not needing high-powered lasers, IRS employs very low energy radiation which does not affect surface structures, as SFG might. (iv) IR absorbance is usually easy to relate to surface concentration, whereas the relationship between SFG signal and concentration is much more complex. (v) While simple selection rules apply to IRS, while those for SFG are also more complex. (vi) On the other hand, signal intensity is sometimes higher with SFG than with IRS, low intensities of which can be limiting factors in examining single crystal surface. (vii) IRS detects both gaseous and surface species (subject to the selection rules), and so the gas pressure needs to be low, whereas as noted above SFG does not suffer this constraint. I am grateful for the assistance of Professor F. Zaera in drafting this paragraph.

The technique of choice therefore depends upon the nature of the questions posed, and on the limitations of cost and theory, which have been outlined above for one particular field. Very precise structural information is obtainable by LEED, and particularly by *photoelectron diffraction* (*PED*, Table 4.1): this last method is, unlike LEED, applicable to disordered adlayers.[17,20] The effect is due to the diffraction of locally generated photoelectrons by other atoms within the same adsorbed species: variation of the diffraction intensity with the photoelectron energy using synchrotron radiation provides structural information.[17,21,22]

A major advance in technique has permitted the calorimetric determination of heats of adsorption of hydrocarbons on thin single-crystal surfaces of metals.[23,24] Films of about 200 nm thickness are grown epitaxially by vapour deposition on the surface of a sodium chloride crystal cut to expose the desired face; it is then dissolved in water, and the film, the surface of which mimics that of the crystal, is then rescued. Its small thermal mass means that the heat liberated is rapidly equilibrated throughout the metal, and the temperature rise can be measured by a remote infrared detector. By using a pulsed molecular beam, simultaneous estimates of sticking probability can be made. Heats of adsorption of hydrocarbons on supported metals at various temperatures have also been reported[25] (Section 4.5).

Direct structural information on hydrocarbon species on supported metals (e.g. C—C bond distances) is hard to come by, but can be obtained with some difficulty using NMR in the SEDOR mode[17,26-28] (Section 2.42 and Table 4.1). NEXAFS, through the polarisation dependence of absorption resonances and selection rules for photoabsorption, can determine the average orientation of adsorbed chromophores with respect to the surface[5,17] (see also Section 2.4.2 and Table 4.1).

In the early days of investigating the structures of chemisorbed hydrocarbons on metal single crystals, much use was made of *ultraviolet photoelectron spectroscopy (UPS)*, especially by Demuth and his associates.[29-36] UV photoemission spectra of valence orbitals were obtained using He I (24.4 eV) or He II (40.8 eV) as the exciting radiation, and comparison with the spectre of the free molecules, combined with calculation of the Hartree-Fock ground-state orbital energies by the self-consistent-field linear-combination-of-atomic-orbitals (SCF-LCAO) method, enabled structural information to be obtained for the principal adsorbed species.[29] The limitation of this technique lay in the fact that the observed spectra were the sum of those due to *all* species present, and for this reason the combination of UPS with its X-ray cousin XPS,[37-39] and with LEED,[33,40] was more informative. In a further development, the angle of incidence of the exciting radiation was changed, giving *angle-resolved UPS (ARUPS)*;[41] this helped to identify the orbitals from whence the photoemitted electrons came. Additional references to these techniques are given in Table 4.1.

While little use has been made of Raman spectroscopy,[42] somewhat more has been done with Surface-Enhanced Raman Spectroscopy (SERS),[43] especially in the study of adsorbed aromatic hydrocarbons.[9]

4.4. IDENTIFICATION OF ADSORBED HYDROCARBON SPECIES

4.4.1. The Catalogue – or 'The Organometallic Zoo'[8,16,17,44]

We noted when discussing hydrogen chemisorption that the old idea of a hydrogen atom forming only a single covalent bond to some other atom was inadequate to describe its interaction with metallic surfaces. A similar problem arises when trying to formulate the structures of chemisorbed hydrocarbons: simple-minded notions of tetravalent carbon and hybridisation of s and p orbitals as sp^3 and sp^2 and sp are simply not sufficient. The difficulty however is not entirely new, as it will certainly have been met when studying organometallic chemistry. We have to face the fact that a molecule such as ethene is able to bond to metals in a large variety of ways, depending on the nature and the extent of the interaction of its orbitals with those of the surface metal atoms. Analogies between chemisorption and co-ordination chemistry have in fact been very helpful,[9,17,40,45-50] although they should not be pushed too far (see later).

Description of the chemisorbed state of a hydrocarbon proceeds through stages, the number that can be accomplished depending on the type of surface and the techniques deployed. There are fewer methods available to the study of molecules on small particles than to single crystal surfaces, and so the information available is more restricted. The first stage requires the composition of the species to be specified, e.g. if the molecule starts as ethene, is its formula still given by C_2H_4? If some C—H bond breaking has occurred, do we have C_2H_3 or C_2H_2 or something else? The second stage enquires into the organisation of electrons within the adsorbed species. Has there been some degree of rehybridisation of the carbon atoms? Or has there just been some engagement of the π orbital lobes with metal orbitals, with little change to the formal sp^2 hybridisation of the carbons? The third and final stage, applicable only to single crystal surfaces, is to establish the new structure in terms of metal-carbon and carbon-hydrogen bond lengths and angles, and to locate its position with respect to the lattice of surface metal atoms. In this last stage in particular we must be ready to find different techniques giving discordant results: a value judgement on the reliability of the method, aided by some chemical common sense and theoretical calculation, then has to be applied.

Table 4.2 sets out the names and structures of a number of adsorbed C_2 species, for the existence of which there is some evidence. To simplify reference to these and other structures and compounds in the text, the following symbolism is introduced; M=C_1 species; E=C_2 species; P=C_3 species, and the pre-superscript gives the number of hydrogen atoms. At the foot of each box there are noted examples of organometallic complexes containing structures of the same type, and which have been used to identify the adsorbed species by their vibrational spectra. The numbering, and much of the other information that follows, have been provided by Sheppard and de la Cruz in their epic reviews.[9,17,51] It is interesting to see how frequently osmium[52–54] and cobalt[44] appear in the cited complexes, the former in particular exhibiting a great variety of coordinated hydrocarbon structures. Higher alkenes form analogous structures, some of which, derived from propene, are shown in Table 4.3, but note that alkylidynes can only be formed from terminal alkenes. With propene and the butenes another adsorbed state becomes possible, namely, the π-alkenyl structure (usually called π-allylic), in which loss of a hydrogen atom results in a delocalised π-bond over three carbon atoms: such structures are often found in co-ordination complexes containing in particular either cobalt or palladium (see also Table 4.4, where some of the structures derivable from the 2-butenes are shown).

Higher alkynes may form structures analogous to numbers **11** to **16** in Table 4.2 but analogues of **16A** to **21** are only formed by terminal alkynes. Molecules containing more than one multiple bond, e.g. 1,2-propadiene, 1,2- and 1,3-butadiene, etc. can have their adsorbed states formulated in many different ways, but in general both C=C bonds are connected to the surface in the same way, i.e. both are either π or di-σ. It will be evident that species containing an odd number of hydrogen

TABLE 4.2. Structures of Chemisorbed Hydrocarbons

4E

2	3	4	5
$H_2C{=}CH_2$ with `*`	$H_2C{-}CH_2$ (bridged to `*`)	$H_2C{-}CH_2$ with two `*`	$H_3C{-}CH$ with two `*`
Ethene	Ethene	Ethene	Ethylidene
4EPtCl[3- (4ERhCl)₂	4EM(CO)₄ 4EMP₂	4EOs₂(CO)₈	4EOs₂(CO)₈

3E

6	6A	7	8	9	10
$H_2C{=}CH{-}$`*`	$H_2C{-}CH$ with `*`	$H_2C{=}CH$ bridged two `*`	$CH_3{-}C{-}$`*` with two `*`	$CH_3{-}C$ with two `*`	$CH_3{-}C{\equiv}[$`*`$]_3$
Vinyl/ethenyl	Ethenyl	Ethenyl	Ethylidyne	Ethylidyne	Ethylidyne
Sn(3E)₄		3EOs(CO)₁₀H	3ECo₃(CO)₉	[3ERu₂(CO)₃Cp₂]⁺	3ECr(CO)₄Br

2E

11	12	12A	13	14
$HC{\equiv}CH{-}$`*`	$HC{\equiv}CH$ with `*`	$HC{=}CH$ two `*`	$HC{=}CH$ three `*`	$HC{=}CH$ two `*`
Ethyne	Ethyne	Ethyne	Ethyne	Ethyne
Various	2ENi		2EOs₃(CO)₁₀	2ECo₂(CO)₆

(continued)

TABLE 4.2. (*Continued*)

	15	16	16A	17	18	19
2E	HC≡CH	HC—CH	H$_2$C—C—*	H$_2$C=C—*	H$_2$C=C—*	H$_2$C=C=[*]$_2$
	Ethyne	Ethyne	Ethylylidyne	Ethenylidene	Ethenylidene	Ethenylidene
	^{2}ECo$_4$(CO)$_{10}$	^{2}ECo$_4$(CO)S		^{2}EM$_2$(CO)$_3$Cp$_2$	^{2}EM$_3$(CO)$_9$H$_2$	^{2}ENi

	20	21	22	23
1E	HC≡C—*	HC≡C—*	HC=C—*	HC—C—*
	Ethenylyl	Ethenylyl	Ethenylidyne	Ethenylidyne
	^{1}E$_4$Sn	^{1}EOs$_3$(CO)$_9$H		

Notes. Bold number indicate the ordinal number; then follows the structure, name, and one or more examples of organometallic complexes containing analogous species. The first column gives the symbol ^{i}E where i is the number of H atoms associated with two carbon atoms. Structures are linearised for ease of presentation, and bond angles are therefore not shown truly. Structures **1** is ^{5}E = H$_3$C CH$_2$*, the ethyl group. The significance of the asterisk is discussed in the text (Section 4.42). Where two or three asterisks are bracketed, they belong to a single metal atom. In **3**, M = Fe and Os in ^{4}EM(CO)$_4$, and Pd and Pt in ^{4}EMP$_2$ (P = PPh$_3$); in **17**, M = Fe and Ru; in **18**, M$_3$ = Os$_3$, Co$_2$Fe and Co$_2$Ru.

TABLE 4.3.   Terminology for Adsorbed C_3 Species

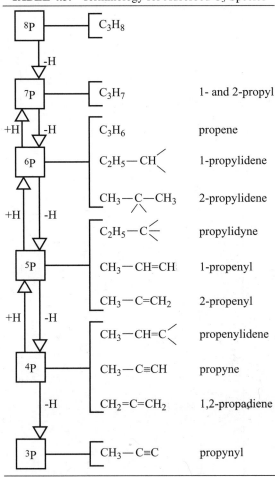

atoms may be formed either by loss of one or more from an alkane or alkene or by gain of one or more from an alkyne or similar molecule. Table 4.3 sets down these processes as applied to C_3 species, showing the structures and names of some of those involved. The reader will be able to suggest some additions to this table (compare Table 4.2). By loss of one or more hydrogen atoms, alkanes can adopt many different adsorbed forms (Table 4.5), but evidence for their existence comes mainly from kinetic/mechanistic studies of their reactions (see Chapters 6, 13, and 14).

TABLE 4.4. Selected Structures formed on Chemisorption of Propene and of 2-Butenes

		π	1,1-di-σ
6P	C_3H_6	$CH_3-CH=CH_2$ ---* Propene	CH_3-CH_2-CH (with two *) 1-Propylidene

		σ	[1]₃-tri-σ	σ	π
5P	C_3H_5	$CH_3-CH=CH$ —* Propenyl	CH_3-CH_2-C-* (with two *) Propylidyne	$H_2C=CH-CH_2$ —* 3-Propenyl	$H_2C\cdots CH\cdots CH_2$ —* π-Propenyl

		π	π	σ
7B	C_4H_7	$H_2C\cdots CH-CH$ with CH_3 —* syn-1-3-Butenyl	$H_2C\cdots CH-CH$ with CH_3 —* anti-1-3-Butenyl	$CH_3-CH=C-CH_3$ —* 2-Butenyl

		2,2-di-σ
8B	C_4H_8	$CH_3-CH_2-C-CH_3$ (with two *) 2-Butylidene

TABLE 4.5. Selected Structures Formed by Dissociative Chemisorption of Propene and n-Butane

$\begin{array}{c} C-C-C \\ | \\ * \end{array}$	$\begin{array}{c} C-C-C \\ | \\ * \end{array}$	$\begin{array}{c} C-C-C \\ / \quad \backslash \\ * \qquad * \end{array}$	$\begin{array}{c} C \qquad C \\ / \backslash \quad / \backslash \\ C \qquad C \\ | \qquad | \\ * \qquad * \end{array}$	$\begin{array}{c} C \qquad C \\ \backslash \quad / \\ C-C \\ / \backslash \\ * \qquad * \end{array}$	$\begin{array}{c} C-C \\ / \backslash \\ C \qquad C \\ | \qquad | \\ * \qquad * \end{array}$
1-absorbed propane	2-absorbed propane	1,2-diadsorbed propane	1,3-diadsorbed propane	2,3-diadsorbed butane	1,4-diadsorbed butane

We may now try to summarise what we have found.

- Simple hydrocarbons form a bewildering variety of adsorbed species: those containing an even number of hydrogen atoms correspond to a parent molecule, although they may have been formed by the loss of a hydrogen molecule or have undergone severe structural reorganisation (see for example species **5** and **15–19** in Table 4.2).
- Molecules containing one or more multiple bonds may chemisorb without the breaking of C—H bonds (see species **2–4** and **11–16** in Table 4.2), but alkanes can only adsorb dissociatively (Table 4.5).
- Adsorbed species having an odd number of hydrogen atoms may be formed either by dissociation of parent hydrocarbon and/or by addition of a hydrogen atom to a more unsaturated species: there is no distinction between, for example, the ethenyl species **6** being formed by loss of a hydrogen from **2**, **3** or **4** and its formation by addition of a hydrogen to **11** or **12**.
- However some of the species shown in these Tables are most unlikely to be intermediates in truly catalytic reactions: thus highly dissociated or reorganised species (e.g. **8** to **10**) are not expected to be intermediates in 'easy' reactions such as hydrogenation. Our task in later chapters will be to try to separate the sheep from the goats.
- Most of the structures shown in Table 4.2 for C_2 species, and some at least of those in Tables 4.3 and 4.4, find analogues in organometallic chemistry, the latter having served importantly in the task of identifying the former, particularly by their vibrational spectral characteristics. Much attention has also been given to the dissociative chemisorption of halogenated molecules (e.g. $H_2C=CHI$;[55] CH_2I_2, CH_3I, C_2H_5I;[56] C_2H_5Cl[57]), which are expected to lead in the first instance to well-defined surface species.[4,57] The correspondence between chemisorbed entities and those occurring as ligands in metal coordination or cluster compounds, and its importance in advancing surface science, has been recognised for many years,[58] although not all the predictions made concerning adsorbed structures have been borne out. Nevertheless the importance of the correlation requires emphasis. *The fact*

that similar structures can often be found both on metal surfaces and in organometallic complexes suggests that surface metal atoms behave rather more as individuals than as part of the throng, and the same theoretical techniques should therefore be applicable to describing both sets of structures.

As hinted above, however, these analogies should be pursued with some care, because other ligands in complexes cannot adequately substitute for metal atoms in surfaces, and as also noted previously the metals that provide the analogous complexes are not always those on the surface of which similar species have been seen. So for example to use a *chromium* complex as a model for an ethylidyne radical triply bonded to a single metal atom is scarcely justifiable, as chromium and other elements having somewhat few d-electrons are well known to form triple bonds with comparative ease, while those near the end of the Transition Series do not. Moreover there are well-established chemisorbed structures that cannot be correlated with that of any conceivable organometallic complex.[21]

- Minimally changed structures are found in complexes containing only one or two metal atoms (e.g. $^4EPtCl_3^-$; $^4EOs_2(CO)_8$), while triatomic complexes can accommodate dissociated species ($^3ECo_3(CO)_9$): extended metal surfaces provide more ample opportunities for dissociation to occur, given the right conditions, and adsorption in the undissociated form is much more restricted.

- There is no mention in Table 4.2 of structures, either on surfaces or in complexes, wherein two or more of the same type of hydrocarbon molecule are attached to the same metal atom or ion. Because of the wide occurrence of carbonyl complexes, where the M—CO bond is similar to that in the alkene complexes, this is somewhat surprising, but in view of their greater size it is likely that steric repulsion would inhibit their formation except perhaps at atoms of low CN. Somewhat unstable complexes of the form $^4E_2M^{II}$ (M=Pt, Pd) are known, that with platinum being the more stable, and a number of complexes denoted as $^4E_2M^0$ and $^4E_3M^0$ (M=Co, Rh, Ni, Pd, Pt and Cu) have been formed by matrix isolation in a solid Group 0 element at very low temperature, as well as $^4EM^0$ species.[59,60] The infrared spectrum of $^4EPd^0$ closely resembles that of the adsorbed structure **2**, but there are additional bands that, because of the metal-surface selection rule, are not visible for the adsorbed species. Moreover the measurement of UV-visible spectra is possible. This fascinating area of π-complex chemistry, which has produced several prophetic insights into catalytic mechanisms, has been sadly neglected for many years: it merits renewed attention.

- There is one other concept that may prove helpful as we come to try to understand what features of a surface determine the type of hydrocarbon structures found on it. *The change in structure consequent upon chemisorption often resembles that caused by electronic excitation.*[46,61,62]

It has been pointed out that the structure of co-ordinated butadiene depends on the π acceptor strength of the *trans*-ligands: if strong, the structure is like that of the free molecule, but if weak it is more like the second excited state, viz. $\cdot CH_2$—CH=CH—$CH_2\cdot$, the terminal atoms forming σ bonds to the metal. On metal surfaces, therefore, the form adopted may be conditioned by the electron donor-acceptor character of the relevant metal atoms, i.e. on their electron concentration and the local density of states.

Finally, those working in this field should constantly remember the advice of William of Occam concerning the unnecessary multiplication of entities. Sir Isaac Newton embodied this as his first 'Rule of Philosophizing'. *No more causes of things should be admitted than are both true and sufficient to explain their phenomena.*

4.4.2. The π and di-σ Forms of Chemisorbed Alkenes

Most of what will be said in the Section will relate to ethene (and ethene-d_4):[63] much of it will also probably apply to higher linear and cyclic alkenes (see 'Further Reading' list), the only additional factor being that activation of allylic methyl or methylene groups adjacent to the double-bond causes easy dissociation of a C—H bond thereon, with the formation of a delocalised three-centre π-bond; the adsorbed state is then a π-alkenylic species (see Table 4.3).

In the σ-*diadsorbed form* of an alkene (Table 4.2, structure **4**), the state of hybridisation of the carbon atoms has changed from sp^2 to sp^3; the π-bond between the carbons has been uncoupled, and two covalent bonds are then formed between them and two metal atoms:

$$H_2C{=}CH_2 + 2^* \rightarrow H_2\underset{*}{C}{-}\underset{*}{C}H_2 \qquad (4.A)$$

The asterisk simply represents a univalent entity corresponding to a single free valence at the surface, with which a covalent σ-bond can be formed; alternatively (as in Tables 4.2 and 4.3) it stands for an atom or an orbital (or orbitals) to which a π-bond may be formed. Except in this latter regard, our understanding of the asterisk has advanced little since an earlier discussion some 40 years ago.[7] Convenient though this device is for the simple depiction of chemisorbed states, we shall shortly encounter structures where it cannot adequately show what is going on.

The composition of the π carbon-metal bond as it exists in complexes such as Zeise's anion $^4EPtCl_3{}^-$ has often been rehearsed in inorganic chemistry texts[64,65] and elsewhere:[66,67] the alkene donates π-electrons into a vacant hybrid orbital on the metal or ion (the σ component) if such is available, and there is a compensating

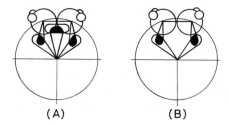

Figure 4.3. (A) Molecular orbital diagram of the bonding ethene to a d^8 metal ion (as in Zeise's salt) or to a surface metal atom having an incomplete d shell. (B) Molecular orbital diagram of the bonding of ethene to a d^{10} or $d^{10}s^1$ metal atom or to a surface metal atom having a filled d shell.

back-donation of electrons from an occupied metal d-orbital into the $2p\pi^*$ antibonding orbital of the alkene (the π component) (see Figure 4.2). This basic model is taken to describe the π-*alkene state* (structure **2** in Table A) on metal surfaces.[58,68] The C—C bond stretches and the molecule becomes non-planar, the hydrogen atoms or other attached groups moving away from the metal (see Figure 4.3). However, when the metal has no low-lying vacant orbitals (e.g. in the d^{10} or $d^{10}s$ states), there can be no σ component, and the structure in complexes of zerovalent atoms then more resembles that of a metallocyclopropane (structure **3** of Table 4.2) (Figure 4.4), in which rotation of the molecule can only be accomplished by the breaking of both covalent bonds and their re-formation, and is therefore much more difficult than with π-bonded molecules.

The evidence for the $\pi\sigma$ *structure* **3** on surfaces rests mainly on a similarity between spectra sometimes observed and those shown by metallocyclopropane complexes: however, it may also be that they are due to a mixture of π-and di-σ-species,[8] or to some experimental artefact.[44] The factor that favours structure

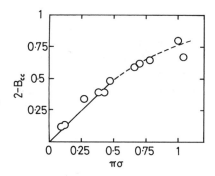

Figure 4.4. Correlation between the $\pi\sigma$ factor and the C—C bond order (B_{CC}) for ethene chemisorbed on various metal surfaces (see Table 4.6 for data).

3 in monatomic complexes, namely the absence of low-lying vacant orbitals, will in the case of metals (which have a high density of states near the Fermi energy) rather lead to the σ-diadsorbed species by the principle enunciated above, that the best process lowers the surface free energy as much as possible: if two metal atoms are involved rather than one, this is a good thing.

The extent to which π or di-σ structures are formed in organometallic complexes depends upon the nature of the metal, its oxidation state, and on the types of other ligands present. With chemisorbed structures, their occurrence (or the appearance of Types B and A spectra that are attributed to them) hinges on the kind of metal, and the size and composition of the site (number and CN of metal atoms, and presence of other adjacent species of the same or quite different type). In both cases, all these factors affect the number and occupancy of the valence orbitals available to take part in the bonding. Additionally the number and nature of the substituent groups about the double bond will influence the character of the π bond in the relevant alkene, and hence the way it interacts with the surface orbitals.

We ought now to confine further discussion principally to metal *surfaces*. Those who use vibrational spectroscopies, LEED, and PED, seek to analyse the structure and composition of surface species present under fixed conditions of temperature and coverage, while those who use techniques that scan over coverage, such as TPD and especially adsorption calorimetry (see Section 4.5.3), recognise species that are labelled π or di-σ at different points in the experiment. With the latter method at close to ambient temperature, the di-σ state prevails at low to medium coverage, perhaps following a small amount of dissociated forms created at particularly active sites such as defects, while the less strongly held π state may be seen at high coverage where only monatomic sites remain.

Much of the experimental information we shall want to review has however been obtained at low temperatures, because only in this way can the first stages of chemisorption be revealed. In general it appears that the π form is the less stable, changing on sometimes only slight warming to the di-σ state:[69,70] this implies that, as suggested by Figure 4.2, that there is little difference in their strengths of adsorption, and only at most a small activation energy for changing one to another. One consequence of this is that spectroscopic criteria often show the two states side by side (e.g., on Fe(111); Pd(100)[71] and (110); Rh(100) and (111);[72] and on Pt(100) and (110)(2 × 1)[8,71,73] and on Pt/Al$_2$O$_3$[72,74,75]). The π form was found at low temperatures (50–200 K) on the Group 11 metals (e.g. Cu(100) and (111),[76] Ag(110) and (111)[77,78]) as expected, on all the low-index faces of palladium,[79] on Pt(111) and Pt(110), and most remarkably on the stepped Pt(210), on which it was stable to room temperature. The di-σ form is stated to occur on the close-packed faces of most metals of Groups 8 to 10 (e.g. Fe(110), Ru(001), Pt(111), (100)(1 × 1) and (5 × 20), and on all three faces of nickel[80,81]), but rarely on palladium.[82,83] STM has revealed that adsorbing ethene on Pt(100)hex.R 0.7° removes the reconstruction, which resumed after its removal.

The main factors determining the prevalence of the π form are therefore (i) degree of occupancy of the d-orbitals, (ii) temperature, (iii) surface roughness, and (iv) surface coverage, and (v) some quality specific to palladium, on which it is distinctly more favoured than on platinum. This last point will be reverted to later.

The frequent observation of the π-bonded structure **2** on stepped surfaces may be understood in the following way. It has been suggested that the electron concentration at the top of a step, i.e. on atoms of low co-ordination number (CN), is *lower* than at the foot, i.e. at atoms of high CN.[8] This allows to $\pi \rightarrow M$ charge transfer (i.e. the σ component) to take place more readily than at a flat surface.[84] By extension of this concept we might expect the π form to occur more on small metal particles than on flat surfaces under equal conditions,[85] and after early unsuccessful attempts[86] it was subsequently recognised to occur[87] on Pt/SiO$_2$ and Pd/SiO$_2$; its existence is no longer in doubt. It has often been seen alongside ethylidyne (structure **10**, Table A). Evidence for the existence of two states of adsorbed ethene has also come from the observation that carbon monoxide displaces the weaker form more readily.[88]

Another significant factor favouring the π form is the presence of other species on the surface. The presence of a partial monolayer of oxygen[89] or carbon atoms, or of caesium,[8] helps, as does hydrogen on Rh(100)[74] and Pd(111):[83,85] this may be because of both steric and electronic effects, as the number of diatomic sites needed for the di-σ form is lowered, and the electron density on the free metal atoms is decreased by withdrawal of charge towards electronegative modifiers. A partial monolayer of hydrogen, often present adventitiously, may be responsible for apparent disagreement between results obtained by different techniques.[83] π-Ethene was reversibly adsorbed on Pt(111) saturated with the di-σ form,[90] with a heat of adsorption of 40 ± 7 kJ mol^{-1}; ethylidyne prevented its formation entirely. The potassium atom (or ion as it probably becomes) inhibited the formation of the di-σ form,[69,91–93] its size also preventing the ethene molecule from approaching closely enough to make a true π-form: an even weaker version took its place.[92] Pre-adsorbed hydrogen on Pt(100)(2 × 1) had the same effect.[94] Bismuth atoms are so large that they stop all adsorption of ethene.[95]

In organometallic complexes having an alkene as one of the ligands, structural studies reveal that the molecule can adopt forms that are intermediate between the π and the metallocyclopropane, the precise structure depending on the variables listed above. In particular the greater the d-electron density on the metal atom, the larger is the di-σ character of the coordinated species. It has been proposed that a similar methodology can be applied to chemisorbed alkenes, and that a C—C *bond order* (B_{CC}) lying between two and one can be estimated[72] from the C—C vibrational frequencies of adsorbed ethene-d$_4$ by first deriving force constants (k_{CC}) using the free molecule and Zeise's salt as the basis of a calibration; these can then be connected to bond lengths using Badger's rule. Bond lengths so obtained are in fair to very good agreement with those measured directly; they of course correlate with bond orders, but it is better to derive orders via an empirical relation with force constants, namely,

$$B_{CC} = (k_{CC}/4.2)^{0.69} \tag{4.1}$$

Values so obtained are listed in Table 4.6.

Another approach has also been tried. In moving from a π to a di-σ structure, there is also a decrease in the C—C vibration frequency and also in the CH_2 scissors (δ) mode.[72,96,97] In hydrogen-containing molecules these are strongly coupled and are difficult to disentangle, so it is preferable to use spectral data for deuterated molecules where coupling is weaker. It is then possible to derive a $\pi\sigma$ *parameter*[96] defined as the sum of the fractional decreases of the two bands divided by 0.366, a factor which normalises the expression so that for free ethene (C_2H_4) it is zero and for $C_2H_4Br_2$ it is unity. Values of this parameter obtained for deuterated molecules ($\pi\sigma$-D) are also given in Table 4.6, and Figure 4.4 shows there to be a reasonable correlation between it and B_{CC}. For many cases there is good agreement between $\pi\sigma$-H and $\pi\sigma$-D[96] (e.g. for Zeise's salt, 0.38 and 0.35 respectively), but in a few cases (e.g. Ni(111) and Fe(110)) there are quite large and unexplained differences. Values of $\pi\sigma$-H shown where necessary in Table 4.6 are in brackets. Ethene co-ordinated to Ag^+ is only weakly perturbed, $\pi\sigma$-H being 0.12. Although the ethene C—C bond orders for palladium and platinum are similar on (110) faces, there is a strong propensity for the di-σ structure to be formed on Pt(111) and for the π structure to arise on Pd(111). Examination of 1-butene and 1,3-butadiene by NEXAFS at 95 K on these surfaces has clearly confirmed this.[98] Recently published values for Rh(100)[74] and Pd(111)[82] are in substantial disagreement with those obtained earlier; like the clock that strikes thirteen, they cast an element of doubt on what has gone before.

The validity of this procedure therefore merits some consideration. While it is accepted that a π-bonded species can be comfortably accommodated on a single metal atom, the di-σ species (4) demands two atoms, and it is difficult to envisage structures that are intermediate between them. Papers that deal with $\pi\sigma$ parameters carefully avoid drawing structures for the intermediate forms, and discussing how many metal atoms might be needed for their chemisorption. Of course if the di-σ form were actually the metallocyclopropane (3), all would be well, because the analogy with alkene *complexes* would apply. However, vibrational spectroscopy recognises the di-σ state (4) more certainly than the alternative (3), as the Type A′ spectrum by which it is defined may be a mixture of π and di-σ forms:[17,51] but it may also be a kind of half-way house between them, occurring when there is an increased opportunity for back-donation of d-electrons to enhance the covalent element of the bonding. A further increase of back-donation, and weakening of the

TABLE 4.6. Bond Orders (B_{CC}) and $\pi\sigma$ Parameters for C_2D_4 (for C_2H_4 in brackets) Chemisorbed on Various Single Crystal Surfaces[8,66,72,96]

Group 8			Group 9			Group 10			Group 11		
Face	B_{CC}	$\pi\sigma$	Face	B_{CC}	$\pi\sigma$	Face	B_{CC}	$\pi\sigma$	Face	B_{CC}	$\pi\sigma$
Fe(110)	1.20	1.00				Ni(111)	1.33	1.04	Cu (100)	1.66	0.27
		(0.55)	Rh(111)	1.52	0.47			(0.80)	Cu*	—	(0.19)
Fe(111)	—	0.86				Ni(110)	1.41	0.66	Ag*	1.88	(0.09)
Ru(001)	1.35	0.78	Rh(100)	1.16	0.62	Pd(100)	1.38	0.70	O...Ag (100)	—	(0.14)
		(0.85)			(0.39)	Pd(110)	1.61	0.38			
O...Ru(001)	—	0.42				Pd(111)	1.61	(0.43)(0.87)			
						O...Pd(100)	—	(0.30)	Au*	—	(0.25)
						Pt(111)	1.13	0.88			
						Pt(110)	1.51	—			
						Pt(210)	1.67	—			

(1) M* represents film or foil.
(2) There are no data yet for cobalt, osmium or iridium.
(3) The symbol O... refers to a surface partly covered by oxygen.

σ-component, may cause the structure to flip over to that of the two-atom di-σ (**4**) as being more comfortable in this way, and making perhaps a greater contribution to the relief of surface stress.

The final word on this problem cannot yet be written, but it seems clear that there are species in which the character of the C—C bond is neither substantially double (as in Zeise's salt) nor single as in ethane. There are indeed considerable variations in both Types A and B vibrational spectra, indicating some flexibility in the kind of structure that can be formed. It seems we must accept that there may be at least two or three different structures co-existing or being capable of being formed at different sites and under different conditions, and that sometimes there may be structures that cannot be reconciled with simple notions of chemical bonding (see Section 4.5.1). This is a situation that does not make the formulation of reaction mechanisms any easier. Two quotations from the literature are apposite to conclude this discussion.

The molecular bonding of ethene is a sensitive probe for the steric and electronic differences between transition metal surfaces.[72]

As all Transition Metals are characterised by their own peculiarities, extreme care has to be exercised when proposing general principles of chemisorptive bonding.[99]

Grateful thanks are due to Professor Norman Sheppard for his advice and assistance in drafting this Section.

There is much less information on hydrocarbon chemisorption on alloy surfaces:[95] two platinum-tin surfaces analogous to Pt(111), namely, PtSn(2 × 2) containing 25% tin in the surface, and PtSn($\sqrt{3} \times \sqrt{3}$)R30° having 33% tin, have however been examined[100] for alkene adsorption. The former has sites of five platinum atoms, and the latter three-atom sites (see Figure 4.5). Tin concentration did not much affect initial sticking probability for either ethene, propene or isobutene, and saturation coverages were similar to those for Pt(111). What was however very marked was the progressive weakening of the adsorption with increasing tin content, due to the loss of trigonal sites, and the extent of dehydrogenation during thermal decomposition was much reduced: in fact all three alkenes desorbed unchanged from the 33% tin surface. There was no sign of

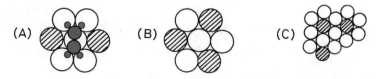

Figure 4.5. Structures of various platinum-tin alloy surfaces having (111) geometry.[206] (A) (111) PtSn(2 × 2), with the probable location of an ethene molecule; (B) (111) PtSn($\sqrt{3} \times \sqrt{3}$)R30°; (C) (111) Pt$_2$Sn($\sqrt{3} \times \sqrt{3}$)R30°.

inductive effects of the methyl substituents on adsorption strength. A structure for ethene on the 25% tin surface was proposed (see Figure 4.5). On silica-supported PtSn bimetallic particles, the π form of ethene was seen at 25% Sn, while at 12.5% tin the diatomic sites needed for the di-σ form occurred more often, and this was then the only structure seen[101] (see also Section 4.5 and Table 4.10).

The adsorption of ethene and ethene-d_4 on Pt$_3$Cu(111) at 95 K gave both the π and di-σ forms, bonding as for Pt(111): they continued to exist in equilibrium and to desorb unchanged.[102] The surface composition was actually Pt$_{85}$Cu$_{15}$, and there was no evidence for any ligand effect. It may however be that the π form resides on the copper and the di-σ on the platinum, as was suggested by similar work on the p(1 × 1)CuRh(111) surface, and on partial overlayers of gold or copper on ruthenium. Peculiar behaviour was shown by the platinum-rich Pt$_{50}$Ni$_{50}$(111) surface, which seemingly failed to adsorb ethene under conditions where Pt(111) does so readily. The behaviour of the PdCu(111) surface has also been compared with Pd(111) using EELS.[103]

To conclude this section, we should note what has not been discussed – and why. The emphasis has been on trying to understand how the nature of the surface metal atoms influences the type of hydrocarbon structure formed, and although there is a great deal of information available on the structures formed by higher and cyclic alkenes, alkadienes, alkynes and benzene, some of which will be touched on below, it is only with ethene that a sufficiently wide range of surfaces has been investigated to make comparisons between them possible. A further omission, to be remedied in Section 4.7, is any attempt to provide a theoretical basis for the observations any more profound than that outlined above. The reason for this delay is that further useful information comes from studying the detailed structures of certain adsorbed molecules (Section 4.5) and the manner of their thermal decomposition (Section 4.6).

4.5. STRUCTURES AND PROPERTIES OF CHEMISORBED HYDROCARBONS[104−110]

In this Section we shall consider (i) detailed structure determinations of ethene, ethyne and benzene, (ii) measurements of the heats of chemisorption of hydrocarbons on single crystals and small particles, and (iii) their spectroscopic characterisation.

4.5.1. Detailed Structures of Chemisorbed Alkenes

The structures displayed in Tables 4.2 and 4.3, and discussed above, are usu-ally qualitatively identified by vibrational spectroscopy using pattern recognition

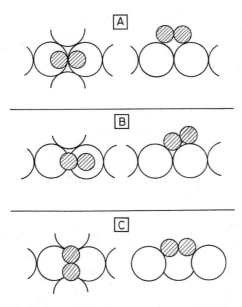

Figure 4.6. Experimentally determined structures of ethene and ethyne chemisorbed on various single crystal surfaces. (A) Ethene on Ni(111) by PED;[20,116] (b) ethene on Ni(110);[21] (C) ethyne on Ni(111).[21,50,147]

involving frequencies and intensities; there is no attempt to define the exact location of the carbon atoms with respect to the lattice of metal atoms with which they are in contact. It is possible to do this by photoelectron diffraction (PED; see Section 4.3) and by LEED intensity analysis.[111] Results are available for ethene on Ni(111)[20] and (110),[100] and Pt(111),[112,113] and for ethyne on Ni(111), Pd(111) and Cu(111), and the important conclusions are shown diagrammatically in Figure 4.6. While these techniques reveal the structures with considerable certainty, the precise C—C bond length is sometimes difficult to pin down, it sometimes appearing to be even longer than the bond in ethane. In the case of ethene on Ni(111) there was a 7% expansion of the Ni—Ni distance between the first and second layers. With ethene on Ni(110)[21] the C—C axis seemed to be not exactly aligned with the rows of nickel atoms, nor were the carbon atoms located directly above the nickel nuclei. There was an interesting long-range order, dictated by the need to minimise overlap of the van der Waals radii. While some structures conform to traditional models, that is, they may safely be described as di-σ states, that found on Ni(110) does not fit easily into the canon of chemisorbed states as shown in Table 4.6, nor would an MO description of its bonding be an easy matter. On Pt(111) there were two non-equivalent sites, depending on whether or not there was a platinum atom below the trigonal hole over which one of the carbon atoms sat, but the two structures are very similar. It appears that atoms in the second layer do not contribute significantly to the chemisorption bonds. Ethene π-adsorbed on Pt(111) saturated with the di-σ form had its plane tilted with respect to the surface, but the C=C bond was parallel to it.[90]

4.5.2. Structures of Chemisorbed Ethyne

The chemisorption of ethyne has commanded somewhat less attention than that of ethene, although there have been numerous studies of it using vibrational spectroscopy, LEED, UPS and other techniques.[40,50,114–118] Table 4.2 shows a number of the structures that have been proposed, together with some organometallic analogues. Much experimental evidence suggests that on Pt(111) the low-temperature form is likely to have the di-σ/π structure **13**; the C—C bond has stretched from 120 pm to about 135 pm, i.e. intermediate between the lengths of single and double bonds, and the linearity of the molecule has been lost, giving a C—C—H angle of approximately 125°.[119] There is also some evidence for the occurrence of the purely π forms **11** and **14**.[50] Further structural input comes from PED;[122] structures have been devised for ethyne on Ni(111),[120] Pd(111)[121] and Cu(111)[122] (see Figure 4.6). Calorimetric work, to be considered in the next section, latter also contributes information on thermal stability, being performed at 300 K. Little ethyne desorbed unchanged from Pt(111) on heating, it mainly decomposing to carbon and gaseous hydrogen, but introduction of tin into this surface diminished the disruption of the molecule, partially in the case of Pt$_3$Sn p(2 × 2) and almost totally with Pt$_2$Sn($\sqrt{3}$ × $\sqrt{3}$)R30°. Quite unexpectedly the major product on this latter surface was then benzene, with some butadiene as intermediate; this may be due to there being a hexagon of platinum atoms surrounding each tin atom[123] (see Figure 4.5). The conversion of ethyne to benzene on palladium surfaces will be considered in Chapter 9. No attempt seems yet to have been made to apply the $\pi\sigma$ formalism in a quantitative way to the forms of chemisorbed ethyne.

RAIRS studies of propyne on Ni(111) and on Cu(110) showed that addition of the methyl group made little difference to the interaction of the triple bond to the surface, the adsorbed states being well described by structure **15** in Table 4.2.

4.5.3. Structures of Chemisorbed Benzene

Although in the earlier sections we have omitted mention of the chemisorption of benzene and other aromatic molecules so as not to muddy the waters unnecessarily, there have been a number of structure determinations of chemisorbed benzene that are conveniently described at this point. On surfaces of trigonal symmetry (i.e. fcc (111) and cph (001)) the molecule sometimes forms disordered overlayers, which can be forced into long-range order by co-adsorbing carbon monoxide[112] or nitric oxide.[124] The general conclusion is that it remains at most only slightly distorted, that is, essentially flat and parallel to the surface,[40] but several studies appear to show an alternation in the C—C bond lengths, i.e. a change towards a cyclohexatriene structure, with a simultaneous buckling of the plane of the ring[125] (but not on Pd(111)[112]). On Pt(100)(2 × 1) this form was seen[125] *at low coverage*, and reacted with hydrogen; a less distorted planar form succeeded it at high coverage, but was not reactive. Similar results were obtained with Pt(111).[126] The existence of different forms at different coverages (as with ethene) may help to explain some of the discrepancies that are found in the literature. Once again however the C—C distances can rarely be found exactly, and the differences in bond lengths are often within experimental error. An exception is the case of Ni(110) where bond lengths are measurable by PED with unusual precision; they were all the same (145 pm) and only a little longer than in the

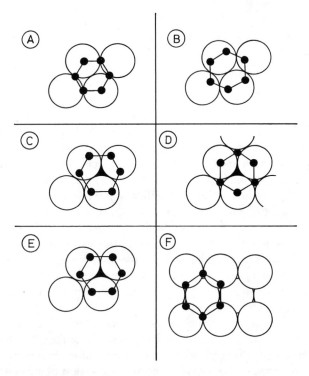

Figure 4.7. Experimentally determined structures of benzene chemisorbed on various single crystal surfaces. (A) On Pt(111) (disordered);[111] (B) on Pt(111) (ordered with co-adsorbed CO);[111] (C) on Rh(111) (ordered with co-adsorbed CO);[112] (D) on Ru(001);[112] (E) on Ni(111) (ordered with co-adsorbed NO: the same structure is found without NO);[112,124,127,128] (F) on Ni(110).[22] In (C), (D) and (E) the molecules are adsorbed over a cph site.

free molecule (139 pm). The structures shown in Figure 4.7 reveal that on the close-packed surfaces of nickel, ruthenium, rhodium and palladium the centre of the benzene molecule lies directly over an hcp hole whereas on platinum it is over a bridge site. There have been three determinations of the structure of benzene on Ni(111), two by LEED[112,127] and one by PED;[128] they agree on the orientation of the molecule (Figure 4.7), but not entirely on the extent of bond length alternation or on the amount of distortion of the nickel surface layer, which is more at low coverage.[128] The very long C—C distances found by LEED are not supported by vibrational spectroscopy measurements. The orientation was different on Ru(001), where the larger atomic radius allowed a mode of packing that is forbidden with the smaller nickel atom.[128] An unusual structure was observed on Co($10\bar{1}0$). Benzene molecules have also been individually resolved by STM: on Rh(111) when co-adsorbed with carbon monoxide at 4 K, they formed a (3 × 3) structure in which two different sites could be distinguished. They have also been observed on Pt(111), Cu(111) and Pd(110).[129,130]

For further references to work on chemisorbed benzene, see the Further Reading section at the end of the chapter.

4.5.4. Heats of Adsorption

The measurement of heats of adsorption has a long and honourable history. The first calorimetric results for hydrocarbons were obtained by Otto Beeck using condensed metal films,[13,14] but the worth of such results is limited if the structure of the adsorbed species is not known and the exact process to which the heat release relates cannot be defined. The recent development of a calorimetric technique applicable to very thin single crystals (Section 4.3) is significant because simultaneous measurement of sticking probability is possible, and plots of differential heat versus coverage often show more or less well-defined plateaux (see Figure 4.8 for an example), which by comparison with other methods (LEED, HREELS etc.) permit identification of the adsorbed structures formed at various coverages with high confidence. This allows estimation of the bond strengths of covalent C—M bonds in each type of species, with the aid of assumed values for C—C, C—H and M—H bond strengths (see Table 4.7) and certain plausible assumptions, at least in those cases where there is no π component.

It was hinted above that one of the factors determining what structure is formed is in fact surface coverage, and the importance of the calorimetric method is that it informs about the sequence in which adsorbed states are formed as coverage increases *at constant temperature*. The most reactive sites are first used to take the most strongly held species, utilising the greatest number of C—M bonds, in line with the principle proposed in Section 4.2; progressively more weakly-held species form as coverage rises (Figure 4.2).[90] The information obtained is thus complementary to that obtained in the study of thermal decomposition (Section 4.6),

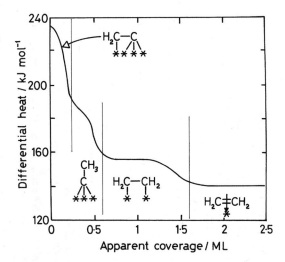

Figure 4.8. Dependence of heat of adsorption of ethene upon coverage for the Pt(110)(1×2) surface; the structures thought to be formed at each stage are shown.[23]

but the order of events is reversed; things seen predominantly at *low* temperature are those found at *high* coverage at room temperature, where only sites of low reactivity remain. Since catalytic reactions usually involved fully occupied surfaces, the relevance of UHV studies at low temperature is reinforced.

Results obtained for the chemisorption of ethene at ~300 K on a number of single surfaces are given in Table 4.8. Initial sticking coefficients are generally high (0.6–0.8), while initial heats are variable between 120 and 305 kJ mol^{-1}; the Pd(100) surface has however been distinguished from the rest by only being able to adsorb ethene reversibly at 300 K, so that no adsorbed structure could be determined and no decomposition took place.[131] In many cases, dehydrogenated species such as ethylidyne (**8**) or ethylylidyne (**16A**) or tetra-σ-ethyne (**16**) are first formed: the ethenylidyne structure (**22**) was also proposed, although the disposition of the bonds about the carbon atoms can only be accommodated by the surface

TABLE 4.7. Average Bond Strengths (kJ mol^{-1}) Used for Calculating the C—M Bond Strengths Shown in Table 4.8

C≡C	962	Ni—H	266
C=C	733	Rh—H	255
C—C	376	Pd—H	270
C—H	412	Pt—H	250

The C—H bond strength depends on its environment.

TABLE 4.8. Initial Sticking Probabilities (σ_0) and Heats of Adsorption ($-\Delta H$) of Ethene at 300 K: Species Identified at Various Stages of Coverage and Corresponding C—M Bond Strengths

Surface	σ_0	$-\Delta H_0$	$-\Delta H_{fin}$	Species proposed	Code	D_{CM}	Species proposed	Code	D_{CM}	Code	D_{CM}	Code	Reference
Ni (100)	0.82	203	100	≡CH	—	204	→ CCH	**21,22**	—	—	—	—	131
Ni (110)	0.78	120	80	CCH	**21,22**	204	→ ?	—	—	—	—	—	133
Rh (100)	0.88	175	100	–HC≡C=	**22**	268	→ ?	—	—	—	—	—	132
Pd (100)	0.75	73	—	—	—	—	—	—	—	—	—	—	131
Pt (100)	0.71	305	135	=CH–CH=	**16**	246				→ **4**	253	**2**	310
Pt (111)	0.67	195	80	≡C–CH$_3$	**8**	238	→ =CH–CH$_3$	**5**	250	→ **4**		**2**	23
Pt (100)hex	0.75	213	130	=CH–CH=	**16**	223	→ ≡C–CH$_3$	**8**	230	→ **4**		**2**	310
Pt (110)(1×2)	0.85	235	140	≡C–CH$_2$	**16A**	229	→ ≡C–CH$_3$	**8**	239	→ **4**		**2**	23
Pt (211)	0.84	180	110	=CH–CH=	**16**	262	→ ≡C–CH$_3$	**8**	274	→ **4**		**2**	311
Pt (311)(1×2)	0.84	220	80	≡C–CH$_2$	**16A**	273	→ ≡C–CH$_3$			→ **4**		**2**	311

$-\Delta H_0$ is the initial and $-\Delta H_{fin}$ the final heat (measured at high coverage). The species named in the first section are those formed first, the arrows printing to species formed at progressively higher coverages. In some cases the final heat is that due to ethene chemisorption in the π-form (**2**) and the penultimate form to the di-σ form (**4**).

Only reversible adsorption occurs on Pd(100), so the adsorbed state cannot be identified: it may be either the π or di-σ form or a mixture of the two. On the nickel surfaces, the species CCH could not be identified certainly; it may be either **21** or **22**. On Ni(110) and Rh(100) no further species could be determined.

(except perhaps in the trough of the fcc(110) surface or at the foot of a step) by straining usual bond angles (and stretching the imagination). It is unfortunate that more attention is not given to the manner of bonding of highly dehydrogenated species to metal surfaces. Only in the case of Ni(100) was it thought likely that the C—C bond broke initially, forming the singly carbon species methylidyne ($\equiv$C—H). Ethylidyne or ethylidene may follow the most strongly held species, and then the σ-diadsorbed state (**4**); finally the π state **2** may appear, but it is by no means certain that the final heats measured (80-140 kJ mol^{-1}) refer just to this state. What is perhaps surprising is that calculated M—C bond strengths lie in so narrow a range, showing only small dependence on surface geometry: for platinum, values lie between about 220 and 270 kJ mol^{-1} (Table 4.8), while those for nickel are a little lower,[23] and for rhodium slightly higher.[132] We must therefore conclude that the *strength* of adsorption depends primarily on the *number* of C—M bonds formed, although on a given surface the bond energy of each bond increases slightly as the number of bonds formed to the surface decreases.[23] Figure 4.2 indicates the energetics of a likely sequence of progressively dehydrogenated species.

Fewer results are available for ethyne (see Table 4.9); initial sticking probabilities are consistently high (0.8), and initial heats fall between 190 and 270 kJ mol^{-1} *except for palladium* which again stands out with a much lower value (112 kJ mol^{-1}) associated with a di-σ (or di-σ/π) structure (**13** or **12a**).[23] On Ni(100) the molecule split at low coverage to form methane, which also seemed to occur together with methylene at higher coverages on Ni(110).[133] On Rh(100)[132] and Pt(211), various rearranged C$_2$ species have been detected. Unfortunately there is little firm information on species existing at high coverage, such as might be intermediates in catalytic processes.

Dumesic and his colleagues have used microcalorimetry[25] to determine heats of adsorption of ethene and ethyne on unsupported platinum black[133] and silica-supported platinum and platinum-containing bimetallic systems:[101,134,135] allocation of observed heats to likely structures was aided by parallel FTIR measurements and bolstered by DFT calculations. Results were obtained at various temperatures between 203 K and ambient, so that, in addition to following the sequence with which adsorbed species were formed as coverage increased, the effects of the temperature dimension were explored, giving results which harmonise with those obtained by LEED and spectroscopic methods. These are summarised in Table 4.10. The values recorded at 203 K represent the molecularly-adsorbed forms **2** and **4**, with the latter presumably predominating; they are similar to the terminal values found at 300 K on single crystal surfaces (Table 4.8). Those at 300 K with pure platinum are due to formation of ethylidyne (**8**) plus a hydrogen atom; progressive addition of tin lowered the number of platinum triplets that could be formed and which are required for ethylidyne, so that the π and di-σ forms could then also appear.[101,136] Addition of gold had the same effect,[137] but there was calculated to

TABLE 4.9. Initial Sticking Probabilities (σ_0) and Heats of Adsorption ($-\Delta H$) of Ethene on Single Crystal Surfaces at 300 K: Species Identified and C—M Bond Strengths (D_{CM}) (kJ mol^{-1})

Surface	σ_0	$-\Delta H_0$	$-\Delta H_{fin}$	Species proposed	Code	D_{CM}	Species Proposed	Code	References
Ni (100)	0.81	264	?	≡CH	—	204	HC≡C	**21**	131
Ni (110)	0.80	190	50	HC≡C—	21	—	=CH₂ + ≡CH	—	133
Rh (100)	0.83	210	80	=CH–CH= or –CH=C=	**16 or 22**	273	?	—	132
Pd (100)	0.83	112	?	–HC=CH–	**12/A**	177	?	—	131
Pt (211)	0.84	270	80	=C=CH₂	**17 or 18**	220, 255	=C=	—	312

$-\Delta H_0$ is the initial and $-\Delta H_{fin}$ the final heat (measured at high coverage). The first-named species are those proposed at low coverage, and the second-named at high coverage.

TABLE 4.10. Initial Heats of Adsorption of Ethene and of *iso*Butene

		Heat of adsorption/kJ mol^{-1}		
Alkene	Catalyst	300 K	203 K	Reference
C$_2$H$_4$	Pt film	148	—	141
	Pt black	160	120^b	133
	Pt/SiO$_2$	157	125	101
	Pt/SiO$_2$	145	—	135
	Pd/SiO$_2$	170	—	135
	Pt-10Au/SiO$_2$^{a}	140	100	137
	Pt$_7$Sn/SiO$_2$^{a}	135	100	101
	Pt$_3$Sn/SiO$_2$^{a}	129	98	101
	PtSn/SiO$_2$^{c}	115	—	135
iso-C$_4$H$_8$	Pt/SiO$_2$	160	—	135
	Pd/SiO$_2$	190	—	135
	PtSn/SiO$_2$^{c}	125	—	135

^{a} The numbers are the *relative* molar amounts.
^{b} 173 K.
^{c} Pt/Sn ratios of 2/1 and 2/3 gave very similar results. Values for 300 K are for alkylidynes.

be a consequential move of charge from the 6sp to the 5d levels of the platinum with a net increase of charge at the expense of the gold, corresponding to the latter's greater electronegativity (see Chapter 1): this caused stronger adsorption of ethylidyne, counteracting the geometric effect. The effect of tin was to increase the population of both the 5d and 6sp levels. Early work gave no evidence for the chemisorption of hydrocarbons on Au(111) or stepped gold surfaces,[138] but more recently weak adsorption of ethene on Au(100) and (111) has been detected,[139] the desorption activation energy being about 40 kJ mol^{-1}. There is also spectroscopic evidence for weak adsorption of ethene on Au/SiO$_2$.[137] Heats of adsorption of ethyne on Pd/SiO$_2$ were higher than for Pt/SiO$_2$, as were the greater values for *iso*butene (Table 4.10). Values for propyne on Pt/Al$_2$O$_3$ decreased slightly with increasing dispersion, but a more marked difference in the same sense was found with Pd/Al$_2$O$_3$, especially at coverages between 0.1 and 0.5, where values were greater than for Pt/Al$_2$O$_3$.[140]

On platinum black at 173 K, ethyne formed a $\pi\sigma$ structure (either **13** or **15**), and at 300 K some ethylylidene (**16A**) was also observed,[133] the heats liberated being respectively 210 and 220 kJ mol^{-1}.

Initial heats of adsorption of ethene reported many years ago[13,14] for condensed metal films were remarkably similar to those given in Tables 4.8 and 4.10 (Fe, 272; Ni, 232; Rh, 210 kJ mol^{-1}). *Plus ça change, plus c'est la même chose.* More recently values have been obtained[141] for a number of hydrocarbons on platinum films (see Table 4.10 for ethene), but adsorbed states could not be identified, and in many cases (e.g. methane, ethane) substantial decomposition must have occurred.

4.5.5. Characterisation by Other Spectroscopic Methods

The UPS technique was briefly introduced in Section 4.3. It was extensively applied to ethene, ethyne and benzene chemisorbed at low temperatures (100–200 K) on single crystal surfaces of nickel, copper, palladium, iridium and platinum (Table 4.1), and, notwithstanding its constraint in observing *all* adsorbed species, it led to the conclusion that under these conditions the molecules were adsorbed without dissociation or rearrangement, i.e. in π or $\pi\sigma$ states. This agrees with the conclusions reached by the other techniques discussed above.

Ethylidyne (**8**) has been recognised on Pt/SiO$_2$ at 300 K using the SEDOR NMR technique applied to heavily ^{13}C-labelled ethene;[27,28] the C—C bond length was 149 pm. This seemed to occur on large platinum particles, where areas of (111) face are most likely;[136] it was also seen by SIMS on platinum black,[142] but on small particles vinylidene (**17**) predominated. Similar SEDOR experiments with ethyne showed 75% vinylidene and 25% ethyne as **12A** or **15**.[143] Adsorbed benzene was shown to rotate freely at 300 K, and cyclopropane was adsorbed, but not strongly, i.e. without loss of hydrogen.[27]

Many other methods have been used to look at chemisorbed states of hydrocarbons, but none of them give direct structural information. The decrease in magnetic susceptibility of paramagnetic metals accompanying adsorption led to estimates of the number of bonds formed between each molecule and the surface,[144,145] and work function changes (Table 4.1) indicated the direction of charge movement at the surface. Techniques such as LEED, SEXAFS, HREELS, PED and FTIR, utilising the interaction of electrons or electromagnetic radiation with the adsorbed layer, remain of paramount importance for structure determination.

4.5.6. C$_6$ Molecules

The reader may be surprised to see so little about higher alkenes, and especially those having six carbon atoms, on which considerable work has been done. Because of the greater number of options for decomposition by C—H bond breaking, their pristine states are harder to access: a few leading references are given at the end of the chapter, and further attention to these molecules will appear in Chapters 12 to 14.

4.6. THERMAL DECOMPOSITION OF CHEMISORBED HYDROCARBONS

The modes of decomposition of chemisorbed hydrocarbon as the temperature is raised from an initial low value to ambient or even higher have been investigated extensively and in very great detail, using all the techniques of structure assessment referred to above. It should by now be quite clear that chemisorbed molecules, especially alkenes, exist at low temperatures in states that will change to more strongly bound configurations given the necessary (often minimal) thermal input (see Figure 4.2), in accordance with the principle of maximum bond formation stated in Section 4.2. Recognition of such events may therefore be necessary in order to find the primeval adsorbed state, but with those techniques currently limited to ambient temperature (e.g. calorimetry on single crystals) understanding the ways of decomposition is unavoidable. More generally the multiplicity of states and pathways has proved an irresistible challenge both to experimentalists and theoreticians, and the subject has acquired a momentum that perhaps outweighs its significance to catalysis. Nevertheless a rich area of surface chemistry has been disclosed, and the chemical explanation of what goes on is by no means accomplished.

And what goes on depends on (i) the molecule, (ii) the metal, and (iii) the surface geometry in the case of single crystals[8,146] (to which our attention will be largely confined). A molecule such as ethene, which is the most widely studied, if chemisorbed at low temperature can on heating undergo either (a) desorption unchanged or (b) dissociation by loss of one or more hydrogen atoms, which may then attack other species with the formation of ethane. These events have been recognised from the time of the work on metal films,[13,14] and the dehydrogenated species, sometimes called 'acetylenic residues', have since time immemorial been the curse of anyone trying to study catalysed reactions of alkenes. Knowing the cause of the problem does not however necessarily make for happiness. It is also evident that dehydrogenation of an alkene may lead to species also formed by the corresponding alkyne, although adsorption *via* the stronger triple bond makes for greater stability. The alkane can also by loss of hydrogen form equivalent species, although usually only at higher temperatures. The formalism of these various species was depicted for C_3 hydrocarbons in Table 4.3.

The following discussion is confined mainly to ethene. A survey of the very extensive literature (see Further Reading section) strongly suggests that the pathway of decomposition depends greatly on the arrangement of atoms in the crystal face used. On faces of trigonal symmetry (fcc(111), cph(0001), fcc(100)−(5 × 20)), and on stepped surfaces exhibiting terraces of this type, ethylidyne (**8**) has usually been observed at room temperature, but on Ni(111) (and the corresponding faces of iron and tungsten) di-σ ethene dehydrogenated to ethyne in the form shown in Figure 4.6: this process has been closely observed by PED[147] (route B in Figure 4.9). On the square (100) plane however the bonding carbon atoms more often form only one or two σ bonds to the surface. The various processes are summarised in Figure 4.9; it is not certain that the π form always precedes the di-σ, although it may be a normal transient state as indicated in Figure 4.2. Recognised routes followed by the π form are shown in Table 4.11, together with some of the surfaces on which they occur: the egregious behaviour of palladium, already noted

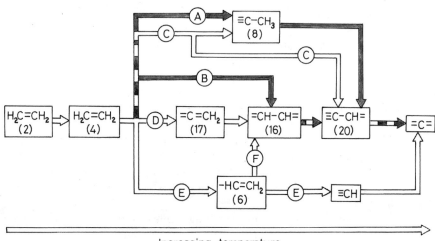

Increasing temperature

Figure 4.9. Reactions of chemisorbed ethene on faces of three-fold symmetry (dark lines) and on faces of four-fold symmetry (open lines): (A) Re, Ru, Rh, Pd, Ir, Pt; (B) W, Fe, Ni; (C) Rh; (D) Pt; (E) Ni, Pd; (F) Ni.

several times, is once again very evident. Rough surfaces such as the fcc(110), either in its normal (1 × 1) or reconstructed (1 × 2) form, and higher index stepped surfaces, have shown additional features (Figure 4.10). Ethylidyne could not form on Pt(110) because of overlap of the van der Waals radii of the methyl group and the metal atoms;[148] this reinforces a point often ignored in the drawing of adsorbed species. Instead a further loss of hydrogen occurred via vinylidene (**17**) to the strange-looking isomeric form ethylylidyne (**16A**), which could be accommodated in (110) troughs or at the foot of steps of the right kind. The tendency to find single carbon fragments is a logical consequence of its formation. The importance of step density in the reactions of adsorbed ethene has however been questioned.[149,150]

Much attention has been given to the way in which di-σ adsorbed ethene (**4**) is transformed into ethylidyne, especially on Pt(111), and every conceivable technique, theoretical as well as experimental, has been applied to this problem (see Further Reading section). The activation energy for the conversion as measured

TABLE 4.11. Reactions of π-Adsorbed Ethene

	$\longrightarrow C_2H_4$ (g)	(i)...Pd (100, 110, 111)
	$\longrightarrow$ di-σC_2H_4	(ii)...Pt (100)
π-C_2H_4	$\xrightarrow{+2H} C_2H_6$	(iii)...Pt (110)
	$\xrightarrow{-H}$?	(iv)...Pt (210)

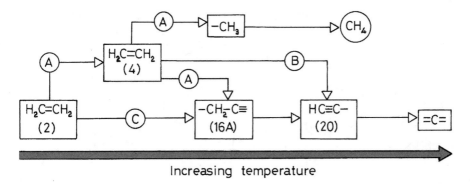

Increasing temperature

Figure 4.10. Additional reactions of chemisorbed ethene on stepped surfaces: (A) Pt(110); (B) Ni(110); (C) Pd(110), Pt(210). Note, reactions depicted in the previous figure may also occur on terraces.

by the accompanying hydrogen desorption is about 43 kJ mol^{-1}. The arguments based on the observations verge on the theological, but it seems likely[151] that there is first isomerisation to ethylidene (**5**) and then dehydrogenation to ethylidyne. STM revealed that its formation proceeds by island growth on a surface saturated with ethene.[130] LEED structure determinations of ethylidyne have been performed on the (111) faces of rhodium[111,112,152] and platinum.[153] On the former it sat on an hcp trigonal hole, and the atoms to which it was attached moved away from the bulk, as did the underlying atom: the surface was therefore corrugated. On the latter it resided on an fcc hole, a difference for which there is no logical explanation. All stages of the decomposition of adsorbed ethene on Pt(111) from 160 to 800 K have been followed by STM. At 230 K the molecule coexists with ethylidyne, the ethene being ordered and the ethylidyne randomly arranged: its formation during heat proceeds by island growth.[129,130]

Self-hydrogenation of adsorbed ethene to ethane using hydrogen atoms released when some of the molecules decompose is apparently confined to platinum, several of whose surfaces show this reaction: on other metals, the hydrogen atoms desorb as molecules, leaving dehydrogenated species behind; these end up as carbon.[146]

The entire energy profile for conversion of ethane through ethene to ethylidyne on Pt(111) is shown in Figure 4.11.

Formation of ethylidyne is not confined to single crystal surfaces; it has been detected on a number of supported metals, including Pt/SiO$_2$,[28,154] Pd/SiO$_2$,[155,156] Rh/Al$_2$O$_3$[157] and even Ni/SiO$_2$,[158] and it is presumably mainly responsible for the all-too-frequently observed deactivation during the catalysis of hydrocarbon reactions. Nor does it arise only from ethene: alkylidynes have also been formed on Ni/Al$_2$O$_3$ from higher 1-alkenes,[159] but they were unstable, and reverted to ethylidyne at higher temperatures. It is also formed from ethyne, both on supported

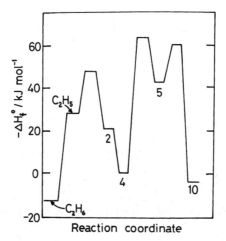

Figure 4.11. Energy profile for the transition of C_2H_6 via C_2H_5 to π and di-σC_2H_4, then to $=CH–CH_3$, and finally to $\equiv C–CH_3$, on Pt(111).[207] Compare with Figure 4.2 which starts with gaseous C_2H_4, but does not include $=CH–CH_3$ as in intermediate.

metals[28,156] and on trigonal-symmetry single-crystal surfaces,[160–164] necessarily with the concomitant formation of other species in which the H/C ratio is less than two.

4.7. THEORETICAL APPROACHES[16,24,165–167]

There is already a very considerable literature on the application of theoretical procedures to our understanding of the chemisorbed states of hydrocarbons, and no doubt its growth is set to continue. At the outset we need to ask ourselves what expectations we have of such work, and what questions and problems we would like to see addressed. There would seem to be two different classes of question demanding attention. The first is a detailed and profound insight into the structure and bonding of individual molecules and their fragments on specific crystal planes or adsorption sites. The second is a more qualitative and broad-brush look over the whole field, to discern what factors may explain satisfactorily the trends we have observed, with particular reference to apparently anomalous behaviour. It may well be that in the course of time the questions of the second type will be answered more precisely by the quantitative methods now being applied to questions of the first type, and indeed one can see already a start in this direction.

Early work on the quantum mechanical analysis of chemisorption and chemisorbed states is admirably covered in Clark's monograph.[165] Application of the extended Hückel molecular orbital method in the hands of A. B.

Anderson[116,117,165,166] gave results which have broadly stood the test of time, and the procedure has been developed by R. Hoffmann to give a detailed account of the bonding of ethene to surfaces of the Group 10 metals;[68] this again points to palladium being unlike the others, the ethene molecule preferring an atop site (presumably in the π-form) rather than a two-fold site. What has revolutionised theoretical chemistry is the use of *density functional theory* (DFT).[24,168] It is unnecessary to attempt to explain it here, as the exposition would only interest its practitioners, and is in any case fully described elsewhere; the experimental chemist, lacking the skills and dare one say the interest in the ritual, has perforce to take its results on trust, while noting its limitations and uncertainties.

It is indeed a very powerful and convincing tool, and is capable of devising the most stable location of an adsorbed molecule or fragment, the type of best bonding, bond lengths and angles, and heats of adsorption[119] to an accuracy of about $\pm$ 20 kJ mol^{-1}. The computational procedure contains a number of disposable parameters, which have to be evaluated and optimised ('calibrated') with reference to experimental facts. Thus for example employing a 'basis set' that treats the valence and outermost core electrons of platinum (i.e. $5s^2 \, 5p^6 \, 5d^9 \, 6s^1$) explicitly, and lumps the inner electrons together as effective core potentials accounting for mass-velocity and relativistic effects, gives very satisfactory results, but including the $5s$ and $5p$ electrons in the core potential gives worse results, because their orbitals are similar in size to those of the valence electrons. Once the calibration is done, the machinery can be applied predictively to systems for which facts are lacking. The power of the method may be shown by calculations made on a ten-atom tetrahedral cluster of platinum atoms, exposing triangular faces;[119] energy minimisation however led to some distortion of the fcc structure, and the normal bond length of 277 pm was reduced to an average of 272 pm. This agrees with experimental findings on small metal particles (Section 2.4). DFT has been further applied to ethene chemisorbed on Pt(111) and Cu(111).[169]

Some values of the calculated parameters are given in Table 4.12; bond lengths and angles agree very well with those found, and many of the heats of adsorption are quite accurate. (The experimental heats in this Table are either taken Tables 4.8 or 4.9 or are those from which the bond energies in Table 4.8 were obtained). The di-σ form of ethene was located at a two-fold bridging site, as observed by PED on Ni(111) (Figure 4.6A); the π-form adopted an atop position on an edge atom, with the C—C axis either parallel or at right angles to the edge. This conclusion was supported by other calculations showing that this form is the more stable when an atom of co-ordination number less than seven is available,[119] and supports the experimental finding of its stability on stepped surfaces (e.g. Pt(210)). In practice this species is likely to rotate freely except at very low temperatures. Ethyne was predicted to sit over a trigonal hole, making σ bonds to two atoms on one side and a π bond to an atom on the other; this is similar but not identical to the structure predicted by vibrational spectroscopy and found by PED on Ni(111). Structures

TABLE 4.12. Estimated Bond Lengths (C—C) and Calculated Heats of Adsorption $(-\Delta H_{ads})$[24,119,187,305,343,354] and Comparison with Experimental Values[23,101,133]

Surface	Structure	Code	C-C/pm	$-\Delta H_{ads}$/kJ mol^{-1} Calc.	23	101,133
Pt(111)	di-σ-C$_2$H$_4$	4	150,148	171,122	136	120
Pt(111)	π-C$_2$H$_4$(i)	2	141	103	120	—
Pt(111)	π-C$_2$H$_4$(ii)	2	141	137		
Pt(111)	CH$_3$—C≡	8	152	127	125	115
Pt(111)	Di-σ,π-C$_2$H$_2$	13	141	209	—	210
Pt(111)	Di-σ,π-C$_2$H$_2$	18	144	278	290	—
Pt(111)	H$_2$C=C=	17	134	262	—	—
Pt(111)	π- C$_2$H$_2$	11	125	88	—	—
Pt(111)	H$_2$C=CH—	6	134	49	—	—
Pt(111)	—CH$_2$—CH=	6A	150	78	—	—
Pt(111)	CH$_3$—CH=	5	151	87	—	—
Pd(111)	Di-σ-C$_2$H$_4$	4	150	48,60	—	—
Pd(111)	π- C$_2$H$_4$	2	142	34,30	—	—

The two forms of π-C$_2$H$_4$ vary in the orientation of the C—C axis (see text). Experimental heats are corrected for the adsorption of hydrogen atoms where dissociation occurs, assuming $-\Delta H_{ads}$(H$_2$) = 90 kJ mol^{-1}.

for vinylidene (di-$\sigma\pi$ (**18**) and di-σ (**17**)), for ethylidyne (**8**) and for C$_1$ species have been advanced, with characteristics close to those observed (see Table 4.12). Such is the confidence now attending these calculations that the probable accuracy of a LEED structure for ethene on Pt(111) can be queried.[24]

While a Pt$_{10}$ cluster of this type is a not unreasonable model for a small metal particle, since particles of this size or even smaller do indeed show catalytic behaviour, it may not be quite large enough to simulate a single crystal surface. The procedure only examines one species at a time, and so the manner of their packing, and possible overlap of van der Waals radii as a factor in determining structure,[124] are not considered. The latter was shown to be important in the behaviour of ethene on Pt(110).[8] It also identifies the most stable site, although not ignoring others that may be relevant in catalysis. A triangular six-atom face is also not large enough to accommodate species such as **16** (di-σ, di-π ethyne). A further problem for the chemist is the translation of the results into a molecular orbital description, as the orbital structure of the cluster lies buried deep within the computational procedure and is reluctant to emerge (see however references 68, 99). These comments do not demean the value of DFT in these systems, and present limitations may well be overcome in future developments.

Recent work on DFT has concentrated on the metals of Group 10 (Ni, Pd, Pt) and on C$_1$ and C$_2$ species (see Further Reading section). Di-σ ethene on Pd(111) is reported as showing a low binding energy, that for the π-form being even lower (see Table 4.12), but the distinction appears to be less marked at low coverage. Other metals have been less studied (for CH$_3$ . . . Rh$_6$, see reference 170) and faces other than fcc(111) are not so popular (for Ni(100) see references 171, 172). These limitations, which will also doubtless be rectified in time, will lead us to consider the broader overview that students of catalysis need. Calculations have also been performed[101] with a ten-atom Pt_6Sn_4 cluster analogous to the Pt_{10} but having tin at each apex, and with $Pt_{16}Sn_3$ in which the triangular face has tin at each corner. The tin atoms weaken somewhat the binding of di-σ and π ethene to platinum atoms held in the same way as before, but they affect that of ethylidyne much more. The results show electron transfer from tin to platinum.

It is noteworthy that both experimental and theoretical studies suggest that only *one* σ C—M bond can be formed and that C=M bonds do not exist; species such as ethylidene (**5**) and ethylidyne (**8**) therefore require respectively two and three metal atoms: this is in keeping with the principle of maximum utilisation of surface free valencies. $\alpha\alpha$-Diadsorbed species such as ethylidene have frequently been postulated as intermediates in reaction of alkanes (see e.g. Chapter 6). Group theoretic arguments indicated that a C—M double bond if formed would have σ and π components.[173]

We return now to the broader prospect, and re-visit the question of the conditions under which the π and di-σ orms of ethene appear. DFT calculations are not yet sufficiently numerous to reveal the variations in C—C bond order and vibration frequency that are observed in practice (Section 4.4.2). The basic model suggests that the di-σ form will occur when there is substantial π back-bonding, and when this is only moderate the π form will result. This led Yagasaki and Masel[8] to propose a correlation between *interstitial electron density* (i.e. that outside the atomic cores) and the occurrence of either form: this quantity emerges from basic theories of the metallic state, which start with the jellium model,[108] and varies with Periodic Group number somewhat as sublimation heat (Section 1.11), although strangely its values in Groups 6 to 9 increase in the sequence: 2nd row<1st row <3rd row. Elements in these Groups (and nickel and platinum) have values greater than four, and generally favour formation of the σ state, while palladium and the Group 11 metals, having values between two and four, favour the π state. This correlation is however based on somewhat limited experimental evidence, and the unusual behaviour of palladium may find alternative explanations. This metal is in so many respects *sui generis*. The concept of electron density also helps to understand why the π state occurs along the ridges of stepped surfaces such as Pt(210); it is thought to be lower there than on terraces. A linear relation between both bond order and C—C vibration frequency with interstitial electron density for ethene on several single crystal surfaces and films has also been observed.

The first and very pictorial attempt to invoke the ideas of co-ordination chemistry and to apply them to chemisorption at metal surfaces[58,174] was introduced in Section 1.23: it was based on identifying the character of orbitals emerging from various crystal planes, assuming that the arrangement within the bulk was maintained at the surface. This notion has been quite properly ridiculed as being

overly simplistic ('it is not accepted'[8]), but it has had its uses[175,176] ('Nevertheless there is a good correlation between the model and the experiment'). This
last statement is based on a compilation of the available orbitals for each type of
structure, and finding that these correlate with the tendency of several platinum
surfaces to form bonds with them.[8] Once again, however, the way in which the
calculation is made remains quite unclear. The principal observation from all experimental studies of hydrocarbon chemisorption is however so obvious that it
escapes attention. *Adsorbed molecules and fragments occupy clearly defined crystallographic sites on metal surfaces; they form σ and π bonds to specific atoms,
having well-defined strengths and lengths, and their spectroscopic characteristics
are well simulated by organometallic complexes.* Exceptions to this generalisation
are rare, and are open to doubt.[21] The rules of chemical bonding apply equally to
chemisorption, and we are therefore dealing with chemistry in two dimensions. In
some cases, species occupy sites that might have been predicted using chemical
common sense; it does not need refined theory to see that ethylidyne, wanting to
form three σ bonds to the surface, will bond at trigonal sites on the fcc(111) plane
(except, curiously, and inexplicably, on Ni(111), where it does not form), rather
than elsewhere. Ethene and ethyne in their adsorbed states are usually more or less
in register with the surface lattice (the asymmetrical di-σ ethene on Pt(111) has
been queried[113]), but there are differences in the orientations of the C—C axes and
in the locations of the carbon atoms with respect to the Ni(111) surface lattice (Figure 4.6) that are not replicated by the DFT calculations on the Pt_{10} cluster referred
to above. It will be interesting to see whether DFT calculations on Ni(111) produce
the correct result. As noted in Section 1.2.2, surfaces having kinks possess chi-
ral sites, and calculations[177] have suggested that enantiomeric hydrocarbons will
bond to kinks of different form with slightly different strengths. There is still a
need for a sound molecular orbital account of all these interactions, one that offers
explanations for the structures adopted, phrased in language understandable by
chemists.

A recurring theme throughout this chapter has been the anomalous characteristics shown by palladium. Specifically, in comparison to its neighbours (Rh,
Ni, Pt) it chemisorbs ethene much more weakly: both the DFT-calculated and
experimental heats of adsorption (Tables 4.8 and 4.12) are lower, the π form is
seen in all low-index faces, bond order estimates confirming this, and it desorbs
unchanged on heating. Ethyne is also weakly adsorbed on palladium (Table 4.9).
One other characteristic that palladium shares with nickel, but which is not highlighted in fundamental studies, is its propensity to form π-alkenyl species when
three or more carbon atoms are present. The question now arises as to whether it
is palladium or platinum in Group 10 that shows unexpected behaviour. It might
be thought that the general similarities shown by metals in the second and third
rows of the Periodic Classification would lead a near-identity of behaviour in palladium and platinum, but the curious discontinuous changes seen in Groups 11

TABLE 4.13. Calculated Hydrocarbon-Metal Bond Energies
(kJ mol^{-1}) for Ethene and Ethynyl Complexes (P = PPh$_3$)

Metal	M^{0}P$_2$ (C$_2$H$_4$)	M^{0}P$_2$ (C$_2$H$_2$)	M^0(CO)$_4$ (C$_2$H$_4$)
Ni	159	189	163
Pd	83	91	128
Pt	88	104	164

and 12,[178] which have been attributed (Section 1.1) to the operation of relativistic effects caused by heavy nuclei, may suggest that it is platinum that is the odd man out.

The importance of relativistic phenomena both in coordination complexes and in chemisorption has been reviewed.[178,179] For complexes containing coordinated ethene or other unsaturated hydrocarbons, comparable quantitative information on all the Group 10 metals is extremely hard to come by, but calculations[180] on various ethene and ethyne complexes (Table 4.13) performed by the non-local quasi-relativistic DF method are instructive. For each complex the bond energy is in the sequence: Ni > Pt > Pd; marked differences in the stabilities and reactivities of complexes of the type M^{II}P$_2$(CH$_3$) (M = Pd, Pt; P = PPh$_3$) were also noted. In this context, it is never remarked that *nearly all reactions homogeneously catalysed by metal salts or complexes, and metal-mediated reactions, involve elements from the first and second rows, and very rarely a third row element.* Ruthenium, rhodium and palladium feature often; osmium, iridium and platinum hardly at all. This is because very generally the complexes of the third row elements are too stable to be reactive.

It was concluded[178] that this is caused by the stabilisation of the 6s electron level relative to that of the 5d level, permitting the latter to be involved in chemical bonding. In the case of alkene and carbene complexes, and chemisorbed molecules, platinum therefore has a greater facility for π-back-bonding than with palladium, resulting in preference for di-σ rather than π bonding, with a consequential greater stability.[181] This simple factor underscores much of the catalytic chemistry of these metals, and will be a recurring theme in subsequent chapters. The discontinuity that arises between palladium and platinum (and somewhat less clearly between rhodium and iridium) has often been noted, and interpreted in various theoretical frameworks: for example, 'The enhanced stability of the π-bound complex on Pd(111) is associated with the smaller orbital exponent of the d-bands in palladium',[8] and 'Relativistic augmented plane-wave calculations have shown the palladium d-band to have lower energy, and therefore overlap with the hydrocarbon's π orbitals is poorer'.[178] Which metal is out of line is therefore a matter of personal choice: what is important is to understand the causes of their characteristics (*rerum cognoscere causas*).

4.8. CHEMISORPTION OF ALKANES[182-186]

This is mercifully a relatively straightforward matter compared with the complexities attending the chemisorption of unsaturated hydrocarbons, at least as far as adsorbed structures go. At low temperatures, alkanes are physically adsorbed with a strength that depends on the size of the molecule, i.e. on the number of contacts it makes with the surface when lying flat. The adsorbed state of n-butane on Pt(111) at 98 K has been studied in some detail using LEED and molecular beam methods, changes in adsorption energy and mobility with coverage being noted.[187] Frennet spoke[182,184,185] of a non-dissociative chemisorption of methane that occurs below 273 K without evolution of hydrogen, but it is not clear how it would be bound to the surface and whether it would be in truly molecular form. With metal films above a critical temperature (Rh, 323 K; Ni, 373 K; Pd, 398 K), hydrogen evolution begins, and equilibrium is established between gaseous hydrogen and methyl, methylene and methylidyne radicals. Above 473 K, equilibrium is reached quickly if carbide formation is impossible, but with tungsten for example it may require more than two days. Detailed studies[185] of the effects of methane and hydrogen pressures on rates and positions of equilibrium have led to the conclusion that the most likely mechanism for chemisorption is:

$$CH_4 + H^* \rightarrow CH_3{}^* + H_2 \qquad (4.B)$$

the reverse of which is an Eley-Rideal step. The forward process seems however also to need a number of vacant 'sites' around the hydrogen atom in order to accommodate the methyl radical, these forming a 'free potential site' or 'landing site'. This mechanism has however not been translated into an easily visualised picture. Methane is in many ways an atypical alkane, and the general picture that emerges is of an initial breaking of a C—H bond with the formation of an alkyl radical; this may be followed, at a rate dependent on the temperature, by the loss of further hydrogen atoms, leading to an alkylidyne,[187] or at higher temperatures to the complete disruption of the molecule.[188]

It is very much harder to chemisorb an alkane than hydrogen. The process is activated,[189] and sticking coefficients are low, but this is attributed to an unfavourable entropy of activation.[184] In atomic terms the electrons in the C—H bond to be broken are shielded from the metal atoms by the other hydrogen atoms, thus forbidding easy reaction. Higher alkanes chemisorb more easily because the stability of their precursor state is greater.

Much effort has been devoted to trying to decide the number of metal atoms that comprise a landing site for an alkane. As with so many mechanistic questions in heterogeneous catalysis, a definite answer cannot be obtained. As to the first question, it had been thought that use of alloy surfaces might provide an answer since dilution of the active metal by an inactive should lead to decreases in the size of various ensembles of the former in a way that can be calculated, so that the dependence of rate or amount of adsorption upon composition should identify the minimum ensemble size required. However, this calculation is fraught with difficulties, which may explain the improbable numbers that the early work produced.

There have been a number of studies of the mechanism of chemisorption on single crystal surfaces, with particular attention being made to the type of energy needed to secure the breaking of the first C—H bond.[183] Work performed during the 1970s revealed the importance of surface geometry: close-packed planes $(Ni(111)^{190}$ and $Ir(111)^{191})$ were unreactive, while the more open planes[190,192] (e.g. Ni(100)) and stepped surfaces[193] (e.g. Ir(110)(1 × 2)) were more reactive. Use of molecular beams[194,195] has shown that energy stored in the surface is relatively unimportant, and that the more critical factor is the rate of vibrational activation of the C—H bond to be broken, using translational and other vibrational modes.[4,196]

On platinum and rhodium surfaces, however, sticking coefficients increase with temperature when the kinetic energy of the molecule is low, and they show activation energies of some 20–40 kJ mol^{-1}. Sticking coefficients for methane were[4,184] as low as 10^{-10} to 10^{-12}. Impact-induced adsorption of methane and ethane produced by hammering physically-adsorbed molecules with other impinging molecules on $Pt(111)$ and $Pt_{25}Rh_{75}(111)$ surfaces suggested that two vibrational quanta in the C—H deformation mode were needed for adsorption to occur.[197] Recent molecular beam work with methane on Ir(110), Ir(111) and Pt(111) has identified a new low translational energy pathway;[198] adsorption of isobutane and of neopentane on Ir(110) has also been followed by this technique.[199]

Nickel and ruthenium 'model' catalysts have also been used. Adsorption of methane on small palladium particles supported on alumina is photo-assisted: UV radiation effects dissociation as well as desorption of physically-adsorbed molecules, the former process requiring larger particles than the latter.[200]

There have been a number of DFT calculations of C_1 species on metal surfaces (see Further Reading section), producing quantitative estimates of binding energies, but few surprises.

4.9. THE FINAL STAGE: CARBONACEOUS DEPOSITS[201–203]

This termed is frequently deployed in the literature to describe ill-defined and unwelcome substances deposited during hydrocarbon reactions on metals, leading to loss of catalytic activity. The work on hydrocarbon adsorption which we have reviewed has shown that, in almost every case, adsorbed species when heated decompose first to C_1 species (e.g. $=CH_2$ or $\equiv CH$), and that these then lose their hydrogen atoms to form carbon atoms, which on Ir(100) have been shown by LEED (and DFT) to reside in fourfold sites in a (2 × 2) overlayer. Further heating may lead to their polymerisation to graphite or to a diamond-like structure.[204] Carbon

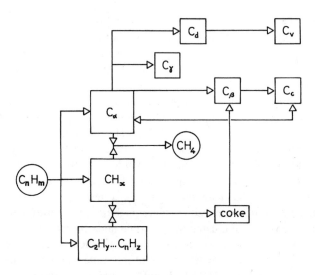

Figure 4.12. Forms of carbon involved in the decomposition of the hydrocarbon C_nH_m and modes of their transformation: C_α, carbidic (atomic) carbon; C_d; dissolved carbon; C_β, amorphous carbon; C_c, graphitic carbon; C_γ, metal carbide; C_v, filamentary carbon.

atoms on Pd(100) affected the reactivity of the surface towards other molecules such as ethene.[49]

A number of techniques have been used to examine the structure and re-activity of carbon deposits on technical catalysts (e.g. ^{13}C NMR[205]). Hydrogen treatment may convert surface carbon to methane if it is not too polymerised, but oxidation is usually more successful in removing it, although it may have other undesired consequences. Several different forms of 'carbon' are recognised (see Figure 4.12). Carbon atoms may dissolve into the surface and into the bulk, form-ing non-stoichiometric interstitial compounds,[11] or, especially with the base metals of Groups 8 to 10, stoichiometric carbides. These metals, particularly nickel, will catalyse the formation of filamentary whiskers if exposed continuously to a hy-drocarbon vapour at high temperature: they constitute carbon 'nanotubes' which have a buckminsterfullerene-type structure. Homologation of C_1 species to higher alkanes will be mentioned in Chapter 13.

REFERENCES

1. H. Hopf, *Classics in Hydrocarbon Chemistry*, Wiley-VCH: Chichester, 2000.
2. J.H. Sinfelt, *Surf. Sci.* **500** (2002) 923.
3. F. Zaera, *Surf. Sci.* **500** (2002) 947.
4. F. Zaera, *Chem. Rev.* **95** (1995) 2654.

5. B.E. Bent, *Chem. Rev.* **96** (1996) 1361.
6. V. Ponec and G.C. Bond, *Catalysis by Metals and Alloys*, Elsevier: Amsterdam (1995).
7. G.C. Bond, *Catalysis by Metals*, Academic Press: London (1962).
8. E. Yagasaki and R.I. Masel in: *Specialist Periodical Reports: Catalysis*, Vol. 11 (J.J. Spivey and S.K. Agarwal, eds.), *Roy. Soc. Chem.* (1994), p. 16.
9. N. Sheppard and C. de la Cruz, *Adv. Catal.* **42** (1998) 181.
10. G.A. Somorjai and K.R. McCrea, *Adv. Catal.* **45** (2000) 385.
11. Z. Karpiński, personal correspondence.
12. R.N. Reifsnyder and H.H. Lamb, *Catal. Lett.* **40** (1996) 155.
13. O. Beeck, A.E. Smith and A. Wheeler, *Proc. Roy. Soc.* **A177** (1941) 159.
14. O. Beeck and A.W. Ritchie, *Discuss. Faraday Soc.* **8** (1950) 159.
15. G.A. Somorjai, *J. Molec. Catal. A: Chem.* **107** (1996) 1.
16. N. Sheppard, *Ann. Rev. Phys. Chem.* **39** (1988) 589.
17. N. Sheppard and C. de la Cruz, *Adv. Catal.* **41** (1996) 1.
18. N. Sheppard and J. Erkelens, *Appl. Spectroscopy* **38** (1984) 471.
19. H.A. Pearce and N. Sheppard, *Surf. Sci.* **59** (1976) 205.
20. A.M. Bradshaw, *Surf. Sci.* **331–333** (1995) 978.
21. T. Gießel, R. Terborg, O. Schaff, R. Lindsay, P. Baumgärtel, J.T. Hoeft, K.-M. Schindler, S. Bao, A. Theobald, V. Fernandez, A.M. Bradshaw, D. Chrysostomou, T. Cabe, D.R. Lloyd, R. Davis, N.A. Booth and D.P. Woodruff, *Surf. Sci.* **440** (1999) 125.
22. J.-H. Kang, R.L. Toomes, J. Robinson, D.P. Woodruff, O. Schaff, R. Terburg, R. Lindsay, P. Baumgärtel and A.M Bradshaw, *Surf. Sci.* **448** (2000) 23.
23. W.A. Brown, R. Kose and D.A. King, *Chem. Rev.* **98** (1998) 797.
24. Q.-F. Ge, R. Kose and D.A. King, *Adv. Catal.* **45** (2000) 207.
25. N. Cardona-Martinez and J.A. Dumesic, *Adv. Catal.* **38** (1992) 149.
26. C.D. Makowka, C.P. Slichter and J.H. Sinfelt, *Phys. Rev.B* **31** (1985) 5663.
27. C.P. Slichter, *Ann. Rev. Phys. Chem.* **37** (1986) 25.
28. K. Sakaie, C.P. Slichter and J.H. Sinfelt, *J. Magnet. Res. A* **119** (1996) 235.
29. J.E. Demuth, H. Ibach and S. Lehwald, *Phys. Rev. Lett.* **40** (1978) 1044.
30. J.E. Demuth, *IBM J. Res. Develop.* **22** (1978) 265.
31. J.E. Demuth, *Phys. Rev. Lett.* **40** (1978) 409.
32. J.E. Demuth and D.E. Eastman, *J. Vac. Sci. Technol.* **13** (1976) 283.
33. T.E. Fischer, S.R. Kelemen and H.P. Bonzel, *Surf. Sci.* **64** (1997) 157.
34. J.E. Demuth, *Chem. Phys. Lett.* **45** (1977) 12.
35. J.E.Demuth and D.E. Eastman, *Phys. Rev. B* **13** (1976) 1523.
36. J.E. Demuth, *Proc. 7th. Internat. Vac. Congr.* and *3rd. Internat. Conf. Solid Surfaces*, Vienna (1977), p. 779.
37. W. Ranke and W. Weiss, *Surf. Sci.* **465** (2000) 317.
38. R. Mason, M. Textor, Y. Iwasawa and I.D. Gay, *Proc. Roy. Soc. A* **354** (1977) 171.
39. I.D. Gay, M. Textor, R. Mason and Y. Iwasawa, *Proc. Roy. Soc. A* **356** (1977) 25.
40. G. Brodén, T. Rhodin and W. Capehart, *Surf. Sci.* **61** (1976) 143.
41. D.R. Lloyd, C.M. Quinn and N.V. Richardson, *Solid State Comm.* **23** (1977) 141.
42. J. Grewe, Ü. Erfürk and A. Otto, *Langmuir* **14** (1998) 696.
43. W. Krasser and A.J. Renouprez, *Solid State Commun.* **41** (1982) 231.
44. R.L. Burwell Jr., *Chem. Eng. News* **44** (1966) 56.
45. E.L. Muetterties, T.N. Rhodin, E. Band, C.F. Brucker and W.R. Pretzer, *Chem. Rev.* **79** (1979) 91.
46. S. Carrà and R. Ugo, *Inorg. Chim. Acta Rev.* **1** (1967) 49.
47. C.E. Anson, N. Sheppard, D.B. Powell, B.R. Bender and J.R. Norton, *J. Chem. Soc. Faraday Trans.* **90** (1994) 1449.

48. C.E. Anson, N. Sheppard, D.B. Powell, J.R. Norton, W. Fischer, R.L. Keiter, B.F.G. Johnson, J. Lewis, A.K. Bhattcharrya, S.A.R. Knox and M.L. Turner, *J. Am. Chem. Soc.* **116** (1994) 3058.
49. I. Ratajczykowa, *Surf. Sci.* **152/153** (1985) 627.
50. L.L. Kesmodel, R.C. Baetzold and G.A. Somorjai, *Surf. Sci.* **66** (1977) 299.
51. N. Sheppard and C. de la Cruz, *Catal. Today* **70** (2001) 3.
52. J. Evans and G.S. McNulty, *J. Chem. Soc. Dalton Trans.* (1983) 639.
53. J. Evans and G.S. McNulty, *J. Chem. Soc. Faraday Trans.* **80** (1984) 79.
54. C.E. Anson, B.F.G. Johnson, J. Lewis, D.B. Powell, N. Sheppard, A.K. Bhattacharya, B.R. Bender, R.M. Bullock, R.T. Hembre and J.R. Norton, *J. Chem. Soc. Chem. Comm.* (1989) 703.
55. F. Zaera and N. Bernstein, *J. Am. Chem. Soc.* **116** (1994) 4881.
56. F. Solymosi, *Catal. Today* **28** (1996) 191.
57. K.C. McGee, M.D. Driessen and V.H. Grassian, *J. Catal.* **157** (1995) 730.
58. G.C. Bond, *Discuss. Faraday Soc.* **41** (1966) 200.
59. G.A. Ozin, *Acc. Chem. Res.* **10** (1977) 21.
60. R.M. Atkins, R. Mackenzie, P.L. Timms and T.W. Turney, *J. Chem. Soc. Chem. Comm.* (1975) 764.
61. R. Ugo, in: *Proc. 5th. Internat. Congr. Catal.* (J.W. Hightower, ed.), North Holland: Amsterdam (1972) 19.
62. T.E. Felter and W.H. Weinberg, *Surf. Sci.* **103** (1981) 265.
63. M.A. Chesters, C. De La Cruz, P. Gardner, E.M. McCash, J.D. Prentice and N. Sheppard, *J. Elec. Spec. Rel. Phen.* **54/55** (1990) 739.
64. N.N. Greenwood and A. Earnshaw, *Chemistry of the Elements*, 2nd edition, Butterworth-Heinemann: Oxford (1997).
65. F.A. Cotton and G. Wilkinson, *Advanced Inorganic Chemistry*, 4th edition, Wiley: Chichester (1980).
66. G.C. Bond, *J. Molec. Catal. A: Chem.* **156** (2000) 1.
67. G.C. Bond, *Platinum Metals Rev.* **44** (2000) 146.
68. Y.-T. Wong and R. Hoffmann, *J. Chem. Soc. Faraday Trans.* **86** (1990) 4083.
69. A. Cassuto, Mane Mane, J. Jupille, G. Tourillon and P. Parent, *J. Phys. Chem.* **96** (1992) 5987.
70. J.-F. Fan and M. Trenary, *Langmuir* **10** (1994) 3649.
71. E.M. Stuve and R.J. Madix, *J. Phys. Chem.* **89** (1985) 105.
72. B.E. Bent, C.M. Mate, C.-T. Kao, A.J. Slavin and G.A. Somorjai, *J. Phys. Chem.* **92** (1988) 4720.
73. G.H. Hatzikos and R.I. Masel, *Surf. Sci.* **185** (1987) 479.
74. C. Egawa, *Surf. Sci.* **454–456** (2000) 222.
75. S.B. Moshin, M. Trenary and H.J. Robota, *J. Phys. Chem.* **92** (1988) 5229.
76. R. Raval, *Surf. Sci.* **331–333** (1995) 1.
77. B. Krüger and C. Benndorf, *Surf. Sci.* **170** (1976) 704.
78. D. Stacchiola, G. Wu, M. Kaltchev and W.T. Tysoe, *Surf. Sci.* **486** (2001) 9.
79. M. Nishijima, J. Yoshinobu, T. Sekitani and M. Onchi, *J. Chem. Phys.* **90** (1989) 5114.
80. L. Hammer, T. Hertlein and K. Müller, *Surf. Sci.* **178** (1976) 693.
81. E. Cooper and R. Raval, *Surf. Sci.* **331–333** (1995) 94.
82. D. Stacchiola, L. Burkholder and W.T. Tysoe, *Surf. Sci.* **511** (2002) 215.
83. D. Stacchiola, S. Azad, L. Burkholder and W.T. Tysoe, *J. Phys. Chem.* **105** (2001) 11233.
84. E. Yagasaki, A.L. Blackman and R. I. Masel, *J. Phys. Chem.* **94** (1990) 1066.
85. Sh. Shaikhutdinov, M. Heemeir, M. Baumer, T. Lear, D. Lennon, R.J. Oldman, S.D. Jackson, and H.-J. Freund, *J. Catal.* **200** (2001) 330.
86. N. Sheppard, N.R. Avery, B.A. Morrow and R.P. Young, in: *Chemisorption and Catalysis*, Institute of Petroleum: London (1971), p. 135.
87. J.D. Prentice, A. Lesiunas and N. Sheppard, *J. Chem. Soc. Chem. Comm.* (1976) 76.

88. M.K. Ainsworth, M.R.S. McCoustra, M.A. Chesters, N. Sheppard and C. de la Cruz, *Surf. Sci.* **437** (1999) 9.

89. H. Steiniger, H. Ibach and S. Lehwald, *Surf. Sci.* **117** (1982) 685.

90. J. Kubota, S. Ichihara, J.N. Kondo, K. Domen and C. Hirose, *Surf. Sci.* **337–338** (1996) 634; *langmuir* **12** (1996) 196.

91. R.D. Cortright and J.A. Dumesic, *J. Catal.* **157** (1995) 576.

92. R.G. Windham, M.E. Bartram and B.E. Koel, *J. Phys. Chem.* **92** (1988) 2862.

93. R.G. Windham, B.E. Koel and M.T. Paffet, *Langmuir* **4** (1988) 1113.

94. E. Yagasaki and R.I. Masel, *J. Am. Chem. Soc.* **112** (1990) 8746; *Surf. Sci.* **222** (1989) 430.

95. C.T. Campbell, *Ann. Rev. Phys. Chem.* **41** (1990) 775.

96. E.M. Stuve and R.J. Madix, *J. Phys. Chem.* **89** (1985) 3183.

97. M. Bowker, J.L. Gland, R.W. Joyner, X.-Y. Li, M.M. Slinko and R. Whyman, *Catal. Lett.* **25** (1994) 293.

98. J.C. Bertolini, A. Cassuto, Y. Jugnet, J. Massardier, B. Tardy and G. Tourillon, *Surf. Sci.* **349** (1996) 88.

99. G. Papoian, J.K Nørskov and R. Hoffmann, *J. Am. Chem. Soc.* **122** (2000) 4129.

100. Y.-L. Tsai, C. Xu and B.E. Koel, *Surf. Sci.* **385** (1997) 37.

101. J.-Y. Shen, J.M. Hill, R.M Watwe, B.E. Spiewak and J.A. Dumesic, *J. Phys. Chem. B* **103** (1999) 3923.

102. C. Becker, T. Pelster, M. Tanemura, J. Breitbach and K. Wandelt, *Surf. Sci.* **433–435** (1999) 822.

103. N.R.M. Sassen, A.J. den Hartog, F. Jongerius, J.F.M. Aarts and V. Ponec, *Faraday Discuss. Chem. Soc.* **87** (1989) 311.

104. G.A. Somorjai and G. Rupprechter, *J. Chem. Educ.* **75** (1998) 171.

105. G.A. Somorjai, *Ann. Rev. Phys. Chem.* **45** (1994) 721.

106. R.W. Joyner in: *Specialist Periodical Reports: Catalysis*, Vol. 5 (G.C. Bond and G. Webb, eds.), *Roy. Soc. Chem.* (1982), p. 1.

107. S. Titmuss, A. Wander and D.A. King, *Chem. Rev.* **96** (1996) 1291.

108. R.I. Masel, *Principles of Adsorption and Reaction on Solid Surfaces*, Wiley: New York (1996).

109. C. Kemball, *Catal. Rev.* **5** (1971) 33.

110. G. Webb in: *Specialist Periodical Reports: Catalysis*, Vol. 2 (C. Kemball and D.A. Dowden, eds.), *Roy. Soc. Chem.* (1978), p. 145.

111. G.A. Somorjai, *Surface Science: An Introduction*, Butterworth-Heinemann: New York (1992).

112. M.A. Van Hove and G.A. Somorjai, *J. Molec. Catal. A: Chem.* **131** (1998) 243.

113. R. Döll, C.A. Gerken, M.A. Van Hove and G.A. Somorjai, *Surf. Sci.* **374** (1997) 151.

114. H. Ibach, H. Hopster and B. Sexton, *Appl. Phys.* **14** (1977) 21.

115. R.C. Pitkethly in: *Chemisorption and Catalysis*, Inst. Petroleum: London (1971), p.98.

116. A.B. Anderson, *J. Chem. Phys.* **65** (1976) 1729.

117. A.B. Anderson, *J. Am. Chem. Soc.* **100** (1978) 1153.

118. N. Sheppard and J.W. Ward, *J. Catal.* **15** (1969) 50.

119. R.M. Watwe, B.E. Spiewak, R.D. Cortright and J.A. Dumesic, *J. Catal.* **180** (1998) 184.

120. S. Bao, Ph. Hofmann, K.-M. Schindler, V. Fritzsche, A.M. Bradshaw, D.P. Woodruff, C. Casado and M.C. Ascencio, *Surf. Sci.* **307–309** (1994) 722.

121. C.J. Baddeley, A.F. Lee, R.M. Lambert, T. Gießel, O. Schaff, V. Fernandez, K.-M. Schindler, A. Theobald, C.J. Hirschmugl, R. Lindsay, A.M. Bradshaw and D.P. Woodruff, *Surf. Sci.* **400** (1998) 166.

122. S. Bao, K.-M. Schindler, Ph. Hoffman, V. Fritsche, A.M. Bradshaw, and D.P. Woodruff, *Surf. Sci.* **291** (1993) 295.

123. Chen Xu, J.W. Peck and B.E. Koel, *J. Am. Chem. Soc.* **115** (1993) 751.

124. S. Bao, R. Lindsay, M. Polcik, A. Theobald, T. Giessel, O. Schaff, P. Baumgärtel, A.M. Bradshaw, R. Terborg, N.A. Booth and D.P. Woodruff, *Surf. Sci.* **478** (2001) 35.
125. F.S. Thomas, N.S. Chen. L.P. Ford and R.I. Masel, *Surf. Sci.* **486** (2001) 1.
126. N.C. Chen, L.P. Ford and R.J. Masel, *Catal. Lett.* **56** (1998) 105.
127. G. Held, M.P. Bessent, S. Titmuss and D.A. King, *J. Chem. Phys.* **104** (1996) 11305.
128. D. Schaff, V. Fernandez, Ph. Hofmann, K.-M. Schindler, A. Theobald, V. Fritsche, A.M. Bradshaw, R. Davis and D.P. Woodruff, *Surf. Sci.* **348** (1996) 89.
129. J. Winterlin, *Adv. Catal.* **45** (2000) 131.
130. S. Chang, *Chem. Rev.* **97** (1997) 1083.
131. L. Vattone, Y.Y. Yeo, R. Kose, and D.A. King, *Surf. Sci.* **447** (2000) 1.
132. R. Kose, W.A. Brown and D.A. King, *Chem. Phys. Lett.* **311** (1999) 109.
133. W.A. Brown, R. Kose and D.A. King, *J. Molec. Catal. A: Chem.* **141** (1999) 21.
134. B.E. Spiewak, R.D. Cortright and J.A. Dumesic, *J. Catal.* **176** (1998) 405.
135. M.A. Natal-Santiago, S.G. Podkolzin, R.D. Cortright, and J.A. Dumesic, *Catal. Lett.* **45** (1997) 155.
136. F.B. Passos, M. Schmal and M.A. Vannice, *J. Catal.* **160** (1996) 118.
137. J.-Y. Shen, J.M. Hill, R.M. Watwe, S.G. Podkolzin and J.A. Dumesic, *Catal. Lett.* **60** (1999) 1.
138. M.A. Chesters and G.A. Somorjai, *Surf. Sci.* **52** (1975) 21.
139. K.A. Davis and D.W. Goodman, *J. Phys. Chem. B* **104** (2000) 8557.
140. A. Guerrero, M. Reading, Y. Grillet, J. Rouquerol, J.P. Boitiaux and J. Cosyns, *Z. Phys. D* **12** (1989) 583.
141. S. Pálfi, W. Lisowski, M. Smutek and S. Černý, *J. Catal.* **88** (1984) 300.
142. Z. Paál, E. Fülöp and D. Marton, *React. Kinet. Catal. Lett.* **38** (1989) 131.
143. Po-Kang Wang, C.P. Slichter and J.H. Sinfelt, *Phys. Rev. Lett.* **53** (1984) 82.
144. G.A. Martin, *Catal. Rev.-Sci. Eng.* **30** (1988) 518.
145. P.W. Selwood, *J. Am. Chem. Soc.* **79** (1957) 3346.
146. G.A. Somorjai, *Adv. Catal.* **26** (1977) 2.
147. A. Bao, Ph. Hofmann, K.-M. Schindler, V. Fritsche, A.M. Bradshaw, C. Casado and M.C. Asensio, *J. Phys. Condens. Matter* **6** (1994) L93.
148. E. Yagasaki and R.I. Masel, *J. Am. Chem. Soc.* **112** (1990) 8746.
149. L.P. Ford, H.L. Nigg, P. Blowers and R.I. Masel, *J. Catal.* **178** (1998) 163.
150. L.P. Ford, P. Blowers and R.I. Masel, *J. Vac. Sci. Technol. A* **17** (1999) 1705.
151. C.-H. Hwang, C.-W. Lee, H. Kang and C.M. Kin, *Surf. Sci.* **490** (2001) 144.
152. G.A. Somorjai, *Catal. Lett.* **12** (1992) 17.
153. U. Starke, A. Barbieri, N. Materer, M.A. Van Hove and G.A. Somorjai, *Surf. Sci.* **286** (1993) 1.
154. Po-Kang Wang, C.P Slichter and J.H. Sinfelt, *Phys. Rev. Lett.* **55** (1985) 2731.
155. D. Stacchiola, H. Molero and W.T. Tysoe, *Catal. Today* **65** (2001) 3.
156. T.P. Beebe Jr., M.R. Albert, and J.T. Yates Jr., *J. Catal.* **96** (1985) 1.
157. G. Blyholder and L. Orji, *Ads. Sci. Technol.* **4** (1987) 1.
158. N. Sheppard, personal correspondence.
159. P.R. Marshall, G.S. McDougall and R.A. Hadden, *Topics in Catal.* **1** (1994) 9.
160. I. Jungwirthová and L.L. Kesmodel, *Surf. Sci.* **470** (2000) L39.
161. P.A.P. Nascente, M.A. Van Hove and G.A. Somorjai, *Surf. Sci.* **253** (1991) 167.
162. J.A.Gates and L.L. Kesmodel, *Surf. Sci.* **124** (1983) 68.
163. L.L. Kesmodel, G.D. Waddill and J.A. Gates, *Surf. Sci.* **136** (1984) 464.
164. N.R.M. Sassen, A.J. den Hartog, F. Jongerius, J.F.M. Aars and V. Ponec, *Faraday Discuss. Chem. Soc.* **87** (1989) 311.
165. A. Clark, *The Chemisorption Bond*, Academic Press: New York (1974).
166. P.J. Hiett, F. Flores, P.J. Grout, N.H. March, A Martino-Rodero and G. Senatore, *Surf. Sci.* **140** (1984) 400.

167. A.B. Anderson and R. Hoffmann, *J. Chem. Phys.* **61** (1974) 4545; A.B. Anderson, *J. Am. Chem. Soc.* **99** (1977) 696.
168. R.A. van Santen and M. Neurock, *Cat. Rev.-Sci. Eng.* **37** (1995) 557.
169. G.W. Watson, R.P.K. Wells, D.J. Willock and G.J. Hutchings, *J. Phys. Chem. B* **104** (2000) 93; *Surf. Sci.* **459** (2000) 93.
170. M. Chen, C.M. Friend and R.A. van Santen, *Catal. Today* **50** (1999) 621.
171. J.L. Whitten and H. Yang, *Catal. Today* **50** (1999) 603.
172. F. Mittendorfer and J. Hafner, *Surf. Sci.* **472** (2001) 133.
173. J. Erkelens and L.P. van't Hof, *J. Catal.* **8** (1967) 103.
174. G.C. Bond, *Platinum Metals Rev.* **10** (1966) 87.
175. R.L. Augustine, *Heterogeneous Catalysis for the Synthetic Chemist*, Dekker: New York (1996).
176. E.W. Müller and Z. Knor, *Surf. Sci.* **10** (1968) 21.
177. D.S. Sholl, *Langmuir* **14** (1998) 862.
178. G.C. Bond, *J. Molec. Catal. A: Chem.* **156** (2000) 1.
179. G.C. Bond and E.L. Short, *Chem. and Ind.* (2002, 3rd June), p. 12.
180. Li Jian, R.M Dickson and T. Ziegler, *J. Am. Chem. Soc.* **117** (1995) 11483.
181. J. Li, G. Schreckenback and T. Ziegler, *Inorg. Chem.* **34** (1995) 3245.
182. A. Frennet in *Hydrogen Effects in Catalysis* (Z. Paál and P.G. Menon, eds.), Dekker: New York (1988), p. 399.
183. W.H. Weinberg, *Langmuir* **9** (1993) 655.
184. A. Frennet, *Catal. Rev.* **10** (1974–5) 37.
185. A. Frennet, *Catal. Today* **12** (1992) 131.
186. A. Frennet, G. Lienard, A. Crucq and L. Degols, *J. Catal.* **53** (1978) 130.
187. J.F. Weaver, M. Ikai, A. Carlsson and R.J. Madix, *Surf. Sci.* **470** (2001) 226.
188. G.-A. Martin, J.A. Dalmon and C. Mirodatos, *Bull. Soc. Chim. Belg.* **88** (1979) 559.
189. D.J. Oakes, H.E. Newall, F.J.M. Rutten, M.R.S. McCoustra and M.A. Chesters, *Chem. Phys. Lett.* **253** (1996) 123.
190. F.C. Schouten, O.L.J. Gijzeman and G.A. Bootsma, *Bull. Soc. Chim. Belg.* **88** (1979) 541.
191. J.M. Derochette, *Bull. Soc. Chim. Belg.* **88** (1979) 549.
192. F.C. Schouten, E.W. Kaleveld and G.A. Bootsma, *Surf. Sci.* **63** (1977) 460.
193. K. Klier, J.S. Hess and R.G. Hermann, *J. Chem. Phys.* **107** (1997) 4033.
194. R.C. Egeberg, S. Ullmann, I. Alstrup, C.B. Mullins and F. Chorkendorff, *Surf. Sci.* **497** (2002) 183.
195. S. Paavvilaien and J.A. Nieminen, *Surf. Sci.* **486** (2001) L489.
196. R.W. Verhoef, D. Kelley, C.B. Mullins and W.H. Weinberg, *Surf. Sci.* **291** (1993) L719.
197. H.E. Newell, D.J. Oakes, F.J.M. Rutten, M.R.S. McCoustra and M.A. Chesters, *Faraday Discuss. Chem. Soc.* **105** (1996) 193.
198. C.T. Reeves, D.C. Seets and C.B. Mullins, *J. Molec. Catal. A: Chem.* **167** (2000) 207.
199. Jungqi Ding, V. Burghaus and W. H. Weinberg, *Surf. Sci.* **446** (2000) 46.
200. H.-J. Freund, M, Bäumer and H. Kuhlenbeck, *Adv. Catal.* **45** (2000) 334.
201. J.L. Whitten and H.Yang, *Surf. Sci. Rep.* **218** (1996) 55.
202. A.T. Bell in: *Structure and Reactivity of Surfaces*, (C. Morterra, A. Zecchina and G. Costa, eds.), Studies in Surface Science and Catalysis', Elsevier: Amsterdam, **48** (1989) 91.
203. G.C. Bond, *Appl. Catal. A: Gen.* **149** (1997) 3.
204. R.I. Kvon, S.K. Koscheev and A.I. Boronin, *J. Molec. Catal. A: Chem.* **158** (2000) 297.
205. M. Pruski, J. Kelzenberg, B.C. Gostein and T.S. King, *J. Am. Chem. Soc.* **112** (1990) 4232.
206. Chen Xu and B.E. Koel, *Surf. Sci.* **304** (1994) 249.
207. F. Zaera, T.V.W. Janssens and H. Öfner, *Surf. Sci.* **368** (1996) 371.
208. A. Guerrero, M. Reading, Y. Grillet, J. Rouquerol, J.P. Boitiaux and J. Cosyns, *Z. Phys. D* **12** (1989) 583.

209. P.D. Holmes, G.S. McDougall, I.C. Wilcock and K.C. Waugh, *Catal. Today* **9** (1991) 15.
210. H. Ibach and S. Lehwald, *J. Vac. Sci. Technol.* **15** (1978) 407.
211. J.-C. Bertolini, J. Massardier and G. Dalmai-Imelik, *J. Chem. Soc. Faraday Trans. I* **74** (1978) 1720.
212. G.A. Somorjai, *Cat. Tech.* (5) (1999) 84.
213. G.A. Somorjai, Zingcai Su, K.R. McCrea and K.B. Rider, *Topics in Catal.* 8 **(1999) 23.**
214. P. S. Cremer, X. Su, G.A. Somorjai and Y.R. Chen, *J. Molec. Catal. A: Chem.* **131** (1998) 225.
215. P.S. Cremer, X.-C. Su, Y.R. Shen and G.A. Somorjai, *J. Phys. Chem. B* **101** (1997) 6474.
216. P. Cremer, C. Stanners, J.W. Niemantsverdriet, Y.R. Shen and G. Somorjai, *Surf. Sci.* **328** (1995) 111.
217. K.R. McCrae and G.A. Somorjai, *J. Molec. Catal. A: Chem.* **163** (2000) 43.
218. X.-C. Su, Y.R. Shen and G.A. Somorjai, *Chem. Phys. Lett.* **280** (1997) 302.
219. C. Egawa, H. Iwai and S. Oki, *Surf. Sci.* **454–456** (2000) 347.
220. D. Arvantidis, U. Döbler, L. Wezel and K. Bakerschke, *Surf. Sci.* **178** (1976) 686.
221. F. Zaera, D.A. Fisher, R.A. Carr and J.L. Gland, *J.Chem. Phys.* **89** (1988) 5335.
222. R.J. Koestner, J. Stohr, J.L. Gland and J.A. Horsley, *Chem. Phys. Lett.* **105** (1984) 332.
223. M.A. Chesters and N.D.S. Canning, *Vacuum* **31** (1981) 695; N.D.S. Canning, M.D. Baker and M.A. Chesters, *Surf. Sci.* **111** (1981) 441.
224. Z. Hlavathy and P. Tétényi, *Surf. Sci.* **410** (1998) 39.
225. T.E. Fischer, S. Kelemen and H.P. Bonzel, *J. Vac. Sci. Technol.* **14** (1977) 424.
226. Y. Nicolou, J.J. Ehrhardt, R. Ducros and A. Cassuto, *Japan. J. Appl. Phys., Suppl. 2*, Part 2 (1974) 521.
227. H.H. Farrell and E.S. Strauss, *J. Vac. Sci. Technol.* **14** (1977) 427.
228. Po-Kang Wang, C.P. Slichter and J.H. Sinfelt, *J. Phys. Chem.* **89** (1985) 3606.
229. I.D. Gay, *J. Catal.* **108** (1987) 15.
230. T. Shibanuma and T. Matsui, *Surf. Sci.* **154** (1985) 215.
231. P.K. Wang, C.P. Slichter and J.H. Sinfelt, *Phys. Rev. Lett.* **53** (1984) 82.
232. M.R. Albert, L.G. Sneddon, W. Eberhardt, F. Greuter, T. Gustavsson and E. W. Plummer, *Surf. Sci.* **120** (1982) 19.
233. J.P. Candy, H. Jobic and A.J. Renouprez, *J. Phys. Chem.* **87** (1983) 1227.
234. R.R. Kavangh, J.J. Rush, R.D. Kelley and T.J. Udovic, *J. Chem. Phys.* **80** (1984) 3478.
235. P.E.C. Franken and V. Ponec, *J. Catal.* **42** (1976) 398.
236. J.L. Gland and G.A. Somorjai, *Adv. Coll. Interface Sci.* **5** (1976) 205.
237. M. Abon, J. Billy and J.C. Bertolini, *Surf. Sci.* **171** (1986) 387.
238. G.-A. Martin and B. Imelik, *Surf. Sci.* **42** (1974) 157.
239. G.A. Martin, J.A. Dalmon and C. Mirodatos, *Bull. Soc. Chim. Belg.* **88** (1979) 559.
240. J.-A. Dalmon, G.-A. Martin and B. Imelik, *C.R. Acad. Sci. Paris, Ser. C* **278** (1974) 1481.
241. S. Tsuchiya and N. Yoshioka, *J. Catal.* **87** (1984) 144.
242. G. Szulczewski and R.J. Lewis, *J. Am. Chem. Soc.* **118** (1996) 3521.
243. R.L. Palmer, *J. Vac. Sci. Technol.* **12** (1975) 1403.
244. R.A. Zuhr and J.B. Hudson, *J. Vac. Sci. Technol.* **14** (1977) 431.
245. B.J. McIntyre, M. Salmeron and G.A. Somorjai, *J. Catal.* **164** (1996) 184.
246. M. Doering, J. Buisset, H.-P. Rust, B.G. Briner and A.M. Bradshaw, *Faraday Discuss. Chem. Soc.* **105** (1996) 163.
247. M. Doering, H.-P. Rust, B.G. Briner and A.M. Bradshaw, *Surf. Sci.* **410** (1998) L736.
248. K.C. Campbell and S.J. Thomson, *Prog. Surf. Membrane Sci.* **9** (1975) 163.
249. S.D. Jackson, B.M. Glanville, J. Willis, G.D. McLellan, G. Webb, R.B. Moyes, S.Simpson, P.B. Wells and R. Whyman, *J. Catal.* **139** (1993) 221.
250. S.M. Davis, F. Zaera, B.E. Gordon and G.A. Somorjai, *J. Catal.* **92** (1985) 240.
251. M. Barber, J.C. Vickerman and J. Wolstenholme, *J. Catal.* **42** (1976) 48.

252. N. Sheppard, D.I. James, A. Lesiunas and J.D. Prentice, *Izv. Khim.* **17** (1984) 95; *Commun. Dept. Chem. Bulgarian Acad. Sci.* **17** (1984) 95 (in English).
253. J. Ryczkowski, *Catal. Today* **68** (2001) 263.
254. *Spectroscopy for Surface Science* (R.J.H. Clark, ed.), Wiley: Chichester (1998).
255. C. De La Cruz and N. Sheppard, *J. Chem. Soc. Chem. Comm.* (1987) 1854.
256. C. De La Cruz and N. Sheppard, *J. Mol. Struct.* **247** (1991) 25.
257. C. de la Cruz and N. Sheppard, *J. Catal.* **127** (1991) 445.
258. C. de la Cruz and N. Sheppard, *J. Chem. Soc. Faraday Trans.* **93** (1997) 3569.
259. M. Primet and N. Sheppard, *J. Catal.* **41** (1976) 258.
260. B.A. Morrow and N. Sheppard, *Proc. Roy. Soc. A* **311** (1969) 391, 415.
261. R.P. Eischens, *Z. Elektrochem.* **70** (1956) 782.
262. A.J. Roberts, S. Haq and R. Raval, *J. Chem. Soc. Faraday Trans.* **92** (1996) 4823.
263. M.A. Chesters, C. De La Cruz, P. Gardner, E.M. McCash, P. Pudney, G. Shahid, and N. Sheppard, *J. Chem. Soc. Faraday Trans.* **86** (1990) 2757.
264. G. Shahid and N. Sheppard, *Canad. J. Chem.* **69** (1991) 1812.
265. G. Shahid and N. Sheppard, *Spectrochim. Acta* **46A** (1990) 999.
266. G. Shahid and N. Sheppard, *J. Chem. Soc. Faraday Trans.* **90** (1994) 507, 513.
267. N.R. Avery and N. Sheppard, *Proc. Roy. Soc. A* **405** (1986) 1, 27.
268. R. Kleyna, D. Borgmann and G. Wedler, *Surf. Sci.* **433–435** (1999) 205.
269. A. Ravi and N. Sheppard, *J. Phys. Chem.* **76** (1972) 2699; *J. Catal.* **22** (1971) 389.
270. R.D. Cortright, E. Bergene, P. Levin, M. Natal-Santiago and J.A. Dumesic, *Proc. 11th. Internat. Congr. Catal.*, (J.W. Hightower, W.N. Delgass, E. Iglesia and A.T. Bell, eds.), Elsevier: Amsterdam **B** (1996) 1185.
271. L.M. Ilharco, A.R. Garcia, E.C. Hargreaves and M.A. Chesters, *Surf. Sci.* **459** (2000) 115.
273. R.B. Moyes and P.B. Wells, *Adv. Catal.* **23** (1973) 121.
274. S. Gönenç and Z.I. Önsan in: *Recent Advances in Chemical Engineering* (D.N. Saraf and D. Kunzru, eds.), Tata McGraw-Hill: New Delhi (1989), p. 266.
275. M. Primet and M.-V. Mathieu, *J. Chim. Phys.* (1975) 30.
276. J.C. Bertolini, G. Dalmai-Imelik and J. Rousseau, *Surf. Sci.* **67** (1977) 478.
277. R. Duschek, F. Mittendorfer, R.I.R. Blyth, F.P. Netzer, J. Hafner and M.G. Ramsay, *Chem. Phys. Lett.* **318** (2000) 43.
278. W. Braun, G. Held, H.-P. Steinrück, C. Stellwag and D. Menzel, *Surf. Sci.* **475** (2001) 18.
279. D.A. King, Rideal Conference, Manchester, 2002.
280. K. Pussi, M. Lindroos and C.J. Barnes, *Chem. Phys. Lett.* **341** (2001) 7.
281. M.C. Tasi and E.L. Muetterties, *J. Phys. Chem.* **86** (1982) 5067.
282. A. Wander, G. Held, R.Q. Hwang, G.S. Blackman, M.L. Xu, P. de Andres, M.A. Van Hove and G.A. Somorjai, *Surf. Sci.* **249** (1991) 21.
283. S. Tjandra and F. Zaera, *J. Catal.* **164** (1996) 82.
284. A. Sárkány, *J. Chem. Soc. Faraday Trans. I* **84** (1988) 2267.
285. N. Vasquez Jr. and R.J. Madix, *J. Catal.* **178** (1998) 234.
286. D.E. Hunka, T. Picciotto, D.M. Jaramillo and D.P. Land, *Surf. Sci.* **421** (1999) L166.
287. J.M. Davidsen, F.C. Henn, G.K. Rowe and C.T. Campbell, *J. Phys. Chem.* **95** (1991) 6632.
288. F.C. Henn, *J. Phys. Chem.* **96** (1992) 5965.
289. D.P. Land, W. Erley and H. Ibach, *Surf. Sci.* **283** (1993) 236.
290. X. Su, K. McCrea and G.A. Somorjai, *Chem. Phys. Lett.* **280** (1997) 302.
291. X. Su, Y, R. Shen, K. Rider and G.A. Somorjai, *Topics in Catal.* **8** (1999) 23.
292. G.A. Somorjai, *Chem. Rev.* **96** (1996) 1223.
293. F. Zaera, *Prog. Surf. Sci.* **69** (2001) 1; *Acc. Chem. Res.* **35** (2002) 125.
294. C. Minot, M.A. Van Hove and G.A. Somorjai, *Surf. Sci.* **127** (1983) 441.
295. G.H. Hatzikos and R.I. Masel, *Surf. Sci.* **185** (1987) 479.

296. G.H. Hatzikos and R.I. Masel in: *Catalysis 1987* (J.W. Ward, ed.), Studies in Surface Science and Catalysis', Elsevier: Amsterdam, **38** (1987) 895.

297. H. Öfner and F. Zaera, *J. Phys. Chem. B* **101** (1997) 396.

298. T.V.W. Janssens and F. Zaera, *Surf. Sci.* **344** (1995) 77.

299. D.R. Lloyd and F.P. Netzer, *Surf. Sci.* **129** (1983) L249.

300. D. Stacchiola and W.T. Tysoe, *Surf. Sci.* **513** (2002) L431.

301. G.A. Somorjai, *Phys. Rev. Lett.* **67** (1991) 626.

302. R.J Koestner, M.A. Van Hove and G.A. Somorjai, *Surf. Sci.* **121** (1982) 321.

303. R.M. Watuwe, H.S. Bengaard, J.R. Rostrop-Nielsen, J.A. Dumesic and J.K. Nørskov, *J. Catal.* **189** (2000) 16.

304. A. Michaelides and P. Hu, *Surf. Sci.* **437** (1999) 362.

305. J.-F. Paul and P. Sautet, *Proc. 11th Internat. Congr. Catal.*, (J.W. Hightower, W.N. Delgass, E. Iglesia and A.T. Bell, eds.), Elsevier: Amsterdam **B** (1996) 1253.

306. V. Pallassana and M. Neurock, *J. Catal.* **191** (2000) 301.

307. Q. Ge and D.A. King, *J. Chem. Phys.* **110** (1999) 4699.

308. K. Johnson, Q. Ge, B. Sauerhammer, S. Titmuss and D.A. King, *Surf. Sci.* **478** (2001) 49.

309. D.J. Klinke II, S. Wilke and L.J. Broadbent, *J. Catal.* **178** (1998) 540.

310. Y.Y. Yeo, A. Stuck. C.E. Wartnaby, R. Kose and D.A. King, *J. Molec. Catal. A: Chem.* **131** (1998) 31.

311. W.A. Brown, R. Kose and D.A. King, *Surf. Sci.* **440** (1999) 271.

312. R. Kose, W.A. Brown and D.A. King, *J. Am. Chem. Soc.* **121** (1999) 4845.

313. Y.Y. Yeo, A. Stuck, C.E. Wartnaby and D.A. King, *Chem. Phys. Lett.* **259** (1996) 28.

314. L. Vattuone, Y.Y. Yeo, R. Kose and D.A. King, *Surf. Sci.* **447** (2000) 1.

315. F. Zaera, *Langmuir* **12** (1996) 88.

316. H. Ibach, M. Balden and S. Lehwald, *J. Chem. Soc. Faraday Trans.* **92** (1996) 4771.

317. I. Jungwirthová and L.L. Kesmodel, *Surf. Sci.* **470** (2001) L39.

318. B.E. Bent, C.M. Mate, J.E. Crowell, B.E. Koel and G.A. Somorjai, *J. Phys. Chem.* **91** (1987) 3249.

319. F. Zaera and R.B. Hall, *J. Phys. Chem.* **91** (1987) 4318.

320. B.E. Koel, J.E. Crowell, C.M. Mate and G.A. Somorjai, *J. Phys. Chem.* **88** (1984) 1988.

321. P. Légaré and G. Maire, *J. Chim. Phys.* (1971) 1206.

322. N. Freyer, G. Pirug and H.P. Bonzel, *Surf. Sci.* **126** (1983) 487.

323. E. Janin, S. Ringler, J. Weissenrieder, T. Åkerman, U.O. Karlsson, M. Göthelid, D. Nordlund and H. Ogawasara, *Surf. Sci.* **482** (2001) 83.

324. T. Tysoe, G.L. Nyberg and R.M. Lambert, *Surf. Sci.* **135** (1988) 128.

325. W.T. Tysoe, G.L. Nyberg and R,M. Lambert, *J. Phys. Chem.* **88** (1984) 1960.

326. Z. Hlavathy and P. Tétényi, *React. Kinet. Catal. Lett.* **62** (1997) 163.

327. J. Gaudioso, H.J. Lee and W. Ho, *J. Am. Chem. Soc.* **121** (1999) 8479.

328. M. Rønning, E. Bergene, A. Borg, S. Ausen and A. Holmen, *Surf. Sci.* **477** (2001) 191.

329. F. Zaera and D. Chrysostomou, *Surf. Sci.* **457** (2000) 71, 89.

330. T.V.W. Janssens, D. Stone, J.C. Hemminger and F. Zaera, *J. Catal.* **177** (1998) 284.

331. W. Akemann and A. Otto, *Langmuir* **11** (1995) 1196.

332. 322. G. Blyholder, D. Shikabi, W.V. Wyatt and R. Bartlett, *J. Catal.* **43** (1976) 122.

333. K. Föger and H.L. Gruber, *Naturwiss.* **11** (1975) 528.

334. D.B. Powell, J.G.V Scott and N. Sheppard, *Spectrochim. Acta* **28A** (1972) 327.

335. S.B. Moshin, M. Trenary and H.J. Robota, *J. Phys. Chem.* **92** (1988) 5229.

336. S.D. Jackson and N. Casey, *React. Kinet. Catal. Lett.* **49** (1993) 231.

337. S.D. Jackson, N. Hussain, and S. Munro, *J. Chem. Soc. Faraday Trans.* **94** (1998) 955.

338. B.J. Bandy, M.A. Chesters, D.I. James, G.S. McDougall, M.E. Pemble and N. Sheppard, *Phil. Trans. Roy. Soc. London A* **318** (1986) 141.

339. C. de la Cruz and N. Sheppard, *Phys. Chem. Chem. Phys.* **1** (1999) 329.
340. A.P. Graham, M.F. Bertino, F. Hoffmann and J.P. Toennies, *J. Chem. Soc. Faraday Trans.* **92** (1996) 4749.
341. W. Erley, A.M. Baro and H. Ibach, *Surf. Sci.* **120** (1982) 273.
342. F. Zaera, *J. Am. Chem. Soc.* **111** (1989) 4240.
343. Q.-F. Ge and D.A. King, *J. Chem, Phys.* **110** (1999) 4699.
344. P. Chen, K.Y. Kung, Y.R. Shea and G.A. Somorjai, *Surf. Sci.* **494** (2001) 259.
345. K. Matusek and P. Tétényi, *React. Kinet. Catal. Lett.* **62** (1997) 171.
346. K.L. Street, V. Fiorin, M.R.S. McCoustra and M.A. Chesters, *Surf. Sci.* **433–435** (1999) 176.
347. M.J. Weiss, C.J. Hagedorn and W.H. Weinberg, *Surf. Sci.* **426** (1999) 154.
348. S. Roke, J.M Coquel and A.W. Kleyn, *Chem. Phys. Lett.* **323** (2000) 201.
349. T.V. Choudry and D. W. Goodman, *J. Molec. Catal. A: Chem.* **163** (2000) 9.
350. Chen Xu, B.E. Koel and M.T. Paffett, *Langmuir* **10** (1994) 166.
351. X.-C. Su, K.Y. Kung, L. Lahtinen, Y.R. Shea and G.A. Somorjai, *J. Molec. Catal. A: Chem.* **141** (1999) 9.
352. D. Stacchiola, M. Kaltchev, G. Wu and W.T. Tysoe, *Surf. Sci.* **470** (2000) L32.
353. G.M. Bernard and R.E. Wasylishen in: *Unusual Structures and Physical Properties in Organometallic Chemistry* (M. Gielen, R. Willem and B. Wrackmeyer, eds.), Wiley: Chichester (2002).
354. M. Neurock and R.A. van Santen, *J. Phys. Chem. B* **104** (2000) 11127.
355. D. Stacchiola, A.W. Thompson and W. Tysoe, *Surf. Sci.* **391** (1997) 145.
356. H. Hoffmann, F. Zaera, R.M. Ormerod, R.M. Lambert, L.P. Wang and W.T. Tysoe, *Surf. Sci.* **268** (1992) 1.
357. L.P. Wang, W.T. Tysoe, R.M. Ormerod, R.M. Lambert, H. Hoffmann and Z. Zaera, *J. Phys. Chem.* **94** (1990) 4236.
358. M.A. Doyle, Sh.K. Shaikhutdinov and H.-J. Freund, *J. Catal.* **223** (2004) 444.
359. H. Öström, A. Fölisch, M. Nyberg, M. Weinelt, C. Heske, L.G.M. Petterson and A. Nilsson, *Surf. Sci.* **559** (2004) 85.

FURTHER READING

1 **Use of vibrational spectroscopy and related techniques** 16, 17, 51, 76, 81, 252–262, 315, 332–339, 355–359

2 **Higher linear and cyclic alkanes** 9, 233, 263–271

3 **Chemisorbed benzene** 43, 125, 126, 273–282

4 **Studies on C_6 molecules** 271, 274, 281, 283–291, 351

5 **Decomposition of alkenes** 5, 8, 23, 84, 94, 203, 292–296, 340–342

6 **Formation of ethylidyne from ethene Pt(111)** 4, 15, 24, 70, 89, 98, 130, 151, 153, 207, 222, 297, 298, 343, 344; **Pd(111)** 82, 83, 161, 162, 299, 300, 352; **Rh(111)** 151, 301, 302.

7 **Applications of Density Functional Theory Ni**, 171, 172, 303, 304; **Pd**, 305, 306, 354; **Pt**, 99, 101, 119, 207, 307, 308; **C_2 species**, 99, 101, 119, 171, 306, 307, 354; **C_1 species**, 99, 170, 303–305, 308, 309.

8 **Chemisorption of alkanes** 4, 182–186, 321, 326, 339, 346–350.

5

INTRODUCTION TO THE CATALYSIS OF HYDROCARBON REACTIONS

PREFACE

The impatient reader may be wondering when we are going to get around to the main subject of the book, namely the reactions of hydrocarbons catalysed by metals. The material of the earlier chapters will be found of use in what follows, and of course the first encounter with catalysis has already occurred when we considered reactions undergone by hydrogen alone (Chapter 3). It is now necessary to look at the essential nature of the catalytic process; it is a kinetic phenomenon and susceptible to quantitative description, although the application of the principles of chemical kinetics as developed for reactions in a single phase is not without its difficulties. A careful and critical evaluation of the experimental and computational procedures used for extracting mechanistic information is therefore in order, and indeed the very concept of reaction mechanism in a catalysed reaction is something demanding attention. At the heart of this discussion is the notion that for every reaction there is a characteristic active centre, the identification of which has been a kind of Holy Grail for catalysis research. Its nature and structure in a particular case may be sought by systematic variation of parameters such as crystal face and particle size, and assessing the effect on the kinetic features of the reaction; and the size of the grouping of metal atoms which the reaction needs can, it has been argued, be deduced by diluting the active metal with an inactive one in an alloy or bimetallic particle.

The purpose of this chapter is to construct a framework of concepts and procedures that will be employed repeatedly in the ensuing chapters.

5.1. THE ESSENTIAL NATURE OF CATALYSIS[1−6]

5.1.1. A Brief History of Catalysis

The basic idea of catalysis can be traced to the writings of J. J. Berzelius,[7] who in 1836 reviewed a number of curious occurrences in which traces of certain substances seemed to have an effect on chemical reactions disproportionate to their amounts. In a passage often quoted, but bearing repetition, he wrote:[8,9]

> *I shall therefore call it the* catalytic power *of substances, and the decomposition by means of this power* catalysis, *just as we use the word* analysis *to denote the separation of the component parts of bodies by means of ordinary chemical forces. Catalytic power actually means that the substance is able to awake affinities which are asleep at this temperature by their mere presence and not by their own affinity.*

We should not mock this first attempt to impart scientific rigour to such an elusive concept, but rather seek to place it in its historical context, remembering how rudimentary were chemical ideas at that time; the metaphor of awaking sleeping affinities is one which sticks in the memory.

It is not absolutely clear why he selected the word *catalysis*, which is formed from two Greek words, namely, the prefix *cata-* meaning 'down' and *lysein*, meaning 'to break'. Both of these words make frequent appearance in the English language (catastrophe, catalepsy; hydrolysis, photolysis). Most probably Berzelius thought that substance causing the effect was breaking down the normal constraints that prevent reactions occurring, and it is in this sense the Chinese language selects the same word for catalysis as for marriage broker. It is somewhat ironic however that in journalistic use the word has come to mean 'a bringing together', which at first sight is quite the opposite to its original meaning, but on reflection we see that the coming together is an inevitable consequence of the abolition of the inhibiting barrier.[8] Although not used in archaic Greek, the term catalysis has sometimes been used before the scientific era to mean a riot or a breaking down of social constraints.

The early work of Sir Humphrey Davy,[10] of Döbereiner,[11] and many others, that led Berzelius to coin the word and apply it to a diverse range of phenomena, has often been described,[1−4] as have the ensuing but somewhat slowly developed practical applications. These may be of interest to the historian of science, but do not greatly help our present purposes. The next significant development was a much clearer definition of what a catalyst is and does, namely, *a catalyst is a substance that increases the rate at which a chemical system attains equilibrium, without being consumed in the process.* This form of words follows that suggested by F. W Ostwald;[2] numerous modifications of the wording have been suggested, mainly in the final phrase, to allow for physical deterioration and deactivation during use. We need not pursue these efforts, as precise definition is the task of the

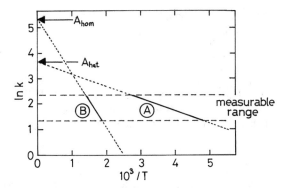

Figure 5.1. Schematic Arrhenius plots of (**A**) catalysed and (**B**) non-catalysed reactions showing how the lower activation energy increases the rate at a given temperature and lowers the temperature needed to achieve a given rate.

pedant not the scientist. As Humpty Dumpty said, *When I use a word it means just what I choose it to mean – neither more nor less* (a sentiment incidentally which erodes the basis of a fair part of philosophy).

The implications of this definition have also been frequently explored.[8] By increasing the rate of a reaction at some fixed temperature, a catalyst can also have the effect of lowering the temperature at which a given rate is achieved (Figure 5.1), and for many practical purposes this is its chief advantage. It can only assist reactions that are thermodynamically favourable, and the position of the equilibrium has to be the same as that which would have resulted, albeit in a much longer time, without it. This also implies that rates of forward and reverse reactions must be helped by the same factor if the equilibrium constant is to remain unaltered. Just how a catalyst manages to exert its influence will be considered in the next section.

5.1.2. How Catalysts Act

It would be useful to know by exactly how much a catalyst speeds up a given reaction, but a comparison of rates obtained with and without its help depends upon the chosen conditions. If we consider a gas-phase reaction, the rate will be proportional to the volume, and will be some function of reactant pressures and temperature; the rate of the catalysed reaction will depend on the surface area of the catalyst, i.e. on the number of points at which the reaction occurs, and on the surface coverages by the reactants, and also on temperature. A precise comparison is therefore clearly impossible, but if it is attempted on the basis of the number of potentially effective collisions between reactants, this will be very much larger (perhaps[12] by as much as 10^6) in the homogeneous case than in the catalysed case.

The rate at each reacting centre must therefore be very much faster with the latter, so that if we now adopt the Arrhenius equation

$$k = A \exp(-E/RT) \qquad (5.1)$$

as a framework for the discussion, A_{cat} will be very much smaller than A_{hom}, and therefore the exponential term must be correspondingly larger, i.e. E_{cat} must be much smaller than E_{hom}. Remembering that in elementary kinetic theory

$$A = PZ \qquad (5.2)$$

where Z is a collision number and P a 'steric factor', there may be some relief in that P_{cat} may exceed P_{hom} by 10^2 or 10^3: but the conclusion is that $(E_{hom} - E_{cat})$ must be *at least* 65 kJ mol^{-1} at a typical temperature just to compensate for the difference in the A terms, and should exceed 100 kJ mol^{-1} for efficient catalysis. There is much experimental evidence to show that this is indeed the case,[12] and it has become an article of faith that *the principal way in which a catalyst acts is by lowering the activation energy of a reaction.* However as we shall see (Section 5.6) it does not automatically follow that the member of a set of catalysts showing the lowest activation energy is necessarily the most active.

To lower the activation energy, the reactants must have found an easier path-way across the potential barrier separating them from the product state (Figure 5.2) The nature of this new route depends upon the reaction and the catalyst, and per-haps also on the experimental conditions, but certainly it is the act of chemisorption that prepares molecules for reaction and it evades the sometimes large energy input needed to secure this preparation in the gas phase. In case of hydrogen-deuterium

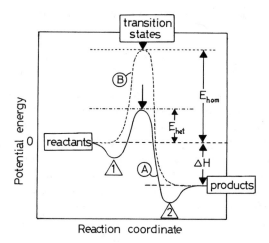

Figure 5.2. Potential energy profile for (**A**) catalysed and (**B**) non-catalysed reactions.

equilibration (Section 3.4.2) dissociative chemisorption removes the necessity to find the 434 kJ mol^{-1} of energy required for breaking the molecule in the gas phase. The reaction can then proceed at a measurable rate well below room temperature, compared to the 800 K or more that the homogeneous reaction needs. There are of course other ways of speeding up this reaction, e.g. by photolysis, but the energy resident in the free valencies of the metal surface is the most convenient way of activating hydrogen molecules. This concept is readily extended to all reactions in which hydrogen is a reactant so that for example a similar comparison can be drawn for hydrogenation of ethene, which defies study as a homogeneous reaction because other parasitic reactions intervene at the necessary high temperature. Catalysed hydrogenation was widely use to measure heats of reaction at about ambient temperature, safe in the knowledge that, following the accepted definition of catalysis, the answer would be the same as that which would have resulted without a catalyst had that experiment been possible. Chemisorption of the hydrocarbon reactant will also help the progress of reaction by its partial conversion towards the configuration of the product: this is a matter that will command out attention in later chapters.

5.1.3. The Catalytic Cycle[13]

The total requirements for a successful and sustained catalysed reaction can be summarised in the catalytic cycle (Figure 5.3) which provides a kind of check-list of points to which we shall return for closer examination. There must exist an *active centre* (Section 5.5) at which reactants can chemisorb in the appropriate for with an optimum energy for conversion to the desired products. Particularly with hydrocarbons there may be inappropriate chemisorbed forms, such as we have met in the last chapter; the optimum adsorption energy is that which just secures

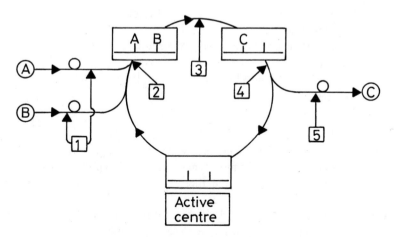

Figure 5.3. The Catalytic Cycle (see text for explanation).

full surface coverage, because stronger adsorption implies a lower reactivity and weaker adsorption means incomplete surface coverage, and inefficient use of the catalyst. These important considerations are enshrined in catalytic science as the *Sabatier Principle* or the *Volcano Plot*.

It is then necessary to bring the reactant molecules to the active centre by a process of *mass transport* which can sometimes be rate limiting (Section 5.2.1). Having been chemisorbed they must then react at the *reacting centre*, and how they do this will occupy us for much of the rest of the book. The product molecules remain on the surface or may be ejected from the surface as it is formed; in the former case it its desorption may be the slow step, and if it does not desorb quickly its further reaction may lead to undesired products. Speedy removal of the product from the neighbourhood of the surface by another mass transport step is often important if it is an intermediate product that is wanted. Conversion of reactants or products into strongly held residues, or adventitious poisons in the feedstock, can block the active centre and lead to deactivation. In their absence the reaction should continue indefinitely.

5.2. THE FORMULATION OF KINETIC EXPRESSIONS[14–16]

5.2.1. Mass Transport versus Kinetic Control[1,2,8]

If it is intended to study the reaction kinetics, that is, the dependence rate upon variables such as reaction pressures or temperature, it is important to be sure that the surface reaction is the slowest of all those forming the catalytic cycle, and that therefore mass transport to and from the surface are not rate limiting. Symptoms of mass transport limitation include the following.

1. The rate is proportional to the catalyst weight W or surface area A_s to a power less than unity.
2. The conversion is not accurately proportional to the inverse of the reactant flow rate F.
3. Thus for kinetic control the rate r should be proportional to the contact time τ so that

$$r = kW/F = k\tau \tag{5.3}$$

4. The temperature coefficient is low, corresponding to an apparent activation energy of $10\text{--}15 \text{ kJ mol}^{-1}$; gaseous diffusion processes do not in fact obey the Arrhenius equation, their rates being proportional to $T^{1/2}$.

If the rate is thought to be governed diffusion of reactants or products within the body of a porous catalyst, it will be increased by diminishing the particle size, i.e., the average length of pore that has to be traversed. Mass transport limitation will arise when the rate of the surface reaction is very fast, for example, at high temperature; a decrease in the apparent activation energy therefore betokens its onset. It will also tend to occur at large contact times and high conversions, where reactant concentrations are low.

Although under conditions of mass transport control the catalyst is inefficiently used, it is sometimes beneficial to operate under these conditions, as for example when in a sequential reaction such as

$$A \rightarrow B \rightarrow C \tag{5.A}$$

it is desired to optimise the yield of the intermediate product B. In what follows it will be assumed that kinetic control is operative, to which region the term *microkinetics* is applied; mass transport control is described in greater detail in textbooks having a chemical engineering flavour.[1,15,17]

5.2.2. The Purpose of Kinetic Measurements[18]

The manner in which the rates of reactant consumption and of formation of each individual product varies with reactant concentrations and temperature affords information that is useful in two ways: first, as providing the basis for the reactor design if the reaction is to be operated on a significant scale, and second, and more to the immediate point, to give a framework within which a reaction mechanism can be formulated. Whatever the practical difficulties of obtaining this information, and they can be considerable with microporous catalysts and those undergoing rapid deactivation, it is *essential* for mechanistic analysis. It cannot be stressed to strongly that *no formulation of a reaction mechanism can be accepted as plausible until shown to be consistent with experimentally determined kinetics.* The corollary is however equally important: *the mechanism cannot be deduced from kinetic measurements alone, because many different mechanisms can lead to the same kinetic expression.*

Kinetic analysis therefore proceeds into two convergent directions. Experimental results are often first fitted to an empirical rate expression in which rate is proportional to the product of the pressure P of each component raised to the power of its *order of reaction*: thus for the process

$$A + B \rightarrow C \tag{5.B}$$

The rate expression will be

$$r = k P_A^{\ a} P_B^{\ b} P_C^{\ c} \tag{5.4}$$

where a, b and c are the orders of reaction, and may be positive, negative or fractional. Such a rate expression is termed a *Power Rate Law* and although results are often quoted in this form it is of limited value, as the exponents may apply only over a limited pressure range, and give only the vaguest idea as to what the mechanism might be.

The second approach starts with an idea of possible mechanism, leading to a theoretical kinetic equation formulated in terms of concentrations of adsorbed reactants and intermediate species; use of the steady-state principle then leads to an expression for the rate of product formation. Concentrations of adsorbed reactants are related to the gas-phase pressures by adsorption equations of the Langmuir type, in a way to be developed shortly: the final equation, the form of which depends on the location of the slowest step, is then compared to the Power Rate Law expression, which if a possibly correct mechanism has been selected, will be an approximation to it. A further test is to try to fit the results to the theoretical equation by adjusting the variable parameters, mainly the adsorption coefficients (see below). If this does not work another mechanism has to be tried.

5.2.3. Measurement and Expression of Rates of Reaction[16,19]

Straightforward laboratory reactors are of two kinds: (i) the dynamic or flow reactor, in which the reactants are forced through a catalyst, and (ii) the static, constant-volume reactor, in which however the reactants can be made to circulate around a closed-loop or 'race-track' and so through the catalyst bed. Where the reactants move, fewer problems with mass transport and with keeping a constant catalyst temperature are met. With the static reactor, the rate follows directly from the changes of pressure or concentration with time and the orders of reaction from the dependence of initial rate on reactant pressure, or the variation of rate with time, according to the traditional precepts of chemical kinetics. For the dynamic reactor, constant conversion is obtained (in the absence of deactivation) by maintaining a fixed flow-rate, and the time dimension is only accessed by changes in this rate.[13] The rate of reaction is then obtained as follows.

A single reactant enters a cylindrical bed of catalyst of length x and cross-sectional area A at a flow-rate of F^0 and emerges at a flow-rate F: the conversion α is then given by

$$F/F^0 = 1 - \alpha \tag{5.5}$$

In a differential slice of the bed of volume $dV = (A dx)$, the change of flow-rate due to reaction is αF, and the rate r is given by

$$r = -dF/dV \tag{5.6}$$

$$= d\alpha/d(V/F^0) \tag{5.7}$$

or if the rate is to be expressed per unit weight of catalysts

$$r = d\alpha/d(W/F^0) \tag{5.8}$$

The average time a molecule spends in the bed is the apparent contact time τ, which is proportional to F/W. Thus

$$r = d\alpha/d\tau \tag{5.9}$$

The conversion equates to the fractional change in concentration

$$\alpha = (c^0 - c)/c^0 \tag{5.10}$$

and for a first-order reaction

$$r = k_1 c = k_1 c^0 (1 - \alpha) \tag{5.11}$$

and

$$\tau = \int d\alpha/k_1 c^0 (1 - \alpha) \tag{5.12}$$

Similar equations may be derived for reactions of other orders.

Information relevant to determining reaction mechanism is obtained by working at low conversion, preferably less than 10%, i.e. by operating the reactor in the *differential* mode. The use of high conversion (>90%) gives information of more practical usefulness; the reactor is then in the *integral* mode.

We thus arrive at a rate per unit weight of catalyst, but for a supported metal we need the rate per unit weight of metal, although we cannot assume its metal content is what we think it to be; accurate chemical analysis is essential. This gives the *specific rate*, but if the metal dispersion or average particle size is known (as it should be) we may then get the rate per unit area of metal (the *areal* rate).[20] So far, so fairly good, but difficulties may arise in studying the effect of rate on particle size (a popular pastime, see Section 5.4) if the size distribution is binodal, as it often is. The catalyst may contain both large and small particles, in amounts that vary with total loading[19] (Section 2.5.6); small particles may escape detection (e.g. by XRD and TEM), and the H/M_s stoichiometry may therefore change with loading. Great care must therefore be exercised in looking for particle size effects on rate or product formation. For a time it was usual to express the rate per unit atom of exposed metal, as determined for example by hydrogen chemisorption; this is termed the *turnover frequency*,[21–23] the units of which are mol (mol surface atom)$^{-1}$s^{-1}. It is however now appreciated that for many reactions the size of the ensemble of atoms needed for reaction is quite large, and that therefore the turnover frequency is not a good basis of comparison.[21,24] It does however represent a way of normalising the performance of a given metal if its dispersion changes from one preparation to another. It must be understood that a TOF is simply a *rate*, and that it only has the units of t^{-1} because the mass units have been cancelled. Its value therefore depends on all operating variables (reactant pressure, temperature etc.), and these must be cited if its value is to have any meaning. This discussion is continued in Section 5.4.

5.2.4. The Langmuir-Hinshelwood Formalism[25]

The essential problem in formulating a theoretical rate expression for a catalysed reaction is that we do not in general have direct access to concentrations of the adsorbed reactants: we have to assume a relationship between them and their pressures in the gas phase. The simplest and most widely used way of doing this is to suppose that the Langmuir adsorption equation (Section 3.2.4) adequately describes the connection: we must however remember that the basic postulates underlying this type of equation should always apply, and much of the discussion concerning the proper derivation of rate equations concerns the definition of site (or sites) involved (see Section 5.5). Whatever the uncertainties and ambiguities associated with this approach, the resulting rate expressions first applied by C. N. Hinshelwood[26] in the 1920s, have proved extraordinarily useful in understanding and interpreting the kinetics of catalysed reactions.

For the simplest possible reaction, namely A $\rightarrow$ X, the mechanism can be formulated as follows

$$A + {}^* \rightarrow A^* \rightarrow X + {}^* \tag{5.C}$$

where A is chemisorbed at an active centre designated by *, and is changed into X which is simultaneously returned to the gas phase. Thus

$$r = k_1 \theta_A \tag{5.13}$$

and by introducing the Langmuir adsorption equation for θ_A

$$r = k_1 b_A P_A / (1 + b_A P_A) \tag{5.14}$$

The dependence rate on P_A will thus follow the dependence of θ_A upon P_A (Figure 5.4). The form of this curve is consistent with the conclusion that when the P_A is slow and/or b_A is small, the rate in simply proportional to P_A, but when either or both is large it is independent of P_A, i.e. the reaction is of zero order. In

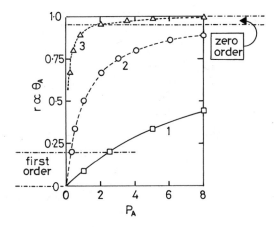

Figure 5.4. Dependence of rate on reactant pressure P_A for a unimolecular reaction: **1**, $b = 0.1$; **2**, $b = 1$; **3**, $b = 10$.

between, the order will seem to be a positive fraction, but a single exponent of the pressure can only describe the change in rate over a very limited range.

When there are two reactants A and B, and the product X is not absorbed, the mechanism becomes

$$A + * \longrightarrow A^*$$
$$B + * \longrightarrow B^* \longrightarrow X + 2^*$$

(5.D)

and the rate expression is

$$r = k_2\theta_A\theta_B$$

(5.15)

If A and B adsorb on equivalent sites

$$\theta_A = b_A P_A/(1 + b_A P_A + b_B P_B)$$

(5.16)

$$\theta_B = b_B P_B/(1 + b_A P_A + b_B P_B)$$

(5.17)

so that

$$r = k_2 b_A P_A b_B P_B/(1 + b_A P_A + b_B P_B)^2$$

(5.18)

The form of the dependence of rate on P_A depends on the size of the term $b_B P_B$; when this is about the same as $b_A P_A$, curves of the type shown in Figure 5.5 result. The highest rate occurs when θ_A equals θ_B, and since P_B is known the position of the maximum reveals the value of b_A/b_B. There are two important limiting cases.

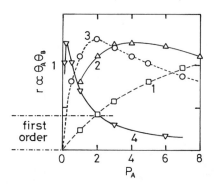

Figure 5.5. Dependence of rate on reactant pressure P_A for a bimolecular reaction: $P_B = 1$; **1**, $b_A = 0.1$; **2**, $b_A = 0.5$; **3**, $b_A = 1$; **4**, $b_A = 10$.

(i) When both reactants are only weakly chemisorbed, the denominator approaches unity, so that

$$r = k' P_A P_B \tag{5.19}$$

where $k' = k_2 b_A b_B$; the reaction is first order in each reactant. (ii) If A is much more strongly adsorbed than B and/or P_A greatly exceeds P_B, then $b_A P_A > 1 > b_B P_B$ and so

$$r = k'' P_B / P_A \tag{5.20}$$

where $k'' = k_2 b_B b_A$; the reaction is first order in B and inverse first in A. Once again, when neither limiting situation applies, the rate dependence upon reactant pressures may be approximate by the Power Rate Law expression

$$r = k_2 P_A^x P_B^y \tag{5.21}$$

where the orders x and y may be either positive or negative fractions. Note however that either exponent may change from positive to negative as the pressure which it qualifies increases (see Figure 5.5).

The principles underlying this treatment are capable of extension to cover a greater number of reactants, inhibition by products, poisoning by adventitious impurities, dissociation of reactants upon adsorption (Section 3.2.4) and many other situations.[27] The relevant rate expressions were collected and comprehensively evaluated many years ago by O. A. Hougen and K. M. Watson,[28,29] and monographs on chemical kinetics[2,15,22,30] often contain a fuller presentation than is thought necessary here.

Contemplation of Figure 5.5 raises another concern. If the curves **1** to **4** are given experimentally by four reactants whose adsorption coefficients b_A have the values shown, and if the rates are measured at various values of P_A, then very clearly the sequence of 'activities' will depend on the value chosen: in particular the reactivity of reactant **4** will depend critically on the pressure of A adopted. This simple observation deprives the concept of 'catalytic activity' of any meaning whatsoever, if the rates are measured only at a single pressure of each reactant.[31]

5.2.5. Effect of Temperature on Rate and Rate Constant

Precise determination of the temperature dependence of the rate of a catalysed reaction is beset by difficulties. We have already met one possible complication, namely, that, as temperature increases, mass transport limitation may set in: there may be a change in the rate limiting step, and in the fraction of surface not covered by strongly-held by-products, and hence available for reaction. We shall turn to the most important source of uncertainty in a moment.

According to the *Arrhenius equation*, which was of course devised to describe a homogeneous gas-phase reaction, the *rate constant k*, which is the rate at unit pressure of the reactants, i.e.

$$r/P_A{}^a P_B{}^b = k \qquad (5.22)$$

is the product of two terms, one of which is independent (or only weakly dependent) upon temperature, and the other of which is exponentially dependent on temperature. In the formalism of the Transition State (or Absolute Rate) Theory, the first of these, the pre-exponential factor A, becomes the product of an entropy term $(\exp(\Delta S^{\neq}/R)$, a frequency factor $(k_B T/h)$ and perhaps a transmission coefficient (κ) for safety's sake. The second term contains only the activation energy, i.e. $\exp(-\Delta H^{\neq}/RT)$. Exact treatments of this important theory will be found all physical chemistry texts and elsewhere. Those who find its logic difficult to follow should take the advice of the mathematician d'Alembert: *Allez en avant et la foi vous viendra*. Its application to chemisorption and catalysed reactions has often been discussed; it requires the identification of the rate-limiting step and the inclusion of the unknown concentration of active centres, but we shall not be able to make a great deal of use of it. There are comparatively few systems to our understanding of which it has made an important contribution. It is however proper to appreciate that 'activation energy' can be regarded in two complementary ways: (1) by applying the Arrhenius equation to experimental results, it is given by

$$E = RT^2.d\ln k/dT \qquad (5.23)$$

and (2) through Transition State Theory it is the energy $\Delta H^{\ddagger}$ required to raise the reactants to the point of almost no return at the top of the potential energy barrier, that is to say, to the transition state.[32]

Now for a homogeneous gas-phase reaction, the two are more or less equivalent, but for heterogeneous reactions the rate constant k is somewhat elusive: as we have used it, it rests on the validity of the Langmuir-Hinshelwood model, and its numerical value depends upon making the best (if not correct) choice of mechanism. Even so, to evaluate it at a sufficient number of temperatures involves considerable work, made easier now by the availability of microprocessor-controlled reactors and on-line computers. Nevertheless in much of the prior literature quite understandably the *rate* is used instead of the rate constant to obtain what must be called an *apparent activation energy* E_a: but changing temperature affects not only the rate of each reacting unit (i.e. k), but also the concentrations of the adsorbed reactants θ_A and θ_B. Consider first a unimolecular reaction (equation (5.13)). Simple chemisorption is always exothermic, and so θ_A will decrease as temperature rises; since the adsorption coefficient b_A is an equilibrium constant, by applying the Van't Hoff isochore we obtain

$$b_A = C_A \exp\left(-\Delta H_A^{\ominus}/RT\right) \qquad (5.24)$$

where $\Delta H_A^{\ominus}$ is the standard heat of adsorption and C_A an integration constant. If θ_A remains close to unity over the whole temperature range, the increase in rate will simply be due to its effect on k: thus

$$k = A_t \exp\left(-E_t/RT\right) \qquad (5.25)$$

E_t being the *true activation energy*. If however the adsorption is weaker, θ_A will decrease significantly, and by joining equations (5.23) and (5.24) we find

$$k = A_t C_A \exp\left[(-E_t - \Delta H_A^{\ominus})/RT\right] \qquad (5.26)$$

and

$$E_a = E_t + \Delta H_A^{\ominus} \qquad (5.27)$$

The true activation energy will therefore be *larger* than E_a, $\Delta H_A^{\ominus}$ having a negative value since chemisorption is exothermic.

By the same token, for a bimolecular reaction when θ_A and θ_B are both small in the temperature range used the order in both is first and so

$$E_a = E_t + \Delta H_A^{\ominus} + \Delta H_B^{\ominus} \qquad (5.28)$$

But when θ_A is high and almost invariant, and θ_B is small

$$E_a = E_t - \Delta H_A^{\ominus} + \Delta H_B^{\ominus} \qquad (5.29)$$

This is because the value of the product $\theta_A\theta_B$ is decreased by increasing P_A and the reaction is inverse first order in A (see Figure 5.5). Equations of this type can be written to describe a variety of situations. The first systematic analysis was undertaken by M. I. Temkin in 1935;[33] he arrived at the general expression

$$E_a = E_t + x\Delta H_A^{\ominus} + y\Delta H_B^{\ominus} \qquad (5.30)$$

where x and y are the respective orders (equation (5.21)); this clearly invokes Power Rate Law concepts, but as we have seen these harmonise with the Langmuir–Hinshelwood formalism in limiting cases.

It will also be apparent that any other factor besides temperature that changes the coverages by adsorbed species will alter the measured activation energy: changing the reactant pressures will do this (Figures 5.4 and 5.5), so that measuring at high pressures will give a value close to E_t, while at low pressures it will be governed by equation (5.28). Indeed one has only to look at the curves in Figure 5.5, imagining that it is temperature that is responsible for causing b_A to change, to realise immediately that the activation energy will depend upon the value of P_A that is selected.[33] While this idea is expressed for a bimolecular reaction, it applies equally to a unimolecular reaction (Figure 5.4).

The distinction between true and apparent activation enquiries is important to draw for several reasons. (1) In trying to understand how catalyst structure and composition affect activity, there are two factors to consider: a thermochemical factor determining the concentration of reacting species, and a kinetic factor controlling their reactivity. E_a contains both, and only when E_t and the relevant heats of adsorption are separated can their individual contributions be assessed. (2) E_a is not a fundamental characteristic of a catalytic system, because its value may depend on the reactant pressures used.[33,34] As we shall see in Section 5.5, there are very helpful correlations to be drawn between kinetic parameters, reactant pressures and orders, and structure sensitivity in the field of hydrocarbon reactions.

5.2.6. Selectivity[12]

In our mission to understand how and why the structure and composition of a catalytic surface determine performance, it is the nature and amounts of the products formed that provide more useful information about the adsorbed intermediates and mechanism than does the rate of reactant removal. The latter is not without its significance, but product analysis is a richer source of inspiration.

A catalytic system is characterised by the *degree of selectivity* with which each product is formed; the term 'selective' is however used very loosely in the literature, sometimes being applied when the product in question is only a small fraction of the total. Strictly speaking, only when that product is unique can the process properly be called 'selective'.

With a single hydrocarbon molecule, there may be formed *either* two or more products simultaneously *or* two or more products in sequence, viz.

$$A \underset{\longrightarrow Y}{\overset{\longrightarrow X}{\rule{1.5cm}{0pt}}} \tag{5.E}$$

$$A \to X \to Y \tag{5.F}$$

In catalytic hydrogenation and hydrogenolysis, both schemes are often encountered: they are easily distinguished because Y is not an initial product in the second scheme, whereas the concentration of the product X will pass through a maximum. Very frequently X is the desired product and conditions have to be sought to maximise its yield, but the two schemes often occur together, viz.

$$A \underset{\longrightarrow Y}{\overset{\longrightarrow X}{\rule{1.5cm}{0pt}}} \tag{5.G}$$

and this exacerbates the problem of obtaining X in high yield. Figure 5.6 shows some examples of how amounts of product may vary with conversion. Here we are only concerned with products detectable in the fluid phase; fuller development of these schemes defining the role of adsorbed intermediates will be undertaken later. Selectivity is then expressed simply as the fraction that one product forms of the total, or

$$-dP_A/dP_X = S_X \tag{5.31}$$

Product ratios, e.g. P_X/P_Y, are often cited, but are generally less useful.

There is a further type of selectivity that arises when two reactants are present, viz.

$$A \to X$$
$$B \to Y \tag{5.H}$$

The selectivities of X and Y are then determined by the relative strengths of adsorption of A and B as well as their inherent reactivities.

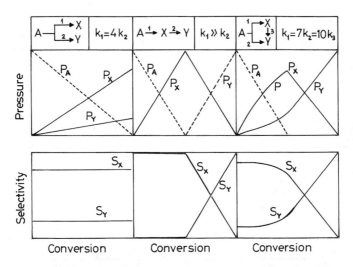

Figure 5.6. Dependence of product pressures P_x and P_y, and of selectivities S_x and S_y on conversion for the reaction schemes 5.E, 5.F and 5.G.

Mathematical treatments of product composition as a function of conversion have been presented.[12,35]

5.2.7. Kinetic modelling[26,36-39]

Kinetic modelling is the art of deducing the best possible rate expression by comparing its predictions with the experimental results, and hence inferring a likely mechanism. This careful definition is predicated on Karl Popper's precept that it is impossible to prove that a theory or model is correct, because there may be a better one around the corner: it is only possible to negate an hypothesis, by showing that it fails to accord with observation. With modern computational facilities it is easy to compare the merits of various theoretical rate equation based on different mechanisms with experimental results, and for a given equation to find the values of the constants giving best fit, by calculating the mean standard deviation or other statistical parameter. Several words of caution are however necessary before blind faith in what the machine says dulls our critical facilities.

First, it is necessary to examine by eye a plot of the calculated curve of rate versus the pressure of a reactant, and superimpose the experimental points. The reason for doing this is because a fairly good overall fit may result from a very good fit over a part of the range, tending to a very poor fit in another (Figure 5.7). In deciding in which range a good fit makes most sense, variable experimental error must be kept in mind, e.g., very fast and very slow rates may not be measured as precisely as moderate ones. A computer will not know about experimental error

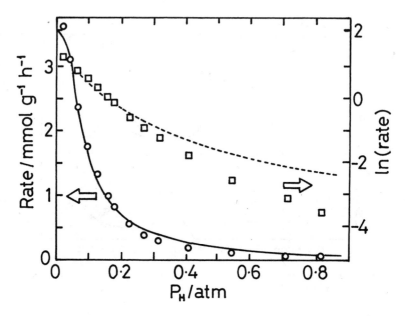

Figure 5.7. Hydrogenolysis of n-butane on Rh/TiO$_2$ at 429 K: comparison of direct (O) and semi-logarithmic (□) plots of rate versus hydrogen pressure. In this and the next figure, the curves are calculated by the rate expression ES5B to be derived in Chapter 13.

unless specifically told. Comparison at low rates is made easier if semi-log or log-log (Figure 5.8) plots are used.

Second, the mechanism on which the rate expression is based should be both plausible and comprehensive, that is, it should take note of other relevant observations in the literature and should not omit any reasonable possibilities. There is a potential conflict here, because the greater the number of disposable parameters in the equation, the better automatically becomes the fit with the results. As the great mathematician Cauchy said: *With five constants I can draw an elephant; with six I can make it wag its tail.* It is desirable to restrict the number to three even it entails some over-simplification.

Third, the values of the disposable constants must be sensible, and their temperature dependence must be rational, giving linear plots of $\ln k$ or $\ln b$ vs. T^{-1}, and afford heats of adsorption having positive values of $-\Delta H_a$ and true activation energies of reasonable magnitude. These criteria are of extreme importance for this reason. One of the surprises for the novice modeller is the great variety of mathematical expressions, and therefore mechanisms, that can generate curves giving good fits to the points (Figure 5.9): but the value of a constant such as the adsorption coefficient of a reactant can vary astronomically with the form of the

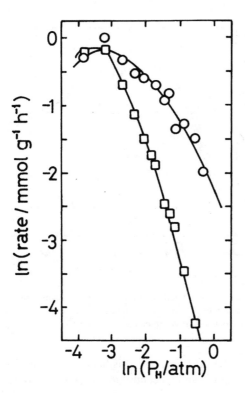

Figure 5.8. Double-logarithmic plots of the rates of hydrogenolysis and of isomerisation of n-butane as a function of hydrogen pressure at 533 K with Pt/SiO$_2$ as catalyst.

expression.[36,40] Clearly therefore the statistical quality of the fit is by itself valueless in selecting a mechanism, and not all best values of that constant are equally acceptable. The precious gift of common sense based on experience has to be applied if mathematical modelling is to have any significance. Further illustrations of the use of the procedure will appear in Chapter 13.

5.3. THE CONCEPT OF REACTION MECHANISM[16,41]

The term 'mechanism' has been freely used in the foregoing sections, but without definition: this is because it admits of no simple description, there being many formulations considered by their begetters to be adequate and satisfactory. Mechanistic discussion is like peeling an onion: it is possible to go through a never-ending series of ever more profound analyses without ever reaching the end,

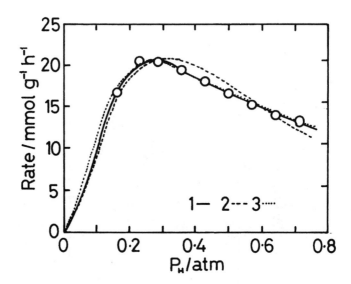

Figure 5.9. Dependence of rate of hydrogenolysis of *n*-butane on hydrogen pressure at 609 K using PtRe/Al$_2$O$_3$; experimental points fitted to three rate expressions. ES5B, ——; ES2, - - -; ES3, $\cdots$. The formulation of these rate expressions is explained in Chapter 13.

and tears may be shed in the process. This Section is based on a part of an article written many years ago,[42] and its central theme runs as follows.

> *Any mechanistic analysis must be made in terms of a model, and every model is limited by its frame of reference. The kind of answer we get depends upon the language in which the question is framed; the value of the answer is determined by the care that has gone into defining the nature of the conceptual model and by the symbolism employed to express it.*

So for example we may ask, 'Is ethene associatively adsorbed during its metal-catalysed hydrogenation?' and we may hope to obtain a straight 'yes or no' answer; but if the question is 'How is ethene adsorbed ?' we have to expect a more discursive reply, as we express our answer in terms of the many structural formula considered in Chapter 4.

In that article[42] it was suggested that at an elementary level a mechanism is understood if the following are established beyond reasonable doubt:

- The nature of all the participating species.
- The qualitative modes of their interaction contributing significantly to the total reaction.
- Quantitative aspects of these interactions expressed on a relative but not absolute basis.

These represent the absolute minimum of information required, and in some cases it is possible to move on from them, at least in a speculative fashion. Modern experimental methods, to be mentioned briefly in Section 5.7, sometimes reveal the structures of possible or probable adsorbed intermediates, and theoretical techniques have a role as well. The concept of the *most abundant surface intermediate* (MASI)[22] has proved useful for some systems, but caution is necessary because that species is not necessarily the vital reactive intermediate; indeed it may only be abundant and observable because it is *unreactive*. Such appears to be the case with ethylidyne in the hydrogenation of ethene; the much less abundant π-adsorbed form may well be the reactive state. It is also necessary to consider, in the case of supported metals, exactly where the reaction is taking place. We shall meet at least one example of a reaction occurring on the support using hydrogen atoms arriving by spillover from the metal (Chapter 10). Formulating a rate expression when this situation occurs needs great care.

Reaction schemes such as 5.G are very incomplete accounts of mechanism. A fuller statement will show symbolically the adsorbed intermediates as in 5.C and 5.D; the existence of such intermediate species may have to be inferred, because they do not necessarily lead to an observable product. Thus, in the reaction sequence 5.I, X and Y *may* be seen in the gas phase, depending on the relative values of the

$$A \longrightarrow A^* \longrightarrow X^* \longrightarrow Y^* \longrightarrow Z^* \qquad (5.I)$$

rate constants, or Z may be the only detectable product. These possibilities occasion much discussion as to whether the straight A to Z process occurs at a different kind of site, and perhaps by a different mechanism, from that giving X and Y as well. This question arises when discussing the hydrogenation of ethyne (Chapter 9) and the hydrogenolysis of *n*-butane (Chapter 13). While it is perfectly possible that a reaction may proceed differently at distinct sites, and may follow different kinetics, the Principle of Economy of Hypothesis, otherwise known as Occam's Razor, requires us first to explore the simplest option, which is the 'rake' mechanism shown in Scheme (5.I).

5.4. THE IDEA OF THE ACTIVE CENTRE[43–48]

The term 'active centre' (or 'active site', which is the same thing) was first employed by Sir Hugh Taylor in his far-sighted papers published in 1925 and 1926.[49] They should be required reading for all students of heterogeneous catalysis, because their thinking has guided generation after generation of researchers in the field, and their message is still very relevant. The central theme is that the surface of a typical metal contains atoms of various co-ordination number, and that each class of reaction will only proceed at a place named the 'active centre' where there is an atom, or a group of atoms, of the appropriate type. Each reaction or reaction class proceeding on a given surface will have its own specific requirement, the stringency of which may vary from one extreme, where only a small fraction of the surface is active, to the other, where all sites are suitable and the whole surface is active. In Taylor's words,[49] *The amount of surface that is catalytically active depends on the reaction catalysed.* The first type of reaction was originally called 'demanding' and the latter 'facile',[19,50,51] but 'facile' has a pejorative connotation, and these terms were later replaced by the still not entirely satisfactory 'structure-sensitive' and 'structure-insensitive'.[19,24,52] The usefulness of these names, and their possible refinement, will be considered again in a moment.

Taylor's picture of a typical surface, containing a highly unstable arrangement of atoms having mainly low co-ordination numbers,[49] is one to which we would not now subscribe, but considering the simple case of atoms on the surface of a cube helps to make the point. The corner atoms might be the locus of a structure-sensitive reaction, while those in the sides (excluding edge atoms), being the majority except for the smallest cubes, could be responsible for a structure-insensitive reaction. An active centre may therefore be regarded as having a specified *number* of atoms having *defined co-ordination numbers,* and held in an arrangement peculiarly effective for the reaction in question.[53]

The number of active centres is therefore generally less than the total number of surface atoms,[54] the ratio of the two being termed the *Taylor fraction* F_T (or Taylor ratio). A.A. Balandin attempted to make Taylor's ideas more precise by proposing a *Multiplet Hypothesis*[55,56] by which each type of reaction required a 'multiplet' of several atoms to be an active centre. So for example to chemisorb a benzene molecule either for its hydrogenation or in its formation by dehydrogenation, a hexagonal multiplet with one central atom as found in the fcc(111) plane would be needed. If the number of atoms composing the multiplet (or *ensemble* as it was called by Kobozev[57,58]) is the Balandin number N_B, then

$$N_B \times F_T = 1 \qquad (5.32)$$

Frennet's *free potential site*[58–60] is a closely related concept to that of the active centre, as is Campbell's *true ensemble requirement.*[61]

It should now be clear that the attribute of 'sensitivity' belongs to a *catalytic system* comprising reactants *plus catalyst*: it cannot be safely assumed that because a reaction appears structure-insensitive on one type of catalyst it will necessarily be so on another. Thus for example we might expect alkene hydrogenation to be 'insensitive' on nickel, but 'sensitive' on copper, where only a few surface atoms are able to assist. It should therefore be possible to classify metals accordingly to their structure sensitivity for a given reaction, but there has been little progress in this direction to date.

We have noted elsewhere a strong desire on the part of scientists to categorise an observation by placing it in one of only two boxes, feeling that thereby they have advanced their understanding. The affixing of the labels 'structure-sensitive' or 'structure-insensitive' to a system is in fact an oversimplification, and conflicts with Taylor's idea[49] of an infinite variability of sensitivity with reaction type. We may conceive the typical surface as having atoms belonging to one of a number of sub-sets containing atoms of specified co-ordination number, or of a defined mix of different co-ordination numbers (e.g. one corner and one edge). We might therefore more usefully speak of a *degree of structure sensitivity*, depending on the number of the sub-sets that qualify to contribute active centres. This concept has been given quantitative expressions in the context of particular size variation by David Avnir[24,62,63] using fractal analysis. A reaction system is assigned a *reaction dimension D_R* defined by the equation

$$\ln (r/t^{-1}) = \ln k + (D_R - 2)\ln R \qquad (5.33)$$

where (r/t^{-1}) is the turnover frequency and R the mean particle radius. This equation affords convincingly linear plots for a number of reactions, and the values of D_R are often nearly integral: they can be interpreted in terms of the involvement of a specific class of surface atom (corner, edge, plane etc.) according to the manner of their expected occurrence as a function of size, as expounded by van Hardeveld and Hartog.[64-66] Fractional numbers denote the involvement of more complex atomic groupings. While the validity of this approach has been seriously questioned,[24] its value lies in attaching shades of grey to a concept that is otherwise seen only in black and white.

Assessment of sensitivity and, for a 'sensitive' reaction, identification of the active centre and estimation of their number, have been prime objectives of much research, but definitive answers are unfortunately elusive. One contributing factor to the difficulties encountered is the likely mobility of surface atoms, especially those on small particles and at high temperatures, where the surface may be in a semi-molten state (Section 2.5.3). Mobility may occur to a lesser extent even on single crystal surfaces.[67] Somorjai has advanced the idea that active centres do not pre-exist, but are formed by adaptation of the surface to the reaction being catalysed, suitable sites being created by reconstruction. This process undoubtedly occurs on single crystals, and there is evidence that strongly adsorbing molecules such as alkynes can withdraw atoms from their normal lattice positions (see Chapter 9). Structure sensitivity was first recognised by a decrease in activity for hydrogenolysis of neopentane relative to its isomerisation after heat treatment of a Pt/C (Spheron 6) catalyst, although little change in particular size was seen.[50,51]

It was thought that active centres were removed by annealing, i.e. by smoothing an initially rough surface, although other explanations are possible.

For single crystal surfaces, a reaction is deemed 'insensitive' if its rate is about the same on all low Miller index planes, but since these differ from small metal particles in not having atoms of very low co-ordination number, the term *face sensitivity* should be used in this case. Two further approaches to the general problem have been tried: (1) systematic variation of particle in supported metal catalysts, and (2) alteration of the composition of the surface of bimetallic catalysts, either supported or unsupported (Section 5.7). These lead respectively to *particle size sensitivity* and *ensemble size sensitivity*, but the three types are not necessarily exactly the same.

The literature is full of attempts to deduce the nature of an active centre by systematically changing the particle size in a supported metal catalyst.[44,68–70] The argument runs as follows. Alteration of the mean particle size, for example, by changing the metal content, the method of preparation, or the thermal pre-treatment, will alter the numbers of atoms having a particular co-ordination number, or the numbers of atoms forming specific groupings, in a way that can be predicted by the use of models such as those advanced by van Hardeveld and Hartog.[64] The so-called B_5 site, comprising five atoms in one of two configurations at a step, has claimed particular attention.[29,65] Oles Poltorak[70] also contributed to early work in this field, coining the term *mitohedrical region* to describe the size range where proportions of edge and corner atoms change quickly. This has also been termed the *mesoscopic region*, as being intermediate between the microscopic and the macroscopic. Correlation with the corresponding change in specific rate may lead to some idea as to the type of active centre responsible for the reaction. It is undoubtedly true that specific rates of many catalysed reactions do change in a regular manner with mean particle size, even if this is only expressed as an H/M_s ratio; the specific rate may either increase or decrease or pass through a maximum (hardly ever a minimum) with increasing size (see Figure 5.10).

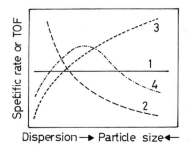

Figure 5.10. Possible forms of dependence of TOF or areal rate on particle size: **1**, no dependence; **2**, negative/antipathetic dependence; no dependence; **3**, positive/sympathetic dependence; 4, TOF passes through maximum.

There are however a number of difficulties and dangers in this procedure.[19,71] (1) As noted in Sections 2.4 and 5.2.3, the particle size distribution may be binodal, a fact only revealed by TEM; in this case, *mean* size as determined by selective chemisorption has no meaning, and the observed activity will be the sum of those given by two quite different kinds of particle. (2) Additionally, the H/M$_s$ ratio is a function of size for very small particles;[72] hydrogen chemisorption is therefore not a reliable method for size estimation. (3) Most importantly, in a supported metal catalyst there will inevitably be a size distribution, even if it is mononodal, and the population of surface atoms of a selected co-ordination number will vary greatly across the distribution. As noted earlier (Section 2.41), only the *mean* surface co-ordination number changes smoothly with size, and use of numbers derived from regular model particles having a complete outer layer can be very misleading. (4) Methods of preparation designed to effect size changes may bring about other differences, notably to the concentration of residual impurities such as chloride ion. (5) Many other properties alter with size besides surface structure as discussed earlier (Section 2.5). (6) Finally, as we have seen, surface mobility may vitiate entirely the possibility of a strictly geometric factor in catalysis. Thus while we may speak of a *particle-size* (or *dispersion*) *sensitivity*, it would be unwise to attribute this to a specific structural property. This strongly negative note is regretted, but it is essential: the concept of sensitivity will be re-visited in the next Section with a somewhat more positive result.

A further cautionary word must be added. Conclusions concerning the particle size sensitivity of a reaction are often based on a quite inadequate amount of experimentation, usually just the rate measured under a single set of conditions. There is therefore no knowing how dependent the conclusion is on the choice of those conditions, and whether the use of other reactant pressures or temperatures would have led to other conclusions. This frequent economy of effort is hard to understand, as the necessary physical characterisation is much more expensive and time-consuming, and is quite inexcusable now that micro-processor-controlled equipment is so freely available. A salutary example is provided by benzene hydrogenation over Ni/SiO$_2$, where with increasing particle size the rate at room temperature increases, but above 453 K it decreases.[58] There are other indications of a change in rate-controlling step between these temperatures (see Chapter 10).

Accepting the existence of 'active sites' implies also that there are also 'inactive centres' or perhaps 'overactive centres' that are quickly inactivated by the destructive chemisorption that occurs so easily with hydrocarbons. The accidental or deliberate removal of such sites by autogenic toxins (i.e. 'carbonaceous deposits', ethylidyne etc.), or by other poisonous species such as sulfur compounds, allows the remaining 'active centres' to exhibit reactions of a structure-insensitive type that could not take place while the overactive centres were in existence. This situation may arise not only through surface heterogeneity, but also on a plane surface by operation of the Principle of Maximum Occupancy (Section 4.2) by which the preferred first reaction is that which utilises the largest size of active centre by forming the greatest possible number of bonds to the surface. There remain in between these species, which are usually dehydrogenated forms of the reactant hydrocarbon (Section 4.6) but may be other toxins, small sites on which other reactions can proceed; so for example ethene hydrogenation is thought to proceed typically on a relatively few sites not occupied by 'spectator species' such as ethylidyne (Chapter 7). Active centres isolated by strongly-held species or other site-blocking species may also catalyse reactions that are impossible on clean surfaces; thus ruthenium catalysts, which are normally very active for hydrogenolysis of alkanes, can when suitably poisoned give very high selectivities for skeletal isomeration (Chapter 14).

According to the Catalytic Cycle (Figure 5.1), an active centre is a place where the reactants are adsorbed adjacently, ready to react. We may therefore attempt a working definition of an active centre as *a location on the surface of a catalyst where reactants may be adsorbed in the right way and with the best strengths to give the desired products efficiently*. It is not necessary for the reactants to adsorb directly onto the active centre; one or both may adsorb elsewhere and then diffuse to the active point; this is what happens in spillover catalysis. There are without doubt cases where two reactants do not adsorb in competition on the same type of site; each has its own requirement, and it may be more closely restricted for one reactant than for the other. The active centre will then contain two sites of different character, on which each reactant is adsorbed non-competitively. Kinetic analysis should reveal when this situation arises.

Correct estimation of the number of reacting centres is essential for the accurate measurement of turnover frequency. Although this is very hard to do in practice, a mental picture of what constitutes a reacting centre may be helpful. Imagine a reaction in a flow system proceeding under quite steady conditions, and freeze the motion of time: the number of reacting centres is the number of places at which at that instant of time the reactants have progressed so far along the reaction co-ordinate through the transition state that formation of products is inevitable. The literature records only one attempt to estimate this number by a transient response method: using ^{14}C-labelled ethene, the fraction of exposed metal atoms constituting reactive centres during its hydrogenation was for Pt/SiO_2 50% and for Ir/SiO_2 17%.[51]

5.5. THE USE OF BIMETALLIC CATALYSTS

In the search for information on the composition of active centres, and for materials of improved catalytic performance, very much use has been made of bimetallic catalysts (see Further Reading at the end of the chapter). The term is preferred to 'alloy' as in many cases the degree of intimacy of the components is uncertain, and in some cases interesting behaviour is found with systems exhibiting only very limited mutual solubility: it implies however that both components are in the zero-valent state during use, although this requirement does not prevent discussion of systems where this is not fully proved, or where perhaps one component is only partially metallic. There is indeed a very narrow and somewhat arbitrary dividing line between 'promotion' and the effects shown in bimetallic catalysts. We should note too that although multi-element formulations often appear in the patent literature, the presence of two components poses quite sufficient difficulties for the academic scientist.

Bimetallic catalysts have had outstanding success in industrial applications, most notably in petrochemistry and in petroleum reforming, where the combinations Pt-Sn, Pt-Ir and Pt-Re have found widespread use, and have been pervasively studied in both industrial and academic laboratories. Their success cannot be assigned to a single cause, but rather to a number of favourable factors working in concert; and it is largely to this success that we owe our extensive knowledge of other bimetallic systems through a kind of scientific spillover.

Early research on bimetallic catalysts[12] used mainly simple reactions such as para- hydrogen conversion and formic acid decomposition, and mainly unsupported forms such as wires, films, foils and powders, since there was no certain way of making supported bimetallic catalysts and no known means of characterising their surfaces. Its motivation was to detect an electronic factor in catalysis, and favoured systems were those in which there was a continuous range of solid solutions; and therefore, it was thought (mistakenly) that there would be a monotonic change in the electron: atom ratio of all component atoms. Significant changes in activity were indeed observed and correlated with composition (see for example Section 3.4.2), but unfortunately the interpretation placed on them was not right. Revision of these views was made necessary by the findings made by electron spectroscopy (Section 2.5.4), which has complicated rather than simplified the task of understanding the results. This phase of research was followed in the period 1970–1990 by the use of supported bimetallics or more complex reactions such as those of hydrocarbons that modelled those occurring in petroleum reforming: very notable contributions were made by John Sinfelt, Wolfgang Sachtler, Vladimir Ponec, John Clarke and many others. One of Sinfelt's major discoveries was that extended mutual solubility is not a prerequisite for a useful bimetallic system, because a surface 'alloy' formed for example by copper atoms on the surface of a ruthenium or osmium particle was stable and usable, and this has led to the widespread use of surface alloy formed by vapour deposition of atoms of one metal onto a single crystal surface of another (Section 1.3.1).

The critical question now addressed was whether the results could be understood simply in terms of the effect of the 'inert' metal on the mean surface ensemble size of the active partner, or whether there was some movement of electrons between the components, modifying the properties of the more active one. Once again, effort was concentrated on bimetallics formed from Groups 10 and

11, because the additional electron brought about a catastrophic decrease in activity, and because many of the combinations showed complete miscibility. It has been the aspiration of many workers in catalysis to find an unambiguous answer to the straightforward question: ensemble size effect or ligand effect? The answer is not however always so simple as the question, and unfortunately most scientists have failed to appreciate that the best answer may be 'a little one and a lot of the other.' We may anticipate the conclusion of a great body of experimental work by saying that, when a bimetallic system is formed by two metals differing greatly in catalytic activity but only slightly in electronic constitution (e.g. Ni-Cu; Pd-Ag), ensemble size is the dominant factor; but when the two metals differ significantly in electronic structure (eg. Pt-Sn; Pt-Re) then the ligand effect may also be significant.

Calculation of the number of pairs, triplets or larger ensembles of one kind of atom randomly dispersed on a plane surface containing two kinds is a simple application of binomial theorem.[13] Use of the results in real systems is however predicated on a number of assumptions, and conditions that have to be met.[13] These may be enumerated as follows.

- The bimetallic system must be homogeneous, comprising a single phase: phase separation may occur below a critical temperature as with the Ni-Cu/SiO_2 catalysts.[58] Ordered superlattices may occur at certain compositions (e.g. Pt_3Cu).
- There must be no short-range ordering, i.e. no preferred formation of clusters of one component in the bulk or on the surface. This condition is quite well met with Ni-Cu and perhaps other Group 10–11 bimetallics, but the greater the disparity in electronic structure the more likely it is to occur.
- If segregation of one component to the surface is thought likely, and in theory it is nearly always possible, the surface composition must be measured, for example, by XPS.
- The possibility of reaction- or chemisorption-induced heterogeneity has to be recognised, as the component interacting most strongly with the reactants may be drawn to the surface. The surface composition has therefore to be checked *after* reaction, as well as before.
- In the case of small particles and stepped or kinked surfaces, the component of lower surface energy should segregate preferentially to the site of the *lowest* co-ordination number, where it makes greatest impact on the total energy; and as its concentration is increased, it will occupy progressively edges, rough surfaces and open planes (e.g. fcc(100)), and finally close-packed planes.[13] Relevant calculations have been performed for a number of systems and these have shown that the distinction between each class of site is not clear cut, but depends on the difference between the energies of energetically adjacent sites.

- Finally it must be remembered that, as particle size is decreased, there is a rising minimum concentration of one component needed to cover the surface, even if its tendency to surface segregation is total. Thus for example at 50% dispersion, less than 50% of either component will not be sufficient to cover the surface fully. For this reason, if there is a broad size distribution the surface composition of all particles may not be the same.

The quantitative significance to be attached to observations of changes in rate with composition is therefore somewhat problematic, and indeed there are further points of uncertainty to be considered. Enhanced activity is sometimes seen on adding a Group 11 element to an active one, and this may well result from a reduced rate of deactivation by elimination of the larger ensembles on which toxic deposits originate. It is therefore important to record *initial* activities and deactivation rates if proper comparison is to be made. It is also possible that atoms of the Group 11 metal adjacent to an ensemble of active atoms can join in the reaction, i.e. hetroatomic sites may be initiated, and in the limit spillover to areas of the 'inactive' component may contribute to the reaction. These possibilities are however not easily confirmed, but the fall in the heat of hydrogen chemisorption on Ni-Cu/SiO$_2$ catalysts with increasing copper content may be one indicator of it.[12] Alternatively, hydrogen atoms may be forced into less energetic sites involving only nickel. Failure to recognise surface heterogeneity and its effect on the location of Group 11 atoms may have important consequences: if for example the active centre demands an active atom of low co-ordination number (CN), and if these are quickly replaced in preparations containing some Group 11 metal, then activity will quickly decrease as the active sites are annihilated, and the size of the active ensemble may be thus greatly over-estimated. The same effect arises when say a single nickel atom is replaced by a copper atom, if several other nickel atoms are thereby eliminated from participating in an active centre.[73] These considerations may explain the unreasonably large values of Balandin number reported in the work of G.-A. Martin and his associates.[58] This tendency for low CN sites to be first occupied by atoms that lower the particles energy has however been used constructively by Bernard Coq and his colleagues to synthesise catalysts in which such sites are absent: marked changes in reaction specificity occur in consequence.

The use of bismuth as an inert site-blocking atom on a single crystal surface (Pt(111)) has produced interesting results.[61] Increasing its surface concentration lowers the chance of dehydrogenation or decomposition of cyclic hydrocarbons as against their chance or desorbing unchanged upon heating: cycloalkanes are affected most, then aromatics, and cycloalkenes least. However the point is strongly urged that one has to distinguish between the number of atoms upon which a molecule can adsorb unchanged and the number needed to produce the state which will upon desorption give the dehydrogenated or decomposed products plus gaseous hydrogen. This latter number, the *true ensemble requirement*, of course includes the number to accommodate the hydrogen atoms that are formed; it is of the order of five to ten for cyclopentene. However such numbers are only reliably derived from the effect of site-blocking upon the dehydrogenation/desorption ratio when this is very small. Campbell concludes[61] with the useful caveat that

> One cannot *in general assess even qualitatively which reactions require a larger number of free sites simply by observing which reaction are poisoned more rapidly by the addition of a site-blocking agent such as bismuth. One must therefore seriously question even the qualitative conclusions about relative ensemble requirements that have previously been obtained in this manner.*

It has often been suggested that a single atom of an active metal in a matrix of an inactive one can perform catalytic functions;[74,75] the thought originated with N.I. Kobozev's concept of the active ensemble,[57] and we shall meet one or two cases when it seems very likely. Such an atom is most likely to suffer a ligand effect from its neighbours, but this does not necessarily imply an actual transfer of charge; it is quite possible that it simply experience a change in the occupancy of its electron energy levels: indeed the broadening of energy bonds as revealed by UPS[58] strongly supports the idea. An analogous effect is seen with the 'giant' magnetic moments shown by iron and cobalt atoms when in the midst of palladium or platinum atoms.[58] Charge transfer in the Pt-Re, Pt-Mo and Pt-W systems is much more likely, and a model of the $Pt-Re/Al_2O_3$ catalyst[76] in which a core of rhenium atoms modifies a surface layer of platinum atoms carries a good deal of conviction. Bimetallic catalysts formed by metals of the same Group rarely produce any surprises.

The point made earlier about the need adequate experimentation before drawing conclusions is just as relevant here as when considering particle size effects. The existence of a true ligand can only be detected if there is a distinct change in the kinetic parameters, particularly the activation energy: extensive work[58] on Ni/SiO_2 and $Ni_{65}Cu_{35}/SiO_2$ shows this is not so in this case and the nature of the active centre remains the same. We shall find other examples of such behaviour. If however the activation energy changes with composition, then the shape of the rate dependence on it will be temperature-dependent, and the shapes at the high and low temperature may support conflicting theoretical models.

It would be quite wrong to conclude this discussion on a totally negative note. The use of bimetallic catalysts has indeed drawn a very clear qualitative distinction between structure-sensitive and structure-insensitive reactions, as was shown clearly by Sinfelt's comparison[77–81] of the rate dependence of cyclohexane dehydrogenation and ethane hydrogenolysis on the copper content of Ni-Cu catalysts: the former was surprisingly constant over a wide range of concentration, while the latter fell steeply.[13] This marked difference in behaviour was confirmed by the work of Clarke,[82] and of many others, using various bimetallic compositions. There is thus general similarity between a reaction's sensitivity to ensemble size and to particle size, since the availability of suitable large ensembles of active atoms to a reaction requiring them diminishes as a size is decreased. However the requirements for ensemble size and for atoms of low co-ordination number are in conflict, as they move in opposite directions as size increases. One cannot predict which will be the dominant factor, the various possibilities being shown in Figure 5.10: the existence of a minimum in the TOF vs. size plot is unlikely. Where TOF increases with dispersion, the terms 'positive' or 'sympathetic' sensitivity have been used (curve 3); for the converse (curve 2), the terms 'negative' or 'antipathetic' are applied.[24] But whatever misgivings one may have concerning the quantitative interpretation of the results, their qualitative meaning is clear; and in reactions of hydrocarbons allowing multiple products, variation of product yields with surface composition reveals further aspects of ensemble-size sensitivity.

One last point: little consideration is given to the co-ordination number requirements of active ensembles in bimetallic catalysts. Statistical calculations have assumed an infinite flat surface, although the Monte Carlo treatment by King places active atoms preferentially in high co-ordination sites,[83] because in those positions they minimise the surface energy. The non-equivalence of ensemble-size sensitivity and particle-size sensitivity must be kept clearly in mind.

5.6. THE PHENOMENON OF 'COMPENSATION'[27,84-89]

It is said that 'In argument there is much heat but little light', and its truth is amply verified by the extensive literature on compensation phenomena: for on no other subject in the field of catalysis has so many words been expended to such little purpose. The experimental observation is in essence very simple: in a series of catalysts or reactions sharing a common feature there is often observed a linear correlation between activation energy and pre-exponential factor, of the form

$$\ln A = mE + c \qquad (5.34)$$

and its occurrence is so frequent, and its precision sometime so great, that theoretical attention to it is inevitable. In most cases the values of E and $\ln A$ have been obtained by plotting the logarithm of the *rate* versus T^{-1} according to the Arrhenius equation in the form (Section 5.25)

$$\ln r = \ln A_a - E_a/RT \qquad (5.35)$$

$\ln A_a$ being the intercept when T^{-1} is zero (Figure 5.1). In such cases the compensation equation should be written

$$\ln A_a = mE_a + c \qquad (5.36)$$

This relation was first observed by F.H. Constable in 1925, and was subsequently noted by G.-M. Schwab,[90] who was the first to use the term 'compensation', and who also devised the name 'Theta Rule', because of a supposed dependence of activation energy on reduction temperature. E. Cremer was also one of the earliest students of the phenomenon. The use of the term 'compensation' is apposite, since an increase in activation energy, which causes a decrease in rate, is *compensated* by an increase in the pre-exponential factor, and indeed exact obedience to equation (5.34) requires there to be a point at which the rates of all members of the set are identical.[84] This is the *isokinetic point*, and at temperatures lower than the *isokinetic temperature* T_i the member having the *lowest* activation energy displays the *fastest* rate; *the reverse is however true above* T_i. By simple algebra, the reciprocal of the slope m equates to RT_i. Thus depending on the location of T_i with respect to the experimental results, one may observe either compensation, negative compensation or no compensation (Figure 5.11). Negative compensation is rare

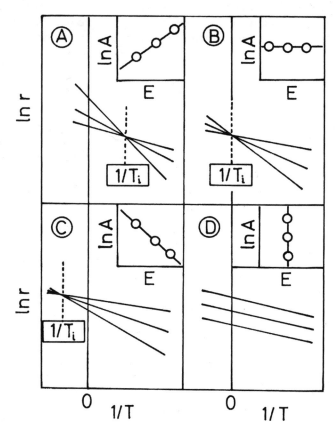

Figure 5.11. Various types of compensation produced by progressive increase of T_i: A, positive compensation; **B** and **D**, no compensation; **C**, negative compensation.

and of doubtful occurrence; cases of no compensation where rate is determined solely by ln A permit a quite straightforward explanation. Would that this could be said of 'compensation' itself.

It is also important to note that the most active catalyst of a group is only that with the lowest activation energy when measurements are made below T_i: above this temperature, the opposite is true, thus negating the commonly held (and taught) correlation of activity with a lowering of activation energy, which is (as the Arrhenius equation shows) not the sole determining factor.

The phenomenon of compensation is not unique to heterogeneous catalysis: it is also seen in homogeneous catalysts, in organic reactions where the solvent is varied and in numerous physical processes such as solid-state diffusion, semiconduction (where it is known as the Meyer-Neldel Rule), and thermionic emission (governed by Richardson's equation[12]). Indeed it appears that kinetic parameters of any activated process, physical or chemical, are quite liable to exhibit compensation; it even applies to the mortality rates of bacteria, as these also obey the Arrhenius equation. It connects with parallel effects in thermodynamics, where entropy and enthalpy terms describing the temperature dependence of equilibrium constants also show compensation.[88] This brings us the area of linear free-energy relationships (LFER), discussion of which is fully covered in the literature, but which need not detain us now.

Much trouble has been taken to find statistical criteria that will establish the validity of an *isokinetic relationship* (IKR)[84,91,92]. The compensation equation (5.35) is not statistically sound, because both slope and intercept are derived from the same results and are therefore not independent: error in one determines the error in the other. The existence of an isokinetic point can only be established by showing that the individual Arrhenius plots have a common solution. This entails the somewhat fruitless discussion of how accurate is 'accurate', and also involves careful assessment of experimental errors.[92,93] This is however a minefield into which we need not enter: it is sufficient to know there are very many cases where activation energy varies over a wide range, much greater than can be excused by experimental error, and even though the points do not lie exactly on a compensation plot (sometimes called a *Constable plot*[89]) we know that there is a correlation between the two kinetic parameters that requires our attention (see Figure 5.12 for an example). It has been suggested that the term 'compensation' be used in such cases, the name IKR being retained for those in which a proper statistical analysis confirms the existence of an isokinetic point.

It would be tedious and of no great value to review the many and varied attempted explanations of the phenomenon of compensation (see Further Reading), because as we shall see shortly, they may be wide of the mark at least as far as reactions of hydrocarbons go. There are however a number of review articles that may be consulted if desired.[86–89,94] On reflection however it seems more fruitful to focus attention on the reason for the variation in the activation energy, and perhaps

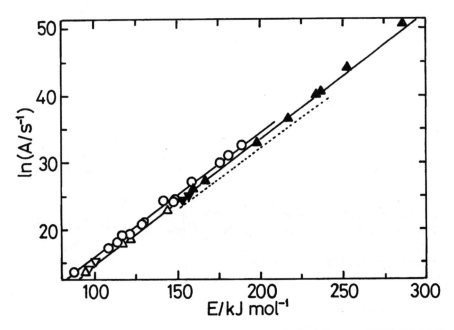

Figure 5.12. Compensation (Constable) plot for the hydrogenolysis of alkanes on EUROPT-1 (6.3% Pt/SiO$_2$).
O Total rates of reaction for linear C$_3$—C$_6$ alkanes and neopentane; $\triangle$ hydrogenolysis rates only; $\blacktriangle$ isomerisation rates only; broken line, results for ethane.

to try to connect this with some attribute of the catalyst or reaction, the variation of which is causing it. It is necessary to remind ourselves that the Arrhenius equation properly applies to the *rate constant* (equation (5.1)), and that when *rate* is used in an Arrhenius plot, the activation energy must be regarded as apparent, i.e. the *true* value moderated by the appropriate heats of adsorption, unless and until it can be shown that the surface concentrations do not change significantly over the range of measurement, when orders of reaction will be zero (Section 5.2.5). This simple truth is unfortunately overlooked by many of those seeking an explanation of compensation. There is a very strong possibility that, at least within the area of our interest, *compensation is simply an inevitable consequence of the use of apparent kinetic parameters*, and *that a major cause of the variation of* E_a *lies in the varying contributions made by the heats of adsorption terms* (Section 5.2.5).

In the case of the hydrogenation of alkenes (Chapter 7), surface coverage by reacting species is high and not very temperature-dependent: activation energies are often in the region of 40 kJ mol^{-1} and compensation is rarely seen. The same applies to the hydrogenation of alkynes (Chapter 9), where activation energies are often about 60 kJ mol^{-1}. It therefore seems likely that, following the Temkin equation (5.30), these are the true activation energies for those processes, and that as expected they do not respond much to changes in catalyst structure or composition. They are therefore widely regarded as being *structure-insensitive* reactions. Some refinement of these statements may be proved necessary in due course, because although orders in the hydrocarbon are often close to zero, those in hydrogen are usually about first. This matter will be taken up again in Chapters 7 to 9. The hydrogenation of benzene and other aromatics is also a special case to be considered in Chapter 10.

Entirely different behaviour is found with the hydrogenolysis of alkanes (Chapters 13 and 14). Here activation energies are high, typically 100–250 kJ mol^{-1}, but sometimes reaching the astronomic value of 400 kJ mol^{-1}. Orders in alkane are about unity, but the rate as a function of hydrogen pressure passes through a maximum, the position and sharpness of which depend upon the alkane and upon temperature.[89] Most significantly, apparent activation energy increases with hydrogen pressure[95] (Figure 5.13), as model calculations confirm[33] (Section 5.2.4), *and the pre-exponential factor varies sympathetically,*[95] as is also the case with the model calculations.[33] Thus *compensation is observed within a single*

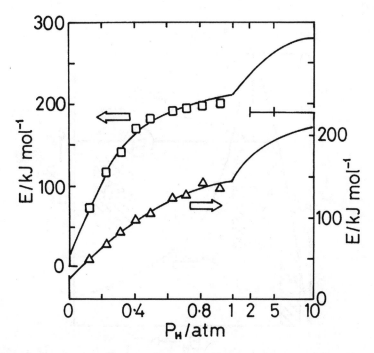

Figure 5.13. Dependence of E_{app} on hydrogen pressure for hydrogenolysis of propane and of n-butane on EUROPT-3 (0.3 % Pt/Al$_2$O$_3$, curves are calculated by eqn. ES5B (see Chapter 13) using constants of best-fit.

system of catalyst and reactants, and its origin is thereby proved to lie in variations of surface coverage caused by differences in heats of adsorption. The high activation energies are attributed to the existence of an endothermic pre-equilibrium in which the alkane first loses several hydrogen atoms; the concentration of this critical dehydrogenated intermediate therefore *increases* with temperature, and E_t is therefore less than E_a, unlike the classic situation where coverage by the adsorbed intermediate *falls* because it is exothermically adsorbed. The often-noted decrease of E_a with the alkane's chain length also receives a ready explanation, as coverage by hydrogen atoms decreases simultaneously. Comparison of alkane reactivities therefore needs to be made at equivalent surface concentrations. Adsorbed hydrocarbon species are present in comparatively small amounts, requiring somewhat specific sites: hydrogenolysis of the C—C bond is therefore regarded as being *structure-sensitive*, and we have therefore succeeded in connecting sensitivity, reaction kinetics, activation energy and compensation in one reasonably satisfying picture (Figure 5.14). In a bimolecular process, it is the adsorption requirement of the more weakly adsorbed reactant that determines the degree of structure sensitivity. Additional flesh will be added to this skeleton in later chapters.

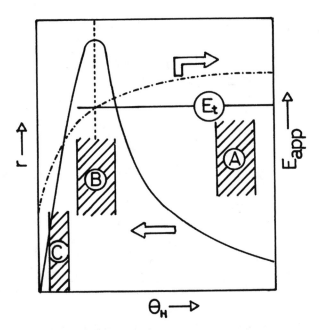

Figure 5.14. Schematic dependence of rate (full line) and of E_{app} (broken line) on θ_H, showing the locations of three different values of K_H.

An interesting, important, but rarely discussed question which now arises is the source of the energy needed to activate adsorbed species for reaction. The translational energy available in a homogeneous gas-phase reaction is lacking in species that are set in place on a surface, although the collection of vibrational quanta into a critical bond remains a possibility. However, just as a solid can act as an energy sink in the recombination of hydrogen atoms (Section 3.4.3), so it may act as an energy source by transfer of phonons (lattice vibrational energy quanta) to adsorbed species. This concept is at the heart of one of the suggested explanations of compensation.[94,96,97]

Finally it may be said that a compensation plot has a number of uses. (1) A trivial use is to detect and then to suspect experimental data points lying well away from the main line.[85] (2) It has been suggested that members of the set defined by a line on a compensation plot enjoy the same basic mechanism, differing only in the energetic profile of the reaction[85] (Figure 5.12). (3) When therefore the data points appear to require two or more separate lines (see Figure 5.15 for an example), it may be thought that two or more different mechanisms are operative.[19]

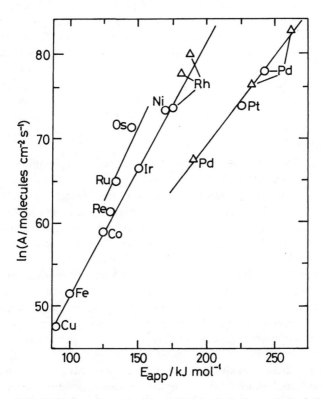

Figure 5.15. Multiple compensation (Constable) plot for hydrogenolysis of ethane.

5.7. THE TEMKIN EQUATION: ASSUMPTIONS AND IMPLICATIONS

We have seen in the last Section how distinguishing between true and apparent activation energies may resolve the long-standing debate over the significance of compensation phenomena, and in Section 5.2.5 how these distinct quantities are linked through the heats of adsorption of the reactants by the Temkin equation (5.31). It is now necessary to explore the assumptions underlying this equation, and to pursue certain implications that arise from it.

The terms that moderate the heats of adsorption in this equation are the orders of reaction x and y, so clearly the Power Rate Law formalism is being employed. While this is adequate to describe limited cases where x or y is unity or zero, a single exponent of the pressure cannot account for the variation of rate with pressure over a wide range, and the use of the Langmuir-Hinshelwood formalism then becomes obligatory (Section 5.2.4). Because in general surface coverages by the reactants will decrease with rising temperature, values of x and y in fixed ranges of pressure will tend to increase, and where tis occurs the measured activation energy should *decrease* as temperature is increased, because according to the Temkin equation the terms containing the heats of adsorption become more significant (the heats of adsorption are of course negative). There are few cases where the operation of the Temkin equation has been observed for certain, part of the difficulty being that a change in activation energy can have other causes, such as the onset of mass-transport limitation or decreasing rate due to deactivation. One possible example is the hydrogenation of aromatic compounds, where above a certain critical temperature the rates start to decrease, and negative activation energies result: the exact interpretation of this effect however remains uncertain (Section 10.2.4).

The fact that Arrhenius plots for hydrocarbon transformations are so often quite linear over the whole of the measured range raises a doubt that the Temkin equation can or should be applied to them, even although the orders of reaction would seem to demand it. The problem may be illustrated as follows. It has been suggested in the previous section that the somewhat low activation energies habitually found with the hydrogenation of alkenes, alkadienes and alkynes are *true* values, because it appears from the limited evidence on the temperature-dependence of the orders of reaction that surface concentrations are not changing much over the temperature range used (see Chapter 7, 8 and 9). Unthinking application of the Temkin equation would require the unknown heat of adsorption of hydrogen to be *added* to the apparent activation energy, since orders in hydrogen are often first (or greater). If, however, as seems possible, the slow step in these cases involves the collision of an undissociated hydrogen molecule with some chemisorbed hydrocarbon species, it would clearly be wrong to do this.

A related problem arises with the hydrogenation of aromatic molecules, where it is well documented that orders with respect to both reactants increase with temperature, but the observed activation energy remains unchanged (Section 10.2.2); so according to the Temkin equation this would imply a variable *true* activation energy (E_t), which seems unlikely: and application of this equation, using heats of adsorption derived by kinetic analysis, led to astronomic values for E_t, which if not impossible are at least improbable. We must therefore conclude that *the Temkin equation* (or a modification of it containing adsorption coefficients rather than reaction orders) *should only be used where the mechanism is established, where reversible adsorption of both reactants is known to occur, and where surface concentration change significantly over the temperature range employed.* An attempt to apply the equation to alkane hydrogenolysis, where *pressure*-dependence of activation energy is well established, will be discussed in Chapter 13 (see also Section 5.6).

We must conclude that the effect of increasing temperature is two-fold: (i) through diminishing surface concentrations of reactants in line with their (coverage-dependent) heats of adsorption, and (ii) through the operation of E_t on each reacting centre. Now the height of the potential barrier to be overcome is also likely to depend on the strengths of adsorption, so the heat terms govern *both* coverage *and* reactivity. This suggests that we might look for some correspondence between E_t and adsorption heats, and that it might be possible for E_t to vary even when surface coverage remains close to unity in a series of related catalysts, because while coverage cannot exceed unity the heat terms can be higher to variable extents than needed to secure complete coverage. It is worth noting that the effect of *decreasing* heats of adsorption with increasing surface coverage, whether due to induced heterogeneity or to the arrival of weaker chemisorbed states, works to reinforce the effects of changing temperature and pressure on rates. Thus if an increase in temperature leads to a lower rate than expected because coverage has decreased, the resulting *increase* in adsorption heat will increase E_t and give an even slower rate. Similarly if an increase in pressure gives a faster rate because there are now more reacting species, this increase in coverage leads to a lower heat of adsorption, a smaller E_t and an even faster rate.

5.8. TECHNIQUES

5.8.1. Reactors[98]

The main types of laboratory reactor were introduced in Section 5.2.3, and, since (as the reader is constantly reminded) this is not intended as a handbook of catalytic practice, all that is necessary now is to add a little further detail, and briefly allude to other types of reactor that may be encountered.

First of all, it is not proposed to deal at all with pilot plant or industrial-scale reactors, as this is a very specialised area, adequately covered in existing texts,[14–16,99] nor is it necessary to deal with three-phase reactors, although some results obtained with their use will be shown; this is an area where great care has to be taken to avoid mass-transport limitation.[8] Most of our concern will be with gas-phase reactions, where continuous-flow microreactors using 0.1 to 1 g catalyst are in common use. There are obvious limits to the rates that can be obtained at fixed temperature by altering flow-rate: the upper limit is dictated by possible entrainment of catalyst in the gas, and the lower limit by the difficulty of measuring very slow flow-rates. Meaningful results can however be obtained at conversions as low as 0.1% in favourable cases. The great advantage of constant-volume reactors is that they can be operated for long periods for very slow rates of reaction.

In the *pulse-flow mode*,[13] shots of a hydrocarbon are injected periodically into a flow of hydrogen, and the whole mix of products and unchanged reactants analysed. In the short residence time, formation of carbonaceous residue is minimised, and the surface is cleansed between shots, but the actual composition of the reacting mixture is unknown, and the method is unsuited for obtaining quantitative kinetics. It is better to inject a hydrocarbon and hydrogen mixture into a hydrogen stream, and with care accurate kinetics are got by using times as short as 1 min: cleansing occurs between reaction periods as before, but if deactivation is unavoidable the regular use of standard compositions still allows viable results to emerge.[95] Problems arise however with microporous catalysts (e.g. zeolite supports) because hydrocarbons are strongly retained and are not quantitatively recoverable in the short term.[100] Microprocessor-controlled reactors allow lengthy sequences of varied conditions to be pre-programmed, and with data processing of product analysis by computer all that remains to do is to write the paper.[101]

Very useful information is obtained by the method of *transient kinetics*.[40,102–106] Here an abrupt change is made to the conditions in a flow-reactor, and the temporal consequences of that change analysed. Most simply the flow of one reactant is stopped, and the change in product yield with time followed: thus, for example, if in hydrocarbon-hydrogen reaction the hydrocarbon flow is stopped, the integrated amount of produce subsequently formed measures the amount of intermediate species on the surface (Figure 5.16). A step change made to the

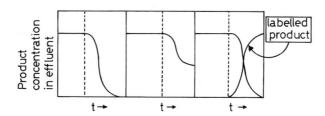

Figure 5.16. Transient kinetics: **A**, flow of reactant stopped; **B**, flow of reactant altered; **C**, isotopic variant of reactant introduced.

flow-rate of either reactant will have consequences that can give kinetic/ mechanistic information, and the replacement of one reactant by an isotopic variant is perhaps the best way of all because then there is no change in surface conditions, so that the emergence of a labelled product reflects the actual surface concentrations (Figure 5.16). Deuterium and tritium are not however suitable labels, as their atoms are very mobile as we shall see in the following chapter. ^{13}C is much to be preferred.

A further advance is the *TAP (Temporal Analysis of Products) reactor*, where very fast reaction steps can be followed, but the equipment needed is sophisticated, and has not been much used for reactions of interest to us. Catalytic reactions can also be run in an infrared spectrometer cell, so that adsorbed species can be inspected after if not during reaction. Finally reactions can now readily be performed inside UHV apparatus, so that detailed knowledge of the state of the surface single crystals during and after reaction is accessible.[13,107]

Heterogeneous catalysis is a subject full of surprises, and one of these, long known but never completely explained, is the difference sometimes observed between the kinetic laws derived from constant volume reactors, (a) by following the change of rate with time, and (b) by observing the effect of changing reactant pressures on initial rate. A specific case in point is the hydrogenation of ethyne and other alkynes, where with hydrogen/alkyne ratios exceeding two, the time-order is accurate zero, i.e. the rate stays the same for much of the reaction, whereas the pressure-order is approximately first;[108,109] and moreover a sudden change in hydrogen pressure alters the rate.[110,111] This does not happen when the hydrogen/alkyne ratio is below two. It seems possible that in the former condition a self-sustaining surface chain reaction is set up (see Chapter 10).

5.8.2. Use of Stable and Radioactive Isotopes[112–115]

The discovery of deuterium by H.C. Urey in 1932 was seized upon by physical chemists interested in kinetics and mechanism, and indeed proved a godsend to students of heterogeneous catalysis, where it quickly revealed a wealth of unexpected subtlety and complexity in the mechanism of hydrocarbon reactions catalysed by metals. H.S. Taylor's laboratory at Princeton was the first to prepare heavy water (D_2O) by electrolysis,[116] but the first publications on the use of deuterium in hydrogenation came in 1934 from Manchester and Cambridge, where it was shown that in the reaction with ethene the deuterium content of the non-condensable fraction fell, so that an *exchange reaction* was taking place[117,18] (Chapter 7). Product analysis was at this time limited to thermal conductivity, and recognition of the composition of the ethene and the ethane had to await the arrival of analytical mass-spectrometry in about 1950. Infrared spectroscopy also made a useful contribution, for example in analysing mixtures of C_2H_2, C_2HD and C_2D_2, and of the positional isomers of $C_2H_2D_2$.[119,120] Taylor's laboratory also made the first measurements of the metal-catalysed exchange of alkanes with deuterium[121] (Chapter 6).

The early analytical use of mass spectrometry was difficult and tedious, as when examining the reaction of ethene with deuterium it was necessary to remove ethene by chemical absorption from a small sample of product in order to obtain the mass spectrum of the ethane fraction, and then to derive the spectrum of the ethene by difference between this and the untreated sample.[122] Obtaining the actual composition from the raw data was also troublesome because the ethane parent ion $C_2X_6^+$ (X = H or D) readily loses X_2 to form $C_2X_4^+$, and before the advent of the computer there was a tiresome calculation to be performed, involving correction for ^{13}C natural impurities and the presence of fragment ions. The extent of fragmentation (less with ethene and with propane than with ethane) decreases at low ionising voltage as the appearance potential of the fragment is approached, but in the case of ethane this is actually below the ionisation potential of the parent ion, so fragmentation is unavoidable.[123] Operating a mass-spectrometer at the necessary low ionising voltage of 12V was at that time a somewhat risky undertaking. The advent of preparative gas chromatography simplified the analytical procedure greatly, and then it was possible to observe the inclusion of deuterium atoms into each of the three n-butene isomers, as well as into the n-butene, in the reaction of 1-butene with deuterium[124] (Chapter 7).

Mass-spectrometry does not however report the position of a deuterium atom in a hydrocarbon molecule (i.e. it cannot distinguish CH_2D-CH_2D from CH_3-CHD_2) except by analysis of the fragment ions, which is neither easy nor reliable.[125] NMR has however come to the rescue and has been of great use in providing further detail of mechanical routes.[126] Microwave spectroscopy has distinguished the isotopomers of propene and of 1-butene and E-2-butene.[13,125] High-resolution mass-spectrometry is however now able to separate ions of nominally the same mass, thus avoiding the problem of subtraction of fragment ion currents: so for example separation of the ions $C_2H_6^+$, $C_2H_4D^+$, $C_2H_2D_2^+$ and $C_2D_4^+$ in the mass range 30.040–30.048 has been achieved.

We end with a word on nomenclature. First, as to reactions: where deuterium is substituted for hydrogen, as say in its reaction with methane, we may speak of exchange or better of *equilibration* of the mixture. In the reaction with ethene, deuterated ethenes are formed by *exchange* or deuteriation because there is no time for them to equilibrate: the process of addition to the C=C bond is sometimes called deuteriumation (logically correct, but inelegant) or deuterogenation (logically incorrect and even more unpleasant). The location of deuterium atoms can be expressed by using the Boughton convention, by which for example CH_2D-CH_2D becomes ethane-1,2-d_2 and CH_3-CHD_2 is ethane-1,1-d_2. This terminology has certain merits that will appear later.

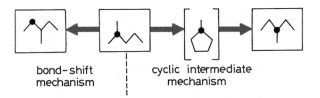

bond–shift
mechanism

cyclic intermediate
mechanism

Figure 5.17. Use of ^{13}C-labelled 2-methylpentane to distinguish bond-shift and cyclic intermediate mechanisms for isomerisation to 3-methylpentane.

Tritium is a weak β^- emitter[127–130] (E_{max}, 18.6 keV) and has seen some use in chemisorption and catalysis of hydrocarbon reactions: its location in a hydrocarbon molecule can be formed by NMR spectroscopy.[131] The ^{14}C carbon isotope (E_{max},155 keV) has been successfully used to monitor chemisorption of labelled hydrocarbons and to assess the importance of final product formation from an intermediate product.[132] Thus for example in the reaction Scheme 5.1, addition of labelled X during the reaction and measurement of the activity in the product Z allows the importance of the re-adsorption of gaseous X to be estimated.[131] ^{13}C has played an important role in the work of François Gault and his associates in determining mechanisms of skeletal isomerisation of hydrocarbons[75,133] (see Chapter 14). The power of this technique can be simply illustrated by comparing the products formed by isomerising 2-methylpentane by alternative mechanisms (see Figure 5.17). 3-Methylpentane is formed by each route, but only the position of the label reveals the route involved. Very often both mechanisms operate simultaneously, and for details of the elegant but tedious way in which the product is degraded chemically to show how much of the label is in each place, references 75 and 133 should be consulted.

REFERENCES

1. J.M. Thomas and W.J. Thomas, *Introduction to Surface Chemistry and Catalysis,* Wiley: New York (1994).
2. R.A. van Santen and J.W. Niemantsverdriet, *Chemical Kinetics and Catalysis,* Plenum: New York (1995).
3. M. Temkin, *Acta Physicochim. URSS* **13** (1940) 1.
4. *Catalysis from A to Z: a Concise Encyclopedia,* (B. Cornits, W.A. Hermann, R. Schlögl and C.-H. Wong, eds.). Wiley-VCH: Weinheim (2000).
5. M. Boudart in; *Handbook of Heterogeneous Catalysis,* Vol. 3 (G.Ertl. H. Knözinger and J. Weitkamp, eds.), Wiley-VCH: Weinheim (1997), p. 958.
6. B.H. Davis in: *Handbook of Heterogeneous Catalysis,* Vol. 1 (G.Ertl. H. Knözinger and J. Weitkamp, eds.), Wiley-VCH: Weinheim (1997), p. 13.
7. J. Trofast, *Chem. Britain* (1990) 432.

8. G.C. Bond, *Heterogenous Catalysis: Principles and Applications,* 2nd Edn., Oxford U.P: Oxford (1987).

9. J.J. Berzelius, *Anals Chim. Phys.* **61** (1836) 146.

10. A.J.B. Robertson, *Platinum Metals Rev.* **19** (1975) 64.

11. D. McDonald and L.B. Hunt, *A History of Platinum,* Johnson Matthey: London (1982).

12. G.C. Bond, *Catalysis by Metals,* Academic Press: London (1962).

13. V. Ponec and G.C. Bond, *Catalysis by Metals and Alloys,* Elsevier: Amsterdam (1995).

14. R.J. Farrauto and C.H. Bartholomew, *Fundamentals of Industrial Catalytic Processes,* Chapman and Hall: London (1997).

15. C.N. Satterfield, *Hetreogeneons Catalysis in Practice,* 2nd. Edn., McGraw-Hill: New York (1991).

16. J.A. Dumesic, D.F. Ruda, L.M. Aparicio, J.E. Rekoske and A.A. Trevino, *The Microkinetics of Heterogeneous Catalysis,* Am. Chem. Soc.: Washington DC (1993).

17. R.J. Wijngaarden, A. Kronberg and K.R. Westerterp, *Industrial Catalysis,* Wiley-VCH: Weinheim (1998).

18. J.W.E. Coenen in: *Perspectives in Catalysis*, (R. Larsson, Ed.), C.W.K. Gleerup: Lund (1981), p. 19.

19. G.C. Bond, *Chem. Soc. Rev.* **20** (1991) 441.

20. IUPAC Manual of Symbols and Terminology for Physicochemical Qunatities and Units, App. II— Part II, Heterogeneons Catalysis, (R.C. Burwell Jr., ed), *Pure and Appl. Chem.* **14** (1976) 71.

21. M. Boudart, *Chem. Rev.* **95** (1995) 661.

22. M. Boudart and G. Djega-Mariadassou, *Kinetics of Heterogeneous Catalytic Reactions,* Princeton Univ. Press: New York (1984).

23. F.H. Ribeiro, A.E. Schach von Wittenau, C.H. Bartholomew and G.A. Somorjai, *Catal. Rev.- Sci. Eng.* **39** (1997) 49.

24. C.O. Bennett and M. Che, *J. Catal.* **120** (1989) 293.

25. H.S. Taylor, *Chem. Rev.* **9** (1931) 1.

26. C.N. Hinshelwood, *Kinetics of Chemical Change,* Clarendon Press: Oxford (1940).

27. *Contact Catalysis* (Z.G. Szabó and D. Kalló, eds.) Elsevier: Amsterdam (1976): vol. 1, #2.2.

28. O.A. Hougen and K.M. Watson, *Chemical Process Principles,* Wiley: New York (1948).

29. R.H. Yang and O.A. Hougen, *Chem. Eng. Prog.* **46** (1950) 146.

30. K.J. Laidler, *Chemical Kinetics,* Harper and Row: New York (1987).

31. G.C. Bond, R.H. Cunningham and J.C. Slaa, *Topics Catal.* **1** (1994) 19.

32. R.I. Masel and Wei Ti Lee, *J. Catal.* **165** (1997) 80.

33. G.C. Bond, A.D. Hooper, J.C. Slaa and A. O. Taylor, *J. Catal.* **163** (1996) 319.

34. G.A. Martin, *Bull. Soc. Chim. Belg.* **105** (1995) 131.

35. J.M. Hermann, *Appl. Catal. A: Gen.* **156** (1997) 285.

36. G.C. Bond, *Ind. Eng. Chem. Res.* **36** (1997) 3173; *Catal. Today* **17** (1993) 399.

37. V.P. Zhdanov and B. Kasemo, *J. Catal.* **170** (1997) 377.

38. V.P. Zhdanov and B. Kasemo, *Surf. Sci. Rep.* **39** (2000) 25.

39. P. Stolze, *Prog. Surf. Sci.* **65** (2000) 65.

40. S.B. Shang and C.N. Kenney, *J. Catal.* **134** (1992) 134.

41. W.L. Holstein and M. Boudart, *J. Phys. Chem. B* **101** (1997) 9991.

42. G.C. Bond and P.B. Wells, *Adv. Catal.* **15** (1964) 92.

43. R. Burch, *Catal To-day* **10** (1991) 233.

44. M. Che, C.O. Bennett, *Adv. Catal.* **36** (1989) 55.

45. R. Burch in *Specialist Periodical Reports: Catalysis,* Vol. 7 (G.C. Bond and G. Webb, eds.), *Roy. Soc. Chem.* (1985), p. 149.

46. *Active Centres in Catalysis*, (R. Burch, ed.), *Catal. Today* **10** (1991) 233.

47. G.A. Somorjai, *Langmuir* **7** (1991) 3176.

48. G.A. Somorjai, *Adv. Catal.* **26** (1977) 2.

49. H.S. Taylor, *Proc. Roy. Soc. A* **108** (1925) 105; *J. Phys. Chem.* **30** (1926) 145; **31** (1927) 277.
50. M. Boudart, *Adv. Catal.* **20** (1969) 153.
51. M. Boudart, A.W. Aldag, J.E. Benson, N.A. Dougharty and C.G. Harkins, *J. Catal.* **6** (1966) 92.
52. A.D. O Cinneide and J.K.A. Clarke, *Catal. Rev.* **7** (1972) 213.
53. G.C. Bond, *J. Catal.* **136** (1992) 631.
54. S.V. Norval, S.J. Thomson and G. Webb, *Appl. Surf. Sci.* **4** (1980) 51.
55. E.K. Rideal, *Concepts in Catalysis,* Academic Press: London (1968).
56. B.M.W. Trapnell, *Adv. Catal.* **3** (1951) 1.
57. A.N. Mal'tsev, *Russ. J. Phys. Chem.* **61** (1987) 1504.
58. G.A. Martin, *Catal. Rev. - Sci. Eng.* **30** (1988) 519.
59. A. Frennet, *Catal. Rev.* **10** (1974–5) 37.
60. A. Frennet, *Catal. Today* **12** (1992) 131.
61. C.T. Campbell, *Ann. Rev. Phys. Chem.* **41** (1990) 775.
62. D. Farin and D. Avnir, *J. Am. Chem. Soc.* **110** (1988) 2039; *J. Catal.* **120** (1989) 55.
63. J.J. Carberry *J. Catal.* **114** (1988) 277.
64. R. van Hardeveld and F. Hartog, *Adv. Catal.* **22** (1972) 75.
65. R. van Hardeveld and F. Hartog, *Surf. Sci.* **17** (1969) 90.
66. R. van Hardeveld and F. Hartog, *Surf. Sci.* **15** (1969) 189.
67. L.P. Ford, P. Blowers and R. I. Masel, *J. Vac. Sci. Technol. A* **17** (1999) 1705.
68. J. R. Anderson, *Sci. Prog., Oxford.* **69** (1985) 461.
69. J.R. Anderson, *Structure of Metallic Catalysts,* Academic Press: New York (1975).
70. E.G. Schlosser, *Ber. Bunsenges. Phys. Chem.* **73** (1969) 358.
71. G.C. Bond and R. Burch in: *Specialist Periodical Reports: Catalysis,* Vol. 6 (G.C. Bond and G. Webb, eds.), *Roy. Soc. Chem.* (1983), p. 27.
72. B.J. Kip, F.B.M. Duivenvoorden, D.C. Koningsberger and R. Prins, *J. Catal.* **106** (1986) 26; *J. Am. Chem. Soc.* **108** (1986) 5633.
73. I. Alstrop and N.T. Anderson, *J. Catal.* **104** (1987) 466.
74. W.R. Patterson and J.J. Rooney, *Catal. Today.* **12** (1992) 113.
75. Z. Schay and P. Tétényi, *J. Chem. Soc. Faraday Trans. I* **75** (1979) 1001.
76. R.W. Joyner and E.S. Shpiro, *Catal. Lett.* **9** (1991) 239.
77. J.H. Sinfelt, *Catal. Rev. - Sci. Eng.* **9** (1974) 147.
78. J.H. Sinfelt in: *Catalysis—Science and Technology'* (J.R. Anderson and M. Boudart, eds.) Springer-Verlag, Berlin: (1981). p. 257.
79. J.H. Sinfelt and J.A. Cusumano, *Adv. Mater. Catal.* (1977) 1.
80. J.H. Sinfelt, *Acc. Chem. Res.* **20** (1987) 134.
81. J.H. Sinfelt, *Internat. Rev. Phys. Chem.* **7** (1988) 281.
82. J.K.A. Clarke, *Chem. Rev.* **75** (1975) 291.
83. J.K. Strohl and T.S. King, J. Catayl. **116** (1989) 540; **118** (1989) 53.
84. W. Linert and R.F. Jameson, *Chem. Soc. Rev.* **18** (1989) 477.
85. G.C. Bond, *Appl. Catal. A: Gen.191* (2000) 23.
86. A.K. Galwey, *Thermochim. Acta* **294** (1997) 205; *Adv. Catal.* **26** (1977) 247.
87. J.J. Rooney. *J. Molec. Catal. A: Chem.* **96** (1995) L1.
88. A.K. Galwey, *J. Catal.* **84** (1983) 270.
89. G. C. Bond, M.A. Keane, H. Kral and J.A. Lercher, *Chem. Rev. - Sci. Eng.* **42** (2000) 323.
90. G.M. Schwab, *Adv. Catal.* **2** (1950) 251.
91. W.C. Conner and W. Linert, *Orient. J. Chem.* **5** (1989) 204.
92. K. Héberger, S. Kemény and T. Vidóczy, *Internat. J. Chem. Kin.* **19** (1987) 171.
93. Z. Karpiński and R. Larsson, *J. Catal.* **168** (1997) 532.
94. G.C. Bond, *Z. Phys. Chem. NF* **144** (1985) 21.
95. G.C. Bond and R. H. Cunningham, *J. Catal.* **166** (1997) 172.

96. W.C. Conner Jr., *J. Catal.* **78** (1982) 238.
97. W.C. Conner Jr. and W. Linert, *Orient. J. Chem.* **5** (1989) 304.
98. L.K. Doraiswamy and D.G. Tajbl, *Catal. Rev.* **10** (1974–5) 177.
99. J.J. Carberry, *Catal. Rev.* **3** (1970) 61.
100. G.C. Bond and Xu Lin, *J. Catal.* **169** (1997) 76.
101. G.C. Bond and A.D. Hooper, *React. Kinet. Catal. Lett.* **68** (1999) 5; A.D. Hooper, PhD thesis, Brunel University, (2000).
102. C. O. Bennett, *Adv. Catal.* **44** (1999) 330.
103. H. Kobayashi and M. Kobayashi, *Catal. Rev.* **10** (1974–5) 139.
104. P. Biloen, *J. Molec. Catal.* **21** (1983) 17.
105. A. Frennet and C. Hubert, *J. Molec. Catal. A: Chem.* **163** (2000) 163.
106. C.O. Bennett, *Am. Chem. Soc. Symp. Ser.* **178** (1982) 1.
107. C.T. Campbell, *Adv. Catal.* **36** (1989) 2.
108. J. Sheridan, *J. Chem. Soc.* (1945), 133, 301.
109. G.C. Bond and J. Sheridan, *Trans. Faraday Soc.* **48** (1952) 651.
110. G.C. Bond, *J. Chem. Soc.* (1958) 2705.
111. G.C. Bond and R. S. Mann, *J. Chem. Soc.* (1958) 4738.
112. K.C. Campbell and S.J. Thomson, *Prog. Surf. Membrane Sci.* **9** (1975) 163.
113. R.L. Burwell Jr., *Langmuir* **2** (1986) 2.
114. F.G. Gault, *Adv. Catal.* **30** (1981) 1.
115. R.L. Burwell Jr., *Catal. Rev.* **7** (1972–3) 25.
116. C. Kemball, *Biograph. Mem. Fellows Roy. Soc.* **21** (1975) 517 (H.S. Taylor).
117. M. Polanyi and J. Horiuti, *Trans. Faraday Soc.* **30** (1934) 1164.
118. A. Farkas, L. Farkas and E.K. Rideal, *Proc. Roy. Soc. A* **146** (1934) 630.
119. G.C. Bond, J. Sheridan and D.H. Whiffen, *Trans. Faraday Soc.* **48** (1952) 715.
120. G.C. Bond, *J. Chem. Soc.* (1958) 4288.
121. K. Morikawa, W. A. Benedict and H.S. Taylor, *J. Am. Chem. Soc.* **57** (1935) 592; **58** (1936) 1795.
122. G.C. Bond, *Trans. Faraday Soc.* **52** (1956) 1235.
123. G.C. Bond and J. Turkevich, *Trans. Faraday Soc.* **49** (1953) 281.
124. G.C. Bond, J.J. Phillipson, P.B. Wells and J.M. Winterbottom, *Trans. Faraday Soc.* **60** (1964) 1847.
125. R.L. Burwell Jr., *Catal. Rev.* **7** (1973) 25.
126. A. Loaiza, M-D. Xu and F. Zaera, *J. Catal.* **159** (1996) 127.
127. L. Guczi and P. Tétényi, *Acta Chim. Acad. Sci.* **71** (1972) 341.
128. L. Guizi, A. Sárkány and P. Tétényi, *Z. Phys. Chem. NF* **74** (1971) 26.
129. L. Guczi, K.M. Sharan and P.Tétényi, *Monatshefte Chem.* **102** (1971) 187.
130. L. Guczi, and P. Tétényi, *Z. Phys. Chem.* **237** (1968) 356.
131. Z. Paál and S.J. Thompson, *J. Catal.* **30** (1973) 96.
132. G.F. Berndt in *Specialist Periodical Reports: Catalysis,* Vol. 6 (G.C. Bond and G. Webb, eds.), *Roy. Soc. Chem.* (1983), p. 144.
133. F. Gault, V. Amir-Ebrahimi, F. Gavin, P. Parayre and F. Weisang, *Bull. Soc. Chim. Belg.* **88** (1979) 475.
134. H.C. de Jongste and V. Ponec, *Bull. Soc. Chim. Belg.* **88** (1979) 453.
135. H. Charcosset, *Internat. Chem. Eng.* **23** (1983) 187, 411.
136. J.H. Sinfelt, *J. Phys. Chem.* **90** (1986) 4711.
137. R. Burch, *Acc. Chem. Res.* **15** (1982) 24.
138. V. Ponec, *J. Molec. Catal. A: Chem.* **133** (1998) 221.
139. E.G. Allison and G.C. Bond, *Catal. Rev.* **7** (1973) 233.
140. R.L. Moss in: *Specialist Periodical Reports: Catalysts,* Vol. 1 (C. Kemball and D.A. Dowden, eds.), Roy. Soc. Chem. (1977), p. 37; Vol. 4 (C. Kemball and D.A. Dowden, eds.), *Roy. Soc. Chem.* (1981), p. 31.

141. D.A. Dowden in: *Specialist Periodical Reports: Catalysis*, Vol. 2 (C. Kemball and D.A. Dowden, eds.), *Roy. Soc. Chem.* (1978) p. 1.
142. L. Guczi and A. Sárkány: *in Specialist Periodical Reports: Catalysis*, Vol. 11 (J.J. Spivey and S.K. Agarwal, eds.), *Roy. Soc. Chem.* (1994), p. 318.
143. V. Ponec, *Appl. Catal. A:* **222** (2001) 31.
144. K.A. Fichthorn and W.H. Weinberg, *Langmuir* **7** (1991) 2539.
145. R. Larsson, *Z. Phys. Chem. Leipzig* **268** (1987) 721.
146. R. Larsson, *Catal. Lett.* **11** (1991) 137; **16** (1992) 273.
147. R. Larsson, *J. Molec. Catal.* **55** (1989) 70.

FURTHER READING

EXCHANGE OF ALKANES WITH DEUTERIUM

PREFACE

The replacement of hydrogen atoms by deuterium (or tritium) atoms is the simplest catalytic change that can happen to a hydrocarbon; it occurs with surprising ease on a number of metal surfaces, and it focuses on the reactivity of C—H bonds, ignoring the C—C bonds. It has been a popular field of study, and has revealed (as all isotopic labelling studies have) an unsuspected wealth of detail, and a number of mechanistic questions have arisen that have required the study of carefully selected large hydrocarbon molecules whose structures allow only certain pathways to operate. It is one of the paradoxes of heterogeneous catalysis that reactions of larger molecules sometimes give more direct and unambiguous information about reaction mechanisms than smaller molecules are capable of. Studies of such molecules do however illuminate the stereochemical principles governing the chemisorption and reactions of hydrocarbon molecules of all sizes; and the reactions singled out for examination here will be found to occur as part of, or in parallel with, the more profound processes to be met in the following chapters. They therefore deserve careful attention.

6.1. INTRODUCTION[1,2]

The process whereby hydrogen atoms in an alkane are replaced by deuterium or tritium atoms in the presence of a metal catalyst is the simplest way of making recognisably different products, but ones that nevertheless are almost chemically identical to the reactants. The simplicity of these reactions, which proceed at conveniently modest temperatures on many metals, and which are generally (but not necessarily[3]) separable from those such as dehydrogenation and hydrogenolysis, has made them an attractive field of study since the earliest work on the lower linear alkanes using a nickel/kieselguhr catalyst in the 1930s.[4] Apart from its intrinsic interest, the behaviour of adsorbed alkyl radicals revealed by its study is relevant to other reactions of hydrocarbons such as hydrogenation and hydrogenolysis in which they are also intermediates: and recently the discovery of homologation of adsorbed carbon species with its potential for utilisation of the lower alkanes has added a further use for the information that exchange reactions provide. Although the reaction may be performed with amounts of alkane and deuterium containing equal numbers of hydrogen and deuterium atoms,[5] it has been usual to employ a large excess of deuterium (up to alkane/deuterium = 10) in order to minimise reverse reactions and so to identify the true initial products: under such conditions, deactivation due to 'carbon' deposition is rarely a problem with methane (although time-dependent changes have sometimes been seen[1,6]), although the larger the alkane the more noticeable it usually becomes, to an extent depending on the type of catalyst being used, and on the plane exposed.[7–9]

The term 'exchange' has been widely used, although strictly speaking it is only the hydrogen atoms that are exchanged or substituted, and 'equilibration' has much to commend it to describe what occurs. Plainly in the absence of any side reactions a position of equilibrium will be reached, the composition of which can be calculated at least approximately by the binomial theorem, knowing the numbers of hydrogen and deuterium atoms present. So, for example, when methane reacts with an excess of deuterium, the final product will contain mainly methane-d_4 and -d_3 (CD_4 and CHD_3). Exchange reactions are almost but not quite thermoneutral because of zero-point energy effects, and so equilibrium compositions are not exactly the statistical ones, although these are approached more closely with rising temperature. Kinetic isotope effects are also observed: thus the reaction of methane-d_4 with hydrogen is slower than that of methane with deuterium, and the activation energy for n-hexane-d_{14} plus hydrogen on palladium film was some 4 kJ mol^{-1} greater than for n-hexane plus deuterium.[8]

A common feature of the reactions of alkanes with deuterium is the occurrence of two simultaneous but distinguishable processes: in one of these, termed *single* or *stepwise exchange*, only one atom is substituted during a single residence of the molecule on the surface, while in the other, termed *multiple exchange*, all or almost all are substituted. The rate of each can be measured by analysis of the products formed, and each can be assigned its characteristic kinetic parameters:[2] multiple exchange usually shows the more negative order in deuterium and the higher activation energy. In the case of methane the stepwise mechanism predominates at low temperatures on many metals (Section 6.2.1). The separate character of the two processes is shown most clearly by (i) a minimum in the centre of the deuterated product distribution, and (ii) changes in product concentration with conversion when equal numbers of hydrogen and deuterium atoms are used: thus for example with methane + deuterium, where multiple exchange predominates, methane-d_3

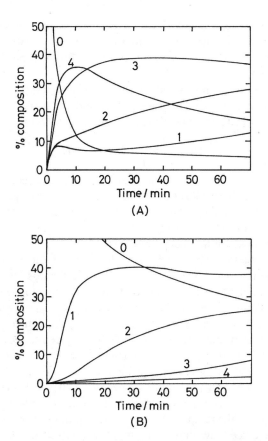

Figure 6.1. Formation of exchanged methanes over (A) rhodium powder at 429 K; (B) $Rh_{25}Pd_{25}$ powder at 418 K: multiple exchange predominates in (A) and stepwise exchange in (B).[23] $0 = CH_4$; $1 = CH_3D$; $2 = CH_2D_2$; $3 = CHD_3$; $4 = CD_4$.

and -d_4 pass through maxima (Figure 6.1a), while when stepwise exchange is the main process methane-d_1 is initially formed in excess of its equilibrium proportion[5] (Figure 6.1b). In a flow-reactor, the two processes are completely separated on rhodium film, no methane-d_2 or -d_3 being formed at low conversion;[1] when natural convection is relied on, re-adsorption of initial products (CH_3D, H_2, HD, CD_4) before they can diffuse away from the surface can obscure the initial product mix. Methane-d_1 and -d_4 are also the sole products formed on (111) and (100) orientated nickel films at 520–540 K.[8,10]

The total process is therefore characterised by (i) the rate of removal of the light (reactant) alkane, (ii) the separate rates of stepwise and multiple exchange (but only for methane, (iii) the total rate of introduction of deuterium atoms into the molecule, and (iv) the mean number M of such atoms initially entering a molecule.

Assigning orders of reaction and activation energises to the formation of individual products is not especially helpful.[11] A first-order kinetic equation for a reversible process is often observed;[12] mechanisms proposed for the types of exchange will be considered in due course.

A question that must receive attention is the way in which alkanes chemisorb and desorb in the presence of hydrogen or deuterium; their atoms will cover at least part of the surface, and a possible mechanism is therefore[1]

$$CH_4 + D^* \leftrightarrow CH_3^* + HD \tag{6.A}$$

The forward process also requires a number of 'free' sites because of the methane molecule's size, and these constitute the 'landing site' or 'free potential site' mentioned in Sections 4.8 and 5.4. The quantitative development of this concept[1,13] is however rendered difficult by uncertainty as whether the sites for the atoms are in fact the same as those for the methyl radicals.

A further problem that has received much attention is the mechanism by which cycloalkanes undergo exchange. With cyclopentane there is an energy barrier to transferring the exchange process from one side of the ring to the other, so that the -d_5 molecule is dominant at low temperatures and the -d_{10} only at high temperatures. The barrier is smaller with cyclohexane and cycloheptane, and absent from larger rings. Several mechanisms for this switch have been suggested (Section 6.3) and a number of molecules having quite complex cyclic structures have been devised or selected, and their exchange characteristics studied, with the intention of discriminating between the possibilities. There have also been a number of studies using bimetallic catalysts, especially with methane, to see whether there is an ensemble size effect on the relative rates of single and multiple exchange. Strangely enough there have been few studies of particle size effects on exchange reactions.

Activities of metals for alkane exchange increase with metal-metal bond strength (i.e. with sublimation energy), presumably because of stronger bonding of the radicals formed when the alkane dissociates, and their consequent higher coverages.[8]

Where substitution of deuterium atoms into a molecule occurs in parallel with their hydrogenation (e.g. alkenes (Chapter 7), aromatics (Chapter 10)) the matter is discussed in the relevant chapter.

6.2. EQUILIBRATION OF LINEAR AND BRANCHED ALKANES WITH DEUTERIUM[14]

6.2.1. Methane[1,2,8,15,16]

Most studies of the methane-deuterium reaction have been performed in static systems, which do not clearly reveal changes of activity with time: in a flow-system, however, films of metals in Groups 5 and 6, and nickel, increase in activity with time-on-stream, due it is thought to the slow build-up of reaction intermediates.[1] A flow-system also enables ready measurement of the effects of varying conversion, and hence the use of very low conversions at which the true initial product distribution is revealed. Frennet[1] has argued convincingly that this is hard to find in a static system, where sufficient hydrogen is released by chemisorption of methane to confuse the picture, even at conversions below 1%. He believed certainly with rhodium, and probably with all the metals studied,[8] that *only* methane-d_1 and -d_4 were formed initially, and that the appearance of the others resulted from an incomplete multiple exchange caused by the unwanted presence of some hydrogen atoms. Nevertheless the procedures used for separating the kinetics of stepwise and multiple exchange provide conclusions that are probably not much in error.

The principal features[2,4] of the kinetics are easily summarised. (i) Orders in methane are similar for stepwise and multiple exchange, ranging from +0.1 to 1.0 depending on the metal and the coverage of the surface with hydrocarbon radicals. (ii) Orders in deuterium are more negative for multiple exchange (-0.6 to -1.4) than for stepwise exchange (-0.1 to -0.9); multiple exchange is therefore suppressed by raising deuterium pressure. (iii) Multiple exchange always has the higher activation energy, so that its importance increases with rising temperature.[1,2,17,18] (iv) Stepwise exchange predominates on tantalum, molybdenum, tungsten[1] and (most notably) palladium,[2,19,20] but multiple exchange is substantial on nickel,[10] ruthenium, rhodium, iridium and platinum.[1,2,17,18] (v) There are no *major* differences in activity between the metals when the same forms are compared; thus, of the metals examined by Kemball, tungsten was most active, and palladium least, but very considerable differences in specific activity have been shown by a given metal in various physical forms (powders, blacks, films: nothing on single crystals) (Table 6.1). Powders of palladium and of platinum showed higher rates than films (rates calculated at 423 K using measured Arrhenius parameters) and films sintered at 773 K were especially low in activity.[19] Powders and blacks have highly defective surfaces, so it is difficult to avoid the conclusion that the reaction is structure-sensitive, although its degree depends upon the temperature selected for the comparison, since activation energies differ enormously (Table 6.1). As the footnote to this table states, rates were measured with greatly different reactant pressures, and so structural effects cannot be clearly separated from kinetic effects. The effect of particle size has not been widely studied, but with Pt/Al$_2$O$_3$ the TOF for stepwise exchange decreased with size above 1.5 nm.[21]

Compensation plots (Section 5.6) represent a convenient way of comparing activities measured under a variety of conditions where Arrhenius parameters are available. A suitable starting point is the work of McKee on ruthenium-palladium bimetallic powders[22] (Figure 6.2A); data points for stepwise and multiple exchange, and for methane removal, lie mainly close to a single line, the range of apparent activation energies being from about 40 to 190 kJ mol^{-1}. Points along the line have almost the same rates at 423 K, differing from one end of the line to the other only about two-fold; small differences from the line do however result in

TABLE 6.1. Rates of Methane Exchange with Deuterium at 423 K (r^{423}) and Apparent Arrhenius Parameters (E, ln A) for Metals in Various Physical Forms

Metal	Form	D_2/CH_4	E/kJ mol^{-1}	ln A	ln r^{423}	References
Ru	Black	10	97	62.9	35.32	17
	Powder	0.5	113	67.94	35.81	22
	/SiO$_2$	8	85	60.69	36.62	26
Rh	Black	10	94	63.77	37.05	17
	Powder	0.5	46	50.4	37.22	5, 23
	Film	1	—	—	40.49	2
	Film	~1	—	—	36.64	1
	/TiO$_2$	10	97	63.08	35.50	25
	/SiO$_2$	10	103	62.80	33.51	23
Pd	Black	10	111	63.56	32.00	20
	Powder	0.5	90.4	63.35	37.65	12
	Powder	0.5	119	70.01	36.17	22
	Powder	0.5	131	70.9	33.54	5, 23
	Film	1	98	62.64	34.69	2
	Thick Film	7	157	72.4	27.76	19
Pt	Black	10	104.5	66.2	36.4	18
	Powder	0.5	86	63.6	39.21	23
	Film	1	—	—	36.64	2
	Thick Film	7	113	56.9	24.77	19
Ir	Black	10	111	68.19	36.63	20

Notes:
1. The rates are those for methane removal and have been measured at various reactant pressures, so that they and the apparent Arrhenius parameters on which they are based have no absolute significance.
2. A and r in molecules m^{-2}s^1.
3. Activation energies are generally rounded to two or three significant figures, but rate may have been calculated from more accurate values.

significant rate changes. Figure 6.2B has results for rhodium-platinum-palladium bimetallic powders,[23] and 6.2C shows values for palladium-gold powders:[12,24] values for single metal films[2,19] and blacks[17] are contained in 6.2D. There are few values for supported metals.[25,26] Lines of identical slope are used in all parts of this Figure. The positions of the points for palladium are highlighted; they are sometimes (but not always) well below the line, signifying low activity: it may be that the possibility of poisoning by dissolved deuterium atoms has not always been thought of; this might also determine this metal's reluctance to catalyse multiple exchange.

The chief purpose of studies with bimetallic systems, either supported[11,27–29] or unsupported, has been to establish the size of the ensemble or 'landing site' necessary for the reactions and to see whether there was variation in the multiplicity of the exchange. We must therefore look at effects of composition on *rate* that are not visible in the compensation plots. Addition of a Group 11 metal to a Group 10 metal[11,12,28,30] has led generally to a progressive loss of activity, but in the palladium-gold system[12,30] there was sometimes a small increase in rate at low gold concentrations, as is seen on other reactions catalysed by this combination.

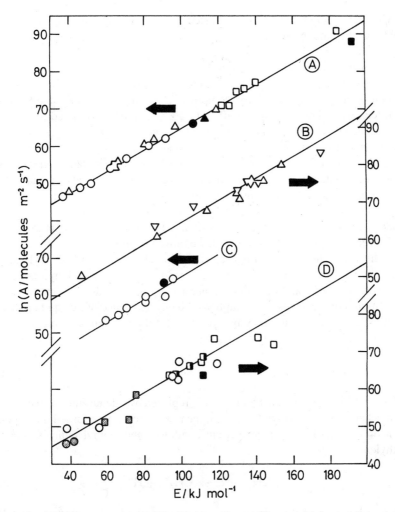

Figure 6.2. Compensation plots of Arrhenius parameters for methane exchange with deuterium: O stepwise exchange; □ multiple exchange; ▲ total exchange based on reactant removal (see Table 6.1). (A) Ru-Pd powders;[22] (B) Rh-Pd powders;[23] (C) Pd-Au powders;[12] (D) films (open, ref. 2; shaded, ref. 1) and blacks (half-filled) of single metals. Identical lines are used in each part: $T_i = 461$ K; points for palladium are filled.

Unfortunately much of this work was done before surface analysis became commonplace, so interpretation of activity changes in terms of ensemble size has not been possible. Use of bimetallics containing metals drawn from Groups 8 to 10 does not contribute much of interest: in most cases a continuous range of solid solutions is formed (Pd-Pt;[19,23] Pd-Rh;[23] Pt-Rh;[5] Pt-Ir;[29] Pd-Ni[31]), and the low activity of palladium has been manifested in systems containing it, as rates rose rapidly as the amount of the more active partner increased, before a constant rate was obtained. The likely surface composition could thereby be inferred. A distinct activity maximum was seen in the platinum-rhodium system[5] at about $Pt_{60}Rh_{40}$, the cause of which is not clear. The ruthenium-palladium system[22] is complicated by their limited mutual solubility; compositions from pure palladium to $Ru_{21}Pd_{79}$ were fcc, while those containing 40 and 62% ruthenium had mixed fcc and cph phases, each with the maximum amount of the metal of opposite structure. There was a marked maximum in rate at 473 K and in M at $Ru_{40}Pd_{60}$, but again the reason is not evident: this composition showed very low activation energies, not fully compensated by the ln A terms. It is worth stressing however that all the rate variations are accommodated by motion along the compensation line, with only minor divergences: the sole exception is with high concentrations of gold in palladium[12] (Figure 6.2C). This observation will contribute usefully to the discussion of reaction mechanisms, to which we must now turn.

This is a matter of some importance, as the release of adsorbed alkyl radicals to the gas phase by reaction with hydrogen is the final step in both hydrogenation and hydrogenolysis, and of course alkane adsorption initiates the latter process. A discussion of mechanism is complicated by the existence of two quite radically different proposals, which may be outlined as follows. Kemball[2] proposed that methane is first dissociatively adsorbed as

$$CH_4 + 2^* \rightarrow CH_3^* + H^* \qquad (6.B)$$

and, with the hydrogen atom being quickly replaced by a deuterium atom by one of the mechanisms set down in Chapter 3, the reverse reaction afforded methane-d_1. Then on certain sites or under certain conditions the methyl radicals becomes methylene,

$$CH_3^* + 2^* \rightarrow CH_2^{**} + H^* \qquad (6.C)$$

and so reiteration of this process and its reverse, the hydrogen atom each time being changed to deuterium, leads to methyl-d_3 and hence in all likelihood to methane-d_4. If the process does not proceed to completion, the other intermediate methanes would be formed. Further dehydrogenation to methylidyne (HC^{***}) to speed the process has been sometimes suggested.[32] This mechanism provides a logical explanation for the more negative order in deuterium for multiple exchange, as more free sites are needed, and the higher apparent activation energy possibly follows from the creation of more free sites as temperature rises, thus facilitating the further exchange.

A different mechanism suggested by Frennet[1,6,13,15,33,34] merits serious consideration, as it is based on extensive studies of the chemisorption both of methane and of hydrogen, and of the exchange kinetics, performed mainly in rhodium films. Attention is focused first on multiple exchange, because (as noted above) he believes with good evidence that methane-d_1 and -d_4 are the *only* proper initial products. Methane chemisorption is formulated as

$$CH_4 + H^* + Z \rightarrow CH_3{}^* + H_2 + Z \qquad (6.D)$$

(see Section 4.8), where Z is the number of uncovered metal atoms constituting the 'landing site' ($Z \simeq 7$), although their role after adsorption is accomplished is not made clear: furthermore, the films being polycrystalline, the exact geometry of this ensemble cannot be determined. All sites are assumed to be energetically equivalent, and no distinction is drawn between those occupied by hydrogen atoms and those on which methane adsorbs. Sites represented by the asterisk seem to be equated to metal atoms (i.e. atop sites), notwithstanding the preference of hydrogen atoms for three- or four-fold sites (Section 3.2). The rate of methane-d_4 formation r_m is given by

$$r_m = k\theta_c P_D \qquad (6.1)$$

where θ_c is the coverage by carbon-containing species (in fact methyl radicals) and P_D is the deuterium pressure. Methane-d_4 therefore results from the complete exchange of the hydrogens in a methyl radical, presumably by repetition of a process such as

$$CH_3{}^* + 2^* \rightarrow CH_2{}^{**} + H^* \qquad (6.E)$$

$$CH_2{}^{**} + D^* \rightarrow CH_2D^* + 2^* \qquad (6.F)$$

The methyl radical is finally released to the gas phase as

$$CD_3{}^* + D_2 \rightarrow CD_4 + D^* \qquad (6.G)$$

which is an Eley-Rideal step conforming to the kinetic equation. The strongly negative dependence of rate upon deuterium pressure[2,4] is explained by the need for a large landing site for the methane molecules. Frennet noted[1] that methane-d_4 formation occurred best on those metals from which hydrocarbon radicals are most easily displaced by hydrogen.

Methane-d_1 was thought to be formed by a totally different process, which involves only a fleeting interaction of methane with the surface, viz.

$$CH_4 + D^* \rightarrow [CH_4D] \rightarrow CH_3D + H \qquad (6.H)$$

Evidence for the formation of CH_5^+ (e.g. in a mass-spectrometer) was adduced in support; methane-d_1 is formed chiefly on metals on which it is *difficult* to displace hydrocarbon species with hydrogen.

Although clearly based on sound experimental evidence, and an unusually thorough independent assessment of surface coverages by reactants, this mechanistic scheme leaves several questions unanswered. (1) The rate of methane-d_1 formation should depend on θ_D and not on a negative power of deuterium pressure, unless the adsorption of the methane has a smaller landing site requirement than for multiple exchange. (2) Rates are not correlated with any physical properties of the metals, nor are the underlying characteristics of methane chemisorption, and the apparent structure sensitivity of the process is not discussed. (3) The conclusions are based on studies of considerable depth on only a small number of metals, and palladium, noted for its limited ability for multiple exchange, was not examined. (4) Changes in rate and reaction character with bimetallic systems have not been analysed. (5) Finally, for what it is worth, Arrhenius parameters for methane-d_1 formation and for multiple exchange both lie generally close to the same compensation line (Figure 6.1), which can be attributed to there being common features in the mechanisms of the two processes. Thus although the classical mechanism originally proposed by Kemball,[2,4] and since adopted by others, does not fully harmonise with Frennet's data, there remain a number of issues which neither scheme wholly resolves.

While there has been no study of methane exchange on any single crystal surface because of insuperable practical difficulties, the reaction of methyl radicals formed by thermal decomposition of methyl iodide with co-adsorbed deuterium has been followed on Pt (111) using TPD and RAIRS:[32,35] this technique which has been widely used to observe the reactions of adsorbed alkyl radicals,[36] avoids the difficult dissociative chemisorption of the alkyl radical. In TPD, all possible methanes from -d_1 to -d_4 vacated the surface below 300 K, and perdeuterated methyl and methylene radicals were observed by RAIRS; interconversion of methylene and methylidyne was suggested to account for the multiply exchanged products. The importance of this works lies in the recognition that the same set of sites can accommodate both single and multiple exchange, thus answering one of the questions posed by both the conflicting theories discussed above.

Lest it be thought that these results and their interpretation are archaic and of little current interest, we must note that the catalytic activation of methane and its consequent use to make hydrogen, oligomers, or oxygenated products such as methanol, has been and continues to be a field of great practical interest, and studies of methane exchange throw light on the molecule's adsorption and decomposition. On 5%Ni/SiO$_2$ the kinetic isotope effect mentioned previously[22] has been confirmed,[37] as has the formation of methane-d_4 as the sole product when deuterium is in large excess; the self-exchange process (see also Section 6.4)

$$CH_4 + CD_4 \rightarrow CH_3D + CHD_3 \qquad (6.I)$$

was not observed. Further attention to processes dependent on the dehydrogenation of methane will be given in Chapter 12.

6.2.2. Ethane and Higher Linear Alkanes[4,8,15,36,38,39]

The exchange of ethane and higher alkanes with deuterium has been the subject of fewer investigations than that of methane, with which there are some similarities but many differences; these may be summarised as follows. (i) The reaction occurs at somewhat lower temperatures, due to the easier activation of ethane. (ii) Tungsten, molybdenum and tantalum films are highly active (at 193–273 K) but give only stepwise exchange: in this respect their behaviour is like that of tungsten in methane exchange. (iii) On other metals of Groups 4 to 10, but especially on cobalt, rhodium and palladium, multiple exchange giving mainly ethane-d_6 accompanies stepwise exchange, initial deuterium numbers M being 2.3 to 5.4;[2,4] their values usually increase with temperature[17,22] (not however on Pt(111)[40]), but separate kinetic expressions for the two processes have not been derived. (iv) Unlike methane exchange, where satisfactory (and satisfying) compensation plots are shown (Figure 6.2), the Arrhenius parameters for ethane exchange when plotted in this way are somewhat scattered[8] (Figure 6.3).

Arrhenius parameters and selected values of M are given in Table 6.2: in conjunction with Figure 6.3, the following comments can be made. (i) Sintered films of palladium and platinum again have low activities and show high activation energies: this appears to be the only bimetallic system looked at. (ii) Certain

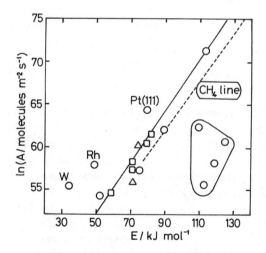

Figure 6.3. Compensation plot of Arrhenius parameters for ethane total exchange with deuterium:[2,19] (see Table 6.2): O foils, films, Pt(111); □ blacks, powders; △ supported Pt. The enclosed points are for Pt and Pd (see text).

TABLE 6.2. Apparent Arrhenius Parameters (E, ln A), Rates at 423 K (r^{423}) and Multiplicity Parameters (M) for Exchange of Ethane with Deuterium on Metals of Groups 8 to 10

Metal	Form	D_2/C_2H_6	E/kJ mol^{-1}	ln A	ln r	M at T/K	References
Ni	Film	8	75.3	57.11	35.70	3.1/~450	1
Ru	Black	10	59	54.51	35.82	4.37/356	17
Rh	Black	10	71	58.17	35.68	4.56/386	17
	Film	8	49	57.80	43.86	5.0/~343	4, 38
Pd	Film	8	89.5	61.95	33.60	4.8/~473	4, 38
	Film^a	9	125	60.57	21.07	3.23/546	19
	Black	10	71	57.20	37.01	3.46/428	20
Pt	Foil	10	113	71.16	35.26	4.02/625	40
	(111)	10	79.5	64.15	38.97	3.96/555	41
	Film	8	52.3	54.12	37.55	3.5/~463	2, 38
	Black	10	79.4	60.36	35.23	2.61/~400	18
	Film^a	9	109.6	62.18	27.46	2.12/573	19
	/SiO$_2$	10	74.1	60.08	39.01	—	88
	/Cab-O-Sil	10	71.6	55.84	35.48	—	88

^{a}Sintered film. See also footnotes to Table 6.1.

surfaces appear abnormally active, at least in comparison with other physical forms (W (see preceding text; Rh film;[38] Pt(111)[41]); with these exceptions, activities do not vary greatly (Table 6.2). Rates on films of other Transition Series metals have been reported,[10] but only on the basis of unit weight of metal.[1,8,38] Orders of reaction were in the range: for ethane, 0.8 to 1.2, and for deuterium, −0.55 to −0.8 (excluding tungsten).

There is a readier means of achieving multiple exchange with ethane than the methane; this involves the repeated inter-conversion of ethyl radicals with ethane, thus:

$$C_2H_5{}^* \xrightarrow[+2^*]{-H} C_2H_4{}^{**} + H^* \xrightarrow{+2D^*} C_2H_4D^* + 4^* + HD \qquad (6.J)$$

Once again it is possible that ethanes-d_4 and -d_5 result from incomplete multiple exchange caused by intervention of hydrogen atoms not efficiently removed from the surface during the single residence of the ethyl radical.[15] On rhodium film, ethane-d_6 was the main product at *low* deuterium pressure, but all ethanes appeared as primary products at higher pressures. Stepwise or single exchange might as with methane occur either by reaction of an ethyl radical at a site, or by a mechanism that does not allow formation of ethene: the high activity and low multiplicity shown by tungsten, molybdenum and tantalum suggest that the mechanism proposed by Frennet[1] for single exchange of methane may apply here also. While with platinum the temperature ranges in which exchange and hydrogenolysis occur are well separated, with rhodium the two ranges overlap[1] because of

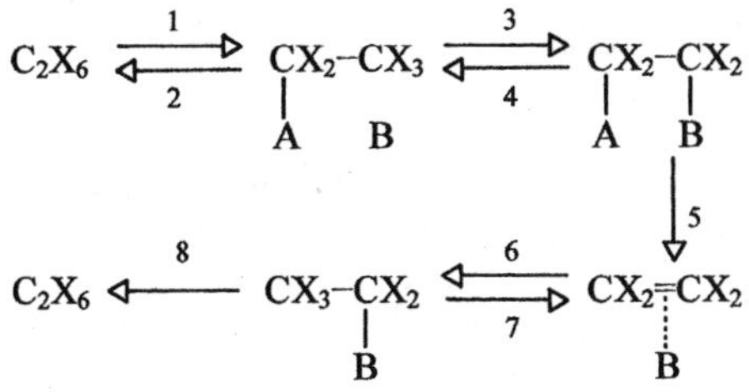

$$C_2X_6 \underset{2}{\overset{1}{\rightleftharpoons}} \underset{A\ \ \ B}{CX_2\text{--}CX_3} \underset{4}{\overset{3}{\rightleftharpoons}} \underset{A\ \ \ B}{CX_2\text{--}CX_2} \quad\Big\downarrow 5$$

$$C_2X_6 \overset{8}{\longleftarrow} \underset{B}{CX_3\text{--}CX_2} \underset{7}{\overset{6}{\rightleftharpoons}} \underset{B}{CX_2\text{=}CX_2}$$

Note: $p = k_3/k_2$; $q = k_5/k_4$; $r = k_7/k_8$; X represents either H or D.

Figure 6.4. Hegarty and Rooney scheme for the exchange of ethane with deuterium.[42]

this metal's much higher activity for C—C bond breaking; ethane exchange for this case will therefore be re-visited in Chapter 14.

Deuteroethane distributions have been interpreted in terms of a parameter **P**, which is the quotient of the rate constants for ethyl to ethene and ethyl reverting to ethane.[2,4,38] For molybdenum, tantalum, rhodium and palladium films, a single value of **P** (respectively 0.25, 0.25, 18 and 28) sufficed to reproduce the observed distribution, assuming that a further deuterium atom is acquired at every opportunity. With other metals, however, *two* simultaneous values of **P** appeared to operate, one contributing 30 to 50% of the reaction having a high **P** value (13.5–18) and another having a much lower **P** value (0.36–2). This analysis has not however been accorded an interpretation in terms of the metals' physical properties or of ensemble sizes and structures responsible for each participant.

An alternative way to explain product distributions has been suggested[42] (Figure 6.4). This involves two sites denoted **A** and **B**, on the first of which the ethyl radical is formed; di-σ ethene adsorption needs both. This then transforms to π-ethene (there is independent evidence that removal of hydrogen from ethyl gives the di-σ and not the π-ethene[41]), thence to ethyl on the **B** site and finally to ethane. Assigning numerical values to the three rate constant quotients produces calculated distributions in fair agreement with those observed. There are supposed to be few type **A** sites and many type **B**. This scheme also necessitates three independent parameters, and one wonders whether all the niceties of Microscopic Reversibility are accommodated by it.

Frennet[6,15] has discussed his observations concerning ethane exchange on rhodium film by analogy with methane exchange, namely, a dual mechanism in which an ethyl radical is heavily exchanged before desorbing, and an ethane molecule is monosubstituted during a brief encounter with the surface. This analysis was again supported by detailed analyses of hydrocarbon coverage.

It has long been a goal of research on alkane exchange to identify isomers of the same mass, e.g. ethane-1^2-d_2 (CH$_3$-CHD$_2$) and ethane-1,2-d_2 (CH$_2$D-CH$_2$D); this has at last been accomplished by NMR of ^{13}C-containing ethane,[40] and the

TABLE 6.3. Analysis of the Products of Reacting ^{13}C-Labelled Ethane with Deuterium on Platinum Foil at 675 K ($P_D/P_E = 10$)

Method	$-d_1$	$-d_2$	$-d_3$	$-d_4$	$-d_5$	$-d_6$
Mass-spec	0.283	0.029	0.044	0.068	0.180	0.395
NMR	0.151	0.054	0.043	0.078	0.205	0.470

/	↓	\
$CH_2D\text{-}CH_2D$	$CH_2D\text{-}CHD_2$	$CHD_2\text{-}CHD_2$
0.038	0.038	0.038
$CHD_2\text{-}CH_3$	CH_3CD_3	$CH_2D\text{-}CD_3$
0.016	0.005	0.038

results found for products formed over platinum foil at 625 K contain some mild surprises (Table 6.3). The $-d_2$ molecule was 70% symmetric, as would be expected for $\alpha\beta$-exchange (equation 6.J), but the 30% of the asymmetric isomer is something of a puzzle: the $-d_4$ molecule had equal amounts of ethane-1,2^3-d_4 and ethane-1^2,2^2-d_4, but there was much more ethane-1,2^2-d_3 than ethane-1^3-d_3. The $-d_2$ product was definitely initial, and not formed by duplicate exchange. It was pointed out that the U-shaped distribution is characteristic of platinum in all physical forms *including the single crystal Pt(111),* and so it is difficult to defend a mechanism that assumes two different kinds of site, and which invokes multiple ethyl-ethene interconversions under conditions where ethane is highly favoured. The alternative is to recall the findings of surface science experiments, which show that ethlidyne is very likely to be formed under the conditions where exchange proceeds. Zaera has therefore proposed a mechanism wherein multiple exhange occurs by repeated interconversion of diadsorbed ethene and ethylidyne:

$$C_2H_6 \underset{+X}{\overset{-X}{\rightleftharpoons}} C_2H_5{}^* \underset{+X}{\overset{-X}{\rightleftharpoons}} C_2H_4{}^{**} \underset{+X}{\overset{-X}{\rightleftharpoons}} X_3C\text{-}C^{***} \tag{6.K}$$

The ethene is supposed to be in the di-σ form, and the ethylidyne largely exchanged, perhaps by interconversion to vinyl and ethylidene (Figure 6.5), before reappearing mainly as ethane-d_6, a process for which there is much independent evidence and which is faster than its hydrogenation to ethane.[36] In this formulation, X may be either hydrogen or deuterium. (The different ways of using asterisks

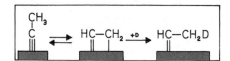

Figure 6.5. Mechanism of exchange of hydrogen atoms in ethylidyne.

to symbolise adsorption sites in processes, figures and in the text ought not to cause confusion; the linear form is used in the text for simplicity). The generally observed increase in the amount of multiple exchange with temperature is then explained by the decrease in the surface concentration of deuterium atoms, which favours the dehydrogenation of ethene into ethylidyne. Dehydrogenation beyond ethene has been rejected in other mechanisms, but Zaera's proposal merits serious consideration.

There remain a number of observations that do not admit of easy explanation by the mechanisms so far presented. The deposition of nickel films in UHV under various conditions produced very varied product distributions: thus films created on a hot substrate (573 K) and orientated so that either (111) or (100) faces were preferentially exposed both gave only ethane-d_2, -d_6 and -d_4 decreasing in that sequence ($M = 3.8$), while a randomly oriented film deposited at 273 K but sintered at 613 K gave mainly ethane-d_1 and -$d_6(M = 3.3)$.[8,10] Temperatures were 443–463 K, but at 273 K, on an unsintered film, ethene-d_1 was almost the only product. Very large amounts of ethane-d_2 and -d_6 were also formed on nickel powder ($P_D/P_E = 9$, $T = 420$–540 K).[43] The means whereby ethane-d_2 is formed, sometimes to the exclusion of the -d_1 and -d_3 species, has not been much discussed, but it would appear that on certain sites loss of two hydrogen atoms gives ethene (either π- or di-σ) and reversal with two deuterium atoms occurs without the intrusion of stepwise exchange.

A synthesis of mechanistic opinion is made difficult by the profound studies by Frennet,[15] Zaera[40] and others having been made on different surfaces. The invocation of ethylidyne as the means of achieving complete exchange is as perfectly reasonable for rhodium as for platinum, but unfortunately the scheme of equation 6.K and Figure 6.5 has not received a quantitative treatment, so its validity is not yet assured. What is very noticeable however is that palladium behaves quite 'normally' in ethane exchange, unlike its failure to give multiple exchange in the reaction of methane. Nothing could point more clearly to the differences in the mechanisms by which these alkanes exchange with deuterium.

6.2.3. Higher Linear Alkanes

With the introduction of the third and further carbon atoms into the linear alkane chain, the number of conceivable mono- and di-adsorbed structures increases rapidly (see Tables 4.3–4.5), and quantitative modelling of the distribution of exchanged molecules becomes progressively more difficult. Thus exchange may proceed not only by the easy $\alpha\beta$ mechanism so favoured by ethane (process 6.J), but also (e.g. on rhodium) through reversible formation of $\alpha\gamma$ species (i.e. 1-propyl $\leftrightarrow$ 1,3-diadsorbed propane (Table 4.5)):[44] further possibilities exist when there are four or more carbon atoms present (e.g. an $\alpha\delta$ process[26]). In the case of propane, early work indicated[4,45] that on nickel film the hydrogen atoms on the secondary carbon exchanged faster than those on the primary carbons, as might be expected on the basis of the relative bond strengths, but more recent work[44] with Rh/SiO$_2$ using deuterium NMR as the analytical tool showed that exchange of *both* secondary hydrogens was somewhat hindred. Apart from this work, there has been no attempt to identify the numerous isotopomers formed in higher alkane exchange with deuterium.

An additional feature of the reaction of the higher alkanes is that, depending on the temperature and the metal used, it is sometimes difficult to separate exchange from hydrogenolysis, as the relevant temperature ranges overlap:[26,46,47] this is particularly so with nickel, ruthenium and rhodium, all of which have high activities for hydrogenolysis, but some workers have turned this to advantage, using exchange to illuminate the process of C—C bond breaking. Exchange reactions on platinum are however usually untroubled by hydrogenolysis.[48]

For each metal the general features of the distributions of exchanged products for all linear alkanes resemble that shown by ethane. Multiple exchange occurs widely, and is perhaps even more significant than with ethane: it increases in important with rising temperature,[4,26,49,50] and with decreasing P_D/P_C ratio, but is inhibited by carbon deposition and by adsorbed oxygen atoms.[4,20] Comparison of metals' tendencies in stepwise and multiple exchange is complicated by their different activities and the consequent need to use different temperature ranges; thus, rhodium film at 250 K showed much less multiple exchange ($M = 5.7$) than palladium film at 320 K ($M = 7.6$),[4,51] but rhodium catalysts have given very complete exchange of n-hexane and n-heptane at moderate temperatures.[52,53] Various forms of nickel catalyst all show some degree of multiple exchange; with Ni/SiO$_2$, increase in particle size by sintering gave more multiple exchange. These results, and those concerning the effect of poisoning (mentioned above), suggest[4,50] that the course of alkane exchange is both particle-size-sensitive and ensemble-size-sensitive, and that a larger active centre is needed for multiple than for stepwise exchange. There is usually some stepwise exchange, but with the larger alkanes such as n-pentane and n-hexane it is difficult to estimate amounts in the centre of the distribution with high accuracy.[48] Hexane-d_2 was however a significant product of the exchange of n-hexane on rhodium film.[4,49]

With pumice-supported metals, there was apparently a third process causing a peak in the centre of the distribution,[4] but an interpretation needing five disposable parameters to account for eight quantities was rightly criticised as having too many variables. There was no significant difference between the products formed from n-hexane on Pt(111) and the kinked Pt(10,8,7), showing the reaction is not face-sensitive; active centres for multiple exchange must be freely available on both. On these surfaces however multiple exchange did not rise with temperature.[48]

Table 6.4 shows Arrhenius parameters, rates and multiplicity factors for higher alkane exchanges with deuterium, mainly on metal films: rates are generally somewhat higher than for ethane, and the temperature ranges used correspondingly

TABLE 6.4. Apparent Arrhenius Parameters (E, $\ln A$), Rates at 423 K (r^{423}) and Multiplicities (M) for the Exchange of Higher Linear Alkanes with Deuterium

Alkane	Metal	Form	D_2/HC	$E/kJ\,mol^{-1}$	$\ln A$	$\ln r^{423}$	M at T/K	References
Propane	W	Film	8	37.7	59.19	48.47	1.73/226	4
	Ni	Powder	8.7	75	59.19	46.82[a]	7.0/448	54
	Ni	Film	8	43.5	59.19	48.47	1.50/302	4
	Rh	Film	8	55.6	66.79	50.98	5.7/249	4
	Pd	Black	10	58	54.83	38.34	4.00/356	20
	Pd	Film	6	99.2	69.09	40.88	7.6/419	49
	Pt	Black	10	79.4	60.68	38.10	5.0/390	18
n-Butane	Pd	Black	10	54	55.00	39.65	8.18/41	20
n-Pentane	Rh	Film	8.4	—	—	—	7.1/273	49
	Pd	Film	8.4	60.7	60.79	43.53	11.8/356	49
	Pt	Black	10	107	75.91	45.60	1.31/~330	18
n-Hexane	Rh	Film	10	—	—	—	6.9/266	49
	Pd	Film	10	70.3	63.79	50.30	13.6/333	49
	Pt	(111)	10	—	—	—	9.2/573	48
n-Heptane	Pt	Kinked	10	—	—	—	10.0/573	48

Notes: [a]Exchange of secondary hydrogen atoms only.
See also footnotes to Table 6.1.

lower. There are insufficient data to create a meaningful compensation plot. Orders of reaction are usually positive in alkane (0.5–1) and negative in deuterium,[4,54] becoming positive at high temperatures as desorption sets in.

Variously structured nickel films also produced some very peculiar product distributions,[4,10] as they did with ethane: at 393–433 K on (111) and (100) oriented films gave propane-d_8 and (usually) -d_2 as the major products, but randomly orientated but sintered film at 400–435 K yielded propane-d_1 and -d_8, and an unsintered film at 273 K gave only the -d_1 isomer. These results closely resemble those obtained with ethane, and lend stress to the importance of surface structure in deciding what intermediates are formed and how they react. The absence of results for single crystal surfaces of nickel is sorely felt.

6.2.4. Branched Alkanes[55]

Provided a branched alkane has only primary, secondary and tertiary carbon atoms, all hydrogen atoms can be exchanged by the $\alpha\beta$ mechanism; the carbon skeletons (devoid of hydrogen atoms) of some such molecules are shown in the first row of Table 6.5. The presence of a quaternary carbon atom as in *neo*pentane (2,2-dimethylpropane) prevents the formation of an $\alpha\alpha$-diadsorbed species, so that multiple exchange, if it occurs, must of necessity proceed through either $\alpha\alpha$- or $\alpha\gamma$-diasorbed structures, or where possible through an $\alpha\delta$ structure.[26,56–58] Carbon skeletons of molecules of this type are shown in the lower part of Table 6.5, but

TABLE 6.5.   Structure of Branched Alkanes

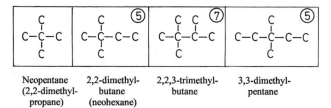

Branched alkanes having tertiary carbon atoms

Isobutane (2-methyl- propane)	Isopentane (2-methyl- butane)	Isohexane (2,3-dimethyl- butane)	3-methylbutane

Branched alkanes having quaternary carbon atoms

Neopentane (2,2-dimethyl- propane)	2,2-dimethyl- butane (neohexane)	2,2,3-trimethyl- butane	3,3-dimethyl- pentane

2,2,3,3-tetramethyl- butane

in some cases the $\alpha\beta$ mechanism can operate in part of the molecule,[54,59,60] and where it can the number of exchangeable hydrogen atoms is shown. Where no secondary or tertiary carbon atom is present, stepwise exchange predominates at low temperatures,[26,61] but multiple exchange by the $\alpha\alpha$ or $\alpha\gamma$ process becomes more evident with rising temperature, clearly with rhodium[44] but even more so with ruthenium.[26] Prohibition of the passage of exchange past a quaternary atom is in itself evidence of the difficulty of forming $\alpha\gamma$-diadsorbed structures, the ease of occurrence of exchange therefore being summarised as $\alpha\beta > \alpha\alpha > \alpha\gamma$. It is however difficult to find *quantitative* expressions for the relative energetics of these processes. There are similarities between *neo*pentane and methane, in for example the non-appearance of multiple exchange in both.[2,20,62,623] Variation of the extents of exchange by the $\alpha\beta$ and $\alpha\gamma$ processes on Pd/SiO$_2$ and Pt/SiO$_2$ having various particle sizes did not however accord with the expectation that the latter should become easier as size increased.[64] Some rates and Arrhenius parameters are given in Table 6.6; palladium's very high activation energy stands out.

There is one further interesting and useful feature shown by branched alkanes.[1] 3-Methylhexane (see Table 6.5) has a centre of optical activity, and the (+) form is observed to racemise during exchange.[50,52] The process over

TABLE 6.6. Apparent Arrhenius Parameters (E, $\ln A$) and Rates at 423 K (r^{423}) for the Exchange of Branched Alkanes with Deuterium

Alkane	Metal	Form	P_D/P_A	E/kJ mol^{-1}	$\ln A$	$\ln r^{423}$	References
Isobutane	W	Film	8	33	55.50	46.12	4
Isobutane	Ni	Film	8	38	56.42	45.70*	4
Isobutane	Pt	/Nb$_2$O$_5$	10	45	54.82	42.05	62
Neopentane	Pd	Black	10	85	57.53	33.36	20
Neopentane	Pd	Film	10	138	78.30	39.03	4
Neopentane	W	Film	10	43	55.73	43.47	4
Neopentane	Rh	Film	10	60	63.10	44.16	4

Notes: *Exchange of tertiary hydrogen atom only; see also footnotes to Table 6.1.

a nickel/kieselguhr catalyst had similar kinetics to exchange, and almost every molecule suffering exchange at its tertiary C—H bond also inverted its configuration. The mechanism by which this occurs has not been fully established.[1] At some stage the conformation about the tertiary carbon atom must become planar, and the deuterium atom must add at the opposite side from that at which the hydrogen atom was removed. One possibility[1] is that adsorption of the alkane occurs by Frennet's mechanism,[2] and its Eley-Rideal reversal may allow inversion to occur, thus:

$$R_1R_2R_3CH + D^* \rightarrow R_1R_2R_3C^* + HD$$

$$R_1R_2R_3C^* + D_2 \rightarrow DCR_1R_2R_3 + D^* \tag{6.L}$$

The existence of exchange in such a way prompts the speculation that it may also occur with other molecules, but remain undetected because no optically active centre is present.

6.3. EQUILIBRATION OF CYCLOALKANES WITH DEUTERIUM[8,65]

It will not have escaped notice that adsorbed species implicated in the exchange of alkanes with deuterium are formally the same as those invoked in the hydrogenation of alkenes: indeed the reiteration of the alkyl-alkene transformation (process 6.J) held responsible for multiple exchange in linear and branched alkanes, and designated the $\alpha\beta$ exchange mechanism, is on the face it of identical with the old and well-tried *Horiuti-Polanyi mechanism*[41] for alkene hydrogenation. This will be discussed further in the next chapter (sections 7.1 and 7.21), but briefly it supposes the sequential addition of two hydrogen atoms to some adsorbed form of the alkene, e.g.

$$C_2H_4 \xrightarrow{+2^*} C_2H_4^{**} \underset{-H}{\overset{+H^*}{\rightleftharpoons}} C_2H_5^* + 2^* \xrightarrow{+H^*} C_2H_6 + 4^* \tag{6.M}$$

The similarity of this scheme with that for multiple exchange is quite evident, and indeed the $\alpha\beta$ exchange mechanism is sometimes called the Horiuti-Polanyi mechanism:[66] the only difference (which turns out to be a point of some importance) is that in exchange the adsorbed alkene is formed from an alkyl radical, whereas in hydrogenation it comes from the gaseous alkene. It may also be wondered whether it is logical to speak of adsorbed alkene being formed under exchange conditions that in many cases are very different (e.g. much higher in temperature) from those in which hydrogenation occurs. It is true that the alkyl-alkene equilibrium set up during exchange will be heavily on the alkyl side, and the alkene surface concentration very low, but it is sufficient in most cases to sustain multiple exchange, and no other proposal has been made to account for it.

The original discovery by Kemball, since frequently confirmed, was that, while five hydrogen atoms on one side of the cyclopentane ring are easily exchanged, there is a considerable energy barrier to accessing the five on the other side of the ring: this is manifested by a higher activation energy, the importance of complete exchange always increasing with temperature. Thus in a typical case at lower temperature there will be maxima at cyclopentane-d_1 and -d_5 (sometimes at -d_2 as well) with no more extensively exchanged molecules, but rising temperature will cause the maximum at -d_5 to diminish and finally disappear, with cyclopentane-d_{10} becoming the dominant product. This apparently simple observation has however created a crop of problems, and efforts to resolve them have revealed hitherto unsuspected depths and subtleties in the mechanisms of metal-catalysed reactions of hydrocarbons. Central to the ensuing discussion has been the way in which the molecule inverts to enable the second set of hydrogen atoms to exchange. Careful consideration of the results obtained under a variety of conditions has led to the conclusion that there are four (or perhaps five) distinct mechanisms at work, each proceeding on a site appropriate to it, but the research has focussed more on the nature of the intermediates than on the form of sites required. It was perhaps unfortunate that this work was undertaken at a time (mainly in the 1960s and 1970s; little since 1980) when single crystal surfaces were not commonly used. The important observation that stepwise and multiple exchange of ethane proceed side by side on the highly uniform Pt(111) surface showed that radically different types of site were not needed, and attention in that case therefore moved to the nature of the intermediate species. It is surprising to say the least that the problems posed by the complex nature of cyclopentane exchange have not been addressed by studies using single-crystal surfaces, which would surely have removed many of the still existing uncertainties.

Before attempting to summarise the debates on mechanisms, a little further background is needed. Cycloalkane exchange is faster than for linear and branched alkanes, and with metal films[2,4,67] and powders[68] it is frequently possible to work at subambient temperatures, even as low as 77 K. Deactivation by 'carbon deposition' is however often a problem; its removed by treatment with hydrogen shows that on Pt/SiO_2 it is an over-dehydrogenated form of the reactant.[69] Tungsten,[2,4] rhodium[25,43] and iridium[67] film have shown exceptional activity, tungsten featuring principally stepwise exchange of one set of cyclopentane's hydrogen atoms, in conformity with its behaviour with other alkanes. Those metals of Groups 8 to 10 that have been studied all show some amount of multiple two-set exchange leading to the fully exchanged product;[70,71] this is particularly evident with palladium in any physical form, and it is on this metal that most of the mechanistic studies have been performed. Once again a quantitative comparison of the tendencies of metals to catalyse the several types of exchange process cannot be made because of their various activities and the temperature ranges in which they have been studied. The available information on Arrhenius parameters and rates is given in Table 6.7. Part of the evidence for the existence of several types of mechanism (and site?) comes from the observation that their relative rates as mentioned by the products to which they give rise are not the same on palladium catalysts differing in physical form.[72]

The mechanisms that have been clearly identified as occurring in cyclopentane exchange are those responsible for forming the following products.

A. Cyclopentane-d_1 by a single stepwise act.
B. Cyclopentane-d_2 by a single operation of the $\alpha\beta$ mechanism or perhaps by dissociate adsorption to give the 1,2-diadsorbed molecule.
C. Cyclopentanes-d_3 to -d_5 by multiple $\alpha\beta$ exchange on one side of the ring.
D. Cyclopentanes-d_6 to -d_{10} by inversion of the molecule, followed by multiple exchange on the other side.

Four different approaches have been used to elucidate the mechanisms: (i) variation of the structure of the alkane, (ii) kinetic analysis, (iii) examination of particle size

TABLE 6.7. Apparent Arrhenius Parameters (E, ln A), Rates at 423 K (r^{423}) and Multiplicities M for the Exchange of Cycloalkanes with Deuterium over Unsupported Metals

Alkane	Metal	Form	E/kJ mol^{-1}	ln A	ln r^{423}	M at T/K	References
Cyclopentane	Rh	Black	51.0	61.03	46.93	3.56/297	4, 43
Cyclopentane	Pd	Film	59.4	61.49	44.60	4.47/273	43
Cyclohexane	Ni	Film	45.2	59.65	46.80	1.2/~273	4, 39
Cyclohexane	Ni	Powder	67.0	58.36	39.31	6.73/523	81, 82
Cyclohexane	Rh	Film	43.5	61.03	48.66	3.8/273	4, 39
Cyclohexane	Rh	Black	63.0	—	—	4.03/265	43
Cyclohexane	Pd	Film	54.4	58.50	43.03	6.5/291	4, 39
Cyclohexane	W	Film	46.0	63.10	50.02	1.5/225	4, 39
Cyclohexane	Pt	Film	50.2	61.03	46.76	2.1/273	4, 39
Cyclohexane	Pt	Powder	79.4	65.69	43.11	—	81

Notes: See also footnotes to T0ble 6.1.

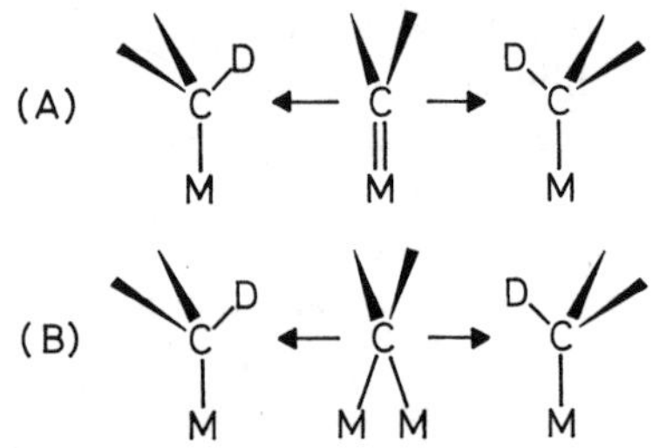

Figure 6.6. Inversion of chemisorbed cyclopentane molecule via $\alpha\alpha$-diadsorbed species: (A) the attached carbon atom is sp^2 hybridised; (B) it is sp^3.

effects, and (iv) use of bimetallic catalysts. By far the most work has been done with (i), so this will be reviewed first.

The cyclopentane molecule is almost planar, and the C—H bonds are held in eclipsed configuration: it is therefore quite reasonably deemed impossible to form a 1, 2-diadsorbed state by removal of hydrogen atoms from opposite sides of the ring. The cyclohexane ring is more flexible, but in its exchange at low temperatures a maximum at $-d_6$ is still observed, although qualitatively it is less marked than the $-d_5$ maximum in cyclopentane: inversion to allow exchange of the second set is therefore easier, but it is not until the ring is enlarged to seven or eight carbon atoms[2,73] that the barrier disappears and uninhibited total exchange occurs. One means of securing inversion of an adsorbed cyclopentyl radical is to assume that an $\alpha\alpha$-diadsorbed species is formed; if this is planar about the attached carbon atom, as has been assumed, then reversion to cyclopentyl may occur by addition of deuterium at either side, opening the way to total exchange (Figure 6.6). Difficulty of forming the $\alpha\alpha$ species accounts for the higher activation energy shown by two-set exchange, and this of course parallels the behaviour of methane (Section 6.2.). The argument remains unaffected if, as seems probable on the basis of information presented in Chapter 4, the attached carbon atom retains its sp^3 state, forming bonds to two different metal atoms or sites (Figure 6.6).

At least three other modes of inversion have received consideration. (1) The diadsorbed alkene may just desorb before responding to the thermodynamic urge to re-adsorb, which after flipping over will allow second-set exchange to proceed: the high activation energy might reflect the endothermic nature of the desorption. (2) The diadsorbed alkene in the di-σ form might 'roll over' to another pair of sites without actually desorbing (Figure 6.7). (3) The adsorbed alkene in the π-form might lose a further hydrogen atom, becoming a π-alkenyl species (see Table 4.2); topside addition of a deuterium atom by an Eley-Rideal step would then lead to inversion and complete exchange as shown in Figure 6.8.

The trouble with all these four proposals is that each can be (and has been) defended on chemical grounds, but can also be criticised. We may note that

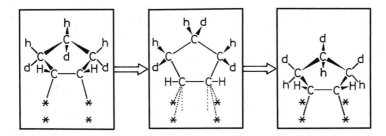

Figure 6.7. Representation of the 'roll-over' process through a planar intermediate.

ready alkene desorption *and* formation of π-alkenyl species are characteristic of palladium and of nickel, on which two-set multiple exchange is prevalent: but it also occurs on platinum and iridium, where these features are less clearly exhibited. Topside addition as shown in Figure 6.8 also implies (by Microscopic Reversibility) topside *removal*, which seems less likely than under side removal. Clearly some further input is needed; and allied with this is the vexed question of whether the state of the alkene is best represented as the σ-diadsorbed or as the π-adsorbed form (structure **2** or **4** in Table 4.2).

The scope and limitations of one-set exchange are clearly illustrated by methylcyclopentane.[74,75] Its exchange has been investigated on Pt/SiO$_2$, Pd/SiO$_2$ and Rh/SiO$_2$, using NMR to assist product identification;[74] adsorption at the unhindered side of the ring permitted exchange of five hydrogen atoms, and the process

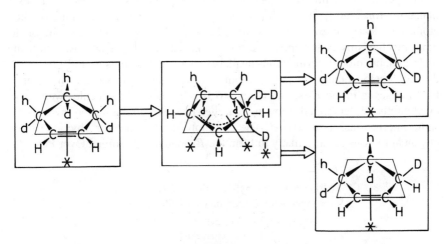

Figure 6.8. Inversion of π-adsorbed cyclopentane by top-side addition of a deuterium atom to a π-alkenyl intermediate Eley-Rideal step: note that Microscopic Reversibility requires that the π-alkenyl can be formed by topside *removal* of an atom.

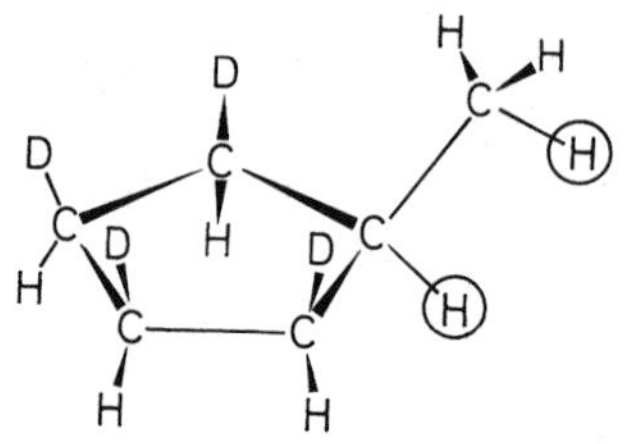

Figure 6.9. Transfer of exchange to the second set in cyclopentane via the methyl group: the σ-diadsorbed form is made by loss of the two indicated hydrogen atoms.

could extend to the methyl group through the formation of an eclipsed diadsorbed structure (Figure 6.9), giving maxima at the d_8-species. However this latter process was harder than the one-side exchange, and had a higher activation energy. Ring inversion was easier for molecules that first adsorbed on the unhindered side, because interference between the methyl group and the surface encouraged formation of the intermediate responsible for inversion. Understanding of the subtleties of exchange in polymethylcyclopentanes may be tested by attempting to predict the maxima in product distributions formed from the structures shown in Table 6.8 over palladium film at 313 to 353 K.[4,76]

There is a further reaction that proceeds under exchange conditions and requires a mechanism similar to that for cycloalkane inversion: this is *epimerisation,* exemplified by conversion of Z-1,2-dimethylcyclohexane to the E-isomer[72,77] (Figure 6.10). As with racemisation of branched alkanes, this also demands a planar intermediate, or some other mode of reconfiguration of a substituted carbon atom, but clearly an $\alpha\alpha$ species cannot be formed at this point, and one of the other mechanisms must apply. These reactions can of course be observed using hydrogen, but more information is available when deuterium is used. With Z-1,2-dimethylcyclopentane, eleven of the fourteen hydrogen atoms were exchangeable, but only seven in the E-isomer (Figure 6.10). Epimerised molecules were however mainly perdeuterated, showing that the change of configuration involves the same intermediate as inversion. The common intermediate for obvious stereochemical reasons is more likely to form the E- than the Z-isomer. Over Pt/Al$_2$O$_3$

TABLE 6.8. Methyl-Substituted Cyclopentanes Exchanged with Deuterium over Palladium Films

1,1,3-Trimethyl -
E-1,1,3,4-Tetramethyl -
Z-1,1,3,4-Tetramethyl -
1,1,3,3-Tetramethyl -
1,1,3,3,4-Pentamethyl -

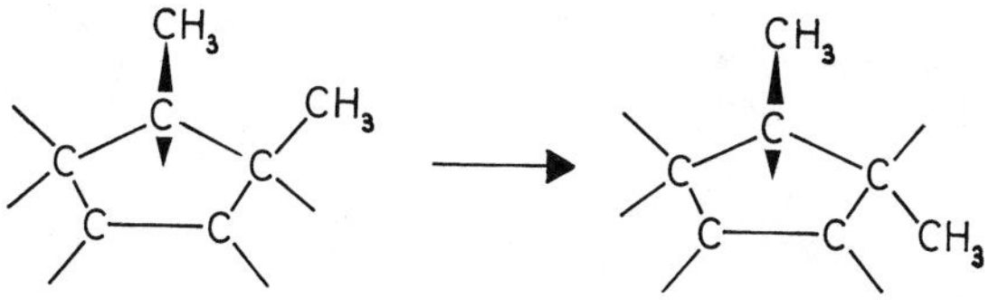

Figure 6.10. Epimerisation of Z- to E-1, 2-dimethylcyclopentane.

catalysts having dispersions between 5 and 65%, the TOF for epimerisation of Z-1,2-dimethylcyclohexane at 453 K was essentially constant, i.e. the reaction was structure-insensitive.[77]

Further attempts to resolve mechanistic ambiguities required the use of poly-substituted cycloalkanes[78] and polycyclic molecules of Byzantine complexity, the synthesis of which must have taxed the ingenuity of research students, and helped to fill many a PhD thesis. One example is 1,1,3,3-tetramethylcyclohexane which contains an isolated trimethylene group in which on palladium film only five of the six hydrogen atoms are exchangeable;[65] this is good evidence for a π-alkene-π-alkenyl mechanism. The study of fused-ring cycloalkanes has also been a fertile field for mechanism discrimination. The argument here revolves around the ability of the alkene formed in the exchange to adopt a π structure that is almost coplanar, and thence a π-alkenyl or mono-adsorbed alkene, as against its retaining the di-σ form and tetrahedral stereo-chemistry, in which 'roll over' can operate. It has to be remembered that these debates were held before the spectroscopic and other structural studies mentioned in Chapter 4 were available: these would suggest that a π-alkene would be the preferred form at *low* coverage and *low* temperature, but only at *high* coverage at *high* temperature. This would suggest that under most of the conditions used in exchange experiments the di-σ form would be stabler than the π form.

It is however worth citing a few of the results obtained with fused-ring cycloalkanes. In norbornane (bicyclo[2.2.1]heptane) there are only two pairs of hydrogen atoms in eclipsed confirmation, but these are isolated by the bridge-head methyne groups, and so $\alpha\beta$ exchange gives only the -d_2 product (Figure 6.11A). With bicyclo[3.3.1]nonane, however, if one of the cyclohexane rings adopts the boat conformation, the bridgehead hydrogen atoms eclipses with an adjacent methylene group and hence the exchange of eight hydrogen atoms becomes possible[72,79] (Figure 6.11B). Further maxima at -d_{10} and -d_{12} may be due to exchange of two atoms in each of the trimethylene groups. Only step-wise exchange is possible with adamantane, because the two cyclohexane rings are fused in the chair conformation (Figure 6.11C). With bicyclo[3.3.0]octane, however, there is no hindrance to inversion, and -d_8 and -d_{14} maxima have been seen[80] (Figure 6.12D); replacement of one of the bridge-head hydrogen

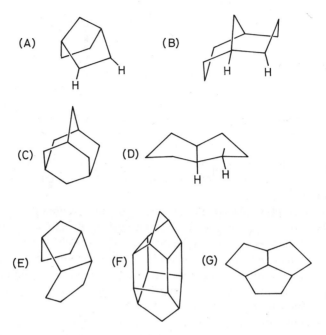

Figure 6.11. (A) Norbornane (bicyclo[2.2.1]heptane): one of the pairs of eclipsed and therefore exchangeable hydrogen atoms is shown.
(B) Bicyclo[3.3.1]nonane: one pair of eclipsed hydrogen atoms is shown, and exchange of eight atoms becomes possible.
(C) Adamantane: all methylene groups are isolated, and only stepwise exchange can take place.
(D) Bicyclo[3.3.0]octane: one hydrogen in each of the methylene groups adjacent to the bridge-head hydrogen is eclipsed with it, so all hydrogen atoms are exchangeable.
(E) *Endo*-trimethylenenorbornane: exchanges only five hydrogen atoms. Find them!
(F) Heptacyclotetradecane: shows only stepwise exchange.
(G) Tricyclo[2.2.0]decane: exchange ten hydrogen atoms.

atoms by a methyl group limits the extent of the exchange to seven and eleven atoms, a result which clearly favours the di-σ adsorbed intermediate. *Endo*-trimethylenenorbornane (Figure 6.11E) exchanges only five hydrogen atoms, not more. Further attempts to distinguish between di-σ-adsorbed alkene and π-alkene intermediates hinged on finding molecules that should have a pair (or pairs) of eclipsed C—H bonds but which could not by any stretch of the imagination allow the relevant part to adopt the nearly flat configuration demanded of an adsorbed π-alkene. Such weird and wonderful molecules as heptacyclotetradecane and tricyclodecane (Figure 6.11E and F) have also been obtained and used. The former only showed stepwise (not $\alpha\beta$) exchange, perhaps because of steric interference by the methylene bridge; the latter exchanged up to ten hydrogen atoms, but it is difficult to envisage any of the carbon atoms changing to the sp^2 state or anything close to it. The unbiased observed might conclude that if there is really a conflict of interpretation (a point we come back to shortly), the idea of the diadsorbed alkane intermediate as presented by Bob Burwell has defeated by π-alkene intermediate mechanism favoured by John Rooney–but only on points.

Having exhausted the inspiration obtainable by tinkering with molecular structure, we must turn for further insights to the use of other approaches. Little use has been made of kinetic analysis, as has been done so extensively (and profitably) with ethane and methane, orders of reaction when determined being expressed just as exponents of reactant pressures.[50,81] Exchange distributions observed with cyclohexane have been interpreted as for ethane,[4] but dependence of the variables on physical properties of the metals used has not been pursued. It has often been noted that above some critical temperature around 373 K there is hydrogenolysis of the cycloalkane, which above about 473 K leads to deactivation by 'carbon' deposition.[43,68,69,81,82] This is more quickly suppressed by addition of an inert metal (Cu to Ni;[68] Sn or Au to Pd[30]), and many (but not all) deliberately introduced poisons affect multiple exchange more than stepwise[77] (the exceptions are $SnCl_2$ and $PbEt_4$). While the inert metals quickly depress the rates of exchange processes[30,68] (although a small addition of gold to palladium can cause an increase, as is seen with other reactions[63]), there has been no attempt to quantify these results in terms of active ensemble size. One is however left with the impression that the needful size decreases as

'carbon' deposition > multiple exchange > stepwise exchange.

This sequence does not however point uniquely to any of the conflicting mechanisms.

The tendency to inversion (i.e. two-set exchange) with palladium is unaffected by addition of gold, but is actually increased by adding tin and is decreased by dissolved hydrogen:[30] it was concluded that raising the energy of the palladium $4d$ states weakens the metal-alkene bond, while lowering it had no effect. Where the bimetallic pair comprises two active metals (Pt-Ir;[29] Pd-Ni[31]), the results approximate to the weighted average of the two components: thus iridium favoured stepwise and pairwise exchange of cyclopentane to $-d_1$ and $-d_2$, and platinum preferred multiple exchange to $-d_4$ and $-d_5$, while nickel tended to give more pairwise and less total exchange than palladium. High -temperature pre-reduction (HTR) of Rh/TiO_2 created the SMSI state (Section 3.53), but while this lowered the rate it did not affect the multiplicity of the exchange, which showed much cyclopentane-d_1 and $-d_2$, and little $-d_{10}$. It did however reduce to multiplicity shown by Pt/TiO_2.[25]

Cyclopentane exchange has been used to 'count' the number of exposed metal atoms in alumina-supported metals: rates at 373 K were about the same for Pt/Al_2O_3 and Re/Al_2O_3, but no exchange past $-d_4$ was found with the latter.[83]

Rates were however about twice as fast with $PtRe/Al_2O_3$, and the exchange pattern was intermediate between those of the pure metals. In the case of Pt/Al_2O_3, specific (areal) rates of stepwise exchange of cyclopentane were almost independent of particle size, increasing only three-fold as it increase from 1.3 to 17.5 nm;[77] rates of multiple exchange increased by a factor of about 20. The areal epimerisation rate for 1,2-dimethylcyclohexane was also independent of particle size with this catalyst.[52] Similar results have been found with Pt/SiO_2[69,84] and Ni/SiO_2,[50] although the type of pre-treatment has a significant effect with the former.

We have therefore only a very incomplete matrix of isolated sets of results from which to deduce anything about site requirements for exchange reactions. It seems that stepwise and perhaps the single $\alpha\beta$ steps need no special site, a conclusion that does not rule out Frennet's mechanism for stepwise methane exchange (Section 6.2.1), but that multiple exchange and inversion appear to need a larger site. It is however very difficult to disentangle the effects of support, particle size and reaction temperature arrive at any generally valid conclusions.

More recent work by Rooney's group[85,86] and others[87] has however revealed further complexities to those which already exist in the behaviour of supported metal catalysts for exchange reactions. Low-temperature reduction (LTR) in the reactants after pre-oxidation completely removed inversion of cyclopentane on palladium film and on $Pd/\gamma-Al_2O_3$, but HTR at 573 K restored it, at least in part; it was thought that inversion was somehow prevented by electron-deficient sites such as Pd^{2+} ions near palladium atoms, or low co-ordination number atoms (Table 6.9): but HTR of palladium on other aluminas or on titania did *not* show the $-d_{10}$ products of inversion. Similar effects were shown by platinum on acidic supports, although Pt/MgO showed much more total exchange than Pt/SiO_2 or Pt/Al_2O_3: the rate was however much smaller. The difficulty with observations of this sort is that the effects cannot be definitively connected with any physical property of the catalysts (electron deficiency, presence of adventitious poisons, (S)MSI etc.) because of the lack of appropriate characterisation. They do however appear to operate with rhodium, as well as palladium and platinum; and unusual catalytic behaviour of metals on magnesia or influenced by other basic components will be encountered again in a later chapter.

Small-ring cycloalkanes have not escaped attention. Cyclopropane underwent exchange as well as hydrogenation on rhodium film at 173 K,[4] as did cyclobutane[82] and ethylcyclobutane[4] on nickel catalysts at about 430 K, although in both these cases ring-opening occurred at the same time. All hydrogen atoms were exchanged on nickel, a fact having some mechanistic significance, as it shows that inversion is possible. On palladium it did not occur with methylcyclobutane, however, a maximum of six hydrogen atoms being exchangeable, and with 1,1-dimethylcyclobutane multiple exchange appeared not to occur at all;[8] these last observations are not mutually consistent. Endocyclic alkenes are most unlikely to be formed in C_4 rings, so other mechanisms of exchange must apply.

TABLE 6.9.   Parameters of Cyclopentane Exchange with Deuterium over Palladium
Catalysts Variously Reduced, and over Supported
Platinum Catalysts

Metal/%	Form/Support	D_2/C_5H_{10}	T_{red}/K	T/K	$-d_1$	$-d_5$	$-d_{10}$	M	References
Pd/-	Film	10.7	573	323	27.7	47.1	11.3	4.7	85
Pd/-	Film	10.7	333	333	31.3	14.2	0	3.8	85
Pd/5	γ-Al$_2$O$_3$(BDH)	10.7	573	345	23.8	3.9	32.4	6.5	85
Pd/5	γ-Al$_2$O$_3$(BDH)	10.7	343	343	18.9	25.2	1.0	3.4	85
Pd/5	γ-Al$_2$O$_3$(Laporte)	10.7	573	318	29.1	21.6	0	3.0^a	85
Pd/5	AlO(OH)^b	10.7	573	273	30.0	17.9	0	2.6^a	85
Pd/5	AlO(OH)^b	10.7	723	343	4.3	15.7	61.0	7.8	85
Pd/14	γ-Al$_2$O$_3$	2	723	313	12	26	25	5.23	4, 52
Pt/2	MgO	13.6	723	343	1.5	12.6	55.0	8.5	86
Pt/0.5	SiO$_2$	10	573	340	12	50	5	4.75	69, 84
Pt/0.5	SiO$_2$	13.6	723	343	14.2	32.0	14.2	5.1	86
Pt/0.35	Al$_2$O$_3$	25	773	373	21.5	21.0	12.0	4.90	83
Pt/0.5	Al$_2$O$_3$	13.6	723	343	12.8	34.2	16.0	5.5	86

Notes:
^{a} Similar results were obtained by reduction at the reaction temperature.
^{b} Boehmite but may transform to Al$_2$O$_3$ under reaction conditions.

6.4. INTERALKANE EXCHANGE

There have been occasional reports that exchange can take place between
normal and deuterated molecules in the absence of hydrogen or deuterium, thus:

$$CH_4 + CD_4 \rightarrow CHD_3 + CH_3D \rightarrow 2\,CH_2D_2 \qquad (6.N)$$

The process took place on Ni/Cr$_2$O$_3$ between 373 and 525 K and must require the
dissociative chemisorption of the methanes followed by random recombination:
not surprisingly, deactivation was found at all temperatures.[4] The reaction between
normal and deuterated n-butane has also been studied on a number of supported
platinum catalysts.[4] Little or no exchange was seen between light methylcyclohexane and deuterated n-octane (or the reverse) over Pt/SiO$_2$ at 755 K either in the
presence or absence of hydrogen or deuterium because of rapid conversion to other
molecules; reaction did however take place between them at lower temperatures,
and on Pt/Al$_2$O$_3$ it occurred both on the metal and the support.[88]

6.5. CONCLUSIONS

Almost all the work mentioned in this Chapter was performed before about
1975, so that either metal films or powders ('blacks') or supported metals were
of necessity used. Thus attempts to assess the sensitivity of exchange processes

either to particle size, or to ensemble size in bimetallic systems, have yielded only results of qualitative significance, and measurement of particle size by TEM and the possible presence of accidental poisons have been largely ignored. Quantitative kinetic analysis has also been omitted (except for methane and ethane), so that it is difficult to connect exchange multiplicity and tendency to inversion to any independently measured property of the catalysing metal. The only study using a single crystal surface (Pt(111)) has been rewarding in the sense that distinct processes are shown to occur through different adsorbed structures on a uniform surface rather than through a variety of sites. It is strange that, for example, cyclopentane exchange has not yet been examined on the low-index faces or stepped surfaces of palladium or platinum (or any other metal): such a study would surely be most informative.

When two very intelligent people inspect the same body of experimental facts and proceed to draw quite different conclusions as to their cause, one might think that the truth lies somewhere in the middle or that both are partly right and partly wrong. We must therefore ask whether the alternative views expressed particularly mechanisms of cycloalkane exchange represent a genuine conflict or whether they are essentially synthetic. Rooney's ideas were originally based on analogies with organometallic complexes, which as we have seen in Chapter 4 are indeed useful models for adsorbed states that are formed *under certain conditions*. However the fact is that a particular form, say the π-alkene, exists in mono-nuclear complexes such as Zeise's salt does not necessarily mean it will also be formed on metal surfaces, where there are many atoms crying out to have the free valences neutralised. The regions of temperature and surface coverage in which this form are stable is restricted. We now realise too that the σ-diadsorbed and the π-adsorbed alkene are only extreme representations of a whole range of intermediate forms, the tendency towards one or the other depending on the chemistry of the surface and probably on the nature of the adsorbed molecules as well. Some of the recent very precisely determined structures of adsorbed species (Section 4.5) are not easily rationalised in terms of emergent molecule orbitals, and theoretical explanations for the structures adopted are still awaited. Applications of this knowledge to reaction mechanisms require these to be inspected on similar surfaces, and it is of little use to the traditional forms of metal catalyst mainly used for alkane exchange, with their evident proliferation of types of surface site.

In conclusion, we may feel that it has been an error to try to force *all* the observations of cycloalkane exchange into a single mould, and to require *all* molecules (and surfaces) to conform to a single type of mechanism. Many (but not quite all) the results are explicable by π-alkene, π-alkenyl and mono-adsorbed alkene structures; opposition to the view that σ-diadsorbed alkane structures predominate seems to have been based on intuition rather than on hard fact, and the lack of any attempt to identify adsorbed structures by spectroscopic methods (perhaps difficult when surface concentrations are low) has not helped to resolve the difference of opinion. The mechanisms suggested by Frennet for single and multiple exchange of methane and ethane differ radically from those originated by Kemball, and which form the basis of all the discussion concerning the behaviour of cycloalkanes, but unfortunately the effect of operating conditions on rates and multiplicities can be explained at least qualitatively by both. The safest conclusions to reach concerning mechanisms of cycloalkane exchange are that either (i) all reactions involve intermediates that are somewhere between the di-σ and π forms or (ii) that some molecules prefer to adopt a form more like one extreme, while others prefer another. Perhaps alkane exchange is like politics—the art of the possible.

The reader wishing to be informed of the intensity of the debate on these reaction mechanisms as it was exercised in the mid-1960s should read the relevant part of the discussion in *Discussions of the Faraday Society,* Vol. 42 (1966).

REFERENCES

1. A. Frennet, *Catal. Rev.* **10** (1974) 37.
2. C. Kemball, *Adv. Catal.* **11** (1959) 223.
3. L. Guczi, B.S. Gudkov and P. Tétényi, *J. Catal.* **24** (1972) 187.
4. G.C. Bond, *Catalysis by Metals*, Academic Press: London (1962).
5. D.W. McKee and F.J. Norton, *J. Catal.* **4** (1965) 510.
6. A. Frennet, A. Crucq, L. Degols and G. Lienard, *Acta Chim. Hung.* **111** (1982) 499; *J. Catal.* **53** (1978) 159.
7. L. Guczi, A. Sárkány and P. Tétényi, *Z. Phys. Chem. NF* **74** (1971) 26.
8. J.R. Anderson and C. Kemball, *Proc. Roy. Soc.* A **226** (1954) 472; J.R. Anderson and B.G. Baker in: *Chemisorption and Reactions on Metal Films*, (J.R. Anderson, ed.), Academic Press: London, Vol. 2 (1971), p. 64.
9. G.C. Bond, *Appl. Catal. A* **149** (1997) 3.
10. J.R. Anderson and R.J. Macdonald, *J. Catal.* **13** (1969) 345.
11. J.A. Dalman and C. Mirodatos, *J. Molec. Catal.* **25** (1984) 161.
12. D.W. McKee, *J. Phys. Chem.* **70** (1966) 525.
13. A. Frennet, G. Lienard, A. Crucq and L. Degols, *J. Catal.* **53** (1978) 150.
14. F.G. Gault, *Adv. Catal.* **30** (1981) 1.
15. A. Frennet in: *Hydrogen Effects in Catalysis*, (Z. Paál and P.G. Menon, eds.), Dekker: New York (1988), p. 399.
16. G.A. Martin, *Bull. Soc. Chim. Belg.* **105** (1996) 131.
17. A. Sárkány, K. Matusek and P. Tétényi, *J. Chem. Soc. Faraday Trans. I* **73** (1977) 1699.
18. L. Guczi, A. Sárkány and P. Tétényi, *J. Chem. Soc. Faraday Trans. I*. **70** (1974) 1971.
19. L. Guczi and Z. Karpiński, *J. Catal.* **56** (1979) 438.
20. A. Sárkány, L. Guczi and P. Tétényi, *Acta Chim. Acad. Sci. Hung.* **96** (1978) 27.
21. N. Khodakov, N. Barbouth, Y. Berthier, J. Oudar and P. Schulze, *J. Chem Soc. Faraday Trans.* **91** (1995) 569.
22. D.W. McKee, *Trans. Faraday Soc.* **61** (1965) 2273.
23. D.W. McKee and F.J. Norton, *J. Catal.* **56** (1964) 252.
24. G.C. Bond and E.G. Allison, *Catal. Rev.* **7** (1973) 25.
25. A. da Costa Faro Jr. and C. Kemball, *J. Chem. Soc. Faraday Trans.* **91** (1995) 741.
26. R. Brown and C. Kemball, *J. Chem. Soc. Faraday Trans.* **89** (1989) 2159.

27. G.A. Martin, J.A. Dalmon and C. Mirodatos, *Bull. Soc. Chim. Belg.* **88** (1979) 559.
28. I.H.B. Haining, C. Kemball and D.A. Whan, *J. Chem. Res. (S),* (1982) 296.
29. D. Garden, C. Kemball and D.A. Whan, *J. Chem. Soc. Faraday Trans.* I **82** (1986) 3113.
30. J.K.A. Clarke and J.F. Taylor, *J. Chem. Soc. Faraday Trans. I* **72** (1976) 917.
31. J.L. Vlasveld and V. Ponec, *J. Catal,* **44** (1976) 352.
32. F. Zaera, *Proc. 10th Internat. Congr. Catal.* (L. Guczi, F. Solymosi and P. Tétényi, eds.), Akadémiai Kiadó: Budapest **B** (1993), 1591.
33. A. Frennet and C. Hubert, *J. Molec. Catal. A: Chem.* **163** (2000) 163.
34. A. Frennet, *Catal. Today* **12** (1992) 131.
35. F. Zaera, *Catal. Lett.* **11** (1991) 95.
36. F. Zaera, *Chem. Rev.* **95** (1995) 2651.
37. K. Otsuka, S. Kobayashi and S. Takenaka, *J. Catal.* **200** (2001) 4.
38. J.R. Anderson and C. Kemball, *Proc. Roy. Soc.* A **223** (1954) 361.
39. P. Tétényi, L. Guczi and A. Sárkány, *Acta Chim. Acad. Sci. Hung.* **97** (1978) 221; L. Guczi and A. Sárkány, *J. Catal.* **68** (1981) 190.
40. A. Loaiza, Mingde Xu and F. Zaera, *J.Catal.* **159** (1996) 127.
41. F. Zaera and G.A. Somorjai, *J. Phys. Chem.* **89** (1985) 3211.
42. B.F. Hegarty and J.J. Rooney, *J. Chem. Soc. Faraday Trans. I* **85** (1989) 1861.
43. A. Sárkány, L. Guczi and P. Tétényi, *J. Catal.* **39** (1976) 345.
44. R. Brown, C. Kemball, J.A. Oliver and I.H. Sadler, *J. Chem. Res. (S)* (1985) 274.
45. G.C. Bond and J. Addy, *Trans. Faraday Soc.* **53** (1957) 383, 388.
46. L. Guczi. B.S. Gudkov and P. Tétényi, *J. Catal.* **24** (1972) 187.
47. H. Öz and T. Gaumann, *J. Catal.* **126** (1990) 115.
48. S.M. Davis and G.A. Somorjai, *J. Phys. Chem* **87** (1983) 1545.
49. F.G. Gault and C. Kemball, *Trans. Faraday Soc.* **57** (1961) 1781.
50. R.L. Burwell Jr. and R.H. Tuxworth, *J. Phys. Chem.* **60** (1956) 1043.
51. E.F. Meyer and C. Kemball, *J. Catal.* **4** (1965) 711.
52. R.L. Burwell Jr., B.K.C. Shim and H.C. Rowlinson, *J. Am. Chem. Soc.* **79** (1957) 5142.
53. T. Chevreau and F. Gault, *J. Catal.* **50** (1977) 124.
54. L. Guczi, A. Sárkány and P. Tétényi, *Bulg. Acad. Sci. (Comm. Dept. Chem.)* **6** (1973) 349.
55. R. Burch and Z. Paál, *Appl. Catal. A : Gen.* **114** (1994) 9.
56. A.C. Faro Jr. and C. Kemball, *J. Chem. Soc. Faraday Trans.* I **82** (1986) 3125.
57. R. Brown and C. Kemball, *J. Chem. Soc. Faraday Trans.* **86** (1990) 435.
58. R. Brown, C. Kemball and I.H. Sadler, *Proc. 9th Internat. Congr. Catal.* (M.J. Phillips and M. Ternan, eds.), Chem. Inst. Canada: Ottawa **3** (1988) 1013.
59. R.L. Burwell Jr., *Langmuir* **2** (1986) 2.
60. R.L. Burwell Jr., *Catal. Rev.* **7** (1973) 25.
61. R. Brown and C. Kemball, *J. Chem. Soc. Faraday Trans.* **92** (1996) 281.
62. R. Brown, C. Kemball and I.H. Sadler, *Proc. Roy. Soc. A* **424** (1989) 39.
63. Z. Karpiński, *J. Catal.* **77** (1982) 118.
64. E. Eskinazi and R.L. Burwell Jr. *J. Catal.* **79** (1983) 118.
65. E.H. van Broekhoven and V. Ponec, *J. Molec. Catal.* **25** (1984) 109.
66. J.K.A. Clarke and J.J. Rooney, *Adv. Catal.* **25** (1976) 125.
67. Z. Karpiński and J.K.A. Clarke, *J. Chem. Soc. Faraday Trans. I* **75** (1975) 2063.
68. V. Ponec and W.M.H. Sachtler, *J. Catal.* **24** (1972) 250.
69. Y. Inoue, J.M. Herrmann, H. Schmidt, R.L. Burwell Jr., J.B. Butt and J.B. Cohen, *J. Catal.* **53** (1978) 401.
70. S.M. Augustine and W.M.H. Sachtler, *J. Phys. Chem.* **91** (1987) 5953.
71. J. Barbier, A. Morales and R. Maurel, *Nouv. J. Chim.* **4** (1980) 223.
72. R.L. Burwell Jr. and K. Schrage, *Discuss. Faraday Soc.* **41** (1966) 215.

73. R.L. Burwell Jr., *Chem. Eng. News* **44** (1966) 56.
74. R. Brown, A.S. Dolan, C. Kemball and G.S. McDougall, *J. Chem. Soc. Faraday Trans.* **88** (1992) 2045.
75. A. Roberti, V. Ponec and W.M.H. Sachtler, *J. Catal.* **28** (1973) 381.
76. J.J. Rooney, F.G. Gault and C. Kemball, *Proc. Chem. Soc.* (1960) 407.
77. J. Barbier, A. Morales, P. Marécot and R. Maurel, *Bull. Soc. Chim. Belg.* **88** (1979) 569.
78. F.G. Gault and J.J. Rooney, *J. Chem. Soc. Faraday Trans.* I **75** (1979) 1320.
79. R.L. Burwell Jr. and K. Schrage, *J. Am. Chem. Soc.* **87** (1965) 5253.
80. J.J. Phillipson and R.L. Burwell Jr., *J. Am. Chem. Soc.* **92** (1970) 6125.
81. A. Sárkány, L. Guczi and P. Tétényi, *J. Catal.* **39** (1975) 181.
82. L. Guczi and P. Tétényi, *Acta Chim. Acad. Sci. Hung.* **77** (1973) 417.
83. S.M. Augustine and W.M.H. Sachtler, *J. Catal.* **106** (1987) 417.
84. J.B. Butt, *Appl. Catal.* **15** (1985) 161.
85. G. Fitzsimmons, C. Hardacre, W.R. Patterson, J.J. Rooney, J.K.A. Clarke, M.R. Smith and R.M. Ormerod, *Catal. Lett.* **45** (1997) 187.
86. T. Baird, E.J. Kelly, W.R. Patterson and J.J. Rooney, *J. Chem. Soc. Chem. Comm.* (1992) 1431.
87. D. Łomot and Z. Karpiński, *Catal. Lett.* **60** (2000) 133.
88. Buchang Shi and B.H. Davis, *J. Catal.* **147** (1994) 38.

7

HYDROGENATION OF ALKENES
AND RELATED PROCESSES

PREFACE

It is now time to consider a catalytic reaction in which the product is distinctly different from the reactant hydrocarbon – the hydrogenation of an alkene (they used to be called 'olefins'; some people do still call them that). As is typically the case, the first member of the series – ethene – has received more than its fair share of attention; this is understandable up to a point, because when it reacts with hydrogen ethane is always the main product, and usually the only one. Also it cannot compete with ethene for the surface, a fact that simplifies the kinetics somewhat. However the reaction with deuterium reveals a level of complexity unsuspected when only hydrogen was used, and the same Horiuti-Polanyi mechanism, which so satisfactorily explains what occurs, also accounts easily for the occurrence of reactions shown by the higher linear alkenes, in which the location and stereochemistry of the double bond changes concomitantly with the process of hydrogen addition. Related stereochemical subtleties are shown in the hydrogenation of substituted cycloalkenes. Just about every catalytic chemist who worked with metals will at some time have hydrogenated ethene, as its apparent simplicity and ease of performance commends it as an attractive way of assessing whether and how the physicochemical nature of the catalyst determines its activity. Many papers are content to record just that: but there have also been profound studies of its mechanism on a limited range of catalysts, which result in this being one of the best understood of catalysed hydrocarbon transformations.

7.1. INTRODUCTION[1-6]

The metal-catalysed hydrogenation of ethene was discovered by Paul Sabatier in the early years of the last century, after which it received little attention until the 1930s when the application of deuterium started to reveal the inner complexities of this apparently simple reaction.[7-9] These studies and those that followed (Section 7.2.4) demonstrated that the reaction is much more than the simple addition of a molecule of hydrogen or of two hydrogen atoms to the carbon-carbon double bond, but this has not prevented its use as an easy way of seeing how changes to the nature of catalysts affect their catalytic behaviour. However most of this extensive work, and the numbers of papers describing it must run into hundreds, records only the conversion or rate under a single set of operating conditions, and, because the response to changes in these variable has not been explored, it is of limited value: nevertheless we will have to note the conclusions reached by using catalysts having different particle sizes (Section 7.2.3). I fear that this apparent want of extended and systematic examination of the dependence of catalytic power on chemical composition and physical structure will be a recurring litany, because it can be said—and it needs to be said—of every class of reaction. I just hope it will not bore the reader.

The attraction of the reaction when performed with hydrogen lies in the singular nature of the final product, because with ethene, and indeed the higher linear alkenes also, the activity of many metals is such that reaction occurs in a range of temperature where the position of equilibrium greatly favours the alkane. Ethene hydrogenation can be followed on metal films even at 173 K. It is only with cyclohexenes, where further dehydrogenation is facilitated by the resonance stabilisation of the aromatic product, that products other than the alkane are observed at moderate temperature. The only other slight complication is that homologation is found in some circumstances (Section 7.3.1). Rates are usually faster than those of alkadienes and alkynes (see Chapters 8 and 9) because the reactive form is less strongly adsorbed; a 'volcano' relation between reactivity and adsorption strength has been proposed.[10]

Alkene hydrogenation is significantly exothermic, and it is not always easy to keep the catalyst isothermal except at low rates. Heats of hydrogenation for a number of alkenes were measured many years ago (Table 7.1),[2] the use of a catalyst ensuring that calorimetry could be conducted at ambient temperature. The values are similar to but perhaps more accurate than those derived from heats of combustion, where subtraction of two large numbers is involved; they reflect the extents to which the π electrons interact with the electrons in the C—H bonds by hyperconjugation. This electron delocalisation is also reflected in the relative stabilities of alkene complexes with Ag^+ cations.[2]

Measurements of the specific rate of ethene hydrogenation made under fixed conditions on catalysts of the same physical form reveal a systematic variation

TABLE 7.1. Heats of Hydrogenation of Alkenes (kJ mol^{-1})[2,64,249,250]

Linear Alkenes	$-\Delta H_H$	Branched Alkanes	$-\Delta H_H$	Cycloalkenes	$-\Delta H_H$
Ethene	137	2-Methyl-propene	119	Cyclopentene	109
Propene	124	2-Methyl-2-butene	112	Cyclohexene	113
1-Butene	126	2,3-Dimethyl-2-butene	111	Cycloheptene	108
E-2-Butene	115	4-Methyl-1-pentene	121	Cyclo-octene	96
Z-2-Butene	119	2-Methyl-1-pentene	112	1,2–Dimethylcyclo-	112.5 (Z)
				hexene*	120.5 (E)
1-Pentene	109	4-Methyl-E-2-pentene	110		
E-2-Pentene	114				
Z-2-Pentene	118				

*Calculated by DFT by Dr. E.L. Short for conversion respectively to Z- and E-dimethyl-cyclohexane.

in activity with the metals' electronic structure,[2] and hence with the strengths of adsorption of *both* of the reactants, which according to Beeck's results[11,12] vary in the same way with Periodic Group number: this is as expected if the slow step requires the breaking of M—H and M—C bonds. Thus in accordance with the Sabatier or Volcano Principle, the activity maximum is located where the best compromise between coverage and bond strength is to be found (Figure 7.1). This occurs at Group 9 or Group 10 (Ir is unaccountably 10^2 less active than Rh according to these old results), and the general level of activity of the base metals is less than that of the noble metals in Group 8 to 10.[3] Measuring rates at constant reactant pressures and temperature is no guarantee that the number of reacting

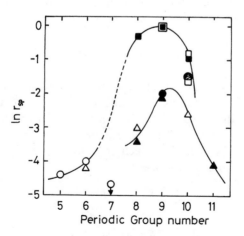

Figure 7.1. Dependence of the activity of metals for ethene hydrogenation on Periodic Group number.[2,3,11] $\triangle$, First row metals; $\square$, second row metals; O, third row metals. Open points, condensed metal films; filled points, silica-supported metals.

centres will be the same, and the lower activities of the base metals may simply reflect smaller areas of active surface resulting from their greater tendency to cause extensive disruption of the alkene.

Correlations between activity, adsorption strengths, and fundamental properties of the metals, both measured and derived, have however long been canvassed,[12] and have pedagogic value for discussing the physical basis of catalytic activity, at least in this class of reaction.[3] However the revelation of the more complex nature of the reaction by the use of deuterium in place of hydrogen[13] provides an additional level for analysing these connections. It will therefore be helpful to anticipate the discussion of the results of isotopic tracer experiments (Section 7.2.4) by setting down the basic mechanisms and terminology at this point.

Numerous works have sought to define *the* mechanism of alkene hydrogenation, as if this were a universal truth, applicable in all circumstances. Our discussion (Section 5.3) teaches that a more catholic view is appropriate, and that we should regard the mechanistic scenario as a kind of pot containing all conceivable elementary steps, from which we withdraw such items and in such quantities as are requisite to account for the observations in any particular case. We must abandon the idea that a single mechanistic statement will be valid everywhere. The Horiuti-Polanyi mechanism, to which reference has already been made in the context of alkane exchange, therefore represents the *simplest possible* way of accounting for what happens when ethene reacts with deuterium. As we shall see, it has been necessary to elaborate the original idea: this does not however demean its value.

The early work on this reaction disclosed dilution of the deuterium content of the non-hydrocarbon reactant by hydrogen; this could only be explained by the simultaneous occurrence of an *alkene exchange reaction,*[7,14–16] i.e.

$$C_2H_4 + D_2 \rightarrow C_2H_3D + HD \tag{7.A}$$

An inevitable consequence is the subsequent formation of more extensively exchanged ethene and of ethanes containing from zero to six deuterium atoms, but the relative rates of the exchange and addition vary widely with the metal and with operating conditions (Section 7.31). These observations are neatly explained by the Horiuti-Polanyi mechanism,[8] which proposed a reversible stepwise addition of deuterium (or hydrogen) atoms, e.g.

$$C_2H_4 \underset{-D}{\overset{+D}{\rightleftarrows}} C_2H_4D \begin{array}{c} \overset{-H}{\nearrow} C_2H_4D \\ {\scriptstyle +H} \\ {\scriptstyle +D} \\ \underset{-D}{\searrow} C_2H_4D_2 \end{array} \tag{7.B}$$

In this scheme the attachment of the species to the surface is not specified, and in particular the structure of the adsorbed ethene is left open for future discussion, as is the provenance of the deuterium atoms, although it will be evident that dissociative adsorption of the deuterium molecule must occur in some way. A more quantitative treatment of this scheme is deferred to Section 7.33. On some surfaces an intermolecular exchange of isotopes can take place in the absence of hydrogen or deuterium, e.g.

$$C_4H_8 + C_3D_6 \rightarrow C_4H_7D + C_3HD_5 \tag{7.C}$$

The reaction presumably needs atoms liberated by dissociative chemisorption of the alkenes with the accompanying formation of alkyl radicals.[2,17]

A note on terminology: the symbol D represents the ^2H isotope; in the text, D is called a deuterium atom and D_2 a deuterium molecule; the same goes for H and H_2. Terms such as protium (for ^1H) or dideuterium (for D_2) are eschewed. The process giving alkane is described as addition or hydrogenation, irrespective of whether hydrogen or deuterium species are involved. The terms deuterogenation and deuteriumation, being hybrids with all the elegance of a mule, are not employed.

Recognition of the stepwise addition of atoms to a chemisorbed alkene and of the stability of the intermediate alkyl radical provided an immediate explanation for other processes that occur during hydrogenation of higher linear and cyclic alkenes. These are *double-bond migration* and *cis-trans* (or *Z-E*) *isomerisation*:[2,6,18–20] these arise naturally because the atom removed in the alkyl reversal step is not necessarily that which was added, as is quite evident from the existence of alkene exchange. Once added, the atom is indistinguishable from other pre-existing atoms. Thus, when four carbon atoms are present, the reaction of 1-butene with hydrogen gives, besides *n*-butane, *Z*- and *E*-2-butenes, and the *Z*-isomer may transform to the *E*-isomer by rotation about the central bond of the 2-butyl radical (see Figure 7.2). The relative amounts of the various isomers formed do not correspond, at least initially, to those predicted by thermochemical considerations[21,22] (Figure 7.3), because the proportions in which they return to the gas phase are determined by the energetics of their conformations in the adsorbed state and not by those in the free molecule.[6] The isomer yields therefore contain mechanistic information and give valuable insights into the structures of adsorbed species. With catalysts that give much isomerisation, the amounts of the isomeric alkenes will approach or even reach their equilibrium values before all are hydrogenated (Sections 7.3 and 7.4); 3-methylcyclohexene will change to the 1-isomer, where the double bond stays because this is where it is most comfortable. Hydrogenation of disubstituted cyclohexenes has interesting stereochemical consequences

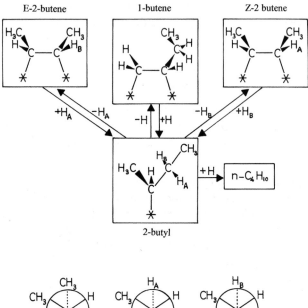

Figure 7.2. Interconversion of n-butene isomers via the σ-adsorbed 2-butyl radical: conformation **I** leads to either Z- or E-2-butene; **II** to Z; and **III** to E.

(Section 7.5). Double-bond migration in propene may be followed by labelling either of the terminal carbon atoms with ^{13}C. The reaction

$$CH_3{-}CH{=}^*CH_2 \rightarrow {}^*CH_3{-}CH{=}CH_2 \qquad (7.D)$$

occurs rapidly on Au/SiO_2 in the presence of hydrogen, and more slowly in Ag/SiO_2;[23] metals of Groups 8 and 10 have not yet been examined.

A besetting problem with the study of alkene hydrogenation is deactivation of the catalyst by strongly-adsorbed 'carbonaceous deposits'[2,24-26] or 'acetylenic residues'[11] that form even well below room temperature; the base metals of Groups 8 and 10 are especially prone to this, but even the noble metals are not immune, and even the most fundamental studies are necessarily made on 'equilibrium' surfaces, much of which is permanently inactivated.[27-29] This raises the vexed question of whether such poisoned surfaces truly reflect the character of the metal, and indeed whether it is not on the carbonaceous overlayer that the catalysis occurs.[30,31] Another idea that has been seriously suggested is that reaction actually occurs, in

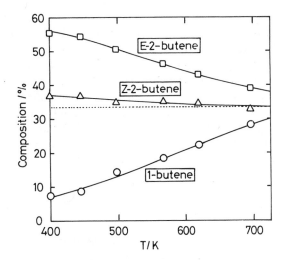

Figure 7.3. Temperature-dependence of equilibrium concentrations of the *n*-butene isomers.[22]

the case of supported metals, on the support, by spillover catalysis;[32-34] the fact that unadulterated metals are also efficient catalysts does not automatically invalidate this hypothesis. These ideas will be considered in Sections 7.2.1 and 7.2.7.

7.2. HYDROGENATION OF ETHENE AND PROPENE[4,6,15,35-37]

It is convenient to treat these two molecules side-by-side because, they are the only ones with which addition and exchange are the only real possibilities. The preponderance of the literature concerns ethene, but where comparison can be made there is little to choose between the ways in which they behave, save for self-evident differences due to the numbers of carbon atoms.

7.2.1. Kinetics of Hydrogenation

The purpose of this section is to provide an overview of the principal kinetic features of the hydrogenation of ethene and of propene, as providing a framework (or at least part of one) within which discussion of mechanisms must be conducted. Their reactions with hydrogen (and with deuterium) are quite comparable; the addition of the methyl group leads to somewhat higher reactivity, due to weaker chemisorption as might be predicted from its lower heat of hydrogenation (Table 7.1). Relative rates for other alkenes will be considered later. The problem of deactivation by 'carbon deposition' has already been mentioned, but quantitative information is scarce, and experience differs: thus various platinum catalysts are reported to lose 25% of their initial activity in about 1 h, before stabilising,[38] but in another case with similar catalysts there appeared to be no problem.[39] Deactivation seems to be more serious with the base metals than with the noble metals, although removing oxygen from the ethene is beneficial,[40] and when working with a constant volume system the order of adding the reactants is important; with nickel film, rates are slowest when ethene is introduced first and fastest and least reproducible when hydrogen is added first, intermediate values being found with simultaneous addition.[41] This is easily understood in terms of the formation of ethylidyne or other decomposition products being inhibited by pre-adsorbed hydrogen atoms, the effect of which is only slowly removed.

It has been seriously suggested that the structure insensitivity of ethene hydrogenation[42] and the constancy of the apparent activation energy finds an explanation if the reaction actually proceeds on top of a layer of carbonaceous species, which act as a bridge by which hydrogen atoms are carried from the metal to the ethene.[43] This concept has however never been quantitatively modelled, nor does it account for the marked differences in behaviour between palladium and platinum that will appear shortly. A similar concept was adopted by Somorjai, who suggested[44] that ethylidyne was the intermediate concerned (on Pt(111)); and although this species is now firmly relegated to the status of a mere spectator,[45–47] its role and that of carbon in general to facilitate movement of hydrogen atoms has never been totally eliminated. It must therefore be remembered that, in the absence of information to the contrary, the metal will be at lease partly covered by strongly retained dehydrogenated species that may (or may not) play some role in the reaction.

By far the greater part of published work has been performed in flow systems, which is in some ways unfortunate because the dependence of rate upon time (i.e. real time as opposed to contact time) is not then accessible: the time course kinetics do not always correspond to the initial rate orders because of a slow response of surface concentrations to pressure changes, and mechanism may alter as the reactants are consumed.[41] Orders of reaction are generally expressed as simple exponents, and only rarely (and usually only for ethene) is Langmuirian formalism used.[2,48] Sometimes the former method only describes the results over a very limited pressure range.[38] All of the most detailed experimental studies have been made with platinum catalysts of various descriptions, and from a broader viewpoint, considering also the reaction with deuterium, this is understandable, because ethene exchange is slight and its occurrence is indeed often ignored in otherwise comprehensive theoretical treatments.[38] By comparison what we know of the reaction on palladium, nickel and other metals is much less detailed.

Alkene hydrogenation is considerably exothermic (Table 7.1), and maintaining isothermal conditions in a catalyst bed is, except at lowest conversions, not a

TABLE 7.2. Kinetic Parameters for the Hydrogenation of Ethene on Platinum Catalysts[a]

Form/Support	Wt. % Pt	$E/\text{kJ mol}^{-1}$	P_H/kPa	P_E/kPa	x	y	T/K	Notes	References
/SiO$_2$	6.3	67	6.7	14.7	1	0	253	b	39
/SiO$_2$	1.2–11.5	29–50	97.5	2.5	—	—	—	c	55
/SiO$_2$	0.04	34.5	20	3.3	~0.5	−0.17	223		38
/SiO$_2$	—	35	—	—	0.77	−0.25	233		51
/SiO$_2$	12	88	6.6	6.6	1.0	−0.2	333	d	29
/SiO$_2$	0.05	67	20	30	0.5	~0	317		251
/Al$_2$O$_3$	5	63	13	13	1.0	−0.35	423	e	103
/Al$_2$O$_3$	1	41	8	6.5	1.2	−0.5	273	f	54
(111)	—	45	2.6	1.3	1.3	−0.6	315	g	44
Foil	—	42	13	2.6	1.3	−0.8	323		2, 44
Wire	—	34.5	20	3.3	0.95	−0.33	336		38
Wire	—	42	13	2.6	1.2	−0.5	323		233

[a] See ref. 2 for pre-1960 results.
[b] Similar results were reported for 0.75% Pt/SiO$_2$, and for Pt/Al$_2$O$_3$and for Pt/MoO$_3$.
[c] No systematic variation of E with either metal content or dispersion (see Figure 7.5).
[d] Catalyst deliberately deactivated.
[e] Experiments conducted with D$_2$.
[f] Similar orders were reported for Pt foil and Pt on other ceramic supports.

TABLE 7.3.   Kinetic Parameters for the Hydrogenation of Ethene on Other Metals*

Metal	Form	$E/\text{kJ mol}^{-1}$	x	y	T/K	References
Fe	/SiO$_2$	35	0.91	−0.04	303	51
Fe	Film	30.5	0.87	−0.6	305	41
Co	/SiO$_2$	35	0.55	−0.19	213	51
Ni	/SiO$_2$	35	0.67	−0.08	233	51
Ni	/SiO$_2$-Al$_2$O$_3$	50	1.09	0.21	401[a]	40
Ni	Film	42	1→0.7	−0.4	305	41
Cu	/SiO$_2$	35	0.69	0.06	353	51
Ru	/γ-Al$_2$O$_3$	36	1	−0.2	327	108
Ru	/SiO$_2$	35	0.95	−0.59	203	51
Rh	/SiO$_2$	35	0.85	−0.74	197	51
Pd	/SiO$_2$	35	0.66	−0.03	243	51
Pd	Al$_2$O$_3$	—	∼0.9	−0.3	298	47
Os	/γ-Al$_2$O$_3$	35	1	—	297	108
Ir	/Al$_2$O$_3$	58	1.4	−0.6	333	103

[a] E falls to 27 kJ mol^{-1} above 408 K, y increases to 0.36.
*See 2 for pre-1960 results.

trivial matter. Ignoring it will vitiate kinetic measurements, and lead to unjustifiable conclusions.[40]

Notwithstanding the many differences of metal, dispersion, surface structure, cleanliness etc, there is a remarkable uniformity in the kinetic parameters observed in the region of room temperature. Of course even with a single metal this uniformity is not complete, and there are occasional cases of the rate (expressed per unit of active area) being badly out of line,[49] perhaps due to unsuspected poisoning of the catalyst, and of activation energies being anomalously high[50] or low.[2] A selection of some of the more recent values (and a few older ones) is given in Tables 7.2 and 7.3, but in general those published before about 1960 are omitted; for these an earlier summary must be consulted.[2] Turnover frequencies given for platinum in older work have been shown as an Arrhenius plot, giving a mean activation energy of 36 kJ mol^{-1}. These provide a useful norm against which subsequent results can be compared; thus the TOF at 273 K should be[49] about 1 s^{-1}, but a tabulation of values for rates or TOFs is not particularly useful because of the variety of circumstances in which they have been measured, and the uncertainty of correcting them to any standard condition.

Activation energies are very frequently between 30 and 50 kJ mol^{-1}, and for different metals in comparable physical form it is sometimes said that the same value suffices for all.[12,51] Orders in hydrogen are always positive and very often are unity, while orders in ethene are usually either zero or sometimes negative. So general is this behaviour that attention is naturally drawn to the few exceptions. With platinum, orders of reaction are essentially independent of the physical form[38] and the support,[52] but are temperature-dependent (Table 7.4), the order in

TABLE 7.4. Temperature-Dependence of Orders of
Reaction for Ethene Hydrogenation on a Platinum Catalyst
$(0.04 \% \text{ Pt}/ \text{ Cab-O-Sil})$[38]

	$r \propto P_H{}^x P_E{}^y$	
T/K	x^a	y^b
223	0.48	−0.17
273	0.67	−0.17
336	1.10	−0.43

$^a P_E = 25$ Torr
$^b P_H = 150$ Torr

hydrogen becoming less positive and that of ethene less negative with falling temperature. These are important factors to be accommodated by the reaction model, but they may be specific to platinum, because for various silica-supported metals the orders in ethene, being necessarily measured at various temperatures, tend to *more* negative values the lower the temperature.[51] Orders in hydrogen however show no such systematic trend (see Tables 7.2 and 7.3).

The limited value of expressing the rate dependence in a reactant pressure as a simple exponent in Power Rate Law formalism has been noted before (Section 5.2) and will doubtless be mentioned again. The concept of an 'order of reaction' has been borrowed from homogeneous gas-phase kinetics, where integral and semi-integral orders are the norm, and only with complex reactions such as those proceeding by chains is this mode of expression inadequate. It is however optimistic to think that a single 'order' can describe the behaviour of a reaction over any significant range of conditions: at best it is an approximation to a more meaningful expression relating rate to the concentration of reacting entities. Ethene hydrogenation provides several convincing examples of the limitations of the 'order' paradigm: (i) 'orders' in both reactants are temperature-sensitive and (ii) the 'order' in ethene on Pt/SiO_2[38] at 273 K is negative below about 50 Torr pressure, but accurately zero above this pressure. Furthermore the Arrhenius equation is not always fully obeyed; activation energy will be pressure-dependent if reaction orders are temperature-sensitive. On fresh Pt/SiO_2 there appear to be two different states of activity, having different activation energies, and between which it is possible to move reversibly by changing temperature; carbon deposition removes this effect, however.[39] Multiple steady states are reported[53] under isothermal conditions on rhodium film simply by altering reactant concentrations, but no details have been given. There have also been many reports of activation energy decreasing[40] and ultimately becoming negative[2] as temperature is raised; this has been reasonably attributed on the onset of ethene desorption, in support of which is the movement of ethene order to higher (i.e. more positive) values.[2]

This then summarises the kinetic framework within which mechanistic discussion has to be conducted (Section 7.2.6), but much other relevant information has to be collected before this can be done; a preliminary consideration of certain features is however in order. The high positive orders in hydrogen carry the implication that there is much scope for further increase of rate if yet higher pressures were used, and hence that the concentration of reacting centres is quite small. First order in hydrogen tells us that a molecule, or probably two atoms, appear in the rate-determining step, most probably indirectly through an ethyl radical: therefore as a working hypothesis for a typical catalyst we may imagine the equilibrium

$$H + H_2C—CH_2 \rightleftharpoons CH_2—CH_3 + 2^* \qquad (7.E)$$
$$* \quad * \quad * \qquad *$$

lying far to the right, but since ethene is so much more strongly adsorbed than hydrogen the surface layer will usually comprise mainly ethene and ethyl radicals (Figure 7.4). Orders in hydrogen greater than first are not easily explained. There has been much discussion as to whether the adsorption of the reactants is competitive for the same sites, or non-competitive if there are sites that only hydrogen can use:[28] an early proposal[52,54] was that both can operate at the same time. Negative orders in ethene suggest competition, but zero orders do not require it.

The general near-constancy of activation energy at about 35 kJ mol^{-1} may lead us to think that this is a *true* and not an *apparent* quantity, because there is no evidence that it depends on reactant pressures[40] (although this can be inferred) or on dispersion.[55] There is of course no reason for its value to be always exactly the same, as minor differences may be caused by variations in surface cleanliness etc; much higher values (in the region of 60–90 kJ mol^{-1}, see Tables 7.2 and 7.3) do however require special thought. True activation energies are associated with constancy of surface coverage, and they are normally *higher* than apparent values (Section 5.25): we must therefore suppose that (i) the rate-determining step is the

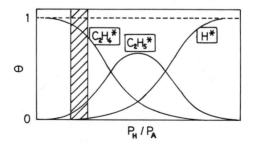

Figure 7.4. Schematic variation of surface coverages by ethene, ethyl and hydrogen atoms as a function of hydrogen pressure/ethene pressure (P_H/P_A). The likely range of coverage of various metals for $P_H/P_A = 1$ is within the shaded area.

same for all catalysts showing an activation energy of 30–50 kJ mol^{-1}; and (ii) within the usual range of measurement, increase of temperature does not so effect the concentration of adsorbed reactants that the rate is thereby depressed, and so there is maximum utilisation of surface throughout.

An interesting application of ethene hydrogenation is its use to estimate the number of pre-adsorbed hydrogen atoms by passing ethene over the catalyst in either a continuous[56] or a pulse mode,[57] and measuring the ethane produced. The method has been applied to a number of supported platinum catalysts, and gives good agreement with other methods (e.g. hydrogen–oxygen titration), providing sulfur is absent (see also Section 3.31). The procedure was first described using 1-pentene[58] (see Section 7.4).

Dr Samuel Johnson remarked of a dog walking on its hind legs, *it is not surprising it is done badly; it is surprising it is done at all.* The same may be said of the hydrogenation of alkenes by the metals of Group 11. Ag/SiO$_2$ and Ag/TiO$_2$ catalysts activated by oxidation and reduction were active[59] for the hydrogenation of ethene and of propene between 313 and 413 K, with activation energies of about 40 kJ mol^{-1}. For the reaction of ethene with deuterium on Au/SiO$_2$, see Section 7.24; for the hydrogenation of 1-pentene on gold catalysts, see Section 7.4.

7.2.2. Structure Sensitivity[50]

The hydrogenation of alkenes has long been recognised as being 'structure-insensitive',[45] by which is meant rates per exposed metal atom (i.e. TOFs) are essentially independent of particle size, the number of such atoms being estimated before reaction, and usually by hydrogen chemisorption. The general truth of this view is not in doubt, but it is necessary to test its veracity and to examine apparent exceptions.

The observation that specific (areal) rates of ethene hydrogenation on platinum catalysts of very different dispersion when measured under fixed conditions are closely similar[38,55] would seem to imply that every surface atom or site capable of adsorbing a hydrogen atom, or a constant fraction thereof, provides an 'active centre' of about the same activity. The logic of this conclusion rests upon the near equivalence sometimes observed between the number of surface atoms estimated (i) by physical techniques and (ii) by hydrogen chemisorption using a 1:1 H/Pt$_s$ stoichiometry, the limitations of which have already been discussed (Section 3.3.1). All surface atoms, whatever their co-ordination number, are equally capable of being or participating in an 'active centre', and so this reaction obeys one of Taylor's limiting conditions that *nearly all surface atoms are active.*[60] It is not only specific rates that show this uniformity; in many cases activation energies also fall within a narrow range (Tables 7.2 and 7.3). There are however alternative explanations of this apparent insensitivity to particle size[38,55] and surface structure.[61] It is common experience that metal surfaces rapidly become partly covered by 'carbonaceous deposits',[38,41,62] and that rates measured under steady state conditions therefore utilise only a fraction of the total surface. It is therefore hypothesised that these deposits occupy the more reactive areas and that reaction occurs only on those remaining sites, uniform in character, on which ethene is adsorbed weakly and reversibly.[62] It is not however obvious that such 'active centres' should be the fixed fraction of the total, as is implied by correlating the number of surface sites *before* reaction with the rate *during* reaction.

We should now examine some studies that apply shades of grey to the black and white picture painted above. A number of careful examinations of particle size effects have been made using 'model' catalysts prepared by depositing metal atoms from the vapour phase onto oxide supports. With both Pt/SiO_2[63,64] and Pt/Al_2O_3,[65] rates at 373 K using a 10:1 hydrogen:ethene ratio show maxima at a size of about 0.6 nm; they then decrease three-fold to reach values which above about 1.7 nm are independent of size. The rates are however at least a factor of 10 less than expected at 373 K on the basis of results[49] for other forms of platinum. Ni/SiO_2 shows similar behaviour, but lower rates at sizes below 0.6 nm were seen only with the platinum catalysts.[66] These papers deserve to be consulted for their extended erudite discussion of the possible geometric basis of catalytic activity, but there is no recording of kinetic parameters or of carbon deposition, and it cannot be assumed that the observed rates refer to equal coverages by reactants: and remembering that the size of a single platinum atom is about 0.28 nm, the significance of particle sizes less than 0.6 nm is not quite clear. No variation in activity has been found using palladium particles between 1 and 3 nm size.[67] R. L. Moss and his associates[55] have made detailed observations on Pt/SiO_2 catalysts containing 1.2 to 11.5% metal, having mean particle sizes between 3 and 6 nm; rates at 193 K varied erratically with size within a factor of four, but Arrhenius plots showed distinct (but again irrational) variations of activation energy with metal loading, and exhibited compensation (Figure 7.5). If as has been suggested above the activation energies reported for this reaction are 'true' values, such variation is not to be expected, but it is clearly beyond experimental error. There are two possible explanations: either (i) each *is* a 'true' value, being slightly influenced by other variable factors than size, e.g. by contaminants, or (ii) they are not *quite* 'true' values, because surface coverages are not equivalent. In the absence of further information such as reaction orders, these possibilities cannot be distinguished. The fact[40] that activation energy is dependent of ethene pressure on Ni/SiO_2-Al_2O_3 does not help, as each case is *sui generis*, and must be considered separately.

There are some indications of particle size dependence with propene hydrogenation:[35] with Pt/SiO_2, rates at 220 K increased less than two-fold as dispersion rose from 5 to 80%, but they also depended on the type of pre-treatment applied.[68,69] With Pd/SiO_2,[70] rates passed through a maximum at about 60%

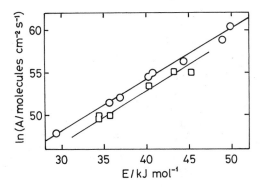

Figure 7.5. Compensation plot of Arrhenius parameters for ethene hydrogenation on Pt/SiO$_2$ reduced at either 483 K (O) or 353 K ($\square$).[55]

dispersion, and again varied with pre-treatment the difference between the two metals found no ready explanation, and the absence of kinetic measurements (and of hydrogen dissolved in the palladium catalysts) did not help. TOF values for Ni/Al$_2$O$_3$ catalysts increased with increasing dispersion.[71]

There is limited information available on structure dependence of rates shown by unsupported metals. TOFs have been reported for ethene hydrogenation on low Miller index faces of nickel, the (100) face being quite inactive, perhaps due to 'carbon' deposition.[36] Step-density on platinum single crystals does not determine activity for this reaction, but does affect its hydrogenolysis.[61] Surface defects introduced into nickel and platinum by argon ion bombardment when removed by annealing caused respectively 100- and 10-fold decreases in activity.[72] Hydrogen-deuterium equilibrium was not however affected. Most interestingly, nickel powder having the abnormal cph structure has been prepared by decomposition of the acetylacetonate, and at 313 K showed a rate (4.1 mol m^{-2} h^{-1}) for propene hydrogenation that was ten times larger than that shown by the usual fcc form.[73] Nickel boride showed very low activity. Linear nickel 'nanostructures' prepared by photolithography showed a maximal rate for ethene hydrogenation at a size of 4 nm.[74]

To conclude, it is evident that there are a number of loose ends that need to be tidied up before the last word on the structure sensitivity of ethene hydrogenation can be written. It is worth noting that many studies use a large excess of hydrogen and a low pressure of ethene,[55,59,64,66] which may bring the reaction into the region of positive order in ethene (Figure 7.4). It has to be continually stressed that valid comparisons of activity can only be made when the concentrations of reacting species are about the same, a condition that is by no means always met.

7.2.3. Ethene Hydrogenation on Bimetallic Catalysts[75,76]

The hydrogenation of ethene was used in a number of laboratories between about 1950 and 1980 to assess the importance of the electronic theory in heterogeneous catalysis. Unfortunately, as we now know, this work was motivated by the concept of d-band filling by electron-rich metals that is incorrect (Section 1.32). Moreover, the existence of a miscibility gap in the nickel-copper system, which was particularly selected for study, was not always appreciated, and in many cases the composition of the surface, and its degree of contamination by carbonaceous deposits, were unknown quantities. In some of the work, great pains were taken to try to obtain homogeneous bimetallic films,[77-79] but X-ray diffraction, often used as the criterion, does not of course disclose the surface composition, which in any case may be altered by the presence of the reactants. In view of these and many other areas of uncertainty, it is not surprising that results from different laboratories were lacking in consistency. We must therefore regretfully review this work only briefly, as it does not add much to our understanding of mechanisms: we shall however find some additional support for the structure-insensitive nature of the reaction.

The bimetallic systems used fall into four classes:

(i) Group 8, 9 or 10 + Group 11 or 14 or RE metal (e.g. Ni-Cu;[2,78-89] Ni-Au;[79] Pd-Cu;[2,82] Pd-Ag;[2] Pd-Au;[75] Ru-Cu;[90] Ni-Sn and Pd-Sn;[91] Pt-Sn;[92] Pt-Cu;[2,82] Pd-Eu and Pd-Yb[93]).

(ii) Group 10 (or 9 +10) metals (e.g. Ni-Pd;[77,86,94] Pd-Pt[95]).

(iii) Group 11 metals or Group 11 + RE metal (e.g. Cu-Ag;[2] Ag-Yb[96]).

(iv) Intermetallic compounds.[97]

In the work of Rienäcker and his associates, using metal foils, rates were often only slightly affected by adding copper or silver to a Group 10 metal, although activation energies changed, and exhibited compensation; the Arrhenius parameters describing the much lower rates found for materials rich in the Group 11 metals fell on a quite separate line, perhaps due to a step change in the concentration of active centres.[2] With nickel-copper films, rates usually increased with nickel content, in one case the activation energy falling from 50 to 34 kJ mol^{-1} at 80% nickel,[88] but in another study they also rose as copper was added to nickel, so that there was a marked minimum in the centre of the series.[78,79] Both studies showed approximate compensation between the Arrhenius parameters, as did values found using nickel-copper foils cleaned under UHV conditions;[89] in one instance the value of the plot was proved by highlighting one questionable value that lay well off the line. With silica-supported nickel-tin and palladium-tin, the reduction of carbon deposition caused by the presence of tin did not overcome the loss of activity due to an electron shift, which was indicated by an increase in both the Ni $3d$ and Pd $4d$ binding energies. This confirms the observations made with metal-tin systems mentioned in Section 3.32. NMR work with Ru/SiO$_2$ catalysts showed that adsorbed hydrogen exists in strong and weak states, the ratio H_w/H_s doubling as dispersion increases only from 20 to 30%: the weak form is associated with atoms of low co-ordination number, and these are the ones that are preferentially replaced when copper is introduced.[90] The consequential decrease in activity for ethene hydrogenation is therefore caused by loss of this form. This adds a further factor to the matrix of problems that have to be resolved when trying to understand how the activities of bimetallic catalysts depend on structure and composition.

Results obtained with mixtures of Group 10 metals do not unfortunately provide information of great interest; they have already been reviewed in depth.[76,77,82,86,94]

It is well known that bimetallic compounds of the type $(RE)M_5$, where RE is a rare-earth element and M is either cobalt or nickel, absorb large amounts of hydrogen, so it is of interest to know whether it is able to hydrogenate alkenes. Compounds of this type not containing hydrogen catalysed ethene hydrogenation at 190 to 230 K at rates proportional to hydrogen pressure and independent of ethene pressure.[97] The hydrides reacted quickly with ethene at 195 K, but rates varied 100-fold with composition, the $LaNi_5$ compound being the most active: activation energies were within error constant at about 24 kJ mol^{-1}, which was also that for hydrogen desorption, so diffusion of hydrogen atoms to the surface was the slow step. On reacting a mixture of hydrogen and ethene over a hydride, hydrogenation occurred without change of pressure until late in the reaction.[97]

7.2.4. Reactions of Ethene and of Propene with Deuterium[13,98,99]

This section is concerned with the reactions of ethene and of propene with deuterium on forms of metal catalyst other than single crystals, which are covered in the next section. A fuller discussion of reaction mechanisms is reserved to Section 7.2.6.

The revelation of the extensive but unsuspected processes occurring during alkene hydrogenation was mentioned in Section 5.7.2, and the principal observations were summarised in the introduction to this chapter. These may be developed as follows. An *alkene exchange reaction* by the addition of a deuterium atom to chemisorbed alkene and the later abstraction of a hydrogen atom (process 7.B) occurs to very different extents on various metals. It is for example very marked with nickel and palladium, and much less with platinum (see later). Now, if the liberated hydrogen atom is very quickly returned to the gas phase as a molecule of hydrogen deuteride, there will be seen a *hydrogen exchange reaction*: if the number of hydrogen atoms released then equates to the number of deuterium atoms appearing in the alkene, the mean number of deuterium atoms in the alkane will be exactly two. If however the hydrogen atoms are employed in the process of forming the alkane, the rate of the hydrogen exchange will be less than that of alkene exchange, and the alkane will necessarily contain less than two deuterium atoms. The characteristics of the reaction that are of chief interest are therefore the

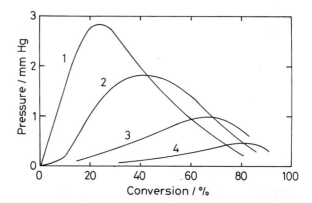

Figure 7.6. Reaction of ethene with deuterium on nickel wire at 363 K; variation of pressures of deuteroethenes with conversion.[100] $1 = C_2H_3D$; $2 = C_2H_2D_2$; $3 = C_2HD_3$; $4 = C_2D_4$.

rates of these exchange processes (and their dependence on operating conditions), the mean number of deuterium atoms entering the alkane, the breadth of the distribution of the deuteroalkanes, and the way in which it changes with conversion.

These features are best illustrated by reference to the reaction of ethene with deuterium on various nickel catalysts[2,6,7,19,100,101] and on supported platinum catalysts.[2,6,28,52,103] On nickel there was seen the stepwise formation of al the deuterated ethenes (Figure 7.6), in consequence of which (hydrogen exchange being minimal) the deuterium number of the ethane rose progressively: thus ethane-d_0 was the major initial product, but all deuterated ethanes were seen, and the formation of ethane-d_6 was more marked towards the end of the reaction, when most of the ethene was ethene-d_4 (Figure 7.7). The stepwise character of the

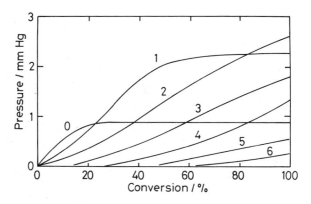

Figure 7.7. Reaction of ethene with deuterium on nickel wire at 363 K; variation of pressures of deuteroethanes with conversion.[100] $0 = C_2H_6$; $1 = C_2H_5D$; $2 = C_2H_4D_2$; $3 = C_2H_3D_3$; $4 = C_2H_2D_4$; $5 = C_2HD_5$; $6 = C_2D_6$.

ethene exchange implies a rapid desorption (and adsorption) of ethene molecules. The reaction over platinum shows quite different behaviour. Although the results were somewhat support-dependent,[52,102] in general alkene exchange was slight when near-stoichiometric reactant ratios were used, and the deuteroethane distribution did not change much with conversion (Figure 7.8). Hydrogen exchange was more noticeable than with nickel, and there was evidently a less ready interchange between gaseous and adsorbed ethene. Even in the absence of detailed information on the extents of alkene and hydrogen exchange, it is possible to assess their importance by the value of the deuterium number of the alkane, and how it changes with conversion.

The form of the distribution of the deuteroalkanes merits further consideration. It was noticed that the amounts of species having more than one or two deuterium atoms declined logarithmically with increasing number of such atoms:[52] the ratio of [alkane-d_{n+1}]/[alkane-d_n] ($n > 2$ or 3) is denoted by σ, and this relation holds, often with surprising accuracy, not only for ethanes and propanes,[2,6] but also for higher alkanes (Sections 7.31 and 7.4), providing the concentration of the deuteroalkane falls within the range of precise measurement. Some examples of this behaviour are shown in Figure 7.9; values of σ are most usually between 0.4 and 0.6, and are often close to 0.5. A constant σ will arise is the ratio of the rates of alkyl reversal to alkene and of alkyl conversion to alkane is independent of deuterium content, and it has been argued that a value of 0.5 suggests that alkene and alkane are formed at the same time by disproportionation of two alkyl radicals. However, it is often the case that the amount of the alkane-d_2 exceeds expectations,[2,6,52] sometimes by a large amount (see Figure 7.9), and the difference between the observed and expected values (given by back-extrapolation of the logarithmic plot) is attributed to a 'direct addition' (DA) process that is independent of that giving the other products. It has been shown[52] that experimental distributions are very well reproduced simply by selecting values for the two parameters σ and the amount of DA. Further attention to mechanism and the source of atoms in the formation of alkane will be given below (Section 7.2.6).

Before considering how operating conditions affect the progress of alkene-deuterium reactions, it is desirable to introduce an alternative and somewhat more profound way of interpreting product distributions.[104] This proceeds by assigning probabilities to each of the four steps in process 7.B as in Scheme 7.1: here X stands for either H or D and the asterisk simply identifies the species that are adsorbed; the probabilities are assumed to be independent of the isotopic composition of the reacting species, and C—H and C—D bonds are supposed to have equal strengths. There are six possible adsorbed ethenes and twelve ethyls (including positional isomers), and one may then write eighteen simultaneous equations defining their concentrations in terms of p, q and r: the ethyl radical profile is then converted into an ethane distribution by the parameter s. This procedure actually affords more information than is available experimentally, as it says how much ethane-d_0 is

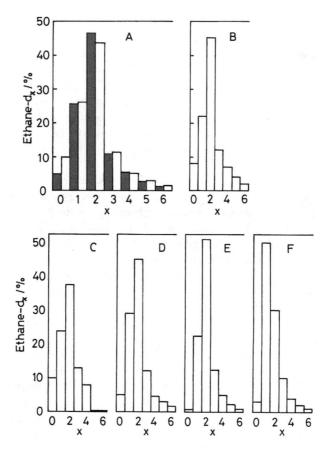

Figure 7.8. Histograms of deuteroethane distributions obtained on various Pt/SiO$_2$ catalysts (A - E) and on Pt(111) (F) under the following conditions.

	Wt.% Pt	D^a	T/K	P_D/P_E	M	% Conv.	References
A	1	—	273	1.1	1.93^b,1.89	45^b,100	52
B	0.04	1.0	273	2	2.08	Low	28
C	2.4	0.3	263	1	1.85	Low	48
D	0.1	—	293	1	1.90	15	102
E	6.3	0.6	293	1	1.96	15	102
F	—	—	333	2	1.57	—	44

$^a D$ = dispersion. ^{b} Shaded columns.

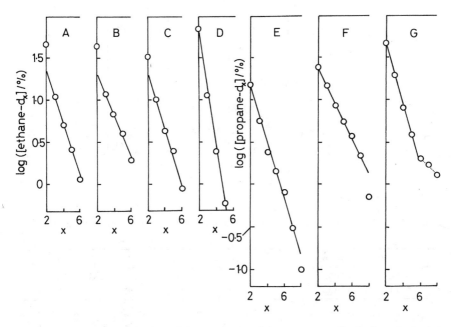

Figure 7.9. Semi-logarithmic plots of deuteroethane and deuteropropane distributions obtained on various platinum catalysts under the following conditions.

	Alkane	Support	Wt.% Pt	P_D/P_E	T/K	% Conv.	σ	References
A	C_2H_4	SiO_2	1	~ 1.1	273	45	0.49	52
B	C_2H_4	SiO_2	0.04	2	273	Low	0.58	28
C	C_2H_4	Al_2O_3	1	1.67	255	39	0.45	103
D	C_2H_4	Al_2O_3	1	125	273	100	0.21	103
E	C_3H_6	Pumice	5	0.03	291	100	0.46	54
F	C_3H_6	Pumice	5	1.23	291	100	0.61	54
G	C_3H_6	Pumice	5	42.2	291	100	0.43	54

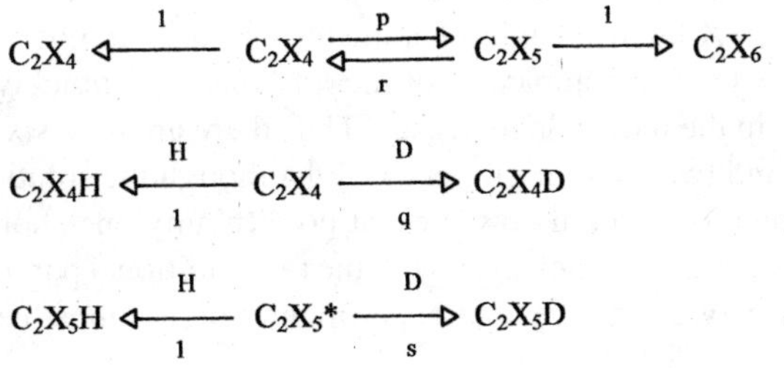

$$C_2X_4 \xleftarrow{\quad 1 \quad} C_2X_4 \underset{r}{\overset{p}{\rightleftarrows}} C_2X_5 \xrightarrow{\quad 1 \quad} C_2X_6$$

$$C_2X_4H \xleftarrow{\quad H \quad} C_2X_4 \xrightarrow{\quad D \quad} C_2X_4D$$
$$\text{\small 1} \qquad\qquad \text{\small q}$$

$$C_2X_5H \xleftarrow{\quad H \quad} C_2X_5{}^* \xrightarrow{\quad D \quad} C_2X_5D$$
$$\text{\small 1} \qquad\qquad \text{\small s}$$

Scheme 7.1. Parameters describing the reaction of ethane with deuterium (after Kemball).[104]

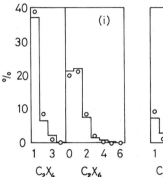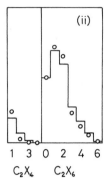

Figure 7.10. Histograms of deuteroethene and -ethane distributions obtained under the following conditions,[125] with best-fit calculated values by Kemball's method[104] (see Table 7.4 and text) shown as ○.

	Metal	Support	P_D/P_E	T/K	% conv.	P'	q'	r'	s'
(i)	Pd	Al_2O_3	1	237	5	75	33	90	20
(ii)	Pt	Al_2O_3	1	298	10	98	33	99	50

returned to the gas phase, and quantifies all positional isomers of the deuteroethanes where these exist. It may also give the ratio of $C_2X_4^*/C_2X_5^*$, or some quantity proportional to it. Experimental results were matched to calculated ones by trial-and-error, aided by intuition and experience. A few examples are shown in Figure 7.10; agreement is often good but never perfect, as the selected parameter values are not the optima, because minimisation routines of the type now commonly used were not then available. The method[104] does however serve to identify which parameter is most in play when some alteration to experimental conditions is made, and for this reason is more informative than the logarithmic approach described above. A weakness of this analysis, however, is its tendency to underestimate amounts of the more heavily deuterated products; this is partially overcome by a modification[105] that allows the C—X bond-breaking parameter r to have a much larger value for X = H than X = D. The complexity of the analysis rises quickly as the number of carbon atoms in the molecule increases. Thus there are fifty-six distinguishable propyl radicals and twenty-four propenes, and although the solution of the eighty simultaneous equations would now present no difficulty one should logically include allylic species as well; however, with the two additional parameters involved, unique matches between theory and experiment would almost certainly become impossible.

We are now in a position to consider how the reactant ratio and temperature affect product distributions. In the simplest manifestation of the Horiuti-Polanyi scheme,[8] deuterium is adsorbed dissociatively, and ethene non-dissociatively: their interaction will lead in the steady state to surface coverages by ethene, ethyl and deuterium or hydrogen atoms that will vary with reactant ratio in the manner shown in Figure 7.4. The curves drawn relate to some specific but arbitrary value for the equilibrium constant of the alkene-alkyl-H,D atom system, and this will depend upon temperature; the value of the reactant ratio that provides a chosen surface composition is likely to depend on the metal, support and other features of the catalyst. The effect of temperature is harder to predict as it depends on the energetics of all the participating processes; concentrations of adsorbed species will of course tend to decrease with increasing temperature.

When alkene is in excess, the reaction in a constant volume system stops when the deuterium is used up, and the deuterium: alkene ratio decreases continuously as the reaction proceeds. Unreacted but partially exchanged alkene remains at the end, and the deuterium number of the alkanes M is less than two, the alkane-d_o becoming a major product (see Figure 7.9 for the propene-deuterium reaction on Pt/pumice[54]). Contrarily when deuterium is in excess, the final alkane deuterium number is greater than two, the alkene-d_0 falls to near zero, and the alkane-d_2 becomes the major product (see also Figure 7.9). Similar but less complete results were seen with propene[103] and ethene[28,52,103] on a variety of supported platinum catalysts.

These trends are readily understandable in terms of Figure 7.4 and Scheme 7.1. The rate of alkene exchange will be proportional to the concentration of adsorbed alkene, and will therefore increase as the ratio P_A/P_D (A = alkene) increases; M decreases because deuterium atoms are entering the alkene rather than going to make alkane. As P_D/P_A increases, hydrogen exchange becomes more important because its rate depends on θ_x^2 or $\theta_x\theta_{ethyl}$, and alkene exchange diminishes: the alkane distribution sharpens, alkane-d_2 becomes the major product, and M increases to a limiting value of about 2.5 (Figure 7.11). The value of σ decreases (compare **C** and **D**, and **F** and **G**, in Figure 7.9) because addition to alkyl is favoured over alkyl reversal to alkene, there being plenty of adsorbed atoms available, but few empty sites (or ethene molecules) to act as atom acceptors. Higher values of p and lower values or r are found at high P_D/P_A. Lower values of σ are also found at low P_D/P_A ratios, because here alkene desorption removes deuterium atoms from the surface and progressive substitution in alkyl radicals is inhibited (compare **E** and **F** in Figure 7.9). Under these conditions, equilibration of propene-d_o and -d_6 proceeds in parallel with alkane formation.[19]

On nickel catalysts with equimolar reactant mixtures, ethene exchange increases in importance with rising temperature, being undetected at 195 K and 223 K, but very significant at 403 K;[106] however, with a ten-fold excess of deuterium it is absent up to 323 K.[19] On platinum catalysts the effects are comparatively minor:[28,52,103] in general the yield of alkane-d_2 and values of M decrease with rising temperature (Figure 7.12), this signifying that here too alkene exchange is becoming relatively more important. These effects may be in part because alkyl

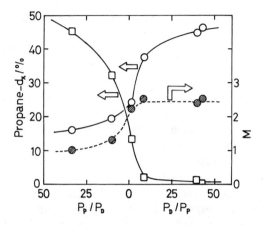

Figure 7.11. Dependence of propane-d_0 (O) and -d_2 (□) yields, and of M, on ratios of reactant pressures (T = 298 K).[54]

reversal becomes easier as more free sites are created by thermal desorption of the reactants; this is confirmed by the tendency of r to increase and p to decrease.[6]

Although it is commonly thought[28,48] that alkene hydrogenation is structure-insensitive, as we have seen (Section 7.2.2), this is not quite true: moreover the belief is based on the rate of hydrogenation, but there is ample evidence to show that rates of [52,102]*alkene exchange* depend on the form of the catalyst (in the case of platinum), and this inevitably affects the shape of the deuteroalkane profile. It is therefore incorrect to think that results obtained with one particular platinum

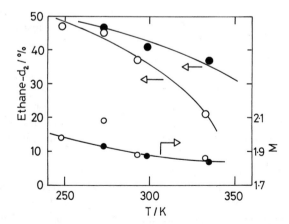

Figure 7.12. Dependence of ethane-d_2 yield and of M on temperature O 1% Pt/SiO$_2$, P_D/P_E ~1.1[52] ● 0.04% Pt/SiO$_2$, $P_D/P_E = 2$[28]

catalyst epitomise all such catalysts. An early study[52] of the behaviour of this metal unsupported and on a number of supports showed significant differences in the relative rate of alkene exchange: it was quite high on platinum foil and on Pt/Al$_2$O$_3$, and low on Pt/SiO$_2$, results which have subsequently been confirmed by other workers[28,44,48,102] (see Figure 7.9). Particle size is unlikely to be the course of the differences, as in one case[52] the catalysts were made by depositing a platinum sol (size ~2 nm) onto the various supports; also sintering of 6.3% Pt/SiO$_2$ (EUROPT-1) made little change to the product distribution.[102]

The variability of the rate of alkene exchange relative to that of hydrogenation to ethane is also clearly seen with results for Pt/TiO$_2$ and Pt/MgO, where the exchange rate is more prominent than for Pt/Al$_2$O$_3$ and Pt/SiO$_2$.[102] Application of Kemball's modified scheme showed[105] this to be due to enhanced chances of ethane desorption (respectively 40 and 20% compared to ~10%) rather than to greater alkyl reversal. The Pt/TiO$_2$ may have been partially in the SMSI condition, and similar effects have been shown elsewhere by magnesia, although the cause is unlikely to be the same (Section 3.3.4). This study[102] emphasises the importance of the theoretical analysis of the component steps[105] in identifying the effects of variables; from the similarity between the values of the s and q parameters it appeared that both ethene and ethyl species drew on the same source of hydrogen and deuterium atoms.

The discussion thus far has been mainly about platinum, but extensive results are also available for nickel wire,[54,101,107] for supported ruthenium, rhodium, palladium, osmium and iridium,[2,103,108,109] with a few for nickel, iron and tungsten films,[104] and for Au/SiO$_2$.[110] Each metal is unique in the values it shows for the relative rates of the component processes, and their dependence on operating variables: their behaviour is most economically described by reference to the four parameters p, q, r and s that give approximate fits to the experimental results; a selection of these is given in Table 7.5. Because these parameters are defined as ratios (see Scheme 7.1), differences in their values over-emphasise changes in relative rates; they are therefore re-defined as

$$i' = i/(1 + i) \qquad (7.1)$$

($i = p, q, r, s$). Inspection of this table at once shows that platinum and iridium distinguish themselves by showing very high values of p' and r' (all over 90%), this accounting for their small propensity for alkene exchange; for all the other metals they are substantially lower, indicating that this process is much more favoured. This is particularly so with Ru/Al$_2$O$_3$ and Os/Al$_2$O$_3$. Thus for example with palladium, with both ethene and propene, the extensive alkene exchange required the mean number of deuterium atoms in the alkene (M) to be often less than unity, and alkane-d_0 the major product (Figure 7.10); hydrogen exchange was minimal.[2,109] Alkene and hydrogen exchange was also very marked in the reaction

TABLE 7.5.   Values of Parameters Describing the Ethene-Deuterium Reaction*

Metal	Form	P_D/P_E	T/K	p'	q'	r'	s'	Notes	References
Ni	Film	—	173	75	67	92	50	a	104
Ru	$/Al_2O_3$	0.54	326	17	90	71	80	b	6
Rh	$/Al_2O_3$	1	255	75	50	80	80		6
Rh	$/Al_2O_3$	1	349	37	67	90	80		6
Pd	$/Al_2O_3$	1	340	67	33	90	10		6
Os	$/Al_2O_3$	0.56	317	44	90	50	80		6
Ir	$/Al_2O_3$	1	257	99	33	90	90		103
Pt	$/Al_2O_3$	1	327	98	33	95	33	c	103

* The parameters i' give the percentage of reaction in the i direction (Scheme 7.1) : values are selected to give
   approximate fits to results obtained under the conditions given in columns 3 and 4.
^{a} An iron film under these conditions is fitted by the same parameters.
^{b} Re/SiO$_2$ gives products similar to those of ruthenium and osmium although its activity is very low.
^{c} These parameters also fit the product description given by Au/SiO$_2$ at 459 K.

of propene on Rh/Al$_2$O$_3$ at 363 K, so that lightly deuterated propanes were major products, but these processes were somewhat less evident with ethene, especially at low temperature (255 K); the main effect of temperature in this case was on the ethene desorption parameter p' (Table 7.4).[111] Gold showed results similar to those for platinum, but of course rates were much lower;[110] tungsten film at 193 K afforded mainly ethane-d_2 with little ethene exchange.[104]

We must now consider the sources of the atoms that are added to adsorbed alkenes and alkyls. While the simplest form of the Horiuti-Polanyi scheme would suggest that only deuterium atoms should be available, it is quite clear that hydrogen atoms can also be used; this is at once apparent from the appearance of alkane-d_0 and -d_1, and is confirmed by the values of the parameters q' and s' (Table 7.4). Thus for example with all the quoted metals (except Ru and Os) q' is only in the range 33–67%, and where q' and s' take similar values there may be a common process of atom addition;[102] but with rhodium and iridium s' is much greater than q' (80–90%, against 33–67%), suggesting that a molecule of deuterium may be involved in alkane formation. The most obvious sources of hydrogen atoms are adsorbed atoms released in alkyl reversal, and hydrogen-rich alkyl radicals; the two are almost equivalent and are kinetically indistinguishable.

There is evidence with nickel[112] and platinum[52] that molecular addition is unimportant: virtually the same alkane profile is given by non-equilibrated and equilibrated mixtures of hydrogen and deuterium.[2] Unfortunately the measurements with platinum were made with propene where 'direct addition' as shown by semi-logarithmic plots is absent (it appears with ethene but not propene, Figure 7.9). In addition to the evidence provided by the high values of q' and s' for ruthenium and osmium, which incidentally overcome the extensive alkene exchange to make ethane-d_2 the major product,[6] the orders in deuterium are often somewhat above unity (Table 7.2), which is hard to explain without some contribution from molecular addition.

Iwasawa and his associates have made a number of studies of the changes to metal particles brought about by high temperature reduction, using supports that have been claimed to induce the Strong Metal-Support Interaction (SMSI, see Section 3.3.5). Some of these have employed XPS[48] or XANES[50] to characterise the effects, and the ethene-deuterium and ethene-hydrogen + deuterium reactions have been used as tests. With Rh/Nb_2O_5 and Ir/Nb_2O_5, activation energies for these reactions between 210 and 321 K were not affected by reduction temperature, but values were low (16–23 kJ mol^{-1}) and exceedingly low for Ir/Nb_2O_5 after HTR.[113] Both catalysts after LTR gave ethane-d_2 as the main product of the ethene-deuterium reaction, but little or no more fully deuterated products were found, so that values of M were very low: this implies the substantial occurrence of alkene exchange (a fact not remarked on), but unfortunately there was no analysis of the ethenes to confirm this. After HTR, both catalysts showed peculiar distributions in which ethane-d_0 and -d_2 were major products, but ethane-d_1 was smaller; again no product contained more than two deuterium atoms, and hydrogen deuteride formation was small. These results were interpreted by a model involving metal sites, peripheral sites influenced by the support, and sites on the support itself.

In a further study, catalysts comprising platinum on yttria, zirconia, titania, niobia and vanadium sesquioxide were examined similarly:[50] changes in the rates of addition and of equilibration caused by HTR, and activation energies for the former, are listed in Table 7.6. The addition rate at 233 K was unaffected by HTR in the case of Pt/Y_2O_3 and actually raised with Pt/ZrO_2: with both Pt/V_2O_3 and Pt/TiO_2 the rates as expected were severely depressed. Deuteroethane profiles were not much affected by HTR, except in the case of Pt/V_2O_3, where deactivation gave products containing 86% ethene-d_2. In all other cases, values of M were well below two, and ethanes-d_5 and -d_6 were never seen, but analysis of the ethenes receives only a passing mention: conversions at which the analyses were made are also not

TABLE 7.6. Effect of High-Temperature Reduction on the Rate of Ethane Hydrogenation and on the Rate of the Simultaneous $H_2 + D_2$ Reaction over Various Platinum Catalysts.

Support	Hydrogenation		Equilibration
	Δr	E/kJmol^{-1}	Δr
Y_2O_3	0	~40	0.22
ZrO_2	2.2	119	0.86
V_2O_3	0.02	—	0.10
TiO_2	0.04	~37	0.09

Δr is the factor by which rate changes on altering reduction temperature from low to high: this has little effect on E.

mentioned. Although the P_D/P_E ratio was changed 50-fold, its effect on product distributions was not stated. Changes in the electronic structure of the metal brought about by HTR, and then by the presence of the reactions, were revealed by XANES measurements, which assessed the density of unoccupied $5d$ states. This was generally decreased by HTR, except with V_2O_3, which has metallic character, so that easy exchange of electrons between metal and support is possible; this change occurred even with yttria and zirconia, which do not exhibit the SMSI. The sense of the electron shift between metal and ethene was also support-dependent. The multiple-site model was again used to explain the deuteroethane distributions: it is otherwise very hard to account for the highly unusual distributions shown by platinum in the SMSI state; it is possible for one site to show massive ethene exchange, so that ethane-d_0 > $-d_1$ > $-d_2$, and for another to give only direct addition, but identification of the sites responsible is speculative. The high activation energies shown by Pt/ZrO$_2$ await explanation. Effects on ethene hydrogenation of adding niobia, vanadia, molybdena and manganese oxide to Rh/SiO$_2$ have also been discussed.[114]

Addition of sodium ion to Pt/SiO$_2$ increased the rate of ethene deuteration but not that of the simultaneous hydrogen-deuterium equilibration;[115] activation energy of the former increased from 33 to 40 kJ mol^{-1} and the Pt $4f_{1/2}$ binding energy increased. This implies the acquisition of negative charge, although it is hard to see how sodium cations can provide electrons. Deuteroethane profiles on the sodium containing catalysts showed M values well over two, showing that hydrogen exchange must have been important: there were sharp cut-offs after ethene-d_4. The multiple site model was again invoked.

Lest it be thought that the Horiuti-Polanyi scheme, or some modification of it, provides all the answers to the problems of alkene hydrogenation and exchange, it must be said that there is considerable evidence that the latter process may involve a quite different mechanism, involving the *dissociative* chemisorption of the alkene,[16] as for example

$$C_2H_4 + 2^* \rightarrow C_2H_3{}^* + H^*$$

$$C_2H_3{}^* + H^* + D_2 \rightarrow C_2H_3D + HD + 2^* \qquad (7.F)$$

Rapid exchange of ethene and propene on iron and nickel films appears to proceed in this way.[36] Very detailed studies by Japanese scientists using microwave spectroscopy have identified the structure of propene-d_1 formed in reaction of propene with deuterium over metals of Groups 10 and 11, either supported on silica[2,23,36] or as powders.[82] Interpretation of the results is somewhat difficult because although addition and exchange show very similar kinetics, and are therefore thought to have the same intermediates, the locations of the deuterium atom in the propene-d_1 are not entirely as expected by the alkyl reversal mechanism. Except on palladium and platinum, the major initial product was propene-2-d_1:[1,116] this could arise if addition of the first deuterium atom gave chiefly the n-propyl radical. On platinum however E-propene-1-d_1, was initially preferred, while on palladium exchange occurred at all possible positions. Further intra-molecular processes suffered by the primary products made for additional complications in the interpretation.[23,82] On Au/SiO$_2$ and Ag/SiO$_2$,[23] exchange took place faster than addition. Over nickel, palladium and rhodium powders, intermolecular exchange of propene-3-d_1 occurred in the absence of hydrogen or deuterium; π-propenyl species were first formed, and the liberated atoms formed propyls from which the exchanged propenes were formed.[1,82] If these processes were also to occur during the normal course of alkene hydrogenation, the methodology devised by Kemball would be seriously compromised.

There are few reports of alkene-deuterium reactions on bimetallic catalysts, but those few contain some points of interest. On very dilute solutions of nickel in copper (as foil), the only product of the reaction with ethene was ethene-d_1;[117] it is not clear whether the scarcity of deuterium atoms close to the presumably isolated nickels inhibits ethane formation, so that alkyl reversal is the only option, or whether (as with nickel film, see above) the exchange occurs by dissociative adsorption of the ethene. Problems also arise in the use of bimetallic powders containing copper plus either nickel, palladium or platinum. Activation energies for the exchange of propene were similar to those for the pure metals[82] (33–43 kJ mol^{-1}) and rates were faster than for copper, but the distribution of deuterium atoms in the propene-d_1 clearly resembled that shown by copper. It was suggested that the active centre comprised atoms of both kinds. On Cu/ZnO, the reaction of ethene with deuterium gave only ethane-d_2, as hydrogens in the hydroxylated zinc oxide surface did not participate by reverse spillover.[82]

Exchange of alkenes with tritium (as hydrogen tritide, HT) is more sensitive bur less informative than exchange with deuterium.[118]

7.2.5. Reactions on Single Crystal Surfaces[119]

It is convenient to collect information on the interactions of alkenes with hydrogen or deuterium on single crystal surfaces into a separate sub-section because the application of methods unsuited to supported metals and powders isolates and identifies species and elementary steps, the existence of which can only be surmised by conventional kinetic studies: this makes an invaluable contribution to a final discussion of mechanism.

Although most of the work so far reported has used the Pt(1 1 1) surface,[120–122] a recurring theme of all the more recent publications is the pivotal role played by the weakly-adsorbed π-alkene in hydrogenation. Sum-frequency generation[123–128] (SFG) (Section 4.3) clearly shows this to be the case with both ethene[62,120,127,129] and propene[31] at room temperature, this form being more reactive than the more strongly held σ-form "by several orders of magnitude". This had long been suspected, but positive confirmation had to await the development of SFG. Theoretical work using DFT on the Pd(111) surface gas however concluded that only the di-σ form is able to react with hydrogen atoms.[130] The surface coverage of the π-form under reaction conditions may be quite low (~ 0.04; it cannot be detected by IRAS[122]), so that turnover frequencies based on the whole surface could be 25 times too small.[62] A dynamic molecular beam study of the ethene-hydrogen reaction also implicates this form as the key intermediate, the rate of ethane formation being proportional to its coverage and that of hydrogen atoms.[131]

The ethene-deuterium reaction has been studied over Pt(111) between 300 and 370 K;[44,46,121] ethane-d_1 was the chief product and the mean deuterium number of the ethanes at 333 K was only 1.57 (Figure 7.8): this increased with time, so ethane exchange was obviously occurring, but as so often happens the deuteroethene composition was not reported. The kinetic parameters obtained are included in Table 7.2. An important feature of this work was the observation that after reaction the surface was coated by 'carbonaceous deposits' which later work identified as ethylidyne groups. The reaction did not however exhibit self-poisoning, a second reaction proceeding at the same rate as the first: it was therefore concluded that reactions went above the ethylidynes, which therefore acted as a bridge for transfer of hydrogen atoms from metal to π-ethene.[43] This notion, previously suggested by Thomson and Webb to explain the apparent structure insensitivity of this reaction, has not however been applied quantitatively to account for either the differences between reactions on various metals as outlined in the last section, or the kinetic parameters. The idea that ethylidyne radicals were themselves intermediates in ethene hydrogenation has not been sustained by further investigations, and they are now relegated to a spectator status. However, the possibility that they may act as bridges in the transfer of hydrogen atoms to the reactant alkane has not been entirely eliminated. The reaction of ethene with a mixture of hydrogen + deuterium has been followed on Pt(100), (110) and (111) between 318 and 423 K; rate dependences on temperature and on 'hydrogen' pressure were determined, but no effects due to 'carbon' formation were seen or discussed.

The propene-deuterium reaction has been examined more recently on the same surface using TPD spectroscopy; the extensive results obtained are described in a lengthy paper.[132] Above 230 K, all deuteropropenes and propanes were found, yields of the latter decreasing logarithmically with increasing number of deuterium atoms. An unexpectedly small amount of propane-d_7 was explained by the reluctance of the hydrogen on the central carbon atom to exchange. The traditional mechanism was confirmed, and adsorption of propene as a vinylic or π-allylic species was definitely ruled out.

Finally, as concerns Pt(111), the reaction of co-adsorbed ethene-d_4 and hydrogen has been studied using a combination of laser-induced thermal desorption, mass-spectrometry and RAIRS:[133] exchange occurred above 215 K with an activation energy of 46 kJ mol^{-1}, this being below the point at which conversion to ethylidyne was seen. This latter reaction had the higher activation energy of about 70 kJ mol^{-1}. As is often found,[134] pre-adsorbed hydrogen favoured the formation of the π-state of ethene; both forms reacted to give an ethyl radical, through which exchange took place, but the π-form reacted faster.

Less, but no less interesting, work has been done with single crystal surfaces of the other metals of Groups 8 to 10. With films of iron and cobalt grown epitaxically on Ni(100), and having fcc structure, ethene reacted with pre-adsorbed hydrogen to give (once again) the π-state, the binding energies being respectively 5 and 8 kJ mol^{-1} greater than for nickel.[135] Platinum deposited on Ni(100) has also been examined.[136] On Rh(111) as on Pt(111), hydrogenation of ethylidyne to ethane and exchange in its methyl group were some 10^6 times slower than hydrogenation of π-ethene.[45] A HREELS study of ethene's reaction with adsorbed hydrogen atoms on Rh(100) at 110 K also gave the π-state with a $\pi\sigma$ factor of 0.39: ethyl radicals were also formed, and were converted to ethane below 200 K.[137] TPD measurements on Rh(111) led to much the same conclusions.[138] Similar work on Pd(100)(1 × 1) with ethene and propene showed that exchange with adsorbed deuterium atoms started at 120 K, and that alkanes were formed at 250–300 K.[139] On Pd(100) the initial adsorption of ethene on an ordered overlayer of hydrogen or deuterium atoms at 90 K gave the π-state plus another even more weakly-held form; they inhibited further adsorption of hydrogen, but ethene-d_4 exchanged via ethyl radicals.[140] On Ir(111) and Ir(110)(2 × 1) above 400 K, the reaction of propene with hydrogen afforded lower hydrocarbons in small amounts that grew as temperature was raised; coverage by strongly-held carbon species followed suit.[141] The earliest work on single-crystal faces was performed using nickel.[142] Work on Ni(111) led to the curious conclusion that dissolved hydrogen atoms emerging from within were more effective than adsorbed atoms because they were more energetic.[143]

Model catalysts (Section 2.3) permit the use of the same techniques of examination as single crystals. Palladium particles formed on alumina-coated NiAl(100) adsorbed ethene in both the π- and σ-forms, but the latter was favoured with increase in particle size: this could indeed be the basis for the weak size dependence noted previously (Section 7.2.2) in alkene hydrogenation. Hydrogen adsorbed more strongly on small particles, and in a now familiar way shifted the adsorbed states of ethene towards the π-form: this reacted with weakly-held hydrogen atoms to form ethane.[62]

7.2.6. The Reaction Mechanism: Microkinetic Analysis, Monte Carlo Simulation, and Multiple Steady States

The nature of the mechanism by which ethene is hydrogenated and the related reactions take places has excited the interest (and sometimes the emotions) of scientists for three-quarters of a century. It is humbling—not to say humiliating—to find that the questions being discussed in detail some 50 years ago have not yet been definitively answered; new questions have also arisen to muddy the water further. It is typical of heterogeneous catalysis that reactions of great formal simplicity admit of so many and varied interpretations. In addressing the First International Congress on Catalysis, the late Sir Eric Rideal remarked, with characteristic modesty as follows. *A great number of workers in the field of catalysis from Sabatier onward have given explanations of the mechanism of the reaction; I myself have advanced three. At least two must be erroneous and judging by the fact that no fewer than three communications are to be made on this subject during this week, it is quite likely that all three of them are wrong.*[144] As we shall shortly see, the number of candidate mechanisms has grown considerably since 1956.

An attempt was made to list the items of knowledge required in making a statement of mechanism (Section 5.3), but this list was minimal, and is capable of extension. There are several suggestions in the previous sections that the simple form of the Horiuti-Polanyi mechanism[8] as expressed in process 7.B and Scheme 7.1 is not entirely adequate; this has long been suspected, and the evidence for a vinylic exchange mechanism needing the dissociative adsorption of the alkene,[36] and the recognition of the likely role of π-alkenylic species on some metals, has only added to the feeling of inadequacy. We may focus on two particular worries. First, there are a number of other possible and plausible reaction steps that can lead to the same range of products as would the Horiuti-Polanyi mechanism; these include for example the reaction of a hydrogen or deuterium *molecule* with either ethene or ethyl, and the disproportionation of two ethyl radicals to give either two ethenes plus a hydrogen or deuterium molecule, or ethene + ethane, without the intervention of hydrogen or deuterium atoms. This by no means exhausts the possibilities. Second, it has several times been proposed that hydrogen or deuterium molecules can chemisorb dissociatively either in competition with sites acceptable to the alkene, or on sites at which the alkene cannot itself adsorb. This is a distinction that did not occur to Horiuti and Polanyi, but has troubled many since their day. It is most convenient to review the current state of mechanistic theory by outlining three quite different approaches, and summarising their conclusions. Those desiring an historical perspective are urged to read the relevant papers and discussional sections in *Discussions of the Faraday Society,* Vol. 8, (1950); they will find much of interest.

The term *microkinetic analysis* has been applied[1,27,144] to attempts to synthesise information from a variety of sources into a coherent reaction model for the hydrogenation of ethene. The input includes steady-state kinetics (most importantly the temperature-dependence of reaction orders[38]), isotopic tracing,[28] vibrational spectroscopy and TPD; it uses deterministic methods, i.e. the solution of ordinary differential equations, for estimating kinetic parameters. It selects a somewhat eclectic set of elementary reactions, and in particular the model provides for both competitive and non-competitive adsorption of hydrogen, as well as further 'activation' of both sorts of atom before they can react. It ignores minor products of the ethene-deuterium reaction such as ethanes-d_3 to -d_6, deuterated ethenes and hydrogen deuteride. It coalesces results obtained on Pt(111) with those for small platinum particles; and so by assuming structure-insensitivity it represents its conclusions as applicable to all platinum catalysts.

A major conceptual difficulty with this model is the lack of definition of the different sorts of adsorption sites proposed. The results presented in Chapters 3 and 4 indicate that there are indeed several possible locations for hydrogen atoms, one of which perhaps could be involved with alkene adsorption and thus be 'competitive'; but the possibility of interconversion of the two types is not considered, nor is the physical nature of 'activation' discussed. Site requirements for the hydrocarbon species clearly allow for site-blocking, as ethyl radicals are taken to use two sites rather than one; the π- and σ-forms of adsorbed ethene are not differentiated. These models shown in Chapter 4 reveal how hard it is to relate observed structures to specific locations on the metal atom lattice that may appear to be logical adsorption 'sites'. Kinetic parameters for the seven forward and seven reverse reactions have been evaluated, and the model with these values is claimed to describe the results adequately. Orders in hydrogen greater than first were not however reproduced. It is important that work of this kind should be undertaken, but we must guard against thinking that the model is wholly valid and its conclusions totally reliable.

Monte Carlo simulations[145–150] claim to avoid some of the above problems by using a stochastic model. Duca and colleagues considered[145,146] a square array of sites (i.e. the fcc (100) plane), a set of adsorption and reaction steps, and the probabilities of their occurrence. This set contains some unusual (and unnecessary?) features; ethane is formed in a physically adsorbed state using two 'sites' and is in equilibrium with gaseous ethane, while ethyl radicals take up three sites, i.e. the two on which ethene was adsorbed and one for a hydrogen atom. The ethyl radical is regarded as a kind of virtual species, and again π- and σ-forms of ethene are not distinguished: the reaction set is somewhat simpler than that used in microkinetic analysis. A 'steric hindrance parameter' defining the number of carbon atoms allowed to be adjacent to a given atom is included; if its value is set low, non-competitive hydrogen adsorption is possible. Hydrogen is 'activated' if it is on a site next to a carbon atom, and ethane is formed when two hydrogen atoms are adjacent to an ethene molecule.

The model was operated[146] by starting with a bare surface, and evaluating the probabilities of all possible events in a 'time-slice': this was repeated about 30 million times after which steady-state behaviour was found. The probabilities were related to real-time by collision numbers based on kinetic theory, and experimental sticking coefficients values of variable parameters were fixed to give best agreement with the experimental dependence of TOF on hydrogen pressure at 298 K; these values then reproduced the order in ethene very satisfactorily, including the change from negative to zero order as its pressure is increased. This followed from the necessarily assumed low value of the steric hindrance parameter, which allows non-competitive hydrogen adsorption (and hence reaction) to occur even when the surface is as fully covered by ethene as possible. While achieving striking success, an unfortunate limitation of this model is its failure to consider the results of isotopic tracer experiments. It is able however to 'study' the reaction under conditions not yet examined experimentally, for example, where the ethene coverage becomes so low that the rate is proportional to its pressure.

A further application[151] of the Monte Carlo procedure has been directed to accounting for sudden transitions from a low to a high activity state in ethene hydrogenation as either temperature or ethene pressure is changed; they are capricious, in that they are sometimes very large and sometimes quite small, and frequently do not appear at all. They have been observed with a fresh Pt/SiO$_2$ catalyst,[39] but not with Pt/MoO$_3$ or EUROPT-1; they were more noticeable with small platinum particles (5–6 nm) than with large ones (16–19 nm), but particle size dependence does not seem to be a uniformly satisfactory explanation. The Monte Carlo simulation used only the basic Horiuti-Polanyi reaction set, and assumed the π-state of ethene, and competitive adsorption of reactants: the transition was observed because, unlike the earlier model, adjacent adsorbed ethenes were allowed, and coverage by ethyls was specifically included. In the low activity state, hydrogen atoms were isolated in a sea of mainly ethene molecules whereas in the high activity state most were present as adjacent pairs. The slow step was therefore thought to change from ethyl formation to ethane formation, thus explaining the change in activation energy. This would require a change in reaction orders, but this has not been tested experimentally.[39] The procedure predicted the occurrence of rate discontinuities as ethene pressure was changed, but the calculated size of the rate change was very large, and much greater than that found in practice in the same study. It also attempted to account for the products of the reaction of ethene with deuterium, and the variation of activity with composition of palladium-gold bimetallic catalysts; the effect of particle geometry has also been addressed.

The group originating this methodology has continued to expand and refine its procedures with thermodynamic and quantum mechanical inputs.[152] The same reaction set and steric hindrance parameter were retained, and the activity transitions referred to in the last paragraph were again targeted;[39] a quantitative treatment of this concept was developed. The rate transitions do however require there to be an order-disorder transition of the adsorbed hydrogen atoms. The arguments deployed are both complex and subtle, and are not easily summarised: the papers cited in this section deserve careful reading as showing the depth of theoretical reasoning that can be applied to this simple reaction.

The third theoretical paradigm that should perhaps be mentioned is the use of *advanced deficiency theory*,[153] (a sub-set of chemical reaction network theory) to interpret the multiple steady states,[154] which have been observed in ethene hydrogenation over rhodium film in a flow system.[53] These alternative states of high and low activity were apparently achieved by alteration of the direction of temperature or hydrogen pressure change, and although they may be related to those described by Jackson et al.[39] it reads more as if the plots of rate vs. variable showed hysteresis; in the absence of actual results one cannot be sure. The theoretical armoury of chemical engineers was then deployed[53] to consider thirteen basic sets of unit reactions, including three variants of the Horiuti-Polanyi mechanism, and some that are only available in American doctoral theses. If this were not enough, these thirteen schemes were then combined in every possible way to allow for duplicate pathways for ethene formation, to give 80 candidate mechanisms, of which 71 were rejected. Unfortunately the limited intelligence of the author prevents any attempt at describing the process by which this conclusion was reached, but one must suppose that it is an advance of some kind if experiments based only on a simple measurement of rate can be shown to be consistent with only nine reaction mechanisms.

DFT has been applied to unravel the means by which hydrogen atoms add to ethene molecules chemisorbed on the Pd(111) surface in the opening step of the hydrogenation sequence.[130] Binding energies of the π and σ forms were similar at low coverage, but the former had to transform into the latter before it could react; computed activation energies were similar to those found experimentally, and references to earlier theoretical work were given.

7.2.7. Catalysis by Hydrogen Spillover and the Reactivity of Hydrogen Bronzes

It has been blithely assumed in what has gone before that the reaction occurs solely on the metal in the case of supported metal catalysts. The phenomenon of *hydrogen spillover* was introduced in Section 3.34, and the idea of *spillover catalysis*, i.e. reaction on the support induced in some way by the presence of the metal, has been noted (Section 5.4) as a possible complicating factor when studying particle size effects. The possibility that spillover catalysis contributes to alkene hydrogenation was considered some years ago as a means of accounting for different rates shown by platinum on various supports,[36,59] but this is now thought very unlikely: one of the strongest contrary arguments is the general similarity, not only of TOF but also of internal kinetic parameters (alkene exchange etc.), shown by single crystals, wires and supported metals. There are however some somewhat special and limited conditions under which alkene hydrogenation can occur in a spillover mode, and these will now be briefly examined, not so much because reaction in this way can qualify as being 'metal-catalysed' as because, being metal-induced, it is an ever-present threat to the rational interpretation of results. So one must always be on one's guard against it, and indeed in Chapter 10 it will emerge as being more than likely.

The various ways in which hydrogen spillover catalysis can emerge may be classified as follows. (1) Hydrogen atoms may migrate from metal particles to a ceramic oxide support where they may react with the alkene; the possibility that the observed reaction is taking place on the metal is overcome by physically removing or isolating the metal-containing catalyst. (2) In the presence of both reactants, continuing reaction can be found on the support after removal of the metal catalyst. (3) Hydrogen atoms may dissolve into the lattice of certain transition metal oxides (V_2O_5, MoO_3, WO_3, see Section 3.3.4), and these may react with alkene. (4) The hydrogen bronze so formed may catalyse hydrogenation in the absence of the metal. Numbers 1 and 3 are examples of spillover catalysis; numbers 2 and 4 might be termed *induced catalysis* because, although the metal played a role, it was (probably) not itself the catalyst. Catalysis by *reverse spillover* occurs when the metal remains but the hydrogen gas is removed: spillover hydrogen (hydrogen spillage) returns to the metal by diffusion, where it reacts with the alkene. This process is recognised when the amount of product vastly exceeds what might be formed by reactants retained on the metal. Short summaries of experimental findings follow.

In work with ceramic oxides, the oxide to be activated is initially in contact with a supported metal catalyst, and the two are subjected to defined treatments before the activated oxide is separately examined. One standard procedure involved heating in 1 atm hydrogen at 573 K for 8 h, lowering the temperature to 383 K and leaving for a further 8 h.[34,155] Two methods have been used for eliminating the catalyst from the subsequent examination. In the first, the catalyst (Ni/Al_2O_3 or Pt/Al_2O_3) was placed in a bucket immersed in the oxide; it was then raised up by a windlass, and isolated.[155] In the second, the catalyst as pellets rested at the foot of the reactor containing the oxide through which hydrogen flowed; it was then replaced by nitrogen and after purging, ethene was introduced above the pellets under conditions such that its back-diffusion to the metal was unlikely.[156] These experiments led to the following conclusions.

Hydrogen spillover to γ- or δ-Al_2O_3 affords atoms that can react with ethene, and it generates sites at which continuous but slow hydrogenation can occur. The activation is either activated or endothermic, because hydrogen spillage increases with temperature. A chain reaction such as

$$H\bullet + C_2H_4 \longrightarrow C_2H_5\bullet \xrightarrow{H_2} C_2H_6 + H\bullet \tag{7.G}$$

may describe what happens, as reaction is inhibited by nitric oxide, which may either react with the hydrogen atoms or displace them.[36] No such effect was shown by silica, and silica-alumina improved with repeated reductions.[156] Rates were slow and the number of reacting centres small (<1 per 10^{-18} m^2). The location of the hydrogen atoms has never been firmly established: they may have been attached to oxide (O^{2-}) or hydroxyl ions,[158] but 'carbon' deposition has been observed after reaction, so this may in fact be where reaction occurs.

Little interest has been shown in the phenomenon in recent years, although the role of hydrogen spillover in controlling carbon deposition during petroleum reforming,[159] or even in the reforming process itself,[160,161] has been canvassed. This early work may therefore not have been entirely in vain.

Hydrogen bronze formation with vanadia,[162,163] molybdena[164,165] and tungsta[166] is achieved by depositing a small amount of an activating metal (usually Pd or Pt) and exposing the material to hydrogen; alternatively the oxides can be admixed with a catalyst (e.g. Pt/Al$_2$O$_3$) before hydrogen treatment.[164] In the latter case, use of a bucket reactor allows study of the reactivity of the bronze by itself; in the former case the effects of metal and bronze cannot be distinguished. With 0.1% Pt (or Pd)/MoO$_3$, successively longer exposures (up to three days) and higher temperatures (up to 333 K) gave H$_x$MoO$_3$, where x is 0.34, 0.9, 1.6 and 2: a mixture of molybdena with Pt/Al$_2$O$_3$ at 433 K gave only H$_{1.6}$MoO$_3$. The bronze phases were stable in ambient air, but were oxidised at 333 K, regenerated molybdena from which the bronze could again be formed; H$_{1.6}$MoO$_3$ reacted with ethene above 373 K, giving in sequence the lower bronzes. This phase in the absence of metal began to react with ethene above 353 K; at 453 K as the ethene pressure was raised, the order was initially positive but became zero, and the activation energy was 54 kJ mol^{-1}. This is about that for proton migration in the bulk. The H$_{1.6}$MoO$_3$ itself acted catalytically for ethene hydrogenation: at 453 K at low ethene pressure (<10 kPa) the order in ethene was first, becoming zero at higher pressures however the time-dependent rate was always zero order, so the ethene pressure in some way determined the state of the surface. The initial rate varied only slightly with hydrogen pressure; the activation energy was 38 kJ mol^{-1}. Pt/MoO$_3$ has sometimes been used for catalytic reductions without the possibility of bronze formation being recognised.

A similar detailed study has been reported[162,167] of the reactivity of vanadium hydrogen bronze H$_x$VO$_{2.5}$ ($x = 1.7$–1.9)[168] formed by depositing platinum on vanadia and exposure to hydrogen at 335 K. This phase has interesting physical properties,[168] and its structure has been examined by inelastic neutron scattering,[169] which shows the hydrogen atoms to have reacted with the V=O bond to give H$_2$O that remains bonded to the vanadium atom; no V—H bonds were detected, although its IR spectrum confusingly showed no bands due to hydroxyl ions either.[162] NMR spectroscopy provides evidence both for[162] and against[169] the presence of V—H bonds, so the structure is still a matter for discussion. Proton mobility is however at least 10^2 slower than in H$_{1.6}$MoO$_3$. Its reactivity towards ethene is greater than that of H$_{1.6}$MoO$_3$ and the kinetics are similar, but it retains some oxidising character, and more readily loses water on outgassing. Reactivity is clearly not simply determined by proton mobility, and unlike $H_{1.6}M_0O_3$, which has metallic character because of electron transfer from the hydrogen to the conduction band, $H_xVO_{2.5}$ is merely a semi-conductor. This difference was thought to explain its greater reactivity.[162] The sesquioxide V_2O_3 oxidises on storage,[170] but is reduced back by hydrogen spillover on application of platinum.

Detailed studies have also been performed[165] on the formation of the tungsten hydrogen bronze H_xWO_3 ($x \leq 0.6$) and of reverse spillover with it and with H_xMoO_3.

7.3. REACTIONS OF THE BUTENES WITH HYDROGEN AND WITH DEUTERIUM[6,36]

7.3.1. The n-Butenes

The further reactions that are readily observed when the alkene contains four or more carbon atoms were introduced in Section 7.2, so that now it is only necessary to recall the main features that characterise each metal before entering into a summary of the results and a discussion of mechanism. The three observable processes are (i) addition (hydrogenation) leading to n-butane, (ii) isomerisation (double-bond migration or Z-E isomerisation), and (iii) exchange in the reactant alkene if deuterium is used. The number of deuterium atoms entering isomerised butenes, and the Z/E ratio in 2-butene formed from 1-butene, are also matters of interest. These reactions, using each of the three n-butenes as reactant, have been widely studied for the extra insights they provide into the form of the adsorbed intermediates and the ways in which they react. All of the metals of Groups 8 to 10 have been examined, those of Group 10 having had the lion's share of attention. The Group 11 metals have been largely neglected, although gold is reported to isomerise 1-butene at 573–673 K in the absence of hydrogen;[171] it will hydrogenate 1-butene if hydrogen *atoms* are supplied to the gold surface.[172]

We may start by considering the results obtained with platinum catalysts: the reactions of the n-butenes with hydrogen and or deuterium have been studied using Pt/Al$_2$O$_3$,[103] on catalysts prepared from reverse micelles[173,174] (Section 2.32), and on platinum foil and various single crystal surfaces.[175] There are a number of common features: (i) orders of reaction, where measured, are either accurately or close to first in hydrogen and zero in the butene; (ii) activation energies for the macroscopic forms are between 33 and 43 kJ mol^{-1} for both 1-butene and Z-2-butene,[175] but lower for the latter on Pt/Al$_2$O$_3$ (21 kJ mol^{-1});[103] (iii) rates of reactant removal for each butene appear to be structure-insensitive[173,175] and all three isomers react at about the same rate.[103] A notable characteristic of platinum catalysts prepared conventionally or in macroscopic form is the slow rate of isomerisation r_i compared to the hydrogenation rate r_h; this is in line with the small amount of ethene exchange seen with this metal (Section 7.3.1), and is again explained by the reluctance of the alkene once adsorbed to desorb (Table 7.4). There is however evidence for some variability of r_i/r_h because isomerisation is relatively more important on the unconventional catalysts, although the cause of this was not identified.[173]

Relative isomerisation rates r_i/r_h naturally vary with the structure of the reactant butene, depending on the variance of the initial condition from that of equilibrium between the butene isomers (Figure 7.3) and on the stabilities σ the adsorbed intermediates. Values of r_i/r_h at room temperature over Pt/Al$_2$O$_3$ were estimated as[103]

$$Z\text{-2-butene}:E\text{-2-butene}:1\text{-butene} = 15:1:5$$

Their interpretation rests on conformational analysis of the adsorbed 2-butyl radical; this can adopt three staggered configurations (Figure 7.2): loss of a hydrogen atom from **I** will give either the Z- or the E-isomer, depending on which goes, **II** will give only E and **III** only Z. Their relative stabilities (discounting interaction of the methyl group with the surface) are **II** > **I** $\cong$ **III**, so that conversion of Z to E should be easier than E to Z, as observed. The medium r_i/r_h shown by 1-butene follows from the fact that it can form both 1- and 2-butyl radicals, the former only reverting to 1-butene or forming butane. The conformational factors also explain why the Z/E ratio is far from the equilibrium value for the free molecules ($\sim$40 at 298 K);[6] in fact it is about 0.6, rising to 1.4 at 379 K as thermal motion overcomes the repulsive interactions between the methyl groups.

When the n-butenes interact with deuterium on platinum catalysts,[103,173,174] there is little exchange of the reactant alkene although the Z- and E-2-butenes formed from 1-butene are substantially and about equally exchanged. The 2-butene-d_0 is however the major product, but, especially on very small particles, products up to 2-butene-d_8 have been observed.[173] Unless the exchange-isomerisation is followed as a function of conversion,[103] and particularly if products are only analysed at a single high conversion,[173] it is difficult to assess the importance of the sequential reaction of the primary products, but the following picture seems feasible. There is a rapid interchange of hydrogen and deuterium atoms (X) between X*, C$_4$X$_8$** and C$_4$X$_9$*, so that each species will contain about the same fraction of each kind. In this way the butene and butyl species acquire deuterium atoms in diminishing numbers up to the possible total, but addition of the second atom to the butyls is faster than butene desorption: those molecules that do desorb are mainly isomerised, but may be quite extensively exchanged. All possible deuterobutanes are formed, butane-d_2 being the major product, but in consequence of the low relative rates of both hydrogen and alkene exchange the mean number of deuterium atoms is close to two.

Microwave analysis of the mono-exchanged 1-butene and Z-2-butene formed by double-bond migration adds some detail to this scenario:[173] in the 1-butene-d_1, the two terminal methylenic hydrogen atoms, taken together, are slightly or significantly less exchanged than that on C2, while in Z-2-butene-d_1 it is one of the C1 atoms that is twice as likely to be exchanged as that on C2. These observations suggest that 1-butyl radicals are somewhat stabler than 2-butyl radicals. The Horiuti-Polanyi mechanism, supplemented by conformational analysis of the intermediate species, therefore provides a reasonably satisfactory explanation of the results, although minor contributions from other mechanisms are not ruled out.[173]

The behaviour of Ir/Al$_2$O$_3$ resembled that of Pt/Al$_2$O$_3$, but relative isomerisation and exchange rates were even lower at 253–293 K.[103] No isomerisation of 1-butene was detected, and only 2.5% of 1-butene-d_1 was detected: mean deuterium numbers of the butane were close to two.

Other metals of Groups 8 to 10 have very different characteristics in respect of reactions of the butenes with hydrogen and deuterium:[6] as might be expected from the way they behave in the ethene- (and propene-) deuterium reactions, nickel, palladium, ruthenium, rhodium and osmium are able under some conditions to exhibit much higher values of r_i/r_h, so that the butenes are able to achieve their equilibrium concentrations before their hydrogenation is finished.

The reactions have been followed on supported nickel catalysts,[2,6,18,19,176,177] on nickel film[176] and wire,[6,18,178] and on 'solvated metal atom dispersed catalysts'.[179] Orders of reaction and activation energies for the possible reactions on all three butenes have been reported, but values of the latter are sometimes very low (8–15 kJ mol^{-1}), this suggesting diffusion limitation. Reactions of 1-butene and Z-2-butene with deuterium have been extensively studied on Ni/Al$_2$O$_3$,[6,19] Ni/SiO$_2$,[177] Ni/pumice,[176] and nickel wire.[6,18,20,178] The work performed by Taylor and Dibeler[18] is especially commendable, as it was undertaken before the advent of gas chromatography; analysis of the butene isomers had to be made by IR spectroscopy. Exchange and isomerisation have higher activation energies than addition, so they become more significant as temperature increases.[6,176] E/Z ratios in 2-butene formed from 1-butene lie between 1.5 and 2, without any marked effect of temperature.[176] With 1-butene, extensive stepwise exchange means that butene-d_0 and -d_1 are major initial products,[18] and the change of product concentrations with conversion recalls that found with the ethene-deuterium reaction (Figures 7.5 and 7.6). Z- and E-2 butenes are equally and substantially exchanged; microwave analysis shows that at low temperature (<273 K) the mono-exchanged 1-butene is mainly 1-butene-2d_1, although exchange at C1 increases with temperature.[176] In the reaction of Z-2-butene, the isomerised 1-butene is more heavily exchanged than the E-2-butene; this is understandable if adsorbed 1-butene and 1-butyl radicals are stabler than the 2-butenes and 2-butyls because of the smaller unfavourable repulsive interactions between methyl groups, allowing more interconversion to

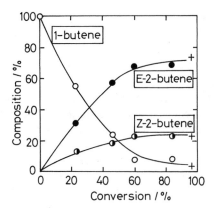

Figure 7.13. Isomerisation of 1-butene on Pd/Al$_2$O$_3$ at 313 K.[181] Equilibrium amounts are indicated by +.

take place before desorption occurs. In line with this view, there is considerable exchange at C3 as well as at C2.[176] It was however argued that the Horiuti-Polanyi mechanism could not account for all the observations, and that other processes involving dissociative chemisorption of the reacting butene were needed.

For supported or unsupported iron, cobalt and nickel catalysts made by the solvated metal atom route[179] (Section 2.2), the values of r_i/r_h in the reaction of 1-butene with hydrogen vary greatly with preparative method; this broad-ranging but somewhat superficial study did not lead to any definitive explanations for the variations.

We must now consider how the butenes react with hydrogen and with deuterium on palladium catalysts.[18,171,173,177,182–185] As with nickel there is very rapid isomerisation and extensive exchange *at low temperatures*,[177,181] so that for example at 273 K, starting with 1-butene, the butenes attain their equilibrium proportions when only half of them have been hydrogenated[181] (Figure 7.13). Under such conditions the butene contains a preponderance of the d_0- and d_1-species.[177,181] However, unlike the situation with nickel, the activation energies for these processes are *smaller* than for addition, so they become *less* important as temperature increases.[181] The opposite effect was found in the ethene-deuterium reaction on Pd/Al$_2$O$_3$. Values of E_h are notably higher than the norm for other noble metals of Groups 8 to 10 (viz. 35–45 kJ mol^{-1}). Again the Z/E ratio for the 2-butenes arising from 1-butene is less than the thermodynamic ratio, being only 1.6 at 273 K, although this value is higher than that found with platinum. The isomerised 2-butenes are however extensively exchanged, to an extent that increases with temperature. There is another significant difference from nickel; microwave analysis shows that in exchanged 1-butene-d_1 nearly all the label is at C1, and in Z-2-butene-d_1 formed by isomerisation it is also mainly at C1.[173] The ratio r_i/r_h shows particle-size sensitivity; its value approaches 100 for particles of mean size 2.8 nm (Pd/α-Al$_2$O$_3$, 273 K),[177] but falls to about unity for particles 4.5 nm in size.[180]

As well as the Horiuti-Polanyi mechanism, two others have been proposed to account for the features seen in the reactions of the butenes. (1) The vinylic dissociative mechanism supposes adsorption of the alkene by breaking a C—H bond at C1 or C2: such a mechanism was indeed suggested many years ago to explain the exchange of ethene with deuterium, and has been resurrected to explain some aspects of reactions on nickel[176] and platinum.[173] We shall meet it again shortly. (2) The allylic dissociative mechanism requires the loss of a hydrogen atom from C3, giving a delocalised π-bond over three carbon atoms (see Section 4.41). Such species are know to be favoured by palladium (and nickel), but the presence of gaseous hydrogen or deuterium is thought to prevent their formation,[173] and reactions over palladium have been interpreted without their aid.[181] This mechanism does not allow the direct E-Z isomerisation; this has to proceed through adsorbed 1-butene. We shall meet this mechanism again also. The unusual temperature coefficients of exchange and isomerisation of the butenes over palladium catalysts may be associated with change in the equilibrium between dissolved and surface hydrogen, the latter becoming more abundant as the β-hydride phase becomes unstable. In support of this view, hydrogen exchange is insignificant at low temperatures.

Rhodium catalysts also show quite peculiar behaviour.[6,173,186] The value of r_i/r_h is very small at and below 273 K, but between 273 and 423 K it increases very rapidly, so that the products at low temperature resemble those given by platinum, while those at high temperature are like those that palladium gives; hydrogen exchange is however much more marked. Increasing temperature therefore favours alkene desorption over addition.

Ru/Al$_2$O$_3$ and Os/Al$_2$O$_3$ have also been investigated.[108] They show striking similarities, both giving rapid isomerisation between 290 and 340 K, to extents that with the 2-butenes decrease slightly with increasing temperature. There is therefore a general resemblance to the character of palladium. At 523 K on Ru/SiO$_2$,[187] both 1-butene and Z-2-butene isomerise rapidly during their hydrogenation, but hydrogenolysis to lower alkanes and homologation to C$_5$ alkenes occur to small extents. These two processes are linked, and involve (i) breaking of a butyl radical into a methylene group and a smaller alkyl radical; and (ii) insertion of the methylene into a M—C bond to give a larger alkyl radical. Mechanisms have been discussed in detail; homologation of 3,3-dimethyl-1-butene has also been observed on Ru/SiO$_2$.[188]

While one has to admire the skill and hard work that has been applied in obtaining the results discussed in this section, it is also necessary to note that many of the conclusions are based on analysis of products obtained only under a single set of conditions (conversion, temperature, reactant ratio etc.); hydrogen exchange is often not monitored, and sometimes[176,177] there are even no butane analyses given.

7.3.2. The Single Turnover Approach[189–92]

We have noted before that one of the principal goals of research in heterogeneous catalysis has been to pin down the identity of the active centre at which reaction occurs. For structure-sensitive reactions this has perhaps a definite physical identity, while for structure-insensitive reactions the problem has been to understand why no such specific site is needed, or why only homogeneous sites are available (Section 5.4). The enquiry demands some kind of conceptual model, and this has been provided by input from homogeneous catalysis and organometallic chemistry, from geometric models of small metal particles, and from an innovative method for performing reactions: this is the *single turnover method*, the virtues of which have been vigorously prosecuted by R. L. Augustine.

This method is based on the following reasoning. If a metal particle is coated with hydrogen atoms under defined conditions and a pulse of an alkene (usually a butene) in stoichiometric amount is passed over it, carried by an inert gas, the observed products will be those that have undergone only a single reaction step, i.e. a single turnover (STO), and thus will reflect directly on the number and nature of the surface sites. With 1-butene as reactant, there is observed some unreacted 1-butene, some *Z*- and *E*-2-butene formed at *isomerisation sites* and some *n*-butane formed at *direct saturation sites*: some butyl radicals remain, and are removed by a hydrogen pulse, giving butane that is formed at *two-step saturation sites*. The original state of the sample is then re-created. This procedure resembles somewhat the 1-pentene titration method[58] for estimating the number of chemisorbed and spiltover hydrogen atoms on platinum catalysts, and indeed for a number of such catalysts the number of hydrogen atoms detected by the STO method correlates very well with that found by conventional selective chemisorption.

Identification of the sites at which these reactions occur is based on an imaginative scheme devised by Samuel Siegel,[193,194] in which, inspired by a knowledge of the mechanisms of homogeneous catalytic reactions involving for example Wilkinson's complex $(PPh_3)_3Rh^ICl$, he assigned specific roles to atoms having a number x of coordinatively unsaturated positions. So for example atoms of type ^xM ($x = 3$ or 2) can react with a molecule of hydrogen to give two adsorbed atoms (processes 7.H and 7.I), and one of these can diffuse to an ^1M atom at which

$$^3M + H_2 \rightarrow {}^3MH_2 \rightarrow {}^1M\,^3MH + {}^1MH \tag{7.H}$$

$$^2M + H_2 \rightarrow {}^2MH_2 \rightarrow {}^1M\,^2MH + {}^1MH \tag{7.I}$$

chemisorption cannot directly occur. This of course ignores the possibility that two ^1M atoms will do the trick. Invoking then the Hartog-van Hardeveld models (Section 2.41), the ^3M sites are reasonably denoted as apical (corner) atoms, ^2M

as edge atoms, and $^1{}_M$ as planar atoms. Augustine has adopted this analysis, and has refined it by recognising two types of direct saturation site, designated $^3{}_{M_I}$ and $^3{}_{M_R}$ at which hydrogen atoms are adsorbed either strongly (*i*rreversibly) or weakly (*r*eversibly). Ledoux and colleagues have used a similar analysis.[17]

Space constraints do not allow a detailed account of how the observations are manipulated to give site densities; activity for hydrogenation and isomerisation is confined by definition to $^3{}_M$, $^3{}_{MH}$ and $^2{}_M$ sites, $^1{}_M$ being assumed inactive. Examination of a number of Pt/SiO$_2$ catalysts (including EUROPT-1) showed the fraction of saturation sites to vary between 27 and 83%, without showing the expected dependence on particle size, although this correlation was nicely established for a series of platinum catalyst having controlled-pore glass (CPG) as support. The problems of interpretation are however highlighted when results for platinum, rhodium and palladium are compared. Analysis of the first pulse of 1-butene showed products in about the same relative amounts as expected from reactions under corresponding conventional conditions, i.e. values of r_i/r_h decreased in the sequence Pd > Rh > Pt. The validity of the method depends critically on the assumption that only one type of reaction proceeds on a given site, and that therefore hydrogenation and isomerisation are independent processes. This is not what we have concluded from the extensive work reviewed in the last section, and it is hardly reasonable to suppose that the fraction of isomerisation sites should vary so much from one metal to another. The method has not been widely applied, and the occasional attempt to use it has not been especially fruitful.

A recurring theme throughout much of the work on reactions of hydrocarbons with hydrogen or deuterium has been the possible existence of two or more distinct forms of adsorbed hydrogen atom, distinguished by their strength of adsorption and reactivity.[190,195] In the hydrogenation of 1-butene on platinum black, an attempt was made[196] to identify the active state as classified by TPD spectroscopy by filling each state with a different isotope (and assuming they did not interconvert); the deuterium content of the resulting butane then pointed to the reactive state, which appeared to be the β-form, i.e. atoms adsorbed atop.

7.3.3. *Iso*butene

*Iso*butene is not nearly so interesting a molecule as the other butenes, as it can undergo only exchange or addition, although with a suitably carbon-labelled molecule the movement of the double bond could in principle be followed, as has been done with propene;[36] such experiments have not however yet been made. Indeed there are very few studies on this molecule to report.[18,197] It was hydrogenated on Pt(111) at 300 K with a P_H/P_C ratio of 10 more slowly than 1-butene,[175] with an activation energy of 49 kJ mol^{-1}, the orders of reaction being 0.7 in hydrogen and -0.2 in *iso*butene. SFG spectra suggested[198] that the reactive form at 295 K is the

π-state; *tert*-butyl groups were also seen, and were less reactive towards hydrogen than isobutyls; only small amounts of dehydrogenated species (e.g. *iso*butylidyne) were present.

Some old work[18,19] showed that the product of its reaction with deuterium on Ni/kieselguhr at 196 K was almost entirely *iso*butane-d_2, so presumably exchange, and movement of atoms between species, does not occur significantly under these conditions. On nickel and rhodium powders, and on Ir/SiO$_2$, microwave spectroscopy has shown addition and exchange to give the mono-deutero species expected if primary and tertiary butyl radicals were involved, but on rhenium powder a π-butenylic species also appeared to contribute.[36] Pd/CaAlO$_4$ was reported to give rapid exchange with a random deuterium distribution, implying a π-butenyl intermediate; Pt/Nb$_2$O$_5$ and Rh/SiO$_2$ gave exchange in the methylene group *via* a vinylic species, while Ir/SiO$_2$ gave little exchange of any kind.[199] A comprehensive kinetic study[200] using PtSn/SiO$_2$ catalysts gave results that were successfully modelled by the Horiuti-Polanyi mechanism, a particular feature being the occurrence of both stepwise and multiple exchange of the *iso*butene, with a sharp cut-off after $-d_7$.

7.3.4. Exchange Reactions between Alkenes

As was mentioned in Section 7.2.4, alkenes are able to undergo reactions between themselves in the apparent absence of hydrogen or deuterium by processes that must involve their dissociative chemisorption or intramolecular atom movement. Thus for example 1-butene is isomerised on a number of supported metals and on platinum black above 373 K.[36] Intermolecular isotopic transfer was also introduced in Section 7.2.4, where work by Hirota and Naito was cited. Extensive studies have been made of exchange and isomerisation caused by the interaction of propene-d_6 with various alkenes,[36] especially 1-butene, the reaction of which has been followed over films of chromium, iron, nickel, rhodium, palladium, iridium and platinum.[17,201] Exchange between ethene-d_0 and ethene-d_4 has been reported to occur rapidly on nickel/kieselguhr at 298 K[202] (*Z-E* isomerisation of ethene-d_2 was even faster), and on palladium single crystal surfaces,[203] but no detailed studies have been reported. The vinylic dissociation mechanism was invoked in most cases, although allylic dissociation contributed over palladium. This work required careful microwave analysis of mono-deuterated products, and extended discussions of mechanism have been given. The relevance of the conclusions to the situation when molecular hydrogen or deuterium is present is however not quite clear.

Hydrogen atoms returning to the metal by reverse spillover can however initiate isomerisation between the butenes and isotopic exchange between labelled and unlabelled alkenes, after which the reaction is self-propagating. Since the amount of the initiator may be immeasurably small, it cannot necessarily be detected by mass-balance measurements. The processes referred to above do not therefore necessarily proceed through vinylic intermediates.[204]

7.4. REACTIONS OF HIGHER ALKENES WITH HYDROGEN AND WITH DEUTERIUM[205–207]

The reactions of linear or branched alkenes containing more than four carbon atoms reveal no types of reaction not already met, but their lower volatility permits their study as liquids or in solution as well as in the vapour phase. Thus for example the relative isomerisation rate r_i/r_h of liquid 1-pentene over Pd/C at 290 K was independent of the conditions of agitation,[208] showing there were no mass-transport effects attributable to the hydrocarbons. The presence of solvents was also without effect.[6]

The hydrogenation of 1-hexene in vapour phase on Pt/SiO$_2$ catalysts was size-insensitive,[209] and on EUROPT-1 (6.3% Pt/SiO$_2$) at 603 K (quite a high temperature for these reactions) 1-hexene and the 2-hexenes all showed extensive double-band migration, although equilibrium ratios were not attained:[210] isomerisation was suppressed by increasing hydrogen pressure, so these results are in line with expectations based on the butenes at lower temperature. The use of alkene titration using 1-pentene to estimate chemisorbed hydrogen has already been noted.[58] In the liquid phase, neither 1-pentene[95,208] nor 1-hexene[6,211] show much isomerisation on platinum at low conversions, but with the latter exchange is ten times faster than isomerisation. The competitive hydrogenation of 1-heptene and 1- and 2-decene on silicalite-coated Pt/TiO$_2$ has also been investigated.[32]

At 378 K, exchange of 1-hexene occurs on Ni/SiO$_2$, causing the value of M in the deuterohexane distribution (Figure 7.14A and B) to increase with conversion.[212] On Raney nickel at 296 K with cyclohexane as solvent it rapidly isomerises, first to the 2-hexenes and afterwards to the 3-hexenes, although complete product separation was not achieved.[213] Similar results were obtained using cyclohexane as the source of hydrogen. Extensive isomerisation is also shown with liquid 1-pentene on Pd/C, Rh/C and Rh/Al$_2$O$_3$, and with Ru/Al$_2$O$_3$, but not with Pt/C, Pt/Al$_2$O$_3$ or Ir/Al$_2$O$_3$, in keeping with what was found in the reaction of 1-butene on these metals.[208,214] At 299 K, 3-methyl-1-butene isomerises on Pd/SiO$_2$ to give the two 2-methyl-butene isomers in approximately their equilibrium amounts, the process being completed after about 40% addition has occurred.[215] The equilibrium proportions at this temperature are 2-methyl-2-butene, 92.0%; 2-methyl-1-butene, 7.8%, 3-methyl-1-butene 0.2%.

Double-bond migration of terminal alkenes forms more of the Z-isomer than corresponds to thermodynamic equilibrium. A study of the exchange and isomerisation of 1-pentene-1,2-d_2 on various types of nickel catalyst has shown that this is due to 'crowding' at the active centre, so that the pentyl radical on losing a

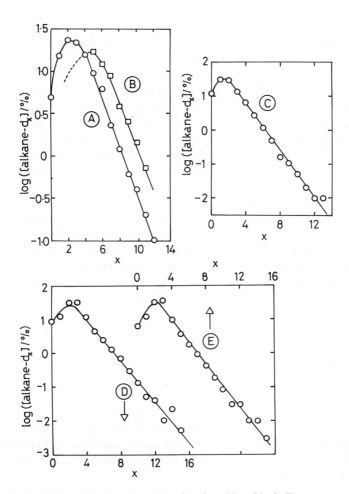

Figure 7.14. Distributions of deuterated products plotted semi-logarithmically.
(A) *n*-Hexanes from 1-hexene + D_2 on nickel, $P_D/P_C = 3.5$.[212]
(B) The same, $P_D/P_C = 70$.[212]
(C) Cyclohexanes formed on Adams PtO_2.[214]
(D) and (E) *Z*- and *E*-decalins formed from $\Delta^{9,10}$-octalin + D_2 on Adams PtO_2.[211]

hydrogen atom preferentially adopts the form of 2-pentene that occupies the smaller volume. Catalysts in which the nickel atoms are widely separately by being in complexes give more of the *E*-isomer.[99]

The glyceride esters of the long-chain fatty acids are not strictly speaking 'hydrocarbons', and the catalytic chemistry and technology of *fat hardening* is therefore not treated in this work: however, the alkyl side-chains behave exactly as if they were hydrocarbons, and the reactions of methyl oleate (where each chain contains one C=C bond) with hydrogen and deuterium have been studied.[216]

We must remember that branched alkenes may contain centres of optical activity; an example is (−)3,7-dimethyl-1-octene, which when hydrogenated on platinum shows little racemisation, but on palladium this happens extensively, to an extent depending upon the form of catalyst and reaction conditions.[205,217] Isomerisation of the double bond to the 2-position negates the optical activity, so that it may return to the terminal position in either the (+)- or (−) form. When tetra-substituted alkenes of the type RR'C=CCRR' are hydrogenated, two centres of optical activity are created; the *E*-form gives the *meso* product, while the *Z*-form gives a racemic mixture.[16]

Gold has the undeserved reputation of being pretty useless as a catalyst, except for oxidising carbon monoxide. However, the specific activity of Au/SiO_2 catalysts for 1-pentene hydrogenation in a large excess of hydrogen increased as the gold content (and hence perhaps particle size) was decreased; active catalysts were mauve in colour, but showed no double-bond migration.[218]

The susceptibility of the hydrogenation of alkenes to sulfur compounds has been reviewed;[98] the effect has been used as a way of estimating the active metal area.[219]

7.5. HYDROGENATION OF CYCLOALKENES

7.5.1. Cyclohexene

On metallic catalysts, cyclohexene can either be hydrogenated or dehydrogenated to benzene, depending on the temperature and hydrogen pressure;[220,221] the first is favoured at low temperatures, but it can also disproportionate to give cyclohexane and benzene:[222]

$$3C_6H_{10} \rightarrow C_6H_6 + 2C_6H_{12} \qquad (7.J)$$

It therefore participates in a somewhat complex set of reactions, which overlap and share common adsorbed intermediates. At near ambient temperatures in the presence of hydrogen, hydrogenation predominates, and the simplicity of product analysis due to the absence of competing reactions such as double-bond migration have made this reaction an attractive one to study.[223,224] Its major drawback is however the speed with which activity is lost through formation of 'carbonaceous residues',[62,220–225] which are probably just highly dehydrogenated forms of the reactant.[220,226] These parasitic reactions are easier with higher alkenes than with ethene, because of the more favourable thermochemisty, and the formation of these strongly-held species raises interesting but difficult questions about the role they may play in the target reaction.

We have already noted the suggestion made some years ago that reactions such as alkene hydrogenation might proceed on top of these species,[43,62] or that they might at any rate act as bridges over which hydrogen atoms might migrate to the reactant adsorbed on the remaining free surface.[227] Since they should occupy and inactivate the more reactive sites, or reside in places where their removal by hydrogenation is least easy, the few remaining sites might be fairly uniform in character, and this would account for the apparent structure insensitivity of this class of reaction.[224] In support of this idea, it has been found that for a series of Pt/SiO$_2$ catalysts, values of the TOF for cyclohexene hydrogenation vary much more when derived from *initial* rates than from the much lower *final* rates. The same conclusion was reached from UHV work on Pt(233), where at low pressure dehydrogenation on clean surface was structure-sensitive, but at high pressures, in the presence of a carbonaceous overlayer, hydrogenation was 'insensitive'.[221] Extensive measurements on supported nickel, palladium[28] and platinum catalysts lend further support to the 'insensitivity' of the catalyst in the steady state. Turnover frequencies are also independent of size for rhodium particles over a very wide range,[229] but for Ru/SiO$_2$ catalysts they show a maximum at about 4 nm. The reaction of cyclohexene with deuterium gives the familiar semi-logarithmic distributions of deuterocyclohexane but there is no discernible break after the -d_6 product (Figure 7.14C)[212,214] showing that there is no barrier to exchange on both sides of the ring.

7.5.2. Other Cycloalkenes

Cycloalkenes other than cyclohexene do not have the same propensity to aromatise, so that the processes of addition and exchange can be followed without complications. The reaction of cyclopentene with deuterium has been studied using various partially deuterated solvents (CH$_3$OD, CH$_3$COOD etc.) to minimise dilution of the deuterium through exchange with the solvent.[214] With platinum there is little alkene exchange, and there is a break in the distribution of the deuteroalkanes after cyclopentane-d_5, which corresponds to the completion of one-side exchange: detailed understanding of the distribution however requires there to be at least three different types of site, varying in their hydrogen: deuterium content and each being responsible for a single aspect of the process. A similar break occurs after cycloheptane-d_7 when cycloheptene is used, but not after cyclo-octane-d_8 in the case of the C$_8$ cycloalkene:[214] this is because the larger ring is so flexible that through an eclipsed conformation the exchange can switch from one side of the ring to the other. These findings exactly parallel those obtained in the exchange of the cycloalkanes with deuterium, although the extent of exchange is much less because of the greater coverage of the surface by adsorbed hydrocarbon species.

The ring in cyclodecene is large enough to allow the existence of E- and Z-isomers having approximately equal stability, although with the E-isomer the ring stops access to one side of the double bond and this is therefore less reactive.[230,231] Their interconversion in solvents has been followed over Pt/C and Pd/C catalysts.[231] As with the n-butenes, isomerisation occurs freely with the latter but not the former; with carbon tetrachloride as solvent, addition is preferentially poisoned with Pd/C (not with Pt/C) due to partial chlorination of the surface. It then becomes clear that in this case the isomerised products formed in the reactions with deuterium contain mainly -d_0 and -d_1 molecules. When considering deuterium contents of isomerised alkenes, it must be remembered that the release of a single hydrogen atoms from an initially formed deuteroalkyl radical can start a chain reaction that will lead to many light isomerised molecules, if alkyl reversal and alkene desorption are easy, and providing the hydrocarbon species are close enough together for this to happen before the hydrogen atom leaves for the gas phase (see process 7.J).

$$
\begin{array}{ccccc}
E\!-\!C_4H_8 & & E\!-\!C_4H_8 & & E\!-\!C_4H_8 \\
\uparrow & & \uparrow & & \uparrow \\
& \xrightarrow{+D} & \xrightarrow[\;Z\!-\!C_4H_8\;]{H} & \xrightarrow[\;Z\!-\!C_4H_8\;]{H} & \\
Z\!-\!C_4H_8 & \quad C_4H_8D & \quad C_4H_9 & \quad C_4H_9 & \cdots
\end{array}
\qquad (7.J)
$$

Heats of hydrogenation decrease progressively as the size of the ring increases,[232] presumably because the greater flexibility minimises the strain in the reactant (see Table 7.1).

7.5.3. Substituted Cycloalkenes: Stereochemical Factors[189,217,233]

Just as the employment of complex cyclic alkanes has proved beneficial in research on alkane-deuterium exchange – at least it has stimulated discussion if not the attainment of conclusions (Section 6.3) – so the hydrogenation of substituted cycloalkenes has served to reveal aspects of mechanism that lie hidden when simpler alkenes are used. The question rarely considered, however, is whether these features are also at work in the simple molecules, or whether they are just operating in the complex cases. Three types of molecules in particular have been used: (i) dialkylcycloalkenes where the double bond is within the ring (*endo-*), (ii) alkylmethylenecycloalkanes where it is adjacent to the ring (*exo-*), and (iii) fused-ring cycloalkenes. Names and structures of some of the compounds used are given in the accompanying Table 7.7.

The hydrogenation of these compounds has attracted much interest for the following reason: 1,2-dimethylcyclohexene (for example) is likely to chemisorb either in the π- or the di-σ form, with the two methyl groups pointing away from the

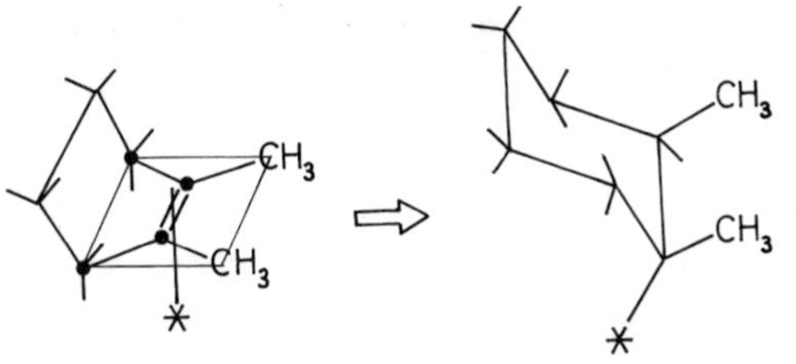

Figure 7.15. Formation of Z-1,2-dimethylcyclohexane from 1,2-dimethylcyclohexene: the latter is shown as the π-adsorbed state, which is converted to the half-hydrogenated state having the same geometry as the final product.

surface; stepwise addition of two hydrogen atoms *from below* therefore inevitably leads by the Horiuti-Polanyi mechanism to Z-dimethylcyclohexane (Figure 7.15). The same should be true for 1,2-disubstituted cyclopentenes and cyclobutenes. However, it is found that in almost every case a significant, sometimes major, amount of the E-isomer is formed; the amount depends on the kind of metal used, the hydrogen pressure, and most importantly the size and shape of the alkyl substituents.[217,234] The two hydrogen atoms must in this case have been added to opposite sides of the ring, and much research has been directed to find out how this can happen; more than half a century's effort has not however provided an unequivocal answer.[235]

In tackling this problem it has been found helpful to look as well at other isomers, especially 2,3-dimethylcyclohexene (**G**), 2- and 4-alkylmethylene-cyclohexanes (**H** and **I**) and the octalins (Table 7.6), and in particular the product of their interaction with deuterium.[236] The exocyclic double bond readily moves to the stablest position within the ring, that is, to where it has most substituents. Relative stabilities are governed by the same considerations as operate with the n-butenes (see Table 7.1), and the equilibrium proportions of **F**, **G** and **H** are as $\sim$85: $\sim$15: 0.3.[237] Isomerisation from *exo-* to *endo-* positions occurs even with platinum catalysts,[238,239] which (together with iridium) are least prone to give double-bond migration, but *exo*-isomers are much more easily hydrogenated, so much of the cycloalkane may arise from this isomer, even although its concentration may be small.

To appreciate how important alkyl reversal is in the reactions being considered, the reactions of monosubstituted C_6 cycles with deuterium are informative: the results obtained[239] with carbon-supported metals after 25% addition are given in Table 7.7 Remembering that deuterium numbers M of the cycloalkane greater than two mean more hydrogen exchange than cycloalkene exchange, and *vice versa*, the results are broadly in line with the characteristics of the three metals as exposed in the earlier sections. A certain amount of alkyl reversal must occur in all cases, but alkene desorption is only important with palladium. With Pt/C

TABLE 7.6.   Structure and Nomenclature of Substituted Cycloalkenes[6]

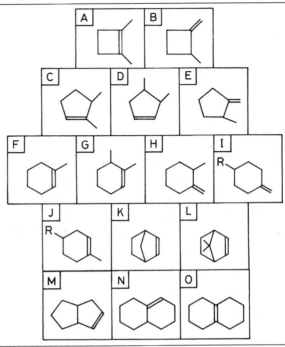

A: 1,2-Dimethylcyclobutene[176]
C: 2,3-Dimethylcyclopentene[176]
E: 2-Methylmethylenecyclopentane[176]
G: 2,3-Dimethylcyclohexene
I: 4-Alkylmethylenecyclohexane
K: Bicyclo[2.2.1]heptene[176]
M: Bicyclo[3.3.0]octene[176]
O: $\Delta^{9,10}$-octalin

B: 2-Methylmethylenecyclobutane[176]
D: 3,4-Dimethylcyclopentene[176]
F: 1,2-Dimethylcyclohexene[237,239,241,242,252-254]
H: 2-Methylmethylenecyclohexane
J: 1-Methyl-4-alkylcyclohexene
L: (+)-Apopinene[243]
N: $\Delta^{1,9}$-octalin[215]

The references cited against **F** also apply to **G–J**.
In most cases the substituent is a methyl group, but in the cyclohexene series *iso*propyl and *tert*-butyl groups have also been used,[6,238,241,252] as have compounds where alkyl groups are in the 1,3-, 1,4- and 2,4-positions.

there must be quite variable amounts of hydrogen exchange to account for these results, and with **Q** the cycloalkane appears to have been considerably exchanged on one side of the ring, leading to the high values of M found with both palladium and platinum. This does not seem to happen with **P** and **R**. In the hydrogenation of **Q**, endocyclic species dominate and lead to the saturated product; the intervention of exocyclic species is only revealed by the use of specifically labelled reactants.[240]

Reverting to the 1,2-dimethylcycloalkenes, we find that on nickel film at 273 K the C_4 and C_5 cycles gave respectively 84 and 66% of the Z-isomer;[176] the

TABLE 7.7. Reactions of Methylcyclohexenes and Methylenecyclohexanes with Deuterium over Carbon-Supported Metals: Product Analysis After 25% Addition.

	P = Methylenecyclohexane	**Q** = 1-Methylcyclohexene	**R** = 4-Methylcyclohexene	
Metal	Reactant	M	M_e	$M_{e,i}$
Pt	P	199	0.22	—
Pt	Q	2.65	0.18	
Pt	R	1.50	0.28	0.41
Rh	P	1.56	0.26	0.84
Rh	Q	—	0.10	
Rh	R	1.48	0.48	0.45
Pd	P	0.65	—	0.59
Pd	Q	2.01	0.90	
Pd	R	1.08	0.60	0.22

M: deuterium number of methylcycloalkane.
M_e: deuterium number of unchanged reactant.
$M_{e,i}$: deuterium number of **Q** formed by isomerisation.

latter gave 75% on Ni/pumice at 298 K,[176] while on platinum it gave 81%, increasing to 95% at high hydrogen pressure.[237] On palladium the Z-isomer formed only 25% of the products.[241] Before trying to understand how Z-addition occurs, we must look at what happens with 2,3-, 1,3- and 1,4-disubstituted cycles, as well as those having an exocyclic double bond (Table 7.6). In molecules of this type, the two sides of the ring are not the same, as one or more of the substituents has already adopted a disposition with respect to the plane of the ring, so that they will be preferentially chemisorbed at the least obstructed side. Addition of hydrogen atoms from below will force a methyl group in the 1- or 2-positions upwards, and this will happen too with the group formed from a methylene substituent (Figure 7.15). However it is not always self-evident which saturated isomer will be formed, as this depends on the conformation adopted (i.e. on non-bonding interactions within the molecule or between the molecule and the surface), either at the chemisorption or some later stage in the reaction path. Table 7.8 contains a small selection of the available results; more are to be found in references.

However, before attempting a short survey of the factors determining these observations, results of a very detailed study of the reactions of $\Delta^{1,9}$- (**N**) and $\Delta^{9,10}$-octalin (**O**) must be presented[6,211,215] (Table 7.10). The product decalin can exist in either the Z- or the E-configuration (Figure 7.16), depending on whether the two hydrogen atoms are on the same or different sides of the common C—C bond. Thus $\Delta^{9,10}$-octalin ought to give Z-decalin, and $\Delta^{1,9}$-octalin either the Z- or the E- form, depending on which of the non-equivalent sides of the molecule faces the surface (Figure 7.17). The double bond is stablest in the 9,10 position, where it is tetra-substituted, but isomerisation (and exchange) of the

TABLE 7.8.(A) Stereochemistry of the Hydrogenation of
Di-alkyl-Substituted Cyclohexenes*: Effect of Hydrogen Pressure[a]

Reactant[b]	Catalyst	%Z	References
F	Adams Pt	$80 \rightarrow 95$	256
F	Pd ?	25	242
G	Adams Pt	$81 \rightarrow 70$	237
H	Adams Pt	$70 \rightarrow 67$	237
I	Adams Pt	$87 \rightarrow 61$	247
I	Pt/C	$45 \rightarrow 47$	239
I	Pd/C	$25 \rightarrow 32$	239
J	Adams Pt	$35 \rightarrow 47$	247
J	Pd/Al$_2$O$_3$	$70 \rightarrow 53$	247
J	Pt/C	$71 \rightarrow 65$	239
J	Pd/C	$39 \rightarrow 45$	239

[a] The two values for %Z are those obtained at low and high (100 atm) hydrogen pressure.
[b] See Table 7.6 : in **I** and **J**, R is the *tert*-butyl group.
*Reference 257 gives results for many other dialkylcyclohexenes.

$\Delta^{1,9}$-octalin was slight except with palladium and rhodium, and $\Delta^{9,10}$-octalin did not isomerise at all.[211,215] The formation of the expected Z-decalin from $\Delta^{1,9}$-octalin was variable (43–74%, Table 7.10) but very low with palladium; larger amounts were formed from $\Delta^{9,10}$-octalin, as expected, palladium again being the exception, and the deuterium content M of the decalins was higher. In all cases the Z-and E-decalins were almost equally deuterated (Table 7.10 records only mean values), and yields declined logarithmically with number of deuterium atoms, all hydrogen atoms being exchangeable (Figure 7.13). There were unusually large differences between the products formed on iridium and on platinum. Because the $\Delta^{1,9}$-octalin is hydrogenated more quickly than the $\Delta^{9,10}$-isomer, it is possible that the E-decalin which was formed from the latter arose in fact from the former,

TABLE 7.8.(B) Stereochemistry of the Reaction of the
Di-substituted Cycloalkenes with Deuterium on Nickel Catalysts[176]

Reactant	Catalyst	% Z	M_Z	M_E
A	Ni film	84	2.2	4.2
B	Ni film	67	2.1	1.9
C′	Ni film	66	3.8	4.9
C′	Ni/pumice	75	3.2	4.2
C	Ni film	30	2.5	2.2
C	Ni/pumice	39	2.5	2.8
E	Ni film	4.8	3.4	3.4

C′ is 1,2-dimethylcyclopentene; M_Z and M_E are respectively the mean deuterium numbers of the Z and E products; experiments with Ni Film at 273 K; with Ni/pumice at 298 K.

TABLE 7.9. Stereochemistry of the Reactions of Octalins with Deuterium on Various Carbon-Supported Metals at 298 K

Reactant	Metal	% Z-decalin	M_e	$M_{e,i}$	$M_{E,Z}$
$\Delta^{1,9}$-octalin[a]	Ru	46	0.10	—	2.3
	Rh	59	1.3	0.10	~2.5
	Pd	17	1.35	0.36	~2.6
	Ir	74	0.05	—	2.55
	Pt	43	0.13	—	~2.0
$\Delta^{9,10}$-octalin[b]	Ru	94	0.16	—	~2.9
	Rh	84	0.59	—	~4.3
	Pd	16	0.14	—	~3.3
	Ir	98	0.21	—	~3.7
	Pt	67	0.14	—	~3.3

M_e is the deuterium number of the reactant; $M_{e,i}$, that of the isomerised reactant; $M_{E,Z}$ is the *mean* deuterium number of the E- and Z-decalin isomers (the individual values are generally very similar).
[a] Conversions between 64 and 97%.
[b] Conversions between 37 and 77%.

notwithstanding that isomerisation in that direction is unfavourable, and was not actually observed.

A thorough explanation of all these results cannot be given in a short paragraph (or perhaps at all); the cited papers contain much more detailed discussion of mechanisms than is possible here. A cardinal question has been the point at which

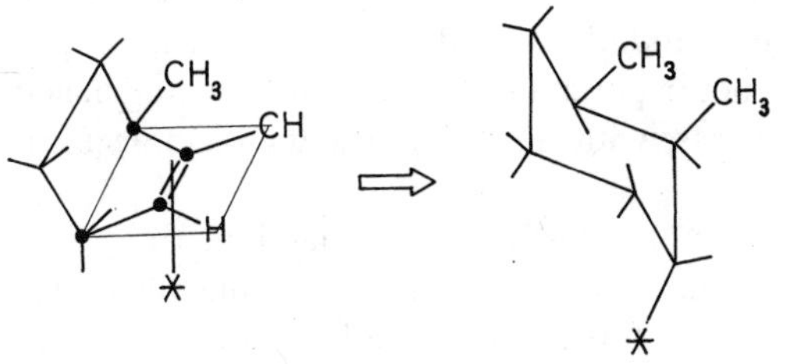

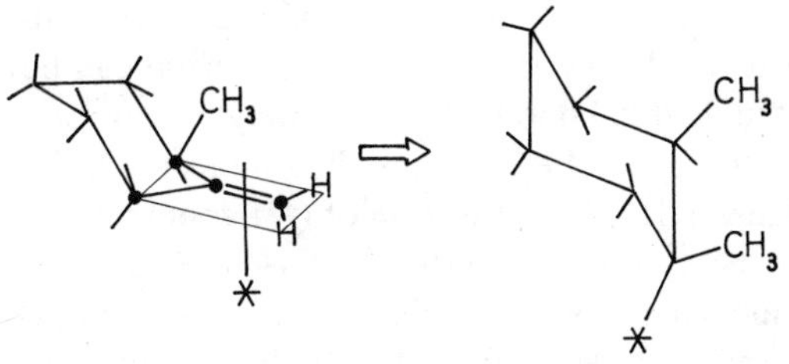

Figure 7.16. Formation of Z-1,2-dimethylcyclohexane from (A) 2, 3-dimethylcyclohexane and (B) 2- methylmethylencyclohexane, after the manner of Figure 7.14.

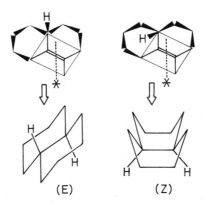

Figure 7.17. Formation of Z- and E- decalin from two different orientations of $\Delta^{1,9}$-octalin shown in the π-adsorbed forms.

the stereochemistry of the product is determined. Assuming the cycloalkene to be adsorbed as the π state,[217] in which the geometry of the free molecule is only slightly distorted (Section 4.4.2), where there is little or no isomerisation (e.g. with platinum, and where the double-bond is already in a fairly stable position) product structure should be fixed at this point (Figures 7.15 and 7.16). Where there is more likelihood of isomerisation (e.g. with palladium, and when the double-bond is exocyclic), then extensive cycloalkene-cycloalkyl interconversion may lead to product structure being decided by interactions within the cycloalkyl species. The nature of the product is fixed when the last hydrogen atom is added, but isomerisation is not important if the reactant is hydrogenated much faster than its isomer: this is the case with $\Delta^{1,9}$- octalin and with molecules containing exocyclic double bonds.

The use of molecules forming enantiomeric pairs (e.g.(+)- and (−)-4-methylcyclohexene and (+)- and (−)-p-menthene (1-methyl-4-isopropyl-cyclohexene)) is helpful because isomerisation creates a distinguishable product that has the same strength of adsorption and reactivity as the reactant, unlike the dialkylcyclohexenes. Isomerisation of (+)-α-pinene can however occur through movement of the double bond first to the *exo*cyclic position, giving β-pinene, so attention has focussed on the corresponding molecule lacking the side chain, i.e. apopinene (Figure 7.18). Isomerisation of (+)-apopinene occurs readily, even on platinum catalysts:[242] this has been attributed to the rigidity of the carbon skeleton, which allows movement of the double bond without alteration of the geometry, unlike cyclohexene, where conformational change giving flattening of the ring is needed. While this is easy with palladium, *via* a π_3-species, it is more difficult with platinum, where such species are not favoured. In the presence of deuterium, the double-bond migration appears to involve a 1,3-sigmatropic shift of a hydrogen

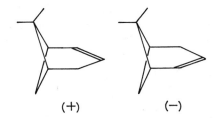

Figure 7.18. Structures of (+)- and (−)- apopinene.

or deuterium atom, although a π_3-intermediate would serve just as well. Use of palladium[230] and platinum[243] catalysts of various dispersions showed that the reactions of (+)-apopinene are mildly structure-sensitive on both, with isomerisation and addition peaking at 60% dispersion. The effect of inhibitors has also been examined; both thiophene and carbon tetrachloride showed minima in plots of k_i/k_h vs. inhibitor concentration.

Further evidence of the stereochemical constraints that determine the course of reactions of cycloalkenes with hydrogen or deuterium is obtained by studying the racemisation of (R)-(−)-10-methyl-$\Delta^{1,9}$-octalin.[244] The paper describing this work, and the review of G.V. Smith's contributions,[242] admirably summarise the current status of the subject.

A number of explanations have been offered for the formation of the stabler E-1,2-dialkylcyclohexanes from the corresponding cycloalkene: these can be summarised as follows. (1) The molecule transiently isomerises to the less stable 2,3-isomer or to the *exo*-isomer, from which the E-cycloalkane is more easily formed, because adsorption with the more hindered face towards the metal is not impossible, especially if the σ-form is adopted. In support of this possibility, small amounts of the 2,3-isomer have been seen in the products of the reaction of 1,2-dimethylcyclopentene.[217] It also accounts for the larger amounts of the stabler but unexpected E-products given by palladium (Tables 7.8 and 7.9), and the larger extent of exchange in the reactant with this metal.[239] E-products are also the more extensively exchanged on nickel catalysts, but only when formed from the 1,2-dialkylcycloalkene.[176] (2) The product is formed in an Eley-Rideal step[245] by a hydrogen molecule from the gas phase attacking the top side of an adsorbed cycloalkene. This possibility is hard to evaluate, and there is no direct evidence for it. (3) A hydrogen atom may move within an adsorbed cycloalkene to a top-side position by a sigmatropic shift; this has been suggested as the way in which (+)-apopinene isomerises to the (−)-form. (4) Vinylic dissociative adsorption of a cycloalkene would give a planar intermediate to which addition of a hydrogen atom from either side would be equally easy:[234,246] this type of adsorption has been implicated in the exchange of cycloalkenes on nickel[176] and platinum.[247] (5)

Carbonaceous residues could act as bridges for top-side addition to adsorbed cycloalkenes, or in solution adsorbed solvent molecules could perform the same role. (6) Finally – a possibility not previously considered, but one which follows logically from (5) – a hydrogen or deuterium atom from an alkyl radical may add *top-side* to a chemisorbed alkene as one step in the alkyl-alkene-alkyl chain shown in process 7.K, or by alkyl disproportionation. This is most likely to happen when species are packed closely together, i.e. when they are small or at low hydrogen pressure.

Robert Augustine[16,189,239,246,248] and others[237,247] have made extensive studies of the effect of varying hydrogen pressure on product stereochemistry chiefly using palladium catalysts. For example, the yield of Z-isomer from 1,2-dimethylcyclohexene increases with hydrogen pressure from 80 to 95%, while that of the 2,3-isomer falls from 80 to 40%. It was believed that the same product composition at low pressure was due to the 2,3-isomer changing to the stabler 1,2-isomer the E-product resulting from the former in equilibrium with the latter. As hydrogen pressure is increased, isomerisation in the sense 1,2- to 2,3- is less favoured than addition so the amount of Z-isomer increases; but the same factor when applied to the 2,3-isomer explains the *increase* in the E-yield. Hydrogenation of the methylmethylenecyclohexanes favours the more unstable isomer (2-methyl-, Z; 3-methyl-, E; 4-methyl-, Z) and their yields increase with hydrogen pressure.[246,248]

A clear example of the repulsive forces that exist between alkyl substituents in adsorbed cycloalkenes is provided by the observation[248] that 1,2-dimethylcyclopentene gives much more of the Z-product than does the corresponding cyclohexene: the lack of flexibility in the C_5 ring means that adjacent Z-substituents experience a strong repulsion, while in the more flexible C_6 ring this is lessened. This effect, felt in the cycloalkyl intermediates, is sufficient to persuade the C_5 reactant to give the stabler isomer, perhaps via isomerisation to the 2,3-isomer.

The processes considered in this section – hydrogenation, disproportionation, and isomerisation – are frequently encountered in the chemistry of the terpenes and steroids, but even with the simpler cycloalkenes many problems remain to be solved. It is unfortunate that no LEED or vibrational spectroscopy seems to have been performed on substituted cycloalkenes, so that stereochemical preferences cannot be related to adsorbed structure as defined for example by the $\pi\sigma$ factor (Section 4.42).

REFERENCES

1. V. Ponec and G.C. Bond, *Catalysis by Metals and Alloys,* Elsevier: Amsterdam (1995).
2. G.C. Bond, *Catalysis by Metals,* Academic Press: London (1962).

3. G.C. Bond, *Heterogeneous Catalysis: Principles and Applications,* 2nd edn, Oxford U.P.: Oxford (1987).
4. G.A. Somorjai, *Introduction to Surface Chemistry and Catalysis,* Wiley: New York (1994).
5. E.K. Rideal, *Concepts in Catalysis,* Academic Press: London (1968).
6. G.C. Bond and P.B. Wells, *Adv. Catal.* **15** (1964) 92.
7. G.H. Twigg and E.K. Rideal, *Proc. Roy. Soc.* **A171** (1939) 55; G.H. Twigg, *Trans. Faraday Soc.* **35** (1939) 934.
8. M. Polanyi and J. Horiuti, *Trans. Faraday Soc* **30** (1934) 1164..
9. A. Farkas, L. Farkas and E.K. Rideal, *Proc. Roy. Soc. A* **146** (1934) 630.
10. J.P. Boitiaux, J. Cosyns and E. Robert. *Appl. Catal.* **49** (1989) 235.
11. O. Beeck, *Adv. Catal.* **2** (1950) 151.
12. O. Beeck, *Discuss. Faraday Soc.* **8** (1950) 118.
13. R.L. Burwell Jr., *Catal. Rev.* **7** (1973) 25.
14. G.C. Bond, *Quart. Rev. Chem. Soc.* **8** (1954) 379.
15. D.D. Eley in: *Catalysis* (P.H. Emmett, ed.), Reinhold: New York, **5** (1955) 49.
16. A. Farkas, *Trans. Faraday Soc.* **35** (1939) 906.
17. M. Ledoux, *J. Catal.* **70** (1981) 375; **60** (1979) 15.
18. T.I. Taylor in: *Catalysis* (P.H. Emmett, ed.), Reinhold: New York, **5** (1957) 257.
19. C.D. Wagner, J.N. Wilson, J.W. Otvos and D.P. Stevenson, *J. Chem. Phys.* **20** (1952) 338, 1331; *Ind. Eng. Chem.* **45** (1953) 1480.
20. G.H. Twigg, *Proc. Roy, Soc.* **A 178** (1941) 106.
21. K.W. Egger and S.W. Benson, *J. Am. Chem. Soc.* **88** (1966) 236.
22. A. Maccoll and R.A. Ross, *J. Am. Chem. Soc.* **87** (1965) 1169.
23. S. Naito and M. Tanimoto, *J. Chem. Soc. Faraday Trans. I* **84** (1988) 4115.
24. G.C. Bond, *Appl. Catal. A: Gen.* **147** (1997) 3.
25. A.T. Bell in: Structure and Reactivity of Metal Surfaces. (C. Morterra, A. Zecchina and G. Costa, eds.), Studies in Surface Science and Catalysis', Elsevier: Amsterdam, **48** (1989) 91.
26. A. Amariglio and H. Amariglio, *J. Catal.* **78** (1982) 44.
27. *The Microkinetics of Heterogeneous Catalysis* (J.A. Dumesic, D.F. Rudd, L.M. Aparicio, J.E. Rekoske and A.A. Treviño, eds). Am. Chem. Soc.: Washington, DC (1993); J.E. Rekoske, R.D. Cortright, S.A. Goddard, S.B. Sharma and J.A. Dumesic, *J. Phys. Chem.* **96** (1992) 1880.
28. S.A. Goddard, R.D. Cortright and J.A.Dumesic, *J. Catal.* **137** (1992) 186.
29. D. Briggs and J. Dewing, *J. Catal.* **28** (1973) 228.
30. L. Pálinkó, F. Notheisz and M. Bartók, Structure and Reactivity of Metal Surfaces. (C. Morterra, A. Zecchina and G. Costa, eds.), Studies in Surface Science and Catalysis', Elsevier: Amsterdam, **48** (1989) 729.
31. P.S. Cremer, X.-C. Su, Y.R. Shen and G.A. Somorjai, *J. Phys. Chem.* **100** (1996) 16302.
32. N. van der Puil, E.J. Creyghton, E.C. Rodenberg, T.S. Sie, H. van Bekkum and J.C. Jansen, *J. Chem. Soc. Faraday Trans.* **92** (1996) 4609.
33. A.B. McEwen, W.F. Maier, R.H. Fleming and J.M. Baumann, *Nature* **329** (1987) 531.
34. D. Bianchi, G.E.E. Gardes, G.M. Pajonk and S.J. Teichner, *J. Catal.* **38** (1975) 135.
35. R.L. Burwell Jr., *Langmuir* **2** (1986) 2.
36. G. Webb in: *Specialist Periodical Reports: Catalysis* Vol. 2 (C. Kemball and D.A. Dowden, eds.), *Roy. Soc. Chem.* (1978), p. 145.
37. J. Horiuti and K. Miyahara, *Hydrogenation of Ethylene on Metallic Catalysts,* National Standards Reference Data Series 13, National Bureau of Standards: Washington DC (1968).
38. R.D. Cortright, S.A. Goddard, J.E. Rekoske and J.A. Dumesic, *J. Catal.* **127** (1991) 342.
39. S.D. Jackson, G.D. McLellan, G. Webb, L. Conyers, M.B.T. Keegan, S. Mather, S. Simpson, P.B. Wells, D.A. Whan and R. Whyman, *J. Catal.* **162** (1996) 10.
40. H.-P. Koh and R. Hughes, *J. Catal.* **33** (1974) 7.

41. K.J. Laidler and R.E. Townshend, *Trans. Faraday Soc.* **57** (1961) 1590.
42. M. Che and C.O. Bennett, *Adv. Catal.* **36** (1989) 55.
43. S.J. Thomson and G. Webb, *J. Chem. Soc. Chem. Comm.* (1976) 526.
44. F. Zaera and G.A. Somorjai, *J. Am. Chem. Soc.* **106** (1984) 2288.
45. J.C. Schlatter and M. Boudart, *J. Catal.* **24** (1972) 482; G.A. Somorjai, M.A. Van Hove and B.E. Bent, *J. Phys. Chem.* **92** (1988) 973.
46. A.L. Backman and R.I. Masel, *J. Vac. Sci. Technol. A* **6** (1988) 1137.
47. T.P. Beebe Jr. and J.T. Yates Jr. *J. Am. Chem. Soc.* **108** (1986) 663.
48. H. Yoshitake and Y. Iwasawa, *J. Phys. Chem.* **96** (1992) 1329; *J. Catal.* **131** (1991) 276.
49. F.H. Ribeiro, A.E. Schach von Wittenau, C.H. Bartholomew and G.A. Somorjai, *Catal. Rev.- Sci. Eng.* **39** (1997) 49.
50. H. Yoshitake and Y. Iwasawa, *J. Phys. Chem.* **96** (1992) 1329.
51. G.C.A. Schuit and L.L. van Reijen, *Adv. Catal.* **10** (1958) 243.
52. G.C. Bond, *Trans. Faraday Soc.* **52** (1956) 1235.
53. P. Ellison, M. Feinberg, M.H. Yue and H. Saltsburg, *J. Molec. Catal. A: Chem.* **154** (2000) 169.
54. G.C. Bond and J. Turkevich, *Trans. Faraday Soc.* **49** (1953) 281.
55. T.A. Dorling, M.J. Eastlake and R.L. Moss, *J. Catal.* **14** (1969) 23.
56. G. Leclercq, J. Barbier, C. Betizeau, R. Maurel, H. Charcosset, R. Frety and L. Tournayan, *J. Catal.* **47** (1977) 389.
57. E. Choren, J. Hernandez, A. Arteaga, G. Arteaga, H. Lugo, M. Arráez, A. Parra and J. Sanchez, *Catal. Lett.* **1** (1988) 283.
58. G.C. Bond and P.A. Sermon, *J. Chem. Soc. Faraday Trans. I* **72** (1976) 745.
59. A. Sárkány and Zs. Révay, *Appl. Catal. A: Gen.* **243** (2003) 347.
60. H.S. Taylor, *Proc. Roy. Soc. A* **108** (1925) 105.
61. L.P. Ford, H.L. Nigg, P. Blowers and R.I. Masel, *J. Catal.* **179** (1998) 163.
62. P.S. Cremer, Xingcai Su, Y.R. Shen and G.A. Somorjai, *Catal. Lett.* **40** (1996) 143; *J. Am. Chem. Soc.* **118** (1996) 2942; A. Fási, J.T. Kiss, B. Török and I. Pálinkó, *Appl. Catal. A: Gen.* **200** (2000) 189.
63. A. Masson, B. Bellamy, G. Colomer, M. M'Bedi, P. Rabette and M. Che in: *Proc. 8th Internat. Congr. Catal.,* Verlag Chemie: Weinheim **IV** (1984) 333.
64. A. Masson, B. Bellamy, Y. Hadj Romdhane, H. Roulet and A. Dufour, *Surf. Sci.* **173** (1986) 479.
65. K. W. Huang and J.G. Ekerdt, *J. Catal.* **92** (1985) 232.
66. Y. Hadj Romdhane, B. Bellamy, V.de Gouveia, A. Masson and M. Che, *Appl. Surf. Sci.* **31** (1988) 383.
67. Sh. Shaikhutdinov, M. Heemeier, H. Bäumer, T. Lear, D. Lennon, R.J. Oldman, S.D. Jackson and H.-J. Freund, *J. Catal.* **200** (2001) 330.
68. J. B. Butt, *Appl. Catal.* **15** (1985) 161.
69. E. Rorris, J.B. Butt Jr. and J.B. Cohen in: *Proc. 8th Internat. Congr. Catal.,* Verlag Chemie: Weinheim **IV** (1984) 321.
70. P.O. Otero-Schipper, W.A. Wachter, J.B. Butt, R.L. Burwell Jr. and J.B. Cohen, *J. Catal.* **50** (1977) 494.
71. Y. Takai, A. Ueno and Y. Kotera, *Bull. Chem. Soc. Japan* **56** (1983) 2941.
72. H. E. Farnsworth and R.F. Woodcock, *Adv. Catal.* **9** (1957) 123.
73. G. Carturan, S. Enzo, R. Ganzerla, M. Lenarda and R. Zanoni, *J. Chem. Soc. Faraday Trans.* **86** (1990) 739.
74. I. Zuburtikudis and H. Saltsburg, *Science* **258** (1992) 1337.
75. E.G. Allison and G.C. Bond, *Catal. Rev.* **7** (1973) 233.
76. D.A. Dowden in: *Specialist Periodical Reports: Catalysis* Vol. 2, (C. Kemball and D.A. Dowden, eds.), *Roy. Soc. Chem.* (1978), p. 1.

77. R.L. Moss, D. Pope and B.J. Davis, *J. Catal.* **62** (1980) 161.
78. M.K. Gharpurey and P.H. Emmett, *J. Phys. Chem.* **65** (1961) 1182.
79. J.S. Campbell and P.H. Emmett, *J. Catal.* **7** (1967) 232.
80. W.K. Hall, L.G. Christner and J.G. Larson, *Preprints ACS Mtg., Divn. Petroleum Chem.* **11** (1966) paper A9.
81. J. Tuul and H.E. Farnsworth, *J.Am. Chem. Soc.* **83** (1961) 2247.
82. S. Naito and M. Tanimoto, *J. Chem. Soc. Chem. Comm.*(1987) 363; G. Ghiotti, F. Boccuzzi and A. Chiorino, *SSSC* **48** (1989) 415.
83. T. Takeuchi, M. Sakaguchi, I. Miyoshi and T. Takatabate, *Bull. Chem. Soc. Japan* **35** (1962) 1390; T. Takeuchi. *Bull. Chem. Soc. Japan* **38** (1965) 322.
84. R.J. Best and W.W. Russell, *J. Am. Chem. Soc.* **76** (1954) 838.
85. J. Tuul and H.E. Farnsworth, *J. Am. Chem. Soc.* **83** (1061) 2253.
86. E.A. Alexander and W.W. Russell, *J. Catal.* **4** (1965) 184.
87. W.K. Hall and P.H. Emmett, J. Phys. *Chem.* **68** (1959) 1102.
88. T. Takeuchi, Y. Tezuka and O. Takayasu, *J. Catal.* **14** (1969) 126.
89. H.E. Farnsworth, *Adv. Catal.* **15** (1964) 31.
90. S. Bhatia, X. Wu, D.K. Sanders, B.C. Gerstein, M. Pruski and T.S. King, *Catal. Today* **12** (1992) 165.
91. M. Masai, K. Honda, A. Kubota, S. Ohnaka, Y. Nishikawa, K. Nakahara, K. Kishi and S. Ikeda, *J. Catal.* **50** (1977) 419.
92. Yong-Ki Park, F.H. Ribeiro and G.A. Somorjai, *J. Catal.* **178** (1995) 66.
93. H. Imamura, K. Igawa, Y. Kasuga and S. Tsuchiya, *J. Chem. Soc. Faraday Trans.* **90** (1994) 2119.
94. R.L. Moss, D. Pope and H.R. Gibbens, *J. Catal.* **46** (1977) 204.
95. G.C. Bond and D.E. Webster, *Ann. New York Acad. Sci.* **158** (1969)540.
96. H. Imamura, K. Fujita, Y. Sakata and S. Tsuchiya, *Catal. Lett.* **28** (1996) 231.
97. K. Soga, H. Imamura and S. Ikeda, *J. Catal.* **56** (1979) 119.
98. J. Barbier, E. Lamy-Pitara, P. Marécot, J.P. Boitiaux, J. Cosyns and F. Verna, *Adv. Catal.* **37** (1990) 279.
99. P.B. Wells in: *Surface Chemistry and Catalysis* (A.F. Carley, P.R. Davies, G.J. Hutchings and M.S. Spencer, eds.), Kluwer: Dordrecht (2003).
100. J. Turkevich, F. Bonner, D.O. Schissler and A.P. Irsa, *Discuss. Faraday Soc.* **8** (1950) 352.
101. J. Turkevich, D.O. Schissler and A.P. Irsa, *J. Phys. Coll. Chem.* **55** (1951) 1078.
102. D. Briggs, J. Dewing, A.G. Burden, R.B. Moyes and P.B. Wells *J. Catal.* **65** (1980) 31.
103. G.C. Bond, J.J. Phillipson, P.B. Wells and J.M. Winterbottom, *Trans. Faraday Soc.* **60** (1964) 1847.
104. C. Kemball, *J. Chem. Soc.* (1956) 735.
105. C. Kemball and P.B. Wells, *J. Chem. Soc. A* (1968) 444.
106. D.O. Schissler, S.O. Thompson and J. Turkevich, *Adv. Catal.* **9** (1957) 37.
107. V.H. Dibeler and T.I. Taylor, *J. Chem. Phys.* **16** (1948) 1008.
108. G.C. Bond, G. Webb and P.B. Wells, *Trans. Faraday Soc.* **64** (1968) 3077.
109. G.C. Bond and J. Addy, *Trans. Faraday Soc.* **53** (1957) 377.
110. P.A. Sermon, G.C. Bond and P.B. Wells, *J. Chem. Soc. Faraday Trans. 1* **75** (1979) 385.
111. G.C. Bond, J.J. Phillipson, P.B. Wells and J.M. Winterbottom, *Trans. Faraday Soc.* **62** (1966) 443.
112. G.H. Twigg, *Discuss. Faraday Soc.* **8** (1950) 152.
113. H. Yoshitake, K. Akasura and Y. Iwasawa, *J. Chem. Soc. Faraday Trans. 1* **84** (1988) 4337.
114. T. Ichijima, *Catal. Today* **28** (1996) 105.
115. H. Yoshitake and Y. Iwasawa, *J. Phys. Chem.* **95** (1991) 7368.
116. K. Tamaru and S. Naito in: *Handbook of Heterogeneous Catalysis,* Vol. 3 (G. Ertl, H. Knözinger amd J. Weitkamp, eds.), Wiley-VCH: Weinheim (1997), p. 1005.
117. Z. Schay and P. Tétényi, *J. Chem. Soc. Faraday Trans. 1* **75** (1979) 1001.

118. Z. Paál and S.J. Thomson, *J. Catal.* **30** (1973) 96. .
119. F. Zaera, *Langmuir* **12** (1996) 88.
120. K.R. McCrea and G.A. Somorjai, *J. Molec. Catal. A: Chem.* **163** (2000) 43.
121. F. Zaera, *J. Phys. Chem.* **94** (1990) 5090.
122. T. Ohtani, J. Kubota, J.N. Kondo, C. Hirose and K. Domen, *J. Phys. Chem. B* **103** (1999) 4562.
123. Y. Chen, *Nature* **337** (1989) 519.
124. G.A. Somorjai, *Cat. Tech.* (5) (1999) 84.
125. G.A. Somorjai and K.R. McCrea, *Adv. Catal.* **45** (2000) 385.
126. G.A. Somorjai, *Chem. Rev.* **96** (1996) 1223.
127. G.A. Somorjai, *Appl. Surf. Sci.* **121/122** (1997) 1.
128. G.A. Somorjai, *J. Molec. Catal. A: Chem.* **107** (1996) 39.
129. P.S. Cremer and G.A. Somorjai, *J. Chem. Soc. Faraday Trans.* **91** (1995) 3671.
130. M. Neurock and R.A. van Santen, *J. Phys. Chem. B* **104** (2000) 11127.
131. H. Öfner and F. Zaera, *J. Phys. Chem. B* **101** (1997) 396.
132. F. Zaera and D. Crysostomou, *Surf. Sci.* **457** (2000) 89.
133. T.V.W. Janssens, D. Stone, J.C. Heminger and F. Zaera, *J. Catal.* **177** (1998) 284.
134. P. Berlowitz, C. Megins, J.B. Butt and H.H. Kung, *Langmuir* **1** (1985) 206.
135. C. Egawa, H. Iwai and S. Oki, *Surf. Sci.* **454–456** (2000) 347.
136. C. Egawa, S. Endo, H. Iwai and S. Oki, *Surf. Sci.* **474** (2001) 14.
137. C. Egawa, *Surf. Sci.* **454–456** (2000) 222.
138. M. Bowker, J.L. Gland, R.W. Joyner, X.-Y. Li, M.M. Slin'ko and R. Whyman, *Catal. Lett.* **25** (1994) 293.
139. Xingcai Guo and R.J. Madix, *J. Catal.* **155** (1995) 336.
140. T. Sekitani, T. Takaoka, M. Fujisawa and M. Nishijima, *J. Phys. Chem.* **96** (1992) 8462; M. Nishijima, J. Yoshinobu, T. Sekitani and M. Onchi, *J. Chem. Phys.* **90** (1989) 5114.
141. J. R. Engstrom, D.W. Goodman and W.H. Weinberg, *J. Phys, Chem.* **94** (1990) 396.
142. R.E. Cunningham and A.T. Gwathmey, *Adv. Catal.* **9** (1957) 25.
143. S.P. Daly, A.L. Utz, T.R. Trautman and S.T. Ceyer, *J. Am. Chem. Soc.* **116** (1994) 6001.
144. E.K. Rideal, *Adv. Catal.* **9** (1957) 8.
145. G. Barone, D. Duca and G. La Manna, *Recent Res. Devel. Quantum Chem.* **2** (2001) 71.
146. D. Duca, L. Botár and T. Vidóczy. *J. Catal.* **162** (1996) 260.
147. A.S. McLeod in: *Catalysis in Application* (S.D. Jackson, J.S.J. Hargreaves and D. Lennon, eds.), Roy. Soc. Chem.: London (2003), p. 86.
148. D.H. Mei, E.W. Hansen and M. Neurock, *J. Phys. Chem. B* **107** (2003) 798.
149. A.S. McLeod and L.F. Gladden, *J. Chem. Phys.* **110** (1999) 4000.
150. A.S. McLeod and L.F. Gladden, *J. Catal.* **173** (1998) 43.
151. A.S. McLeod and L.F. Gladden, *Catal. Lett.* **43** (1997) 189.
152. D. Duca, G. La Manna and M.R. Russo, *Phys. Chem. Chem. Phys.* **1** (1999) 1375.
153. P. Ellison and M. Feinberg, *J. Molec. Catal. A: Chem.* **154** (2000) 155.
154. O.N. Temkin, A.V. Zeigarnik, R.E. Valdés-Pérez and L.G. Bruk, *Kinet. Catal.* **41** (2000) 298.
155. D.A. Dowden: in *Specialist Periodical Reports: Catalysis* (C. Kemball and D.A. Dowden, eds.), *Roy. Soc. Chem.* **3** (1980), p. 136.
156. M.S.W. Lau and P.A. Sermon, *J. Chem. Soc. Chem. Comm.* (1978) 791.
157. W.C. Conner Jr. in: *New Aspects of Spillover Catalysis* (T.Inui, K. Fujimoto, T. Uchijima and M. Masai, eds.) Studies in Surface Science and Catalysis, Elsevier: Amsterdam **77** (1993) 61.
158. J.F. Cevallos-Candau and W.C. Conner Jr., *J. Catal.* **106** (1987) 378.
159. E. Kikuchi and T. Matsuda in: *New Aspects of Spillover Catalysis* (T. Inui. K. Fujimoto, Y. Uchijima and M. Masai, eds.) Studies in Surface Science and Catalysis, Elsevier: Amsterdam **77** (1993) 53.

160. F. Roessner, V. Roland and T. Braunschweig, *J. Chem. Soc. Faraday Trans.* **91** (1995) 1539.
161. F. Roessner and V. Roland, *J. Molec. Catal. A: Chem.* **112** (1996) 401. .
162. J.P. Marcq. G. Poncelet and J.J. Fripiat, *J. Catal.* **87** (1984) 339.
163. N.I. Il'chenko, *Kinet. Katal.* **8** (1967) 215.
164. R. Benali, C. Hoang-Van and P. Vergnon, *Bull. Soc. Chim. Frnace* (1985) 417.
165. J.P. Marck, G. Poncelet and J.J. Fripiat, *J. Catal.* **87** (1984) 339.
166. G.C. Bond and P.A. Sermon, *J. Chem. Soc. Faraday Trans. I* **72** (1976) 233.
167. J.P. Marcq, X. Wispenninckx, G. Poncelet, D. Keravis and J.J. Fripiat, *J. Catal.***73** (1982) 309. 1
168. D. Tinet and J.J. Fripiat, *Rev. Chim. Minérale* **19** (1982) 612.
169. G.C. Bond, P.A. Sermon and C.J. Wright, *Mater. Res. Bull.***19** (1984) 701.
170. G.C. Bond and M.A. Duarte, *J. Catal.* **111** (1988) 189.
171. S.H. Inami, B.J. Wood and H. Wise, *J. Catal.* **13** (1969) 397.
172. R.S. Yolles, B.J. Wood and H. Wise, *J. Catal.* **21** (1971) 66.
173. M. Boutonnet, J. Kizling, V. Mintsa-Eya, A. Choplin, R. Touroude, G. Maire and P. Stenius, *J. Catal.* **103** (1987) 95.
174. M. Boutonnet, J. Kizling, R. Touroude, G. Maire and P. Stenius, *Appl. Catal.* **20** (1986) 163.
175. Cheonho Yoon, M.X. Yang and G.A. Somorjai, *J. Catal.* **176** (1998) 35.
176. V. Mintsa-Eya, L. Hilaire, A. Choplin, R. Touroude and F.G. Gault, *J. Catal.* **82** (1983) 267.
177. V. Mintsa-Eya, L. Hilaire, R. Touroude, F.G. Gault, B. Moraweck and A. Renouprez, *J. Catal.* **76** (1982) 169.
178. T.I. Taylor and V.H. Dibeler, *J. Phys. Chem.* **55** (1951) 1036.
179. K.J. Klabunde and Y. Tanaka, *J. Molec. Catal.* **21** (1983) 67.
180. J. Goetz, M.A. Volpe and R. Touroude, *J. Catal.* **164** (1996) 369.
181. G.C. Bond and J.M. Winterbottom, *Trans. Faraday Soc.* **65** (1969) 2779.
182. J.P. Boitiaux, J. Cosyns and S. Vasudevan, *Appl. Catal.* **6** (1983) 41; **15** (1985) 317.
183. J.P. Boitiaux, J. Cosyns and E. Robert, *Appl. Catal.* **35** (1987) 193; **32** (1987) 145.
184. S. Hub, L. Hilaire and R. Touroude, *Appl. Catal.* **36** (1988) 307.
185. K.S. Sim, L. Hilaire, F. Le Normand, R. Touroude, V. Paul-Boncour and A. Percheron-Guegan, *J. Chem. Soc. Faraday Trans.* **87** (1991) 1453.
186. J. I. McNab and G. Webb, *J. Catal.* **10** (1965) 19.
187. E. Rodrignez, M. Leconte and J.M. Basset, *J. Catal.* **131** (1991) 457.
188. J. Toyir, M. Leconte, G.P. Niccolai and J.M. Basset, *J. Catal.* **152** (1995) 306.
189. R.L. Augustine, *Heterogeneous Catalysis for the Synthetic Chemist,* Dekker: New York (1996), Chs. 3 and 4.
190. R.L. Augustine, D.R. Baun, K.G. High, L.S. Szivos and S.T. O'Leary, *J. Catal.* **127** (1991) 675; R.L. Augustine, *Catal. Today* **12** (1992) 139.
191. R.L. Augustine and R.W. Warner, *J. Catal.* **80** (1983) 358.
192. R.L. Augustine and M.E. Lenczyk, *J. Catal.* **97** (1986) 269.
193. S. Siegel and M. Dunkel, *Adv. Catal.* **9** (1957) 15.
194. S. Siegel, J. Outlaw Jr. and N. Garti, *J. Catal.* **52** (1978) 102.
195. A.F. Carley, H.A. Edwards, B. Mile, M.W. Roberts, C.C. Rowlands, F.E. Hancock and S.D. Jackson, *J. Chem. Soc. Faraday Trans.* **90** (1994) 3341.
196. S. Tsuchiya and N. Yoshioka, *J. Catal.* **87** (1984) 144.
197. R.D. Cortright, E. Bergene, P. Levin, M. Natal-Santiago, in: *Proc. 11thInternat. Congr. Catal.* (J.W. Hightower, W.N. Delgass, E. Iglesia and A.T. Bell, eds.), Elsevier: Amsterdam **B** (1996) 1185.
198. P.S. Cremer, Xingcai Su, Y.R. Chen and G. Somorjai, *J. Chem. Soc. Faraday Trans.* **92** (1996) 4717.
199. R. Brown, C. Kemball and I.H. Sadler, *Proc. Roy. Soc. A* **424** (1989) 39.

200. R.D. Cortright, E. Bergene, P. Levin, M. Natal-Santiago and J.A. Dumesic, *Proc. 11th. Internat. Cong. Catal.* (J.W. Hightower, W.N. Delgass, E. Iglesia and A.T. Bell, eds.), Elsevier: Amsterdam **B** (1996) 1185.
201. V. Mintsa-Eya, R. Touroude and F.G. Gault, *J. Catal.* **66** (1980) 412.
202. R.L. Arnett and B.L. Crawford Jr., *J. Phys. Chem.* **18** (1950) 118.
203. T.M. Gentle and E. L. Muetterties, *J. Phys. Chem.* **87** (1983) 2469.
204. P.B. Wells and G.R. Wilson, *Discuss. Faraday Soc.* **41** (1966) 237.
205. P.N. Rylander, *Catalytic Hydrogenation over Platinum Metals,* Academic Press: New York (1967).
206. M. Bartok, *Sterochemistry of Heterogeneous Metal Catalysis,* Wiley: New York (1983).
207. M. Kraus, in Handbook 3p.1051.
208. G.C. Bond and J.S. Rank, *Proc. 3rd. Internat. Congr. Catalysis,* **II** (1964) 1.
209. O.M. Poltorak and V.S. Boronin, *Russ. J. Phys. Chem.* **39** (1965) 781, 1329; **40** (1966) 1436.
210. Z. Paál, M. Räth, B. Brose and W. Gombler, *React. Kinet. Catal. Lett.* **47** (1992) 43.
211. G.V. Smith and R.L. Burwell Jr., *J. Am. Chem. Soc.* **84** (1962) 925.
212. R.L. Burwell Jr. and R.H. Tuxworth, *J. Phys. Chem.* **60** (1956) 1043.
213. L. Horner, *Ann. New York Acad. Sci.* **158** (1969) 456.
214. J.J. Phillipson and R.L. Burwell Jr., *J. Am. Chem. Soc.* **92** (1970) 6125.
215. A.W. Weitkamp, *Preprints ACS Mtg. New York* 1967, **11** (4), paper A.13; *J. Catal.* **6** (1966) 431.
216. E. Selke et al., *Preprints ACS Meeting, Divn. Petroleum Chem.,* **11** (4) (1966) A63.
217. S. Siegel, *Adv. Catal.* **16** (1966) 123.
218. G.C. Bond and D.T. Thompson, *Catal. Rev.- Sci. Eng.* **41** (1999) 319.
219. L.G. Tejuca and J. Turkevich, *J. Chem. Soc. Faraday Trans. I* **74** (1978) 1064.
220. X.-C. Su, K.Y. Kung, J. Lahtinen, Y.R. Shen and G.A. Somorjai, *J. Molec. Catal. A: Chem.* **141** (1999) 9.
221. S.M. Davis and G.A. Somorjai, *J. Catal.* **65** (1980) 78.
222. M.A. Aramendía, V. Boráu, I.M. García, C. Jiménez, A. Marinas, J.M. Marinas, and F.J. Urbano, *J. Molec Catal. A: Chem.* **151** (2000) 261.
223. A. Farkas and L. Farkas, *Trans. Faraday Soc.* **35** (1939) 917.
224. M.S.W. Vong and P.A. Sermon in: *Catalyst Deactivation 1991*(C.H. Bartholomew and J.B. Butt, eds.), Studies in Surface Science and Catalysis Elsevier: Amsterdam **68** (1991) 235.
225. D.E. Gardin, Xingcai Su, P.S. Cremer and G.A. Somorjai, *J. Catal* **158** (1996) 193.
226. M. Boudart and W.C. Cheng, *J. Catal.* **106** (1987) 134.
227. A. Fási, I. Pálinkó, T. Katona and M. Bartók, *J. Catal.* **167** (1997) 215.
228. E.E. Gonzo and M. Boudart, *J. Catal.* **52** (1978) 462.
229. M. Boudart and D.J. Sajkowski, *Faraday Soc. Gen. Discuss.* **92** (1991) 57.
230. G.V. Smith, O. Zahraa, A. Molnár, M.M. Khan, B. Rihter and W.E. Brower, *J. Catal.* **83** (1983) 238.
231. G.V. Smith and M.C. Menon, *Ann. New York Acad. Sci.* **158** (1969) 501.
232. I. Jardine and F.H. McQuillin. *J. Chem. Soc. (C)* (1966) 458.
233. L. Hilaire and F.G. Gault, *J. Catal.* **20** (1971) 267.
234. R.L. Burwell Jr., *Chem. Rev.* **57** (1957) 895.
235. G.V. Smith, *J. Catal.* **181** (1999) 302.
236. R.L. Augustine and P. Techasauvapak, *J. Molec. Catal.* **87** (1994) 95.
237. S. Siegel and G.V. Smith, *J. Am. Chem. Soc.* **82** (1960) 6082, 6087.
238. J. T. Kiss, I. Pálinkó and A. Molnár, *J. Molec. Struct.* **293** (1993) 373.
239. R.L. Augustine, F. Yaghmail and J.F. Van Peppen, *J. Org. Chem.* **49** (1984) 1865.
240. H.H. Kung, R.J. Pellet and R.L. Burwell Jr., *J. Am. Chem. Soc.* **98** (1976) 5603.
241. R.L. Augustine and J. van Peppen, *Ann. New York Acad. Sci.* **158** (1969) 482.
242. G.V. Smith in: *Catalysis of Organic Reactions* (R.E. Malz Jr., ed.), Dekker: New York (1996), p. 1.

243. F. Notheisz, M. Bartók, D. Ostgard and G.V. Smith, *J. Catal.* **101** (1986) 212; G.V. Smith, F. Notheisz, A.G. Zsigmond and M. Bartok, *Proc. 10th. Internat. Congr. Catal.* (L. Guczi, F. Solymosi and P. Tétényi, eds.), Akadémiai Kiadó: Budapest C (1992) 2463.

244. G.V. Smith, Y. Wang, R. Song and M. Jackson, *Catal. Today* **44** (1998) 119.

245. A. Couper and D.D. Eley, *Discuss. Faraday Soc.* **8** (1950) 172.

246. R.L. Augustine and H.P. Bentelman, *J. Catal.* **97** (1986) 59.

247. G.V. Smith and J.R. Swoap, *J. Org. Chem.* **31** (1966) 3094.

248. G.C. Bond, *Chem. Soc. Rev.* **20** (1991) 441.

249. *Physical Properties of Hydrocarbons* (R.W. Gallant and C.L. Yaws, eds.), 2nd edn. Gulf. Co.: Houston (1992).

250. G.H. Aylward and T.J.V. Findlay, *S I Chemical Data,* 2nd edn., Wiley: Milton, Queensland (1991).

251. J.H.. Sinfelt, *J. Phys. Chem.* **68** (1964) 856.

252. S. Siegel and M. Dunkel, *Adv. Catal.* **9** (1957) 15.

253. J-F. Sauvage, R.H. Baker and A.S. Hussey, *J. Am. Chem. Soc.* **82** (1960) 6090.

254. S. Siegel and B. Dmuchousky, *J. Am. Chem. Soc.* **84** (1962) 3132.

HYDROGENATION OF ALKADIENES AND POLY-ENES

PREFACE

In this and the next chapter the principal theme will be the selective *partial* reduction of an unsaturated hydrocarbon to a desired intermediate product, usually an alkene, and, connected with this, its selective reduction in the presence of an excess of the alkene; this represents an important industrial problem, to the solution of which much academic effort has been devoted. The formal framework for selectivity in heterogeneous catalysis has been presented in Chapter 5; our task is now to see how those principles apply, first, to the hydrogenation of dienes, and then in the next chapter to the hydrogenation of alkynes. In the case of dienes, different modes of addition of hydrogen atoms result in different products, and these modes are sensitive to the metal used and to other experimental variables. The task of defining the mechanisms by which products are formed, and of identifying probable intermediates has been assisted by the application of isotopic labelling by the use of deuterium, and a number of studies have sought to illuminate the role of the surface composition in determining the origin of selectivity by using bimetallic and selectively poisoned catalysts. From this work there has been obtained a quite high state of understanding of the ways in which the hydrogenation of dienes proceeds.

8.1. INTRODUCTION

8.1.1. Types of Unsaturation

The principal subject of this chapter is the hydrogenation of molecules containing two carbon-carbon double bonds. These may exist in a variety of situations: thus in a linear chain of carbon atoms they may be adjacent as in propadiene

TABLE 8.1. Heats of Hydrogenation of Multiply-
Unsaturated Hydrocarbons (kJ mol^{-1})*

Alkynes	$-\Delta H_{\text{H}}$
Ethyne→ethene	174
Propyne→propene	163
1-Butyne→1-butene	165
2-Butyne→Z-2-butene	153
2-Butyne→E-2-butene	157

Alkadienes	$-\Delta H_{\text{H}}$
Propadiene→Propene	172
1,2-Butadiene→1-butene	162
1,2-Butadiene→Z-2-butene	169
1,3-Butadiene→1-butene	110

Cycloalkenes and cycloalkanes	$-\Delta H_{\text{H}}$
Cyclopropene→cyclopropane	224
Cyclopropane→propane	157
Cyclobutene→cyclobutane	129
Cyclobutane→ n-butane	153

*See *Physical Properties of Hydrocarbons* (R.W. Gallant and C.L. Yaws, eds.),
2nd. edn., Gulf Publ. Co.: Houston, Texas (1992) and ref. 25.

(allene), conjugated as in 1,3-butadiene or separated by one or more methylene groups as in 1,4-pentadiene. The closer they are together, the more they interact;[1] thus the heat of hydrogenation of allene to propene is greater than that for propyne to propene (Table 8.1), while that of butadiene to 1-butene is typical of the values for alkenes (see Table 7.1). So when one or more methylene groups separates the double bonds, we may expect them to act independently in chemical terms, although it is of course possible for both to be chemisorbed at the same time, and on metals that are effective for double bond migration (such as palladium) they may move into the stabler conjugated state, so that the major products will be the same as those given by the 1,3-diene. Pairs of double bonds of any of these types may exist in cyclic molecules (Sections 8.2.2, 8.2.3 and 8.4.3). In terpene and steroid chemistry, cyclic dienes are frequently met,[2,3] and their hydrogenation follows complex pathways, especially when isomerisation is possible; we shall not enter far into this minefield.

Triplets or quartets of double bonds can also be found in linear hydrocarbons and in larger ring systems (e.g. C_{10} or C_{12}); the principles outlined above also apply to them (Section 8.4.3). Compounds in which three or more double bonds exist adjacently, either in linear or cyclic systems, are referred to as *cumulenes*; they show some interesting features when being hydrogenated.[3,4]

8.1.2. Practical Applications of Selective Hydrogenation: Outline of Mechanisms[5-7]

Various dienes and alkynes are significant components of the products formed by the steam-cracking of naphtha designing to give streams of the lower alkenes (C_2 to C_4) for further petrochemical manipulation, such as selective oxidation or polymerisation. The main type of alkene made, and the nature and concentrations of the polyunsaturates, depends on the severity of the steam-cracking. The presence of these polyunsaturates is highly undesirable, as they interfere with subsequent processing of the alkenes by reason of either their ready oxidation or oligomerisation (in the case of dienes) or their toxicity towards catalysts (in the case of alkynes). It is therefore essential to treat these streams to remove them to very low limits (generally to ppm), ideally by reducing them to useful alkenes, but without hydrogenating any of the alkene, which is present in great excess, or causing its isomerisation. This is a taxing practical matter of very great industrial importance,[7,8] which has spawned an enormous patent literature, and has attracted the attention of a number of academic laboratories, motivated by the desire to understand the basics of selective hydrogenation and to discover ever more effective catalysts. In laboratory work, 1,3-butadiene has been undoubtedly the model compound of choice, and there is very extensive literature on its hydrogenation;[9-14] the term 'butadiene' therefore is taken to mean the 1,3-isomer so often will the word be used. Corresponding studies on the selective hydrogenation of alkynes will be considered in Chapter 9.

The formal representation of selective processes was introduced in Chapter 5 (Section 5.2.6). In the present context the successful selective hydrogenation of a diene to an alkene demands that (i) the alkene formed from the diene should vacate the surface rather than awaiting reduction to the alkane, and (ii) the stronger chemisorption of the diene should totally exclude the alkene from the surface, so that it emerges unscathed. When a diene is used by itself, the reaction will cease entirely when it is completely hydrogenated by one mol of hydrogen per mol of diene if the catalyst is perfectly selective. These are counsels of perfection, but catalysts based on palladium have a long history of successful use in this process. The extent of their success is measured by the residual content of the diene and the loss or gain in the concentration of the alkene,[15] as well as by conventional criteria such as longevity. In the broader context, i.e. away from alkene manufacture, selectivity can take two other forms:[2,3] (i) *chemoselectivity*, where a C=C bond is hydrogenated rather than some other potentially reducible function such as >C=O or -NO_2, and (ii) *regioselectivity*, where the reduction of one C=C bond is favoured over that of another in different surroundings: so for example a terminal bond is often more easily reduced than an internal one.

One of the major themes of basic research having a practical orientation has been the use of bimetallic catalysts for improved selectivity. Palladium is the predominant active metal that features in almost all industrial uses, and numerous additives have been tried, with some success. Organic molecules, particularly those containing nitrogen[16] or sulfur,[17] act as selective poisons, being less strongly adsorbed than dienes but more strongly held than alkenes.

One very major industrial application of selective hydrogenation is *fat hardening*.[18–21] This subject will not be dealt with in detail, as the molecules being used are triglycerides, that is, esters of glycerol and unsaturated long-chain fatty acids (mainly having 18 carbon atoms); such a chain does however behave much like a hydrocarbon. The object of the game, put simply, is to reduce selectively two of the three C=C bonds which occur in each chain of many animal and vegetable oils, without causing the remaining one to change from *Z* to *E*. This process is carried out under conditions of mass-transport limitation by supported nickel catalysts, although the use of palladium has been proposed.[19,21] It alters materials that are easily autoxidisable to stable products that are edible (e.g. margarine) or have culinary use. High selectivity requires that the double bonds, not initially conjugated, rapidly become so, otherwise full hydrogenation would result; this determines the choice of catalytic metal. Copper also gives selective reduction, but is unsuitable because trace amounts of copper in the product destabilise it.

Our main task in this chapter is to try to understand how, when and why dienes and their relatives chemisorb to the exclusion of alkenes (the reasons are not always straightforward), and to explore the mechanisms that, additional to those revealed in the last Chapter, account for the formation of the observed products.[10,14,22,23]

8.2. HYDROGENATION OF 1, 2-ALKADIENES (ALLENES)

8.2.1. Hydrogenation of Propadiene

Little is known by direct experimentation of the manner in which propadiene is chemisorbed at metal surfaces.[24] At low temperature (90 or 130 K) it is held on Cu(110) and on silver film as a molecular species in which the orthogonal π-systems are oriented so that one methylene group is normal to the surface and the other is parallel to it:[25,26] on Ni(111) however it isomerises to di-σ-di-π-propyne[27] (see structure **15** in Table 4.2). This arrangement of the two sets of π-orbitals at right-angles to each other means that compounds of the type $R_1R_2C=C=CR_3R_4$ have non-super imposable mirror images (Figure 8.1) and are therefore optically active:[28] the C_3 unit therefore behaves just as a single carbon atoms does. It would therefore appear that simultaneous chemisorption of both bonds as π-species at a flat surface is improbable, although strong bonding through one, and weak bonding through the other, is not impossible: but a tetra-σ species is unlikely to be seen except possibly at steps. Bonding through just one of the two double-bonds is therefore most likely: study of propadiene's chemisorption on single crystal

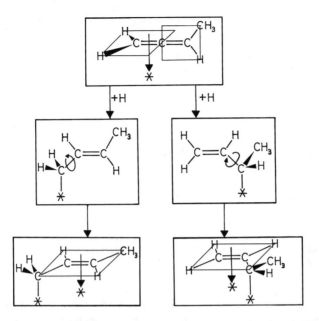

Figure 8.1. Formation of $\pi\sigma$ half-hydrogenated states in the hydrogenation of 1,2-butadiene.

surfaces, and of its coordination in complexes of the chlorides of Groups 9 and 10 metals, are lacunae that deserve to be filled.

The propadiene molecule is clearly quite highly strained, because the stabilisation that follows the opening of one of the two double bonds is very considerable. Its heat of hydrogenation to propene is greater than that for propyne (Table 8.1), and an early study[29,30] showed that it is hydrogenated to propene on pumice-supported metals of Group 10 with high selectivity, especially on nickel and palladium: kinetic parameters are summarised in Table 8.2. In admixture with either ethyne or propyne, it was hydrogenated at similar rates, and neither propene nor cyclopropane interfered with its reduction.[31] Kinetic parameters for the reaction over all the Group 8 to 10 metals as powders are available[32−34] (Table 8.2): selectivities to propene were always high (>0.95), even for iridium. For nickel-copper bimetallic powders, the rate was maximal at 10–20% copper, but the activation energy was constant at ∼25 kJ mol^{-1} for 0–50% copper, then rising to 49 kJ mol^{-1} for pure copper.[35] Temperature-programmed reaction on Pd(100) pre-covered with hydrogen has confirmed its selective reduction to propene,[36] but on Pd/Al$_2$O$_3$ at 293 K the reaction was much less selective.[37]

During the reaction, part of the propadiene is converted into products of higher molar mass;[30,32] these are linear and branched C$_6$ products.[37] Their formation has been explained[30] by isomerisation of the half-hydrogenated state

TABLE 8.2. Kinetic Parameters for the Hydrogenation of Propadiene on Pumice-Supported Metals $r \propto P_H{}^x P_P{}^y$

Metal	x	y	$E/\text{kJ mol}^{-1}$	References
Fe	1.5	—	39	34
Co	1	—	45	34
Ni	1	~0	54	29,30
//	1	—	33	34
Ru	1.4	—	13	34
Rh	0.8	—	40	34
Pd	1	>0	51	29,30
//	1	—	54	34
Pd*	1.1	−0.4	29	37
Os	0.9	—	19	34
Ir	1	—	22	34
Pt	1.1	<0	71	29,30
//	1	—	73	34

Note: references 32 and 33 give somewhat different activation energies for the reaction on metal powders.
*Reaction with D_2 on Pd/Al_2O_3.

(2-propenyl) gives a free-radical form that can attack a neighbouring adsorbed propadiene molecule to give a C_6 species, but further chain-propagation does not occur.

Propadiene reacted with deuterium over Pd/Al_2O_3 at 293 K to give deuterated propenes and propanes, but neither exchanged propadiene nor hydrogen deuteride was observed;[37] the concentration of adsorbed hydrogen and deuterium atoms was therefore low, and propadiene must have been strongly adsorbed. Burwell's N *profile method*[36–39] was applied to the distribution of deuteropropenes. This procedure is based on the assumption that there is a 'pool' of hydrogen and deuterium atoms, and that each position in the hydrocarbon equilibrates with this pool: from the observed distribution, the deuterium content of the pool a and the N profile (where N_i is the fraction of molecules in which i positions have equilibrated) can be deduced. In this case, a for propene formation was $83 \pm 2\%$, this figure being confirmed by determining by NMR spectroscopy the ratio of the two isotopes on the central carbon atom. If the direct transfer of atoms between adsorbed species is possible, as discussed in Chapter 7, then the ratio $a/(1-a)$ represents the integral over all contributing processes. The value of a for propane formation is smaller (77%), suggesting that different processes are involved in its formation.

8.2.2. Hydrogenation of Substituted 1, 2-Alkadienes

The reaction of 1,2-butadiene with hydrogen and with deuterium has been followed on palladium[25,37,40] and nickel[26,40] catalysts, with results that are summarised in Table 8.3. Those for the two palladium catalysts differ somewhat, e.g.

TABLE 8.3.   Reaction of 1,2-Butadiene with Hydrogen and Deuterium: Product
Composition and D Content of 'Pool'

Catalyst	1-Butene	Z-2-Butene	E-2-Butene	D/%	References
Pd/Al$_2$O$_3$	40	53	7	94	25
Pd/Al$_2$O$_3$	40	60	0	81	26,40
Ni/SiO$_2$*	38	57	5	~84	26,40
Ni/Al$_2$O$_3$	26	44	30	~93	26,40

*Ni powder gave similar results.

in respect of the yields of the d_2-butenes and in the values of a; some effect of metal particle size or other characteristic of the catalyst was likely responsible. Although only a small fraction of the reactant was lost by dimerisation, in the case of the nickel catalysts[26] no less than half disappeared in this way, and the different types exhibited two quite different behaviours. In each case there was little exchange in the reactant, but some isomerised giving mainly 1,3-butadiene-d_0 as the result of an intramolecular hydrogen atom transfer. Nickel powder and Ni/SiO$_2$ however gave 1-butene as the major product (Type A mode), while Ni/Al$_2$O$_3$ gave mainly E-2-butene (Type B mode, Table 8.3). A similar dichotomy has been seen in the reaction of 1,3-butadiene (Section 8.3.4), and the eccentric behaviour of Ni/Al$_2$O$_3$ is now known to be a consequence of its being contaminated with sulfur. It is believed that this encourages the conversion of $\pi\sigma$ half-hydrogenated states into π-alkenylic species, and that this opens an additional channel through which E-2-butene can be formed. Of the various possible ways in which the reactant might chemisorb, the most probable is that shown in Figure 8.1. An alternative explanation of the effect of sulfur is that for some reason it encourages the insertion of the first hydrogen atom at the central unsaturated carbon atom, and that the σ-adsorbed butenyl species can then rotate freely as shown in Figure 8.1, adopting the preferred E-conformation before the remaining π-bond engages the surface and the σ C—M bond is broken.

The importance of alkyl substituents about the pair of adjacent double bonds in determining preferred orientations in the chemisorbed state (or *attitudes of presentation*[41]) is well illustrated by the extensive work of Crombie and his colleagues, using 5% Pd/BaSO$_4$ as catalyst. Their results can be illustrated by reference to 3,4-dimethyl-1,2-pentadiene(**I**), 4-methyl-2,3-hexadiene (**II**) and 3,4,4-trimethyl-1,2-pentadiene (**III**). Formation of the final alkane was usually slight until the diene had been wholly removed, although with **III**, which has an isobutyl substituent, it started earlier, probably because the reactant molecules could not pack so tightly, and the concentration of adsorbed hydrogen atoms was in consequence higher. Similarly, isomerisation of the initially formed alkenes was delayed, except with **III**, where for the same reason some Z-E isomerisation started early.

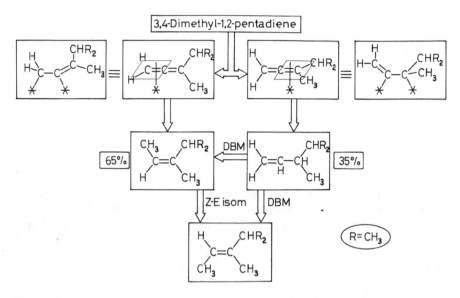

Figure 8.2. Reaction scheme for the hydrogenation of 3,4-dimethyl-1,2-pentadiene: configurations of the adsorbed reactant are shown both as π- and di-σ-structures (alkenes only isomerise after the reactant has disappeared).

Favoured orientations of **I** at the surface and the products to which it gives rise on Z-addition of two hydrogen atoms are shown in Figure 8.2. The structure in which the least substituted double bond is adsorbed is preferred, but only by a factor of about two over the alternative: similar results were obtained with **II** and **III**, so the size and number of the substituents seems to make little difference. This is the more easily understood when it is recalled that even in the π-adsorbed state planarity is modified, a tendency that is even more marked if the di-σ-adsorbed state is envisaged, although this is unlikely with palladium. However, the way in which the first-formed alkenes are created by Z-addition of hydrogen is perhaps more easily visualised if the reactant structures are written as di-σ states (see Figure 8.2). Isomerisation of the initial products occurs when they can regain access to the surface, giving the stabler structures in which the number of substituents is maximised and their steric repulsions minimised. A very dramatic increase in regioselectivity occurs if a substituent is a functional group that is capable of anchoring the molecule to the surface, since $H_2C{=}C{=}CH$ ($COOCH_3$) is reduced with 100% selectivity to $CH_3{-}CH{=}CH(COOCH_3)$.[42]

Allenic dienes having four different substituents possess chirality, as noted above: the various possible modes of adsorption and the structures of products resulting from their hydrogenation have been explored.[42]

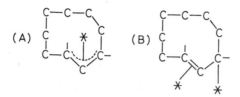

Figure 8.3. The half-hydrogenated state of cycle-1,2-nonadiene shown in plan as (A) a π_3 structure and (B) a $\pi\sigma$ structure.

If an alicyclic ring be large enough, it can accommodate a 1,2-diene group, but the ring must have at least nine carbon atoms for this to be possible. Hydrogenation of cyclo-1, 2-nonadiene and -decadiene over Pd/C poisoned with pyridine afforded[12,43] the corresponding Z-cycloalkene exclusively or predominantly, although the E-isomers were converted to Z-isomers, the latter being stabler due to the constraints imposed by the ring. However E-isomers constituted respectively 17 and 32% of the initial products, and it was argued that, since application of the foregoing principles should lead only to Z-isomers, the reaction must have proceeded through a π-allylic half-hydrogenated state, the stablest configuration of which (Figure 8.3) permits the two isomers to be formed with equal probability. However, the planar $\pi\sigma$-structure (Figure 8.3B) would answer just as well.

8.2.3. Hydrogenation of Cumulenes

It is possible for more than three carbon atoms to be connected by double bonds: compounds are known in which four or six atoms are so conjoined, and are known as *cumulenes*. Their probable modes of chemisorption are as shown in Figure 8.4, because the tetra-phenyl compounds have been hydrogenated on the Lindlar catalyst (Pd-Pb/CaCO$_3$) to respectively the 1,3-diene or to the 1,3(Z), 5-triene.[2,4]

8.3. HYDROGENATION OF 1,3-BUTADIENE

8.3.1. General Characteristics of Butadiene Hydrogenation[3,22,44]

The principal parameters that characterise this reaction are (i) the selectivity to the sum of all three butenes S_{tot}, and (ii) the fraction of the butenes that

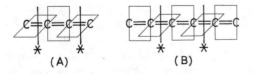

Figure 8.4. Di-π chemisorbed states of (A) C$_4$- and (B) C$_6$-cumulenes.

each isomer constitutes (S_1, S_{Z2} and S_{E2}). Over many of the metals of Groups 8 to 10, S_{tot} is close to unity and (as with 1,2-butadiene) the composition of the isomers is unchanged until the reactant has been consumed: it is therefore more strongly adsorbed than the products. In harmony with this, orders of reaction are frequently about zero in butadiene and first in hydrogen. On platinum, however, S_{tot} is usually much less than unity. The composition of the butenes varies widely: sometimes the main product is 1-butene, with significant E-2-butene but little Z-2-butene (Type A behaviour), but occasionally the E/Z ratio is very high (~8; Type B behaviour); intermediate situations are not unknown. Mechanisms attempting to account for these situations, and for the results of experiments using deuterium, will be considered in Section 8.3.6. Activation energies are typically about 40 kJ mol^{-1}, a value that is typical for a *true* activation energy where the surface composition does not change much with temperature; this is consistent with the observed near-zero orders in butdiene.

Since the reaction was first studied in depth, in the early 1960s, it has proved an attractive means of exploring how catalyst structure and composition determine behaviour: unlike, say, the hydrogenation of ethyne, where there is only a single selectivity to be measured, butadiene hydrogen affords the additional flexibility provided by the formation of the three butenes, and it was hoped, with some justification as it turned out, that these would reveal additional depths of mechanistic understanding. The reaction often shows a lack of response of rate and of selectivities to particle size, and is therefore deemed to be structure-insensitive. Effects attributable to other variables such as deliberate or accidental additives or to the nature of the support can be however quite marked. The industrial relevance of the reaction,[8] noted above, has led to a number of works in which bimetallic catalysts have been used, because it has long been know that the good selectivity shown by palladium could be made excellent by the addition of another metal, especially one from Group 11. This work is summarised in Section 8.3.7.

We begin by looking at the limited work carried out on single crystals (Sections 8.3.2 and 8.3.3), and then more to supported metals (Section 8.3.4) and finally to other unsupported forms (Section 8.3.5). Larger linear, branched and cyclic molecules containing two or more conjugated double bonds are dealt with in Section 8.4.

8.3.2. Chemisorbed States of 1, 3-Butadiene[45,46]

This molecule has not enjoyed the popularity of ethene and ethyne, but it has been subjected to a number of studies on supported metals:[45] unfortunately these do not lead to any very firm conclusions, partly because polymerisation or decomposition intervene to cloud the picture. Two studies on single crystal surfaces do however provide firmer and quite important conclusions. On Pt(111) at 95[46] or

170 K,[45] butadiene was adsorbed in a di-σ form, which may have been 1,2-di-σ as suggested by VEELS, or 1,4-di-σ (i.e. *CH$_2$—CH=CH—CH$_2$*) as indicated by NEXAFS:[46] this latter technique identifies the structure as being parallel to the surface, which the 1,2-di-σ form would not be. The methods agreed that a more strongly held form occurred at 300 K, and this may have been the 1,2,3,4-tetra-σ state. On Pd(111) however NEXAFS measurements[46] also showed a form in which the plane of the species was parallel to the surface, but it was clearly a di-π structure, which is in keeping with what is known about alkenes (Chapter 7): it also changed to a more firmly attached but unidentified form at 300 K. The low temperature structures may well have some relevance to those involved in hydrogenation above about 273 K, but the intermediate species may not be bonded to the surface in the same way as the reactant.

8.3.3. Hydrogenation of 1,3-Butadiene on Single Crystal Surfaces

There have been correspondingly few papers describing this reaction on single crystal surfaces of metals.[47–50] Four concern the reaction on various surfaces of platinum, but their value is unfortunately somewhat limited; two of them[49,50] used both Pt(100) and Pt(111), but TOFs were given at quite different temperatures; with another pair[47,48] it is only possible to compare results on Pt(110). While activation energies (43 $\pm$ 10 kJ mol^{-1}) and order of reaction where determined were broadly comparable (Table 8.4), values of TOF sometimes showed a marked variation between different surfaces, and it is not possible to define a unique hierarchy of activity. Most disappointingly there is little information on product selectivities apart from S_{tot} (Table 8.4): in one publication[49] only partial information was

TABLE 8.4. Kinetic Parameters for the Hydrogenation of 1,3-Butadiene on Various Single Crystal Faces of Platinum, and on Platinum Foil

Face/Form	T/K	H$_2$/C$_2$H$_4$	TOF/s^{-1}	x	y	E/kJ mol^{-1}	S_{tot}	References
(100)	300	10	0.27	—	—	—	—	49
	298	—	2	—	—	$\sim$42	54	47
	373	25	11	0.5	0	33,53	52	50
(110)	298	—	4	—	—	$\sim$42	64	47
	373	25	28	2 → 1*	0	39	—	50
(111)	300	10	85	—	—	—	—	49
	373	25	8	1	1 → 0*	38,39	40,67	50
(755)	300	10	0.71	1.12	−0.13	41	62	49
Foil	300	10	0.9	1.16	−0.10	41	60	49

Note: (i) Differences in *TOF* at $\sim$300 K may be partly due to use of different pressures of reactants.
(ii) Temperatures are those at which *TOFs* were measured; the orders may have been found at slightly different values.
*Orders change as shown as pressure of variable increases.

provided, and in the others there were major gaps or only a qualitative statement. Failure to record the most pertinent and important results is to be deeply regretted.

When hydrogen + deuterium mixtures were used in the reaction,[48,50] the rate of isotope equilibration was more than 10 times faster than butadiene hydrogenation, and had an activation energy of 17 kJ mol^{-1}.[48] Apart from showing that equilibration can proceed in parallel with hydrogenation, this mode of experimentation has little to commend it. S_{tot} decreased with increasing 'hydrogen' pressure below 125 Torr, but then levelled out: neither it nor the activation energy was affected by adsorbed carbon formed by deep dehydrogenation of adsorbed hydrocarbon species, and the rate depended only on the area of clean surface.[48] On Pt(100), coverages by potassium and sodium up to 40% had only slight positive effects on rate and S_{tot}.[47] Poisoning of the (110) surface by sulfur caused the rate to decrease linearly with coverage, one sulfur atom blocking one site for hydrogen dissociation.[51] Activation energy and mechanism were unchanged, and there was no indication of an effect on product selectivities.

On Pd(100)(1 × 1) precoated with either hydrogen or deuterium atoms, temperature-programming showed that butadiene was hydrogenated with 100% selectivity to butenes.[36] The Pd(110) face was almost ten times more active than the (111) face at 300 K;[15,52] species intermediate in the hydrogenation have been examined on the (110) surface.[53]

Work with bimetallic single crystal surfaces will be mentioned in Section 8.3.7.

8.3.4. Hydrogenation of 1, 3-Butadiene on Supported and Unsupported Metals

This reaction has been examined in great depth on a number of supported and unsupported metals in various laboratories,[19,14,54–58] and extensive use of deuterium for isotopic labelling has led to very detailed mechanistic schemes.[10,38,59] In this section we focus on the products formed with hydrogen, and the effect of catalyst composition and structure thereon: studies bearing directly on detailed mechanism are deferred to the next section.

As noted in Section 8.3.1, measurements of orders of reaction and activation energies indicate strong adsorption of butadiene to the exclusion of butenes, and little if any competitive adsorption of hydrogen; activation energy values are usually low, as befits *true* values, the surface concentrations being invariant with temperature. Any departure from these norms is therefore noteworthy and Table 8.5 contains a selection of results for alumina-supported metals, some of which are 'usual' and a few that are not. Of the latter kind, we may note that hydrogen order x may be significantly greater than unity, while simultaneously the orders in butadiene y are somewhat negative (Co, Pd, Pt). Somewhat higher activation energies have sometimes been reported (Pd, Pt), while a low value was associated

TABLE 8.5. Kinetic Parameters for the Hydrogenation of 1,3-Butadiene

Metal	Form	x	y	T/K	E/kJ mol^{-1}	References
Fe	/Al$_2$O$_3$	—	—	—	46	59
Co	/Al$_2$O$_3$	1.0	0	433	36	59
Co	Powder	1.5	−0.5	441	51	59
Ni	Powder	1.3	0.8	473	52	66
Ni	/Al$_2$O$_3$	1.0	−0.2	351	63	59
Ru	/Al$_2$O$_3$	1.0	0	287	52	65
Rh	Sponge	1.0	—	383	31	96
Rh	/Al$_2$O$_3$	1.0	0.1	273	46	65
Pd	/Al$_2$O$_3$	1.7	−0.7	289	70	65
Pd	/Al$_2$O$_3$	1	0	295	85	129
Pd	/Pumice	0.9	0	293	44	63
Pd	Sponge	1.2	—	383	48	96
Ir	/Al$_2$O$_3$	0.8	0	273	19	65
Pt	/Al$_2$O$_3$	1.3	−0.5	289	81	65,75
Cu	/Al$_2$O$_3$	1.0	0	378	55	65
Ag	/SiO$_2$	—	—	345	38	60
Au	/Al$_2$O$_3$	1.0	0	473	36	130

with a low S_{tot} in the case of iridium. The base metals (Fe, Co, Ni, Cu) together with palladium, silver[60] and gold[54,61,62] showed values of S_{tot} that were very close if not exactly equal to unity (Table 8.6). Such values were generally not temperature dependent,[63] nor did they vary much with hydrogen pressure. With the other metals, high values of S_{tot} were approached as hydrogen pressure was lowered or temperature raised. Both these trends would have given *lower* concentrations of hydrogen atoms on the surface; we must therefore deduce that (notwithstanding the values of the orders) some hydrogen is adsorbed alongside the butadiene on the noble metals of Groups 8 to 10 (except palladium, which always appears exceptional).

Values of kinetic parameters and S_{tot} are also available for films,[54] wires[64] and powders:[59,64] the principal difference in the case of wires lies in the necessarily much higher temperatures that had to be used. With rhodium, iridium and platinum, this had the consequence that very much greater values of S_{tot} were found, while orders and activation energies were generally similar to those in Table 8.6. Reasons for these high selectivities will be considered later. In the case of metal films, where values for S_{tot} are available for the majority of metals in Groups 3 to 11, mostly at 273–293 K; nearly all exceed 95%, exceptions being the metals of Group 6 (Cr, 90%; Mo, 84%; W, 90%), rhenium (80%) and most notably platinum (59%).

We come now to the matter of the relative amounts of the butene isomers formed, expressed as S_1, S_{Z2} and S_{E2}. Except for the base metals of Groups 8 to 10, where special factors apply, it is fairly easy to quote typical values for each metal, because in general they were only weakly dependent on temperature (unless

TABLE 8.6. Hydrogenation of 1,3-Butadiene: Product Selectivities for Various Forms of Metal Catalysts

Metal	Form	S_{tot}	S_1	S_{Z2}	S_{E2}	Type	T/K	References
Fe	/Al$_2$O$_3$	0.98	23	32	45	B	471	59
	Film	0.97	69	12	19	A	273	54
Co	/Al$_2$O$_3$	1.00	28	8	64	B	348	59,63
	/Al$_2$O$_3$	—	74	17	9	A	650	100
	/Al$_2$O$_3$	—	82	7	11	A	333	14
	/Al$_2$O$_3$	0.98	72	10	18	A	573	131,132
	Powder	~1	72	10	18	A	573	59
	Film	~1	70	10	20	A	297	54,66
Ni	/Al$_2$O$_3$	1.00	26	8	66	B	373	40,59
	/Al$_2$O$_3$	0.95	64	11	25	A	296	133
	/SiO$_2$	—	59	12	29	A	373	40
	/SiO$_2$	0.96	61	14	25	A	353	133
	Powder	—	62	13	25	A	373	14,40,59,66
	Film	~1	69	9	22	A	273	54
Ru	/Al$_2$O$_3$	0.74	69	12	19	—	273	14
	/Al$_2$O$_3$	0.90	57	22	21	—	293	56
Rh	/Al$_2$O$_3$	0.74	51	17	32	—	289	65
	/Al$_2$O$_3$	0.89	51	21	28	—	293	56
	Film	0.92	59	10	31	—	265	65
	Wire	0.99	38	25	37	—	488	64
Pd	/BaSO$_4$	0.94	49	11	40	—	261	44
	/Al$_2$O$_3$	1.00	65	2	33	—	273	14,65
	/Al$_2$O$_3$	1.00	67	5	28	—	273	8
	/Al$_2$O$_3$	1.00	53	5	42	—	308	25
	/Al$_2$O$_3$	1.00	60	5	35	—	298	132
	/Al$_2$O$_3$	0.98	58	10	32	—	293	56
	/Pumice	1.00	48	4	48	—	292	63
	/Pumice	0.99	78	3	19	—	273	93
	Film	0.98	61	6	33	—	198	54
	Wire	1.00	41	16	43	—	408	64
	Sponge	~1	56	10	34	—	458	94
	h.f.*	0.995	99	0.5	0.5	—	313	134
Os	/Al$_2$O$_3$	0.43	65	16	19	—	297	135
Ir	/Al$_2$O$_3$	0.25	59	22	19	—	297	135
	Wire	0.92	58	22	20	—	465	54
Pt	/Al$_2$O$_3$	0.50	72	10	18	—	273	135
	/Al$_2$O$_3$	0.60	65	13	22	—	273	75,136
	/Al$_2$O$_3$	0.78	82	6	12	—	298	76
	/Al$_2$O$_3$	0.53	68	21	23	—	293	56
	/SiO$_2$	0.68	78	8	14	—	290	75,136
	Film	0.59	72	9	19	—	273	54
	Wire	0.99	46	28	26	—	538	64
	Powder	—	77	10	14	—	273	97
	(111)	—	77	8	15	—	393	50
	Foil	—	67	13	20	—	300	49

(*continued*)

TABLE 8.6. (*Continued*)

Metal	Form	S_{tot}	S_1	S_{Z2}	S_{E2}	Type	T/K	References
Cu	/Al$_2$O$_3$	1.00	87	7	6	—	333	59,135
	/SiO$_2$	1.00	82	9	9	—	295	129
	Film	0.99	78	14	8	—	383	54
Ag	Film	0.96	50	20	30	—	415	54
	/SiO$_2$	1.00	66	10	24	—	345	60
Au	/Al$_2$O$_3$	1.00	59	28	13	—	443	10,135
	Film	1.00	73	11	16	—	431	54

Note: The above results have been obtained at various H$_2$/butadiene ratios and degrees of conversion, but neither has a major effect.
*Metal inside hollow fibre.

unusually high, as for wires), and on conversion or hydrogen pressure,[54] save for those cases where S_{tot} was significantly below unity. There the isomers interconvert and move towards their equilibrium proportions at rates that parallel the change in the hydrogen/butadiene ratio that occurs with increasing conversion if their ratio is initially more than unity.[65] This occurs *a fortiori* the higher the temperature, and confirms the presence of some competitively adsorbed hydrogen. Product selectivities observed at conversions where these complications are absent are shown in Table 8.5; some values from different laboratories are included to emphasise the universality of what is found, but the cited publications contain many more values for which we have no space. Butene selectivities for the metal films are plotted against Periodic Group number in Figure 8.5.

Although it may seem at first sight that the values of the selectivities are almost random, closer inspection reveals some important regularities. (1) The Z-2-butene selectivity is almost always the smallest; in the case of palladium at about room temperature (under which conditions the β-hydride phase may be formed) it is 5% or less. (2) 1-Butene is almost always the major product, and with copper, manganese and cobalt (Type A) it is more than 85%. Only with zirconium, and with iron, cobalt and nickel as Type B, is E-2-butene formed as major product. (3) Except for the base metals of Groups 8 to 10, product distributions under comparable conditions show little significant dependence on physical form, as befits a structure-insensitive reaction.

With these base metals, the dependence of the form of product distribution on the kind of support used (if any) (Table 8.6), and on reduction temperature, was for a long time a major puzzle:[26,40,63] it was finally decided that the trouble with alumina was due to sulfate ions in the support being reduced by hydrogen spillover to hydrogen sulfide, which partially covered the metal with sulfur atoms, causing Type B behaviour.[14,66] Similar effects had been seen with 1,2-butadiene.[26] The mechanism by which this happened is mentioned in the next section. This idea has been comprehensively confirmed by depositing sulfur atoms onto metal films although the obvious ploy of using alumina made by a sulfate-free route has

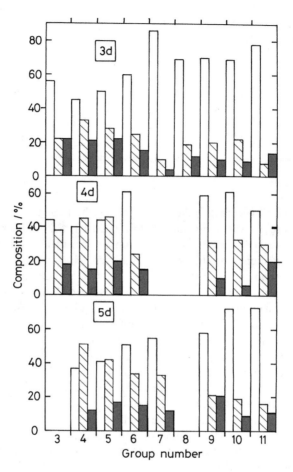

Figure 8.5. Product selectivities in the hydrogenation of 1,3-butadiene on metal films (data for Ir are for wire):[54] 1-butene, open bar; E-2–butene, hatched bar; Z-2-butene, filled bar.

not been tried. With manganese and nickel films, there was a progressive trend from Type A to Type B distributions as sulfur coverage was increased, and fully sulfided films of a number of Groups 6 to 10 metals all (except Pt) showed Type B behaviour, and very high values of S_{tot}. On platinum it appeared to act as a non-selective poison;[48,51,54] iron and cobalt films showed typical Type A behaviour. Other electronegative additives (P, As, Se) applied to Co/Al$_2$O$_3$, and halogens applied to nickel powder had similar effects.[14,66] Hydrided films of metals in Groups 3 to 5 also mainly showed Type B characteristics,[14,54] because hydrogen atoms tend to withdraw charge from the metal atoms.

The high selectivity that is characteristic of palladium and certain other metals (Table 8.6) appears to be a consequence of the rapid formation of strongly adsorbed oligomers (C_8, C_{12} etc) that probably act by decreasing the mean size of the active ensembles.[15,67–69] Maintenance of high selectivity therefore depends upon keeping to conditions where they are retained, and using too high a hydrogen pressure is therefore not helpful. Too much oligomer formation is however also inimical to high butene selectivity, because three-dimensional carbonaceous islands are formed, and these restrict the movement of reactant and product to and from the active centres, leading to loss of selectivity. Butene isomer ratios were also changed,[68] and activation energies decreased to 20–25 kJ mol^{-1} under these circumstances. Non-stoichiometric PdC_x can also be formed with palladium black;[68] this is less likely to occur with small particles, as is hydride formation. The role[47,70,71] of added alkali metals has also been explored.

To prepare the ground for discussion of mechanisms, some further regularities may be noted: these are most clearly seen in the results for films (Table 8.6 and Figure 8.5), where complications due to adventitious poisons, support and particle size effects are likely to be circumvented. Two trends in particular have been noted. (1) Values of S_1 tend to increase on passing from left to right across the Periodic Table, although they are generally larger for the $3d$ metals than for the $4d$ and $5d$ metals. For these last two series, values of S_1 have been related specifically to the Pauling electronegativity. (2) There is a kind of inverse correlation between S_1 and S_{E2}/S_{Z2}; at least this ratio is less than three for all 'pure' metals, while those that are affected by sulfur show higher values. The exception is again palladium, where room temperature values can be as high as 16 (Table 8.6). It therefore seems that, because S_{Z2} is uniformly low, it is the ratio of S_1/S_{Z2} that is the principal variable; they are mechanistically linked, and one can substitute for the other.

As noted above, the total selectivity to butenes S_{tot} increases with temperature for those metals where it is not already near unity,[64] increasing values approaching unity when wires are used at high temperatures (Table 8.6). Now for alumina–supported metals of Groups 8 to 10, the value of S_{tot} at 298 K in each group (Table 8.6) falls within increasing atomic number, viz.

$$\text{Fe} > \text{Ru} > \text{Os}; \quad \text{Co} > \text{Rh} > \text{Ir}; \quad \text{Ni} \cong \text{Pd} > \text{Pt}.$$

Serendipitous use of measurements of hydrogen retained by metal powders formed by hydrogen reduction of salts has provided a likely explanation for these trends: except for iron (no datum) and palladium (exothermic absorption), the trends were the opposite of those above. The amounts *occluded* are however greater by large factors than those due to simple dissolution, and the metals in question (unlike those in Groups 4 and 5) do not form salt-like hydrides. It was therefore supposed that the hydrogen is held in cavities that result from the failure of atoms formed in reduction of the salt to relax into the perfect metal lattice, the more so as their mobility at the reduction temperature T_R decreases because their melting temperature T_M rises. The amount of occluded hydrogen should therefore increase with T_M or with Huttig temperature, which is one-third of T_M and the point at which surface mobility becomes significant. This is qualitatively what is observed, and it was then believed that this extra hydrogen became activated by some means not fully explained, and contributed to the non-selective reduction of butadiene. This would also explain the effect of temperature on S_{tot}, assuming that part of the occluded hydrogen is released as temperature is increased. This model received strong support from work on iridium powders,[72,73] which gave S_{tot} more than 95% even at 273 K when the H/Ir ratio was not more than 0.03. Furthermore the value of S_{tot} rose as the dispersion of Ir/SiO$_2$ and Ir/Al$_2$O$_3$ *decreased*; the extent of cavitation is expected to vary inversely with particle size. This is a subtle manifestation of a particle size effect that may apply in other systems; but platinum films and single crystals also show quite low values of S_{tot} *even at low temperatures.*[49,54]

There have been several studies aimed at detecting support effects. With a series of supports of increasing acidity (MgO, Al$_2$O$_3$, SiO$_2$-Al$_2$O$_3$), S_{tot}, S_1 and the E/Z ratio all decreased somewhat,[74] but the effects were not large. It is of course possible that they were due to other causes such as a particle size effect or partial sulfur poisoning; it is easy but unwise to draw too hasty conclusions. Some unconventional supports have been used. Pt/MoO$_3$ gave almost the same results as Pt/SiO$_2$;[75] Pt/Nb$_2$O$_5$ after reduction at 573 K or 773 K was less active than Pt/Al$_2$O$_3$, but product distributions were similar:[76] there is thus no real evidence for a strong metal-support interaction affecting the reaction.[77] With Pd/ZnO, reduction above 423 K led to lower rates but an increase in S_1:[78] some partial reduction of the support, followed by surface or bulk alloy formation, would occur under these conditions. To obtain high values of S_{tot} it is necessary for there to be no constraint on the diffusion of the butenes away from the active sites; the location of the metal within a porous catalyst particle is therefore of critical importance.[79,80]

This point has been reinforced dramatically by work using palladium introduced into hollow fibres of cellulose acetate or polysulfone (Table 8.6); not only was S_{tot} close to 100%, but the predominant product (96–99%) was 1-butene, with small and roughly equal amounts of the 2-butene isomers. It is hard to understand this result (which passed without comment or reference to the literature) in terms of diffusional effects; it seems more likely that the palladium was in a state such that only one of the two double bonds could be chemisorbed at a time. A palladium-cobalt catalyst made similarly behaved likewise, but was unable to isomerise 1-butene, which the palladium catalyst could.

Although in general product selectivities are much the same for small supported particles as for large unsupported ones, there is much evidence to show that small palladium particles are less *active* than large ones.[70,81–84] This conclusion applies both to particles made by atomic beam deposition or by plasma sputtering onto silica,[81] silicon carbide[82] or graphite[83,85] and to those made chemically:[84,86] particles less than 1 nm in size were found to be inactive, while those larger than about 2 nm had the same order of activity as single crystal surfaces.[81,83] Those made by the atomic beam method appeared however to be more susceptible to deactivation by carbon deposition than those made chemically, and there have been indications that high S_{tot} is a consequence of this, or even of carbide formation. The low activity of very small particles has been blamed on the excessively strong adsorption of butadiene on low coordination number atoms: remembering that such small particles may not show their metallic character (Section 2.5), we expect their atoms to behave more as individuals, and hence to be able to form complexes with alkadienes analogous to those they can form with single atoms and ions. The increase in the Pd $3d$ binding energy as size decreases is another symptom of this effect.[52,70]

8.3.5. The Reaction of 1, 3-Butadiene with Deuterium: Reaction Mechanisms

The material presented in the last section provides ample scope for a serious discussion of reaction mechanisms, but there is one remaining input that needs to be addressed: as always, the use of deuterium in place of hydrogen and identifying the locations of deuterium atoms in the products provides an additional constraint on the possibilities, and often suggests unsuspected pathways. The reaction of butadiene with deuterium has been studied on alumina-supported cobalt, nickel, copper, rhodium, palladium and platinum.[25,38,59] Product analyses have been made at different conversions, but the exigencies of the work have precluded detailed investigation of variables such as temperature and reactant pressures.[59] Each analysis required the separation of five hydrocarbons by preparative gas chromatography and the mass-spectroscopic analysis of each one; the tedium of this procedure readily excuses the absence of more extensive work. What has been reported is however quite enough to be helpful; perhaps when sanity returns to the funding of scientific research, some of the many remaining gaps will get filled.

The main conclusions are similar for all the metals studied. There was limited but variable exchange of the reactant's hydrogen atoms (Table 8.7), and only small amounts of hydrogen appeared in the deuterium,[38,59] showing that the concentration of atoms on the surface under the conditions of the experiments was small. The distributions of deuterium in each of the butenes were broadly the same, showing that each was an initial product, and not formed by the isomerisation of any other: thus both 1,2- and 1,4-addition processes operate. The mean deuterium number M for each butene isomer increased with conversion to an extent and in a manner that reflected how exchange in the butadiene exceeded that in the deuterium: a selection of initial values is shown in Table 8.7. Little purpose is served

TABLE 8.7.　Reaction of 1,3-Butadiene with Deuterium on Metals of Groups 9 to 11: Butadiene Exchange and Deuterium Content of Butenes[38,59]

| Metal | Butadiene Exchange | | Initial M^a of Butenes | | | |
	M^a	T/K	1-b	Z-2-b	E-2-b	$%D^b$
Co	0.020	374	1.0	1.4	1.4	68
Ni	0.005	341	1.6	1.8	1.5	65
Cu	0.042	393	0.5	0.5	0.5	66
Rh	0.06	373	1.8	2.0	1.9	72
Pd	0.05	290	1.7	1.6	2.0	72
Pt	0.01	293	1.7	2.0	1.9	6.2

[a] M = mean number of D atoms per molecule (for butadiene, values after similar conversions).
[b] Deuterium content of 'pool' of atoms, derived from the N-profile procedure.

by quoting detailed deuterobutene distributions; they have been fully analysed by the N-profile procedure, which yields the deuterium content D of the 'pool' of atoms with which N positions in the hydrocarbon have equilibrated. We proceed to consider the mechanisms that have been advanced to account for the calculated profiles.

Before entering into detail, a few remarks of a somewhat philosophical nature are in order. Mechanistic statements for C_4 species are inevitably more complex than those for C_2 and C_3 species, and structures become possible that have no analogues in the smaller ones (Section 4.4). In particular, some of them may be represented in alternative but operationally equivalent ways, the choice between them only being resolved by reference to the relevant organometallic chemistry. For example, the adsorbed state of the butadiene molecule may be formulated (see Figure 8.6) as either di-σ or π (**II**) or tetra-σ or di-π (**I**) or as hybrid $\pi_3\sigma$ (**IA,IIIB**) or π_4 (**IIIA**) forms. [Note, the subscript to π shows the number of carbon atoms in the π-alkenylic bond]. The way in which the reaction proceeds, i.e. the structure of the half-hydrogenated state, then depends on the structure selected for the butadiene. Following precedent, C=C bonds interacting with the surface are shown in Figure 8.6 just as π-bonds; di-σ equivalents are not represented, nor are the tri-σ versions of **IV** and **VII**, nor is **I** as π-di-σ.

An additional and very important dimension is provided by the fact that many of the species shown in Figure 8.6 can exist in two different conformations designated *syn* and *anti*. This applies to butadiene whether as the di-π forms **I** and **III** or the $\pi_3\sigma$ forms **IA** and **IIIB** (**IIIA** can only adopt the *syn* form), and to the half-hydrogenated butenyls formed therefrom. Now in the gas phase the *anti* conformation of butadiene is preferred to the *syn* by a factor of about 20 at room temperature,[14,54] so *direct* chemisorption into any of these structures (not going via **II**) automatically favours the former, and this geometry is maintained after the first hydrogen atom is added. The π_3-butenyls cannot interconvert, except via **IV**, **V** and **VII**: a temporary detachment of the π-bond from the surface, followed by

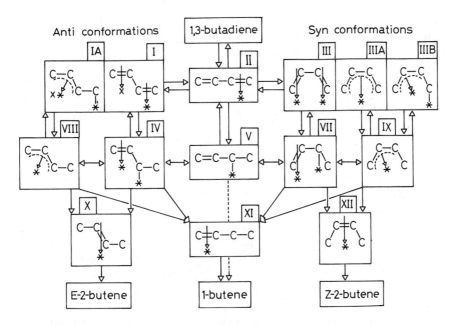

Figure 8.6. Mechanistic scheme for the hydrogenation of 1,3-butadiene.

rotation about the central C—C bond, can accomplish this. Thus adding the second hydrogen atom to an *anti*-π-butenyl must give *E*-2-butene (or 1-butene), and a *syn*-π-butenyl must lead to *Z*-2-butene (or 1-butene). The cards appear to be stacked in favour of *E*-2-butene and 1-butene from the outset, although a mechanism leading to *Z*-2-butene can be envisaged.

It is absolutely necessary to appreciate the limitations of conventional symbolism for representing the bonds between atoms in a molecule and even more so for showing bonds between adsorbed species and the metal atoms to which they are attached. Even such a simple molecule as carbon monoxide presents insuperable difficulty in this regard. Our discussion of the extensive measurements made on chemisorbed ethene (Section 4.42) have shown that a range of bonding interactions between the pure di-σ and the pure π forms is possible, the former being closely shown by platinum and the latter by palladium. These may therefore be regarded as canonical structures from which intermediate forms are constructed by mixing. The same distinction between these metals has been shown by the NEXAFS study[46] referred to in Section 8.3.2. Unfortunately, no more extensive work has been done on butadiene, and in particular there are no structural studies by LEED or PED to show how the molecule might be affixed to the surface: it would be useful to have independent confirmation of the presumed prevalence of the *anti* conformation. In writing mechanisms, therefore, we have to remember that our symbolic structures only reflect the predominant character of the species in question; the truth may either be that hybrid structures exist on uniform sites, or that a range of forms between the limiting ones reside on sites of different character, depending on the coordination numbers of the contributing atoms and the presence of other adsorbed species in the neighbourhood. The literature makes frequent reference to the need to invoke different sites to explain certain features of the results,[14,59] although direct evidence for their existence is rarely attainable.

Wells and his associates have devised comprehensive schemes of mechanism that embrace all features of the experimental results:[10,14,22,38,59] these are combined in Figure 8.6 in a slightly different format, but retaining their numbering of the adsorbed species. On the first line are five representations of chemisorbed butadiene: on the second are shown the consequences of adding a hydrogen atom to the *right-hand terminal carbon atom*. Species **I**, **III** and **IIIA** are symmetrical, but addition to the *left-hand* terminal carbon atom of **IA** and **IIIB** leads respectively to *E*- and *Z*-2-butene by somewhat different routes to those shown. Addition to the *second* carbon atoms counting from the right leads in all cases to 1-butene; addition to an internal carbon atom in a π-butenyl or π_3-butadiene will give respectively a 1,3-di-σ or a 1,3,4-tri-σ species, which may be the route by which *n*-butane is formed. However, in most cases S_{tot} is high, so this last possibility is usually ignored. This scheme provides a framework within which mechanistic considerations may be debated; of course not all processes occur on all catalysts.

The full mechanism provides a plethora of routes to the observed products, and by adjustment of the positions at which hydrogen atoms are added, and the relative rates of the steps, it should be possible to account for every conceivable blend. To facilitate discussion of how the nature of the metal and the state of its surface determines the choice of elementary steps, the complete scheme has been factorised into three parts.[14,54] In mechanism 1, the sequence is simply **II** $\rightarrow$ **V** $\rightarrow$ 1-butene. In mechanism *2*, species **I** and **III** can equilibrate via **II**, and lead respectively through **IV** or **VIII** to 1-butene + *E*-2-butene, and through **VIII** or **IX** to 1-butene + *Z*-2-butene. The existence of species **VIII** and **IX** is not however an absolute requirement to account for the products. Provided the constraint that favours the *anti-* conformation is relaxed somewhat in the adsorbed state, and the structures **I** and **III** (and their counterparts) are of comparable stability, the occurrence of *Z*/*E* ratios of about unity can be explained. Indeed on the very simple-minded assumption that they are equally probable, and that the two ways of adding the second hydrogen atom are also equally balanced, we should obtain the three isomers in the ratio 50:25:25, which is not so far from the observed with several metals (e.g. Sc, V, Ir). The *Z*/*E* ratio is said to be fixed by the relative stabilities **IV** and **VII**, equilibrating via **V**: but equilibration of **I** and **III** via **II** will have the same effect.

Concerning mechanisms 1 and 2, it has been suggested that there ought to be a negative correlation between S_1 and the E/Z ratio; in fact, there is only an approximate connection with the pure metals.[14,54] The tendency of S_1 to increase as the d-band is filled is clearer, and has been assigned to a decreasing tendency to form π_3 and π_4 intermediates. As a generalisation this may have some truth, but there are important vertical differences as well, seen most clearly in Group 10. It may well be that the state of hybridisation of the C—M bonds in butadiene (measured by an unknown $\pi\sigma$ factor as for ethene (Section 4.4.2)), and hence by inference in the butenyls, is a dominant factor. The simple mechanism 1 is clearly important for platinum, and for copper and manganese which have respectively filled and half-filled d-shells. The behaviour of nickel resembles that of platinum, with palladium ($3d^{10}$ in the free state) as always being the odd man out: gold also resembles platinum more than silver or copper. The low S_{tot} usually shown by platinum may have its origin in the preponderance of structure **III**, which from the data for heats of hydrogenation cannot be much more strongly adsorbed than the butenes.

Because of the impossibility of conformational interconversion in π_3 species, if they can be formed from gaseous butadiene without the intervention of **II**, and if the route is then solely through **VIII** and **IX**, then we may expect a high Z/E ratio, reflecting the relative stabilities of the *syn-* and *anti-* conformations in the gas phase. But the route **IA** $\rightarrow$ **VIII** can give *both* 1-butene *and* E-2-butene, and their sum sometimes exceeds 90% and can approach the 95% expected on the basis of the gaseous stabilities (e.g. with Mn, Pd, and especially the Type B Fe, Co and Ni, see Table 8.6). Mechanism 3 therefore advocates *only* states in which there is a delocalised π_3 bond attaching intermediates to the surface. This mechanism is pre-eminently adopted by palladium and by surfaces showing Type B characteristics: the effect of sulfur and other electronegative species is said to favour the formation of π_3 and π_4 species, but for reasons not entirely clear also to inhibit mechanism 1,[14,54,66] which certainly contributes in the case of palladium.[38] It is one of the surprising conclusions of the isotopic labelling work, which has been pursued with great diligence, that metal surfaces must exhibit heterogeneity, or at least various adsorbed species must be possible, although the overall reaction shows all the symptoms of structure-*insensitivity*.

8.3.6. Hydrogenation of 1, 3-Butadiene by Bimetallic Catalysts[5,22,87]

There have been numerous studies of this reaction catalysed by bimetallic catalysts, of which palladium has been the predominant component. The objective of much of this work has been to discover catalysts giving superior selectivity to butenes (S_{tot}), especially in the presence of excess butenes in the reactant flow. The choice of palladium as the constant partner[88] needs no explanation, as it stands out by its good activity and selectivity; nevertheless it is not perfect, and to seek to improve it further is not a waste of time (unlike *gilding refined gold*).

Three types of work have been undertaken using (1) single crystal and polycrystalline alloys and surface alloys, (2) intermetallic compounds, and (3) conventional supported bimetallic catalysts.

An early study[89] using nickel-copper films afforded Arrhenius parameters that showed good compensation over a range of activation energies from 21 to 62 kJ mol^{-1}, corresponding to almost constant activities at the median temperature; rates were however some ten times greater for pure nickel and somewhat greater than for pure copper. This observation harmonises with what is known of the structure of nickel-copper systems made in this way (Section 1.32). Product distributions were also almost constant, but with copper the 1-butene was initially more than 90% of the whole.

There have been several studies of palladium-nickel alloys, ranging in composition from Pd_1Ni_{99}[88] through Pd_5Ni_{95}[88] to Pd_8Ni_{92}:[90] the (111) surface of this last gives a high value of S_1 (>84%). Examination of polycrystalline Pd_1Ni_{99} and Pd_5Ni_{95} by LEIS and XPS show surface enrichment by palladium (respectively 20 and 50%),[90] the latter in particular giving faster rates than pure palladium: values of S_{tot} were always 100%. Pd 4d binding energies in the alloys were greater than for the pure metal, due it was thought to hybridisation of the two sets of valence orbitals; shifts were small (0.1 eV) for surface atoms, but larger (0.9 eV) for those inside. The active site was considered to comprise two palladium atoms. Depositing palladium on Ni(110) and (111) at 295 K gave rates only a little faster than those for the corresponding palladium faces, but annealing at 475 K, which induced formation of a superstructure, resulted in a further 6- to 10-fold increase.[52] Amorphous $Pd_2Ni_{50}Nb_{48}$ ribbon became highly selective even at 100% conversion after oligomer formation; butadiene adsorbed competitively with the oligomers, but the butenes did not.[69]

The (111) and (110) surfaces of platinum-nickel and platinum-iron alloys have also been studied. The (111) surface of $Pd_{50}Cu_{50}$ gave rates that were very much faster than that for palladium alone.[22] By evaporation of silver onto palladium deposited on amorphous silica, bimetallic particles about 3 nm in size and containing 17% silver were formed: they were about 15 times more active than palladium alone,[91] and the second stage of the reaction was entirely inhibited. This is one of the few significant observations to have been made with these model systems, and suggests an important way in which selectivity in the presence of excess alkene can be achieved. 'Microfabricated' square or cubic particles of palladium-gold have been formed on silica,[92] but butene selectivities (~75% 1-butene, ~25% E-2-butene) were independent of gold content; Arrhenius parameters showed approximate compensation. The lack of more detailed kinetic studies with these surfaces so lovingly and laboriously constructed is much to be regretted.

There have been several studies in which *intermetallic compounds* have been used as catalysts, but care is necessary because in some cases structural changes can occur during pre-treatment.[93] Palladium-rare earth intermetallics MPd_3 (M = La,

Ce, Pr, Nd, Sm) were active at 273 K, giving mainly higher values of S_{tot} with lower values of S_1 and of the E/Z ratio.[93] Detailed studies have also been reported on CePd$_3$ and ZrPd$_3$ at higher temperatures;[94] product selectivities resembled those shown by palladium sponge, but reactions stopped soon after the reactant was exhausted. With CePd$_{3-x}$Au$_x$ ($0 < x < 1$), selectivities were like those for CePd$_3$,[95] while for CeRh$_{3-x}$Pd$_x$ and ZrRh$_{3-x}$ there were complex variations that have been described in detail.[96]

The platinum-germanium system has six intermetallic phases, and Pt$_3$Ge, Pt$_2$Ge and PtGe have been examined; as the Pt/Ge ratio decreases, starting from pure platinum, the rates decreased, but S_{tot} rose to unity; there were only minor changes in product selectivities.[97] Other systems that have been studied include Co$_x$Ge, Co$_x$Sn and Co$_x$Al.[97]

Available results for the hydrogenation of butadiene on *supported* bimetallic catalysts are surprisingly fragmentary; they are summarised in Table 8.8 for those catalysts where palladium is the major or active component. In view of the variety of methods of preparation and supports, and the general absence of surface analysis, definite conclusions are hard to detect, but it appears that addition of a Group 11 metal often improves S_{tot},[98] and increases the conversion to which high selectivity is maintained. With palladium-silver particles formed by evaporation,[99] it also inhibits the hydrogenation of excess butene; this advantage is however offset by lowered activity.[68,71,79,98] The beneficial effect of Group 11 additives may thus be ascribed to a decrease in coverage by oligomer through removal of ensembles of the right size and disposition to allow their formation, while at the same time retaining small ensembles of palladium atoms that would otherwise need oligomers to create. Whether there is a loss or gain of activity therefore depends on whether lowering the number of active centres by the additive outweighs the gain by lesser oligomer

TABLE 8.8. Hydrogenation of 1,3 Butadiene: Effect of Metallic Additive on Supported Palladium Catalysts

| Additive | Physical Form | Effect | | | | Note | References |
		r	S_{tot}	S_1	E/Z		
Cr	Cr added to Pd/SiO$_2$	—	+	—	—	a	138
Co	Pd added to Co/Al$_2$O$_3$	—	~	+	—		100,131
Ni	(Pd + Ni)/Nb$_2$O$_5$	?	?	~	?		132
Cu	(Pd + Cu)/Al$_2$O$_3$	?	+	?	?	b	98,129
Ag	(Pd + Ag)/Al$_2$O$_3$	—	+	?	?	b,c	68,71
Au	(Pd + Au)/pumice	—	—	~	—		63
Zn	Pd/Zn (T$_{red}$ > 473 K)	—	?	+	?		78
Sn	(Pd + Sn)/Al$_2$O$_3$	?	+	?	?	d	139

[a] Cr added in various forms (Cr$_2$O$_7$$^{2-}$, CrO$_4$$^-$, Cr(CO)$_6$; not all as Cr^0).
[b] Excess 1-butene in reactant flow.
[c] Unique use of ^{13}C-labelling to study selectivity.
[d] Various Sn precursors used.

formation. These effects will be seen again, and discussed further, in the following chapter. The most detailed investigation has been on the palladium-gold system,[63] and this dates from 1966: initial values of S_{tot} were above 98%, and S_1 showed a maximum ($\sim$60%) between $Pd_{40}Au_{60}$ and $Pd_{30}Au_{70}$. In this region the activation energy also passed through a maximum (62 kJ mol^{-1}) before falling to 37 kJ mol^{-1} at Pd_5Au_{95}. Reaction orders were 0.7–0.9 in hydrogen and zero in butadiene. The results were discussed in great detail, and importance was attached to the dissolved hydrogen, and especially the β to α phase transition that occurred between 353 and 393 K. The butene isomer distribution characteristic of palladium (see Table 8.6) seems always to be shown by bimetallic catalyst in which palladium is a member.

The addition of tin to platinum caused a loss of activity when alumina was the support, but not when niobia was used; in this latter case, reduction at 773 K resulted in a very high value of S_1 (89%).[76] The palladium-cobalt[100] and palladium-tin[101] combinations have also been examined.

8.4. HYDROGENATION OF HIGHER ALKADIENES

8.4.1. Linear Alkadienes

Extension of the linear carbon chain to five or six atoms opens up additional possible reaction paths. Higher 1,3-alkadienes are expected to behave in a similar fashion to 1,3-butadiene,[102] except the two double bonds may differ in reactivity under the influence of the alkyl substituent, and their reduction naturally affords distinguishable products. Thus for example 1,3-hexadiene can give 1- and 3-hexenes by respectively 3,4- and 1,2-addition as well as 2-hexenes by 1,4-addition or isomerisation.[103] A new feature in the higher 1,3- alkadienes is the existence of geometrical isomerism: Z- and E-1,3-pentadienes have been examined separately.[102,104]

It was mentioned in Section 8.1.1 that increasing the separation between the pair of double bonds enabled them to react more independently of each other. Comparison of the behaviours of 1,3- and 1,5-hexadienes,[103,105] and the reactions of even larger α,ω-dienes,[106,107] illustrates this well. With 1,2-alkadienes, their strong adsorption is a consequence of the mutual strain induced by the proximity of the double bonds, but with 1,3-dienes it is due to the simultaneous interaction of both with the surface,[108] because the heat of hydrogenation of 1,3- butadiene to 1-butene is similar to that for 1-butene to butane (Table 7.1). When the separation is greater, the probability of both double bonds being accommodated by the surface is reduced, and the character of the reaction is changed. A possible complication, however, is the occurrence of double bond migration during reaction, to give the more stable conjugated 1,3-isomer, or at least an equilibrium mixture in which it predominates. However, with Pd/C catalyst, 1,5-hexadiene gave 1-hexene as the

TABLE 8.9. Hydrogenation of 1,3-Pentadiene: Product Selectivities[102,104]

Metal[a]	Form	Isomer	State[b]	S_{tot}	S_1	S_{E2}	S_{Z2}
Co[B]	/Al$_2$O$_3$	E	V	~1	90	5	5
Co[B]	/Al$_2$O$_3$	Z	V	1	88	6	6
Co[A]	/Al$_2$O$_3$	E	V	~1	31	55	14
Co[A]	/Al$_2$O$_3$	Z	V	0.99	39	36	25
Ni[B]	/Al$_2$O$_3$	E	V	~1	29	60	11
Cu	/Al$_2$O$_3$	E[c]	V	1	69	19	12
Ru	/C	E	L	0.70	10	67	23
Rh	/C	E	L	0.80	20	67	13
Pd	/C	E	L	0.98	32	61	7
Pd	/Al$_2$O$_3$	E	V	~1	40	52	8
Pd	/Al$_2$O$_3$	Z	V	~1	38	44	18
Ir	/C	E	L	0.46	17	65	18
Pt	/C	E	L	0.68	38	49	13
Pt	/Al$_2$O$_3$	Z	V	0.91	52	19	29

Notes: Product selectivities are those found at low conversions, but change with conversion is usually small.
[a] Cobalt and nickel catalysts are designated Type A or Type B, depending on reduction temperature: the latter are likely to be sulfur-contaminated.
[b] V, reactant as vapour; L, as liquid.
[c] The Z-isomer gives a similar distribution.

major initial product, followed by the E- and Z-isomers of 2-hexene, equilibrium proportions being found after 60% conversion.[106]

The hydrogenation of Z- and E-1,3-pentadienes has been studied in great detail on a number of metals, using the reactant as vapour,[102] liquid or solution:[104] a small selection of the available results is shown in Table 8.9. In the gaseous reactions, selectivities to pentenes were usually very high (but not with platinum), while in the liquid phase only palladium gave selectivities close to unity. Z-E isomerisation in the reactant was very marked with Cu/Al$_2$O$_3$ and Type B Co/Al$_2$O$_3$ giving an equilibrium mixture containing about 75% of the E-isomer at 433 K.[102] Extended discussion of the mechanisms, structures and processes contributing in every case is not warranted, because the principles elaborated for the reactions of 1,3- butadiene must apply here also, with the additional effect of the methyl substituent. What is however quite clear is that 3,4-addition is favoured over the other modes, so that the methyl group must, by an electronic effect, cause greater reactivity in the adjacent double bond, and not an inhibiting effect due to steric interference.[102] It is however possible to chart the reaction paths assuming the intervention of π_4 or $\pi_3\sigma$ alkenylic intermediates (Figure 8.7).

The four possible orientations of the two reactants are shown, and they are then represented as the possible $\pi_3\sigma$ structures (π_4 structures are only possible in the *syn* conformations). We then assume that in forming the alkene the σ C—M bond is first broken and the second hydrogen atom is added at one or other end of the delocalised bond. Assuming values for the relative stabilities of *syn* and *anti* conformers, and

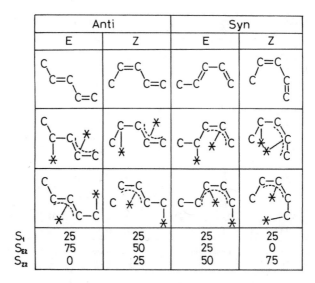

	Anti		Syn	
	E	Z	E	Z
S_1	25	25	25	25
S_{E2}	75	50	25	0
S_{Z2}	0	25	50	75

The selectivities at the bottom are obtained by (i) breaking the σ
C–M bond by H atom addition, (ii) adding a second H atom with
equal probability to either end of the π_3 bond.

Figure 8.7. *Syn* and *anti* conformers of *Z*- and *E*-1,3-pentadiene, and their representation as alternative $\pi_3\sigma$ structures.

for the effects of alkyl groups on the point of addition of the second hydrogen, product distributions can be calculated. It is important to minimise the number of arbitrary assumptions, and to keep assumed values within reasonable limits. The results shown in Table 8.10 take the ratio of *anti:syn* conformers to be four, and the positive weighting to the 3,4-addition mode to be 1.5 or 2.0. Good agreement is obtained with experimental results for the reaction of *E*-1,3-pentadiene over palladium catalysts, but it is not so good for the *Z*-isomer, which probably is partly isomerised during the reaction. To support this view, the calculation for the equilibrium mixture (*Z*, 83; *E*, 17) agrees quite well with observation. The two isomers interconverted in the absence of hydrogen on cobalt catalysts.[109]

The reactions of 1,3- and 1,5-hexadienes with hydrogen have been carried out with Pd/Al_2O_3 catalysts of various dispersions; the reaction was not structure-sensitive nor were product distributions affected.[103] With the 1,3-isomer, selectivities were very high until at least 95% conversion, and product selectivities were not greatly conversion-dependant. Representative values are shown in Table 8.11. The *Z/E* ratio of the reactant was not specified, so that model calculations cannot be made with certainty, but the predominance of Z-isomers in the product argues for the preference for *anti*-conformations in the adsorbed states. $Pd_{5.2}Sn_{0.8}/Al_2O_3$

TABLE 8.10. Comparison of Calculated and Observed Product Selectivities in the Hydrogenation of 1,3-Pentadiene on Palladium Catalysts[102,104]

Isomer	a/s^b	$f_{3,4}{}^c$	S_1	S_{E2}	S_{Z2}
E	4	1	25	65	10
E	4	1.5	33	58	9
E	—	—	32	61	7
E	4	2	40	52	8
E	—	—	40	52	8
Z	4	1	25	40	35
Z	4	1.5	33	36	31
Z	—	—	39	40	22
$(Z + E)_{eq}{}^a$	4	1.5	33	54	13
$(Z + E)_{eq}{}^a$	—	—	38	51	11

Notes: ^{a}Assumed to have Z/E=17:83.
^{b}Assumed ratio of *anti:syn* conformers.
^{c}Assumed weighting of the 3,4-addition mode.

gave similar results. With 1,5-hexadiene, however, hydrogenation over Pd/Al_2O_3 and catalysts also containing tin or silver, total selectivity fell progressively with conversion, and product selectivities moved in the direction of equilibrium concentrations, indicative the relatively weaker adsorption of this reactant. The more extensive formation of the 2-hexenes reflects the occurrence of double-bond migration before desorption.

1,7-Octadiene on a palladium catalyst was hydrogenated at 323 K to a mixture of octenes and isomerised to other octadienes, but S_{tot} was somewhat low.[110]

Pt/TiO_2 hydrogenated 1,3-hexadiene with a total selectivity of 90%, with a broad mix of other products (see Table 8.4) but when deposited on a ceramic membrane the product was 100% 1-hexene;[111] the suggestion that this was due to lack of back-mixing implies that re-adsorption of 1-hexene was somehow prevented, but this does not normally take place, and a full explanation still needs to be found.

TABLE 8.11. Product Selectivities for the Hydrogenation of Hexadienes on Al_2O_3- (or TiO_2)- Supported Catalysts[103,111]

Metal	Isomer	S_1	S_{E2}	S_{Z2}	S_{E3}	S_{Z3}	Note
Pd	1,3-	35	31	3	31	0	
$Pd_{3.8}Ag_{5.2}$	1,3-	27	31	5	35	0	a
Pd	1,5-	30	26	12	6	26	
$Pd_{5.2}Sn_{0.8}$	1,5-	48	18	10	10	12	
Pt	1,3-	29	29	2	29	11	b

^{a}This is the distribution at 90% conversion: the others are approximate values obtained by extrapolating graphical results to zero conversion.
^{b}TiO_2 support (363 K).

8.4.2. Branched Alkadienes[44]

Isoprene (2-methyl-1,3-butadiene) is the precursor to a great range of naturally occurring molecules, including terpenes, di- and triterpenes, and steroids, the skeletons of which are composed of branched C_5 units: it is also of course the origin of natural rubber. The addition of a methyl group at C2 to butadiene creates a molecule the hydrogenation of which has certain interesting features, because 1,2- and 3,4-additions are now distinguishable, and the products are 2-methyl-2-butene, 3-methyl-1-butene and 2-methyl-1-buene. No geometrical isomerism is possible in this system.

The reaction has been studied mainly on palladium catalysts of various types,[110,112–116] including the bimetallic Pd-Ag/SiO$_2$ and Pd-Au/SiO$_2$ systems.[115] Values of S_{tot} were generally high (>0.96), but the conversion-independent product distributions varied slightly but significantly with composition, in an interesting manner (Table 8.12). Once again we may chart the progress of the reaction in terms of $\pi_3\sigma$ and π_3 intermediates (Figure 8.8). The two half-hydrogenated π_3 states differ in the disposition of the methyl substituents, and if they have no effect on the chances of adding the final hydrogen atom to the termini of the delocalised bond, the consequence will be

$$[3,1] : [2,2] : [2,1] = 1 : 2 : 1 \tag{8.1}$$

This ratio is indeed close to that often found[115] (Table 8.12), but close examination has revealed small variations. These have been attributed to the different effects of the electron-releasing methyl groups on the locus of the addition of the final hydrogen atom. If the carbon atom carrying n methyls is nf times more able to accept a hydrogen atom than that having none, the ratio then becomes

$$[3,1] : [2,2] : [2,1] = (1 + 2f) : 2 : (1 + f) \tag{8.2}$$

TABLE 8.12. Hydrogenation of Isoprene (2-Methyl-1,2-Butadiene): Product Selectivities Obtained with Palladium Catalysts,[110,113,115] and Matching Calculations[115,117]

Form/Composition	S_{tot}	S_{22}	S_{31}	S_{21}	f
/BaSO$_4$	>95	45.4	29.9	24.7	—
		44.9	29.2	25.8	0.15
/SiO$_2$	>95	49.1	26.8	24.1	—
		48.5	26.2	25.2	0.04
Pd$_{20}$Ag$_{80}$//SiO$_2$	98	42.3	30.5	27.2	(0.225)
Pd$_{20}$Au$_{80}$//SiO$_2$	97	48.2	25.5	26.3	(0.05)
/C+Φ_3Bi	86	56.6	20.0	23.5	—
		56.0	20.0	24.0	−0.14
/δ-Al$_2$O$_3$	—	78	15	7	(−0.48)

Note: Three examples of matching calculations are shown; values of f in brackets give approximate fits to the experimental distributions shown.

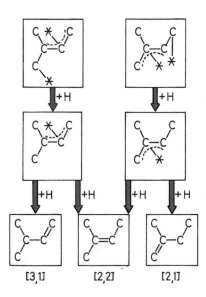

Figure 8.8. Adsorbed states of isoprene (2-methyl-1,3-butadiene) as alternative $\pi_3\sigma$ structures and their half-hydrogenated states, leading to the three isomeric 2-methyl butenes.[115,117]

Thus by matching observed distributions to those calculated by varying f, the value of this parameter can be estimated; it varies from about $+0.24$ to -0.75. Illustrative examples of calculated distributions are given in Table 8.12. A refinement of the procedure to allow different reactivities of the two π_3 intermediates has only a marginal effect. The source of the effect of *increasing* electron density in proportion to the number of methyl substituents has been confirmed by quantum mechanical calculations,[115] from which it must be concluded that when f is positive the hydrogen atom must be $\delta+$, and when it is negative it must be $\delta-$. We therefore have a potentially useful method for identifying the polarity of the H—M bond involved in the reaction, and hence perhaps for studying metal-support interactions.

TOFs for palladium black and various supported palladium catalysts vary enormously, probably due to differences in surface cleanliness rather than particle size, and values of f lie between $+0.225$ (Pd/CaCO$_3$) and -0.015 (Pd/SiO$_2$): they were mainly positive.[115] With increasing amounts of gold or silver added to Pd/SiO$_2$, they became more positive: values of S_{tot} were greater than 96%, and values of TOF (based on surface Pd atoms titrated by CO) were larger (markedly so in some cases) than for Pd/SiO$_2$.

The analysis has been extended to cover many of the published product selectivities found when hydrogenating isoprene on palladium catalysts.[117] When

f is less than -0.5, the 3,1-isomer is not formed, and when it is -1 the 2,2-isomer is the only possible product: such results are sometimes encountered. A particularly interesting set of results was obtained by selective poisoning of Pd/C with Ph_3M (M = N, P, As, Sb, Bi):[113] toxicity increased and values of f decreased with rise in atomic number of M, and it was thought that increasing electron donation counteracted the positive charge on the hydrogen atoms, so that f had to become more negative. Other systems analysed included one where marked solvent effects were revealed. In one extensive study of the effect of adding Group 14 elements (and Sb) to Pd/Al_2O_3, full product analyses were very unfortunately not recorded:[112] constant values of TOF and of activation energy were (55 kJ mol^{-1}) were observed, and S_{tot} was increased, especially by lead. Changes with time-on-stream have been followed with eggshell $Pd/\delta-Al_2O_3$ catalysts of various dispersions: conversion decreased and S_{tot} increased: as observed elsewhere, faster rates were found with catalysts of *lower* dispersion.[110]

Platinum catalysts have given product distributions that were not consistent with the mechanism that seems to apply to palladium.[118,119] Isoprene hydrogenation has also been performed on Pt(111);[50] at 423 K, S_{tot} was very low (37%). Raney nickel has also been used.[120]

The method of analysis described above can also be used on the homogeneously-catalysed reaction: a number of organometallic complexes have given product mixes that are understandable on the basis of π_4 or $\pi_3\sigma$ intermediates.[117]

8.4.3. Cycloalkadienes[2,44]

Cyclopentadiene has received remarkably little attention:[44] this is strange, because the molecule is planar and rigid, and should be invaluable for mapping the geometry of adsorption sites. On various copper catalysts, it is hydrogenated with 100% selectivity to cyclopentene with an activation energy of 54 kJ mol^{-1}:[121] its chemisorption is however weak, because the rate equation is

$$r = k P_H P_C{}^L \qquad (8.3)$$

where $P_C{}^L$ indicates a dependence on cyclopentadiene pressure described by the Langmuir adsorption equation.

Reactions of cyclohexadienes will be considered in Chapter 12 as intermediates in the dehydrogenation of C_6 cycles to benzene. There is no information on cycloheptadienes.

1, 3-Cyclo-octadiene is a conjugated diene, so that selective reduction to cyclo-octene should occur readily: it has also been only little studied. On Pd/pumice catalysts of various dispersions, TOFs decreased sharply as particle size fell in the

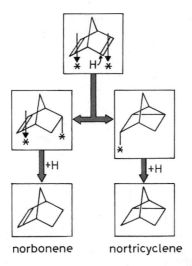

norbornene   nortricyclene

Figure 8.9. Addition of a hydrogen atom to norbornadiene (tricyclo[2.2.1]heptadiene) leading to norbornene and nortricyclene.

2 to 3 nm region (30% dispersion), maximum selectivity being found at about 50% dispersion.[70,87] Sodium ion in pumice increased the rate of electron transfer *to* the metal, thus weakening the chemisorption bond.[122] On Cu/TiO$_2$ (Degussa P-25), preparation by slow deposition of hydroxide from an ammine solution by dilution with water gave a much more active catalyst than conventional impregnation.[123] FeCu/SiO$_2$ catalysts gave fully selective reduction of the diene.[124]

On an unspecified supported palladium catalyst at 323 K, 1,5-cyclo-octadiene isomerised to the 1,3-form at about the same rate as it was hydrogenated; S_{tot} was about 90%, and some mathematical modelling was attempted.[107] Other cyclic dienes have also been examined.[125]

Norbornadiene (bicyclo[2.2.1]octadiene) also has its two double bonds in a fixed geometry. On supported copper and gold catalysts above 353 K, and on Co/pumice and Pt/pumice above 333 K, nortricyclene and its isomer norbornene (Figure 8.9) were formed simultaneously, but not on Pt/MgO.[126] Silver catalysts were inactive, as were all Group 11 metals for norbornene hydrogenation. A 1,3-dipolar addition of a hydrogen molecule was proposed, although the products are equally well accounted for by the mechanism shown in the Figure. Molecular addition may however be important in cases where the concentration of adsorbed hydrogen atoms is low, as with the Group 11 metals and Pt/TiO$_2$ in the SMSI state. The activity of gold for this reaction passed without substantial comment.

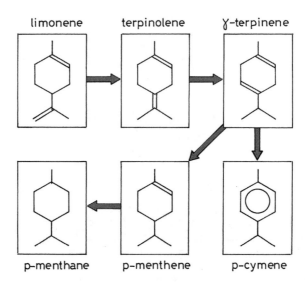

Figure 8.10. Hydrogenation of limonene (1-methyl-4-isopropenyl-cyclohexene).[127]

Extremely complex reaction paths are encountered in the hydrogenation of terpenes, by reason of the diversity of isomeric mono-enes and dienes that are possible: the presence of the C_6 ring also allows dehydrogenation to an aromatic molecule to occur. A recent detailed study of the hydrogenation of limonene (1-methyl-4-isopropenyl-1-cyclohexene) illustrates well the complexity that can be found;[127] it supplements many qualitative observations to be found in the literature.[2] The principal pathway on Pd/C or Pd/Al$_2$O$_3$ at 273 and 323 K involved progressive double-bond migration as shown in Figure 8.10 to the 1,4-cyclohexadiene, which then (following the tendency of such molecules) suffered disproportionation to the mono-ene and aromatic molecule.

1, 5, 9-cyclododecatriene was hydrogenated progressively to a diene and the mono-ene on Pd/Al$_2$O$_3$ at 423 K, with maximum selectivities of respectively about 50 and 65%.[128]

REFERENCES

1. K.J. Klabunde, S.C. Davis, H. Hattori and Y. Tanaka, *J. Catal.* **54** (1978) 254; P. Herman, D. Simon, P. Sautet and B. Bigot, *J. Catal.* **167** (1997) 33.
2. P.N. Rylander, *Catalytic Hydrogenation over Platinum Metals,* Academic Press: New York (1967).
3. P.N. Rylander, *Catalytic Hydrogenation in Organic Synthesis,* Academic Press: New York (1979).
4. R. Kuhn and H. Fischer, *Chem. Ber.* **92** (1959) 1849; **93** (1960) 2285.
5. L. Guczi and A. Sárkány in: *Specialist Periodical Reports: Catalysis*, Vol. 11 (J.J. Spivey and S.K. Agarwal, eds.), *Roy. Soc. Chem.* (1994), p. 318.

6. R.J. Farrauto and C.H. Bartolomew, *Fundamentals of Industrial Catalytic Processes*, Chapman and Hall: London (1997).
7. J.P. Boitiaux, J. Cosyns, M. Derrien and G. Léger, *Hydrocarbon Proc.* (1985) 51.
8. J. Goetz, D.Yu. Murzin and R.A. Touroude, *Ind. Eng. Chem. Res.* **35** (1996) 703.
9. R.L. Moss in: *Specialist Periodical Reports: Catalysis*, Vol. 1 (C. Kemball and D.A. Dowden, eds.), *Roy. Soc. Chem.* (1977), p. 37; Vol. 4 (C. Kemball and D.A. Dowden, eds.), *Roy. Soc. Chem.* (1981), p. 36.
10. G. Webb in: *Specialist Periodical Reports: Catalysis*, Vol. 2 (C. Kemball and D.A. Dowden, eds.), *Roy. Soc. Chem.* (1978), p. 145.
11. D.A. Dowden in: *Specialist Periodical Reports: Catalysis*, Vol. 2 (C. Kemball and D.A. Dowden, eds.), *Roy. Soc. Chem.* (1978), p. 1.
12. S. Siegel, *Adv. Catal.* **16** (1966) 123.
13. R.L. Augustine, *Heterogeneous Catalysis for the Synthetic Chemist*, Dekker: New York (1996).
14. P.B. Wells, in: *Surface Chemistry and Catalysis* (A.F. Carley, P.R. Davis, G.J. Hutchings and M.S. Spencer, eds.), Kluwer: Dordrecht (2003).
15. T. Ouchaib, J. Massardier and A. Renouprez, *J. Catal.* **119** (1989) 517.
16. J.P. Boitiaux, J. Cosyns and S. Vasudevan, *Appl. Catal.* **15** (1985) 317; J.P. Boitiaux, J. Cosyns and G. Martino in: *Metal-Support and Metal-Additive Effects in Catalysis,* (B. Imelik, C. Naccache, G. Coudurier, H. Praliaud, P. Meriaudeau, P. Gallezot, G.A. Martin and J.C. Védrine, eds.), Studies in Surface Science and Catalysis, Elsevier: Amsterdam, **11** (1982) 355.
17. J. Barbier, E. Lamy-Pitara, P. Marécot, J.P. Boitiaux, J. Cosyns and F. Verma, *Adv. Catal.* **37** (1990) 279.
18. G.C. Bond, *Heterogeneous Catalysis – Principles and Applications,* 2nd. edn., Oxford University Press: Oxford (1987).
19. B.I. Rosen, *US Patent* 4479902 (1984) to UOP Inc.
20. L.F. Albright, *Chem. Eng.* (1967) 249.
21. V.I. Savchenko and I.A. Makaryan, *Platinum Metals Rev.* **43** (1999) 74.
22. Á. Molnar, A. Sárkány and M. Varga, *J. Molec. Catal. A: Chem.* **173** (2001) 185.
23. G.C. Bond and P.B. Wells, *Adv. Catal.* **15** (1964) 92.
24. B.E. Bent, C.M. Mate, J.E. Crowell, B.E. Koel and G.A. Somorjai, *J. Phys. Chem.* **91** (1987) 3249.
25. E.F. Meyer and R.L. Burwell Jr., *J. Am .Chem. Soc.* **85** (1963) 2881.
26. J. Grant, R.B. Moyes, R.G. Oliver and P.B. Wells, *J. Catal.* **42** (1976) 213.
27. L.J. Shorthouse, S. Haq and R. Raval, *Surf. Sci.* **368** (1996) 296.
28. L. Crombie, P.A. Jenkins, D.A. Mitchard and J.C. Williams, *Tetrahedron Lett.* (1967) 4297.
29. G.C. Bond, PhD thesis, Birmingham University 1951.
30. G.C. Bond and J. Sheridan, *Trans. Faraday Soc.* **48** (1952) 658.
31. G.C. Bond and J. Sheridan, *Trans. Faraday Soc.* **48** (1952) 664.
32. R.S. Mann and A.M. Shah, *Canad. J. Chem.* **50** (1972) 1793.
33. R.S. Mann and D.E. Tiu, *Canad. J. Chem.* **46** (1968) 3249.
34. R.S. Mann and D.E. To, *Canad. J. Chem.* **46** (1968) 161.
35. R.S. Mann and A.M. Shah, *Canad. J. Chem.* **48** (1970) 3324.
36. Xing-Cai Guo and R.J. Madix, *J. Catal.* **155** (1995) 336.
37. R.G. Oliver and P.B. Wells, *J.Catal.* **47** (1977) 362.
38. A.J. Bates, Z.K. Leszczyński, J.J. Phillipson, P.B. Wells, and G.R. Wilson, *J. Chem. Soc. (A)* (1970) 2435.
39. G.V. Smith and R.L. Burwell Jr., *J. Am. Chem. Soc.* **84** (1962) 925.
40. R.G. Oliver, P.B. Wells and (in part) J. Grant, 5th *Internat. Congr. Catal.* (J.W. Hightower, ed.), North Holland: Amsterdam **1** (1972) 659.
41. L. Crombie, P.A. Jenkins and D.A. Mitchard, *J. Chem. Soc. Perkin Trans. I* (1975) 1081.

42. L. Crombie, P.A. Jenkins and J. Roblin, *J. Chem. Soc. Perkin I* (1975) 517.
43. W.R. Moore, *J. Am. Chem. Soc.* **84** (1962) 3788.
44. B.B. Corson in: *Catalysis*, (P.H. Emmett, ed.), Reinhold: New York, Vol. 5 (1957), p. 59.
45. N. Sheppard and C. de la Cruz, *Adv. Catal.* **41** (1996) 1.
46. J.C. Bertolini, A. Cassuto, Y. Jugnet, J. Massardier, B. Tardy and G. Tourillon, *Surf. Sci.* **349** (1996) 88.
47. J. Massardier, J.C. Bertolini, P. Ruiz and P. Delichère, *J. Catal.* **112** (1988) 21.
48. J. Oudar, S. Pinot and Y. Berthier, *J. Catal.* **107** (1987) 434.
49. C. Yoon, M.X. Yang and G.A. Somorjai, *Catal. Lett.* **46** (1997) 37.
50. C.-M. Pradier and Y. Berthier, *J. Catal.* **129** (1991) 356.
51. J. Oudar, S. Pinot, C.-M. Pradier and Y Berthier, *J. Catal.* **107** (1987) 445.
52. P. Hermann, J.M. Guigner, B. Tardy, Y. Jugnet, D. Simon and J.C. Bertolini, *J. Catal.* **163** (1996) 169.
53. S. Katano, H.S. Kato, M. Kawai and K. Domen, *J. Phys. Chem. B* **107** (2003) 3671.
54. R.B. Moyes, P.B. Wells, J. Grant and N.Y. Salman, *Appl. Catal. A: Gen.* **229** (2002) 251.
55. W.G. Young, R.L. Meier, J. Vinograd, J. Bottinger, L. Kaplan and S.L. Lindin, *J. Am. Chem. Soc.* **69** (1947) 2046.
56. K. Shimazu and H. Kita, *J. Chem. Soc. Faraday Trans. I* **81** (1985) 175.
57. K. Okamoto, K. Fukino, T. Imanak and S. Teranishi, *J. Catal.* **174** (1982) 173.
58. J.P. Boitiaux, J. Cosyns, and E. Robert, *Appl. Catal.* **35** (1987) 193; **32** (1987) 145.
59. J.J. Phillipson, P.B. Wells and G.R. Wilson, *J. Chem. Soc. (A)* (1969) 1351.
60. A. Sárkány and Zs. Révay, *Appl. Catal. A: Gen.* **243** (2003) 347; A. Sárkány, *Appl. Catal. A: Gen.* **165** (1997) 87.
61. M. Okumura in ONRI Report No. 3. Aug. 1999, p. 54.
62. M. Okamura, T. Akita and M. Haruta, *Catal. Today* **74** (2002) 265.
63. B.J. Joice, J.J. Rooney, P.B. Wells and G.R. Wilson, *Discuss. Faraday Soc.* **41** (1966) 223.
64. P.B. Wells and (in part) A.J. Bates, *J. Chem. Soc. (A)* (1968) 3064.
65. G.C. Bond, G.Webb, P.B. Wells and J.M. Winterbottom, *J. Chem. Soc.* (1965) 3218.
66. M. George, R.B. Moyes, D. Ramarac and P.B. Wells, *J. Catal.* **52** (1978) 486.
67. A. Sárkány, *J. Catal.* **180** (1998) 149; *React. Kinet. Catal. Lett.* **68** (1999) 153.
68. A. Sárkány, *Appl. Catal. A: Gen.* **175** (1998) 245.
69. A. Sárkány, Z. Schay, Gy. Stefler, L. Borkó, J.W. Hightower and L. Guczi, *Appl. Catal. A: Gen.* **124** (1995) L181.
70. A.M. Venezia. A. Rossi, D. Duca, A. Martorana and G. Deganello, *Appl. Catal A: Gen.* **125** (1955) 113.
71. J.W. Hightower, B. Furlong, A. Sárkány and L. Guczi, *Proc. 10th·Internat. Congr. Catal.* (L. Guczi, F. Solymosi and P. Tétényi, eds.), Akadémiai Kiadó: Budapest **C** (1993) 2305.
72. A.G. Burden, J. Grant, J. Martos, R.B. Moyes and P.B. Wells, *Faraday Discuss. Chem. Soc.* **72** (1981) 97.
73. P.B. Wells, *J. Catal.* **52** (1978) 498.
74. M. Primet, M. El Azhar and M. Guenin, *Appl. Catal.* **58** (1990) 241.
75. S.D. Jackson, G.D. McLellan, G. Webb, L. Conyers, M.B.T. Keegan, S. Mather, S. Simpson, P.B. Wells, D.A. Whan and R. Whyman, *J. Catal.* **162** (1966) 10.
76. D.A.G. Aranda and M. Schmal, *J. Catal.* **171** (1997) 398.
77. M. Schmal, D.A.G. Aranda, R.R. Soares, F.E. Noronha and A. Frydman, *Catal. Today* **57** (2000) 169.
78. A. Sárkány, Z. Zsoldos, G. Steffler, J.W. Hightower and L. Guczi, *J. Catal.* **141** (1993) 566.
79. H. Miura, M. Terasaka, K. Oki and T. Matsuda, *Proc. 10th Internat. Congr. Catal.* (L. Guczi, F. Solymosi and P. Tétényi, eds.), Akadémiai Kiadó: Budapest **C** (1993) 2379.
80. M. Komiyama, K. Ohashi, Y. Morioka and J. Kobayashi, *Bull. Chem. Soc. Japan* **70** (1997) 1009.

81. V. de Gouveia, B. Bellamy, Y. Hadj Romdhane, A. Masson and M. Che, *Z. Phys.: Atoms, Molecules and Clusters* **12** (1989) 587.
82. A. Berthet, A.L Thomann, F.J. Cadete Santos Aires, M. Brun, C. Deranlot, J.C. Bertolini, J.P Rozenbaum, P. Brault and P. Andreazza, *J. Catal.* **190** (2000) 49.
83. R. Tardy, C. Noupa, C. Leclereq, J.C. Bertolini, A. Hoareau, M. Terilleux, J. F. Faure and G. Nihoul, *J. Catal.* **129** (1991) 1.
84. J.P. Boitiaux, J. Cosyns and S. Vasudevan, *Appl. Catal.* **6** (1983) 41; **15** (1985) 317.
85. J.C. Bertolini, P. Delichere, B.C. Khanra, J. Massardier, C. Noupa and B. Tardy, *Catal. Lett.* **6** (1990) 215.
86. J. Goetz, M.A. Volpe and R. Touroude, *J. Catal.* **164** (1996) 369.
87. G. Deganello, D. Duca, G. Fagherazzi and A. Benedetti, *J. Catal.* **150** (1994) 127.
88. P. Miegge, R.L. Rousset, B. Tardy, J. Massardier and J.C. Bertolini, *J. Catal.* **149** (1994) 404.
89. P.F. Carr and J.K.A. Clarke, *J. Chem. Soc. A* (1971) 985.
90. L.J. Shorthouse, Y. Jugnet and J.C. Bertolini, *Catal. Today* **70** (2001) 33.
91. L. Lianos, Y. Debauge, J. Massardier, Y. Jugnet and J.C. Bertolini, *Catal. Lett.* **44** (1997) 211.
92. A.C. Krauth, G.H. Berstein and E.E. Wolf, *Catal. Lett.* **45** (1997) 177.
93. K.S. Sim, L. Hilaire, F. Le Normand, R. Touroude, V. Paul-Boncour and A. Percheron-Guegan, *J. Chem. Soc. Faraday Trans.* **87** (1991) 1453.
94. A. Bahia and J.M. Winterbottom, *J. Chem. Tech. Biotech.* **60** (1994) 305.
95. A. Bahia, I.R. Harris, C.E. King and J.M. Winterbottom, *J. Chem. Tech. Biotech.* **60** (1994) 347.
96. A. Bahia, I.R. Harris, C.E. King and J.M. Winterbottom, *J. Chem. Tech. Biotech.* **60** (1994) 337.
97. T. Komatsu, S. Hyodo and T. Yashima, *J. Phys. Chem. B* **101** (1997) 5565.
98. A.H. Weiss, S. LeViness, V. Nair, L. Guczi, A. Sárkány and Z. Schay, *Proc. 8th Internat. Congr. Catal.* Verlag Chemie: Weinheim **5** (1984) 591.
99. V. de Gouveia, B. Bellamy, A. Masson and M. Che in: *Structure and Reactivity of Surfaces* (C. Morterra, A. Zecchina and G. Costa, eds.), Studies in Surface Science and Catalysis', Elsevier: Amsterdam, **48** (1989) 347.
100. A. Sárkány, M. Zsoldos, J.W. Hightower and L. Guczi in: *Science and Technology in Catalysis 1994*, Kodanska: Prague (1994), p. 99.
101. S. Verdier, B. Didillon, S. Morin and D. Uzio, *J. Catal.* **218** (2003) 288.
102. P.B. Wells and G.R. Wilson, *J. Chem. Soc. (A)* (1970) 2242.
103. E.A. Sales, M.de J. Mendes and F. Bozon-Verduraz, *J. Catal.* **195** (2000) 96.
104. G.C. Bond and J.S. Rank, *Proc. 3rd. Internat. Congr. Catal.* (W.M.H. Sachtler, G.C.A. Schuit and P. Zwietening, eds.), North Holland: Amsterdam **2** (1965) 1225.
105. R. Brayner, G. Vian, G.M da Cruz, F. Fiévet-Vincent, F. Fiévet and F. Bozon-Verduraz, *Catal. Today* **57** (2000) 187.
106. L. Horner and I. Grohmann, *Liebigs Ann.* **670** (1963) 1.
107. M. Di Serio, V. Balato, A. Dimiccoli, M. Mattucci, P. Iengo and E. Santacesaria, *Catal. Today* **66** (2001) 403.
108. N. Vasquez and R.J. Madix, *J. Catal.* **178** (1998) 234.
109. P.B. Wells and G.R. Wilson, *Discuss. Faraday Soc.* **41** (1966) 237.
110. J.-C. Chang and T.-C. Chou, *Appl. Catal. A: Gen.* **156** (1997) 193.
111. C. Lange, S. Storck, B. Tesche and W. F. Maier, *J. Catal.* **175** (1998) 280.
112. H.R. Aduriz, P. Bodnariuk, B.Coq and F. Figuéras, *J. Catal.* **129** (1991) 47.
113. Y. Fujii and J. C. Bailar Jr., *J. Catal.* **52** (1978) 342.
114. P. Kripylo and D. Klose, *Chem. Tech. (Leipzig)* **29** (1977) 322.
115. G.C. Bond and A.F. Rawle, *J. Molec. Catal. A: Chem.* **109** (1996) 261.
116. J.-R. Chang, T.-B. Lin and C.-H. Cheng, *Ind. Eng. Chem. Res.* **36** (1997) 5096.
117. G.C. Bond, *J. Molec. Catal. A: Chem.* **118** (1997) 333.
118. R.L. Burwell Jr., *Langmuir* **2** (1986) 2.

119. G.C. Bond, F. Garin and G. Maire, *Appl. Catal.* **41** (1988) 313.
120. S. Sane, J.M. Bonnier, J.P. Damon and J. Masson, *Appl. Catal.* **9** (1984) 69.
121. G. Pajonk, M.B. Taghavi and S.J. Teichner, *Bull. Soc. Chim. France* (1975) 983.
122. L.F. Liotta, A.M. Venezia, A. Mortorana and G. Deganello, *J. Catal.* **171** (1997) 177.
123. F. Boccuzzi, A. Chiorino, M. Gargano and N. Ravasio, *J. Catal.* **165** (1997) 129, 140.
124. Y. Nitta, Y. Hiramatsu, Y. Okamoto and T. Imanaka, *Proc 10th Internat. Congr. Catal.* (L. Guczi, F. Solymosi and P. Tétényi, eds.), Akadémiai Kiadó: Budapest **C** (1992) 2333.
125. I. Jardine and F.J. McQuillin, *J. Chem. Soc. (C)* (1966) 458.
126. V. Amir-Ibrahimi and J.J. Rooney, *J. Molec. Catal.* **67** (1991) 339.
127. R.J. Grau, P.D Zgolicz, C. Gutierrez and H.A. Taher, *J. Molec. Catal. A: Chem.* **148** (2000) 203.
128. F. Stüber, M. Benaissa and H. Delmas, *Catal. Today* **24** (1995) 95.
129. B.K. Fourlong, J.W. Hightower, T.Y.L. Chan, A. Sárkány and L. Guczi, *Appl. Catal.* **117** (1994) 41.
130. D.A. Buchanan and G. Webb, *J. Chem. Soc. Faraday Trans.* **71** (1975) 134.
131. A. Sárkány, Z. Zsoldos, G. Steffler, J.W. Hightower and L. Guczi, *J. Catal.* **157** (1995) 179.
132. A. Sárkány, *Appl. Catal. A: Gen.* **149** (1997) 207.
133. S.D. Jackson, J. Willis, G.J. Kelly, G.D. McLellan, G. Webb, S. Mather, R.B. Moyes, S. Simpson, P.B. Wells and R. Whyman, *Phys. Chem. Chem. Phys.* **1** (1999) 2573.
134. C.-Q. Liu, Y. Xu, S.-J. Liao and D.-R. Yu, *Appl. Catal. A: Gen.* **172** (1998) 23.
135. G.C. Bond, *Catalysis by Metals*, Academic Press: London (1962).
136. S.D. Jackson, M.B.T. Keegan, G.D. McLellan, P.A. Meheux, R.B. Moyes, G. Webb, P.B. Wells, R. Whyman and J. Willis, in: *Preparation of Catalysts V* (G. Poncelet, P.A. Jacobs, G. Grange and B. Delmon, eds.) Elsevier: Amsterdam,(1991), p. 135.
137. J.T. Wehrli, D.J. Thomas, M.S. Wainwright, D.L. Trimm and N.W. Cant, *Proc. 10th. Internat. Congr. Catal.* (L. Guczi, F. Solymosi and P. Tétényi, eds.), Akadémiai Kiadó: Budapest **C** (1993) 2289.
138. A. Borgna, B. Moraweck, J. Massardier and A.J. Renouprez, *J. Catal.* **128** (1991) 99.
139. Sun Hee Choi and Jae Sung Lee, *J. Catal.* **93** (2001) 176.

HYDROGENATION OF ALKYNES

PREFACE

The theme of the previous chapter is continued, because once again the reactions to be considered are dominated by the much stronger adsorption of the reactant compared to the products; high selectivities in the hydrogenation of alkynes to alkenes are frequently met. Emphasis is now placed rather on the causes of this high selectivity, rather than as before on the structure of the alkenes produced; the main product is often the one formed by simple addition of two hydrogen atoms to the same side of the alkyne. Because of the importance of selective alkyne hydrogenation in petrochemical operations and in organic synthesis, reaction mechanisms have been deeply researched, and much effort has been devoted to technical improvements to catalyst performance.

9.1. INTRODUCTION

9.1.1. The Scope of the Literature

The hydrogenation of alkynes has been widely studied, and there is a correspondingly large literature. Much of this concerns ethyne, and the vast majority of the papers deal with its reaction on palladium, which is outstanding for its high activity in this reaction, and for the generally high selectivity with which ethene can be formed from it. The motivation for this work has been the necessity in industrial practice to remove small amounts of alkynes (and dienes) from alkene streams produced by steam-cracking of hydrocarbons (see later). A subsidiary factor if the suitability of palladium for semi-hydrogenation of alkynes in the fine chemicals sector. Much of the work has been conducted in flow systems to mimic industrial practice, although the use of static systems has revealed puzzling aspects of the kinetics that still await full explanation.

The emphasis in this Chapter will be on the factors that determine the *degree of selectivity* with which the intermediate alkene is formed; whereas in the previous Chapter, it lay more on the nature of the products formed, this is of lesser importance here, because under most circumstances the major product is that which arises from the addition of two hydrogen (or deuterium) atoms to the *same* side of the chemisorbed alkyne, i.e. Z-addition. Minor amounts of the products of E-addition do occur, and the mechanism by which they are formed merits discussion. However in the case of ethyne itself, this consideration is irrelevant unless deuterium is used, so the overwhelming thrust of the work on its hydrogenation has been towards understanding the mechanism by which ethene is so selectively formed. Further significant aspects of the reaction are best discussed after a brief description of the industrial problems.

9.1.2. Industrial Applications of Alkyne Hydrogenation[1–6]

The petrochemical industry is largely based on the conversion of alkenes containing mainly two to four carbon atoms into products of greater value by selective oxidation, polymerisation and other processes, for which purpose alkenes streams containing very low concentrations of alkynes and alkadienes (preferably <5 ppm) are required. Alkenes are produced by non-selective thermal processes (steam cracking) or catalytic cracking of naphtha fractions, and are separated into cuts containing predominantly molecules of a specified number of carbon atoms. However each cut contains the multiply unsaturated impurities (alkynes and dienes), so the problems of their selective removed by hydrogenation are similar. Much work has been done with ethyne, and a lesser but still significant amount with 1-butyne, where because of the lower volatility of the C_4 fraction the reaction can be conducted in the liquid phase or in solution (Section 9.3.2).[7] Two types of ethene feedstock are used, both containing 0.2 to 2% ethyne, but differing in their hydrogen and ethene contents: in the front-end cut the hydrogen content exceeds 15%, but in the tail-end cut it is only 0.5 to 3%. Typical reactant concentrations used in laboratory work designed to mimic industrial conditions are shown in Table 9.1,[4] although these have been varied widely in explorations of their effects on performance.[8–10]

The technical success of a palladium catalyst is measured by its ability to lower the alkyne or diene content almost to zero without hydrogenating any of the alkene: indeed, ideally its content ought to rise slightly. Thus the alkene selectivity S_e may take either positive or negative values, depending on whether it is being created or destroyed. Since the ethene: ethyne ratio may be as high as 345, this is a very tall order indeed for any catalyst, and in addition, with the C_4 cut, 1-butene is the main component and the desired product, so that double-bond migration must not occur. Although palladium is the best metal it is not ideal for the following reasons. (1) Its selectivity is not perfect, and under some laboratory conditions (e.g. high H_2/C_2H_2 ratio) it can fall catastrophically. (2) The reaction is accompanied by the formation of oligomers, which foul up the plant (Section 9.3.3). (3) As we have seen in earlier chapters, it is among the most active metals for alkene isomerisation. Nevertheless its high activity and intrinsic selectivity make it the metal of choice, and in industrial practice is often used as 0.04% Pd/Al_2O_3; in laboratory work this concentration has been varied (0.005 to ~10%) to examine particle size and related effects. The selection of such a low concentration is, as we shall see, dictated not only by considerations of cost. Much research has been directed to overcoming these limitations, and two lines in particular have been followed: (i) the use of a selective poison added to the feedstock (carbon monoxide is most often used,[11] Section 9.3.4); and (ii) the use of palladium-containing bimetallic catalysts (Section 9.3.5).

In the fine chemicals sector there is frequently the need to hydrogenate a molecule containing one or more $C\equiv C$ bonds with high degrees of stereo- and regioselectivity. Examples of this include the synthesis of insect sex pheromones,[11] and of vitamin A. Here also the inherent high selectivity to partially reduced products shown by palladium is employed, but as before it is beneficially reinforced by selective poisons and appropriate choice of support (Section 9.3.5).

9.1.3. The Chemisorbed State of Alkynes

There have been a number of studies of the chemisorption of ethyne, employing FTIR, RAIRS, LEED, HREELS, STM and PED.[13] The structures discovered have been reviewed in Section 4.4.3, so only a brief recapitulation is needed. These investigations reveal two basic structures, coded types A and B (respectively structure **15** and **13** in Table 4.2). In Type A, the H—C—C—H plane is parallel with the surface, and the molecule is held by two σ and two π bonds (di-σ/di-π): on fcc(111) surfaces, the carbon atoms lie above two adjacent trigonal holes, and so the molecule forms a 'long bridge' as shown in Figure 9.1. In Type B, the plane is tilted as a consequence of only a single π bond at one side (di-σ/π) (Figure 9.1); this should represent a slightly weaker mode of bonding. There is little evident chemical logic in which surfaces show which structure. The low index faces of copper give only Type A, as does the (100) face of all three Group 10 metals: Ni(110) and Pd(110) show Type B, as do the (111) faces of palladium and platinum; Ni(111) gives Type A.[13] There is clearly scope for a thorough theoretical study of the bonding of ethyne to metal surfaces. It has to be remembered however that this work is usually carried out at low temperature, and so does not necessarily tell us the forms that are reactive in hydrogenation at ambient temperature and above. A RAIRS study of propyne on Ni(111) and Cu(110) suggests structures analogous to those shown by ethyne:[14] on Ni(111) the C—C bond order in a Type A structure has been lowered to almost unity, and the structure is stable from 110 to 293 K.

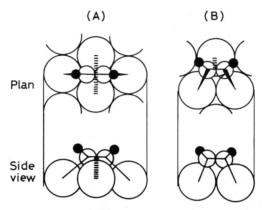

Figure 9.1. Structures of ethyne chemisorbed on fcc(111) surface: (A) long bridge; (B) short bridge.[105]

9.1.4. The Origin of Selectivity in Alkyne Hydrogenation

Work described in the literature can be divided into two broad categories:
(i) the hydrogenation of an alkyne by itself (including the use of deuterium as an
isotopic label, Section 9.2.4), and (ii) its hydrogenation in the presence of a large
excess of the product alkene, in order to simulate industrial practice (including
the use of compounds labelled with radioactive carbon). There are some intermediate cases, where smaller amounts of the alkene have been added.[15-17] Work of
the second type ventures into areas that are met in the first type only at very high
conversions, and are therefore usually ignored: important and difficult questions
are thereby raised that are irrelevant to the hydrogenation of the alkyne alone.

In this latter category, it was long ago suggested that two factors are relevant:
these are illustrated by the simplified reaction scheme shown as Scheme 9.1. Less
than complete selectivity to ethene occurs if either (i) chemisorbed ethene fails to
desorb, and remains on the surface long enough to be hydrogenated to ethane, or
(ii) ethene that has desorbed can re-adsorb in competition with the ethyne. These

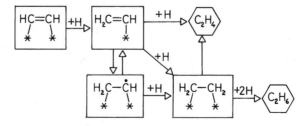

Scheme 9.1. Simplified scheme for the mechanism of ethyne hydrogenation. Note: in this and the
following Schemes no attempt is made to show all interactions of π orbitals with the surface.

two factors are readily distinguished by observing how the selectivity alters with conversion of the ethyne: if it remains constant, then the first explanation applies, whereas if it continually decreases,[18] the second explanation has to be invoked. These two possibilities were termed respectively *mechanistic* and *thermodynamic factors*.[16] Of course, as the Scheme shows, both may operate at the same time, but the thermodynamic factor will assume greater importance towards the end of the reaction, when the ethene/ethyne ratio has become large, or when additional ethene has been deliberately added.

Under industrial conditions the mechanistic selectivity is not of prime importance, because the ethyne concentration is so low; but it is disastrous if the ethene can adsorb in competition with the ethyne, and become hydrogenated to ethane. The amount of ethane arising from the two sources can easily be distinguished by labelling the ethyne with either ^{13}C or ^{14}C. What is therefore needed is a very large thermodynamic factor, i.e. for the ethyne to be much more strongly adsorbed than the ethene. If this situation is secured, not only will access of ethene to the surface be prevented by low concentrations of ethyne, but the ethyne will displace the chemisorbed ethene formed by its hydrogenation. The two selectivity factors are therefore closely linked. Thus if selectivity to ethene in the hydrogenation of ethyne alone is high and independent of conversion until almost the end of the reaction, it is likely that excess ethene will not gain access to the surface. On the other hand,[18] when the selectivity is not very high even at the outset, as sometimes happens with rhodium and iridium, it falls as conversion increases, because the ethene formed is able to compete successfully for a share of the surface. This is not always the case, however.[15]

It will also become clear that the behaviour of alkynes closely resembles that of the 1,2- and 1,3-alkadienes already discussed. Their strong chemisorption was due to the simultaneous interaction of both double bonds in the latter case, and probably to the large release of strain that accompanied the opening of one of the double-bonds (or at least the disengagement of one set of π orbitals) in the latter case. The heat of hydrogenation reflects the magnitude of the release of strain and hence the strength of the interaction with surface atoms: for ethyne to ethene it is 172 kJ mol^{-1} compared to 137 kJ mol^{-1} for ethene to ethane. These reactions are thus quite exothermic, and in industrial use care must be taken to avoid temperature excursions, as these would lead to loss of selectivity.

9.1.5. Interpretation of Results: Some Preliminary Comments

As the results are presented in more detail in the following sections, it will be natural to wonder what further development of the mechanistic framework in Scheme 9.1 will be needed to account for them. Such considerations are postponed to Section 9.3.2, but the following points should be borne in mind. The reaction system palladium-ethyne-hydrogen is in fact extremely complex,[19] and there are many complicating aspects barring the way to simple explanations. It is simplistic to try to assign a particular observation to a single cause, as it is more probable that a number of factors are simultaneously at work. In particular it is notoriously difficult to isolate a single parameter of the system (e.g. metal particle size) and to study this by itself, because other things change at the same time. Taking this variable as an example, we have seen (Section 2.5) that consequential geometric and electronic effects cannot meaningfully be separated, although this is sometimes attempted. More importantly, the tendency to form the β-hydride phase is size-sensitive,[19] and dissolved hydrogen atoms have been implicated in the reaction;[19-21] the likelihood of deactivation by carbonaceous deposits[22-25] or by formation of a carbide phase[26] may also change, and the former have been assigned important roles in determining selectivity. As particle size is lowered, spillover becomes more probable, as does a metal-support interaction. Interpretation of results is therefore a minefield through which we must walk delicately.

9.2. HYDROGENATION OF ETHYNE: 1, IN STATIC SYSTEMS[27-29]

9.2.1. Introduction

The classic work on this reaction, published between about 1945 and 1970,[16-18,30-37] was mainly conducted in constant-volume reactors, the progress of the reaction being followed by the pressure fall. Forced recirculation of the reactants through the catalyst bed was occasionally used, but more usually trust was placed on natural convection, which was usually adequate as rates were comparatively slow. Only rarely was the effect of added ethene noted.[16-17] This *modus operandi* had the advantage that the effect on product yields of a smoothly varying conversion was immediately apparent, and certain interesting and informative kinetic phenomena were observed that were inaccessible to flow systems.[30-32] Its disadvantage is that the rate and kinetic form of the reaction depends somewhat on the initial conditions such as the order in which the reactants are introduced. If the hydrocarbon is added first, the surface may become partially coated with strongly-adsorbed derived species before the hydrogen is introduced; if the reverse sequence is used, some time may elapse before surface concentrations attain their steady state. Work of this type is the concern of this section, which also covers studies made in UHV systems.

Much of the work designed to elucidate or to improve upon palladium catalysts in the industrial context has naturally been performed in continuous-flow systems, sometimes employing a spinning basket;[38] the effect of added ethene has been a recurrent theme. In such work the catalyst is usually allowed to reach its steady level of activity before results are taken, or conditions altered to ascertain their effects. Work of this kind is summarised in Section 9.3.

9.2.2. Kinetic Parameters

In the study of non-catalysed reactions, there are several methods available for determining orders of reaction, and when properly performed they lead to the same results. It is therefore disconcerting to find that this is not always the case with heterogeneously catalysed reactions, the hydrogenation of ethyne being a case in point. In a closed system, the variation of rate with extent of reaction ought to be governed by the remaining pressures of the reactants raised to the power of the order, viz.

$$r = k P_H{}^x P_C{}^y \tag{9.1}$$

the values of x and y being determined from the way in which the *initial* rate depends on the pressure of each reactant varied separately. If one reactant is in large excess, its pressure will not change much as the reaction proceeds, and so the rate is chiefly dependent upon the pressure of the other reactant. These concepts are fully explained in all textbooks of physical chemistry.

It has been observed routinely that when ethyne is admitted first to a vessel containing a supported metal of Group 10, and hydrogen is then added such that the hydrogen: ethyne ratio (P_H/P_C) is two or more, the rate remains constant until at some point it starts to accelerate (Figure 9.2), notwithstanding the initial rate method giving the orders in hydrogen and ethyne as respectively about one and zero (see Table 9.1).[16,17,27,28,30–34,39] The acceleration, which is most marked with nickel and palladium, starts when the ethene/ethyne ratio has risen to the point where ethene can compete with the ethyne for space on the surface and can itself by hydrogenated; and, because its rate of hydrogenation is much faster than that of

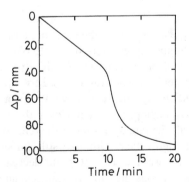

Figure 9.2. Hydrogenation of ethyne over Pd/pumice at 293 K in a constant volume system; pressure fall versus time ($P_E = 50$ Torr; $P_H = 150$ Torr).[33]

TABLE 9.1. Kinetic Parameters for the Hydrogenation of Ethyne in the Absence of Added Ethene: $r \propto P_H^x P_E^y$

Metal	Form	E/kJ mol^{-1}	x	y	T/K	$S_{2,i}$	S_0	References
Fe	/pumice	64	~1	~0	429	91	30	28,36
Co	/pumice	17	~1	~0	470	90	50	28,36
Ni	Powder	—	1	−0.5	403	—	—	28
Ni	/pumice	51	1	0	353	83	60	28,36
Ni	/pumice	43	1	−0.14	371	85	—	30
Cu	/pumice	38	~1	0.3	473	90	60	28,36
Cu	/various	88	1	−0.1	~450	99	—	171
Pd	/pumice	50	1	−0.5	322	92	25	28,36
Pd	/Al$_2$O$_3$	46	1	−0.5	273	97	—	17
Pd	/α-Al$_2$O$_3$	—	1.42	—	293	96	—	16
Pd	/SiO$_2$	71	1	0	387	97	—	39
Pd	/SiO$_2$	—	—	—	293	94	—	15
Pd	Foil	40	1.04	—	300	31	—	43
Ag	/SiO$_2$,TiO$_2$	39	—	—	353	100	0	66
Au	Al$_2$O$_3$	34	0.4	0.1	~360	100	30	67
Pt	/pumice	50	1.2	−0.7	346	82		28,36
Pt	/Al$_2$O$_3$	39	1.5	−0.7	383	90	28	34
Rh	/pumice	65	~1	~0	358	86	25	37
Rh	/Al$_2$O$_3$	38	1.4	~0	403	90	—	18
Rh	/SiO$_2$	—	—	—	293	74	—	15
Ir	/Al$_2$O$_3$	—	~1	−0.3	403	55	15	34
Ir	/SiO$_2$	—	—	—	293	16		15
Ru	/Al$_2$O$_3$	44	1	0	385	90	8	41
Os	/Al$_2$O$_3$	33	1	0	398	65	16	41

E is the activation energy for total reaction: values for the formation of C$_2$ products and of oligomers may differ slightly. T/K is the temperature at which the orders and initial C$_2$ selectivities ($S_{2,i}$) were measured : S_0 is a rounded value for the oligomer selectivity, which *in most cases* was obtained at about T/K. Further details of earlier work (<1960) are to be found in references 27 and 39. See also 16 and 85.

ethyne, acceleration occurs.[16,40] The acceleration point therefore depends on the selectivity to ethene in the early part of the reaction, and for this reason it occurs sooner as either P_H or temperature is raised.[17,27,34] With platinum, however, the acceleration is more gradual, and occurs at least initially without loss of selectivity; it is due to the negative order in ethyne (Table 9.1). Zero-order pressure-time curves have been found in butadiene hydrogenation over the Group 10 metals,[31] and with ethyne hydrogenation over copper,[32] but not over the metals of Groups 8 and 9:[18,32,41] with these metals the rate decreases as the reaction proceeds, in line with the initial rate law, although acceleration can still sometimes be seen.[41] When the initial P_H/P_C ratio is less than two, the rate decreases continuously, being determined by the instantaneous pressures of the reactants;[40] no acceleration is observed if there is insufficient hydrogen left to hydrogenate the ethene. Careful and detailed studies on the base metals of Groups 8 to 10,[32] especially nickel,[30,31]

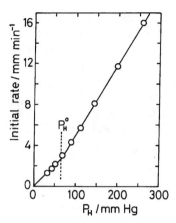

Figure 9.3. Hydrogenation of ethyne over Ni/pumice at 486 K in a constant volume system: dependence of initial rate on P_H ($P_E = 30.5$ Torr).[30]

have revealed sudden slight changes of rate during the zero-order phase when hydrogen was admitted first.

The insensitivity of the rate to the varying hydrogen pressure after prior admission of ethyne was (and is) most puzzling, but there are further surprises to come. Although rates were proportional to hydrogen pressure when P_H/P_C ratio was less than two, in the case of the base metals at higher pressures the zero-order rate constant assumed a greater value (Figure 9.3), so the full rate expression becomes

$$r = k_A P_H + k_B(P_H - P_H^0) \qquad (9.2)$$

where P_H^0 is about twice P_C.[30,31] Variation of temperature showed that the corresponding activation energies E_A and E_B on Ni/pumice were respectively 43 and 100 kJ mol^{-1}. Even more surprising was the observation that the sudden injection of more hydrogen or ethyne during the zero-order phase caused the rate to change in accordance with the initial rate law for the latter and equation 9.2 for the former.[30] These strange findings have received no further attention since they were first disclosed. They were discussed in some detail at the time,[16,30-32] but no firm conclusion was reached: it seems most likely that when a sufficiently high pressure of hydrogen is present a certain concentration of reactive centres is set up, and this is sustained by some kind of chain reaction while reactant pressures decrease slowly. Fragmentary results suggest that the same effect may be seen with other metals,[42] but no systematic studies have been reported. A further comment will be offered later when the mechanism is re-considered, but the mystery remains.

The principal kinetic features shown by the metals of Groups 8 to 10 and copper have marked similarities, which can be summarised as follows: (1) Orders in hydrogen are usually close to first, although with platinum[34] and rhodium,[18] and palladium (especially at higher temperatures),[16,17,43] they are significantly greater, approaching 1.5 (see Table 9.1). (2) Orders in ethyne are most often close to zero, but especially with palladium and platinum they can be substantially negative (see also Table 9.1), suggesting some degree of competition between the reactants for the surface. (3) Activation energies are mainly between 30 and 50 kJ mol^{-1}, with only few exceptions (Table 9.1): this is the range in which *true* values can be expected, so the composition of the reactive layer cannot be very temperature-dependent. (4) Ethene selectivities vary considerably, but are high (>80%) in most cases, especially with palladium (except in the form of foil[43]), but are markedly lower with osmium and iridium[15,44,45] (Table 9.1): they are however somewhat lower than those shown by butadiene[27,46] (Table 8.6). (5) They decrease with increasing hydrogen pressure (Figure 9.4) and decreasing temperature[27] (Figure 9.5), as expected intuitively, since higher concentrations of hydrogen in the reactive layer must favour formation of ethane. (6) All metals catalyse the formation of *oligomers*; their yields vary very much with reaction conditions; the activation energy for their formation is higher than that for C_2 products, and the process is inhibited by increasing hydrogen pressure. Over the base metals of Groups 8 to

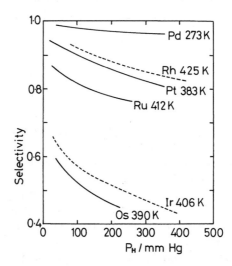

Figure 9.4. Hydrogenation of ethyne over Al_2O_3-supported metals: effect of P_H on selectivity ($P_E = 50$ Torr).[27]

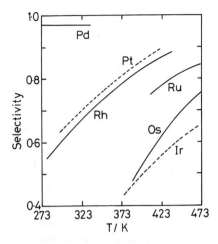

Figure 9.5. Hydrogenation of ethyne over Al_2O_3-supported metals: effect of temperature on selectivity ($P_E = 50$ Torr; $P_H = 200$ Torr).[27]

10 (and copper) it is frequently the major process, but on the noble metals the oligomer yield is 10–30% (Table 9.1). The composition of the oligomers will be considered later (Section 9.3.3).

There has been much discussion in the literature as to whether there is only one type of active site on palladium catalysts, on which ethyne is very much more strongly adsorbed than ethene (a factor of 2200 has been derived from one study), or whether on the other hand there is also another type which is specific to ethene chemisorption and hydrogenation, which is poisoned by carbon monoxide, and which may possibly develop as the surface becomes partially coated with carbonaceous residues. This latter possibility is envisaged from work in which a large excess of ethene has been used in order to simulate industrial conditions, and which frequently employed labelled ethene or a double-labelling procedure (e.g. $^{14}C_2H_4 + C_2H_2 + D_2$). This work will be discussed in the next Section; for the moment we may simply note that in static systems in the absence of added ethene the selectivity to ethene remains remarkably constant on palladium (and most other metals) until the ethyne has almost disappeared, so that under these conditions the second type of site either does not exist or is in very low concentration: it may of course arise or develop under the simulated industrial conditions. However the fact that in static systems the ethyne pressure at which acceleration begins is *increased* by raising the pressure of hydrogen or of added ethene, or by temperature,[17] shows that competitive adsorption occurs, as is indeed shown by the negative orders in ethyne that are sometimes seen[42] (Table 9.1). When the initial P_H/P_C ratio reaches 10, the reaction is non-selective throughout its course,[17] an observation that has also occasioned discussion.[17,47,48]

Most metal catalysts lose activity as they are used for ethyne hydro-genation (platinum may be an exception,[16] and palladium sometimes); this is caused by the formation of strongly-adsorbed derived species, which could in-clude ethylidyne, and which are usually called simply *carbonaceous residues* or *deposits*.[22−24,26,42,49−51] They are not formed on supported metals from ethyne itself,[50] but only during hydrogenation. It has also been shown by simultaneous X-ray diffraction that a carbide phase is also created in palladium,[27,52] and by the same method the relevance of the dissolved hydrogen content has been established.[53−55] The carbide phase $PdC_{0.13}$ is not however capable of forming the hydride phase.[56] The unique ability of palladium in selective hydrogenations has often been ascribed to its marked propensity to dissolve hydrogen;[42,57,58] hydrogen dissolved in nickel is also reactive towards ethyne. It has been firmly established that (1) prolonged exposure to hydrogen poisons its activity for *ethene* hydrogena-tion, thus improving its selectivity,[16] and (2) the α-phase (low H content) is more selective than the hydrogen-rich β-phase[19] (see however reference 59). Thus the better selectivity shown by *small* particles[26,60] (see also Section 9.3.1), the effect of including an inactive metal (Section 9.3.5), and the effects of reaction conditions, may all find their cause in the level of the dissolved hydrogen,[55,61,62] which if high may create a high concentration of *surface* hydrogen atoms (or atoms in sites readily accessible to the surface), thus causing non-selective behaviour. We may recall that the low selectivity shown by iridium in butadiene hydrogenation (Table 8.6) was attributed to *occluded* hydrogen,[44] which participated in the reaction: similar low selectivity is seen in ethyne hydrogenation[26,44,45,63] (Table 9.1), and may have the same cause. Although the phenomenon is not the same as that shown by palladium, the effect on the availability of hydrogen at the surface is the same, and that without doubt determines alkene selectivity.

A kind of transmutation of elements has been achieved by electrochemically-pumping sodium[64] or potassium[65] into a thin platinum film from the electro-active support $A\beta''$-alumina (A = Na or K). Selectivity for ethene rose from a very modest <20% to 90% in the case of potassium; although the rates of hydrogenation of both ethyne and ethene individually were suppressed by increasing alkali metal concentration on the platinum, the latter was the more affected, so selectivity increased. The effects were reversible. By reference to work on Pt(111), it was though that the alkali metal tended to convert strongly-held di-σ ethene to the weaker π form, while at the same time increasing the strength of the hydrogen chemisorption.[65]

Ag/SiO_2 and Ag/TiO_2 after activation by oxidation and reduction were active for ethyne hydrogenation at 353–443 K[66] (Table 9.1); both showed 100% selectiv-ity to ethene and no oligomer formation at the lower temperatures, but selectivity fell and more oligomers were made as temperature increased. Rates were slower than for butadiene; the activation energy on Ag/SiO_2 was 39 kJ mol^{-1}.

Au/Al$_2$O$_3$ was active between 313 and 523 K, and also formed ethene with 100% selectivity;[67] ethyne reacted 2000 times faster than ethene, and the rate was maximal for gold particles of 3 nm size (Table 9.1).

9.2.3. The Formation of Benzene from Ethyne

We have noted that the formation of oligomers usually occurs in parallel with the hydrogenation of ethyne; we shall return to this later, but under certain conditions the main or only product is benzene. This reaction has been intensively studied using single crystal surfaces of palladium,[68,69] chiefly Pd(111),[70-73] although benzene has been detected as a product of the interaction of ethyne with the surfaces of palladium foil[74] and film, Pd/Al$_2$O$_3$ and Pd/C, as well as on nickel and copper single crystal surfaces, and Ni/SiO$_2$.[69] Since much of this work was carried out in static UHV reactors, it is convenient to mention it briefly at this point.

The reaction is clearly structure-sensitive because it occurs most extensively on Pd(111).[68,69] This would have gladdened the heart of A.A. Balandin had he lived to learn of it, but although the hexagonal structure of this face is clearly relevant, the reaction proceeds through an adsorbed C$_4$ species[73] which is tilted with respect to the surface.[70] On Pd(100) the process is inhibited by adsorbed hydrogen,[43] although on foil it is accelerated. Electron donors assist it and electron acceptors inhibit it.[75] Trimerisation of propyne only occurs on the Pd(111) surface.[68] It is suggested that the surface is first covered with a layer of ethylidyne, on which the reaction actually proceeds,[74] although the C$_4$ species may be formed by reaction of ethyne with ethylidene.[43,70] Further insights into the mechanism of trimerisation are obtained by the study of bimetallic systems in which the characteristic properties of palladium are modified by tin[76] (PdSn/SiO$_2$) or gold[77] (Au/Pd(111) and Pd/Au(111)) or by deposition on a single crystal of another metal having a different crystal habit (Ru).[78] With PdSn/SiO$_2$, benzene selectivity was highest at low P$_H$/P$_C$ ratios, n-hexane becoming the major product when this ratio exceeded two; no ethane was apparently observed.[76] Deposition of gold onto Pd(111) greatly enhanced benzene formation, Pd$_6$Au and Pd$_7$ ensembles being responsible; depositing palladium onto Au(111) to give the ($\sqrt{3} \times \sqrt{3}$)R30° structure (equivalent to Pd$_2$Au) was far more effective than unmodified Pd(111).[77] Reactive desorption from Pd/Ru(001) revealed three states of the adsorbed precursor, corresponding to different modes of bonding.[78]

9.2.4. The Reaction of Ethyne with Deuterium[27,79]

As was the case with ethene and other alkenes, replacing hydrogen by deuterium and analysis of reaction products reveals hitherto hidden aspects of reaction mechanism. Early work with platinum[40] and nickel[2,80] indicated that neither

TABLE 9.2. Initial Distributions of Deuteroethenes from the Reaction of Ethyne with Deuterium ($P_D/P_C = 2$)

Metal	Form	T/K	$-d_0$	$-d_1$	$-d_2$	$-d_3$	$-d_4$	M	References
Ni[a]	/Pumice	368	2	20	67	10	1	1.88	81
Pd	/Al$_2$O$_3$	288	2	22	65	9	2	1.87	35
Pt	/Al$_2$O$_3$	362	1	19	66	11	3	1.96	35
Rh	/Al$_2$O$_3$	408	1	12	36	28	23	2.60	35
	/Al$_2$O$_3$	393	2	13	38	27	20	2.50	35
Ru	/Al$_2$O$_3$	443	6	7	30	36	21	2.59	41
Os	/Al$_2$O$_3$	458	3	13	33	35	16	2.48	41

[a] The reaction on nickel was performed with $C_2D_2 + H_2$ (for convenience).

exchanged ethyne nor hydrogen deuteride was returned to the gas phase in significant amounts, and this has been confirmed at least for the metals of Groups 10 in later extensive studies. Nevertheless all possible deuterated ethenes were found[35,41,81] (Table 9.2), so that redistribution reactions must have occurred within the adsorbed layer. Analysis of the reaction scheme (Scheme 9.2) proceeds as for ethene, but is simpler because no provision of ethyne desorption is needed: the important parameter is that for vinyl reversal ($1-p$), as this determines the breadth of the ethene distribution. The parameters s and q fix the chance of adding a deuterium atom to respectively an ethyne or a vinyl (or ethenyl) radical, their values and that of p being taken as independent of deuterium content. C—H and C—D bonds are assumed to be equally strong, so that probabilities of loss of these atoms is determined by simple statistics. Solution of the set of simultaneous equations for formation of the ethenes then allows optimum values of the parameters to be found.[35,81] The ranges within which the observed values fall are shown in Table 9.3; although reaction conditions affect the precise values (see below), these ranges are sufficiently distinct to merit discussion. In most cases the values of s and q may be taken as equal, so this permits a further simplification.

Inspection of Table 9.3 shows that the values of s, q are all between 60 and 90%, suggesting (as noted above) that variation in the vinyl reversal term p is chiefly responsible for differences in the ethene distributions. This is indeed so; its values are high in Groups 8 and 9, accounting for the high deuterium numbers

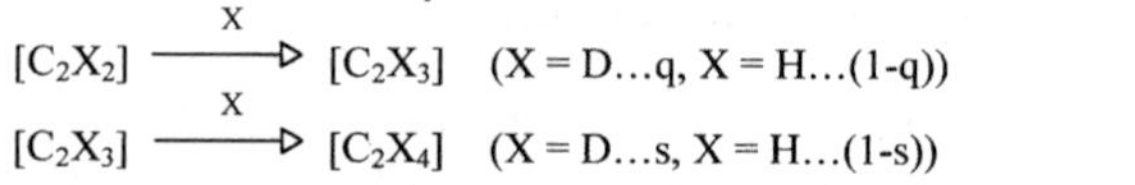

$$C_2X_2 \longrightarrow [C_2X_2] \underset{1-p}{\overset{}{\rightleftarrows}} [C_2X_3] \overset{p}{\longrightarrow} [C_2X_4] \longrightarrow C_2X_4$$

$$[C_2X_2] \overset{x}{\longrightarrow} [C_2X_3] \quad (X = D...q, X = H...(1-q))$$

$$[C_2X_3] \overset{x}{\longrightarrow} [C_2X_4] \quad (X = D...s, X = H...(1-s))$$

Scheme 9.2. Analysis of deutero-ethenes formed from ethyne and deuterium.[35,81] Species in square brackets are in the adsorbed state.

TABLE 9.3. Reaction of Ethyne with Deuterium: Range of Values of s and $(1-p)$

Parameter	Ni	Pd	Pt	Ru	Ir	Ru	Os
s	80	79–86	85–92	75–88	78–88	70–90[a]	60–90[a]
$(1-p)$	~40	20–38	21–37	79–89	60–81	85–96	85–96

[a] $s \geq q$ (for convenience).

M in the ethenes (Table 9.2), but much lower in Group 10, where M is much closer to two. These trends are similar to those of ethyl reversal found in the ethene-deuterium reaction (Section 7.2.4). The procedure allows the estimation of the composition of the alkynes (Table 9.4); the effect of the different extents of vinyl reversal is evident. This is also reflected in the extents to which hydrogen deuteride is returned to the gas phase, since when M exceeds two it must take place, there being little alkyne exchange. In Group 10 it is zero or very slight; in Group 9, moderate; and in Group 8, very marked.[7] To some extent this may reflect the temperature range in which convenient activity is found; this varies in the sequence $10 > 9 > 8$.

Similar trends in product distributions were found with each metal; increasing temperature led to less ethene-d_2 and a broader distribution, and this was also the case when the ratio P_H/P_C was decreased.[82] These two effects are linked, because the parameter p is lowered in each case, this being due to a decrease in the concentration of adsorbed hydrogen atoms and/or vinyl radicals.

Ethene-d_2 exists in three forms, namely, the E, the Z and the asymmetric (a); these are distinguishable by infrared spectroscopy.[80] The results obtained show that the reaction is more complex than simple Z-addition, as significant amounts of the E-isomer were always seen, and especially with the metals of Groups 8 and 9 the a-isomer was also clearly apparent (Figure 9.6). Evidently the process of vinyl reversal allows the formation of the E- and a-isomers; this is made apparent by the systematic decreases in their yields as the total amount of ethene-d_2 increases, this being achieved by an increase in p (Figure 9.6). The points for the noble metals of Groups 8 to 10 lie about the same curves, while those for nickel lie slightly above them. The suggested mechanism by which the E- and a-isomers are formed is shown in Scheme 9.3; this would however predict equal amounts of the two, which is clearly not the case, if as is assumed the atoms are always added

TABLE 9.4. Reaction of Ethyne with Deuterium: Calculated Surface Concentrations (%) of Ethynes

Metal	$-d_0$	$-d_1$	$-d_2$
Pd, Pt	90	9	1
Rh, Ir	70	25	5
Ru, Os	40	40	20

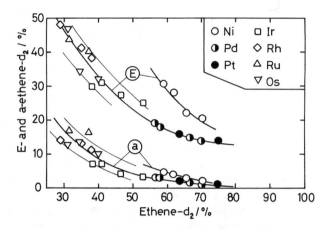

Figure 9.6. Yields of E-$C_2H_2D_2$ and of a-$C_2H_2D_2$ as a function of total yield of $C_2H_2D_2$ on various Al_2O_3-supported metals.[27]

from below. The difficulty may be resolved if there is another way in which the E-isomer can be formed at the expense of the Z-isomer; for example, the vinyl radical may isomerise to a free-radical or a tri-σ form (Scheme 9.4) in which there is a planar carbon atom, to which a hydrogen or deuterium atom can add from either side, giving equal amounts of the two isomers. This free radical is also thought to initiate the process of oligomerisation by attacking other ethyne or vinyl species, or simply dimerising. Analysis of the butadiene formed by reaction of deuterium on Pd/Al_2O_3 showed the d_2 molecule to be the major one (61%), the composition

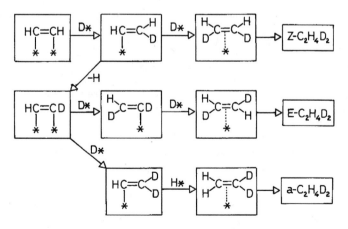

Scheme 9.3. Routes to the formation of Z-, E- and a-ethene-d_2.

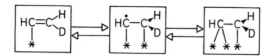

Scheme 9.4. Isomerisation of the ethenyl radical into the free-radical form.

of the hydrogen + deuterium atom pool from which it arose being defined by a Burwell N parameter of 0.85, similar to that for the ethene.[48] Butadiene therefore was made mainly by recombination of two C_2HD^* species. Similar results were obtained with Pd/TiO$_2$, but the N parameter was smaller ($\sim$0.7). A C_4 free-radical intermediate has been detected on palladium by a radical trapping technique,[83] although it may not have been an adsorbed vinylic species. Scheme 9.3 suggests that if p were to become unity and ethene-d_2 the only product, some 10–12% would be in the E-form, so that 20–34% of the vinyls would have passed through the free-radical form; this figure would be larger if some had entered the process of oligomerisation. Calculated relative concentrations of the ethynes on various metals are given in Table 9.4.

Ethyne-d_2 was found to react with ethyne-d_0 over nickel/pumice above 333 K, forming ethyne-d_1 and nothing else;[84] at 392 K equilibrium was reached in 2h, the equilibrium constant (3.4) being close to that calculated by statistical mechanics. The reaction followed a first-order course, the order in total pressure being 0.65 and the activation energy of 45 kJ mol^{-1}. Reproducible rates were found with freshly reduced catalyst, but performing a hydrogenation experiment led to some subsequent loss of exchange activity. Ethyne-d_2 exchanged only with the acidic hydrogen of propyne, with an order in total pressure of 0.47 and an activation energy of 53 kJ mol^{-1}. It seemed unlikely that the ethynes dissociated on chemisorption, and an intermolecular mechanism was suggested. This interesting reaction also occurred at and below room temperature on Ni/kieselguhr,[80] but it has received no further attention since its discovery in 1951; it did however take place on Pd(111) at 298 K.[68]

9.3. HYDROGENATION OF ETHYNE: 2, IN DYNAMIC SYSTEMS WITH ADDED ETHENE[85]

9.3.1. Kinetics and Selectivity

It is no easy task to summarise the extensive work that has been undertaken to understand how ethyne can be selectively removed by hydrogenation in the presence of a concentration of ethene that can by up to at least 400 times greater. Extensive studies have been performed in the laboratories of Guczi (Hungary) and Weiss (U.S.A.),[42,57,82,86,87] of Borodziński (Poland),[8,9,49,52,88] of

Duca (Italy)[4,23,24,89,90] and of Gigola (Argentina):[91,92] many others have made shorter but important contributions.[2,10,15,22,38,93–99] For the most part, standard analytical methods have been used, but the single- and double-labelling techniques (using respectively either ^{13}C-[94] or ^{14}C-labelled reactants,[96] or one of these plus deuterium[82,94]) have provided vital information. Reactant concentrations have been chosen to imitate either front-end or tail-end conditions,[4] although they have also been varied over wide ranges. Palladium has been employed as black,[82] but chiefly supported, on alumina, silica[8] and pumice.[4] Industrial catalysts usually have α-Al_2O_3 as support, but pumice-supported catalysts show very good stability because their macroporous structure is not easily blocked by oligomers.[4] The particle size of the palladium is by no means always measured, but it plays a determining role. The following paragraphs attempt to draw together results that bear in particular on (i) the effect of particle size on rates and selectivity, (ii) kinetic and isotopic-labelling results, leading to ideas on site multiplicity and identity, and (iii) the use of mathematical models to describe the system.

The form of the reactor has a bearing on the significance of the results obtained, especially in the measurement of selectivity at high ethyne conversion. The use of a *gradientless reactor*[52] is strongly recommended, where the catalyst is in the form of a very thin bed, so that there is no concentration gradient through the bed.

Table 9.5 provides information on the effects of varying metal dispersion. Although the temperature in every case was close to 300 K, the supports differ,[42] and only in two cases are the reactant concentrations about the same. It is therefore perhaps not surprising that the results are not in agreement: in three of the cases, the *highest* turnover frequencies were found at the *lowest* dispersions, although their values differ very considerably. This point is not discussed in the literature, and there is no obvious explanation. The selectivity expressed as the *absence* of ethane formation either improved as dispersion increased,[57,60,91] or was independent of it.[4,8,89] This small exercise illustrates vividly the problem so often encountered in heterogeneous catalysis, namely that of perceiving some general framework on which analysis of mechanisms and mathematical modelling can be performed. Each group has perforce to construct a model to fit its own results, and it is rare for any one group to pause to wonder why its findings do not agree with those of others. This is not a good way of achieving progress in science.

We may turn now to consider work in which the composition of the reaction mixture has been changed in order to identify the source of the non-selective product ethane. In the light of the foregoing paragraph, it is worth pointing out that most of these studies have used only one catalyst: fortunately, many have used industrially manufactured catalysts containing 0.04% Pd/α-Al_2O_3. The ICI 38-3 has been reported as having a dispersion of only 18%,[91] which seems low for such a low loading; however, the surface area of the support is also small. The dispersion of the Polish C-31-1A was 28%.[9] However we must first note work performed in static systems using catalysts of quite high metal loading (5%) and relatively low

TABLE 9.5. Particle-Size Sensitivity of Ethyne Hydrogenation in the Presence of Excess Ethene, and Estimations of Turnover-Frequencies, on Palladium Catalysts

Support	[Pd]/%	Variant[a]	D/%	[C₂H₂]	[H₂]	[C₂H₄]	P_{tot}/kPa	T/K	TOF/s⁻¹	S[b]	References
SiO₂	2.6	T_{calc}	29	0.3	0.5	39	101	353	0.11	⊄D	8
	3.1		5						0.70		
Pumice	0.05	T_{red}	65	0.3	0.55	97	101	300	2×10^{-3}	⊄D	4,89
	0.05		12				$''$	$''$	34×10^{-3}		
α-Al₂O₃	0.06	T_{red}	58	0.84	16	72	1520	288	0.28	↑	91
	0.04		5				$''$	$''$	7.3		
Al₂O₃	0.07	—	60	0.15	0.35	99.5	101	300	26×10^{-3}	↑	57
	8.7		6				$''$	$''$	8×10^{-3}		

[a] Procedure used to alter the dispersion: T_{calc} = calcination temperature; T_{red} = temperature of reduction or pretreatment in hydrogen.
[b] Selectivity to ethene or decrease in ethane formation: ⊄D = independent of dispersion; ↑, increases as dispersion increases.
Inlet concentrations of reactants are in %; balance to 100% made up by inert gas (He, Ar, CH₄).

TABLE 9.6. Classification of Active Centres for Ethyne Hydrogenation in the Presence of Excess Ethene, and Their Functions

Type		Notes	References
A_1	C_2H_2 and C_2H_4 adsorbed competitively	$[A_1 + A_2] > [E]$	8,9,49,88
A_2	C_2H_2 and C_2H_4 adsorbed non-competitively		
E	C_2H_2, C_2H_4 and H_2 adsorbed competitively		
I	$C_2H_2 \rightarrow C_2H_4$		15,51
II	$C_2H_2 \rightarrow C_2H_6$		
III	$C_2H_4 \rightarrow C_2H_6$		
X	C_2H_2 and C_2H_4 adsorbed competitively	$[X] > [Y]$	38,94
Y	$C_2H_4 \rightarrow C_2H_6$		
α-PdH	$C_2H_2 \rightarrow C_2H_4 \uparrow \rightarrow C_2H_6$	α-PdH on *large* particles	19,29,104
β-PdH	$C_2H_2 \rightarrow C_2H_6 : C_2H_4 \nrightarrow$	β-PdH on *small* particles	

ethene/ethyne ratios. Under these conditions with Pd/Al_2O_3[17] (and $PdAg/Al_2O_3$[16]) progressive increase in the amount of ethene caused the acceleration in rate to start even earlier, suggesting that it competed with the ethyne for the available surface. Use of $^{14}CCH_4$ showed[15] that most of the ethane (85%) came from the ethyne, but the amount of labelled ethane increased with the $^{14}CCH_4$ pressure, again indicating a contribution from a competitive mechanism. Thermodynamic selectivity was even greater with rhodium (92%) and iridium (97%). Surfaces were however substantially covered by carbonaceous overlayers, and the same amount of $^{14}CCH_4$ was adsorbed in the presence and absence of ethyne. These observations pointed to the existence of three types of site[15] (see Table 9.6), but independent evidence for them is lacking.

Experiments with $^{13}CCH_4$ and an ethene/ethyne ratio of 49 carried out with an ICI catalyst showed[38,94] that ethene adsorbed competitively with the ethyne, but the results required two types of site (see Table 9.6) (or two modes of chemisorption of the ethyne) to explain them. Their properties did not however match any of those proposed by Webb. Type **X**, in the majority, adsorbed both hydrocarbons, but ethene was favoured by a factor of 2200; Type **Y** adsorbed ethene only, perhaps because of its high concentration. The main source of the ethane was confirmed as ethene, since in the reaction with deuterium the main product was ethane-d_2. When the pressure of ethyne was varied in the presence of excess ethene,[9,88] its rate of removal (and that of formation of dimers) passed through a maximum, while that of ethane formation fell to zero at an ethyne pressure of 2 kPa (see Figure 9.7). The ethane rate was almost independent of the ethene pressure. Extensive work by Borodziński and colleagues led[8,9,49] to detailed proposals for the identity of two types of site, designated **A** and **E**, that were thought to be created as the carbonaceous overlayer developed, and a third type (**E**$_s$) that may play a role on certain supports.[100] Type **A** sites, in the majority, were small, so that only ethyne and hydrogen could adsorb on them, the former perhaps as vinylidene ($>C=CH_2$),

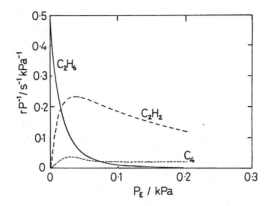

Figure 9.7. Rates of ethyne removal and of ethane and C_4 products formation as a function of P_E over Pd/Al$_2$O$_3$ at 343 K in the presence of excess ethene (outlet pressures, ethene = 40 kPa, hydrogen = 0.6 kPa).[88]

while Type **E** was larger and could adsorb both ethyne and ethene (competitively), as well as hydrogen (Table 9.6). The model has also been found to describe results obtained at 373–498 K with higher reactant concentrations but lower TOFs caused by extensive deactivation.[101] Double-labelling work with ^{14}CCH$_2$ or ^{14}CCH$_4$ and deuterium confirmed that ethene was not hydrogenated above a critical pressure of ethyne.[87] Relative rates of hydrogenation of the two hydrocarbons were however structure-sensitive, ethane formation from ethene being much more marked on *large* palladium particles.[91]

Marked oscillations have been observed[102] in the production of hydrogen deuteride during the reaction of ethyne with a mixture of hydrogen + deuterium on Pd/Al$_2$O$_3$ in the presence of carbon monoxide at 363 K, the amplitude being greatest when its concentration was about 100 ppm; this signifies periodic change to surface coverage by the reactants. No other reaction parameters were however recorded.

9.3.2. Mechanisms and Modelling[85]

We now approach the unenviable task of divining what common elements are to be found in the diverse studies surveyed above; what significant features of the materials and procedures used affect the results obtained; what extensions to the simple scheme (Scheme 9.1) are necessary or desirable; and what reaction models have been used to give quantitative description of the results.

It must be said at once that no single mechanism or model can be sufficiently flexible to apply to all systems. Certain themes do however emerge regularly; these are (i) the probable occurrence of vinylidene ($>$C=CH$_2$) and ethylidyne

($\equiv$C—CH$_3$) as alternatives to symmetrically-bonded species, either in the predominant route to ethene[43] and ethane,[88] or just in conditions of low selectivity;[17,48] (ii) the probable operation of two or three separate types of site during ethyne hydrogenation with excess ethene (Table 9.6); (iii) the likely importance of carbonaceous deposits in determining selectivity or in creating sites at which selective reaction can occur;[8,23,49,51,88,90,101] and (iv) face sensitivity.[103] Other imponderables already noted include the possible formation of carbide and hydride phases in palladium. To add to the misery, we have seen that even the sense of the particle size effect on TOF and selectivity cannot be agreed (Table 9.6), and supports appear to exert an important but poorly understood influence.

As to (i) there appears to be little direct evidence, although the balance of argument favours some role for them;[15,43,48,51] as to (ii), although they are sometimes[2,49] (but not always[4]) employed in mechanistic schemes with apparent success, it is disappointing to see the lack of agreement on their functions (Table 9.6), so it must be concluded that this question is still *sub judice*. It would seem to be a case of *Quot homines, tot sententiae*. In two cases at least it was recognised that different sites and different modes of chemisorption constitute equivalent descriptors of events.[8,49,94] A further possibility, first mentioned many years ago[104] but never fully evaluated, is that there may exist a range of particle sizes, such that some features of the reaction may occur on small particles, and others on large particles. Thus the small-size fraction, being in the α-hydride phase, might catalyse the sequential process of ethyne to ethene to ethane, but with ethene being the major product, while the larger-size fraction in the β-hydride phase might chemisorb ethyne but not ethene, and convert it wholly to ethane (Table 9.6). Unfortunately the very low metal contents of the industrial catalysts makes detailed characterisation (e.g. by TEM) difficult.

Concerning (iii) we are on firmer ground, because there is direct evidence[15,51] for the existence of 'carbonaceous deposits', which are probably adsorbed oligomers, although in the region of ambient temperature they may simply be a dehydrogenated or re-structured form of ethyne (e.g. HC$\equiv$C— or $\equiv$C—CH$_3$). At higher temperatures carbon 'whiskers' are seen growing away from the surface, and on Ni/NiAl$_2$O$_4$ they appear to take the form of nanotubes.[105] In static reactors it is possible that they form quickly when ethyne is admitted first, and that they then affect the rate and the form of the pressure-time curve; but whether they or the hydrogen atoms they contain actually assist the continuous reaction remains uncertain. With respect to (iv), evidence for face sensitivity lies in the claim that palladium particles exposing only (100) and (110) facets are very selective for producing ethene and are not prone to 'carbon' deposition.[103]

Notwithstanding these difficulties, a number of detailed mechanistic schemes have been proposed and evaluated quantitively:[2,4,10,49,98,101] they contain up to eleven unit steps, and provide varying degrees of harmony with experiment. Alternatively, rate expressions based purely on Langmuir-Hinshelwood formalism have been examined.[99] A time-dependant Monte Carlo algorithm has also been used[23,43,90] to study the effects of carbonaceous residues for front- and tail-end mixtures; the involvement of steric hindrance by surface species was said to be essential for successful simulation.

9.3.3. Oligomerisation

A besetting problem with the industrial process to remove traces of alkynes alkadienes from alkene streams using palladium catalysts has been the formation of higher hydrocarbons by oligomerisation.[5] Although in this respect palladium is better than base metals such as nickel[28] (which presumably explains why this cheaper metal is not used), and while the fraction of ethyne that reacts in this manner is small, nevertheless in a continuous operation these higher products accumulate, and cause problems. The carbonaceous deposits, so often mentioned, may be partly C_2 species such as ethylidyne, but they also comprise adsorbed forms of oligomers: in the steady state their formation is followed by release into the fluid phase.

The literature mentions numerous analyses of their composition. On palladium, C_4 products predominate; they contain mainly butenes (chiefly 1-butene) and butadiene:[39] but C_6 products including benzene are also found (Section 9.2.3), but molecules covering as wide range of molar masses have also been found.[2,38] On Ni/pumice the range extends to greater than C_{31}; even-number carbon atom product are favoured.[36,39] It is not however clear what conditions allow the apparently exclusive formation of benzene,[9,69,71,73,75] as discussed in Section 9.2.3. The yield decreases with increasing chain length, rather in the style of Fischer-Tropsch synthesis, suggesting a radical-chain mechanism, initiated by the free-radical form of the vinyl radical (see Scheme 9.1). The formation of oligomers and their strongly bonded precursors only occurs during hydrogenation and not when catalysts are exposed to ethyne alone; and only ethyne, and species derived from it, participate in making them. Thus hydrogenation of ethyne in the presence of excess propene gives C_4 but not C_5 products.[38] Temperature is not an important variable, but oligomerisation is usually made slower by raising the hydrogen/ethyne ratio.[2] As we have seen, the partial blocking of the surface by strongly held oligomers is held to be necessary in order to lower the mean size of the active centres, and thus to prevent further hydrogenation to ethane.

9.3.4. Gaseous Promoters

It has long been recognised that the continuous injection of low levels of carbon monoxide into flow reactors large or small has a beneficial effect on selectivity in the sense of inhibiting ethane formation when ethene is present in large excess.[2,106] Levels of about 1500 ppm are sufficient to stop ethane being

produced.[94] It appears that it is chemisorbed on palladium with a strength intermediate between those of ethyne and ethene, preventing the latter but not noticeably
interfering with the former;[11] indeed ethyne can displace part of the chemisorbed
carbon monoxide. It is likely to occupy sites too small to accept the hydrocarbons, and its main role may be to discourage hydrogen chemisorption;[15] a lower
concentration of hydrogen on (or in) the metal should also improve mechanistic
selectivity. Carbon monoxide is a notorious poison for the diffusion of hydrogen
through palladium devices intended to purify it. Continuous feeding is necessary
because it appears as a component of the liquid oligomers, named *green oil*, the
production of which it inhibits.[94] Sulfur and phosphorus compounds are also effective both in improving selectivity and suppressing oligomerisation.[107] Surface
sulfur also improves the selectivity shown by Pt/kieselguhr catalyst.[38]

9.4. USE OF BIMETALLIC CATALYSTS FOR ETHYNE HYDROGENATION[85,108–110]

The extensive work that has been performed on ethyne hydrogenation using
bimetallic catalysts based on palladium has had as its objective (i) the further improvement of the selectivity with which it can be hydrogenated in the presence of
excess ethene, and (ii) lowering the production of 'green oil'. Work of this type is
not novel; Sheridan briefly examined NiAg/pumice,[35] and some relevant patents
were granted more than 50 years ago.[28] It was natural to believe that diluting
palladium with an inert metal might be beneficial. As we have seen (Section 5.5)
there have been two schools of thought as to how such benefits might arise: (1) by
creating isolated atoms (or small groups of atoms) of the active component, processes requiring large ensembles might be minimised, and (2) by some alteration
in the electron concentration[111] or the orbital energies of the active elements, a
favourable moderation of adsorption energies might arise. Since many bimetallic
catalysts show improvement over palladium in both the targets, it is possible that
both causes are at work. Reduction in the size of the palladium ensembles might
well be expected to inhibit oligomerisation, while weakening the adsorption of the
ethene might well improve selectivity.[112] Furthermore the second component will
certainly affect the solubility of hydrogen and the range of stability of the hydride
phases.

Industrial practice will clearly favour low metal loadings, and it is to be expected that it will be difficult to achieve conjunction of two (or more) components
when their surface concentrations are small;[5] nevertheless there are strong indications that they are successfully made and used, at least in some installations.
Methods for preparing bimetallic catalysts have been reviewed in depth,[111] and the
merits of employing organometallic complexes containing two elements have been
emphasised. The cost of large-scale manufacture by such a method may however

be considerable, and bimetallic particles containing palladium and silver can be made at a total loading of 0.45% by sequential or simultaneous impregnation of precursor solutions.[59] Adequate description of the method used is not however always provided.[111]

An early study using a static system showed[16] that PdAg/α-Al$_2$O$_3$ catalysts containing 10–30% silver gave better mechanistic selectivity (98–100%) than Pd/α-Al$_2$O$_3$; orders in hydrogen (1.7–1.8) were raised, and the activation energy was greater (84 kJ mol^{-1}). Added ethene however competed with the ethyne. No methods of surface analysis were available at that time, but surface enrichment by silver is to be expected, and is indeed found.[111] Silver decreases hydrogen solubility, but raises the temperature at which the β-hydride phase decomposes.[111] However, diffusivity is improved, and a palladium-silver alloy is used for purifying hydrogen by selective diffusion; tendency to failure through hydrogen embrittlement is also reduced (Section 3.1).

The palladium-silver system continues to attract attention.[59,112–115] Highresolution TEM has been applied to PdAg/SiO$_2$ catalysts, the selectivity of which (with excess ethene) depends critically on the type of pretreatment. A sequence of precursor calcination, reduction, re-oxidation and reduction at 773 K gave excellent selectivity and less oligomerisation; this was attributed to a surface reconstruction that lowered the mean size of the palladium ensembles.[59] The addition of nitrous oxide to a palladium-silver catalyst operating under excess ethene gave a further improvement in selectivity,[112] and metal aluminates have been claimed as supports for catalysts containing palladium plus silver or a base metal.[113] The design of the support is important to minimise deactivation due to pore blocking by oligomers; a recent patent[114] recommends the use of 'a moulding of trilobal cross-section with holes through the lobes' as a support for palladium-silver. Copper,[45,112,116] gold[112,117,118] and tin[112] have also been explored as moderators of palladium catalysts.

Other additives or promoters have been advocated. The oxides of cerium, titanium and niobium added to Pd/SiO$_2$ improved selectivity with excess ethene at high conversion of ethyne,[119] and the sodium content of pumice or of synthetic supports also had a beneficial effect,[120–123] due it was thought to some induced change in the electron concentration in the palladium weakening the chemisorption of ethene. Silicon deposited on Pd/SiO$_2$ by CVD of silane, and then oxidised to silica, has the effect of increasing selectivity in the presence of excess ethene by (i) inhibiting chain-growth in oligomerisation, (ii) weakening ethene adsorption and (iii) lowering the amount of hydrogen that can be adsorbed.[124]

The best-known metallic promoter is lead. This has long been known in the shape of the *Lindlar catalyst*,[11,125,126] which is PdPb/CaCO$_3$ together with a nitrogen base (quinoline). It has been widely used in organic synthesis in the liquid phase, where it has excellent selectivity for carbon triple bond reduction, giving Z-alkenes in high yields. The use of PdPb/α-Al$_2$O$_3$ also appears to have been helpful in ethyne hydrogenation,[127] although there is no indication of its being much used in this connection. In the Lindlar catalyst the lead is not in solid solution, but rather it appears to decorate the surface of palladium particles,[11,128] as might be expected from its method of preparation, which involves treating Pd/CaCO$_3$ with lead acetate. This, with Gigola's procedure which used Pb^{n}Bu$_4$ as the lead source, places this type of catalyst in the class of those developed by Figuéras, Coq and their colleagues,[129] who applied alkyls of Group 14 elements to various supported metals to create two-dimensional surface 'alloys' (Section 1.3.1; see also 9.4.5). It has been suggested[130] that in the Lindlar catalyst the reacting hydrogen atoms carry a fractional *negative* charge, thus making the attack on the triple bond *nucleophilic*, whereas on normal palladium it is *positively* charged, and the process is *electrophilic*.

The selective performance of platinum is raised by the addition of rhenium,[131] and admixing with copper or gold helps even that of iridium.[128]

Arrhenius parameters for ethyne hydrogenation on nickel-copper powders are shown as a compensation plot in Figure 9.8. While at 323 K rates decreased progressively as the copper content rose, those at 473 K were maximal at 40 and 60% copper, as indicated by their positions on the compensation plot. As discussed in Section 5.5, this result makes protracted discussion of the relevance of solid-state parameters to catalytic activity of doubtful value, if activity measurements are confined to a single temperature. Results are also available for the nickel-cobalt system. Intermetallic compounds based on cobalt (CoGe, CoGe$_2$, CoAl) were much more selective than cobalt itself.[132]

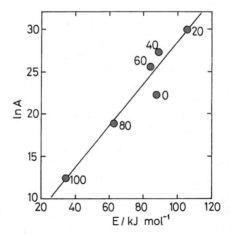

Figure 9.8. Compensation plot for ethyne hydrogenation on nickel-copper powders (A in units of rate of pressure fall m^{-2}).[32]

The poisoning of Pd/Al_2O_3 by mercury vapour is a complex phenomenon that has been fully investigated;[33] activity for ethene hydrogenation was suppressed while that of ethyne remained, so that improved selectivity resulted. Other metals examined in less detail showed similar behaviour.

9.5. HYDROGENATION OF HIGHER ALKYNES

9.5.1. Propyne[133–140]

The hydrogenation of propyne has been much less fully investigated than that of ethyne, with which it shares broadly similar characteristics; orders in hydrogen are close to first, and in propyne zero or slightly negative or positive, tending to become more negative as temperature is raised.[135,139] Activation energies lie in the range 30–70 kJ mol^{-1}, being on average a little greater than comparable values for ethyne (Table 9.1). Addition of the methyl group has no profound consequences, and there is no suggestion that access of hydrogen to the surface is any easier than with ethyne. Alkene selectivities are just as high if not higher, but the striking difference is the much smaller amount of oligomerisation (Table 9.7). This must be due to the greater difficulty of forming carbon-carbon bonds, as the methyl group causes the adsorbed species not to have the necessary propinquity. Propene selectivities and oligomer yields generally decrease with increasing P_H/P_P and decreasing temperature, being largely governed (as in the case of ethyne) by the concentration of adsorbed hydrogen atoms. This may be determined by the size of the active centre remaining after an almost complete layer of strongly adsorbed (perhaps partially dehydrogenated) reactant molecules has been formed; various palladium catalysts retained di-σ propene, $\sigma\pi$ propyne and propylidyne on the surface after hydrogenation at 273 K.[140] A scheme analogous to the second in

TABLE 9.7. Kinetic Parameters for the Hydrogenation of Propyne

Metal	Form	$E/kJ\ mol^{-1}$	T/K	$S_{3,i}{}^a$	$S_o{}^b$	References
Fe	powder	34	452	98	—	136
Co	powder	31	318	83	—	136
Ni	powder	51	340	83	—	136
Ni	/pumice	59	364	86	13	137
Ni	powder	38	358	86	3	134
Cu	powder	89	446	100	47	134
Pd	/pumice	69	409	94	7	136
Pt	/pumice	72	348	95	6	136
Pt	/SiO$_2$	—	333	100	—	139

In all cases the order in H_2 was about 1, and in propyne either zero or slightly negative, becoming more negative with increasing temperature.
[a] Initial propene selectivity
[b] Selectivity to oligomers

Table 9.6 was proposed. Addition of copper to nickel in powder form improved selectivity, but oligomerisation also increased;[134] it was also especially marked with cobalt.[63] Simultaneous hydrogenation of ethyne and propyne shows that the former must be somewhat more strongly adsorbed on nickel and platinum; this is understandable, since the C—C single bond in propyne is shorter (146 pm) than that in ethane (154 pm) due to hyperconjugation with the triple bond, which is thereby weakened.[28] The relief of strain upon adsorption is therefore less.

The reaction has been studied on copper catalysts:[133,134,141] on Cu/SiO_2 prepared by ion exchange, reaction with deuterium produces propyne-1-d_1 and propenes containing deuterium only in the vinylic positions (CH_3—CX=CX_2, where X may be H or D).[133]

9.5.2. The Butynes[27,142]

Extending the alkyl substituent from methyl to ethyl makes little difference to the characteristics of the hydrogenation; the only new feature introduced with 1-butyne is the possibility of Z- and E-2-butenes being formed in addition to the expected 1-butene; in fact this constitutes the sole or major product in most cases. On Pd/Al_2O_3,[103,144] and $Pd/BaSO_4$ in ethanol solution,[27] it is 98%, and with other palladium catalysts[7] (including a number of rare-earth intermetallics[145]) the 2-butenes were formed only in traces. The exception seems to be Cu/SiO_2,[146] where 1-butene was only 72% of the products. Total selectivity to butenes was very high on palladium[7,143,144] and copper[146] catalysts, but as expected was lower with platinum. On Pt/Al_2O_3 it was 85% (independent of dispersion, pressure and temperature[147]) or 90%[148] or rising to 80% as deactivation proceeded with a three-fold excess of hydrogen: a much larger ratio gave much lower selectivities.[149] It would appear that the larger alkyl group does not create steric interference between adjacently adsorbed molecules, creating sites for non-competitive adsorption of hydrogen, although with these metals the extent of oligomerisation is slight. With palladium[7,149] and rhodium[147] catalysts, turnover frequencies increased with decreasing dispersion, but there is very little kinetic information to illuminate the cause. The selectivity shown by palladium and platinum has however been ascribed[150] to the existence of a strongly adsorbed form of 1-butyne, which is in equilibrium with the reactive form, but blocks low coordination number sites, thus acting as a selective poison.

Once again the most informative technique for elaborating the mechanism is the use of deuterium as an isotopic tracer. On 0.03% Pd/Al_2O_3 the 1-butene was 72% 1-butene-1,2-d_2 and no molecules contained more than three deuterium atoms;[143] thus only the terminal hydrogen could have exchanged. A small amount of 1-butyne-1-d_1 was observed, and this may have been the precursor to 1-butene-d_3. In another study,[149] 1-butene-d_2 was 70–80% of the product, the rest having been formed by a 'different route'. The amounts of the 2-butenes formed

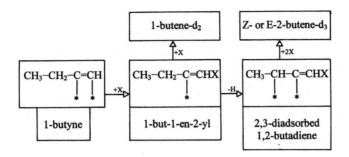

Scheme 9.5A. Reaction of 1-butyne with deuterium.

have been too small for their isotopic composition to be determined, but it is thought they may be formed through isomerisation of adsorbed 1-butyne to 1,2-butadiene (see Scheme 9.5a).

The hydrogenation of 2-butyne presents an even simpler picture. On the base metals (Fe, Co, Ni, Cu)[27,151,152] and palladium[12,153] the total selectivity was unity, and on all other metals examined it was above 90%.[27] Even with iridium with an unfavourable 10/1 hydrogen:butyne ratio it was 96%. Z-2-Butene was the exclusive product on palladium and copper,[27,153] and on the former the reaction with deuterium gave 99% Z-2-butene-2,3-d_2 at 287 K: the yield of other products however rose with temperature.

The cleanliness of the reactions of propyne, butynes and higher alkynes on certain metals, especially palladium and copper, and of ethyne on silver and

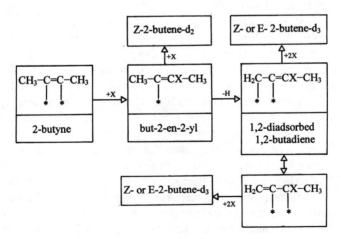

Scheme 9.5B. Reaction of 2-butyne with deuterium. Note: in Schemes 9.5A and 9.5B, all species are shown as only bonded by σ bonds for simplicity; butenyl radicals can be written in free radical forms, and final products assume X = D.

TABLE 9.8. Product Distributions from the Hydrogenation of 2-Butyne over Alumina-Supported Metals, and Certain Activation Energies

Metal	T/K	Z-2-B	E-2-B	1-B	$S_{tot}/\%$	E/kJmol^{-1}	References
Fe	473	76	4	20	700	—	27
Ru	363	79	5	16	97	42	155
Os	393	74	4	22	90	46	155
Co	413	88	7	5	100	28^b	152
Rh^a	427	85	8	7	99	–	152
Ir^a	433	87	8	5	96	–	152
Ni	425	95	5	1	100	38^b	152
Pd	298	100	0	0	100	–	152
Pt^a	431	87	8	5	97	–	152
Cu	397	100	0	0	100	140^b	152

^{a}H$_2$:2-butyne = 4 (Rh); 10(Ir); 7(Pt); :2 in all other cases
^{b}Unsupported powders

gold, requires some consideration. With the Group 11 metals, a low coverage by hydrogen atoms may discourage further reaction of the alkene, which doubtless desorbs readily once formed. In the case of palladium, the difference between ethyne and 2-butyne is quite marked; the former as we have seen can readily succumb to non-selective reaction when the conditions are right, and the apparent failure of 2-butyne to follow suit may be due to its inability to form the analogue of the $\alpha\alpha$-diadsorbed vinylidene. The expectation that the larger molecules would fit less well on the surface, and hence leave more gaps for the non-competitive adsorption of hydrogen, does not seem to be realised in practice. The possibility that molecular hydrogen participates in the rate-determining step finds little support, but the occurrence of hydrogen orders greater than unity still requires explanation.

The reaction on palladium-gold wires also gave results of great simplicity.[123] The activation energy was constant within error across the series (0–94% Au), the rate being determined solely by the pre-exponential factor, which fell smoothly with increasing gold content (Figure 9.9A). The manner of the activity change suggested (reasonably) that an ensemble of four palladium atoms made the active centre. This assumed the absence of surface segregation or island formation. The marked decrease in rate as the gold content was raised necessitated the use of progressively higher temperature ranges, so that it appears that product distributions were more reflecting temperature than surface composition. Below about 400 K the rise in the small amounts of E-2-butene and 1-butene correlated better with gold content than with temperature, but thereafter they correlated well with temperature (Figure 9.9B). Unfortunately the dependence of product yields on temperature for each catalyst was not reported, so the two variables cannot be reliably separated.

A supported palladium-lead intermetallic compound, represented as Pd$_3$Pb/CaCO$_3$, was better than the Lindlar catalyst for the semi-hydrogenation

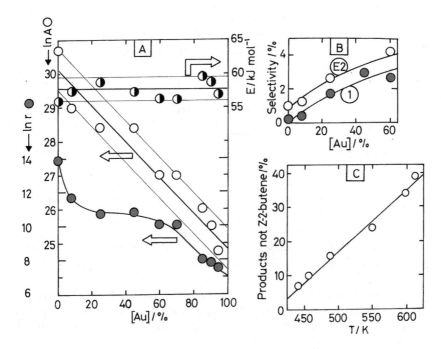

Figure 9.9. Hydrogenation of 2-butyne on palladium-gold wires: (A) activation energy, ln A, ln (rate of 2-butyne removal); (B) selectivities to E-2-butene and 1-butene versus composition; (C) yields of products other than Z-2-butene versus temperature for gold contents above 60%.[123]

of 2-butyne in that the rate of the second stage was slower.[12,128] The properties of the substitutional alloy PdB$_x$ have also been noted.[154]

Product selectivities shown by the metals of Groups 8 to 10 plus copper for 2-butyne hydrogenation are given in Table 9.8:[27,152] the few kinetic parameters known are also given. Ways of forming products other than Z-2-butene are also explained in Scheme 9.5; the occurrence of 1,2-butadiene as an intermediate is supported by its observation as a minor product on rhodium and iridium. Nickel-copper alloy powders showed an activity maximum at Ni$_{90}$Cu$_{10}$, and activation energies between 27 and 38 kJ mol^{-1}: orders were close to unity in hydrogen and zero or a little negative in 2-butyne.[152] The quantity of oligomers formed increased with copper content.

The reactions of 2-butyne with hydrogen and with deuterium have been studied in detail on Ru/Al$_2$O$_3$ and Os/Al$_2$O$_3$ catalysts:[155] orders in hydrogen and in 2-butyne were respectively first and zero on both, and selectivities fell with increasing hydrogen pressure and decreasing temperature as expected (Figures 9.4 and 9.5). Abbreviated results for the deuterated products are given in Table 9.9: there was no exchange in the reactant, but hydrogen exchange occurred to balance the deuterium numbers greater than two. The mechanism has been discussed in detail; the much greater deuterium content of the 1-butene supports the notion of its being formed by reversal of a 2-butenyl species to 1,2-butadiene, followed by addition of two deuterium atoms (Scheme 9.5B).

9.5.3. Alkyl-Substituted Alkynes Having More Than Four Carbon Atoms[125,156]

By now it should be possible to predict what products will be observed in the hydrogenation of alkynes containing larger alkyl substituents. Table 9.10 contains relevant results, and few surprises; the very high stereospecificity shown in the reduction of 1-pentyne even on a metal such as rhodium, where the total selectivity is quite low, is especially noteworthy. Cessation of reaction after reduction of 2-pentyne can be achieved by using dimethylformamide as solvent; this acts as a selective poison, preventing adsorption of Z-2-pentene. 2-Hexyne has been reduced selectively to Z-2-hexene at a Pt/TiO_2 membrane.[157] Binary platinum-palladium Adams oxide has been used to catalyse hydrogenation of 1-octyne; the β-hydride phase was not formed when the platinum content exceeded 10%. 4-Undecyne unusually yielded E-4-undecene as the major product (68%); it was thought the first-formed Z-isomer isomerised before desorbing, but it is unclear why this does not occur with other molecules.

The reaction of 3-hexyne with deuterium on Pd/Al_2O_3 proceeded without exchange either into the alkyne or the deuterium.[153] Z-3-Hexene was the major product, of which 87% was the $-d_2$ isomer; molecules containing more than four deuterium atoms were absent, but the n-hexane was very fully exchanged. It appears that once the reaction proceeds to the n-hexyl stage, the normal $\alpha\beta$-exchange mechanism comes into play, and extensive exchange occurs before the radical is liberated as n-hexane. On the Northwestern series of Pt/SiO_2 and Pt/Al_2O_3 catalysts having dispersions based on H/Pt ratios of between 7 and 109%, 3-hexyne

TABLE 9.9.   Products of the Reactions of 2-Butyne with Deuterium over Alumina-Supported Ruthenium and Osmium Catalysts

Metal	Butene	%d_2-isomer	Other isomers	M^a
Ru[b]	E-2	75	no $-d_0$ or $-d_{\geq 4}$	1.87
	Z-2	87	no $-d_0$ or $-d_{\geq 5}$	1.90
	1-	36	$-d_1$ to $-d_5$ only	2.75
Os[c]	E-2	57	$-d_0$ to $-d_8$	2.41
	Z-2	65	$''$	2.00
	1-	21	$''$	3.27

^{a}M = mean deuterium number
^{b}T = 364 K
^{c}T = 373 K

TABLE 9.10. Hydrogenation of Alkyl-substituted Alkynes in the Liquid Phase: Product Selectivities

Alkyne	Metal	Form	S_{tot}	1*	2*	3*	Reference
1-C$_5$	Pd	/C	97	>99	—	—	172
	Rh	/C	63	>99	—	—	172
	Pt	/C	93	>99	—	—	172
	Ir	/C	55	75	25^Z	—	172
2-C$_5$	Pd	/C	100	98	2^E	—	172
	Rh	/C	81	96	4^E	—	172
4-C$_8$^{d}	Pd	/Zeolite A^a	99	>90	—	—	172
4-C$_{11}$^{d}	Pd	/various	—	—	68^E	—	27
	Pt	/C	90	92	8^E	—	153
	Ir	/C	60	91	9^E	—	162
	Ni	Raney	100	100	—	—	173
3-C$_6$	Pd	/Al$_2$O$_3$	94	92	5^E	3^{E2} + Z	173
1-C$_7$^{b}	FeCu	/SiO$_2$	100	—	—	—	173
1-C$_8$^{c}	Pd	/sepiolite	89	100	—	—	173
3-C$_9$	Pd	/Zeolite A^a	97	97	—	—	27

*1: selectivity to expected alkene by Z-addition without isomerisation
*2,3: superscript letters identify minor product, e.g. for 1-pentyne on Ir, 25% Z-2-pentene.
^{a}Treated with Ph$_2$(EtO)$_2$ Si.
^{b}Solvent C$_2$H$_5$OH
^{c}Solvent CH$_3$OH
^{d}Various solvents.

was hydrogenated with S_{tot} of about 87% and S_{Z3} of about 80%, both independent of dispersion.[158,159] A curious feature of this reaction was the self-activation of the catalyst, more marked with the smallest particles; this was thought due to 'reconstruction' of the particles to give structures more conducive to the reaction: unfortunately this idea was not checked by examination of the catalyst after use.

Considerable insights into the interaction of unsaturated molecules has been obtained by using molecules containing bulky substituents such as the ^{t}Bu group, which might be expected to inhibit or indeed prevent chemisorption; nevertheless di-*tert*-butylethyne (and the corresponding *E*-di-*tert*-butylethene) are easily hydrogenated.[142]

On the same series of Pt/SiO$_2$ and Pt/Al$_2$O$_3$ catalysts just mentioned, the TOF for di-*tert*-butylethyne hydrogenation relative to that of cyclopentene (which was almost structure-insensitive) *fell* by a factor of three as dispersion increased.[158,159] This was quite unexpectedly contrary to the predicted trend, which expected that TOF would be larger on *small* particles, because of the prevalence of surface atoms of low co-ordination number on which the molecule might the more comfortably sit. It was speculated that the potential gain in energy minimisation through chemisorption was sufficiently large to withdraw surface atoms from their normal places, so that the alkynes could interact with them more effectively. If this were to be a general phenomenon, we should have another way of explaining the origin of structure-insensitivity. There were also other interesting effects of dispersion and support. On catalysts of *low* dispersion, isomersation of the alkenes and their hydrogenation to alkane proceeded freely before all the alkyne had reacted, presumably because the alkyne molecules could not pack tightly enough to prevent these reactions occurring: but on catalysts of *high* dispersion this hardly happened until all the alkyne had reacted, values of S_{tot} being about 96% on Pt/Al$_2$O$_3$, because the stronger adsorption of the alkyne prevented re-adsorption of the alkenes. Values of S_{tot} and of S_Z were however consistently greater with Pt/Al$_2$O$_3$ than with Pt/SiO$_2$, for which there was no ready explanation.

'Borides' of palladium[154] and of several base metals,[160] prepared by reduction of metal salts with sodium borohydride, have been successfully used for the selective hydrogenation of alkynes.

Terminal alkynes are more reactive than internal alkynes, presumably because they are more strongly adsorbed; thus 1-octyne has been reduced to 1-octene in the presence of 4-octyne, which remained unaffected.[160]

9.5.4. Aryl-Substituted Alkynes[112,160]

Phenyl- and diphenylethyne have been the subject of a number of studies, no doubt because of the simplicity of the products formed, but the aromatic ring is *much* less reactive than the triple bond, and in most circumstances remains untouched while the triple bond is reduced.[161,162] Only occasionally has a phenylalkylethyne been examined (e.g. 1-phenyl-1-pentyne, or phenylpropyl-ethyne[163]); even then on Pd/montmorillonite with THF as solvent the expected Z-2-pentene was formed with 96% total selectivity and 96% stereoselectivity.[163] Otherwise the complications due to possible double-bond migration and Z-E isomerisation are absent, and this makes life much simpler. The reactions may be easily followed in the liquid phase, so that solvent effects can be looked at. It is even possible for reaction to be obtained by physically mixing the solid diphenylethyne with Pd/C, so that on exposure to hydrogen the reactant is attacked by spillover hydrogen.[164] Selective reduction of phenylethyne has also been achieved using the amorphous alloy $Pd_{81}Si_{19}$ in supercritical carbon dioxide.[165]

Particle size effects have been sought. On Pd/pumice the reaction of phenylethyne was structure-insensitive in the range of dispersion between 14 and 62%,[166] but on Pd/C and Pd/SiO_2-Al_2O_3 in n-hexane both rate and selectivity decreased with increasing dispersion.[167] Bimetallic systems have also been examined. With PdCu/pumice, pre-reduction at 298 K gives rates that were independent of copper content up to 8%, and selectivities that increased slightly;[168] it was thought that under these conditions the copper remained in the $+2$ state, because pre-reduction at 623 K led to rates that decreased with copper content. The same reaction on $PdAu/Al_2O_3$-La_2O_3-Nb_2O_5 gave high selectivities that were independent of gold content,[169] as was found with 2-butyne hydrogenation at the palladium-rich end of the same system. Addition of tin to Pt/Nylon depressed the rate, but did not improve selectivity.[161]

9.5.5. Multiply-Unsaturated Molecules

Further interesting information concerning the relative adsorption strengths and reactivities of various types of carbon-carbon unsaturation is obtained by the use of molecules containing two or more such reactive groups. Much of this treasury lies buried in the literature of organic chemistry, and does not readily come to the notice of the catalytic chemist. It is often concerned just with yields, not with kinetics and mechanism; the Lindlar catalyst is most often used.[112,160] The two functions may be conjugated (e.g. C≡C—C=C) or non-conjugated; they may be the same (e.g. C≡C—C≡C), because the types will diverge upon reaction. When multiple bonds are conjugated, the intervening single bond is shortened due to electron delocalisation, as with 1,3-butadiene (Section 8.12); in 1,3-butadiyne it is only 136 pm long, but in 1,4-hexadiyne it is 138 pm because part of the charge moves to the terminal single bonds. Addition of deuterium to 1,3-pentadiyne on Pd/α-Al_2O_3 gave Z-1,3-pentadiene-d_4 as the major product, but the reaction did not proceed in well-defined stages, and other possible products appeared simultaneously.[170]

1-Ethynylcyclohexene has been reduced selectively to 1-ethenylcyclohexene by the Lindlar catalyst; addition of alkyl groups to give for example the arrangement C=C—C≡C—C—C was without effect, but the grouping C=C—C≡C—C=C led to endless complications.[112] Hydrogenation of 2-methyl-1-buten-3-yne (valy-lene) to isoprene proceeded with 88% selectivity on $Pd/\alpha-Al_2O_3$ reduced at 773 K, better rates being obtained with larger particles; addition of small amounts of antimony, tin or lead from their n-butyl compounds further improved selectivity.[129] Compounds containing either non-conjugated or conjugated triple bonds are usually reduced with Lindlar catalyst successively to the enyne and then to the diene.[112,160]

9.6. CONCLUSION

It may be helpful to try to draw together some of the characteristics that are shared by alkyne hydrogenations in their many manifestations, and to attempt some generalisations: these by definition will be statements that are more or less true most of the time.

Quite evidently alkynes are strongly chemisorbed on all relevant metals, due to the simultaneous engagement of both sets of π orbitals simultaneously with those of the metal surface. The accurate representation of this bonding, and that in intermediates, as in Scheme 9.1, is hard to depict, and at a certain modest level of mechanistic understanding is largely irrelevant. Orders of reaction in hydrogen, being frequently about first, signify that the involvement of two hydrogen atoms or species containing them, e.g. alkenyl + atom, in the slowest step; there is little evidence for the reversible chemisorption of the molecule during reaction, even (and this is somewhat surprising) when disubstituted alkynes are used. Although in many cases there appears to be no competition between the reactants for the sur-face, nevertheless the alkyne blocks the dissociation of the hydrogen. Sometimes however the order in alkyne is somewhat negative, which implies that free sites may be needed for the hydrogen to access the reactive layer. Such negative orders are slightly temperature-dependent, which means that derived activation energies are not *quite* true, although their values are such as to suggest this.

The strength of the alkyne chemisorption is usually blamed for the inverse particle size effect, namely, for small particles being less active per unit area than large ones because the alkyne is less reactive. This supposition does not seem to have been validated by thermochemical or kinetic information, but is logical in that low coordination atoms abundantly present on the surface of small particles should be able to form chemisorption complexes analogous to those known in organometallic chemistry. The strange behaviour of di-*tert*-butylethyne, noted above, supports this contention, and there are even indications when the reactant is in the liquid phase that metal can be eluted from the catalyst, presumably as an alkyne complex. This particular mode of adsorption does however allow highly selective formation of alkene, so a compromise has to be struck in terms of optimum particle size between good selectivity and high activity.

The outstanding quality of palladium in both these respects is well known but not well understood. Foremost among the complexities that have to be addressed are (i) the tendency to form hydride phases, and (ii) the frequent formation, especially with ethyne, of a carbonaceous overlayer, derived perhaps from oligomers, some of which escape into the fluid phase. Both these factors are invoked to account for palladium's remarkable properties, but both are responsive to reaction conditions and especially to particle size; support effects may also operate. So many factors have to be kept in play when discussing mechanisms that one may safely conclude that none so far suggested is wholly satisfactory. The rich literature on alkyne hydrogenation deserves careful attention, and should be a fine source of inspiration for further research.

REFERENCES

1. R.J. Farrauto and C.H. Bartholomew, *Fundamentals of Catalytic Processes*, Chapman and Hall: London (1997), p. 434.
2. A.N.R. Bos and K.R. Westerterp, *Chem. Eng. Process.* **32** (1993) 1.
3. C.N. Satterfield, *Heterogeneous Catalysis in Practice,* 2nd edn., McGraw-Hill: New York (1991).
4. D. Duca, F. Frusteri, A. Parmaliana and G. Deganello, *Appl. Catal. A: Gen.* **146** (1996) 269.
5. J.P. Boitiaux, J. Cosyns, M. Derrien and G. Léger, *Hydrocarbon Proc.* (1985) 51.
6. J.P. Boitiaux, J. Cosyns and S. Vasudevan, *Appl. Chem.* **15** (1985) 317.
7. J.P. Boitiaux, J. Cosyns and S. Vasudevan, *Appl. Catal.* **6** (1983) 41.
8. A. Borodziński, *Catal. Lett.* **71** (2001) 169.
9. A. Borodziński, *Catal. Lett.* **63** (1999) 35.
10. V.A. Men'shchikov, Yu.G. Fal'kovich and M.É Aérov, *Kinet. Catal. (Eng. Edn.)* **16** (1975) 1335.
11. W. Palczewska, I. Ratajczkowa, I. Szymerska and M. Crawczyk in: *Proc. 8th. Internat. Congr. Catal.*, Verlag Chemie: Weinheim **IV** (1984) 173.
12. J. Sobczak, T. Bolesawska, M. Pawowska and W. Palczewska, in: *Heterogeneous Catalysis and Fine Chemicals* (M. Guisnet et al., eds.), Elsevier: Amsterdam (1988), p. 197.
13. N. Sheppard and C. de la Cruz, *Adv. Catal.* **42** (1998) 181; *Catal. Today* **70** (2001) 3.
14. A.J. Roberts, S. Haq and R. Raval, *J. Chem. Soc. Faraday Trans.* **92** (1996) 4822.
15. A.S. Al-Ammar and G. Webb, *J. Chem. Soc. Faraday Trans. I* **75** (1979) 1900.
16. G.C. Bond, D.A. Dowden and N. Mackenzie, *Trans. Faraday Soc.* **54** (1958) 1537.
17. G.C. Bond and P.B. Wells, *J. Catal.* **5** (1965) 65.
18. G.C. Bond and P.B. Wells, *J Catal.* **5** (1966) 419.
19. W. Palczewska in: *Hydrogen Effects in Catalysis* (Z. Paál and P.G. Menon, eds.), Dekker: New York (1988), p. 373.
20. K.L. Haug, T. Bürgi, T.R. Trautman and S.T. Ceyer, *J. Am. Chem. Soc.* **120** (1998) 8885.
21. R.J. Rennard Jr. and R.J. Kokes, *J. Phys. Chem.* **70** (1966) 2543.
22. M. Larson, J. Jansson and S. Asplund, *J. Catal.* **178** (1998) 49.
23. D. Duca, G. Barone and Z. Varga, *Catal. Lett.* **72** (2001) 17.
24. A. Borodziński, *Polish J. Chem.* **72** (1998) 2455.
25. A.J. den Hartog, M. Deng, F. Jongerius and V. Ponec, *J. Molec. Catal.* **60** (1990) 98; J. Houzvicka, R. Pestman and V. Ponec, *Catal. Lett.* **30** (1995) 289.
26. J. Stachurski and A. Frąkiewicz, *J. Less Comm Met.* **108** (1985) 249.
27. G.C. Bond and P.B. Wells, *Adv. Catal.* **15** (1964) 92.

28. G.C. Bond in *Catalysis*, (P.H. Emmett, ed.), Reinhold: New York (1956), p. 109.

29. G. Webb in: *Comprehensive Chemical Kinetics* (C.H. Bamford and C.E.H. Tipper, eds.), Vol. 20, Elsevier: Amsterdam (1978).

30. G.C. Bond, *J. Chem. Soc.* (1958) 2705.

31. G.C. Bond and R.S. Mann, *J. Chem. Soc.* (1958) 4738.

32. G.C Bond and R.S. Mann, *J. Chem. Soc.* (1959) 3566.

33. G.C. Bond and P.B. Wells, *Proc. 2nd. Internat. Congr. Catal.*, Vol. 1 (1961), p. 1139.

34. G.C. Bond and P.B. Wells, *J. Catal.* **4** (1965) 211.

35. G.C. Bond and P.B. Wells, *J. Catal.* **6** (1966) 397.

36. J. Sheridan, *J. Chem. Soc.* (1944) 373; (1945) 133, 301, 305, 470.

37. J. Sheridan and W.D. Reid, *J. Chem. Soc.* (1952) 2962.

38. W.T. McGown, C. Kemball and D.A. Whan, *J. Catal.* **51** (1978) 83.

39. G.C. Bond, *Catalysis by Metals,* Academic Press: London (1962.)

40. A. Farkas and L. Farkas, *J. Am. Chem. Soc.* **61** (1939) 3397.

41. G.C. Bond, G. Webb and P. B. Wells, *J. Catal.* **12** (1968) 157.

42. J.H. Moses, A.H. Weiss, K. Matusek and L. Guczi, *J. Catal.* **86** (1984) 417.

43. H. Molero, B.F. Bartlett and W.T. Tysoe, *J. Catal.* **181** (1999) 49.

44. P.B. Wells, *J. Catal.* **52** (1978) 498.

45. N.R.M. Sassen, A.J. Den Hartog, F. Jongerius, J.F.M. Aarts and V. Ponec, *Faraday Discuss. Chem. Soc.* **87** (1989) 311.

46. V. Ponec and G.C. Bond, *Catalysis by Metals and Alloys*, Elsevier: Amsterdam (1996).

47. P.B. Wells in: *Surface Chemistry and Catalysis* (A.F. Carley, P.R. Davis, G.J. Hutchings and M.S. Spencer, eds.), Kluwer: Dordrecht, (2003).

48. R.G. Oliver and P.B. Wells, *J. Catal.* **27** (1977) 364.

49. A. Borodziński and A. Cybulski, *Appl. Chem. A: Gen.* **198** (2000) 51.

50. A.S. Al-Ammar and G. Webb, *J. Chem. Soc. Faraday Trans. I* **74** (1978) 657.

51. E.A. Arafa and G. Webb, *Catal. Today* **17** (1993) 411.

52. A. Borodziński, *Polish J. Chem.* **68** (1994) 583.

53. A. Borodziński and A. Janko, *React. Kinet. Catal. Lett.* **7**. (1977) 163.

54. J. Zielinski and A. Borodziński, *Appl. Chem.* **13** (1985) 305.

55. A. Borodziński. R. Duś, R. Frąk, A. Janko and W. Palczewska, *Proc. 6th. Internat. Congr. Catal.* (G.C. Bond, P.B. Wells and F.C. Tompkins, eds.), *Roy. Soc. Chem.*: London (1976) paper A7.

56. J. Stachurski, *J. Chem. Soc. Faraday Trans. I* **81** (1985) 2813.

57. A. Sárkány, A.H. Weiss and L. Guczi, *Appl. Catal.* **10** (1984) 369; *J. Catal.* **98** (1986) 550.

58. Y. Inoue and I. Yasumori, *J. Phys. Chem.* **67** (1971) 880.

59. Y. Jin, A. K. Datye, E. Rightor, R. Gulotty, W. Waterman, M. Smith, M. Holbrook, J. Maj and J. Blackson, *J. Catal.* **203** (2001) 292; T.V. Choudhary, C. Sivadinarayana, A.K. Datye, D. Kumar and D.W. Goodman, *Catal. Lett.* **86** (2003) 1.

60. K. Tamaru, *Bull. Chem. Soc. Japan* **23** (150) 64, 180, 184.

61. D.H. Everett and P.A. Sermon, *Zeit. Phys. Chem. (Wiesbaden)* **14** (1979) 109.

62. M. Boudart and H.S. Hwang, *J. Catal.* **39** (1975) 44.

63. A.J. den Hartog, M. Holderbusch, E. Rappel and V. Ponec, *Proc. 9thInternat. Congr. Catal.* (M.J. Phillips and M. Ternan, eds.), Chem. Inst. Canada: Ottawa (1988) 1174.

64. S. Tracey, A. Palermo, J.P. Holgado Vazquez and R.M. Lambert, *J. Catal.* **179** (1998) 231.

65. A. Palermo, F.J. Williams and R.M. Lambert, *J. Phys. Chem. B* **106** (2002) 10215.

66. A. Sárkány and Zs. Révay, *Appl. Catal. A: Gen.* **243** (2003) 347.

67. J. Jia, K. Haraki, J.N. Kondo, K. Domea and K. Tamaru, *J. Phys. Chem. B* **104** (2000) 11153.

68. T.G. Rucker, M.A. Logan, T.M. Gentle, E.L. Muetterties and G.A. Somorjai, *J. Phys. Chem.* **90** (1986) 2703.

69. D. Stacchiola, H. Molero and W.T. Tysoe, *Catal. Today* **65** (2001) 3; D. Stacciola and W.T. Tysoe, *Surf. Sci.* **513** (2002) L431.
70. W.T. Tysoe, G.L. Nyberg and R.M. Lambert, *J. Phys. Chem.* **90** (1986) 3188; *J. Phys. Chem.* **90** (1986) 3188.
71. R.M. Ormerod, R.M. Lambert, D.W. Bennett and W.T. Tysoe, *Surf. Sci.* **330** (1995) 1.
72. C.J. Baddeley, R.M. Ormerod and R.M. Lambert, *Proc. 11th Internat. Congr. Catal.* (J.W. Hightower, W.N. Delgass, E. Iglesia and A.T. Bell, eds.), Elsevier: Amsterdam (1996) 371.
73. D. Stacchiola, G. Wu, H. Molero and W.T. Tysoe, *Catal. Lett.* **71** (2001) 1.
74. M.A. Logan, T.G. Rucker, T.M. Gentle, E.L. Muetterties and G.A. Somorjai, *J. Phys. Chem.* **90** (1986) 2709.
75. A.F. Lee, C.J. Baddeley, C. Hardacre, G.D. Moggridge, R.M. Ormerod, R.M. Lambert, J.P. Candy and J.-M. Basset, *J. Chem. Phys. B* **101** (1997) 2797.
76. C.J. Baddeley, T. Tikhov, C. Hardacre, J.R. Lomas and R.M. Lambert, *J. Phys. Chem. B* **100** (1996) 2189.
77. J. Storm, R.M. Lambert, N. Memmel, J. Onsgaard and E. Taglauer, *Surf. Sci.* **436** (1999) 259.
78. T.I. Taylor in: *Catalysis* (P.H. Emmett, ed.), Reinhold: New York, Vol. 5 (1957), p. 257.
79. R.L. Arnett and B.L. Crawford Jr., *J. Chem. Phys.* **18** (1950) 118; J.E. Douglas and B.S. Rabinovitch, *J. Am. Chem. Soc.* **74** (1952) 2486.
80. G.C. Bond, *J. Chem. Soc.* (1958) 2488.
81. J. Margitfalvi, L. Guczi and A.H. Weiss, *React. Kinet. Catal. Lett.* **15** (1980) 475; *J. Catal.* **72** (1981) 185.
82. F. King, S.D. Jackson and F.E. Hancock, in: *Catalysis of Organic Reactions*, (R. E. Malz Jr., ed.), Dekker: New York (1996), p. 53.
83. G.C. Bond, J. Sheridan and D.H. Whiffen, *Trans. Faraday. Soc.* **48** (1952) 715.
84. A. Molnár, A. Sárkány and M. Varga, *J. Molec. Catal. A* **173** (2001) 185.
85. L. Guczi, R.B. La Pierre, A.H. Weiss and E. Biron, *J. Catal.* **60** (1979) 83.
86. J.M. Moses, A.H. Weiss, K. Matusek and L. Guczi, *J. Catal.* **86** (1984) 417.
87. A. Borodziński and A. Gołębiowski, *Langmuir* **13** (1997) 883.
88. D. Duca, Z Varga, G. La Manna and T. Vidoczy, *Theor. Chem. Acc.* **104** (2000) 302.
89. D. Duca, G. Barone, Z. Varga and G. La Manna, *J. Mol. Struct. (Theochem)* **542** (2001) 207.
90. H.R. Aduriz, P. Bodnariuk, M. Dennehy and C.E. Gigola, *Appl. Catal.* **58** (1990) 227.
91. C.E. Gigola, H.R Aduriz and P. Bodnariuk, *Appl. Chem.* **27** (1986) 133.
92. S. Apslund, C. Fornell, A. Holmgren and S. Irandoust, *Catal. Today* **24** (1995) 181.
93. W.T. McGowan, C. Kemball, D.A. Whan and M.S. Scurrell, *J. Chem. Soc. Faraday Trans.* **73** (1977) 632.
94. Li Zon Gva and Kim En Kho, *Kinet. Katal.* **29** (1988) 381.
95. A. S. Al-Ammar and G. Webb, *J. Chem. Soc. Faraday Trans. 1* **74** (1978) 195.
96. G.C. Battison, L. Dalloro and G.R. Tauszik, *Appl. Catal.* **2** (1982) 1.
97. A.N.R. Bos, E.S. Bootsma, F. Foeth, H.W.J. Sleyster and K.R.A. Westerterp, *Chem. Eng. Processing* **32** (1993) 53.
98. Sh.E. Duisenbaev, M.S. Kharson, Z.T. Baisembaeva and S.L. Kiperman, Abstracts, Europacat II, 1995, p. 736.
99. A. Borodziński, *Polish J. Chem.* **69** (1995) 111.
100. M.J. Vincent and R.D. Gonzalez, *Appl. Catal. A: Gen.* **217** (2001) 143.
101. U. Schröder, *J. Catal.* **146** (1994) 586.
102. World Appl. 98/10,863A to Nissan Girdler Catalyst Co.
103. G. Webb in: *Specialist Periodical Reports: Catalysis,* (C. Kemball and D.A. Dowden, eds.), Roy. Soc. Chem. Vol. 2 (1978), p. 45.
104. J.A. Peña, J. Herguido, C. Guimon, A. Monzón and J. Santamaría, *J. Catal.* **159** (1996) 313.
105. Y.H. Park and G.L. Price, *Ind. Eng. Chem. Res.* **30** (1991) 1700.

106. Phillips Petroleum Co., World Appl. 00/23,403.

107. D.A. Dowden in *Specialist Periodical Reports: Catalysis*, (C. Kemball and D.A. Dowden, eds.), *Roy. Soc. Chem.* Vol. 2 (1978), p. 1.

108. R.L. Moss in *Specialist Periodical Reports: Catalysis*, (C. Kemball and D.A. Dowden, eds.), *Roy. Soc. Chem.* Vol. 1 (1977), p. 37; Vol. 4 (C. Kemball and D.A. Dowden, eds.), *Roy. Soc. Chem.* (1981), p. 31.

109. E.G. Allison and G.C. Bond, *Catal. Rev.* **7** (1973) 233.

110. Q.-W. Zhang, J. Li, X.-X. Lin and Q.-M. Zhu, *Appl. Catal. A: Gen.* **197** (2000) 221.

111. L. Guczi and A. Sárkány, in: *Specialist Periodical Reports: Catalysis*, (J.J. Spivey and S.K. Agarwal, eds.), *Roy. Soc. Chem.* Vol. 11 (1994), p. 318.

112. Piyasan Praserthdam, Suphot Phutanasri and Jumpot Meksikarin, *Catal. Today* **63** (2000) 209; Piyasan Praserthdam, Bongkot Ngamsoma, N. Bogdanchikova, Suphot Phatanasri and Mongkonchanok Pramotthan, *Appl. Catal. A: Gen.* **229** (2002) 310.

113. Phillips Petroleum Co., World Appl. 01/19, 763.

114. Sued-Chemie G, World Appl. 01/58,590.

115. G. Gislason, W.-S. Xia and H. Sellers, *J. Phys. Chem.* A **107** (2002) 767.

116. S. Leviness, V. Nair, A.H. Weiss Z. Schay and L. Guczi, *J. Molec. Catal.* **25** (1984) 131.

117. C. Visser, J.G.P. Zuidwyk and V. Ponec, *J. Catal.* **35** (1974) 407.

118. A. Sárkány, A. Horváth and A. Beck, *Appl. Catal. A: Gen.* **229** (2002) 117.

119. J.H. Kang, E.W. Shin, W.J. Kim, J.D. Park and S.H. Moon, *Catal. Today* **63** (2000) 183.

120. A.M. Venezia, A. Rossi, L.F. Liotta, A. Martorana and G. Deganello, *J. Catal.* **147** (1996) 81.

121. L.F. Liotta, A.M. Venezia, A. Martorana and C. Deganello, *J. Catal.* **171** (1997) 177.

122. L.F. Liotta, A.M. Venezia, A. Martorana, A. Rossi and G. Deganello, *J. Catal.* **171** (1997) 169.

123. H.G. Rushford and D.A. Whan, *Trans. Faraday Soc.* **71** (1975) 3577.

124. Eun Woo Shin, Jung Hwa Kang, Woo Jae Kim, Jae Duk Park and Sang Heup Moon, *Appl. Catal. A: Gen.* **223** (2001) 161; *J. Catal.* **208** (2002) 310.

125. P.N. Rylander, *Catalytic Hydrogenation over Platinum Metals,* Academic Press: New York (1967).

126. W. Palczewska, A. Jabłonski, Z. Kaszkur, G. Zuba and J. Wernisch, *J. Molec. Catal.* **25** (1984) 307.

127. M.A. Volpe, P. Rodriguez and C.E. Cigola, *Catal. Lett.* **61** (1999) 27.

128. W. Palczewska, A. Jabłonski, Z. Kaszkur, G. Zuba and J. Wernisch, *J. Molec. Catal.* **25** (1984) 317.

129. H.R. Aduriz, P. Bodnariuk, B. Coq and F. Figuéras, *J. Catal.* **129** (1991) 47.

130. J. Yu and J.B. Spencer, *J. Am Chem. Soc.* **117** (1997) 5257.

131. R. Pestman, A.J. den Hartog and V. Ponec, *Catal. Lett.* **4** (1996) 287.

132. T. Komatsu, M. Fukui and T. Yashima, *Proc. 11th. Internat. Congr. Catal.*, (J.W. Hightower, W.N. Delgass, E. Iglesia and A.T. Bell, eds.), Elsevier: Amsterdam **B** (1996) 1095.

133. N.J. Ossipoff and N.W. Cant, *J. Catal.* **148** (1994) 125.

134. R.S. Mann and K.C. Khulbe, *Canad. J. Chem.* **46** (1968) 623.

135. R.S. Mann and K.C. Khulbe, *Canad. J. Chem.* **45** (1967) 2755.

136. R.S. Mann and S.C. Naik, *Canad. J. Chem.* **45** (1967) 1023.

137. G.C. Bond and J. Sheridan, *Trans. Faraday Soc.* **48** (1952) 651.

138. G.C. Bond and J. Sheridan, *Trans. Faraday Soc.* **48** (1952) 664.

139. S.D. Jackson and G.J. Kelly, *J. Molec. Catal.* **87** (1994) 275.

140. S.D. Jackson and N. J. Casey, *J. Chem. Soc. Faraday Trans.* **91** (1995) 3269.

141. H.N. Choski, J.A. Bertrand and M.G. White, *J. Catal.* **164** (1996) 484.

142. R.L. Burwell Jr., *Discuss. Faraday Soc.* **41** (1966) 259.

143. E.F. Meyer and R.L. Burwell Jr., *J. Am. Chem. Soc.* **85** (1963) 2881.

144. S. Hub and R. Touroude, *J. Catal.* **114** (1988) 411; S. Hub, L. Hilaire and R. Touroude, *Appl. Catal.* **36** (1988) 307.

145. K.S. Sim, L. Hilaire, F. Le Normand, R. Touroude, V. Paul-Boncour and A. Pecheron-Guegan, *J. Chem. Soc. Faraday Trans.* **87** (1991) 1435.

146. J.T. Werli, D.J. Thomas, M.J. Wainwright, D.L. Trimm and N.W. Cant, *Proc. 10th. Internat. Congr. Catal.* (L. Guczi, F. Solymosi and P. Tétényi, eds.), Akadémiai Kiadó: Budapest C (1992) 2289.

147. J.P. Boitiaux, J. Cosyns and E. Robert, *Appl. Catal.* **32** (1987) 145; **35** (1987) 193.

148. P. Maetz and R. Touroude, *Catal. Lett.* **4** (1990) 37.

149. Ph. Maetz, J. Saussey, J.C. Lavalley and R. Touroude, *J. Catal.* **147** (1994) 48; *J. Molec. Catal.* **91** (1994) 219.

150. P. Maetz and R. Touroude, *Appl. Chem. A. Gen.* **149** (1997) 189.

151. K.C. Khulbe and R.S Mann, *Catal. Rev.-Sci. Eng.* **24** (1982) 311.

152. R.S. Mann and C.P. Khulbe, *Canad. J. Chem.* **48** (1970) 2075.

153. E.F. Meyer and R.L. Burwell Jr., *J. Am. Chem. Soc.* **85** (1963) 2877.

154. W. Palczewska, M. Cretti-Bujnowska, J. Pielaszek, J. Sobczak and J. Stachurski, in: *Proc. 9th. Internat, Congr. Catal.* (M.J. Phillips and M. Ternan, eds.), Chem. Inst. Canada: Ottawa (1988), 1410.

155. G. Webb and P.B. Wells, *Trans. Faraday Soc.* **61** (1965) 1232.

156. S. Siegel, *Adv. Catal.* **16** (1966) 23.

157. C. Lange, S. Storck, B. Tesche and W.G. Maier, *J. Catal.* **175** (1998) 280.

158. H.H. Kung, R.J. Pellet and R.L. Burwell Jr., *J. Am. Chem. Soc.* **98** (1976) 5603.

159. R.L. Burwell Jr., D. Barry and H.H. Kung, *J. Am. Chem. Soc.* **95** (1973) 4.

160. R.L. Augustine, *Heterogeneous Catalysis for the Synthetic Chemist*, Dekker: New York (1996), Chs. 3 and 4.

161. S. Galvagno, Z. Polarzewski, A. Donato, G. Neri and R. Pietropaolo, *J. Molec. Catal.* **35** (1986) 365.

162. Y. Nitta, Y. Hiramatsu, Y. Okamoto and T. Imanaka, *Proc. 10th. Internat. Congr. Catal.* (L. Guczi, F. Solymosi and P. Tétényi, eds.), Akadémiai Kiadó: Budapest C (1992) 2333.

163. Á. Mastalir, Z. Király, G. Szöllösi and M. Bartók, *Appl. Chem. A: Gen.* **213** (2001) 133.

164. I. Pri-Bar and K.E. Koresh, *J. Molec. Catal. A. Chem.* **156** (2000) 173.

165. R. Tschan, R. Wandeler, M.S. Schneider, M.M. Schubert and A. Baiker, *J. Catal.* **204** (2001) 219.

166. D. Duca, L.F. Liotta and D. Deganello, *Catal. Today* **24** (1995) 15.

167. F. Carturan, G. Facchin, G. Cocco, S. Enzo and G. Navazio, *J. Catal.* **76** (1982) 405.

168. L. Guczi, Z. Schay, Gy. Steffler, L.F. Liotta, G. Deganello and A.M. Venezia, *J. Catal.* **182** (1999) 456.

169. D.J. Ostgard, K.M. Crucilla and F.P. Daly in *Catalysis of Organic Reactions* (R.E. Malz Jr., ed.), Dekker: New York, 1996, p. 199.

170. P.B. Wells and G.R. Wilson, *Disc. Faraday Soc.* **41** (1966) 237.

171. M.B. Taghavi, G. Pajonk and S.J. Teichner, *Bull. Soc. Chim. Fr.* **7–8** (1978) I-302.

172. G.C. Bond and J.S. Rank, *Proc. 3rd Internat. Congr. Catal.*, (W.M.H. Sachtler, G.C.A. Schuit and P. Zwietening, eds.), North Holland: Amsterdam **2** (1965) 1225.

173. M.A. Aramendia, V. Borau, C. Jimenez, J.M. Marinas, M.E. Sempere, F.J. Urbano and L. Villar, *Proc. 10th. Internat. Congr. Catal.*, Vol. C (1992), p. 2435.

HYDROGENATION OF THE AROMATIC RING

The aromatic ring is the third type of unsaturated structure, the hydrogenation of which we need to consider. Although it is possible for its hydrogen atoms to be substituted for deuterium, the possible products of the reduction of benzene are effectively but two in number, viz. cyclohexene and cyclohexane: somewhat special conditions of catalyst type and reaction conditions are needed to procure the former, so that commonly cyclohexane is the only product observed. In comparison with say the reactions of the butenes or ethyne, the hydrogenation of benzene is simplicity itself, and for this reason it has been the choice of many who have wished for a quick and easy way of assessing the activity of a catalyst. Very simple analytical techniques are all that one needs, and this helps to explain the attractiveness of this class of reaction before the advent of gas-chromatography.

The straightforwardness of the reaction does not however extend to its interpretation. The mode of chemisorption of the ring has long been debated, and several different structures have been advanced. One is reminded of the student who was required to answer correctly only one of two questions in an oral examination. Question: 'What is the colour of blue litmus?' Answer: 'Red'. This was incorrect. Question: 'What is the structure of benzene?' Answer: 'The Lord only knows'. This was thought to be correct, so the student was passed.

The hydrogenation of alkyl-substituted benzenes leads us into areas of stereochemistry that are related to those we have visited in Chapter 7; and the hydrogenation of fused or condensed aromatic ring systems likewise has fascinating stereochemical consequences also akin to those discussed in Section 7.5.3. The reduction of these molecules has mainly been conducted by organic chemists, and mechanistic aspects have scarcely been examined.

10.1. INTRODUCTION

10.1.1. Scope

The single aromatic ring as it exists in benzene is stabilised by the simultaneous involvement of the three sets of π electrons; this can be represented as *resonance* between canonical structures in which double-bonds are differently located, the net effect being that all the six C—C bonds are the same, and are intermediate in length between the formal single- and double-bonds (0.139 nm). The electron system is therefore hard to disrupt, and much of aromatic chemistry depends upon the greater reactivity of the C—H bonds, which can be manipulated in many different ways without the ring being affected. Hetero-aromatic molecules containing nitrogen or sulfur atoms are also resonance-stabilised, but they will not be considered here. Despite its stability, benzene and other aromatic molecules are quite strongly chemisorbed on metal surfaces, although whether this involves loss of the resonance energy by uncoupling the electron system, or leaving it intact as in for example bis(benzene)chromium, is something that will engage our attention in Section 10.13.

The size of the resonance energy in benzene can be estimated as follows. The heat of hydrogenation of benzene to cyclohexane is 206 kJ mol^{-1}, and that of the hypothetical non-resonating cyclohexatriene can be estimated as three times that of Z-2-butene, namely, 359 kJ mol^{-1}: the resonance energy is thus 150 kJ mol^{-1}. The heat liberated on hydrogenating 1,3-cyclohexadiene is 232 kJ mol^{-1}, so its production from benzene would be endothermic by 24 kJ mol^{-1}; it is therefore an unlikely product. Cyclohexene ($-\Delta H_H = 120$ kJ mol^{-1}) is therefore the only possible product besides cyclohexane. The attachment of alkyl groups to the benzene ring lowers the heat of hydrogenation only very slightly, due to a small delocalisation of the ring electrons.

The temperature-dependence of the thermochemical parameters has important consequences for the study of ring hydrogenation. The free-energy change for forming cyclohexane form benzene ($-\Delta G^{\ominus} = 98$ kJ mol^{-1}) increases rapidly with temperature, so that $\Delta G^{\ominus}$ becomes zero and the equilibrium constant unity at 560 K. At temperatures above about 473 K there is therefore a significant reverse reaction, the occurrence of which complicates kinetic analysis. The position of equilibrium in this sensitive range naturally depends upon the reactant ratio. It thus becomes clear that the *dehydrogenation* of cyclohexane or cyclohexene can be studied at comparatively modest temperatures, these reactions therefore being feasible for industrial-scale use (see Chapter 12). Its analogue with C$_5$ and C$_7$ rings, or indeed with linear alkanes, is hardly possible, because without the pull of resonance stabilisation much higher temperatures are needed to obtain useful amounts of the products, and at these temperatures the molecules will suffer total disruption to carbon and hydrogen.

Since the hydrogenation of benzene under most circumstances affords cyclo-hexane as the only product, the reaction has proved attractive as an easy means of assessing catalytic activity; it can be followed manometrically in a static reactor, or, in a flow-system, where product analysis may be carried out very simply. Many of the studies described in the literature antedate the arrival of gas chromatography, and the reaction is still amenable to study in laboratories deprived of this facility. It may of course be conducted either in the vapour or liquid phase. One of the earliest studies of the behaviour of bimetallic (Ni-Cu) systems was made in 1934 using benzene hydrogenation,[1] and a review by Hilton A. Smith published in 1957 on the hydrogenation of all types of aromatic compounds, including hetero-aromatic, on all kinds of catalyst was supported by no fewer than 465 references.[2] In what follows, therefore, a certain degree of selection has to be applied, and attention will be focussed on more recent publications.

On surveying the literature it becomes clear that certain themes emerge reg-ularly, and that features of the reactions are sometimes common to benzene and its alkyl-substituted derivatives. The hydrogenation of toluene and the xylenes is often studied alongside that of benzene, with beneficial consequences, and it is therefore not appropriate to deal with each molecule individually. After a short in-troduction to the earlier literature (Section 10.2.1), the kinetics and mechanisms of these reactions in the low-temperature regime will be considered (Sections 10.2.2 and 10.2.3), and then attention is turned to the inversion of rate as temperature is raised: this interesting phenomenon has been widely studied (Section 10.2.4). The reaction of the aromatic ring with deuterium leads to simultaneous exchange and addition, so they can also be examined side-by-side (Section 10.2.5). Other general themes can be identified: these include particle-size effects (Section 10.2.2) and the use of bimetallic catalysts (Section 10.2.6). Features specific to alkyl-substituted benzenes are dealt with in Section 10.3, and the hydrogenation of multiple ring systems in Section 10.4.

10.1.2. Industrial Applications of Benzene Hydrogenation[3]

Benzene is the raw material for the two most important routes for the man-ufacture of polymers that are collectively referred to as Nylon. In the first route, cyclohexane is oxidised to adipic acid ($HOOC—(CH_2)_4—COOH$), and for this reason the full hydrogenation of benzene has commanded much attention. Be-cause of the large heat of reaction, the process is best performed in the slurry phase, using an unsupported catalyst such as Raney nickel: the process heat is removed by vaporisation of the cyclohexane, part of which is returned to the feed, which typically contains 20% benzene + 80% cyclohexane. Reaction conditions are 30–50 atm. hydrogen and temperatures of 453–503 K. In the second route, the starting compound is cyclohexene, and for some time efforts have been made to effect the *partial* hydrogenation of benzene to this molecule (see Section 10.2.7).

Separation from unreacted benzene is easier than with cyclohexane, because an azeotrope is not formed. The quality of diesel fuels is improved by removal of aromatics, and hydrogenation has been explored as a means of eliminating them: sulfur-tolerant catalysts are needed in this application.[4–9] The combination of palladium + platinum is sometimes found to be more thiotolerant than either separately,[10] although this is not the case with model (Pt + Pd)/γ-Al$_2$O$_3$ in the hydrogenation of tetralin.[11]

10.2. KINETICS AND MECHANISM OF AROMATIC RING HYDROGENATION

10.2.1. Introduction: Early Work[1,6,12–14]

For the reasons summarised above, the hydrogenation of benzene in particular has commanded immense attention, and this has generated an enormous literature; it is therefore perhaps inevitable, although at first sight disappointing, that the results obtained are so diverse and contradictory. Keane and Patterson[4] have noted that they are characterised by the following features.

(1) Rate maxima between 423 and 473 K.
(2) Product inhibition or easy desorption of cyclohexane.
(3) Competitive or non-competitive adsorption of the reactants.
(4) Temperature dependence of the order in hydrogen either from zero to 0.5 or from 0.5 to 3.
(5) Activation energies between 25 and 94 kJ mol^{-1}.

And that is just with nickel catalysts; the cited paper gives some 20 references to support this summary. What are we to make of such a medley of results? We have to suppose that within the constraints imposed by the manner in which the reaction was conducted, the type of catalyst and the range of conditions explored, that each of the quantitative measurements is correct; only those such as (2) and (3) which depend on the *interpretation* of what was observed may possibly be in error. This situation is by no means unique, as we have seen, but is more obvious here simply because of the greater amount of work that has been performed. Catalysis is indeed like Cleopatra; a thing of 'infinite variety'. How then are these conflicting conclusions to be reconciled? If we have faith in the scientific method we must conclude that somewhere there are logical explanations for these differences; and it is our task—not an easy one—to try to identify their causes. The best help comes from the more comprehensive kinetic studies, and it is these that will receive most attention.

One of the great problems with studying benzene hydrogenation is the pervasive presence of traces of thiophen, which is a notorious poison, and its removal to low levels is essential for obtaining reproducible results. A scientist was once asked if he took much trouble to purify his benzene. 'No', he replied; 'Not *much* trouble: *very much* trouble'.

Early work on this family of reactions has several times been reviewed.[1-3,12,15] That published by Hilton Smith in 1957 gives a particularly full account of work prior to that date, and will be of interest to historians of the science, as it describes *inter alia* several attempts by Russian scientists to establish a particle size effect and to confirm the validity of Balandin's hypothesis that the fcc(111) plane would be particularly adept for the hydrogenation. The other reviews set the tone for what has been amply confirmed by later work, and the following generalisations can be offered. (1) Among the metals of Groups 8 to 10 the base metals (and palladium) are the least active. (2) Orders in hydrogen are larger than those of the aromatic, the former being often close to unity or even greater, and the latter generally between zero and 0.5; the aromatic molecule is therefore the more strongly adsorbed. (3) Activation energies lie mainly in the range 35–50 kJ mol^{-1}, values over 55 kJ mol^{-1} being rare. (4) The rate is often observed to pass through a maximum as temperature is raised; while this is not unique to benzene hydrogenation, it has been much more often seen than with any other system. These observations provide the framework for a more detailed examination of the reactions' characteristics.

10.2.2. Kinetics of Aromatic Ring Hydrogenation

The hydrogenation of aromatic hydrocarbons can be followed either in the vapour or the liquid phase or in solution, depending upon the molar mass and volatility. Much of the older qualitative work was performed in solution, where an additional dimension, namely, the pH, could be beneficially employed: such work does not however allow profound study of kinetics and mechanism, so it will not be dwelt on here. For the more active metals that can be used at moderate temperatures, the reaction ceases with formation of the alicyclic ring, but with the less active metals (Re, Tc) or metals in less active forms such as wire (Ir), the necessary higher temperatures cause deeper-seated changes, and alkanes containing six or fewer carbon atoms are among the products. These may arise from strongly adsorbed and dehydrogenated residues that at lower temperatures[12,16] cause deactivation.

There are no recent comprehensive measurements of the relative activities of metals for benzene hydrogenation, and since most of the detailed work concerns only nickel, palladium and platinum a short summary of what is known about other metals is all that can be made. Quantitative measurements of seven of the metals of Groups 9 to 10 supported on silica were reported many years ago,[1] and the low activity of the base metals has been confirmed by measurements on films.[17] Rhenium in Group 7 also had low activity,[16,18] but technetium is stated, in one of the very few papers mentioning the catalytic activity of this metal, to be as active as palladium.[16] Catalysis by the metals of Group 7 is a very much-neglected area. Osmium had low activity,[18] while tungsten as film was highly active;[17] rhodium and iridium are comparable with platinum.[1,18–20] All who have used palladium agree that it is only moderately active,[4,5,16,19,21] sometimes being placed below nickel and iron in the hierarchy;[17,22] in view of its outstanding activity for the hydrogenation of alkynes, this is somewhat surprising. Ru/C and Rh/C have been assigned highest activities for hydrogenating *liquid* benzene.[2] Ring hydrogenations in solution catalysed by Adams platinum oxide proceed best under acidic conditions, which neutralise the retained sodium.[23]

The manner of performing the reaction appears to have some strange consequences. While most work has been carried out with continuous-flow reactors, the pulse-flow mode has sometimes been used,[24,25] and this has been claimed to account of an observation[26] that rate was a function of catalyst:support ratio in a physical mixture, an effect not observed in continuous flow.[27,28] However as we shall see shortly there is now good evidence for support involvement, at least in some cases. The pulse-flow mode generates very much higher TOFs[29] (by a factor of 10^3), presumably because the surface is thereby kept clean; it is however hard to imagine that only one part in 10^3 of the surface is active when continuous flow is used, although there is evidence to support such a conclusion.[30]

Table 10.1 shows some of the more recent kinetic measurements for benzene hydrogenation; extensive compilations of earlier results are to be found in references 1, 12, 15 and 29, but they do not alter or add to the picture significantly. The following points should be noted. (1) Values of apparent activation energy are remarkably consistent at 50 ± 10 kJ mol^{-1}, although lower values are sometimes reported:[16,31,32] they increase with benzene pressure over rhenium powder, and with hydrogen pressure over Ni/SiO$_2$, but in this case the effect was removed by using rate constants rather than TOFs. Those for toluene and o-xylene are sometimes a little higher (64–80 kJ mol^{-1}).[4,29,33,34] (2) Orders of reaction are almost always expressed in the Power Rate Law approximation (Section 5.2), and only rarely have the reaction orders been interpreted using Langmuir-Hinshelwood formalism.[22,26,35] (3) There is extensive information on the temperature-dependence of the reaction orders, which almost always increase with temperature (the table shows only maximum and minimum values): their significance will be considered further in connection with the inversion of the rate. Once again toluene and o-xylene behave almost exactly as benzene.[4,33,36]

Before proceeding to consider kinetic equations and implied reaction mechanisms, we may note some other pertinent features of these reactions. (A) Benzene hydrogenation was subject to the influence of the Strong Metal-Support Interaction (Section 3.35) when titania and vanadium sesquioxide were used as supports for rhodium, platinum and iridium;[20,37,38] even Pt/SiO$_2$[39] and Ni/SiO$_2$[40] when heated

TABLE 10.1. Kinetics of the Hydrogenation of Benzene on Metals of Groups 8 and 11

$$r \propto P_H{}^x P_B{}^y$$

Metal	Form	E/kJ mol^{-1}	T/K	x	y	Reference
Co	/SiO$_2$[c]	24	338	1.0	0.2	31
			452	1.7	0.5	
Ni	/SiO$_2$	~50	300	0.5[b]	0.05	35
			360	0.65	—	
Ni	/SiO$_2$	49	408	0.8	0	4,53
			523	2.3	0.5	
Ni	/SiO$_2$	52	298	0.5	0.1	60
			473	2–3	0.3–0.5	
Ni	/Al$_2$O$_3$	42	403	1.25	—	25
Ni	/K-Y zeolite	59.5	403	0.7	0.02	122
			523	2.3	0.5	
Cu	/SiO$_2$	58	306	1.5	0.6	62
			400	2.5	0.6	
Pd	various[a]	~50	353	0.5	0	30,34
			573	4	0.8	
Pt	various[a]	42–54	317	0.6	0.1	46
Ru	/Al$_2$O$_3$	50	313	1	0	
Ru	/SiO$_2$	−26	303	1.2	0	19
			400	2.0	0.2	61

[a] Includes supported catalysts and unsupported powders.
[b] Order decreases with increasing P_H
[c] Similar results for Co/Al$_2$O$_3$, Co/SiO$_2$-Al$_2$O$_3$ and Co/C.

in hydrogen to a high enough temperature showed similar effects. (B) Inclusion of tungsten (as oxide) in Pt/Al$_2$O$_3$[41] and of molybdenum in Pt/SiO$_2$ or Pt/Al$_2$O$_3$[42] led respectively to improved stability and greater activity. (C) In a study of the effect of sulfur and its compounds as poisons, it was found that hydrogen sulfide and sulfur dioxide are *non-selective*, as they affect hydrogenation and exchange equally, while sulfur formed by reaction between them is a *selective* poison, suppressing hydrogenation much more than exchange.[43,44]

There have been a number of attempts over the years to define the *structure sensitivity* of benzene hydrogenation. Extensive early work by Russian scientists, reviewed by Smith,[2] covered cobalt, nickel, palladium and platinum on various supports, and alteration in particle size was supposedly achieved by changing the metal loading, in one case down to 0.03%. However when this work was conducted, X-ray diffraction was the only technique available for estimating particle size, although change in the intensity of the (111) reflection did sometimes produce evidence to support Balandin's ideas, as noted above. Some of the more recent observations are summarised in Table 10.2. There is a general perception that the reaction *is* structure-insensitive, or almost so (Figure 10.1), but this conclusion may derive from the somewhat small range of dispersion covered, and the frequent use of

TABLE 10.2. Structure-Sensitivity in the Hydrogenation of Benzene

Metal	Form	Size Range/Dispersion	Conclusion	Reference
Ni	/SiO$_2$	0.5–5 nm	Slight maximum at ~1.3 nm	123
Ru	/SiO$_2$	0.7–9.5 nm	Maximum at ~3.5 nm	45
Rh	/Al$_2$O$_3$	10–150 nm	TOF constant except at size <1 nm	124
Pd	/Al$_2$O$_3$	10–150 nm	TOF constant	124,125
Pd	/Al$_2$O$_3$	12–100%	TOF constant	126
Pd	/SiO$_2$	4–48%	TOF increases with dispersion[a]	126
Pd	/C	2.5–21 nm	Mild sensitivity below 4 nm	127
Ir	/γ-Al$_2$O$_3$	0.5–3.3 nm	TOF increases with size[c]	128
Pt	/Al$_2$O$_3$	7–65%	TOF constant	43
Pt	/Al$_2$O$_3$	7–65%	Almost insensitive[b]	44
Pt	/Al$_2$O$_3$	4–89%	See text	27
Pt	/SiO$_2$	7–81%	See text	27
Pt	/SiO$_2$	4.5–64 nm	TOF increases × 2 as size increases	129

[a] This effect may be due to impurities in the support
[b] See Figure 10.1
[c] Reactant was toluene.

unnecessarily high reduction temperatures (>673 K), the need for which 'seems to be part of the lore of hydrogenation catalysis'.[27] In fact a careful study[27] has shown that the situation is quite complex. With Pt/Al$_2$O$_3$, the TOF was only independent of dispersion between about 5 and 50% dispersion if it had been reduced at 723 K; at higher dispersions, and throughout if lower reduction temperatures were used, it decreased (see Figure 10.2) by as much as a factor of 10. There were differences too between Pt/Al$_2$O$_3$ and Pt/SiO$_2$, especially when reduced at 723 K, where the latter's TOF *increased* with dispersion. 'Differences in the extent of surface reconstruction after the high temperature pre-treatment' were thought to be responsible.

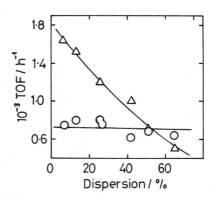

Figure 10.1. Turnover frequencies (h^{-1}) for benzene hydrogenation (O) and exchange with deuterium (△) over Pt/Al$_2$O$_3$ (T = 358 K; P_B = 0.555 Torr; P_D = 205 Torr).[44]

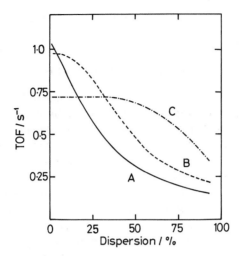

Figure 10.2. Turnover frequencies (s^{-1}) for benzene hydrogenation over pre-oxidised Pt/Al_2O_3 reduced at (A) 373 K; (B) 573 K; (C) 723 K ($T = 353$ K; $P_B = 68$ Torr; $P_H = 692$ Torr).[27]

No evidence was found of an effect of diluting Pt/Al_2O_3 with more support, although in other work a four-fold increase was reported. Ir/Al_2O_3 behaved as Pt/Al_2O_3 when reduced at low temperature,[44] while with Ru/SiO_2 the rate passed through a maximum as dispersion was altered.[45] Extreme care has therefore to be taken before any generalisations about structure-sensitivity are advanced.

Very significant support effects have been seen in the work of Vannice's group.[30,34,46] TOFs for benzene hydrogenation on platinum catalysts were more than five times greater with acidic supports such as silica-alumina and titania, although activation energies and orders of reaction were unaffected. Similar results were obtained with toluene,[29] and with palladium catalysts,[21,34,47] although chlorination of Pt/Al_2O_3 of various dispersions had comparatively little effect.[27] The enhancement was ascribed to the reaction of benzene adsorbed on acidic centres on the support with spiltover hydrogen: this will be referred to again when considering reaction mechanisms.

With metals active in hydrogenolysis (Ru, Re, Ni, Ir, see Chapters 13 and 14), the process of breaking C—C bonds overlaps with that of hydrogenation, and because its activation energy is higher it grows in importance as temperature is raised, and finally becomes dominant. Thus with ruthenium above 405–420 K, technetium above about 440 K and rhenium above about 475 K, the hydrogenolysis rate was easily measurable,[48] activation energies being respectively 125, 121 and 134 kJ mol^{-1}. The process usually goes all the way to methane, although occasionally intermediates have been seen (e.g. with cobalt catalysts[48]). In most cases, however, the occurrence of hydrogenolysis is either not apparent in the range of temperature used, or is ignored.

10.2.3. Rate Expressions and Reaction Mechanisms

In kinetic studies of the hydrogenation of aromatic hydrocarbons, the dependence of rate upon reactant pressures has usually been expressed in Power Rate Law formulations, that is, by orders of reaction that are simple exponents of the pressures. These as we have seen (Section 5.2) are at best approximations to more fundamental expressions based on concentrations of *adsorbed* species,[4,5,14] although they may well represent results over the limited range in which measurements were made. The Langmuir-Hinshelwood formalism has however sometimes been used, and heats of adsorption of the reactants in their 'reactive' states derived from the temperature-dependence of their adsorption coefficients.[4,5,22,30,35]

The next step is to develop a quantitative framework based on an assumed mechanism, and to test this against the experimental results. A particular feature is the dependence of the orders, especially those of hydrogen, upon temperature (Table 10.1). Aspects of the perceived mechanisms, which will now be briefly reviewed, usually ignore any precise description of the reactant hydrocarbon or derived species: these matters will be covered in the section dealing with the exchange reaction.

One particularly vexed question is whether the reaction occurs *only* on the metal particles in supported metal catalysts. Clearly, reaction *can* occur on them, as unsupported metals are effective, but TOFs are lower than on supported metals. This may be in part because metal sites may be deactivated by 'carbonaceous deposits', especially at higher temperatures.[30] Vannice and his associates[34,46–48] have interpreted the larger TOFs found with acidic supports as evidence for a contribution from some spillover catalysis, i.e. reaction of hydrogen atoms moving from the metal to aromatic molecules adsorbed at Brønsted acid centres on the support: but, since the enhancement (that is, the TOF on an acidic support compared to that on a neutral support) can be as much as five-fold, we must conclude that up to 80% of the reaction proceeds by the spillover route—which is no mean contribution. Other workers appear to ignore this possibility,[4,5] which, since neutral or weakly acidic supports are commonly used, may be a reasonable thing to do. A distinction must be drawn between spillover of hydrogen from metal to the support granule on which the metal sits, which is easy, and spillover to another granule of support admixed with the catalyst, which is more difficult (Section 3.34). With Pt/C, however, the latter is possible, as the rate increased to a limiting value as the catalyst was diluted with support.[49]

A further vexed question is whether the reaction proceeds through hydrogen *molecules* arriving from the gas phase or through chemisorbed hydrogen *atoms*. Most workers opt for the latter, and, when the temperature-variation of their coverage is interpreted by the Langmuir equation, values for the entropy and enthalpy of adsorption can be derived.[4,22,0,34,35] A few choose the molecular route,[35,50,51] inaptly termed 'Rideal-Eley', and in one case it is specifically claimed that three pairwise additions are involved. Martin and his associates[35] believed that an ensemble of about four uncovered nickel atoms were needed for adsorption of benzene, this being followed by a molecular hydrogen collision: the rate dependence on hydrogen pressure thus became $(1 - \theta_H)^4 P_H$, and the fractional positive orders (Table 9.1) were seen to result from a competition between these two terms.

There is also disagreement concerning the number of types of site involved and whether adsorption of the reactants is competitive[50] or non-competitive[30,45,48] (or both[50]). Where two different *metallic* sites are invoked, this is because steric constraints to the chemisorption of the aromatic molecule may leave gaps in the adsorbed layer through which hydrogen can enter, to occupy (as atoms) highly coordinated (e.g. trigonal) sites. A recurring theme in discussions of mechanism is the likely multiplicity of states of adsorbed hydrogen, of which only the weakest is reactive. Where the aromatic molecule is supposed to reside on the support, the hydrogen molecule cannot of course compete. The presence of benzene did not inhibit para-hydrogen conversion,[52] as happens with alkenes and alkynes.

The two most thorough interpretations of the kinetics are those of Keane and Patterson[4,5,53] for Ni/SiO$_2$ and of Vannice and associates for various platinum[29,46,54] and palladium[30,44,47,48] catalysts. In the first case, the reactants were taken to adsorb reversibly on separate sites, the addition of the first hydrogen atom being rate-determining; but regarding this step as a 'pre-equilibrium' seems a doubtful benefit. The experimental reaction orders, both of which increased with temperature (Figure 10.3 and Table 10.1), 'reveal the temperature-induced changes in the reactive species'. In the second case, provision was made for inhibition by strongly-held dissociated species such as phenyl and tolyl, the composition of which depended upon reaction conditions; with the xylenes, however, the played no role. Reaction orders varied with temperature much as with Ni/SiO$_2$. The extent of the contribution from the spillover catalysts was not explicitly stated, but it presumably depended on the degree to which metal sites were rendered unavailable to strongly held entities. The large variations in activity as the support was changed originated entirely in the pre-exponential term, so that the energy profile of the transition state did not depend on the contribution from spillover catalysis, an observation not easily explained.

It seems somewhat odd that two sets of results similar in respect of kinetic parameters, including their temperature-dependence, should be described by mechanisms that differ so considerably. It is of course possible for the mechanistic framework demanded by every catalyst to be unique, however improbable this appears. This seemingly straightforward class of reactions is in fact very complex, and none of the mechanistic proposals embraces all the potentially available information. If each group produces a scheme that satisfies its results within the

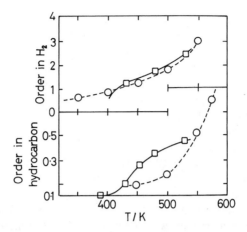

Figure 10.3. Orders of reaction for benzene hydrogenation as a function of temperature. ◯ Pd on various supports (fixed pressures, $P_H = 600$ Torr; $P_B = 50$ Torr).[30] ☐ Ni/SiO$_2$ (fixed pressures, $P_H = 714$ Torr; $P_B = 30$ Torr).[4]

limitation of the range of variables covered, it is not surprising that there is a diversity of opinions.

It is interesting to compare the mechanisms proposed for the hydrogenations of ethyne and benzene. With the former, there is no suggestion of a role for spillover catalysis; with the latter there is no role for 'carbonaceous deposits' in creating active centres or in acting as vehicles for hydrogen atom transfer.

10.2.4. Temperature-Inversion of Rates

When the temperature used for the hydrogenation of benzene or its alkyl-substituted derivatives is raised above a value that is usually between 423 and 437 K, the rate suddenly stops rising and begins to fall. The inversion temperature T_{max} is usually sharply defined and can be estimated accurately, although it is less distinct at low hydrogen pressures than high, perhaps because the effect is obscured by de-activation. The phenomenon has been seen with most of the metals of Groups 8 to 10 (Co, Ni, Ru, Rh, Pd, Pt), as well as with technetium,[16] rhenium,[16] copper,[62] and nickel-copper bimetallics[59] (see also 'Further Reading' list at the end of the chapter). Values of T_{max} are rarely below but sometimes above the 423–473 K range, especially with palladium. They decrease with increasing alkyl substitution by about 10 K per methyl group.[4] They sometimes depend on the hydrogen pressure used,[4,5,60] but not with toluene on Ni/SiO$_2$; with Rh/TiO$_2$ they were higher (485 K) after high temperature reduction than after low temperature reduction (420 K).[37,38]

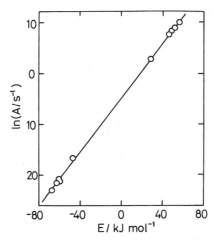

Figure 10.4. Compensation plot for the hydrogenation of aromatics on Ni/SiO$_2$:[4] Arrhenius parameters derived from TOFs, conditions as in Figure 10.3.

Some workers have chosen to ignore this effect, but its analysis can be very informative; in what follows, we examine the methodology of Keane and Patterson.[4,5,14,53]

They have made a detailed study with a Ni/SiO$_2$ catalyst of the temperature-dependence of the reaction order (Figure 10.3 and Table 10.1), and have determined activation energies using benzene, toluene and the three xylene isomers: the change in rate with temperature above T_{max} also conforms to the Arrhenius equation, and affords negative apparent activation energies. All the Arrhenius parameters obtained using TOFs both above and below T_{max} lay about a single compensation line (Figure 10.4), as did those derived from rate constant, for which activation energies were, as expected, higher (e.g. for benzene, 49 compared to 29 kJ mol^{-1}). However the temperature-dependence of the reaction orders immediately implies a dependence of activation energy on reactant pressures, and this was indeed established for hydrogen pressure variation in the reaction of toluene; values of activation energy derived from rate constants k were essentially invariable. Care is taken not to apply the labels 'apparent' and 'true' to these activation energies, as classical theory would suggest. The authors carefully point out that what we might call $E \leftarrow k$ or E_k is not a *true* activation energy, as it still contains a dependence upon heats of adsorption, values of which were obtained from adsorption coefficients extracted from the temperature-dependence of the reaction orders through a Langmuir-type relation. Values, independent of pressure, ranged from 74 kJ mol^{-1} for benzene to 120 kJ mol^{-1} for o-xylene, but for hydrogen they were the same for each aromatic molecule, but coverage-dependent from 31 kJ mol^{-1} ($\theta = 0.97$) to 77 kJ mol^{-1} ($\theta = 0.02$). Using a form of the Temkin equation (eqn.5.29) in which the heat terms were not moderated by the orders of reaction, values of E_t between 155 and 270 kJ mol^{-1} were obtained; this may reflect the obstinacy of the aromatic ring to attack.

The analysis is still not entirely satisfactory for several reasons. (i) Values of $-\Delta H_a(H_2)$ seem somewhat large for 'weakly' adsorbed hydrogen,[50] and others have obtained substantially different values from their results. (ii) The cause of the inversion is not well established. It is probably not due to deactivation by carbonaceous residues,[30] as the effect is reversible.[34,53,58] Although there are inflexions in the order versus temperature plots at about T_{max} (Figure 10.3),[4] the Van't Hoff isochore plots for hydrogen show no breaks at this point,[4,5] and it is illogical to ascribe the negative activation energy to the intrusion of the heat of desorption of the aromatic molecule, as this should operate at all temperatures. It might be thought that the decrease in rate of the forward reaction was a consequence of the growing importance of the reverse reaction, due to the effect of temperature on the position of equilibrium, but at least for benzene and toluene it becomes noticeable only at temperatures above T_{max}. A definitive explanation is still awaited. (iii) Comparison with the reactions of ethene and of ethyne is instructive, if a little worrying. For these reactions there are no comparably detailed measurements of the effect of temperature on orders of reaction, although negative orders in the hydrocarbons ought to become less negative as temperature is raised if they are caused by overly strong adsorption, and on palladium the hydrogen order for ethyne hydrogenation becomes more positive at higher temperatures.[61,63] We have tended to regard activation energies for these reactions of respectively about 45 and 65 kJ mol^{-1} as being *true* in the sense that concentrations of adsorbed reactive intermediates do not change significantly within the range of measurement. Although rate maxima have been detected in the reactions of both ethene[1] and ethyne,[63] they have not been much studied and their cause is obscure. What is certain, however, is that the heats of adsorption of the reactants in their 'reactive' states must be considerable, and the application of the Temkin equation (5.29) without modification would lead to much larger values. We are thus faced with the paradox that estimation of true activation energies only requires correction of the apparent values by addition of the heats of adsorption when adsorbed concentrations change significantly, but not otherwise.

10.2.5. Hydrogenation of Benzene Over Bimetallic Catalysts

Studies of benzene hydrogenation over bimetallic catalysts are of limited value in illuminating reaction mechanism except when conducted with deuterium, when comparison of exchange with addition is informative (Section 10.2.6) otherwise what is found is more relevant to the understanding of how bimetallic systems behave, and how their composition determines their activity.

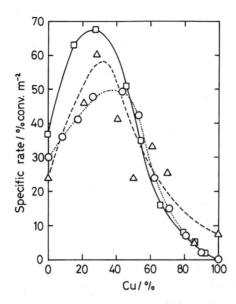

Figure 10.5. Hydrogenation of benzene on nickel-copper powders: rate dependence on composition. ◯, 435K;[65]□ ,435K;[67] △, 463K.[66]

The nickel-copper system has been several times investigated, using powders,[64−67] foils,[68] films[69] and silica-supported catalysts,[64] so it is of some interest to compare the results, especially where Arrhenius parameters are available. We may recall that the equilibrium surface concentration remains fixed at about 23% nickel over a wide range of total composition at moderate temperatures due to the occurrence of a miscibility gap (Section 1.3),[69] although catalysts reduced at low temperatures may not have equilibrated. The form of dependence of activity on composition may therefore be expected to vary with method of preparation. Thus sintered films show constant rates and activation energies in this region,[69] while other forms show distinct maxima at about 30–40% copper[65−67] (Figure 10.5): this may be due in part to a lesser tendency to deactivation of nickel-rich catalysts by 'carbon' deposition.[59] Activation energies are *higher* for the bimetallics, and approximately constant in the mid-composition range (Figure 10.6: the apparent compensation shown in reference 65 may be simply due to experimental scatter); they tend to decrease at high copper concentrations.

Somewhat different behaviour has been found when Group 11 metals are added to platinum[70] and palladium.[58,65,71] Activities fall continuously, and activation energy initially,[65] but the former is determined subsequently by the decrease in the pre-exponential factor (Figure 10.7). Activities were also increased by adding molybdenum,[41] rhenium[18,72] or iridium[18] to platinum. There are conflicting

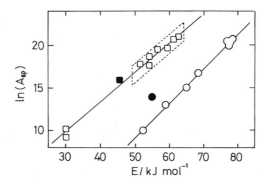

Figure 10.6. Compensation plots for the hydrogenation of benzene on nickel-copper powders: ○ ref. 67; □ ref. 65; filled points, Ni only. Note: in the upper plot, the points within the box almost certainly show compensation because of experimental scatter.

reports on the effect of adding copper to ruthenium.[73,74] The more detailed study[73] showed (unusually) how activities change with time, the bimetallics being stabler than ruthenium alone and thus after 24 h being much the more active. Orders for both reactants became more positive with increasing temperature, as was found with nickel[4] and platinum,[34] but maximum rates occurred at the remarkably low temperature of $\sim$300 K, so that most results were obtained above this value. Other reports concern the nickel-platinum,[75] cobalt-platinum,[76] palladium-rhodium,[77] osmium-iridium[18] and nickel-cobalt and -iron systems.[1]

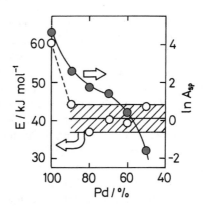

Figure 10.7. Hydrogenation of benzene on palladium-copper powders: activation energy and pre-exponential factor (ln A_{sp}) as functions of composition.[65]

10.2.6. Exchange of Aromatic Hydrocarbons with Deuterium

Substitution of the hydrogen atoms on the aromatic nucleus for deuterium atoms occurs simultaneously with the process of addition, although it is frequently very much faster. It was one of the set of reactions that was used in the 1930s to explore the applications of the newly-discovered deuterium in the laboratories of Polanyi[78] and the Farkas brothers.[52] Much thought has been given to its mechanism, and this has led to the belief that it is distinct from that of the addition process. Evidence for this rests on the following observations. (1) The dependence of TOFs on particle size is quite different for each process. (2) Kinetic parameters are also not the same. (3) Exchange has been observed on surfaces that have no activity for addition (Ag film;[17] PdAu film containing less than 40% Pd[79]).

The exchange reaction of benzene has been studied in much less detail than its hydrogenation. All the metals of Groups 8, 9 and 10 have been examined,[80] but unfortunately only a very short summary is available; more detailed information can only be found for palladium,[17,79] platinum,[17,52,78,81,82] nickel,[69,78,83] and iridium.[83] Rates of exchane on several metal films of Transition Metals ran parallel to those for hydrogenation, and correlated with electrical conductivities.[8] Exchange is not a simple stepwise process, as more extensively exchanged molecules (up to benzene-d_6) appear as *initial* products: the mean number of deuterium atoms in exchanged benzene M depended on the metal and on temperature (e.g. on Pd film, 1.8 at 273 K, and 2.7 at 311 K[17]). On nickel, multiple exchange was thought to be absent because most of the deuterated cyclohexane had only six deuterium atoms.[83] On iridium the products of exchange were mainly benzene-d_1, with a little benzene-d_6. Assuming a dissociative exchange mechanism involving repeated interconversion of phenyl and phenylene species (Scheme 10.1), observed distributions could not be reproduced by a single value of the parameter describing the chance of phenyl $\rightarrow$ phenylene. The extent of multiple exchange is reported[80]

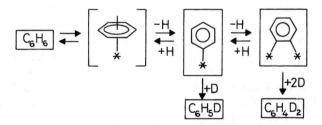

Scheme 10.1. Dissociative mechanism for the exchange of benzene.
Note: reiteration of these steps can account for the formation of molecules containing up to six deuterium atoms as initial products.

to decrease in the following sequence (probably obtained with silica-supported metals):

$$Ru > Rh, Pd, Os > Ni, Ir > Pt, Fe, Co$$

but the chemical logic behind it is hard to discern. The deuterium content of the cyclohexane increased as the reaction proceeded, due to the use of progressively more exchanged benzenes. Cyclohexane-d_{12} constitutes the major product over palladium and palladium-gold film,[79] where the rate of exchange must exceed that of addition. Cyclohexanes do not exchange further after being formed.

Measurements of kinetic parameters for exchange are sparse. Activation energies on platinum and palladium films were much higher than for addition,[79] and on the latter the order in deuterium was negative (-0.5) for exchange but positive (0.8) for addition. Orders of zero for both reactants have been recorded for a number of metal films.[8]

The ways in which the rates of the two processes depend on particle size has been followed with Pt/Al$_2$O$_3$[44] (Figure 10.1) and Ir/Al$_2$O$_3$,[84] and with Ni/SiO$_2$ and iridium probably supported on silica.[83] In each case addition showed a small dependence on size, while the exchange rate increased markedly as the mean size *increased*. However, the form of the distribution of the initially formed exchanged benzenes and cyclohexanes was *not* size-dependent.[83] Strongly adsorbed sulfur-containing molecules may be expected to lower the average size of ensembles of free atoms: molecules such as thiophene and sulfur dioxide affected both reactions equally, but elemental sulfur selectively deactivated exchange.[43]

The two reactions have been examined on bimetallic films. With the nickel-copper system, rates, activation energies and M values were constant in the region of constant surface composition (23% Ni), but activation energies were higher (105 kJ mol^{-1}) than for pure nickel (50 kJ mol^{-1}).[69] With palladium-gold films, activity for addition fell to zero when the gold content exceeded 60%, but the activation energy for exchange stayed in the range 68–90 kJ mol^{-1} up to 82% gold: values for addition were much lower (18–28 kJ mol^{-1}).[79]

The introduction of one or more alkyl groups onto the aromatic nucleus divides the remaining hydrogen atoms into sets depending on their proximity to the substituent, and the rate of exchange in each set may thereby be distinguished, either for steric reasons (e.g. the alkyl substituent(s) may cause the ring to tilt rather than lying perfectly flat) or because of differences in C—H dissociation energies induced by the substituent(s).[12] Exchange may also occur in the alkyl group, the hydrogen atoms of which can also fall into sets depending on how close they are to the ring. Study of the exchange reactions of alkyl benzenes may therefore be expected to illuminate reaction mechanisms. These factors are well illustrated by the exchange of n-propyl benzene with deuterium over nickel film.[12] Its hydrogen atoms may be divided into four sets (Figure 10.8); the hierarchy of exchange rates

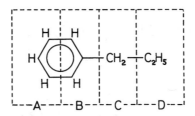

Figure 10.8. Exchange of n-propylbenzene with deuterium: allocation of hydrogen atoms into differentiated groups.

on unsintered and sintered films is given in Table 10.3. What was most remarkable was that the rate for ring set A relative to that for the alkyl hydrogen atoms was lowered by sintering by a factor of 50; it was concluded that the ring exchange mechanism could not therefore be the same as that for the alkyl group, which is well established to proceed by dissociation of a C—H bond.

Exchange of p-xylene with deuterium has been examined on various supported platinum catalysts (Pt/α-Al$_2$O$_3$, Pt/γ-Al$_2$O$_3$, Pt/SiO$_2$) and on platinum film.[82] At 360-373 K, the ratio of the rates of exchange and addition on Pt/γ-Al$_2$O$_3$ was almost constant between 28 and 100% dispersion, higher values of M appearing at dispersions (D) greater than 83%. On Pt/α-Al$_2$O$_3$ however this ratio was much higher ($D = 8$ or 18%), but the value of M was lower (1.1). Ring exchange and addition were much faster on platinum film (T = 273 K), exchange in the side-chain being faster than in the ring. p-Xylene exchange has also been studied on films of palladium and tungsten.

An extensive study has been performed on the exchange reaction between benzene and benzene-d_6 at 273 K, using films of no fewer than 16 metals.[8,12] The mean number of alter-atoms entering each molecule ranged from three for titanium to one for palladium, but for many metals it was close to two. This suggested that under these conditions (i.e. low concentration of either hydrogen or deuterium atoms) the benzene easily lost *two* hydrogen (or deuterium) atoms to form adsorbed phenylene, which then collected two new atoms to form an exchanged benzene. In each Transition Series, rates appeared to decrease linearly with increasing atomic radius.

TABLE 10.3. Exchange of n-Propylbenzene with Deuterium on Nickel Film: Reactivity Sequence of Groups of Hydrogen Atoms as shown in Figure 10.8[130]

Condition	Reactivity Sequence
Unsintered film, 273 K	A = C >> B > D
Sintered film, 303–323 K	C > D > A > B

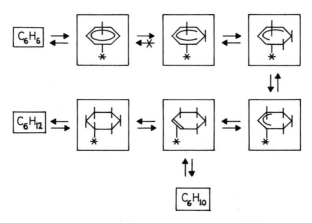

Scheme 10.2. Mechanism for the hydrogenation of benzene by successive addition of hydrogen (or deuterium) atoms.

Note: all steps are shown as reversible, so that the mechanism *could* also describe the exchange: however the loss of resonance energy on addition of the first atom probably makes its reversal very unlikely. The mechanism also describes the formation of cyclohexene (Section 10.27) and the dehydrogenation of cyclohexane.

It remains now to review briefly the various mechanisms that have been proposed to describe the process of ring exchange. There seems to be general agreement, based on the criteria briefly listed in the opening paragraph, and subsequently elaborated, that exchange occurs in a way which is quite separate from the route that leads to addition, for which the first and probably slowest step is the addition of a hydrogen or deuterium atoms to the ring, with loss of the resonance stabilisation. It is therefore unlikely that exchange occurs by reversal of this step, and likely that addition of further hydrogen or deuterium atoms is fast (Scheme 10.2). The dissociative mechanism for exchange, perhaps proceeding through a flat π-adsorbed intermediate, may account for the negative order in deuterium, as two free 'sites' are needed to chemisorb benzene as phenyl, and also the preference that the reaction apparently has for large ensembles of atoms, as it goes most easily on large particles. The consistently higher activation energy for exchange than for addition may also be consistent with this mode of reaction. Detailed argument along these lines has not however been offered. We must conclude that the mechanism of the exchange of aromatic molecules has not yet been firmly established, although as we shall see shortly the nature of the products obtained in the exchange of naphthalene provides some pointers. There are also some notable gaps in our knowledge of aromatic exchange: it does not seem to have been established for example whether rates exhibit a maximum as temperature is raised, or whether spillover to the support plays any role in the case of supported metals.

10.2.7. Hydrogenation of Benzene to Cyclohexene[85]

As noted in Section 10.12, the partial hydrogenation of benzene to cyclohexene is an economically attractive first step in the synthesis of polyamides, because
it is more easily separated from unreacted benzene than is cyclohexane. The reaction is however a difficult one to accomplish, because cyclohexene is normally
much the more easily hydrogenated,[86] so its appearance in any significant amount
depends upon constructing a catalyst which is in effect selectively poisoned for
the hydrogenation of the alkene but not for the more strongly adsorbed benzene.
There have been many attempts to modify metals (mainly ruthenium[87]) to secure
this result, in ways that are now briefly recounted.

Many of the available results have been obtained in three-phase systems, using autoclaves at 423–473K and 10–70 atm hydrogen;[88] unsupported ruthenium
catalysts have often been used, sometimes generated *in situ* by precipitation of the
hydroxide from an aqueous solution $RuCl_3$, followed by its reduction on introduction of hydrogen.[87,89] All agree that the presence of water is essential in securing
reasonable selectivity to cyclohexene, but its role is not well understood.[88] Early
results indicated that iron salts arising from corrosion of the stainless steel of
the autoclave might be beneficial to selectivity, although inimical to rate.[89] Many
salts have subsequently been tried, the sulfates of iron, cobalt and zinc proving
best;[90,91] in these cases, selectivities of about 50% have been obtained, at a sac-
rifice of two-thirds of the rate. The role of the salts may be to render the metal's
surface hydrophilic and thus to secure better access of water to the surface,[90] and
so fulfil its role, whatever that is. Various bases have been used to precipitate the
$Ru(OH)_3$, those of calcium and strontium being best (selectivities respectively
64 and 67%).[87] Various supports for ruthenium have also been tried,[92] those of
ytterbium, zirconium and iron (Fe^{III}), with potassium or calcium hydroxide, all
giving selectivities over 70%.[87]

A practical large-scale process would be better conducted in a continuous
mode, for which vapour-phase reaction is preferable, so this has attracted some
attention. The presence of water is still essential, but other modifiers such as ethylene glycol (1,2-dihydroxy-ethane) and ϵ-caprolactam have proved beneficial.[93]
A good understanding of the role of these various parameters and modifiers is not
yet available.

Alkyl substitution on the benzene ring increases the selectivity for partial
reduction.[2,88.,93,94] With toluene and ethylbenzene, using ruthenium with zinc
oxide, the 2,3- alkylcyclohexene was surprisingly the main product, this being
less stable than the 1, 2- isomer; *m*-xylene however gave chiefly 1,4-dimethyl-1,
2-cyclohexene. The presence of tert-butyl groups is even more effective in
easing the intermediate alkene off the surface: thus even with rhodium 1,2-di-
tert-butylbenzene gave ~40% of the 1,2-dialkyl-2,3-cyclohexene, while 1,3,5-tri-
tert-butylbenzene afforded 65% of the 1,3,5-trialkyl-1,2-cyclohexene.[95]

Cyclohexadiene is very rarely seen as a product of benzene hydrogenation, but inexplicably the 1,3-isomer constituted about 80% of the initial products when the reaction was conducted in isopropanol with a commercial 1% Ru/C catalyst: smaller but still significant amounts were found with 2.5% Ru/TiO$_2$.[96] Metallic impurities in the supports (C, Al$_2$O$_3$, SiO$_2$, TiO$_2$) were found to matter. The 1,4-isomer reacted much more quickly than benzene over Ru/SiO$_2$ and Cu/SiO$_2$, but hydrogenation of a mixture of the 1,3-isomer and ^{14}C-labelled benzene on several metals including ruthenium showed that the label appeared more in cyclohexane than in cyclohexene,[97,98] perhaps because the diene was largely excluded from the surface by the more strongly adsorbed benzene.

10.3. HYDROGENATION OF ALKYL-SUBSTITUTED BENZENES

10.3.1. Kinetic Parameters

Although the work described in Section 10.2 related predominantly to benzene, many of the studies described used toluene and the xylenes as well, and because the hydrogenation of all of these compounds had many features in common it would have been inappropriate to separate them. However there are some further aspects which merit attention, and which if introduced earlier might have interrupted the flow of the argument; and also there are results to be mentioned concerning more highly substituted molecules. Some recapitulation of what has been said before is however inevitable in order to provide a complete picture.

It is now well-established that rates of hydrogenation decrease with the number of methyl substituents,[99] although with the xylenes and more extensively substituted molecules they depend upon their relative positions[1,2,86] (Figure 10.9). For the former the sequences of rates or TOFs on Ni/SiO$_2$,[4,5] Raney nickel and Adams platinum[1,2] is *para* > *meta* > *ortho*. There is general agreement that the cause of the reactivity hierarchy is to be sought in the electron-releasing effect of the substituents, which increases the electron density within the ring in proportion to their number, and thus leads to stronger π-donor bonds to the surface. This effect is demonstrated by a decrease in ionisation potential, and to an increase in stability constants for the formation of charge-transfer complexes,[12] as well as in the heats of adsorption deduced from the kinetic analysis of the reactions on Ni/SiO$_2$.[4,5] Although it might explain a decrease in the rate of exchange (for which quantitative results are lacking), it is by no means clear that it would affect the ease of attack on the ring, and the variations of rate with isomer structure argue for a geometric as well as an electronic effect. Relative rates for *n*-alkyl substituents of increasing chain length reach a lower limit after four carbon atoms,[1,2] and molecules with secondary and tertiary substituents are less reactive than corresponding methyl compounds.[86] Penta- and hexamethyl-benzene are very difficult to hydrogenate.[1]

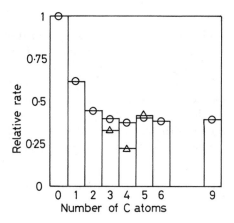

Figure 10.9. Effect of n- ($\ominus$) and sec- ($\triangle$) alkyl substituents on the rate of hydrogenation of the aromatic nucleus on platinum.[2]

Apparent and derived true activation energies increase (and those obtained above T_{max} become more negative) as the number of substituents increases when Ni/SiO$_2$ is used,[4,5] but the effect is not the same with all catalysts.[12] The similarities in the temperature dependence of the reaction orders suggests[4,5] that the tightness of packing of aromatic molecules on the surface is not much affected by inserting one or two methyl groups, and the energetics of binding to the surface is indeed the source of the reactivity differences.

The difference between benzene and toluene has given a group of French scientists the bright idea of using their relative reactivities, or more properly the relative values of their adsorption coefficients (K_T/K_B) derived therefrom, as a way of seeing how the electron-deficient character of surface metal atoms depends upon prevailing circumstances.[100] The more electron-deficient the surface, the more strongly should toluene be adsorbed relative to benzene, and thus its reactivity should be diminished relative to benzene: in other words K_T/K_B should measure the number of unoccupied states in the metal's d-band.[100,101] Thus this ratio has been shown to decrease dramatically on passing from Group 8 to 9 to 10 (Table 10.4), and its values (except for that of ruthenium) run parallel to the electronic specific heat, which in turn depends on the density of states at the Fermi surface.[100] Two examples of the application of this concept may be cited: (1) with

TABLE 10.4. Ratio of the Adsorption Coefficients for Toluene (K_T) and Benzene (K_B) on the Noble Metals of Groups 8 to 10

	Ru	Rh	Pd	Os	Ir	Pt
K_T/K_B	200	10	1	55	24	8

$Pt_{1-x}Zr_x$ supported on carbon or alumina, K_T/K_B is proportional to x, suggesting electron transfer *from* platinum *to* zirconium, as predicted by the Engel-Brewer theory, and (2) chemisorption of sulfur on platinum has been shown to *decrease* electron density of the surface, while carbon has the opposite effect.[102] The ratio K_T/K_B was very large for ruthenium, about 10 for rhodium and about unity for palladium,[85] which may help to explain their different activities in these and other reactions. An extensive kinetic study of the hydrogenation of mixtures of benzene and toluene on Ni/Y zeolite has however revealed a situation of some complexity,[33] and it is not certain that the original simple concept is totally valid.[103]

10.3.2. Stereochemistry of the Hydrogenation of Alkyl-Substituted Benzenes[6,104]

It has been known since 1922[2,105] that the principal product formed by the hydrogenation of dialkylbenzenes is usually the corresponding *Z*-dialkylcyclohexane, although the amount of the *E*-isomer depends upon a number of factors. (1) With the xylenes, it varies with the isomer,[1,53,88,106,107] and with rhodium, ruthenium and platinum the sequence is *para* > *meta* > *ortho*. (2) Its proportion increases with temperature[2] and with hydrogen pressure[36,108] in each case. (3) Rhodium affords more of the *E*-isomer than does ruthenium.[2] These observations relate to experiments performed with liquid or solution, and other factors have also been recognised.[2] In the gas-phase on Ni/SiO$_2$, the *E*-selectivities for each isomer vary with temperature as shown in Figure 10.10;[53] it may be more than coincidence that *o*-xylene achieves the same value as the temperature-independent value shown by *m*-xylene, but no explanation has been suggested. Ru/Al$_2$O$_3$, Ru/SiO$_2$ and Ru/TiO$_2$ all gave 10–20% of the *E*-isomer from *o*-xylene, its amounts decreasing with increasing particle size.[109]

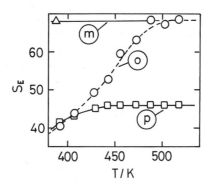

Figure 10.10. Temperature-dependence of the selectivity to *E*-dimethylcyclohexane in the hydrogenation of each of the xylene isomers over Ni/SiO$_2$.[53]

There have been lengthy discussions concerning the significance of these observations, without firm conclusions being reached.[53,106] Epimerisation of the cyclohexane is clearly not a viable general explanation,[53] and the phenomena are obviously related to those seen with alkylcyclohexenes, discussed in Chapter 7. Stereochemistry is presumably decided as the final pair of hydrogen atoms is added, so that Z-addition, as with the cyclohexenes, is not unexpected; but the considerable formation of the E-isomer may require the temporary desorption and re-adsorption of one of the intermediates,[110] as was suggested previously (Section 7.5.3).

10.4. HYDROGENATION OF MULTIPLE AROMATIC RING SYSTEMS

10.4.1. Polyphenyls

This short section is concerned with molecules having two or more benzene rings that are not condensed. The rings therefore have a degree of independence from each other, and so their reduction may show some selectivity.

Biphenyl has been hydrogenated on five of the noble metals supported on carbon, and palladium again distinguished itself by showing 100% selectivity to cyclohexylbenzene: the others gave values between 50 and 65%.[94] It also excelled in reducing diphenylmethane to cyclohexylphenylmethane with 89% selectivity.[111] Over Adams platinum, m-terphenyl suffered selective reduction of the central ring, but with the *para*-isomer it is the terminal ring that was reduced first.[88]

Many years ago, Homer Atkins and his colleagues examined the reduction of triphenylmethane catalysed by nickel powder;[112] it went non-selectively to the tri-cyclohexylmethane. 1,3,5-Triphenylbenzene was reduced to the tri-cyclohexylcyclohexane. With phenylethanes, rates decreased slightly with increasing number of phenyl groups.[113]

10.4.2. Fused Aromatic Rings: (1) Naphthalene[94,114]

This section is concerned with the hydrogenation of naphthalene and of tetralin (tetrahydro-naphthalene) and their alkyl derivatives, and the reactions with deuterium. The focus of interest lies in the stereochemistry of the reactions, in the analysis of the products (sometimes very complicated), and their significance for the understanding of reaction mechanisms.

There is much evidence from early work summarised by H.A. Smith,[2] and amply confirmed by later careful work,[114] that naphthalene can be hydrogenated to tetralin, i.e. one of the rings could be saturated, with a high degree of selectivity on nickel and the noble metals (Table 10.5), although in some cases it declined somewhat at higher temperatures. In further hydrogenation, yields of octalin were

TABLE 10.5. Selectivities in the Hydrogenation of Naphthalene to Tetralin (S_T), Octalins (S_O) and Decalin (S_D) on Metals of Groups 8 to 10[114]

Metal	Support	T/K	S_T	S_O	S_D
Ru	Al_2O_3	303	82.0	1.9	16.1
Rh	Al_2O_3	298	95.1	0.9	4.0
Pd	Al_2O_3	373	99.8	—	0.2
Ir	C	298	86.5	0.4	13.1
Pt	Al_2O_3	473	96.3	0.2	3.5

low because of the fully saturated decalin. It is natural to enquire how the two initially identical rings differ when they interact with the surface to account for this highly selective partial reduction.[115] There are no structural studies to guide us, nor indeed are there any on other compounds having two or more rings (e.g. biphenyl) that might delineate the separation of adsorption sites. It has however been suggested[114] that there are two possible forms, in one of which the interaction is through only one of the rings, and in the other of which it is through one (or perhaps two?) of the C=C double bonds represented in one of the canonical forms (Figure 10.11). It must also be remembered that the total resonance energy in naphthalene (255 kJ mol^{-1}) is less than twice that of benzene, so that the second ring will be harder to reduce than the first.

The presence of a single alkyl substituent in the 1-position has a marked effect on which ring is reduced first: with Pd/C as catalyst, the proportions of reduction of the ring carrying the substituent increase in the sequence: methyl (34%) < ethyl (45%) < isopropyl (68%) < *tert*-butyl (97%).[88] With platinum, however, the unsubstituted ring is the more reactive.[114] The size and position of the substituent also influences its rate of reduction. There are no kinetic studies to differentiate between effects on the adsorption coefficients and on the rates of reaction of the molecules once they have been adsorbed, of the kind conducted with benzene and toluene; a combination of steric and electronic factors may be at work.

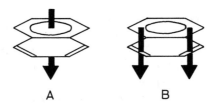

Figure 10.11. Representations of the adsorbed states of naphthalene.[114]

    (A) Interaction with the surface through one ring.

    (B) Interaction through one (or two) C=C double bonds in a frozen canonical form (the likely form on palladium).

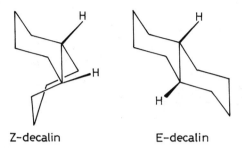

Z-decalin   E-decalin

Figure 10.12. Structures of Z- and E-decalin.

Decalin exists as Z- and E-isomers (Figure 10.12), and if all ten hydrogen atoms were added to the same side of the naphthalene ring Z-decalin would be the sole product. In fact both products are formed;[114] ruthenium showed very high selectivity to the expected Z-product, while palladium gave about 50% of the E-isomer, the other noble metals giving intermediate values. Palladium was also notable for producing more than 80% of the E-isomer from $\Delta^{1,9}$-and $\Delta^{1,10}$-octalins (Section 7.53).[114] Selectivities to the two isomers were about the same when tetralin was the reactant. The mechanism whereby the E-isomer is formed has been the subject of much discussion, and the problem is closely related to that of E-addition to substituted cycloalkenes (Section 7.5.3). The possibility that the E-isomer arises from topside reaction of a hydrogen (or deuterium) molecule has been dismissed for a variety of reasons, so the remaining option, namely, that of desorption, inversion, and re-adsorption of an octalin intermediate, has received careful consideration.[114] This hypothesis requires it to be shown that octalins are indeed formed, that they are sufficiently reactive, and that they in fact do what is expected of them. Of the six possible octalin isomers, the $\Delta^{1,9}$-predominates, and it is Z-addition of this isomer that leads to E-decalin[114] (Scheme 10.3). Table 10.5 indicates that small amounts of octalins are indeed formed and are reactive in the presence of the aromatic compound; in fact they were initially very significant products (15–30%) of the hydrogenation of both naphthalene and tetralin, but especially with the former their concentration quickly decreased as reaction proceeded. Palladium is noted for its ability to hydrogenate alkenes in the presence of aromatics, while ruthenium, rhodium and iridium favour aromatics over alkenes. This has been nicely demonstrated by hydrogenating o-xylene with a mixture of palladium and (say) ruthenium catalysts: the latter produced cycloalkenes, which were then rapidly hydrogenated by the former, so that lower amounts of the cycloalkenes and more of the E-dimethylcyclohexane were found than with ruthenium alone. The distinctive properties of palladium are again evident, but the different behaviours of the various noble metals still need an explanation. With palladium and platinum on (or in?) various zeolites, their acidity affected the ratio

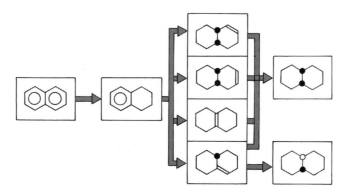

Scheme 10.3. Formation of *Z*- and *E*-decalins by *Z*-addition of hydrogen to octalin intermediates in the hydrogenation of naphthalene or tetralin.

of *Z*- and *E*-decalins formed from naphthalene, high proportions of either being obtainable by appropriate choice of components: metal particle size was not a determining factor.[116]

The consequential studies of the hydrogenation of the octalins[117] have been reviewed in Section 7.5.3. To cut a very long story short, the suggested route to *E*-decalin has been validated, or at least has been found to be satisfactory; of course if we believe Karl Popper there may be an even better explanation around the corner.

The addition of a methyl substituent onto the naphthalene nucleus differentiates the two rings: hydrogenation of 1- and 2-methylnaphthalene has been examined on all the noble metals of Groups 8 to 10 (except osmium),[114] and with Ru/Al$_2$O$_3$ about 80% of reduction occurred at the unsubstituted ring. (Note the numbering of the atoms in the methyltetralins starts on the *reduced* ring). Hydrogenation of the methyltetralins revealed another novel feature, namely, their isomerisation in parallel with their reduction: thus hydrogenation of 1-methyltetralin at 473K and 55% conversion with Pd/Al$_2$O$_3$ gave 38% 5-methyltetralin, 2% octalins and 60% decalins. The other metals gave small amounts (<4%) of the isomer, and similar but less marked effects were seen with 2-methyltetralin. There are four stereoisomers of each of the methyldecalins (see Scheme 10.4 for their structures and naming). Except with palladium, where the *E-anti*-isomer predominated, the *Z-syn*-isomer was the principal one: ruthenium gave 95% of the two *Z*-isomers, but palladium only 55%, the others giving intermediate values. It was suggested that re-adsorption of the methyloctalin in its least hindered orientation was slow and structure-determining with ruthenium, whereas formation of the half-hydrogenated state controlled product structure with palladium. Of course, as isotopic labelling experiments have indicated, the alkene may re-adsorb in the same orientation: it

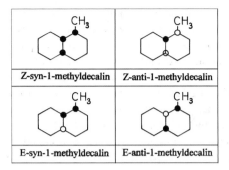

Scheme 10.4. Designation of the isomers of 1-methyldecalin.

appeared that the way the methyloctalin re-adsorbed was more affected by the methyl group in the 1- than in the 2- position.

The use of dimethylnaphthalenes[114] brought additional complications, but no new principles. There are 68 geometric isomers of the dimethyldecalins, but each symmetrically-substituted naphthalene gives only six products, and unsymmetrically-substituted eight: nevertheless their gas-chromatographic analysis required a capillary column 15.2m long, having the equivalent of 10^5 theoretical plates. Once again, ruthenium and palladium (the only two metals used) had clearly differentiated properties: the former hydrogenated cleanly, with a minimum of desorption and re-adsorption, and because Z-addition predominated it was concluded that the transition state for saturation had a high conformational energy. Palladium on the other hand induces in adsorbed molecules 'a state of rather violent agitation', in which hydrogen atoms migrate freely and double-bonds move at will, and desorption-re-adsorption is frequent: on this metal therefore the transition state to the alkane has a somewhat low conformational energy, and so the less stable Z-isomers are major products.

The exchange of naphthalene with deuterium has also been followed on the noble metals of Groups 8 to 10, excepting osmium.[118] Stepwise and multiple exchange were detected, the former dominating on palladium, iridium and platinum at 473 K at high deuterium pressure. Each metal however disclosed its individuality. On palladium, for example, the d_2- and d_4- products were most marked, perhaps because naphthalene, adsorbed in the diene form (Figure 10.11), added two deuterium atoms, then inverted and lost the two hydrogen atoms to revert to naphthalene: reiteration of these steps would give the d_4- molecule. Ruthenium and rhodium showed similar behaviour. With platinum, a sharp discontinuity between d_4- and d_5- molecules suggested that exchange might experience difficulty in proceeding from one ring to the other. The results were interpreted in terms of Burwell's equilibration model (see Section 8.3.5).

It is necessary to offer a short appreciation on the scope and import of the work reported by A.W. Weitkamp in a very substantial paper,[118] a major review[114] and a lengthy conference abstract,[117] concerning the reactions of hydrogen and of deuterium with naphthalene and its homologues, and tetralin and the octalins and their homologues, inadequately summarised in this section and Section 7.5.3. Perhaps because of its intricacy and complexity, this monumental, detailed and comprehensive study has not received the acclamation it deserves, nor has it been integrated into the corpus of knowledge of the mechanisms of metal-catalysed hydrocarbon reactions. This major project was conducted in the R and D Department of the American Oil Company, and its execution required all the resources of an industrial laboratory: it was beyond the reach of any academic institution, and it is unlikely that such a single integrated set of reactions will ever again be subject to such intensive study. As well as its purely catalytic content, which is vast, the work embraced the identification, conformational analysis and thermodynamic properties of the methyl- and dimethyl- homologues of naphthalene and its hydrogenated products. The skills deployed in achieving the outcome are quite remarkable. Much of what was discovered was not fully explicated, and could usefully be re-examined with the aid of now-available computational procedures. Anyone wishing to develop a new area of research in this sector of catalysis would be well advised to scan these publications, but unfortunately there was little physical characterisation of the catalysts reported, and it seems that the stereochemical variations that these reactions enjoy makes them very suitable for further exploring particle size effects and the behaviour of bimetallics. The very clear differences between ruthenium and palladium are only one of the puzzles that this work gives rise to. Undoubtedly there is here a further major area to explore; someone really should examine naphthalene on single crystal surfaces, especially of palladium. It has to be the task of another generation of scientists to exploit this family of reactions, the better to understand how hydrocarbon transformations occur on metal surfaces.

10.4.3. Fused Aromatic Rings: (2) Multiple Fused Rings[1,2,88,111]

The hydrogenation of polycyclic aromatic hydrocarbons has long been a subject of great interest, chiefly to organic chemists, but hardly any fundamental studies of mechanism have been reported. Attention has focussed on the ease with which the various rings can be reduced, and on what intermediate partially reduced products can be formed. The structures of some of these compounds are shown in Scheme 10.5; some of them are notoriously carcinogenic.

It is hard to summarise the older literature,[1,2] because such a variety of catalysts and conditions was used. With anthracene, the central ring **2** was first reduced, but as further hydrogen atoms were added isomerisation occurred, leading to a product in which only the terminal rings **1** and **3** were reduced.[88] Phenanthrene

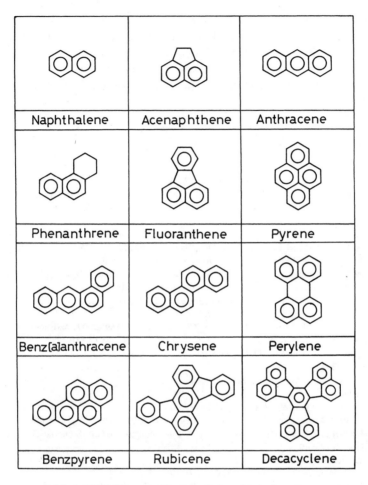

Scheme 10.5. Structures of polycyclic aromatic hydrocarbons.

behaves similarly. With fluoranthene, the sequence is **2** ... **1** ... **3**; with chrysene it is **1** ... **2** ... **4**, the final product containing a C=C bond between rings **2** and **3**. In the case of perylene, the sequence is **1,4**, then **2, 3**, the central ring retaining aromatic character. The middle rings in rubicene and decacyclene not surprisingly resisted reduction. More recently the reduction of some of these molecules has been studied using noble metals supported on carbon or alumina. Rhodium and palladium were the most active, and the relative ease of reducing the various rings was said to be governed by the LUMO electron density, the π-bond order and the stability of the product.[119]

The grandfather of all fused ring systems must be *buckminsterfullerene*, C_{60}, the structure of which is too well known to need depiction: with Ni/Al_2O_3 at 323–423 K it is hydrogenated to $C_{60}H_{36}$, no lighter hydrocarbons being formed.[120] Reduction occurs faster than that of toluene. This paper gives references to previous work with noble metals. The use of $C_{60}H_{36}$ as a means of hydrogen storage has been suggested.[121]

REFERENCES

1. G.C. Bond, *Catalysis by Metals*, Academic Press: London (1962).
2. H.A. Smith in: *Catalysis* (P.H. Emmett, ed.), Reinhold: New York, **5** (1957) 175.
3. R.J. Farrauto and C.H. Bartholomew, *Fundamentals of Industrial Catalytic Processes*, Chapman and Hall: London (1997).
4. M.A. Keane and P.M. Patterson, *J. Chem. Soc. Faraday Trans.* **92** (1996) 1413.
5. M.A. Keane and P.M. Patterson, *Ind. Eng. Chem. Res.* **38** (1999) 1295.
6. A. Stanislaus and B.H. Cooper, *Catal. Rev.- Sci. Eng.***36** (1994) 75.
7. M. Sanati, B. Harrysson, M. Faghihi, B. Gevert and S. Järås, *Specialist Periodical Reports:* Vol. 16 (J.J. Spivey and S.K. Agarwal, eds.), Roy. Soc. Chem. (2000) 1.
8. R.B. Moyes, K. Baron and R.C. Squire, *J. Catal.* **22** (1971) 333; R.G. James and R.B. Moyes, *J. Chem. Soc. Faraday Trans. I* **75** (1978) 1666.
9. A. Corma, A. Martínez and V. Martínez-Corma, *J. Catal.* **169** (1997) 480; R.M. Navarro, B. Pawelec, J.M. Trejo, R. Mariscal and J.L.G. Fierro, *J. Catal.* **189** (2000) 184.
10. J.L. Rousset, L. Stievano, F.J. Cadete Santos Aires, G. Geantet, A.J. Renouprez and M. Pellaria, *J. Catal.* **202** (2001) 163.
11. H. Yasuda and Y. Yoshimura,*Catal. Lett.* **46** (1991) 43.
12. R.B. Moyes and P.B. Wells, *Adv. Catal.* **23** (1973) 121.
13. G.C. Bond and P.B. Wells, *Adv. Catal.* **15** (1965) 92.
14. G.C. Bond, M.A. Keane, H. Kral and J.A. Lercher, *Catal. Rev.- Sci. Eng.* **42** (2000) 323.
15. F.H. Ribeiro, A.E. Schach von Wittenau, C.H. Bartholomew and G.A. Somorjai,*Catal. Rev. - Sci. Eng. 39* (1997) 40.
16. H. Kubicka, *J. Catal.* **12** (1968) 223.
17. J.R. Anderson and C. Kemball, *Adv. Catal.* **9** (1957) 51.
18. G. Leclercq, H. Charcosset, R. Maurel, C. Bertizeau, C. Bolivar, R. Frety, D. Jaunez, H. Mendez and L. Tournayan, *Bull. Soc. Chim. Belg.* **88** (1979) 577.
19. A. Amano and G. Parravano, *Adv. Catal.* **9** (1957) 719.
20. O.H. Ellestad and C. Naccache in: *Perspectives in Catalysis* (R. Larsson, ed.), C.W.K. Gleerup: Lund (1981), p. 95.
21. M.A. Vannice and P. Chou, *Proc. 8th Internat. Congr. Catal.* Verlag Chemie: Weinheim **5** (1984) 99.
22. P.C. Aben, *Rec. Trav. Chim.* **89** (1970) 449.
23. C.W. Keenan, B.W. Giesemann and H.A. Smith, *J. Am. Chem. Soc.* 76 (1954) 229.
24. G. Blanchard and H. Charcosset, *J. Catal.* **66** (1980) 465.
25. A.M. Sica, E.M. Valles and C.E. Gigola, *J. Catal.* **51** (1978) 115.
26. K.M. Sancier, *J. Catal.* **20** (1971) 106.
27. A.F. Flores, R.L. Burwell Jr. and J.M. Butt, *J. Chem. Soc. Faraday Trans.* **88** (1992) 1191.
28. M.A. Vannice and W.C. Meiken, *J. Catal.* **23** (1971) 401.
29. S.D. Lin and M.A. Vannice, *J. Catal.* **143** (1993) 554.

30. Pen Chou and M.A. Vannice, *J. Catal.* **107** (1987) 140.
31. W.F. Taylor and H.K. Staffin *J. Phys. Chem.* **71** (1967) 3314.
32. W.F. Taylor, *J. Catal.* **9** (1967) 99.
33. B. Coughlan and M.A. Keane, *Catal. Lett.* **5** (1990) 101; *Zeolites* **11** (1991) 12; M.A. Keane, *Indian J. Technol.* **30** (1992) 51.
34. Pen Chou and M.A. Vannice, *J. Catal.* **107** (1987) 129.
35. C. Mirodatos, J.A. Dalmon and G.A. Martin, *J. Catal.* **105** (1987) 405.
36. A.K. Neyestanaki, P. Mäki-Arvela, H. Backman, H. Karlm, T. Salmi, J. Väyrnen and Yu. D. Murzin, *J. Molec. Catal. A: Chem.* **193** (2003) 237.
37. You-Jyn Lin, D.E. Resasco and G.L. Haller, *J. Chem. Soc. Faraday Trans. I* **83** (1987) 2091.
38. You-Jyn Lin, D.E. Resasco and G.L. Haller in: *Proc. 6*[th]. *Internat. Congr. Catal.* (G.C. Bond, P.B. Wells and F.C. Tompkins, eds.), *Roy. Soc. Chem.*: London **2** (1976) 855.
39. G.A. Martin, R. Dutartre and J.A. Dalmon, *React. Kinet. Catal. Lett.* **16** (1981) 329.
40. G.A. Martin and J.A. Dalmon, *React. Kinet. Catal. Lett.* **16** (1981) 325.
41. J.L. Contreras and G.A. Fuentes in: *Proc. 11*[th]*Internat. Congr. Catal.*, (J.W. Hightower, W.N. Delgass, E. Iglesia and A.T. Bell, eds.), Elsevier: Amsterdam **B** (1996) 1195.
42. G. Leclercq, S. Pietrzyk, T. Romero, A. El Gharbi, L. Gengembre, J. Grimblot, F. Aïssi, M. Guelton, A. Latef and L. Leclercq, *Ind. Eng. Chem. Res.* **36** (1997) 4015; Yu.I. Yermakov, B.N. Kuznetsov and Yu.A. Rindin, *React. Kinet. Catal. Lett.* **2** (1975) 151.
43. R. Maurel, G. Leclercq and J. Barbier, *J. Catal.* **37** (1975) 324; J. Barbier, A. Morales, P. Marécot and R. Maurel, *Bull. Soc. Chim. Belg.* **88** (1979) 369.
44. D.S. Cunha and G.M. Cruz, *Appl. Catal. A: Gen.* **236** (2002) 55.
45. N. Kitijama, A. Kono, W. Veda, Y. Moro-oka and T. Ikawa, *J. Chem. Soc. Chem. Comm.* (1986) 674.
46. S.D. Lin and M.A. Vannice, *J. Catal.* **143** (1993) 539.
47. M.V. Rahaman and M.A. Vannice, *J. Catal.* **127** (1991) 251.
48. M.V. Rahaman and M.A. Vannice, J. Catal. 127 (1991) 267.
49. S.T. Srinivas and P. Kanta Rao, *J. Catal.* **148** (1994) 470.
50. S. Smeds, T. Salmi and D. Murzin, *Appl. Catal. A: Gen.* **150** (1997) 115; S. Smeds, T. Salmi, L.P. Lindfors and O. Krause, *Appl. Chem.* **144** (1997) 177; S. Smeds, D. Murzin and T. Salmi, *Appl. Catal. A: Gen.* 141 (1996) 207; 145 (1996) 253.
51. A. Parmaliana, M. El Sawi, G. Mento, V. Fedele and N. Giordano, *Appl. Catal.* **7** (1983) 221.
52. A. Farkas and L. Farkas, *Trans. Faraday Soc.* **33** (1937) 827.
53. M.A. Keane, *J. Catal.* **166** (1997) 347.
54. S.D. Lin and M.A. Vannice, *J. Catal.* **143** (1993) 563.
55. J.M. Orozco and G. Webb, *Appl. Catal.* **6** (1983) 67.
56. G.C. Bond, F. Garin and G. Maire, *Appl. Catal.* **41** (1988) 313.
57. R.Z.C van Meerten, A.C.M. Verhaak and J.W.E. Coenen, *J. Catal.* **44** (1976) 217.
58. C.A. Leon y Leon and M.A. Vannice, *Appl. Catal.* **69** (1991) 305.
59. W.A.A. Barneveld and V. Ponec, *Rec. Trav. Chim.* **93** (1974) 243.
60. R.Z.C. van Meerten and J.W.E. Coenen, *J. Catal.* **37** (1975) 37.
61. M.J. Vincent and R.D. Gonzalez, *Appl. Catal. A: Gen.* **217** (2001) 143.
62. M.C. Schoenmaker-Stolk, J.W. Verwijs and J.J.F. Scholten, *Appl. Catal.* **29** (1987) 91; M.C. Schoenmaker-Stolk, J.W. Verwijs, J.A. Don and J.J.F. Scholten,*Appl. Catal.* **29** (1987) 73.
63. D. Stacchiola, H. Molero and W.T. Tysoe, *Catal. Today* **65** (2001) 3; D. Stacchiola, G. Wu, H. Molero and W.T. Tysoe,*Catal. Lett.* **71** (2001)1.
64. G.A. Martin and J.A. Dalmon, *J. Catal.* **75** (1982) 233.
65. D.A. Cadenhead and N.G. Masse, *J. Phys. Chem.* **70** (1966) 3558.
66. G. Rienäcker and S. Unger, *Z. Anorg. Chem.* **274** (1953) 47.
67. W.K. Hall and P.H. Emmett, *J. Phys. Chem.* **62** (1958) 816.

68. E.A. Alexander and W.W. Russell. *J. Catal.* **4** (1965) 184.
69. P. van der Plank and W.M.H. Sachtler, *J. Catal.* **12** (1968) 35.
70. J. Bandiera and P. Meriaudeau, *React. Kinet. Catal. Lett.* **37** (1988) 373.
71. E.G. Allison and G.C. Bond, *Catal. Rev.* **7** (1973) 233.
72. C. Betizeau, G. Leclercq, R. Maurel, C. Bolivar, H. Charcosset, R. Frety and L. Tournayon, *J. Catal.* **45** (1976) 179.
73. M.C. Schoenmaker-Stolk, J.W. Verwijs and J.J.F. Scholten, *Appl. Catal.* **30** (1987) 339.
74. A.J. Hong, A.J. Rouco, D.E. Resasco and G.L. Haller, *J. Phys. Chem.* **91** (1987) 2665.
75. M. Abon, J. Billy, J.C. Bertolini, J. Massardier and B. Tardy in: *Metal-Support and Metal-Additive Effects in Catalysis,* (B. Imelik, C. Naccache, G. Coudurier, H. Praliaud, P. Meriaudeau, P. Gallezot, G.A. Martin and J.C. Védrine, eds.), Studies in Surface Science and Catalysis, Elsevier: Amsterdam, **11** (1969) 269.
76. P. Tétényi and V. Galsán, *Appl. Catal. A: Gen.* **229** (2002) 181.
77. G. del Angel, B. Coq and F. Figuéras, *J. Catal.* **95** (1985) 167.
78. J. Horiuti and M. Polanyi, *Trans. Faraday Soc.* **30** (1934) 1164.
79. A.O Cinneide and J.K.A. Clarke, *J. Catal.* **26** (1972) 233.
80. F. Hartog, *Preprints ACS Mtg., Divn. Petroleum Chem.*, Vol. 11 (1966), Paper A11.
81. G.A. Somorjai, M.A. Van Hove and B.E. Bent, *J. Phys. Chem.* **92** (1988) 973.
82. J.W. Hightower and C. Kemball, *J. Catal.* **4** (1965) 363.
83. R. van Hardeveld and F. Hartog, *Adv. Catal.* **22** (1972) 75.
84. J. Barbier and P. Marécot, *Nouv. J. Chim.* **5** (1981) 393.
85. P. Kluson and L. Červerný, *Appl. Catal. A: Gen.* **128** (1995) 13.
86. F. Hartog and P. Zwietering, *J. Catal.* **2** (1963) 79.
87. L. Ronchin and L. Toniolo, *Catal. Today* **48** (1999) 255; *Appl. Catal. A: Gen.* **208** (2001) 77; *Catal. Today* **66** (2001) 363.
88. R.L. Augustine, *Heterogeneous Catalysis for the Synthetic Chemist,* Dekker: New York (1996).
89. C.U.I. Odenbrand and S.T. Lundin, *J. Chem. Tech. Biotech.* **30** (1980) 677; **31** (1981) 660.
90. J. Struijk, R. Moene, T. van der Kemp and J.J.F. Scholten, *Appl. Chem. A: Gen.* **89** (1992) 77;
91. S.-H. Yie, M.-H. Qiao, H.-I. Li, W.-J. Wang and J.-F. Deng, *Appl. Chem. A: Gen.* **176** (1999) 129; M.M. Johnson and G.P. Nowack, *J. Catal.* **38** (1975) 518.
92. Chin. Petrol. Corp.,*U. S. Patent* 6006423.
93. P.J. van der Steen and J.J.F. Scholten, *Appl. Catal.* **58** (1990) 291.
94. P.N. Rylander,*Catalytic Hydrogenation over Platinum Metals*, Academic Press: New York (1967).
95. H. van Bekkum, H.M.A. Buurmans, G. van Minnen-Pathuis and B.M. Wepster, *Rec. Trav. Chim.* **88** (1969) 777.
96. P. Kluson, L. Červený and J. Had, *Catal. Lett.* **23** (1994) 299.
97. V.I. Derbentsev, Z. Paál and P. Tétényi, *Z. Phys. Chem. NF* **80** (1972) 51.
98. P. Tétényi, and Z. Paál, *Z. Phys. Chem. NF* **80** (1972) 63.
99. R. Gomez, G. Del Angel and G. Corro, *Nouv. J. Chim.* **4** (1980) 219.
100. R. Szymanski, H. Charcosset, P. Gallezot, J. Massardier and L. Tournayan, *J. Catal.* **97** (1986) 366; Tran Thanh Phuong, J. Massardier and P. Gallezot, *J. Catal.* **102** (1986) 456.
101. T. Ioannides, M. Tsapatsis, M. Koussathana and X.E. Verykios, *J. Catal.* **152** (1995) 331.
102. J. Barbier, P. Marécot and L. Tifonti, *React. Kinet. Catal. Lett.* **32** (1986) 269.
103. D. Poondi and M.A. Vannice, *J. Catal.* **161** (1996) 742.
104. M. Bartók, *Stereochemistry of Heterogeneous Metal Catalysis*, Wiley: New York (1983).
105. A. Skita and A. Schneck, *Ber.* **558** (1922) 139.
106. S. Siegel, *Adv. Catal.* **16** (1966) 123.
107. S. Siegel and M. Dunkel, *Adv Catal.* **9** (1957) 15.
108. S. Siegel, G.V. Smith, D. Dmuchovsky, D. Dubbell and W. Halpern, *J. Am. Chem. Soc.,* **84** (1962) 3136.

109. P. Reyes, M.E. König, G. Pecchi, I. Concha, M. Lopez Granados and J.L.G. Fierro, *Catal. Lett.* **46** (1997) 71.
110. S. Siegel, V. Ku and W. Halpern, *J. Catal.* **2** (1963) 348.
111. P.N. Rylander and D.R. Steele, *Engelhard Ind. Tech. Bull.* **5** (1965) 113.
112. H. Atkins, W.H. Zartman and H. Cramer, *J. Am. Chem. Soc.* **53** (1931) 1425.
113. W.H. Zartman and H. Atkins, *J. Am. Chem. Soc.* **54** (1932) 1668.
114. A.W. Weitkamp, *Adv. Catal.* **18** (1968) 1.
115. S.D. Lin and C. Song, *Catal. Today* **31** (1996) 93.
116. A.D. Schmitz, G. Bowers and C.-S. Song, *Catal. Today* **31** (1996) 45.
117. A.W. Weitkamp, *J. Catal.* **6** (1966) 431.
118. A.W. Weitkamp, *Preprints ACS Mtg., Divn. Petroleum Chem.* **11** (1966) Paper A13.
119. K. Sakanishi, M. Ohira, I. Mochida, H. Okazaki and M. Soeda, *Bull. Chem. Soc. Jpn.* **63** (1989) 3994.
120. T. Osaki, T. Tanaba and Y. Tai, *Phys. Chem. Chem. Phys.* **1** (1999) 2361.
121. Natl. Inst. Adv. Ind. Technol., *Japanese Appl.* 2003/012, 572.
122. P. Reyes, M. Oportus, G. Pecchi, R. Frety and B. Moraweck, *Catal. Lett.* **37** (1996) 193.
123. J.W.E. Coenen, W.M.T.M. Schats and R.T.Z. van Meerten, *Bull. Soc. Chim. Belg.* **88** (1979) 435.
124. G.A. del Angel, B. Coq, G. Ferrat and F. Figuéras, *Surf. Sci.* **156** (1985) 943.
125. S. Fuentes and F. Figuéras, *J. Catal.* **61** (1980) 443.
126. S. Fuentes and F. Figuéras, *J. Chem. Soc. Faraday Trans. I* **74** (1978) 174.
127. A. Benedetti, G. Cocca, S. Enzo and F. Pinna, *React. Kinet. Catal. Lett.* **13** (1980) 291.
128. O. Alexeev and B.C. Gates, *J. Catal.* **176** (1998) 310.
129. T.A. Dorling and R.L. Moss, *J. Catal.* **5** (1966) 111.
130. E. Crawford and C. Kemball, *Trans. Faraday Soc.* **58** (1963) 2452. 1983.

FURTHER READING

Temperature inversion of the rate of benzene hydrogenation
Pt 29, 38, 46, 48, 51, 54, 55, 56, 58
Pd 16, 30, 34, 55, 58
Ni 4, 5, 33, 50, 53, 59, 60
Ru 16, 61
Ru, Fe 16
Co 31
Cu 62

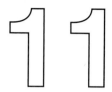

HYDROGENATION OF SMALL ALICYCLIC RINGS

PREFACE

Cyclopropane is a fascinating molecule: although its carbon atoms are linked by what are formally single C—C bonds, it partakes some of the properties of an alkene. It is quite easily hydrogenated, but much less strongly adsorbed than alkenes. As with benzene, its bonding defies simple description, and one might say it is 'neither fish, flesh, fowl or good red herring'. Although only of small practical value, its hydrogenation and particularly that of its alkyl-substituted derivatives have been quite widely used as a way of assessing aspects of catalyst structure and composition that determine activity and specificity: but it needs to be said that the interpretation of what is observed falls short of what might have been hoped for.

The convention is adopted here of calling the processes of converting the cyclopropane and cyclobutane rings to the corresponding alkanes *hydrogenation*, and the breaking of further C—C bonds when it occurs (as it does simultaneously at higher temperature) *hydrogenolysis*. Cyclobutane is more like an alkane and less like an alkene than is cyclopropane; however, further enlargement of the ring to cyclopentane gives a molecule that is undoubtedly a cyclic *alkane*, although as we shall see (Chapter 14) this too has some unusual properties Alkene character diminishes with increasing ring size, rather like the Cheshire cat, which slowly disappeared, starting from its tail, until only the smile was left.

11.1. INTRODUCTION

It might at first sight seem unlikely that a stable molecule could be formed by joining three methylene groups as in a triangle, since distortion of hybrid sp^3 orbitals to allow a C—C—C bond angle of 60° would appear impossible. However,

TABLE 11.1. Some Physical and Structural Properties of Small Alicyclic Rings C_xH_{2x}[1]

x	L(C—C)/nm	D(C—C)/kJ mol^{-1}	$-\Delta H_f^{\ominus}$/kJ mol^{-1}	$-\Delta H_H$/kJ mol^{-1}
3	0.150	238	53.3	157
4	0.156	293	27.7	153
5	0.154	385	77.3	69
6	0.154	397	123.2	44

L(C—C) = bond length; D(C—C) = bond dissociation energy; $-\Delta H_f^{\ominus}$ = standard heat of formation; $-\Delta H_H$ = heat of hydrogenation to the corresponding alkane.

cyclopropane *is* a comparatively stable molecule, and the fact of its existence has taxed the subtlety of the minds of theoreticians for more than half a century. The strain imposed by its geometry gives it some of the characteristics of an alkene, because it undergoes addition with halogens and with hydrogen, this latter process being our particular concern. Its heat of hydrogenation to propane is greater than that of ethene to ethane (Table 11.1); its C—C bond length (0.152 nm) is shorter than that in ethane, but the C—C bond dissociation energy is much less than that in the larger alicyclic molecules.[1] It enters into conjugation with C=C bonds, but is unable to transmit it, and in many other respects its properties are intermediate between those of alkenes and alkanes.[2] In a word we have *a mystery wrapped in an enigma.*

The several attempts[3] that have been made to find theoretical frameworks for understanding these unique properties need not delay us for long. The earliest picture was that proposed by Donald Walsh in 1949:[4] he suggested that the C—C bonds are formed by the intra-annular overlap of one of the sp^2-hybridised orbitals of each carbon atom with three p orbitals. This creates a concentration of charge in the centre of the ring, and leads to one concept of how the molecule can chemisorb (see below). A valence-bond approach[5,6] led to a "bent-bond" model, in which hybridised $sp^{4.12}$-orbital lobes formed angles of 104°, and overlapped outside the equilateral triangle. One study[5] aimed to show that these two approaches were compatible. An *ab initio* SCF-MO calculation[6] did not support Walsh's model, but concluded that the molecule's reactivity was due to a relatively low-lying excited σ-orbital, in contrast to alkenes, where the analogous orbital has π-character. A recent survey[3] concludes that the orbitals directed to the hydrogen atoms have sp^2 character, while those to adjacent carbon atoms are approximately sp^5, the extra p character relieving some of the strain that would feature in sp^3 hybridisation. The electron density is then directed away from the ring at an angle of 21° to the line joining the carbon nuclei, confirming the 'bent-bond' model noted above. Since cyclopropanes can be synthesised by insertion of a methylene diradical (e.g. from diazomethane) into a C=C bond, it has also been possible to regard the bonding as being the result of resonance between the three possible canonical forms (Figure 11.1). The reactivity of and the bonding in the C_3 ring are often discussed in organic

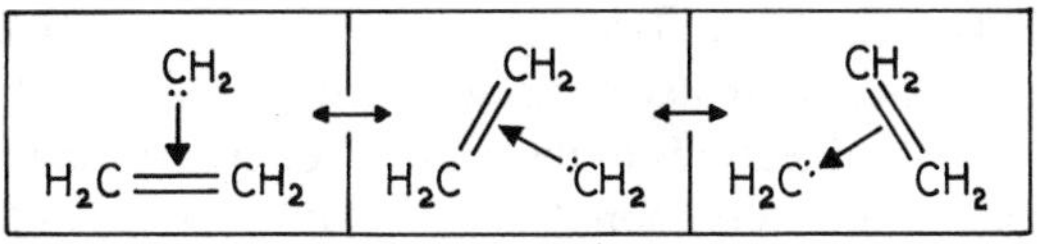

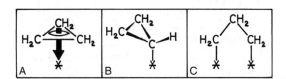

Figure 11.1. Representation of cyclopropane as a resonance hybrid of methylene and ethene.

chemistry texts; for recent experimental and theoretical studies, see references 7 and 8.

Alkenic unsaturation in conjunction with the cyclopropane ring has interesting theoretical and practical effects. Methylenecyclopropane is perhaps better formulated as trimethylenemethane;[9] its hydrogenation[2] and that of derivatives[10] has been examined (Section 11.4). Cyclopropene sounds as if it would be very highly strained, but its methyl derivatives have apparently been made and hydrogenated[11] (see also Section 11.4).

Not much is known about the chemisorption of cyclopropane.[12] Its heat of adsorption on platinum film has been measured[13] ($197\,kJ\,mol^{-1}$), but like other hydrocarbons is suffers variety of fragmentation processes on clean metal surfaces;[14–17] self-hydrogenation leads only to propane on palladium and platinum, while on other metals (Rh, Ni, Mo, Fe), more active in hydrogenolysis, methane and ethane are also formed.[18] Reactions in the absence of hydrogen, brought about only by hydrogen atoms derived from the reactant, run parallel to those in the presence of hydrogen. Three forms have been suggested as intermediates in the hydrogenation to propane (Fig. 11.2). Structure **A** shows the molecule lying flat on the surface, bonded to a metal atom by interaction of delocalised electrons within the ring with vacant orbitals on the metal. UHV studies provide some evidence for this form at low temperatures on single crystals,[14] and it harmonises with Walsh's model, but no detailed description of the bonding has been offered. Structure **B** is a dissociatively-adsorbed state typically shown by alkanes, but it has been dismissed because reaction with deuterium only very rarely leads to cyclopropanes-d_x, and then usually just in small amounts (Section 11.2.2). Structure **C** is clearly destined to lead to propane, and it could arise through **A** as a transitory state. Many other species containing more or fewer hydrogen atoms have been advanced as intermediates in the degradation, hydrogenation or hydrogenolysis, and we shall note some

Figure 11.2. Possible adsorbed states of cyclopropane.

of them presently. The kinetics of hydrogenation (Section 11.2) do however show clearly that cyclopropane is usually much less strongly adsorbed than hydrogen.

One of the significant features of the reaction of cyclopropane with hydrogen is that on metals active for hydrogenolysis, i.e. all the metals of Groups 8 to 10 excepting palladium and platinum, fragmentation to methane and ethane takes place at much lower temperatures than would occur with propane. Ethene and propene are also observed under certain conditions.[15]

Much interest has been shown in the regiospecificity of the hydrogenation of alkyl-substituted cyclopropanes (Section 11.3). In general it is the least obstructed C—C bond that is broken most easily, so for example methylcyclopropane gives chiefly *iso*butane, but the alternative mode giving *n*-butane has a higher activation energy, and thus becomes more important as temperature is raised.[2,19] This reaction has often been deployed in attempts to provide a more detailed view of the structure of a catalyst's surface[19,20] (Section 11.3.1), but it has to be said that the results are somewhat disappointing. While 1,2-dimethycyclopropane gives mainly 2-methylpentane, when adjacent substituents form part of another cycle the intervening C—C bond is strained, and breaks first. This permits one of the few synthetically useful reactions of cyclopropane to be effected: for example, methylene insertion into cyclohexene gives bicyclo[4.1.0]heptane, which is converted by hydrogenation to cycloheptane.

The cyclopropane ring can exist in a variety of environments:[1,10,11] among the more interesting structures the hydrogenation of which has been investigated are spiropentane (bicyclo[2.2.0]pentane), bicyclopropyl, cyclopropylmethanes $((C_3H_5)_x CH_{4-x}, x = 1$ to 4), nortricyclene and phenylcyclopropane (Figure 11.3; see Section 11.3.3). Fusion of a C=C double bond directly onto the cyclopropyl ring entirely changes the mode of ring breaking, so that a bond adjacent to the

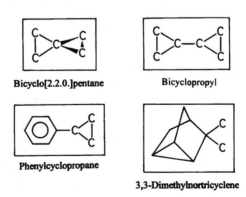

Bicyclo[2.2.0.]pentane   **Bicyclopropyl**

Phenylcyclopropane

3,3-Dimethylnortricyclene

Figure 11.3. Structures of molecules containing the cyclopropane ring.

point of substitution is now the most reactive: the same effect is produces by an ethenyl substituent (Section 11.4).

The cyclobutane ring behaves much more like an alkane:[1] its hydrogenation requires significantly higher temperature than does cyclopropane (Section 11.4). The ring is puckered, with a dihedral angle of 20°, and the C—C bonds are longer than those in ethane or the larger alicyclic molecules; other relevant facts are given in Table 11.1. Its alkyl- and methylene- substituted derivatives have also enjoyed much attention, presumably for the same reason as the corresponding cyclopropanes.

11.2. HYDROGENATION AND HYDROGENOLYSIS OF CYCLOPROPANE

11.2.1. Kinetics

There have been numerous studies of the kinetics of the reaction of hydrogen with cyclopropane, but some record only activation energies and not orders of reaction. Some distinguish between hydrogenation to propane and hydrogenolysis to methane and ethane, the kinetics of these two types of process being quite different (see Tables 11.2 and 11.3): many however have contented themselves with measuring parameters of the overall reaction, or had to of necessity because the means of analysis were lacking. We focus first on hydrogenation.

The difference between cyclopropane and alkenes on the one hand and alkanes on the other is nowhere more clearly seen than in the kinetics of their reactions with hydrogen and in the relative activities of metals for those reactions. The ranking of metals activities for cyclopropane hydrogenation,[21-23] namely,

$$Rh > Ni > Pt > Pd > Ir > Ru \sim Os > Co$$

is quite different from that for alkane hydrogenolysis, but depends on the surface cleanliness temperature and reactant concentrations at which the rates were measured. The earliest kinetic work showed that on the Group 10 metals the order in hydrogen was zero and in cyclopropane first; the former clearly saturated the surface and the latter was at most weakly adsorbed.[24] Extension of this work and the use of other metals refined this simple picture. Rates frequently passed through maxima as hydrogen pressure was raised, the position of the maximum increasing with temperature, and the order above the maximum becoming less negative.[2] The negative values quoted in Table 11.2 were measured at 323 K. Orders in cyclopropane became more positive as the fixed hydrogen pressure was raised, and with temperature at constant hydrogen pressure. This behaviour is typical for the competitive adsorption of the reactants, and the application of Langmuir-Hinshelwood

TABLE 11.2. Kinetics of the Hydrogenation of Cyclopropane ($r \propto P_H{}^x P_{hc}{}^y$)

Metal	Form	$E/\text{kJ mol}^{-1}$	x	y	T/K	Reference
Fe	[a]	56	−1	1	373	39
Co	Powder	45	−0.1	0.7	394	23
Ni	Powder	54	−0.1	0.8	300	23
Ni	Film	31[b]	−0.1	0.6	∼250	27
Ni	/Pumice	44	0	1	∼443	24
Ni	/SiO$_2$	54	−0.1	0.8	300	87
Ni	/SiO$_2$–A1$_2$O$_3$	58	−0.2	0.6	366	87
Ru	(0001)	26	0.75	0	450	14
Ru	(1120)	20	0.9	0	450	14
Rh	Pumice	42	−0.36	∼0.4	323	2
Pd	Film	61[b]	−0.9	0.1	∼250	27
Pd	/Pumice	34	0	1	395	24
Pd	/Pumice	—	−0.8	0.4[c]	323	29
Pd	/Pumice	42	−0.45	∼0.4[c]	323	2
Ir	(111)	41	1 → −0.5	+ve	450	15
Ir	(110)(1 × 2)	36	1	∼0	425	15
Ir	/Pumice	48	−0.2	∼0.4[c]	323	2
Ir	/Pumice	41	0	0.53[c]	323	29
Pt	Powder	51	−0.5	1	352	23
Pt	Film	46[b]	−0.2	0.2	∼223	27
Pt	/Pumice	37	0	1	356	24
Pt	/Pumice	33	−0.35	0.9	473	28
Pt	/Pumice	36	−0.3 → −0.6	0.9[c]	323	2
Cu	Powder	46	−0.1	0.6	419	23

[a] Reduced magnetite with added K.
[b] Reaction with D$_2$.
[c] Measured using the H$_2$ pressure giving maximum rate.

formulism based on a rate-controlling reaction of a hydrogen atom with an adsorbed cyclopropane led to estimates of the adsorption coefficients for hydrogen, which decreased, as they should, with increasing temperature and indeed as the order suggested they would. However, the values depended very much on the method of preparation of the catalyst, and since dispersions were not at that time measured it is likely that their variation was at least partly responsible. Measurement of the hydrogen order for this reaction might be developed into a simple way for estimating metal dispersion.

Other forms of the Langmuir-Hinshelwood method have been tried. Orders in hydrogen more negative than −0.4 seemed to require the participation of a chemisorbed hydrogen *molecule*,[2] while an equation based on a slow reaction of a hydrogen atom with a propyl radical worked for a Pt/SiO$_2$, but not for EUROPT-1 (6% Pt/SiO$_2$).[25,26] Other workers have failed to extract maximum significance from their results by expressing them as exponents of reactant pressures, changing from positive to negative in ways not well explained. Table 11.2 shows a selection of reaction orders expressed in this way: for supported metals,[2] powders[23] and films,[27] hydrogen orders are zero or slightly negative, while those for cyclopropane are generally positive ($0.4 < y < 1$).[2,28,29] Values of activation energies shown in Table 11.2, and others not shown there,[16,18,22,30-33] are mostly in the range 45 ± 10 kJ mol^{-1}. If values close to the mean were found under conditions where the surface concentrations were not changing much, 45 kJ mol^{-1} might be regarded as the 'true' activation energy. Exceptions to these generalisations are shown by iron catalysts (where hydrogenolysis intrudes on powder[16] and film[19]) and by palladium;[22,27] in these cases, hydrogen appears to be *much* more strongly chemisorbed than cyclopropane. Ruthenium catalysts have shown both higher[34] and lower[14,33,34] values. Quite different behaviour has been shown by ruthenium single crystal surfaces[14] and by Ir(110)(1 × 2);[15] *positive* hydrogen orders and *zero* cyclopropane orders were observed, and in the former case activation energies were unusually low (20–26 kJ mol^{-1}). These orders suggest that these surfaces are monopolised by cyclopropane molecules, or species derived from them.

This virtual constancy of activation energy as metal, its form, and method of preparation are varied leads us naturally to consider the hydrogenation reaction's structure sensitivity. It was probably the first hydrocarbon reaction to be declared structure-*insensitive*, by the lack of dependence of TOF on particle size, support and physical form:[31,35,36] but this was for platinum, and later work with Ni/SiO$_2$[37] showed maximum activity and minimum activation energy at a particle size of about 1 nm. Even more dramatic variations were shown by variously supported ruthenium catalysts,[33] where those made from the trichloride had TOFs that increased by 10^5 as dispersion increased in the range 7 to 43% (Figure 11.4): activation energies between 24 and 49 kJ mol^{-1} showed no consistent size-dependence. Structure-sensitivity (or insensitivity) is therefore a function of the whole catalytic system, and not just of the reaction. It is difficult to generalise on such limited observations, but it seems likely that sensitivity is associated with the stronger adsorption of cyclopropane-derived species than hydrogen, and with propensity of metals to engage in simultaneous hydrogenolysis; indeed there may be a causal connection between these two occurrences (see later).

Ru/TiO$_2$ catalysts have unusual character. On first reduction (HTR1) of RuCl$_3$/TiO$_2$, catalysts of only moderate activity were obtained, because they are somewhat poisoned by chloride ion and partially in the SMSI state (Section 3.3.5). Oxidation followed by low-temperature reduction (O/LTR) removed chloride ion and greatly increased particle size; much higher activities for cyclopropane hydrogenation (and alkane hydrogenolysis, see Chapter 13) were obtained.[34] A second high-temperature reduction (HTR2) converted catalysts very largely into the SMSI state, with loss of much of their activity. However, in contrast to the situation with alkanes, with cyclopropane the *most* active catalyst has the *highest* activation energy. Hydrogenation selectivity decreased in the sequence HTR2 > HTR1 > O/LTR, indicating that (not surprisingly) hydrogenolysis requires a larger ensemble

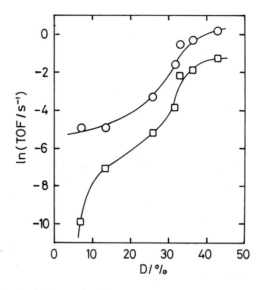

Figure 11.4. TOFs for the hydrogenation (circles) and hydrogenolysis (squares) of cyclopropane on variously supported ruthenium catalysts of different dispersion.[33]

of free atoms than does hydrogenation. Reduction of Rh/La_2O_3 at 673 K or above has led to most of the activity for cyclopropane hydrogenation being lost; Rh/SiO_2 was unaffected.[38]

On metals of moderate to high activity for hydrogenolysis, i.e. on all metals of Groups 8 to 10, except palladium and platinum, the splitting of a second C—C bond accompanies the breaking of the first, methane and ethane being the products.[14–16,18,22,23,27,37] Some ethene may also be formed, but complete conversion to methane requires high temperatures: methane/ethane ratios greater than unity have also been found particularly with *large* (>5 nm) ruthenium particles,[33] and the activation energy was then very high (94 kJ mol⁻¹). Activation energies for hydrogenolysis were in general somewhat larger than for hydrogenation,[14,23,33,34,37], so that the former became more important as temperature was raised. With supported metals, orders of reaction for the two processes did not differ greatly, orders for hydrogenolysis tending to be more negative in hydrogen and less positive in cyclopropane (Table 11.3). The situation was clearest with the single-crystal studies: on ruthenium surfaces, the highest selectivities to propane were found at *low* hydrogen/cyclopropane ratios,[14] but on Ir(111) more extensive cracking occurred at low hydrogen pressure.[15] Hydrogenolysis was also structure-sensitive on nickel[37] and ruthenium[33] catalysts; in the latter case, selectivity to propane improved with decreasing dispersion (Figure 11.4). This may appear to conflict with the consequences of inducing SMSI with Ru/TiO_2 catalysts,

TABLE 11.3. Kinetics of the Hydrogenolysis of Cyclopropane ($r \propto P_H{}^x P_{hc}{}^y$)

Metal	Form	E/kJ mol^{-1}	x	y	T/K	Reference
Co	Powder	78	−0.2	0.6	394	23
Ni	Powder	67	−0.3	0.4	300	23
Ni	/SiO$_2$	67	−0.3	0.4	300	87
Ni	/SiO$_2$-Al$_2$O$_3$	55	0.5	0.6	366	87
Ru	(0001)	43	1.8	−0.5	450	14
Ru	(1120)	43	1.8	−0.6	450	14
Ir	(111)	111	+ve.	∼0	450	15
Ir	(110)-(1×2)	96	∼0	∼0	425	15

but the effects cannot be strictly comparable. Extensive fragmentation of cyclopropane occurred on iron catalysts.[16,27,39,]

Little work has been reported on the use of bimetallic catalysts.[32,40,41] Addition of copper to nickel as powder lowered the overall rate, without major change to the propane selectivity. Activation energies varied irrationally between 40 and 93 kJ mol^{-1}.

11.2.2. The Reaction of Cyclopropane with Deuterium[42]

Early work using Pt/pumice showed[28] that the main initial products at 273 K were propanes-d_x (x = 2 to 4), but unless a large excess of deuterium was used its hydrogen content rose with increasing conversion, and mean deuterium number (M) of the propane fell. The value of M increased from about four to six as temperature increased to 473 K; this was due to a marked increase in the amounts of propane-d_7 and -d_8 formed, their activation energies being significantly higher than those for the lighter molecules. At constant temperatures (273 and 473 K), however, product distributions were independent of the deuterium/cyclopropane ratio. Thus at 473 K, propanes-d_7 and -d_8 were the major products, but a secondary maximum appeared at propane-d_4. Later work[29] confirmed and extended these measurements, and distributions of propanes-d_x were modelled by random extraction of atoms from two 'pools'. Pool A had a deuterium content that rose from 90 to 96% and constituted 23 to 64% of the whole between 323 and 473 K, while pool B had 45–50% deuterium and decreased from 71 to 33% of the whole; a small amount of 'direct addition' to propane-d_2 was also suggested. Similar studies with pumice-supported palladium, rhodium and iridium gave broadly similar results, but yields of propanes-d_7 and -d_8 were higher (Pd, ∼80%; Rh, ∼77%; Ir, ∼80%) and essentially independent of reaction conditions. The same mode of interpretation was adopted.

It very soon became clear that the products of the reaction of cyclopropane with deuterium could be explained by supposing a low concentration of propyl radicals and a high concentration of deuterium atoms on the metal surfaces, their interactions being similar to those experienced by alkanes, with multiple exchange predominating. Parallel measurements[28,29] with propane confirmed this picture, although there were significant differences, as activation energies were higher and activities lower, and a stepwise exchange was also noted. These changes were due to the greater ease of chemisorbing cyclopropane, and to the fact that the opening step created structure **C** (Figure 11.2) and not a propyl radical. Propyl radicals were also formed by dissociation of 1- and 2-chloropropane, and behaved similarly.[29,43]

On films of the metals of Group 10, rhodium, iron and tungsten, the cyclopropane-deuterium reaction gave product distributions in line with those found with the pumice-supported metals where comparison is possible; nickel and particularly iron gave much multiply-exchange propanes.[27] The kinetic parameters are given in Table 11.2.

Very slight exchange of the hydrogens in cyclopropane was noted on Rh/pumice and Ir/pumice;[29] it appeared to be substantial on rhodium film at 173 K,[44] suggesting that it is a low activation energy process. It also occurred significantly on tungsten film.[27]

More recently the reaction has been examined over silica-supported rhodium, palladium, iridium and platinum, with the help of NMR spectroscopy.[45] The results were much in line with those already mentioned, the amounts of propane-d_8 increasing in the sequence: Pt < Ir < Rh < Pd. The propane-d_2 was not surprisingly mainly propane-1,3-d_2. A new feature was observed with Ir/Al$_2$O$_3$; the support made a significant contribution to the overall reaction, and this necessitated a revision of a previous interpretation of the activity of this metal in alkane exchange.[46]

11.2.3. Reaction Mechanisms

Discussion of the mechanisms involved in the reactions of cyclopropane with hydrogen and deuterium is relatively straightforward because of the evident similarity if not identity of unit steps with those encountered in the reactions of alkanes (Chapters 6, 13 and 14). This has not however inhibited scientists from suggesting a variety of possible intermediate adsorbed species.

The exchange process can be swiftly dismissed because of its small importance. Its simplest formulation involves the reversible dissociative chemisorption of the cyclopropane molecule as structure **B** (Figure 11.2). However, the reactivity of the C—H bond is probably not that much different from the bond in alkanes, while cyclopropane exchange appears to go best at low temperature, and has a low (and possibly even negative) activation energy. An alternative route is therefore through the undissociated structure **A**, where exchange occurs as a simple displacement (Scheme 11.1). There is no kinetic information to assist us.

The kinetics of the hydrogenation clearly point to competitive adsorption of the reactants, the sign before the hydrogen order and the magnitude of the

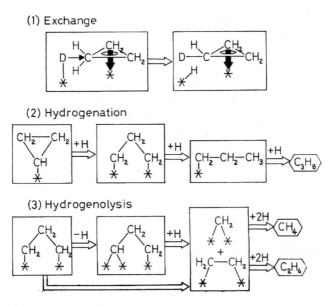

Scheme 11.1. Mechanisms involved in the reactions of cyclopropane with hydrogen and with deuterium.

positive cyclopropane order depending upon the ranges of surface coverage in which the measurements were made: the change in hydrogen order from positive to negative (at low temperatures) as its pressure is increased suggests that the slow step requires both a hydrogen atom *and* a vacant site on which cyclopropane can chemisorb, either as structure **B** or structure **C** (Figure 11.2). Where the orders are respectively unity and zero (Table 11.2), the slow step may be the collision of a gaseous cyclopropane molecule with a fully covered layer of hydrogen atoms. Once the propyl radical stage is reached, the way is open for the mechanisms envisaged for alkane exchange (Chapter 6) to operate. There are however one or two difficulties. (1) It is possible that the rate equation based on the above mechanism is not universally applicable. (2) There is no analogue to the stepwise exchange of alkanes, because the initial propyl radical must have at least *one* deuterium atom, and even although multiple exchange is usually the favoured process there are often subsidiary maxima at propane-d_4 and -d_2. The latter is perhaps easier to account for, but the former is less easily explained. The idea mentioned above (Section 11.2.2) that there was a 'pool' of atoms containing about half of each kind with which the C_3 species equilibrated was severely criticised, with some justification because four or sometimes five disposable parameters were needed to model the whole propane distribution. The concept was however resurrected by

R.L. Burwell Jr. to interpret the way in which unsaturated hydrocarbons interacted with deuterium (Section 8.3.5).

We have already noted that the occurrence of hydrogenolysis of cyclopropane is easier than that of propane, because the initial chemisorption needs a smaller energy input. The fact that methane *and ethane* are the usual products shows that a second C—C bond breaking is followed by the speedy desorption of the C_2 fragment, and the somewhat special and exigent conditions needed for total conversion to methane reflects the known greater stability of the C—C bond in ethane (or its predecessor). There have been several detailed discussions of likely intermediate strucutres; those shown in Scheme 11.1 are as reasonable as any.

11.3. HYDROGENATION OF ALKYLCYCLOPROPANES[47]

11.3.1. Mono-alkylcyclopropanes

The substitution of one of the six hydrogen atoms of cyclopropane for an alkyl group creates differentiated C—C bonds in the ring, so that for example the hydrogenation of methylcyclopropane can give either *n*- or isobutane; small amounts of the corresponding alkenes are also sometimes found. The first work on this reaction showed[1,2,42] that, with a Pt/pumice catalyst, *n*-butane constituted 95% of the products at room temperature, a difference in activation energy of only some 7 kJ mol^{-1} being sufficient to decrease this figure to about 70% at 523 K. It reacted somewhat faster than cyclopropane, and although the kinetics were of similar form the differences in the orders of reaction under equivalent conditions suggested that it was more strongly adsorbed, and reacted faster because it covered more of the surface. This should provide some clue as to the mode of chemisorption of cyclopropane ring. The steric effect of the methyl group on the process of chemisorption was clearly negligible, and its role is more likely to have been as a partial electron donor to the ring, with a consequent stronger engagement to the surface through delocalised electrons. Other explanations are no doubt possible, as the substituent might affect the strength of C—H bonds on the opposite carbon atoms or of the C—C bond between them, thus favouring adsorption as the analogue of structures **B** or **C** (Figure 11.2). The apparent absence of structural information on methylcyclopropane does not help to resolve this question. The presence of the methyl group, or particularly of a longer chain alkyl group, also provides possible further modes of adsorption.

The easier breaking of the bond opposite the substituent has been frequently confirmed,[19,27,47] although as we shall see there are some exceptions. A *steric* effect due to a methyl group may just be sufficient to make the C2—C3 bond more likely to break, but a weakening of this bond relative to the C1—C2 bonds in the adsorbed state is also possible. This question also has not received much attention in the literature,[42] although the reactions of more heavily substituted molecules provide some further insights (Section 11.3.2).

The regiospecificty of the ring opening is not simply a characteristic of the methylcyclopropane molecule; the nature of the catalytic metal is important as well. The percentages of *iso*butane found on metal films are:[27] for platinum, 97% (in harmony with that cited above); for palladium, 85%; for nickel, 75%; and for tungsten about 50%. It may be that different intermediate species are formed (e.g. π-propenyl on Pd and Ni). In the reactions with deuterium, the extents of exchange were not the same in the two isomers. On nickel there was more stepwise and less multiple exchange in the *n*-butane than in the isobutane; with platinum, it was the opposite, while on palladium the two processes overlapped. All ten hydrogen atoms were exchangeable. There was however no exchange in the reactant except on tungsten film, where only the five ring hydrogen atoms were substituted.

The methylcyclopropane-hydrogen reaction has been used in extensive studies to evaluate particle size effects. Series of Pt/SiO_2, Pt/Al_2O_3, Pd/SiO_2 and Rh/SiO_2 catalysts have been prepared in various ways, and their dispersions (D) exhaustively characterised by selective gas chemisorption and X-ray line profile analysis; values of TOF and of *iso*butane selectivity (S_i) for this reaction have been reported.[19,20,30,47−49] Standard pretreatment sequences were applied to the precursor (oxidation, helium flushing, reduction (variable temperature) and further helium flushing (optional)). The ranges of dispersion used and changes in TOF and S_i observed as dispersion was increased are recorded in Table 11.4; the standard reduction temperature (T_{red}) was 623 K and final helium flushing was used. The most startling observation was the extent to which TOF in particular varied with T_{red}. With Pt/SiO_2,[19,20,30,47] good activities were found after hydrogen treatment at ambient temperature, although reductions may not have been complete, but minimal rates were found after reduction at 473 ± 50 K, and rates some

TABLE 11.4. Effect of Degree of Dispersion (D) on the TOF and *iso*Butane Selectivity (S_i) in the Hydrogenation of Methylcyclopropane at 273 K

Metal	Support	D range/%	$\Delta(TOF)^a$	$S_i/\%^b$	$\Delta E/\text{kJ mol}^{-1c}$	References
Pt	SiO_2	6–81	Increase $\times 2^d$	90–95	6	19,20,30,50
Pt	Al_2O_3	4–106	∝ at $D > 30\%$	94	7	20,49
Pd	SiO_2	15–85	Increase $\times 7^e$	75	6	47,50
Rh	SiO_2	10–110	Increase $\times 9$	95–78	3	47,48

^{a}Change in TOF on increasing the dispersion
^{b}Where two figures are given, they refer respectively to minimum and maximum dispersion
^{c}Difference in activation energies for C2—C3 and C1—C2 bond-breaking processes
^{d}The effect is small at D < 40%
^{e}TOF is maximal at D = 65%: the increase given is for 15–65% dispersion, and the factor depends somewhat on T_{red}.
See text for other details

ten times greater than these after reduction at ~773K. A similar trend was found with Pt/Al$_2$O$_3$.[49] Most importantly however the form of the dependence of TOF on D inverted as T_{red} was increased: thus for T_{red} ~300 K, TOF decreased with increasing D; at T_{red} ~473 K it was more or less independent of it; but for T_{red} ~773 K it increased two-fold for dispersions greater than 40%.[30] Thus the apparent structure-insensitivity noted above for cyclopropane hydrogenation may have depended on a fortuitous choice of supports and reduction conditions.

The significance of these results has been discussed at length.[19,20,30,48,49] They are the kind of results that tempt one to give up catalysis and turn to something simpler and more straightforward, such as the philosophical basis of quantum mechanics. The reader of these papers will empathise with Omar Khayyam, who having *Heard great argument, About it and about, but evermore Came out from that same door as in I went*. No detailed interpretation for the effect of T_{red} on TOF or the inversion of the structure sensitivity has been advanced. Other observations also await explanation. For example the activity of Rh/SiO$_2$ catalysts decreased if the final helium purge was omitted,[48] but this may have been due to the retention of strongly chemisorbed hydrogen, although this was not measured. For some reason Pt/Al$_2$O$_3$ (unlike Pt/SiO$_2$) showed almost no dependence of TOF on D,[49] while Pd/Al$_2$O$_3$ showed a quite sharp maximum in TOF at about 65% dispersion. Analogous dependences of TOF on pretreatment conditions were found for propene hydrogenation (Chapter 7).

The various *iso*butane selectivities[50] (Table 11.4) and their dependences on D indicate of the formation of different intermediate species, and the relatively high yield of *n*-butane on Pd/SiO$_2$ may suggest at an exocyclic π-alkenylic species may be formed (Figure 11.5). Activation energies for the harder breaking of the C1—C2 bond were 3 to 7 kJ mol^{-1} greater than for the easier C2—C3 opening, but orders of reaction were unfortunately not measured. It is quite easy to think of many other measurements that might have been made, but then one can't have everything.

There is one more surprise to come. On two single-crystal faces of iridium, n-*butane* was the major product of methylcyclopropane hydrogenation at about 400 K, its selectivity being approximately 99%.[15] It is difficult to be precise because results are only shown graphically, and *n*-butenes were also formed at higher temperatures. Such a complete reversal of normal behaviour is quite astonishing,

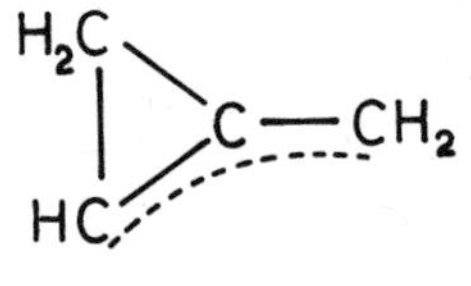

Figure 11.5. The adsorbed state of methylcyclopropane as an exocyclic π-alkenic species.

although it was said that extrapolation of the results on Ir(111) would show the selectivity falling to 50% at about 300 K. The explanation advanced above to explain the results found with palladium may apply, although iridium does not usually form π-allylic species easily. There was one significant point of difference between the two faces; on Ir(111), hydrogenolysis led to methane and propane by breaking of the C1—C2 and C2—C3 bonds, whereas on Ir(110)(1 × 2) methane, ethane and propane were formed in equal amounts, which implies random breaking of pairs of ring C—C bonds. This difference will take some explaining. The reaction does not appear to have been studied on supported iridium catalysts.

The hydrogenation of ethyl- and of n-propylcyclopropane has been studied on Rh/SiO$_2$,[51] Pd/SiO$_2$,[52,53] Pt/SiO$_2$[53,54] and Ni/SiO$_2$;[55] the activity sequence was Pt >> Rh > Pd, and selectivities to the product of ring-opening opposite the substituent were high and independent of hydrogen pressure. They were however markedly lower for the n-propyl compound in the case of Pd/SiO$_2$, and for this molecule much lower for Pd/SiO$_2$ than for Pt/SiO$_2$. Rate dependence on hydrogen pressure took various forms; this matter is treated more fully in the following section, but we may note that the curves for the ethyl compound on Pd/SiO$_2$ at 373 K resembled that in Figure 11.6(b), while on Rh/SiO$_2$ at 318 K it was like that in Figure 11.6(d).

A short comment on expressing product selectivities as ratios is in order. This practice over-emphasises differences, because a change in the ratio of a/b from 10 to 20 only means that the fraction of a increases from 0.91 to 0.95, so that,

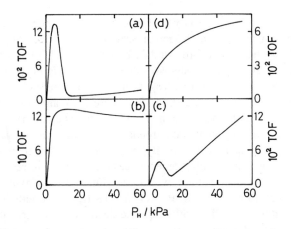

Figure 11.6. Hydrogenation of alkyl-substituted cyclopropanes: dependence of rate of the preferred splitting route on hydrogen pressure for silica-supported metals (curves for fresh and run-in catalysts are usually similar).
1,1-Dimethylcyclopropane: (a) on Pd and Rh at 318 K;[54] (b) on Pt at 318 K;[52] (c) on Rh at 318 K.[55]
Z-1,2-Dimethylcyclopropane: (d) on Ni at 473 K.[55]

since composition and not ratio is measured, values of ratios may be subject to large experimental errors.

11.3.2. Poly-alkylcyclopropanes

The hydrogenation of *Z*- and *E*-1,2-dimethyl-, 1,1-dimethyl- and 1,1,2,2-tetramethylcyclopropanes has been thoroughly investigated on silica-supported nickel,[55] copper,[56,57−59] rhodium,[51,56] palladium[52,56] and platinum[54,56] catalysts by Hungarian scientists at the University of Szeged. The results in the main have been reported graphically as TOFs for individual products at a single temperature as a function of hydrogen pressure. Emphasis was placed on TOF rather than on product selectivity, which in most cases can only be estimated roughly. No attempt has been made to model the results kinetically, and rate dependence on the hydrocarbon pressure was not investigated: this might have thrown light on their strengths of adsorption, which can now only be guessed through the rate dependence on hydrogen pressure. It is usually preferable to report total rates *and* selectivities.

Great variation in the reaction parameters was found, depending on the reactant and catalyst used. Selectivities to the preferred product (i.e. 2-methylbutane; *neo*pentane; 2,3,3-trimethylbutane) in most cases exceeded 90% and often were even higher: in general they were independent of hydrogen pressure, or showed only slight dependence. Rates fell in the sequence

$$Pt >> Rh > Pd > Cu \geq Ni$$

Molecules containing a quaternary carbon atom were especially reactive over Cu/SiO$_2$,[59] which also at low hydrogen pressures afforded quite large amounts of isomerised *alkenes* having the preferred carbon skeleton. 1,2-Dimethylcyclopropane reacted about 2000 times faster over Pt/SiO$_2$ than over Pd/SiO$_2$.

The dependence of rate on hydrogen pressure adopted a variety of forms; some representative examples are shown in Figure 11.6, and some further information relating to *Z*- and *E*-1,2-dimethylcyclopropane is contained in Table 11.5. It is difficult to discern any systematic trends. Rates for alternative processes usually showed the same dependence on hydrogen pressure, but there were several exceptions. There was no regular dependence of reactivity on reactant structure for each metal. In some cases there was no maximum rate as hydrogen pressure was raised (Figure 11.6 (a)); with Pt/SiO$_2$ at 318 K, 1,1-dimethylcyclopropane showed a maximum rate followed by a gentle decline, characteristic of competitive chemisorption[54] (Figure 11.6 (c)), whereas the same reactant on Pd/SiO$_2$ at 373 K showed an extremely sharp maximum[52] (Figure 11.6 (a)). The strangest behaviour was shown by the tetramethylcyclopropane on Rh/SiO$_2$[51] (Figure 11.6 (d)): other reactants performed as in Figures 11.6 (a) and (c). Further examples

TABLE 11.5. Parameters of the Hydrogen Pressure Dependence for the Hydrogenation of Z- and E-1,2-Dimethylcyclopropane on Silica-Supported Metals[51,52,54,55,57]

Isomer	Metal	P_H/kPa at r_{max}	$10^2 TOF/s^{-1}$	$S_{2MB}/\%$	Form
Z-	Ni	~65	17	90	No max.[a]
	Cu	~30	0.6	89	[c]
	Rh	~15	8	92	Slight max.[b]
	Pd	~15	7	~95	[c]
	Pt	~25	4	89	Sharp max.
E-	Ni	~30	6	89	Sharp max.
	Cu	~30	2	67	Slight max.
	Rh	~5	4	84[e]	[d]
	Pd	~40	20	75[f]	Slight max.
	Pt	~15	15	97	Max.

Notes:
1. The figures in columns 3, 4 and 5 relate to the process for the formation of 2-methylbutane, or the point at which this is maximal.
2. Temperatures were: for Ni, 473 K; for Cu, 498 K; the others, 318 K.
3. Forms of the rate vs. pressure curves are generally similar for both processes, and for fresh and run-in catalysts.
4. Similar results for ethyl-, 1,1-dimethyl- and 1,1,2,2-tetramethylcyclopropane will be found in the cited papers.
[a] As in Figure 11.6(b)
[b] As in Figure 11.6(c)
[c] Rate attains maximum value and stays constant.
[d] As in Figure 11.6(d)
[e] At $P_H = 65$ kPa.
[f] S_{2MB} varies with P_H: this value is for low P_H

of irrational behaviour will appear in Section 11.5. This situation calls to mind Ovid's definition of chaos, which he called *rudis indigestaque moles* - a rough and indigestible mass. More recently it has been described as *A state of order that we do not yet understand.*

A clearer picture emerges from limited work on metal films,[60] where the Z-isomer of 1,2-dimethylcyclopropane gave much more of the disfavoured product (*n*-pentane) than did the E-isomer, especially on platinum (see Table 11.6). The disposition of the methyl groups in the E-isomer clearly militates against flat chemisorption of the ring and the fission of the C1—C2 bond, whereas in the Z-isomer this is possible if less likely than C2—C3 bond fission, perhaps because

TABLE 11.6. 2-Methylbutane Selectivity in the Hydrogenation of 1,2-Dimethylcyclopropane Isomers on Metal Films[85]

Metal	T/K	Z-	E-
Ni	296	91.1	99.5
Pd	298	90.9	99.5
Pt	293	80.8	99.6

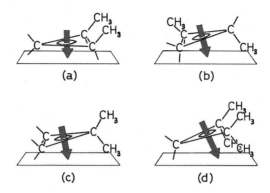

Figure 11.7. Representations of the adsorbed state of (a) Z- and (b) E-1-dimethylcyclopropane; (c) 1,1-dimethylcyclopropane; (d) 1,1,2,2-tetramethylcyclopropane.

of an electronic factor. Steric inhibition due to methyl groups in the chemisorption of the other molecules is represented in Figure 11.7.

The reactions of 1,1-dimethyl- and 1,1-diethylcyclopropanes with hydrogen and with deuterium has also been examined on films of all the metals of Groups 8 to 10 (except Ru and Os);[61] the results have a pleasing simplicity. Ring fission was exclusively at the C2—C3 bond and with deuterium the 1, 3-dideuteroalkanes were the chief products, except on platinum and to some extent on iridium, where -d_3 and -d_4 alkanes were also found. The presence of the quaternary carbon atom forbids the easy $\alpha\beta$ exchange, so $\alpha\alpha$-diadsorbed species must have been responsible for further exchange. Hydrogenolysis on the base metals gave methane and *iso*butane from the dimethyl-compound.[62,63] Further studies[61] with hydrogen-deuterium mixtures led to the conclusion that hydrogenation occurred on platinum by *simultaneous* addition of two atoms to 2,3-diadsorbed species. A study of the reaction of 1,1-dimethylcyclopropane with deuterium on Cu/SiO$_2$ has also been reported.[64]

11.3.3. The Cyclopropane Ring in More Complex Hydrocarbons[1]

The cyclopropane ring can exist within larger molecules. For example the whole family cyclopropylmethanes $(C_3H_5)_xCH_{4-x}$ ($x = 1$–4) has been synthesised, and detailed studies of dicyclopropylmethane ($x = 2$) have been carried out with various platinum and nickel catalysts.[65] The routes whereby it is converted into C$_7$ products on platinum is shown in Scheme 11.2: steps in which bond scission *adjacent* to a substituent occurs are a consequence of dissociative chemisorption on the side-chain. Clearly the product formed by two-fold breaking of the bonds opposite the substituent predominates, although others are significant. This molecule provides an easy way of assessing mass-transport within porous catalysts through

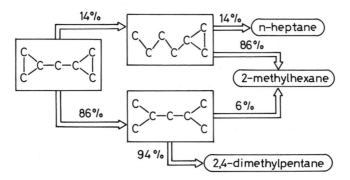

Scheme 11.2. Reactions of dicyclopropylmethane with hydrogen on platinum and nickel catalysts; the numbers relate to the reactions on platinum.

measuring the relative amounts of the intermediate and final products formed. On nickel catalysts there was some demethanation and less of the intermediates, 2-methylhexane forming some 70% of the C_7 products.

Modes of breaking the cyclopropane ring within more complex hydrocarbons give some further insights into the flexibility with which it can chemisorb.[1,10] Thus *spiro*pentane (bicyclo[2.2.0]pentane) is hydrogenated to 1,1-dimethylcyclopropane, although neither ring can easily lie flat on the surface. With bicyclo[2.1.0]pentane, however, the bond between the points of substitution breaks first because more strain is thus released, and the product is cyclopentane. Similarly addition of methylene to 9,10-octalin forms a cyclopropyl ring between the pre-existing rings, and this is hydrogenated to give *E*- and *Z*-9-methyldecalins. In the reaction of 3,3-dimethylnortricyclene to 7,7-dimethylnorbornane, it is hard to see how the reactant can chemisorb so that the relevant bond is broken: considerations of strain-release must again predominate. These processes are illustrated in Scheme 11.3.

11.4. HYDROGENATION OF CYCLOPROPANES HAVING OTHER UNSATURATED GROUPS

Methylenecyclopropane in the free state may also be regarded as trimethylenemethane[9] (Scheme 11.4A), but on chemisorption the former structure is the more appropriate because with this compound hydrogenation gives simultaneously methylcyclopropane and butanes in which the *n-isomer* predominates.[1,2,66] Breaking at the C—C bond opposite the point of substitution also occurs with ethenylcyclopropanes, and prior chemisorption at the C=C bond facilitates the splitting of the nearer C—C bond. However the insertion of a methylene group

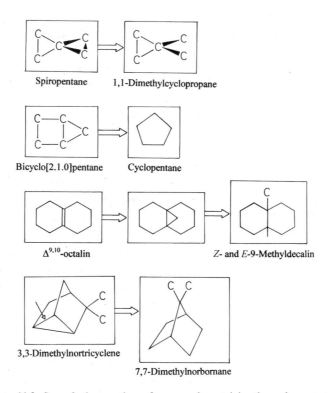

Scheme 11.3. Some further reactions of compounds containing the cyclopropane ring.

between the double bond and the ring insulates the latter, and the normal mode is resumed.

A detailed study of the hydrogenation of methylenecyclopropane (MCPr) on Pt/pumice was reported many years ago.[2] The hydrocarbon was more strongly chemisorbed than hydrogen (and even than *iso*butene, with which it was compared), the rate passing through a maximum at low MCPr pressure. The reactants were competitively adsorbed, but in the reverse order from that seen with cyclopropane and hydrogen. The activation energy was 73 kJ mol^{-1}. At 290 K the butanes were the sole products (*n*-butane selectivity ~70%), but at 373 K the major products were the butenes, all of which were found. Diadsorbed intermediates having had the ring broken were displaced as butenes by MCPr and they only began to be hydrogenated after about 60% of the MCPr had reacted.

Hydrogenation of 3-carene on platinum gave Z-carane, but on palladium migration of the double bond into conjugation with the ring allowed splitting of the strained ring C—C bond to give a cycloheptane[10] (Scheme 11.4B). The strong

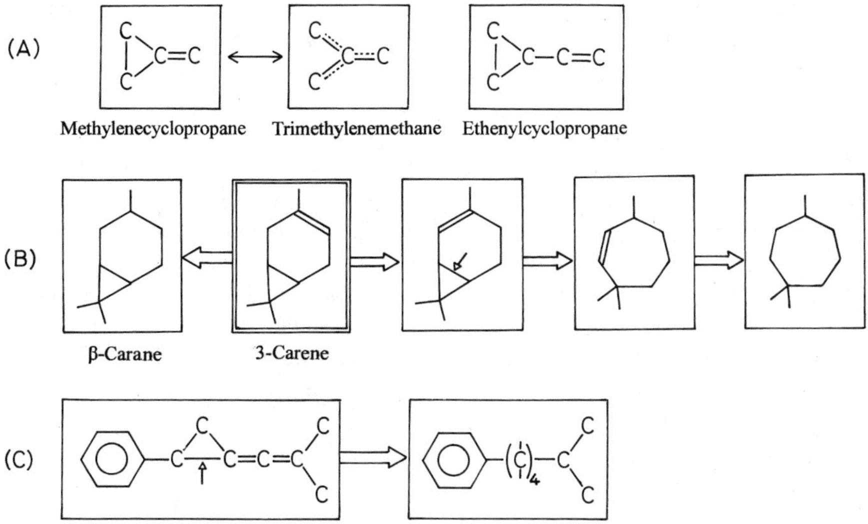

(A) Methylenecyclopropane   Trimethylenemethane   Ethenylcyclopropane

(B) β-Carane   3-Carene

(C)

Scheme 11.4. Unsaturated cyclopropanes and some of their reactions

isomerising propensity of palladium also enables an alkene in which the ring has been broken to desorb, and where there are unsaturated functions at C1 and C2, it is the intervening C—C bond that breaks preferentially (see also Scheme 11.4C).

Cyclopropene is a highly strained system, but loss of a hydride ion gives the cyclopropenyl cation, which is formally an aromatic compound; the tri(cyclopropyl) cyclopropenyl cation is very stable and can even exist in aqueous solution. Notwithstanding the very great strain in the parent hydrocarbon, the hydrogenation of methyl-substituted cyclopropenes has been reported.[11]

Phenyl substituents have the same effect as alkenic unsaturation in promoting ring fission adjacent to the point of substitution, but a study of the hydrogenation with deuterium of a number of phenyl- and phenyl-methyl-substituted hydrocarbons at temperatures between 369 and 443K adds some further refinement to this simple picture.[66] In most cases the products contained only two deuterium atoms, added across a C—C bond adjacent to the phenyl group, but where the carbon atom bearing the phenyl group also carried a hydrogen atom this also was liable to be exchanged *after* ring opening: there was no exchange in any of the reactants. With the 1,2-phenylmethylcyclopropanes, the bond broken was that opposite the methyl group. Placing a methylene group between the phenyl group and the ring again interfered with the conjugative effect and lowered the reactivity.

Consideration of the likely steric interference between the substituents and the surface led to the view that reaction involved arrival of the hydrocarbon at the surface with one of the unsubstituted ring carbons pointing towards it ('corner orientation'),[66] followed by attack by a deuterium atom. It is clear that in some cases 'flat' chemisorption (as in structure **A**, Figure 11.2) would be at best weak and that $\alpha\gamma$-diadsorption by opening of the bond that predominantly broke would be almost impossible. The proposed mechanism therefore is the only remaining alternative; catalysis is, like politics, the art of the possible. It is however risky to believe that conclusions based on heavily substituted rings inform on the mechanism of reaction of the unsubstituted ring, and that product analysis alone can allow a definitive model to be reached. In the absence of kinetic information (reaction orders, activation energy etc.), no inkling of the effect of substituents on chemisorption strength can be obtained. In the interpretation offered, the electronic effects of the substituents worked only on the free molecules, and affected only the disposition of charge within them, and hence their reactivity. In particular no thought was given to the possibility that the aromatic ring might engage with the surface holding it in place while the small ring suffered attack. This concept has neatly explained the effects of alkenic substituents.[2]

11.5. HYDROGENATION OF ALKYLCYCLOBUTANES AND RELATED MOLECULES

Very little is known about the hydrogenation of cyclobutane itself. On Pt/Al$_2$O$_3$ the rate was first order in the hydrocarbon, but the order in hydrogen was expressed indirectly in terms of assumed loss of hydrogen atoms in the adsorbed state.[67] There have however been a number of publications concerning its alkyl derivatives. As we have seen (Section 11.1 and Table 11.1), cyclobutane resembles much more an open-chain alkane than does cyclopropane, so the motivation for studying its alkyl derivatives is to see how the position of ring opening is affected by substituents, and what intermediate species may be formed. Once again however it will be hard to separate steric and electronic effects in deciding what products are formed.

The two modes of ring-opening of methylcyclobutane give n-pentane and 2-methylbutane; on films of nickel, palladium and platinum, the latter predominated, its selectivity S_{2MB} decreasing with increasing temperature as befits the easier path: it also rose with increasing hydrogen pressure on platinum at 323 K, but fell at 403 K (Figure 11.8).[60] Orders of reaction in hydrogen on platinum were negative (-1 to -2), at 323–403 K; nickel gave some hydrogenolysis, platinum a little and palladium none, and activation energies were respectively 33.5, 69 and 82 kJ mol^{-1}. On Pt/Al$_2$O$_3$ model catalysts and on EUROPT-1 (6% Pt/SiO$_2$), orders were positive in both reactants at low hydrogen pressure,[68] but increasing this caused rates to pass through maxima, the position of which increased with temperature but was not affected by dispersion. The reactants were clearly competitively

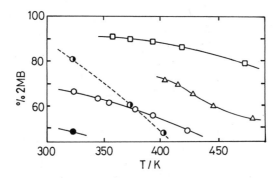

Figure 11.8. Hydrogenation of methylcyclobutane: dependence of 2-methylbutane selectivity on temperature for films of nickel (□), palladium (△), and platinum (○): hydrogen pressures 0.017 atm (filled points), 0.057 atm (open points), 0,17 atm (half-filled points).[60]

adsorbed, and the fact that reaction occurred as low as 373 K suggests that the mode of the ring's chemisorption was by C—C bond breaking and not by C—H dissociation. TOFs increased with hydrogen pressure, and values of S_{2MB} passed through minima; they decreased with rising temperature (as in Figure 11.8), values being between 50 and 70%. Model Rh/TiO$_2$ catalyst began to lose activity due to the SMSI effect when reduced above 373K, although Rh/Al$_2$O$_3$naturally did not: values of S_{2MB} were ~90% independent of temperature.[69]

Extensive studies of the hydrogenation of methylcyclobutane have also been reported by the Hungarian scientists at Szeged.[70–72] Forms of rate dependence for each of the products as a function of hydrogen pressure on fresh and run-in ('working') catalysts comprising silica-supported nickel, rhodium, palladium and platinum at two temperatures have been given, and these display a bewildering variety (see Figure 11.9). In most cases the two rates run closely parallel (see however Figure 11.9(f) for an exception), but the forms do not in general match any of those found in the reactions of the alkylcyclopropanes (Figure 11.6). Only results for Pd/SiO$_2$ at 523 K show[71] sensible behaviour characteristic of competitive adsorption; in most other cases there are clear signs of inhibition at low hydrogen pressure, giving apparent orders in hydrogen greater than unity. This may be due to the formation of hydrogen-depleted species, of which authors were well aware, although in a number of cases the form of rate-pressure curve was similar of both the fresh and used catalysts (Figure 11.8). The very peculiar behaviour of Pt/SiO$_2$at 573 K disappeared on raising temperature by 50 K; it defies rational explanation. Unfortunately not enough detail on experimental procedure has been provided, as it is unclear whether the hydrogen pressure was varied randomly, and whether checks were performed on reproducibility of rates. Values of S_{2MB} were independent of hydrogen pressure (for Pt/SiO$_2$,0.55 ± 0.05 at 573 K; 0.58 ± 0.05 at 623 K; for Pd/SiO$_2$, 0.54–0.61 at 523 K; 0.45–0.60 at 673 K), and were similar

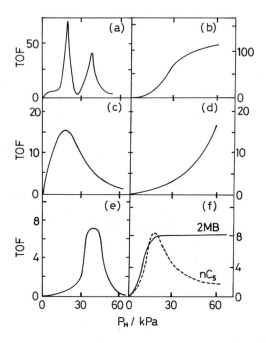

Figure 11.9. Hydrogenation of alkylcyclobutanes: dependence of rate of 2-methylbutane formation on hydrogen pressure for silica-supported metals.
Methylcyclobutane: (a) on Pt at 573 K; (b) on Pt (fresh) at 623 K; (c) on Pd at 523 K; (d) at 673 K; (e) on Pt (run-in) at 623 K, showing rates of both routes.[71,72]
n-Propylcyclopropane: (f) on Ni at 523 K.[70]
Note: except for Pt at 623 K, curves for fresh and run-in catalysts are similar, although rates for the latter are of course much slower.

for fresh and run-in catalysts. Measurements were also undertaken in the pulse reaction mode,[73] which minimises carbon deposition: selectivities, independent of temperature, were close to those just mentioned for Pt/SiO$_2$ and Pd/SiO$_2$ but some hydrogenolysis and ring enlargement was noticed. Rh/SiO$_2$ showed much hydrogenolysis above 473 K and Ni/SiO$_2$ above 523K; values of S_{2MB} were respectively 0.6 to 0.75 and 0.70 ± 0.05.

The number of deuterium atoms in the products of the reaction of methyl-cyclobutane with deuterium gives some indication of the structure of adsorbed intermediates. Thus simple alkyl reversal with $\alpha\beta$-diadsorbed species would give either -d_3 or -d_7 products depending on whether or not ring inversion via the methyl group occurred or not, whereas formation of an exocyclic π-alkenylic species would give alkanes having up to nine exchanged atoms.[74] On palladium film there was stepwise exchange to -d_5 and -d_6 at 298 K, and at 418 K products contained mainly five to nine deuterium atoms. On platinum at 273 K there was again step-wise exchange, but at 415 K most products had two to six deuterium atoms. This distinction between the two metals in terms of their ability to form π-alkenylic forms parallels their behaviour in alkane exchange generally (see Chapter 6).

Over platinum black at 573 to 663 K, ethylcyclobutane showed selectivities to 3-methylpentane that increased from 30 to 60% with rising hydrogen pressure at 663 K, and decreased from 52 to 30% with rising temperature at low hydrogen pressure.[75,76] Extension of the side-chain permits other processes to occur, so that ring enlargement to methylcyclopentane and aromatisation to benzene was seen at low hydrogen pressures. Adsorption through the alkyl group also becomes easier, and breaking the ring close to the substituent more likely.

Results analogous to those for methylcyclobutane are available for the n-propyl compound.[77–81] Kinetic studies showed the same variety of rate as hydrogen pressure curves, but in general the two molecules exhibited some similarity.[80] On Pd/SiO$_2$ it appeared that the n-propyl compound was the one strongly chemisorbed, as the rate maximum occurred at a higher hydrogen pressure. Selectivities to the 'preferred' product (3-methylhexane) were broadly in the same ranges, but were notably lower for Ni/SiO$_2$ (0.30–0.55 at 523 K). Once again Pd/SiO$_2$ showed no hydrogenolysis or ring-enlargement, and Rh/SiO$_2$ was very active for the former (83% at 673 K). Under the same conditions the regiospecific effect of a single alkyl group appears to become smaller the longer the alkyl chain.

A constant and recurrent theme in the study of metal-catalysed hydrocarbon transformations has been the complicating role of carbonaceous deposits, which are recognised as forms of the reactants that are dehydrogenated and form multiple bonds to the surface to an extent depending critically on temperature and hydrogen pressure. Analysis of the behaviour of alkanes and alicyclics[9,82,83] has led to the conclusion that such species are inimical to the reactions of the latter, but are that the former may still be reactive in their presence and may even require their participation. This of course is not a new idea, but it still needs quantitative evaluation.

The phenomenon of bistability was shown in the hydrogenation of n-propylcyclobutane, but curiously only with run-in Rh/SiO$_2$ catalyst.[84]

The 1, 2-dimethylcyclobutanes contain three non-equivalent C—C bonds, the breaking of which gives either n-hexane or 3-methylpentane or 2,3-dimthylbutane; Table 11.7 records the proportions of each formed over films of several metals at the lower of the temperatures used.[85] Not surprisingly, with the E-isomer the breaking of the C1—C2 bond is disfavoured compared to the Z-isomer *except over palladium* where formation of the planar π-alkenyl species may be possible. Fission of the C3—C4 bond is markedly easier in the E-isomer, which cannot lie flat on the surface before bond-breaking occurs; this inhibits C2—C3 breaking, this being especially marked with nickel and rhodium.

TABLE 11.7. Product Yields (%) from the Hydrogenation of Z- and E-Dimethylcyclobutane on Various Metal Films[60]

Metal	Isomer	T/K	n-hexane	3-methylpentane	2,3-dimethylbutane
Ni	E	408	3	17	80
	Z	373	19	29	52
Rh	E	273	1	8	91
	Z	273	6	26	68
Pd	E	433	13	34	53
	Z	428	17	38	45
Pt	E	358	8	22	70
	Z	373	22	38	40

1,1,3,3-tetramethylcyclobutane has only one type of ring C—C bond. Its reaction with hydrogen on films of a number of metal films has been examined.[86] On rhodium and iridium, extensive hydrogenolysis occurred, while on early transition metals (Ti, Zr, Ta, Mo) *iso*butane was the main product. Palladium gave mainly 2,2,4-trimethylpentane, and platinum gave this plus some ring-enlargement to 1,1,3-trimthylcyclopentane. Possible mechanisms were presented.

The hydrogenation of methylenecyclobutane over Pt/pumice at 373 K gave about 90% of methylcyclobutane but above 473 K linear alkenes (especially E-2-pentene) were formed; at 573 K, these appeared to arise through 1-methylcyclo-1-butene.[2] No other studies of this reaction have been reported.

The C_4 ring can exist as part of complex organic compounds, and its fission is aided by the existence of strain and unsaturated substituents;[11] thus for example 1,2-diphenylbenzocyclobutene (Scheme 11.5) was easily reduced to 1,2-dibenzylbenzene. Two fused cyclobutane rings (i.e. bicyclo[2.2.0]hexane) reacted to form cyclohexane, and where there are three fused rings, as in the molecule in Scheme 11.5 that defies simple naming, this reaction occurred, but one ring remained intact.[11]

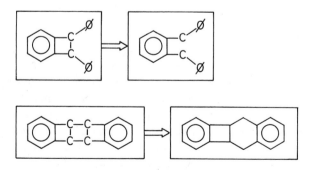

Scheme 11.5. Reactions of molecules containing the cyclobutane ring.

REFERENCES

1. J. Newham, *Chem. Rev.* **63** (1963) 123.
2. G.C. Bond and J. Newham, *Trans. Faraday* Soc. **56** (1960) 1501, 1851.
3. J. March, *Advanced Organic Chemistry: Reactions, Mechanisms and Structure,* 4th ed., Wiley-Interscience: New York (1992).
4. A.D. Walsh, *Trans. Faraday Soc.* **45** (1949) 179.
5. W.A. Bernett, *J. Chem. Educ.* **44** (1967) 17.
6. R.J. Buenker and S.D. Peyerinhoff, *J. Phys. Chem.* **73** (1969) 1299.
7. S. Yamamoto, M. Nakata, T. Fukuyama and K. Kuchitsu. *J. Phys. Chem.* **89** (1985) 3298.
8. J. Gauss, D. Cremer and J.F. Stanton, *J. Phys. Chem. A* **104** (2000) 1319.
9. C.J. Cramer and B.A. Smith, *J. Phys. Chem.* **100** (1996) 9664.
10. R.L. Augustine, *Heterogeneous Catalysis for the Synthetic Chemist*, Dekker: New York (1996).
11. P.N. Rylander, *Catalytic Hydrogenation over Platinum Metals*, Academic Press: New York (1967).
12. W.H. Weinberg, *Langmuir* **9** (1993) 655.
13. S. Černý, M. Smutek, F. Buzek and A. Cuřnova. *J. Catal.* **47** (1977) 159.
14. P. Lenz-Solomun and D.W. Goodman, *Langmuir* **10** (1994) 172.
15. J.R. Engstrom, D.W. Goodman and W.H. Weinberg, *J. Phys. Chem.* **94** (1990) 396.
16. H.F. Wallace and K.E. Hayes, *J. Catal.* **29** (1973) 83.
17. G.F. Taylor, S.J. Thompson and G. Webb, *J. Catal.* **12** (1968) 150.
18. R. Merta and V. Ponec in: *Proc. 4th Internat. Congr. Catal.* Akadémiai Kiadó: Budapest (1968) 53; *J. Catal.* **17** (1970) 79.
19. P.H. Otero-Schipper, W.A. Wachter, J.B. Butt, R.L. Burwell Jr. and J.B. Cohen, *J. Catal.* **53** (1978) 414.
20. J.B. Butt, *Appl. Catal.* **15** (1985) 161.
21. A. Verma and D.M. Ruthven, *J. Catal.* **46** (1977) 160.
22. R.A. Dalla Betta, J.A. Cusumano and J.H. Sinfelt, *J. Catal.* **19** (1970) 343.
23. J.H. Sinfelt, D.J.C. Yates and W.F. Taylor, *J. Phys. Chem.* **69** (1965) 1877.
24. G.C. Bond and J. Sheridan, *Trans. Faraday Soc.* **48** (1952) 713.
25. S.D. Jackson, G.D. McLellan, G. Webb, L. Conyers, M.B.T. Keegan, S. Mather, S. Simpson, P.B. Wells, D.A. Whan and R. Whyman, *J. Catal.* **162** (1996) 10.
26. S.D. Jackson, M.B.T. Keegan, G.D. McLellan, P.A. Meheux, R.B. Moyes, G. Webb, P.B. Wells, R. Whyman and J. Willis in: *Preparation of Catalysts V* (G. Poncelet, P.A. Jacobs, G. Grange and B. Delmon, eds.) Elsevier: Amsterdam (1991), p. 135.
27. J.R. Anderson and N.R. Avery, *J. Catal.* **8** (1967) 48.
28. G.C. Bond and J. Turkevich, *Trans. Faraday Soc.* **50** (1954) 1335.
29. G.C. Bond and J. Addy, *Trans. Faraday Soc.* **53** (1957) 383, 388; *Adv. Catal.* **9** (1957) 44.
30. P.H. Otero-Schipper, W.A. Wachter, J.B. Butt, R.L. Burwell Jr. and J.B. Cohen, *J. Catal.* **50** (1977) 494.
31. D.R. Kahn, E.E. Petersen and G.A. Somorjai, *J. Catal.* **34** (1974) 294.
32. J.M. Beelen, V. Ponec and W.M.H. Sachtler, *J. Catal.* **28** (1973) 376.
33. J. Schwank, J.Y. Lee and J.G. Goodwin Jr., *J. Catal.* **108** (1987) 495.
34. G.C. Bond and R. Yahya, *J. Molec. Catal.* **68** (1991) 243.
35. G.A. Somorjai, *Introduction to Surface Chemistry and Catalysis,* Wiley: New York (1994).
36. M. Boudart, A. Aldag, J.E. Benson, N.A. Dougharty and C.G. Harkins, *J. Catal.* **6** (1966) 92.
37. J.W.E. Coenen, W.M.T.M. Schats and R.Z.C. van Meerten, *Bull. Soc. Chim. Belg.* **88** (1979) 435.
38. G.R. Gallaher, J.G. Goodwin Jr. and L. Guczi, *Appl. Catal.* **73** (1991) 1.
39. K.E. Hayes and H.S. Taylor, *Zeit. Phys. Chem. NF* **15** (1958) 127.
40. C. Visser, J.G.P. Zuidwijk and V. Ponec, *J. Catal.* **35** (1974) 407.
41. P. Baláž, I. Sorták and R. Domanský, *Chem. Zvesti* **32** (1978) 75.

42. G.C. Bond, *Catalysis by Metals,* Academic Press: London,1962.

43. G.C. Bond and J. Addy, *Trans. Faraday Soc.* **53** (1957) 377.

44. J.R. Anderson and C. Kemball, *Proc. Roy. Soc.* **A224** (1954) 272.

45. R. Brown and C. Kemball, *J. Chem. Soc. Faraday Trans.* **86** (1990) 3815.

46. R. Brown, C. Kemball and I.H. Sadler, *Proc. 9th. Internat. Congr. Catal.,* (M.J. Phillips and M. Ternan, eds.), Chem. Inst. Canada: Ottawa **3** (1988) 1013.

47. R.L. Burwell Jr., *Langmuir* **2** (1986) 2.

48. Z. Karpiński, T.-K. Chang, H. Katsuzawa, J.B. Butt, R.L. Burwell Jr. and J.B. Cohen, *J. Catal.* **99** (1986) 184.

49. S.S. Wong, P.H. Otero-Schipper, W.A. Wachter, Y. Inone, M. Kobayashi, J.B. Butt, R.L. Burwell Jr. and J.B. Cohen, *J. Catal.* **64** (1980) 84.

50. R.K. Nandi, F. Molinaro, C. Tang, J.B. Cohen, J.B. Butt Jr. and R.L. Burwell Jr. *J. Catal.* **72** (1982) 289.

51. F. Notheisz, I. Pálinkó and M. Bartók,*Catal. Lett.* **5** (1990) 229.

52. I. Pálinkó, F. Notheisz and M. Bartók, *J. Molec. Catal.* **63** (1990) 43.

53. B. Török, Á. Molnár and M. Bartók, *Catal. Lett.* **33** (1995) 331.

54. I. Pálinkó, F. Notheisz, J.T. Kiss, M. Bartók, *J. Molec. Catal.* **77** (1992) 313.

55. I. Pálinkó, F. Notheisz and M. Bartók, *J. Molec. Catal.* **68** (1991) 237.

56. I. Pálinkó, *J. Catal.* **168** (1997) 543.

57. I. Pálinkó, Á. Molnár, J.T. Kiss and M. Bartók, *J. Catal.* **121** (1990) 396.

58. Á. Molnár, I. Pálinkó and M. Bartók, *J. Catal.* **114** (1988) 478.

59. M. Bartók, I. Pálinkó and Á. Molnár, *J. Chem. Soc. Chem. Comm.* (1987) 953.

60. G. Maire, G. Plouidy, J.C. Prudhomme and F.G. Gault, *J. Catal.* **4** (1965) 556.

61. T. Chevreau and F. Gault, *J. Catal.* **50** (1977) 124.

62. T. Chevreau and F. Gault, *J. Catal.* **50** (1977) 143.

63. T. Chevreau and F. Gault, *J. Catal.***50** (1977) 156.

64. J.T. Kiss, I. Pálinkó and Á. Molnár, *J. Molec. Struct.* **293** (1993) 273.

65. R.L. Burwell Jr. and J. Newham, *J. Phys. Chem.* **66** (1962) 1431, 1438.

66. J.A. Roth, *J. Catal.* **26** (1972) 97.

67. J. Barbier, P. Marécot and R. Maurel, *Nouv. J. Chim.* **4** (1980) 385.

68. C. Zimmermann and K. Hayek, *Proc. 10th Internat. Congr. Catal.,* (L. Guczi, F. Solymosi and P. Tétényi, eds.), Akadémiai Kiadó: Budapest **C** (1992) 2375.

69. G. Rupprechter, G. Seeber, H. Goller and K. Hayek, *J. Catal.* **186** (1999) 201.

70. B. Török and M. Bartók, *J. Catal.* **151** (1995) 315.

71. B. Török, J.T. Kiss and M. Bartók, *Catal. Lett.* **46** (1997) 169.

72. B. Török, M. Török and M. Bartók, *Catal. Lett.* **33** (1995) 321.

73. B. Török and M. Bartók, *Catal. Lett.* **27** (1994) 281.

74. L. Hilaire, G. Maire and F.G. Gault, *Bull. Soc. Chim. France* (1967) 886.

75. Z. Paál, M. Dobrovolszky and P. Tétényi, *React. Kinet. Catal. Lett.* **2** (1975) 97.

76. Z. Paál and M. Dobrovolszky, *React. Kinet. Catal. Lett.* **1** (1974) 435.

77. B. Török, I. Pálinkó, Á. Molnár, and M. Bartók, *J. Molec. Catal.* **91** (1994) 61.

78. B. Török, I. Pálinkó, Á. Molnár and M. Bartók, *J. Catal.* **143** (1993) 111.

79. B. Török, Á. Molnár, I. Pálinkó, and M. Bartók, *J. Catal.* **145** (1994) 295.

80. B. Török, I. Pálinkó and M. Bartók, *Catal. Lett.* **31** (1995) 421.

81. M. Bartók, B. Török, Á. Molnár and J. Apjok, *React. Kinet. Catal. Lett.* **49** (1993) 111.

82. A. Fási, J.T. Kiss, B. Török and I. Pálinkó, *Appl. Catal. A: Chem.* **200** (2000) 189.

83. I. Pálinkó, F. Notheisz and M. Bartók, *Catal. Lett.* **1** (1988) 127.

84. I. Pálinkó and B. Török, *Catal. Lett.* **45** (1997) 193–197.

85. G. Maire and F.G. Gault, *Bull. Soc. Chim. France* (1967) 894.

86. J.K.A. Clarke, B.F. Hegarty and J.J. Rooney, *J. Chem. Soc. Faraday Trans. I* **84** (1988) 2511.

87. W.F. Taylor, D.J.C. Yates and J.H. Sinfelt, *J. Catal.* **4** (1965) 374.

12

DEHYDROGENATION OF ALKANES

PREFACE

We have now concluded our survey of the processes in which hydrogen is added to the various types of carbon-carbon unsaturation, the final products of which are alkanes of some type or other. We must now look at the reverse processes, since in many cases the unsaturated products that can be derived by removing hydrogen from an alkane are more useful and therefore more valuable than the reactant; and hydrogen if it can be recovered is also a desirable product. Dehydrogenation is however usually endothermic and needs high temperatures to secure reasonable yields of products; the exceptions are cyclohexane and related molecules, which convert to aromatic molecules at much lower temperatures, due to the pull of the resonance stabilisation. A further attractive process that has been much studied is the conversion of methane by dehydrogenation to higher alkanes.

The process of dehydrogenation does not enjoy the plethora of mechanistic subtleties that have engaged our attention in earlier chapters. The principal concern has been to discover catalysts that are stable at the necessarily high temperatures, and that are selective towards dehydrogenation, rather than the competing processes of hydrogenolysis and of carbon deposition that bedevil attempts to isolate the target reaction.

12.1. INTRODUCTION

The dehydrogenation of acyclic alkanes is at one and the same time both simpler and more complex than any of the reactions so far considered: it is simpler in the sense that the mechanism for the formation of alkenes involves only adsorbed alkyl radicals, alkenes and hydrogen atoms, and with rare exceptions lacks any of the subtlety so common with hydrogenations; but it is more complex in the sense that it is markedly endothermic (to the extent that the corresponding hydrogenations are exothermic), and therefore requires high temperatures to achieve significant progress towards equilibrium. At such temperatures, parasitic reactions involving excessive dehydrogenation, and formation of carbonaceous deposits, lead to rapid loss of activity, and so the large-scale realisation of the economically attractive production of alkenes from alkanes demands the design of catalysts able to withstand this type of insult. Much of what follows in this Chapter concerns the ways in which this is done.

The success of the petrochemical industry has depended in no small measure on the conversation of C_2 to C_5 alkenes to more useful and valuable products, by processes that include polymerisation, oligomerisation, selective oxidation and epoxidation:[1] these feedstocks were formerly available as by-products of refinery operations such as naphtha cracking, but the supply did not match demand, and so alkane dehydrogenation became economically viable. The processes are however equilibrium-limited, that is, complete conversion cannot be attained at any realistic temperature; so for example the maximum conversions at 800 K and 1 atm pressure are for ethane, 10%; for propane, 28%; and for isobutane, 43%.[2] Another way of expressing the problem is to note the temperatures needed for 50% conversion; these are respectively 983, 863 and 813 K. This limitation can of course be overcome by using oxidative dehydrogenation, where the hydrogen removed is converted to water; this has also led to extensive research to discover oxide catalysts that are selective for this process, but this work is beyond the scope of this book. Another way of defeating equilibrium is to combine dehydrogenation with continuous reaction of the alkene with methanol; this avoids a costly separation step, and produces an ether. Much attention has been given to the sequence of reactions: *iso*butane $\rightarrow$ *iso*butene $\rightarrow$ methyl-*tert*-butylether (MTBE), which has value as a fuel additive.[2] The equilibrium limitation can also be overcome by continuously removing the hydrogen produced by diffusion through a palladium-based permselective membrane.[2,3] With cyclic alkanes, however, the thermodynamics are (as we shall see, Section 12.3) very much more favourable, so that much lower temperatures will suffice.

The higher the temperature used, the more likely are the adsorbed hydrocarbon species to lose more than the two desired hydrogen atoms, and thus to form more than two C—M bonds. These species will polymerise, and form the carbonaceous deposits that have been termed[4] the *bête noire* of all hydrocarbon reactions, and which we have met continually in the preceding five chapters (see Section 12.5). Now one way of inhibiting this unwanted happening would be to add hydrogen to suppress the tendency to over-dehydrogenation, by providing a greater surface coverage by hydrogen atoms (Section 12.2.6). However the process of dehydrogenation results in an increase in pressure at constant volume, so by Le Chatelier's Principle the equilibrium conversion attainable will fall as pressure is increased: theoretically 100% conversion could be obtained at zero pressure.

The need to suppress carbon formation by adding hydrogen is thus in conflict with the desirability of operating at reduced pressure, but to do this entails additional power consumption and construction costs. Each industrial process resolves these conflicting factors in its own way.[2]

There is however one beneficial way of using the tendency of metal surfaces to dehydrogenate alkanes and to form polymeric structures. Methane is abundantly and cheaply available, and it has been found that on certain metals it will do just this, and by alteration of the reaction conditions the hydrogen released can sometimes crack the deposits into mainly C_2 and C_3 fragments that appear as ethane and propane. There will be some hydrogen left over, which is a bonus on top of the formation of the more useful alkanes. The considerable literature on this process is summarised in Section 12.3.

Most of the metallic catalysts used for alkane dehydrogenation are based on platinum, because of its low activity for hydrogenolysis, but by itself it cannot prevent its deactivation by carbon deposits, and so much thought has been given to its promotion, chiefly by combining it with one or more other inert metals. The thinking was that the steps leading to the carbon deposit required a larger ensemble of the active atoms than those that led just to alkene, so that diluting platinum with an inert partner would lead on average to smaller ensembles, and hence to better performance. As well as lessening carbon formation, the promoter ought to increase the rate of dissociative chemisorption of the alkane, improve the tolerance of the catalyst to carbon, and help to gasify the carbon that is formed. Progress in this direction is reviewed in Sections 12.2.2 and 12.2.3. Control of particle size might also be relevant in limiting carbon deposition, and this has also been examined (Section 12.23). Polymerisation of the alkene, once formed, by acid centres on the support is another possible route to carbon deposition, since larger alkanes (and alkenes) dehydrogenate more easily than smaller ones. Supports must therefore be non-acidic and the importance of correct choice is stressed in Section 12.2.2.

The process of converting an acyclic alkane into a cyclic one is also formally a dehydrogenation, and is termed *dehydrocyclisation* (DHC).[5] At one time the superior octane rating of aromatics (see Section 14.1.1) encouraged research into this process, and it was established that n-hexane could cyclise into methylcyclopentane (MCP), which then underwent ring-enlargement to cyclohexane; dehydrogenation gave benzene. It is convenient to separate this last stage, which has been deeply researched, from the earlier steps, which occur alongside the multifarious reactions of C_5 and higher alkanes that constitute the business of petroleum reforming. Dehydrogenation of cyclohexane (and cyclohexene) is discussed in Section 12.3; the other reactions are reserved to Chapter 14.

We may conclude this Introduction by referring to Scheme 12.I, which lays out the main transformations that alkanes can undergo on catalysts having only a metallic function. This portrayal is somewhat simplified, as it omits some of

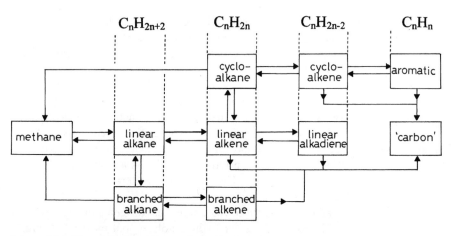

Scheme 12.1. Formal display of the reactions of hydrocarbons.
- Horizontal processes involve addition or removal of hydrogen
- Vertical processes involve isomerisation
- Not all possible processes are included (e.g. isotopic exchange, epimerisation, and some routes to 'carbon'), but all that are shown are possible in practice.

the possible routes for carbon deposition and hydrogenolysis as well as one or two reactions of interest but of minor importance, such as epimerisation and isotopic exchange. As already noted, dehydrocyclisation may proceed in a way not depicted here: this and skeletal isomerisation will occupy us in Chapter 14, and hydrogenolysis in the next chapter.

12.2. DEHYDROGENATION OF ACYCLIC ALKANES

12.2.1. Introduction: Alkane Chemisorption

Most of the publications to be noted in this Section refer to the dehydrogenation of propane or of *iso*butane, and almost all have employed platinum as the active ingredient, for the reason mentioned above: most describe the behaviour of this metal modified by an inert partner, which is usually tin. Other metals have been examined only rarely,[6] although early work highlighted the usefulness of the noble metals of Groups 8 to 10,[4] but the rapid deactivation of metals in the pure state has made them hard to study in a quantitative way. Other added metals besides tin have been tried,[6-8] and occasionally it appears that they may contribute to activity, perhaps through bimetallic sites.[7]

Some reference has already been made to the chemisorption of alkanes on metal surfaces (Sections 4.8 and 6.1); it is a difficult process requiring a significant activation energy, and probably two 'sites', although as we have seen they may not be identical, as the hydrogen atom may be content with a place that would not suit an alkyl radical. In the steady state, the rates of desorption of hydrogen and of alkene, and adsorption of alkane, will determine the concentrations of alkyl, alkene and hydrogen atoms: since these are thought to be low, the rate-determining step is usually reckoned to be the chemisorption of the alkane, the reaction then proceeding by a reversal of the Horiuti-Polanyi mechanism for alkene hydrogenation[9] (Section 7.26). Although alkane chemisorption is faster on rough single-crystal surfaces (e.g. fcc(110)-(1 × 2) than on smooth surfaces (e.g. fcc(111), dehydrogenation has long been recognised as being particle-size insensitive,[2] but for the reasons discussed earlier (Section 5.4) face-sensitivity and particle-size sensitivity are not necessarily equivalent: moreover, particle-size *insensitivity* can be induced by carbon deposition.[10] Alkane chemisorption is inhibited by pre-adsorbed hydrogen,[2] so the rate of dehydrogenation is expected to show a negative order in hydrogen.

12.2.2. Supported Platinum and Platinum-Tin Catalysts

In more recent work, such information as is available concerning platinum itself has appeared in conjunction with work on a bimetallic system, and it is therefore best not to try to separate it. Very much attention has been paid to the method of preparation of platinum-tin catalysts, to their characterisation (by methods that include EXAFS,[7,11] Mössbauer spectroscopy[12,13] and more conventional techniques), and to the choice of support. It is clear that the way in which a catalyst is prepared, and in particular the conditions used for calcination and reduction, determine the structure of the finished article.[14,15] Although the most usual procedure starts with chloroplatinic acid and stannous chloride, many other compounds of each metal have been tried (e.g. $PtCl_2P_2$;[16] Z-$PtCl(SnCl_3)P_2$ (P=PPh_3);[17] Sn^nBu_4 added to Pt/support;[18] aqueous $Sn(OH)^nBu_3$):[11] co-impregnation[14] and successive impregnation[16] have had their adherents. Curiously, however, little use seems to have been made of the $PtCl_2(SnCl_3)_2$ complex which is formed by interaction of chloroplatinic acid with stannous chloride, some of the Sn^{II} being used to reduce Pt^{IV} to Pt^{II}: this complex has an intense blue colour, and was formerly much used for the colorimetric estimation of platinum in dilute solution.

What is very clear indeed is that the degree of interaction between the two metals depends very much on the nature of the support. Both alumina and silica have often been used, and their comparison is instructive. With silica, reduction leads easily to the formation of bimetallic particles[19] (PtSn, $PtSn_2$), as well as platinum and tin, depending on the ratio in which the two metals are present.[14] At low Pt/Sn ratios, tin segregates at the surface; hydrogenolysis and fast deactivation then result. At Pt/Sn = 1, where PtSn predominates, the catalyst has been shown to be more stable, for *n*-hexane dehydrogenation giving high selectivity to alkene, with little benzene or hydrogenolysis. The bimetallic particles were however extremely large (20–30 nm) compared to the platinum particles ($\sim$2 nm). Other reports contradict this, but there is general agreement that smaller activity is the price that has to be paid for improved selectivity.[13,19]

The amount of hydrogen that can be chemisorbed decreased with increasing tin content, and, where particles of different composition are present, temperature-programmed desorption cannot be used to estimate dispersion. Analysis of products formed from *iso*butane extends the conclusions reached with *n*-hexane. A 6/1 Pt/Sn ratio hardly altered the character of 1.2% Pt/SiO$_2$, which gave ~33% isomerisation to *n*-butane and only ~26% selectivity to *iso*butene: smaller alkanes were the other products. With Pt/Sn = 1:1 or 1:3, these almost disappeared and there was no isomerisation: *iso*butene selectivity was ~99% at 673 K. Most significantly, a 0.04% Pt/SiO$_2$ showed almost the same behaviour: these results point very clearly to the need for *larger platinum ensembles for the parasitic reactions, and the sufficiency of a small ensemble for dehydrogenation.* Decomposition of Sn^{n}Bu$_4$ on the surface of Pt/SiO$_2$ gave tin atoms *on* the platinum, and at 823 K a performance similar to that just described.[18] Pt/SiO$_2$ catalysts prepared by the reverse micelle method or by the sol-gel route are reported to dehydrogenate propane with only modest carbon deposition.[20]

The story with alumina is quite different. The usual pre-treatment leaves much of the tin in an oxidised state (probably Sn^{II}) associated with the support, and only a small part in contact with the platinum; nevertheless, there are many beneficial effects. Dispersion and stability in propane dehydrogenation were improved, although the initial activity was the same as in the absence of tin, the early high production of ethane and methane decreasing to 10% of the products with continued use.[21] With *n*-heptane at 773 K, tin suppressed hydrogenolysis and aromatisation somewhat, but isomerisation still occurred: *n*-heptene selectivity was 63%.[22] This system has been very thoroughly studied. The tin either increases or decreases the amount of hydrogen chemisorbed, the very high H/Pt ratios sometimes found being attributed to hydrogen spillover (Section 3.3.4). This seems to be assisted by the tin aluminate layer which forms, it is thought, on the surface of the support, and which modifies the properties of the largely platinum particles that reside on it. The tin is said to promote the mobility of 'carbon' residues from metal to support, although it is not obvious how this is accomplished. It is certain, however, that tin does not diminish the amount of 'carbon' formed, indeed, quite the opposite; but the 'carbon' that is created is less inimical to activity, because much of it is located on the support. Temperature-programmed oxidation of the 'carbon' on PtSn/Al$_2$O$_3$ revealed three distinct types[19] – on the metal (A), on the support (B), and graphitic carbon on the support (C). Types A and B increased with temperature and propane pressure, and were inhibited by hydrogen: Type A reached a limit, and Type B increased continually with time-on-stream.

The theory and practice of preparing bimetallic catalysts has been comprehensively reviewed.[25]

A limited number of other supports have been examined. The need to incorporate basicity makes magnesia an obvious choice,[12,17] but increasing tin concentrations cause a marked growth in particle size, and increased breadth of the size distribution: quite severe sintering also occurred during use. With *iso*butane at 753 K, a catalyst containing 2.7% platinum (Pt/Sn = 1:1) gave only 50% *iso*butane, and considerable isomerisation.[17] Aluminates ($MgAl_2O_4$) are useful supports: the latter is used in the Phillips Petroleum Company STAR process, while alumina promoted with zinc and cobalt (also presumably as $(Zn,Co)Al_2O_4$) is used in the UOP Oleflex process.[2] $MgAl_2O_4$ formed by decomposition of the corresponding hydrotalcite when supporting platinum + tin gave a stable catalyst for propane dehydrogenation.[26] Pt/Nb_2O_5 gave much less aromatisation with *n*-heptane than Pt/Al_2O_3, and Pt_1Sn_1/Nb_2O_5 gave 78% selectivity to *n*-heptene, with little hydrogenolysis or isomerisation. Platinum (1%) with 10 or 20% niobia on alumina gave similar results.[22]

High selectivity to *iso*butene does not depend upon small particle size or the presence of a support. The Pt(111) surface covered to 25% by tin atoms was slightly more active for *iso*butane dehydrogenation at 673 K than Pt(111) itself, and the selectivity, already high (99%), scarcely increased.[27]

12.2.3. Other Metals and Modifiers

Much less work has been done with metals other than platinum or modifiers other than tin. Palladium, rhodium and iridium (as well as platinum) have been examined in conjunction with tin, indium, antimony, titanium, vanadium, molybdenum, manganese and iron for dehydrogenating the butanes.[6] Many of the combinations gave useful selectivities to the butenes, but the best were $PtSn/Al_2O_3$, $PtIn/Al_2O_3$ ($S = 99\%$) and $PtFe/SiO_2$ ($S = 97\%$). The kinetics of *iso*butane dehydrogenation on $PtIn/Al_2O_3$ have been fully explored,[28] this system closely resembling $PtSn/Al_2O_3$.

An especially useful catalyst for dehydrogenating the lower alkanes has been prepared by depositing the complex $[PtMo_6O_{24}]^{8-}$ onto magnesia;[7] after calcination at 773 K, Pt^{4+} ions replaced Mg^{2+} ions in the surface, and Mo^{6+} ions in distorted octahedral coordination also resided on the surface but not in contact with the platinum. Hydrogen reduction at 773 K led to small ($\sim$1 nm) platinum particles and some reduction of the Mo^{6+} ions, which then formed dimers. Such a catalyst was much more active and (below 773 K) stabler for dehydrogenating the butanes than Pt/Al_2O_3 or a PtMo/MgO prepared conventionally; the improvement in the case of propane was somewhat less, but selectivities of 96–97% were obtained. The cause of the better performance was discussed, but, notwithstanding the excellent and extensive use of EXAFS, no firm conclusion was reached. The large-scale manufacture of this catalyst might prove to be very expensive.

It is somewhat strange that a number of potentially useful supports have not been mentioned in the open literature: no use seems to have been made of titania or zirconia, or of $CaAl_2O_4$ or $CaSiO_3$. There may of course be good reasons; perhaps they have been tried and found wanting.

12.2.4. Kinetics and Mechanism

Quantitative studies of the kinetics of dehydrogenation of lower alkanes are few and far between, but those few have considerable significance.

In terms of the value of the conclusion reached, the examination of propane dehydrogenation on platinum-gold powders has perhaps been the most useful.[2] In a deservedly much-quoted paper[8] it was shown that at low platinum content (0–15%) the rate at 633 K was a linear function of it; this clearly suggested that a single platinum atom constituted the active centre.[29] The addition of gold lowered the activation energy only slightly, but increased the order in propane from -1.1 to -0.49 without changing the order in hydrogen. This observation helps to explain the apparent structure-insensitivity of alkane dehydrogenations, although other factors may sometimes contribute (see Section 12.32).

Two independent studies[9,23] of *iso*butane dehydrogenation have reached the same conclusions (an occurrence almost unique in the history of metal catalysis). The mechanism is well described by an inverse Horiuti-Polanyi scheme, for which the constants for the four forward and four backward reactions have been evaluated.[9] Adsorption of *iso*butane is the rate-limiting step, and reaction in the presence of deuterium showed extensive exchange in the *iso*butane, but little into the unreacted *iso*butane. At or above 773 K, the alkyl-alkene interconversion was clearly much faster than alkyl to alkane, but the exchanged *iso*butane had about the same deuterium content as the *iso*butane,[23] showing that both arose from the same alkyl + alkene 'pool'. Rate dependences on reactant pressures have been shown graphically,[9] but precise values of orders (negative in hydrogen, positive in *iso*butane) were not given. The activity of $PtSn/SiO_2$ was increased somewhat by the addition of potassium (Pt:Sn:K = 1:1:3), but a much larger effect ($\times 50$) was produced by using potassium-loaded L zeolite (KL) as support. Kinetic parameters were unaltered, but the cause of the effect was not established with certainty. The role played by hydrogen in dehydrogenations, and its unique character on platinum, will be considered below (Section 12.4).

Although alkane dehydrogenations appear to be structure-insensitive, there is one study[30] that reveals an interesting aspect of particle-size effects. Catalysts containing metal particles with mean sizes of from 2 to 5 nm were prepared by a combination of altering metal content and sintering temperature. The areal rate of dehydrogenation of 2,3-dimethylbutane decreased about four-fold as particle size was increased, but the relative amounts of the two possible products (i.e. 2,3-dimethyl-1- and -2-butenes) also changed, the fraction of the former being formed initially rising from about 0.58 to 0.70, and on platinum wire it was 0.90. Equilibration between the isomers continued as the reaction proceeded. It was thought that the alkene existed in the α,β-diadsorbed form on large particles, the C=C bond then preferring to be in the central position, while on small particles it adopted a π-allylic form, from which both isomers could arise.

The composition of the butene isomers formed in the dehydrogenation of *n*-butane over Pt/Al$_2$O$_3$ has been followed as a function of temperature, space velocity and time-on-stream in the continuous flow mode, and also at 673 K in the pulse mode.[31] In the former case, the initial distribution at 673 K was: 1-butene, 23%; *E*-2-butene, 42%; and *Z*-2-butene, 35%, which approximates to the equilibrium proportions (respectively 21, 42 and 36%), but with increasing time, and loss of activity due to 'carbon' deposition, the 1-butene proportion fell, but the *Z/E* ratio remained constant. This change may have been a consequence of the increased 'crowding' of the active centres as the coverage by 'carbon' increased. At higher temperatures (773 and 873 K), the initial concentration of 1-butene rose (52% at 873 K), apparently exceeding the equilibrium amount, the decrease with time-on-stream continuing. In the pulse mode, product composition was independent of pulse number, but the amount of 1-butene exceeded its equilibrium amount. Some *iso*butane was also formed by isomerisation.

12.3. DEHYDROGENATION OF CYCLOALKANES

12.3.1. Overview

A number of strands converge in this section, because several different kinds of process occur simultaneously, and it is not easy to separate them. Most of the work that concerns us has been carried out with cyclohexane, because not only is the progress of its dehydrogenation easily followed (as with benzene hydrogenation) but also because of its relevance to the catalytic chemistry of petroleum reforming (see Chapter 14). The rich variety of the processes that take place is due entirely to the ease with which, at typical reaction temperatures, hydrogen atoms migrate from one hydrocarbon species to another, so that their surface concentrations depend in a statistical manner on reactant/product concentrations in the gas phase, and on temperature. To illustrate this point, it hardly matters whether one starts with cyclohexane or cyclohexene, because the former will very soon transform to the latter on the surface, and a position of quasi-equilibrium comprising cyclohexyl and cyclohexene, in amounts that depend on the number of available hydrogen atoms, will quickly be set up. As their further dehydrogenation towards benzene proceeds, transitory diene species may be made on the surface, but they do not desorb.[32] Just as benzene is not easily hydrogenated to cyclohexene, so the dehydrogenation of cyclohexane does not stop at it; benzene is usually the sole product (on Pt(111) even at 235 K), although the conditions that favour its selective hydrogenation do not appear to have been tried for selective *de*hydrogenation.

The thermochemistry for the dehydrogenation of cycloalkanes is more favourable than that for acyclic molecules; the free energy change becomes zero at 560 K, and the reaction becomes detectable in the hydrogenation of aromatics (i.e. reactions do not go to completion) above about 473 K. Lower temperatures can therefore be used, and one advantage of this is that carbon deposition, although not negligible, is less serious, so it is easier to work with pure metals, and the use of bimetallic systems has been less common. However, hydrogenolysis (chiefly to methane) also occurs at the same time with the more active metals (Ni,[33] Ru[34]). A further reaction dimension is provided by the fact that at these moderate temperatures the hydrogen atoms released from a chemisorbed molecule of cyclohexene can *add* to another molecule, the nett reaction being

$$3\,C_6H_{10} \rightarrow C_6H_6 + 2C_6H_{12} \tag{12.A}$$

Cyclohexene has sometimes been used for hydrogenation of organic molecules in place of molecular hydrogen. The situation on the surface in the presence of hydrogen is therefore one of considerable complexity;[35,36] in what follows we focus on the dehydrogenation of cyclohexane.

12.3.2. Reaction on Pure Metals

By far the largest amount of work has been performed with platinum catalysts. There have been a number of studies employing single crystals,[32,37–41] and these showed that stepped and kinked platinum surfaces were more active than smooth faces. A recent detailed study[32,40] using Pt(111) and sum-frequency generation started with cyclohexene, which at 200 K was chemisorbed as a π-species, which at 217 K transformed to the di-σ-state, and at 283 K to a three-centre π-allylic C_6H_9 form, and thence to benzene at 383 K. However, experiments with bismuth post-doping after cyclohexane adsorption showed that benzene existed on this surface as low as 235 K, and that no intermediate form could be displaced.[42,43] Dehydrogenation of cyclohexene was faster on Pt(100) than on Pt(111) because only the 1,3-cyclohexadiene (and not the 1,4-isomer) was formed, this being thought the necessary intermediate in benzene formation.[40] Various C_6 cyclic molecules have also been examined on Ni(100).[44]

Little attention has been paid to the kinetics of dehydrogenations. Sinfelt[45] has re-examined early work on methylcyclohexane,[46] and has concluded that at 588 K it reacts eight times more slowly than cyclohexane, the activation energy E $\leftarrow$ k being 138 kJ mol^{-1}. On platinum black, cyclohexane exchanged with deuterium below 373 K, and above 473 K dehydrogenation without hydrogenolysis was observed: the distributions of deuterium atoms in reactant and product were similar, indicating that random loss of both kinds of atom occurred in the dehydrogenation steps. The transient response method has also been applied.[47] Pt/KL zeolite was more active than Pt/SiO$_2$, the addition of potassium to which lowered activity but improved thiotolerance.

Cycloalkane dehydrogenation has been regarded as the archetypal structure-*insensitive* reaction,[48−52] but recent work has qualified this belief. Changes in the free surface area during reaction are due less to particle growth[53] than to 'carbon' deposition; the character of the reaction was consequently found to change from *sensitive* to *insensitive*.[10] This provides the clearest possible indication that the 'carbon' formation homogenises the surface, eliminating high activity sites, the concentration of which is size-dependent, and leaving only low activity sites that are common to all sizes. One other factor has been identified:[53] high-temperature reduction of Pt/α-Al$_2$O$_3$ creates a strongly-held form of hydrogen, particularly on (or in?) very small particles, and this acts as a poison. Its removal by flushing with argon increased the TOF by about 40-fold, and this altered the form of structure-sensitivity: after argon purging, highly dispersed catalysts were the *more* active. However, with freshly reduced catalyst of dispersion less than 50%, the effect of particle size on TOF might well be missed. There have been many other indications that strongly held hydrogen in platinum catalysts is harmful to their activity in hydrocarbon reactions.[54]

If, however, conditions are chosen appropriately, rates are a linear fraction of amounts of hydrogen chemisorbed: thus with platinum and rhodium on ceria, alumina and their mixtures, such a linear plot has been used[55] as a means of estimating dispersion from activity measurement, this being a more sensitive and simple procedure than determining an adsorption isotherm. Thus after reduction at 1273 K, the dispersion of a platinum catalyst was estimated to be 3%.

Modification of ensemble size is achieved not only by introducing an inert or low activity element (see next section), but also by using a support or an oxidic component that can induce the Strong Metal-Support Interaction. Thus with Rh/SiO$_2$ at 733 K, the specific rate for cyclohexane dehydrogenation was independent of reduction temperature, but reduction of Rh/Nd$_2$O$_5$ at 473 and 773 K gave rates that were respectively about 10 and 40 times smaller.[56] The activation energy ($\sim$105 kJ mol^{-1}) was affected, but was smaller than for Rh/SiO$_2$ (151 kJ mol^{-1}). The use of niobia as support also suppressed simultaneous hydrogenolysis.[22,57] Addition of manganese to Rh/SiO$_2$ after calcinations gave the mixed oxide Rh$_2$MnO$_4$/SiO$_2$, which on reduction at 573 K resulted in small rhodium particles modified by inter-action with MnO$_x$;[58] however, after reduction at 473 K, high activity for converting cyclohexane was observed.

Although little work has been published on the other metals of Groups 8 to 10, some of them feature as one terminus of a bimetallic series to be considered presently. In the reaction of cyclohexane in the presence of deuterium on nickel powder, deactivation above 373 K lowered activity 100-fold, without changing activation energy;[33] but above 473 K, the products were benzene, n-hexane and (strangely) some toluene. Exchanged products declined as these other reactions set in, and under these conditions exchanged cyclohexane was not returned to the gas phase. Palladium on sepiolite or $AlPO_4$-SiO_2 caused disproportionation (process 10.B) at low temperature and dehydrogenation when the temperature was raised sufficiently to cause hydrogen to desorb.[59] Pd/Al_2O_3 ($D=40\%$) was 20 times less active than Pt/Al_2O_3($D = 50\%$);[60] this work was supported by calculations using SCF/CNDO methodology, which confirmed C—H bond breaking as the slow step. Methylcyclohexane dehydrogenation was four times faster on Pd(111) than on the (110) or (100) surfaces.[61]

12.3.3. Reaction on Bimetallic Catalysts[4,62]

There is extensive work to report on dehydrogenation of cycloalkanes on bimetallic catalysts, mainly in the supported form. Its motivation is quite clear: to minimise parasitic reactions such as 'carbon' deposition and hydrogenolysis, so as to have a catalytic system capable of working for long periods of time and at high selectivity.

Studies with single-crystal surfaces[27,63] have helped to elucidate the modifying effect of tin on platinum catalysts, as indeed they have in the case of isobutane.[27] Chemisorption of cyclohexene on Pt(111), $Pt_3Sn(111)$ and $Pt_2Sn(111)$ showed that its strength of adsorption decreased with increasing tin content, and that it was unable to form a di-σ bonded species on the last,[63] although adjacent pairs of platinum atoms still existed. The LEED structures of these surfaces were shown in Figure 4.2. This tendency might help to explain the beneficial effect of tin in this reaction;[64] 'carbon' deposition under reaction conditions was also suppressed. Deposition of tin atoms onto Pt(111) increased the TOF for cyclohexane dehydrogenation at 773 K, the maximum enhancement ($\times 1.7$) being at 20% tin coverage.[27] Both these studies imply some electronic modification of platinum by tin.

Experiments with PtPb/Al_2O_3[65] and with PtSb/Al_2O_3[66] have shown that these additives have similar effects to those produced by tin.[65] They decrease activity at low temperature by occupying active centre, but their beneficial effect is shown at high temperatures, where they diminish 'carbon' formation on the metal, and help to maintain activity. In trying to assess the role of the second component, attention must be given to the conditions under which the measurements are made.

The inclusion of tin in an Ir/Al_2O_3 catalyst (5wt.% of each) completely suppressed the hydrogenolysis of cyclohexane, which in its absence gave n-hexane as a major by-product;[67] the rate at 526 K was however decreased, but the activation energy fell from 208 to 125 kJ mol^{-1}. This remarkable and important result seems to have escaped attention: such a catalyst might compete successfully with PtSn/Al$_2$O$_3$.

In the combination PtMo/SiO$_2$, activity for cyclohexane dehydrogenation at 543 K decreased linearly with increasing molybdenum content;[68] no significant difference in rate at 513 K was found between Pt/C and Pt$_3$Zr/C catalysts.[69] Rhenium is much more active than platinum for hydrogenolysis, but when a bimetallic catalyst containing them is pre-sulfided a ReS$_x$ species is formed that acts as a selective site-blocking agent, leaving small ensembles of free platinum atoms available for reaction. With Pt(111) the maximum rate of cyclohexane dehydrogenation occurred when θ_{Re} was 0.5.[70] A thorough kinetic study of the dehydrogenation of methylcyclohexane on Pt/Al$_2$O$_3$ and on PtRe/Al$_2$O$_3$ presulfided showed that the latter was less active and had the higher activation energy (133 as against 196 kJ mol^{-1}), the rate being unaffected by changing hydrogen pressure. The results were modelled using Hougen-Watson methodology, and thermodynamic parameters were derived.[71] The activity of PtRe/Al$_2$O$_3$ at 843 K for the cyclohexane reaction only declined when the rhenium content exceeded about 50%.[72,73] Sulfiding led to better selectivities.[74]

The platinum-ruthenium couple is well known to exhibit synergism in a number of catalytic processes, especially those of an electrochemical character, but the activity of ruthenium for hydrogenolysis greatly exceeds that of platinum.[75] With PtRu/Al$_2$O$_3$, this was a linear function of composition in the cyclohexane-hydrogen reaction, but a maximum rate of dehydrogenation at 570 K was found when the surface contained about 55% platinum.[76]

Some very important conclusions have been obtained from work on the platinum-gold system.[41,77] Deposition of gold atoms onto Pt(100), and of platinum atoms onto Au(100), both increased the rate of cyclohexane dehydrogenation at 373 K, the maximum rate being some five times greater than for Pt(100) alone.[77] In the latter case, the maximum occurred after coating with enough gold atoms to form about a monolayer, but it appeared that island formation took place, leaving some small platinum ensembles uncovered. A similar enhancement was found with PtAu(111) surface alloys at 573 K;[41] in both cases a small amount of cyclohexene was also detected.

It finally remains to note several studies in which metals of Groups 8 to 10 other than platinum have had their reactivity towards cyclohexane dehydrogenation moderated by inert additions. In a very famous paper,[78] extensively cited, a series of nickel-copper powders were prepared and found to be homogeneous by XRD, magnetisation and hydrogen chemisorption measurements. The specific rate of cyclohexane dehydrogenation at 589 K was essentially constant the activation energy being about 220 kJ mol^{-1} for nickel contents between about 20 and 95%. This rate was a little greater than that for pure nickel, and these results were in stark contrast to those for ethane hydrogenolysis, the rate of which decreased dramatically as nickel content fell (see Chapter 13). They provide a telling demonstration that for this reaction at least particle-size and ensemble-size sensitivities are equivalent. This is consistent with the view that a single active atom is sufficient to effect reaction. Nickel and palladium have also been beneficially promoted by addition of tin.[79,80]

This informative work on the nickel-copper system led to a number of related studies. A further important paper by Sinfelt showed that bulk mutual solubility was not an essential precondition for obtaining interaction between active and inactive metals in small particles. Thus silica-supported ruthenium and osmium were both modified by the presence of copper,[34] but for the cyclohexane-hydrogen reaction it was only the rate of hydrogenolysis that was lowered. Benzene selectivities therefore rose to 95% at Ru:Cu or Os:Cu ratios of unity. The interaction of ruthenium and copper depends somewhat on the type of silica used, however.[81] Benzene selectivity and rates were also increased by adding silver to rhodium, the RhAg/TiO$_2$ catalyst at 573 K showing the extremely high selectivity of 99.6%.[82]

12.4. THE CHEMISORPTION OF HYDROGEN ON PLATINUM[2]

Clearly the strength of chemisorption of hydrogen is one of the factors determining activity in dehydrogenation, because if the atoms are reluctant to recombine and desorb the alkane will be unable to find the necessary free sites for its own adsorption: this effect is manifested in negative orders in hydrogen, and ultimately the reaction would grind to a halt. However, the requirement for vacant sites conflicts with the need to maintain a presence of hydrogen on the surface to counteract excessive dehydrogenation, and the consequential development of 'carbon' deposits. A catalytic system that is successful in practice, and stable in the long term will have to adopt a set of conditions in which these conflicting needs are reconciled.

Nevertheless in the steady state on Pt/Al$_2$O$_3$ the transfer of hydrogen atoms between the surface and hydrocarbon species formed from cyclohexane, *and the desorption of benzene*, have been shown to be fast compared to cyclohexane chemisorption and hydrogen desorption, which therefore control the overall rate. Now although our earlier discussion of hydrogen chemisorption on metals (Chapter 3) revealed nothing unusual about platinum, a careful consideration[2] of the literature shows rates of desorption of hydrogen from platinum are uniquely slow, and the strength of adsorption unusually strong. This is shown by (i) very low values for the pre-exponential term associated with the desorption rate constant, and (ii) a very high value for the adsorption coefficient on Pt/SiO$_2$ compared to those for PtRh/SiO$_2$ bimetallic catalysts. No agreed explanation for these fundamentally important observations has been forthcoming, but it seems probable that an important contributing reason for the success of bimetallic catalysts containing platinum is that the additive causes the chemisorption of hydrogen to be weakened.

There are experimental results to support this assertion. First, where orders of reaction in hydrogen can be compared, they are less negative on bimetallic catalysts than on platinum alone.[8,71] Secondly, the temperature-programmed desorption (TPD) profiles of platinum catalysts were changed by the presence of tin, which gave rise to a marked increase in the size of the low temperature (623 K) desorption peak. Corresponding experiments with single crystals however showed the opposite effect with $Pt_3Sn(111)$, and the same difference only with $Pt_2Sn(111)$. There is the additional possible complication with platinum catalysts of very strong held (dissolved?) hydrogen formed in high temperature treatment, the negative effect of which has already been noted: it is quite possible that this might form at temperatures commonly used for dehydrogenation (673–773K), but it seems not to be known whether tin or other additives prevent this.

The kinetic study of *iso*butane dehydrogenation mentioned above[23] shows clearly that with Pt/Al_2O_3 at 900 K increasing the hydrogen/*iso*butane ratio favoured dehydrogenation over carbon formation, but the use of high ratios is uneconomic. Although at the same temperature the beneficial effect of hydrogen was much less marked with $PtSn/Al_2O_3$ because of the weaker adsorption of hydrogen; nevertheless it can operate at lower hydrogen:*iso*butane ratios since the form of 'carbon' deposit is changed, more going to the support and less to the metal. The possibility that the 'carbon' deposit may act as a reservoir for hydrogen in this and other reactions will be considered below.

Bewilderment is sometimes expressed that a catalysed reaction is able to bring a reacting system to that position of equilibrium that would hypothetically be reached without it, and that the catalyst is somehow able to anticipate the state of equilibrium that should pertain only to free molecules. There is indeed evidence in many systems that the products returned to the gas-phase are not in the proportions that meet thermodynamic expectations: this is seen most clearly in the hydrogenation of, for example, 2-butyne, where the less stable Z-2-butene is formed preferentially (see Chapter 9). With cyclohexene and hydrogen at about 473 K the catalyst has to decide whether to make benzene or cyclohexane, utilising chemisorbed species that differ greatly from their gas-phase counterparts. In these cases the initially formed products *are* fixed by factors relating to adsorbed intermediates, but repetition of the adsorption-reaction-desorption cycle eventually allows the proper position of equilibrium to be attained. In the case of the 2-butyne reaction, isomerisation of the 2-butenes has to await the near-total removal of the alkyne in order for them to gain access to the surface. When this can occur, the relative stabilities of the various conformations, determined by repulsive interactions between non-bonded atoms, allow reaction to proceed in the general sense of what equilibrium demands, although those interactions are not identical with those that operate in the free molecules. With cyclohexane dehydrogenation, the effect of rising temperature in decreasing the adsorption coefficients of hydrogen and of benzene enables reaction to go in the expected direction.

12.5. THE FORMATION, STRUCTURE, AND FUNCTION OF CARBONACEOUS DEPOSITS[83,84]

It may be helpful at this point to try to draw together some of the many threads concerning 'carbon' deposition that have appeared many times in the previous chapters: a comprehensive and unifying model is not yet available, and indeed it is doubtful if it ever will be, so many are the factors that determine the state of 'carbon' on metal surfaces. Nevertheless it is possible to make a few generalisations, and an attempt to do so is opportune now because the last section of this chapter concerns a constructive use of surface carbon to create useful products. The term *'carbon'* will be used as an *omnium gatherum* for what has been variously named coke, acetylenic residue, carbonaceous deposit and probably other things as well. The following short survey may be amplified by reference to review articles.[1,25,83,85]

The form and quantity of 'carbon' existing on the surface of a metal catalyst depends *inter alia* upon the following variables: (i) the nature of the metal, (ii) its physical form, i.e. single crystal, powder or 'black', small supported particle etc., (iii) the nature of the support, if any, (iv) the type of hydrocarbon applied, (v) the presence of other molecules, especially hydrogen, and the hydrogen: hydrocarbon ratio, (vi) the time and especially the temperature of exposure. We may briefly consider the importance of each of these factors.

Within Groups 8 to 10, the base metals (Fe, Co, Ni) are distinctly more prone to form 'carbon' than the noble metals: metals to the left of these Groups (except perhaps manganese) are also very susceptible, so there is qualitative correlation between a metal's propensity to form stoichiometric or non-stoichiometric carbides, the strength of C—M bonds that it can form at the surface, and its tendency to deposit 'carbon' under the appropriate conditions. As we have seen, lowering the size of the ensemble of active atoms including an inert partner (e.g. PtSn, PtAu, RhAg) does not always reduce 'carbon' formation, as the additive may have other undesirable effects, but in general this move is usually helpful. This may be because (in the case of tin) it weakens the bonds of hydrocarbon species to the surface, so assisting their migration away from the metal. The same effect is produced by sulfided platinum-rhenium.[70] It may be (in the case of tin) that it weakens the bonds of hydrocarbon species to the surface, so assisting their migration away from the metal.[86]

Somorjai and his colleagues have developed a model[87] for the states of 'carbon' on a platinum surface containing steps and kinks, in which much of the surface was obscured by a 'carbonaceous overlayer' with islands of 3D 'carbon', leaving only a few single atoms or pairs at steps uncovered. It was felt that the higher activity of sites at steps would cause hydrogen if present to break C—M bonds. If this is so, then very small metal particles that expose only atoms of low coordination number should be more resistant to 'carbon' deposition than larger particles, powders or macroscopic forms. Quantitative evidence on a particle-size effect is difficult to find: while supported metals are not immune to 'carbon' formation, highly reproducible activity *can* be found, even in such an unpromising system as ethyne hydrogenation on nickel/pumice.[88] The protracted use of macroscopic forms over a period of days is rarely if ever attempted, so exact comparison with supported metals is impossible, but intuitively one feels they would not last the course.

The process of 'carbon' deposition begins with the removal from the chemisorbed hydrocarbon of more hydrogen atoms than are strictly necessary to achieve the intended process. As we have seen in Chapter 4 (Figure 4.2), the progressive dehydrogenation of ethene proceeds through the sequence[83,84]

$$\text{ethylidyne} = \text{vinyl} \rightarrow \text{vinylidene} \rightarrow \text{ethyne} \rightarrow \text{ethynyl}$$

$$\textbf{(8)} \qquad \textbf{(6)} \qquad \textbf{(17)} \qquad \textbf{(16)} \qquad \textbf{(20)}$$

We shall see in the next chapter, however, that in alkane hydrogenolysis it is probably necessary to remove two or even three hydrogen atoms to form species in which the C—C bond is so strained that it breaks, before the process of hydrogen *addition* can start. Thus *species that are irrelevant to and toxic towards reactions such as alkane exchange and alkene hydrogenation may be the necessary intermediates for reactions such as hydrogenolysis*, where they become reactive at higher temperature. The Pt—C bond is essentially non-polar, since NEXAFS studies of deactivated Pt/Al_2O_3 indicate the absence of electron transfer.

'Carbon' formation can then proceed in various ways. The di-carbon species shown above may break down to mono-carbon species then polymerise, or polymerisation may occur without this. Processes in which free radicals occur, or acidity in the catalyst, both encourage this, and (as noted above) catalysts for dehydrogenation must be neutral or basic to prevent acid-catalysed reactions of the alkene. The more dehydrogenated the reactant hydrocarbon, the greater the tendency to 'carbon' formation: alkadienes[89] and aromatics[90] are notorious in this respect. With alkanes, increase in molar mass (i.e. the number of C—H bonds) also assists degenerate events.

Because of its importance and frequent occurrence, the process of 'carbon' formation and its structure have been widely studied, and many physical techniques[83] (including recently positron-emission tomography[91]) have been deployed. Of these, temperature-programmed methods (oxidation, TPO; reaction with hydrogen, TPRe) are the simplest and most informative. TPO can distinguish between 'carbon' on the metal, which is relatively easily oxidised, from 'carbon' on the support, which is less reactive. Admixture of the sample with a Pd/SiO_2 catalyst ensures that the effluent contains only carbon dioxide, and no monoxide.[92] The use of the TPRe strictly requires estimation of the methane (and possibly other alkanes) that emerges, since because the H/C ratio in the 'carbon' is unknown, the amount of hydrogen consumed is not an accurate indicator of the carbon content. The ratio varies widely and depends upon catalyst composition[2] and especially temperature, since the end products are either amorphous carbon, or (frequently) graphite, or even above 870 K a diamond-like sp^3 form that is difficult to oxidise or reduce.[93] A NEXAFS study has revealed no electron transfer between platinum and carbon, so there can be no *electronic* modification of neighbouring free sites.[94]

Many different manifestations of 'carbon' have been recognised, and an attempt is made to show their interrelations in Scheme 10.2. 'Carbon' may either dissolve into the metal, forming some kind of carbide, or it may cover some or all of the metal surface, or migrate to the support, where it can deactivate by blocking the pores of a microporous catalyst; or even encapsulating the whole particle; or most interestingly it may grow away from the surface as filaments. These were formerly called the *vermicular* form, but are now recognised to be tubular forms (*nanotubes*) having structures related to *buckminsterfullerene*;[95] they grow particularly well on base metals, and usually carry a small metal particle at their head. They are now prepared on a significant scale by the reaction of ethyne with a base metal at high temperature, and are the subject of intensive study because of the uses they may prove to have in solid-state devices.

From the academic point of view, 'carbon' formation is regarded as an unmitigated nuisance by those trying to study hydrocarbon reactions and to obtain quantitative information on them for purposes of mathematical modelling. In practice one either has to work with a catalyst in its stable but highly deactivated state,[96] or to use very short reaction periods interspersed with cleansing times when only hydrogen is passed.[97] Even so, frequent recourse must be had to 'standard' conditions, to ensure constancy of activity, or to monitor deactivation should it occur: in this case, bracketing with 'standard' periods can still lead to usable results. The pulse reaction mode, in which hydrocarbon pulses are injected into a hydrogen stream, is unsuitable for precise work.

Perhaps the most contentions and often-considered aspect of 'carbon' is its possible actual participation in catalysed processes. The idea originated with the suggestion by Thomson and Webb[98,99] that alkene hydrogenation proceeded on top of a 'carbon' layer, utilising hydrogen atoms associated with it, and accounting for the structure-insensitivity and constant activation energy shown by this process. It is certainly true (or as true as anything can be in catalysis) that exposure of metal surfaces to ethene leads to the rapid formation of a layer of ethylidyne species (**8**),[100] but the idea originally canvassed that hydrogenation occurred through the ethylidyne $\rightleftharpoons$ ethylidene equilibration could not be sustained[84] (see Chapter 7). Ethylidyne does not however totally prevent hydrogen chemisorption, and indeed it now appears that it actually promotes di-σ ethene chemisorption on Pd(111).[101] 'Carbon' inhibits the hydrogenation of alicyclic molecules, but has lesser effect on the reactions of alkenes. Unfortunately some obvious and critical experiments with isotopic labels have never been performed, nor has the concept of 'carbon' acting as a hydrogen reservoir ever been quantitatively modelled: only in the case of benzene hydrogenation has it been formally incorporated in a reaction mechanism (Chapter 10). If indeed it were the vehicle by which hydrogenation occurred, it would be hard to explain the observed orders of reaction, and particularly the characteristic patterns of behaviour that distinguish the metals of Groups 8 to 10.

Little attempt seems to have been made to estimate the number of 'free' surface metal atoms in coked catalysts, and hence to find TOFs, assuming these to be the seat of the residual activity. While the use of hydrogen chemisorption might be considered risky, that of carbon monoxide ought to be suitable. Based on the loss of its IR intensity, the active metal area of Pt/Al_2O_3 used for n-heptane reforming was only 8% of its initial value, but its extent of adsorption slowly increased as it displaced some of the 'carbon'.[102]

The spectator of research on 'carbon' might well sympathise with Dogberry's belief that *They that touch pitch will be defiled.*

12.6. THE HOMOLOGATION OF METHANE[85]

Every cloud has a silvery lining, and there is one possibly useful way of employing the process whereby methane after dissociation at a metal surface enters a C—C bond-forming reaction. Although methane is abundantly available, it is of limited value except as fuel or as a feedstock for steam-reforming; the concept of its homologation into high alkanes of great utility was therefore a most attractive one, and its activation under reducing conditions promised better success than has ever attended its oxidative dehydrogenation to methanal or methanol. The reaction was therefore intensively studied through the 1990s and a considerable literature resulted.

Although the first work, and much subsequent effort, employed platinum catalysts[103−109] (especially EUROPT-1, 6.3% Pt/SiO_2), it appeared that better results would be obtained with metals that had a reputation for extensive C—C bond formation, starting from C_1 species as in Fischer-Tropsch synthesis:[110] attention was therefore focused on cobalt[105,107,110−114] and ruthenium.[71,105,106,110,115] Conventional supports have been used, but zeolites have also been investigated: the problem of fully reducing cobalt ions on alumina or in zeolite has been overcome by including a more easily reducible metal (e.g. ruthenium[110,113]) which assists the reduction, probably through hydrogen spillover. Other metals (e.g. Rh,[116] Pd,[117] Cu[118]) have been looked at in a cursory manner.

The amount of methane chemisorbed may be estimated by the amount of hydrogen released, or better by TPO of the retained 'carbon'. Chemisorption at high temperature leads to several distinguishable forms of 'carbon' that differ in reactivity,[116] but the composition of mono-carbon species CH_x formed at lower temperatures (573–723 K) can be assessed by the composition of deuteromethanes formed by their reaction with deuterium.[109] The best chance of C—C bond formation is when x is two or three, but in general the procedure has been to decompose methane at high temperature (up to 973 K) and to hydrogenate the 'carbon' at lower temperature to avoid the hydrogenolysis of any ethane or propane that may be formed. No conditions seem to have been found under which useful yields of higher alkanes have been obtained, and aspirations to found a major chemical process have been disappointed. Interest in this subject has therefore waned.[119] The few references cited should lead any interested reader towards other relevant papers.

REFERENCES

1. G.C. Bond, *Heterogeneous Catalysis: Principles and Applications*, 2nd edn., Oxford Univ. Press: Oxford (1987).
2. D.E. Resasco and G.L. Haller in: *Specialist Periodical Reports: Catalysis*, Vol. 11 (J.J. Spivey and S.K. Agarwal, eds.), *Roy. Soc. Chem.* (1994), p. 379.
3. P. Quicker, V. Höllein and R. Dittmeyer, *Catal. Today* **56** (2000) 21.
4. D.A. Dowden in: *Specialist Periodical Reports: Catalysis*, Vol.2 (G.C. Bond and G. Webb, eds.), *Roy. Soc. Chem.* (1978), p. 1.
5. B. Topsøe, B.S. Clausen and F.E. Massoth, *Hydrotreating Catalysts*, Springer-Verlag: Berlin (1996).
6. Jia Jifei, Xu Zusheng, Zhang Tao and Lin Liwu, *Chin. J. Catal.* **18** (1997) 101.
7. D.I. Kondarides, K. Tomishige, Y. Nagasawa, V. Lee and Y. Iwasawa, *J. Molec. Catal. A: Chem.* **111** (1996) 145; D.I. Kondarides, K. Tomishige, Y. Nagasawa, and Y. Iwasawa in: *Preparation of Catalysts VI*, Elsevier: Amsterdam (1995) p. 141.
8. P. Biloen, F.M. Dautzenberg and W.M.H. Sachtler, *J. Catal.* **50** (1977) 77.
9. R.D. Cortright, E. Bergene, P. Levin, M. Natal-Santiago and J.A. Dumesic in: *Proc. 11th Internat. Congr. Catal.*, (J.W. Hightower, W.N. Delgass, E. Iglesia and A.T. Bell, eds.), Elsevier: Amsterdam **B** (1996) 1185.
10. M.S.W. Vong and P.A. Sermon in: *Catalyst Deactivation 1991* (C.H. Bartholomew and J.B. Butt, eds.) Elsevier: Amsterdam, SSSC, **68** (1991) 235.
11. F.Z. Bentakar, J.P. Candy, J.M. Basset, F. Le Peltier and B. Didillon, *Catal. Today* **66** (2001) 303.
12. W.-S. Yang, L.-W. Lin. Y.I. Fan and J.-L. Zang, *Catal. Lett.* **12** (1992) 267.
13. R.D. Cortright and J.A. Dumesic, *J. Catal.* **148** (1994) 771; R.D. Cortright, J.M. Hill and J.A. Dumesic, *Catal. Today* **55** (2000) 207.
14. S.M. Stagg, C.A. Querini, W.E. Alvarez and D.E. Resasco, *J. Catal.* **168** (1997) 75.
15. P.J.C. Anstice, S.M. Becker and C.H. Rochester, *Catal. Lett.* **74** (2001) 9.
16. J. Llorca, N. Homs, J.-L.G. Fierro, J. Sales and P. Ramírez de la Piscina, *J. Catal.* **166** (1997) 44.
17. J. Llorca, N. Homs, J. Léon, J. Sales, J.L.G. Fierro and P. Ramirez de la Piscina, *Appl. Catal. A: Gen.* **189** (1999) 77.
18. F. Humblot, J.P. Candy, F. Le Peltier, B. Didillon and J.-M. Basset, *J. Catal.* **179** (1998) 459.
19. M. Larsson, M. Haltén, E.A. Blekkan and B. Anderson, *J. Catal.* **164** (1996) 44.
20. A.G. Sault, A. Martino, J.S. Kawola and E. Boespflug, *J. Catal.* **191** (2000) 474.
21. O.A. Barriås, A. Holmen and E.A. Blekkan, *J. Catal.* **158** (1996) 1.
22. M. Schmal, D.A.G. Aranda, R.R. Soares, F.E. Noronha and A. Frydman, *Catal. Today* **57** (2000) 169.

23. Lyu Kam Lok, N.A. Gaidai, B.S. Gudkov, S.L. Kiperman and S.B. Kogan, *Kinet. Catal.* **27** (1986) 1184.
24. S.B. Kogan and M. Herskovitch, *Ind. Eng. Chem. Res.* **41** (2002) 5949.
25. L. Guczi and A. Sarkány, *Specialist Periodical Reports: Catalysis*, Vol. 11 (J.J. Spivey and S.K. Agarwal, eds.), *Roy. Soc. Chem.* (1994), p.318.
26. D. Akporiaye, S.F. Jensen, U. Olsbye, F. Rohr, E. Rytter, M. Rønnekleiv and A.I. Spjelkavik, *Ind. Eng. Chem. Res.* **40** (2001) 4741.
27. Yong-Ki Park, F.H. Ribeiro and G.A. Somorjai, *J. Catal.* **178** (1995) 66.
28. Lyu Kam. Lok, N.A. Gaidai, B.S. Gudkov, M.M. Kostyokovskii, S.L. Kiperman, N.M. Podkletnova, S.B. Kogan and N.R. Bursian, *Kinet. Catal.* **27** (1986) 1190.
29. J.R. Patterson and J.J. Rooney, *Catal. Today* **12** (1992) 113.
30. M. Nakamura, M. Yamada and A. Amano, *J. Catal.* **39** (1975) 125.
31. S.D. Jackson, D. Lennon and J.M. McNamara in: *Catalysis in Application* (S.D. Jackson, J.S.J. Hargreaves and D. Lennon, eds.) Roy. Soc. Chem.: London (2003), p.39; *Catal. Today* **81** (2003) 583.
32. X.-C. Su, K.Y Kung, J. Lahtinen, Y.R. Shen and G.A. Somorjai, *J. Molec. Catal A: Chem.* **141** (1999) 9.
33. A. Sárkány, L. Guczi and P. Tétényi, *J. Catal.* **39** (1975) 181.
34. J.H. Sinfelt, *J. Catal.* **29** (1973) 308.
35. B.E. Koel, D.A. Blank and E.A. Carter, *J. Molec. Catal. A: Chem.* **131** (1998) 39.
36. J.J. Rooney, *J. Catal.* **2** (1963) 53.
37. F.C. Henn, *J. Phys. Chem.* **96** (1992) 5965; X. Su, Y.R. Shen and G.A. Somorjai, *Chem. Phys. Lett.* **280** (1997) 302; X. Su, K. McCrea, K. Rider and G.A. Somorjai, *Topics in Catal.* **8** (1999) 23.
38. D.H. Parker, C.L. Pettiete-Hall, Y.-Z. Li, R.T. McIver Jr. and J.C. Hemminger, *J. Phys. Chem.* **96** (1992) 1888.
39. X. Su, K. Kung, J. Lahtinen, R.Y. Shen and G.A. Somorjai, *Catal. Lett.* **54** (1998) 9.
40. K.R. McCrea and G.A. Somorjai, *J. Molec. Catal. A: Chem.* **163** (2000) 43.
41. J.W.A. Sachtler and G.A. Somorjai, *J. Catal.* **89** (1984) 35.
42. C.T. Campbell, *Ann. Rev. Phys. Chem.* **41** (1990) 775.
43. C.T. Campbell, J.A. Rodriguez, F.C. Henn, J.M. Campbell, P.J. Dalton and S.G. Seimanides, *J. Chem. Phys.* **88** (1988) 6585.
44. S. Tjandra and F. Zaera, *J. Catal.* **164** (1996) 82.
45. J.H. Sinfelt, *J. Molec. Catal. A: Chem.* **163** (2000) 123.
46. J.H. Sinfelt, H. Hurwitz and R.A. Shulman, *J. Phys. Chem.* **64** (1960) 1559.
47. P. Ledoux, Y.S. Hsia and S. Kovenklioglu, *J. Catal.* **98** (1986) 367.
48. J.A. Cusumano, G.W. Dembinski and J.H. Sinfelt, *J. Catal.* **5** (1966) 471.
49. M. Guenin, M. Breysse, R. Frety, K. Tifouti, P. Marécot and J. Barbier, *J. Catal.* **105** (1987) 144.
50. H.E. Swift, F.E. Lutinski and H.H. Tobin, *J. Catal.* **5** (1966) 285.
51. O.M. Poltorak and V.S. Boronin, *Russ. J. Phys. Chem.* **39** (1965) 781; 1329; **40** (1966) 36.
52. M. Guenin, M. Breysse and R. Frety, *J. Molec. Catal.* **25** (1984) 119.
53. A. Rochefort, F. le Peltier and J.P. Boitiaux, *J. Catal.* **138** (1992) 482; **145** (1994) 409.
54. P.G. Menon and G.F. Froment, *Appl. Catal.* **1** (1981) 31.
55. E. Rogemond, N. Essayem, R. Frety, V. Perrichon, M. Primet and F. Mathis, *J. Catal.* **166** (1997) 229.
56. T. Ichijima, *Catal. Today* **28** (1996) 103.
57. D.A.G. Aranda, A.L.D. Ramos, F.B. Passos and M. Schmal, *Catal. Today* **28** (1996) 119.
58. K. Kunimori, T. Wakasugi, Z. Hu, H. Oyanagi, M. Imai, H. Asano and T. Uchijima, *Catal. Lett.* **7** (1990) 337.
59. M.A. Aramendía, V. Boráu, I.M. García, C. Jiménez, A. Marinas, J.M. Marinas and F.J. Urbano, *J. Molec. Catal. A.: Chem* **151** (2000) 261.

60. M.E. Ruiz-Vizcaya, O. Novaro, J.M. Ferreira and R. Gómez, *J. Catal.* **51** (1978) 108.
61. A. Corma, M.A. Martin, J.A. Pajares, J. Perez-Pariente, M. Avalos and M.J. Yacaman, *J. Molec. Catal.* **48** (1988) 199.
62. R.D. Gonzalez, *Appl. Surf. Sci.* **19** (1984) 181.
63. Chen Xu and B.E. Koel, *Surf. Sci.* **304** (1994) 249.
64. F.M. Dautzenberg, J.N. Helle, P. Biloen and W.M.H. Sachtler, *J. Catal.* **63** (1980) 119.
65. J. Völter, G. Lietz, M. Uhlemann and M. Hermann, *J. Catal.* **68** (1981) 42.
66. Hoang Dang Lanh, Ho Si Thoang, H. Lieske and J. Völter, *Appl. Catal.* **11** (1984) 195.
67. R. Frety, B. Benaichouba, P. Bussiére, D. Santos Cunha and Y.L. Lam, *J. Molec. Catal.* **25** (1984) 173.
68. G. Leclercq, S. Pietrzyk, T. Romero, A. El Gharbi, L. Gengembre, J. Grimblot, F. Aïssi, M. Guelton, A. Latef and L. Leclercq, *Ind. Eng. Chem. Res.* **36** (1997) 4015; G. Leclercq, T. Romero, S. Pietrzyk, J. Grimblot and L. Leclercq, *J. Molec. Catal.* **25** (1984) 67.
69. R. Szymanski and H. Charcosset, *J. Molec. Catal.* **25** (1984) 337.
70. Changmin Kim and G.A. Somorjai, *J. Catal.* **134** (1992) 179.
71. V.A. van Trimpont, G.B. Marin and G.F. Froment, *Ind. Eng. Chem. Fundam.* **25** (1986) 544.
72. G. Leclercq, H. Charcosset, R. Maurel, C. Betizeau, C. Bolivar, R. Frety, D. Jaurez, H. Mendez and L. Tournayan, *Bull. Soc. Chim. Belg.* **88** (1979) 577.
73. C. Betizeau, G. Leclercq, R. Maurel, C. Bolivar, H. Charcosset, R. Frety and L. Tournayan, *J. Catal.* **45** (1976) 179.
74. P. Biloen, J.N. Helle, H. Verbeek, F.M. Dautzenberg and W.M.H. Sachtler, *J. Catal.* **63** (1980) 112.
75. S. Engels, Nguyen Phuong Khue and M. Wilde, *Z. Anorg. Allg. Chem.* **463** (1980) 96.
76. H. Miura, M. Osawa, T. Suzuki, K. Sugiyama and T. Matsuda, *Chem. Lett.* (1982) 1803.
77. J.W.A. Sachtler, M.A. Van Hove, J.P. Bibérian and G.A. Somorjai, *Phys. Rev. Lett.* **45** (1980) 1601.
78. J.H. Sinfelt, J.L. Carter and D.J.C. Yates, *J. Catal.* **24** (1972) 283.
79. M. Masai, K. Honda, A. Kubota, Y. Nishikawa, K. Nakahara, K. Kishi and S. Ikeda, *J. Catal.* **50** (1977) 419.
80. H.E. Swift and J.E. Bozik, *J. Catal.* **12** (1968) 5.
81. A.J. Hong, A.J. Rouco, D.E. Resasco and G.L. Haller, *J. Phys. Chem.* **91** (1987) 2665.
82. G.L. Haller, D.E. Resasco and A.J. Rouco, *Faraday Discuss. Chem. Soc.* **72** (1981) 109.
83. A.T. Bell in: (C. Morterra, A. Zecchina and G. Costa, eds.), Studies in Surface Science and Catalysis', Elsevier: Amsterdam, **48** (1989) 91.
84. G.C. Bond, *J. Molec. Catal. A: Chem.* **149** (1997) 3.
85. L. Guczi, R.A. van Santen and K.V. Sarma, *Catal. Rev.-Sci. Eng.* **38** (1996) 249.
86. L.-W. Lin, T. Zhang, J.-L. Zang and X.-S. Xu, *Appl. Catal.* **67** (1990) 11.
87. S.M. Davis, F. Zaera and G.A. Somorjai, *J. Catal.* **77** (1982) 439.
88. G.C.Bond, M.A. Keane, H. Kral and J.A. Lercher, *Catal. Rev.-Sci. Eng.* **42** (2000) 323.
89. A. Sárkány, *Kinet. Catal. Lett.* **68** (1999) 153.
90. A. Sárkány in: *Catalyst Deactivation 1987,* (B. Delmon and G.F. Froment, eds.) Elsevier: Amsterdam (1987), p.125.
91. R.A. van Santen, B.G. Anderson, R.H. Cunningham, A.V.G. Mangnus, L.J. van IJzendoorn and M.J. A. de Voigt, *Angew. Chem. Intntl. Edn.* **35** (1996) 2785.
92. G.C. Bond, C.R. Dias and M.F. Portela, *J. Catal.* **156** (1995) 295.
93. R.I. Kvon and A.I. Boronin, *Catal. Today* **42** (1998) 353.
94. Z. Paál and A. Wootsch in: *Catalysis in Application* (S.D. Jackson, J.S.J. Hargreaves and D. Lennon, eds.), Roy. Soc. Chem.: London (2003), p. 8.
95. E.G. Rakov, *Russ. Chem. Rev.* **70** (2001) 827.
96. R.D. Cortright, S.A. Goddard, J.E. Resoske and J.A. Dumesic, *J. Catal.* **127** (1991) 342.

97. G.C. Bond and R.H. Cunningham, *J. Catal.* **166** (1997) 172.
98. S.J. Thomson and G. Webb, *J. Chem. Soc. Chem. Comm.* (1976) 526.
99. G. Webb in: *Specialist Periodical Reports: Catalysis,* Vol. 2 (C. Kemball and D.A. Dowden, eds.), *Roy. Soc. Chem.* (1978), p.145.
100. G.A. Somorjai, *Introduction to Surface Chemistry and Catalysis,* Wiley: New York (1994).
101. D. Stacchiola and W.T. Tysoe, *Surf. Sci.* **513** (2002) L431.
102. M. Bowker, T. Aslam, C. Morgan and N. Perkins in: *Catalysis in Application*, Roy. Soc. Chem.: London (2003), p.1.
103. M. Wolf, O. Deutschmann, F. Behrendt and J. Warnatz, *Catal. Lett.* **61** (1999) 15.
104. E. Marceau, J.-M. Tatibouët, M. Che and J. Saint-Just, *J. Catal.* **183** (1999) 384.
105. L. Guczi, K.V. Sarma and L. Borkó, *React. Kinet. Catal. Lett.* **68** (1999) 95.
106. A. Amariglio, M. BelgUed, P. Paréja and H. Amariglio, *Nature,* **352** (1991) 789; *J. Catal.* **177** (1998) 113, 121.
107. Yang Zifeng, Xue Jinzhen, Shen Shikong and Wang Hongli, *Chin. J. Catal.* **17** (1996) 249.
108. Z. Hlavathy, Z. Paál and P. Tétényi, *J. Catal.* **166** (1997) 118.
109. S. Monteverdi, A. Amariglio, P. Paréja and H. Amariglio, *J. Catal.* **172** (1997) 259.
110. L. Guczi, K.V. Sarma and L. Borkó, *J. Catal.* **167** (1997) 495.
111. G. Boskovic and K.J. Smith, *Catal. Today* **37** (1997) 25.
112. J.S.M. Zadeh and K.J. Smith, *J. Catal.* **183** (1999) 232.
113. L. Guczi, Z. Koppány, K.V. Sarma, L. Borkó and I. Kiricsi in: *Prog. Zeolites and Micropor. Mater.* (H. Chon, S.-K. Ihm and Y.S. Uh, eds.) Elsevier: Amsterdam (1997), p.861.
114. L. Guczi, K.V. Sharma, Zs. Koppány, R. Sundararajan and Z. Zsoldos in: *Natural Gas Conversion IV* (M. de Pontes, R.L. Espinoza, C.P. Nicolaides, J.H. Scholtz and M.S. Scurrell, eds.) Elsevier: Amsterdam, Studies in Surface Science and Catalysis, **68** (1997) 333.
115. J.N. Carstens and A.T. Bell, *J. Catal.* **161** (1996) 423.
116. N. Matsui, K. Nakagawa, N. Ikenaga and T. Suzuki, *J. Catal.* **194** (2000) 115.
117. F. Solymosi, A. Erdöhelyi, J. Cserényi and A. Felgévi, *J. Catal.* **147** (1994) 272.
118. F. Solymosi, *Catal. Today* **28** (1996) 191.
119. M.C.J. Bradford, *J. Catal.* **189** (2000) 238.

13

REACTIONS OF THE LOWER
ALKANES WITH HYDROGEN

PREFACE

We now approach the final chapter in which reactions achieving drastic restructuring of reactant hydrocarbons are to be considered. Some of these are of immense practical importance, while others, such as hydrogenolysis, are usually regarded as a nuisance, but these have attracted great interest from academic scientists and those in industry with an interest in fundamental studies. Hydrogenolysis is a difficult reaction in more senses than one: it requires somewhat high temperatures, even the most active metals performing only above about 373 K; it is 'demanding' in the sense of being regarded as structure-sensitive; it is (like most hydrocarbon reactions) bedevilled by 'carbon' formation; and its mathematical modelling has taxed the ingenuity of several generations of scientists.

We have met the breaking of C—C bonds by hydrogen already in Chapter 11, but the molecules considered there (cyclopropane and cyclobutane) had some degree of alkene-like character and reacted easily (especially the former). In this chapter we shall be involved with linear and branched alkanes having two, three or four carbon atoms. C—C bond fission is the principal process, but with the butanes skeletal isomerisation is also possible, and dehydrogenation sometimes happens at the same time. Reactions of acyclic and cyclic alkanes having five or more carbon atoms feature in the following chapter, where isomerisation and dehydrocyclisation are the important reactions. Some limited overlap between this chapters and the next is unavoidable.

We start with 'A short philosophical digression' that can be skipped if your interest does not run in that direction.

13.1. INTRODUCTION

13.1.1. A Short Philosophical Digression

Physical scientists are not much given to philosophy. They tend to be down-to-earth people, prepared to believe what their senses tell them, and not very concerned about the wider philosophical implications. This reluctance is very clearly seen in the study of heterogeneous catalysis, where the construction of models and the drawing of inferences are exceptionally difficult but vitally important; the great variation in the vigour with which these actions are pursued reflects the uncertainty that scientists feel on departing from tangible observations.

The difficulties attending the extraction of an acceptable conceptual model for a reaction mechanism from experimental findings has been a recurrent theme of the last six chapters. A suggestion as to what needed to be defined before we could say we understood a reaction mechanism was made in Section 5.3; but this was only a start. Since in this chapter and the next the problems of resolving the nature of adsorbed intermediate and their modes of interaction becomes acute, it may be helpful to preface our discussion with a short reflection on exactly what it is we are trying to do, and how far our efforts are likely to meet with success.

The amount of information needed to start a discussion of mechanism is quite small: speculation can then proceed untrammelled by a superfluity of experimental facts. This was often the case with early works on many systems, but not surprisingly as further observations were made the model had to be refined and extended: and it became more complicated. More assumptions and suppositions had to be introduced, and more uncertainty entered the conclusions.

An example of a complicating refinement lies in the choice of terminology and symbolism used to represent our ideas. One started with statements employing letters as chemicals symbols, e.g. H for the hydrogen atom; but this gives no idea of its size relative to that of other atoms, and it is easy to come to think that a hydrogen atom really *is* a capital aitch. This kind of symbolism is quite unsuited for describing what goes on at metal surfaces or the structure of imagined adsorbed intermediates. Scale drawings on paper or a visual display unit are *essential* for accurately representing adsorbed species, but even this is not quite the end, because there is then the need to show their dynamic interaction both in geometric and energetic dimensions. Dynamic computer simulation is the only way to manage the first of these. There is truly no end to the number of layers of the mechanistic onion!

If correct logic is applied, we must suppose that every mechanistic statement is valid within the confines of the theoretical concepts and symbolism available at the time, and the accessible experimental conditions and techniques. The virtuous circle of experiment → model → experiment, which is supposed to characterise the scientific method, seems to be followed less often than it might, and one rarely reads about predictions that a particular model suggests. A common failure is the disregard of relevant results from analogous systems, and the NIH (not invented here) factor is often perceived in 'Discussion' sections of papers. Under-interpretation is more usual than over-interpretation. Popper's dictum has been mentioned several times: it is necessary to appreciate that one's mechanistic statement is not the last word, but should be used as a springboard to design the experiments to test it. Self-doubt is not a sin, and it is better to be critical oneself of one's conclusions than to await the criticisms of others. Incidentally, Popper's admirable philosophical position was anticipated by Alfred, Lord Tennyson:

> *For nothing worthy proving can be proven*
> *Nor yet disproven; wherefore be thou wise:*
> *Cleave ever to the sunnier side of doubt.*

Scientists and poets throughout the ages have been aware of the conflict between the mass of available information and the understanding of its significance. J.W. von Goethe summed it up succinctly: *We know accurately only when we know little; with knowledge, doubt increases.* This may be paraphrased as: *The more we know, the more we know we don't know.* Somewhat more optimistically, Francis Bacon, Lord Verulam, believed that *If a man begin with certainties, he shall end in doubts; but if he will be content to begin with doubts, he shall end in certainties.*

13.1.2. Alkane Hydrogenolysis: General Characteristics

In the context of alkanes, hydrogenolysis is the breaking of C—C bonds by the action of hydrogen, leading to alkanes of lower molar mass. It is not a reaction that is deliberately practised on a large scale, but it is a parasitic reaction that occurs in parallel with other useful reactions of alkanes to be considered in the next chapter. In order to learn how to avoid or minimise it, it becomes necessary to find out as much as possible about it, and this is most easily done with molecules containing only two to four carbon atoms. Alkanes can also be 'cracked' by an acid-catalysed reaction on solid acids or acidic supports, but in this chapter we are solely concerned with reactions that proceed on purely metallic sites: the cooperation of metallic and acidic sites in petroleum reforming will be briefly considered in Chapter 14.

The reaction has also proved interesting because of its reputation for structure-sensitivity. There have been numerous studies of the hydrogenolysis of alkanes (especially ethane, which has the merit of giving only a single product), examining their response to variations in particle size, and ensemble size in bimetallic systems, to face-sensitivity with single-crystal surfaces and to the Strong Metal-Support Interaction. The reactions are exothermic, but need temperatures that generally exceed 373 K because activation energies are high. It is generally believed that a multi-atom site is necessary in order to accommodate a species in which two or more C—M bonds from *different* carbon atoms are formed, as these are needed to produce the strain in the C—C bond that causes it to break. This of course demands the breaking of several C—H bonds, and sites to receive the hydrogen atoms thus released; this apparently accounts for the very large negative orders in hydrogen that are often observed. The complex history of the attempts to create mathematical models to describe the kinetics of this class of reaction will be surveyed in Section 13.2.

13.1.3. Problems in Studying Reaction Kinetics

The conditions under which hydrogenolysis occurs readily are generally speaking those that favour 'carbon' deposition.[1] This can be limited by the use of high hydrogen: alkane ratios, and many studies have been conducted using ratios of 10 or more. However, every silvery lining must have its cloud, and the disadvantage of doing this is that the negative order in hydrogen means that rates are lower than they might have been if a lower ratio had been used. There is a paradox here that cannot be avoided. The reactive form of the alkane is partially dehydrogenated, and moving to *high* hydrogen/alkane ratios makes its formation *less* likely, but it is also the form that at *low* ratios leads to the formation of 'carbon' and hence to deactivation. You can't win. There are therefore major problems in obtaining meaningful kinetic results, and it is worth emphasising that for purposes of kinetic modelling it is very important to have results over as wide a range of reactant ratios as possible, and particularly to establish the co-ordinates of the rate maximum. It therefore becomes necessary to seek ways of overcoming this problem.

The injection of hydrocarbon pulses into a hydrogen stream has been widely practised, and reveals the activity of metal catalysts and the reactivity of alkanes under standard conditions,[2] but does not allow determination of reaction orders. One procedure that has proved suitable for acquiring good kinetic results is the *short reaction period* (SRP) method,[3,4] in which an alkane-hydrogen mixture of known composition is passed over the catalyst for 1 min and sampled; hydrogen is then passed for 19 min to cleanse the surface, and then another SRP started. This minimises or even eliminates the loss of activity experienced when continuous flow of reactants is used; in the *n*-butane-hydrogen reaction over PtRe/Al$_2$O$_3$ at 582 K, it even leads to some increase in activity.[3] Variations on this standard procedure have been introduced when it has been desired to study the effects of 'carbon' laydown. When deactivation does occur, further improvement in the quality of results may be obtained by having frequent recourse to a standard reactant concentration, and adjusting the rate found with a different reactant ratio by interpolation, remembering that this rate is fixed by the *previous* standard value. Only with zeolite-supported platinum did these procedures not work, because hydrocarbon pulses were totally retained.[5] For further details, the cited references should be consulted.

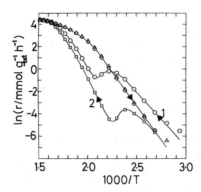

Figure 13.1. Hydrogenolysis of n-butane on 1.5% Re/Al$_2$O$_3$: Arrhenius plots for two successive thermal cycles.[6]

The distinction between lowered activity because of (i) 'carbon' formation and (ii) inappropriate concentrations of *reactive* species is established by the subsequent use of a standard reactant mixture, but sometimes extent of the problem depends unexpectedly on temperature and on the direction of temperature change. Rhenium is a very active metal for hydrogenolysis, and the variation in rate of the n-butane-hydrogen reaction as temperature was raised and then lowered in a stepwise manner is shown in Figure 13.1.[6] Initially some 'carbon' formation occurs, and this leads to a small rate maximum, but then as temperature is raised further it is gradually removed and the reaction continues with a higher activation energy. On lowering the temperature the surface remains in a clean state and a smoothly decreasing rate is observed. This effect may be described as *hysteresis* caused by 'carbon' formation. In a second thermal cycle, 'carbon' formation starts sooner, but the rates found with descending temperature exactly reproduce those in the first cycle.

The tendency to deactivate by 'carbon' formation runs parallel to activity in hydrogenolysis because the same or similar intermediate species are involved. Thus platinum, with which most work has been done and which has been chosen as best for other desirable reactions (dehydrogenation (see Chapter 12), isomerisation etc.) because of its low activity for hydrogenolysis, is much easier to work with than the base metals. On this metal, and probably generally, ease of forming the multiply-bonded intermediates increases with the number of carbon atoms, since as chain-length increases, so does the chance of C—H bond fission, and not all the dehydrogenated species are those needed for hydrogenolysis.[7,8] So with n-butane, a 1,1,2-triadsorbed species may lead to the breaking of the terminal C—C bond, whereas a 1,2,3-triadsorbed species may end up as 'carbon'. Consequentially little deactivation is experienced when ethane reacts over platinum, but problems mount as the chain-length grows.

13.1.4. Ways of Expressing Product Composition

Ways in which *rates* of reaction may be expressed have been considered in Section 5.2.3. In the present context they are usually given in *specific* or *areal* terms, or as *turnover frequencies*, i.e. rate per (presumed) active centre. Where standard catalysts such as EUROPT-1 have been used, the rate per unit mass of catalyst or metal is sufficient, because this can readily be translated into areal units if comparison with other catalysts is desired.

There is much greater variation in the means used to express product selectivities. This is a very important aspect of research in this area, because the nature and amounts of the products formed is determined by the type and reactivities of the adsorbed intermediates, and hence reveals intimate features of the reaction mechanism. The larger and more complex the structure of the reactant alkane, the greater is the need for clarity. Giving the fraction of each product as a percentage of the total[9] is not very useful, neither is the use of ratios of pairs of products,[10] because the overall picture is obscured. The Budapest group have employed a *fragmentation factor* (ζ)[11] defined as

$$\zeta = \Sigma C_i / \Sigma(i/n)C_i \tag{13.1}$$

where C_i is the amount of product containing i carbon atoms formed from a reactant having n carbon atoms. Thus ζ is two for a single C—C bond splitting and n where complete breakdown to methane occurs. This parameter must apply only to initial product yields, as it will increase with reaction progress as first-formed products react further. The products of complex alkanes containing different types of C—C bond have been characterised by a *reactivity factor* (ω) that defines the chance of a particular bond breaking relative to its statistical chance.[12] By far the most useful procedure for the lower alkanes is simply to define the selectivity to the molecule having j carbon atoms (S_j) as

$$S_j = C_j / A \tag{13.2}$$

where C_j is the molar fraction of that product and A the moles of reactant converted.[3,13–16] This leads to the relations[17] ....

For propane:	$S_1 + 2S_2 = 3$	(13.3)
For *n*-butane:	$S_1 + 2S_2 + 3S_3 = 4$	(13.4)

Defined thus, the selectivities can be directly fed into equations derived from a general mechanistic scheme, yielding more fundamental parameters, as will be disclosed shortly. When isomerisation of *n*-butane is observed, the fraction reacting in this way is denoted by S_i, and the other selectivities are based on the remaining fraction.

13.2. HYDROGENOLYSIS OF THE LOWER ALKANES ON SINGLE METAL CATALYSTS: RATES, KINETICS, AND MECHANISMS

13.2.1. The Beginning

The orderly presentation of the very extensive literature falling under this heading presents considerable difficulties. According to Max Planck, *The chief problem in every science is that of endeavouring to arrange and collate the numerous individual observations and details which present themselves, in order that they may become part of one comprehensive picture.* Let us see how this problem might be addressed.

The information available is of two types. (1) First is the dependence of *rate* on the kind of alkane, on the nature of the catalyst, on reactant concentrations and on temperature. (2) Second is the dependence of *product selectivities* or other descriptive factor[18] (for molecules having three or more carbon atoms) on these variables. This separation is somewhat arbitrary, but the first type focuses on the rate-determining step, and leads us into a discussion of mechanism that lacks the refinements needed to understand the origin of selectivities and their variation with conditions. Some people have combined these two features by evaluating orders of reaction and activation energies for the formation of each individual product,[19-22] and assume by implication that each reaction that can be formulated to give these products demands a different sort of site: but the numbers that emerge can only be used in a qualitative way, and the differences in site architecture cannot be defined. This approach is therefore somewhat sterile, and the alternative, which is to treat the rate of product removal and product selectivities separately, is to be preferred and is the one adopted here.

References to relevant reviews are collected in the Further Reading section at the end of the chapter.

13.2.2. Kinetic Parameters

It is convenient to start by considering the results obtained many years ago by John Sinfelt[23] and his associates for the hydrogenolysis of ethane;[24-29] they are models of clarity, and they provide a suitable framework for results on other systems. In terms of activity, metals divide themselves easily into three classes: (i) the very active (ruthenium and osmium); (ii) the moderately active (the majority); and (iii) the least active (palladium and platinum). This distinction is clearly drawn when the Arrhenius parameters are depicted as a compensation plot

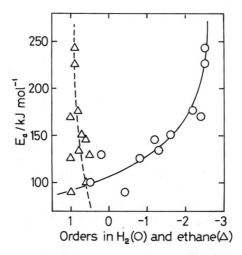

Figure 13.2. Hydrogenolysis of ethane: orders of reaction as function of activation energy for various silica-supported metals.[24,25]

(Figure 5.15);[30] it is however necessary to draw the lines appropriately.[25] It was suggested earlier (Section 5.6) that metals sharing a single compensation line might operate by a common mechanism, or at least have some mechanistic features linking them: by definition there must be some temperature (the *isokinetic temperature*) at which they all show the same rate, and for the majority of metals this should be at about 383 K. However at any other temperature their rates will vary as determined by the Arrhenius parameters between the activation energies and the orders of reaction. The orders in hydrogen, which were in most cases negative, tended to become more positive as the temperature of measurement was necessarily increased due to lessening activity, but palladium and platinum were exceptions. There is a good correlation between activation energy and hydrogen order, and a less good correlation with ethane order (Figure 13.2):[30] the least active metals showed (not surprisingly) highest activation energies, but also most negative orders in hydrogen. This frequently observed inhibition by hydrogen has always been taken to imply that the hydrocarbon has to *lose* a certain number of hydrogen atoms to be activated and thus to need a certain number of free surface sites. Strong hydrogen chemisorption would thus run in parallel with weak alkane chemisorption (Figure 13.2). The results suggested that the adsorption of hydrogen on palladium and platinum must be exceptionally strong, a possibility that was mooted for platinum in the previous chapter in the context of dehydrogenation. This qualitative picture will be quantified in a later section.

Similar if less clear categorisation of metals according to their activities is possible using results obtained with ethane hydrogenolysis by metal blacks,[31] and with *n*-pentane hydrogenolysis on silica-supported metals.[32] However in both these cases the databases are smaller.

Very extensive results are available for the hydrogenolysis of alkanes on many of the metals of Groups 8 to 10. Only iron, cobalt and osmium have been little studied, the first because of its great propensity to destructive adsorption of alkanes. Activation energies have frequently been reported (although their precision is quite rarely given), but reliable pre-exponential factors (i.e. based on specific rates or TOFs) are less common, especially in early work where the metal dispersion was often not determined. Their accurate estimation also demands a precise knowledge of the metal content of supported metals, and assuming the truth of the 'nominal' content is a frequent source of error. Much less often are orders of reaction given, and then usually at only a single temperature; some of these orders are quoted in Table 13.2. More detailed studies employing wider ranges of conditions will be the subject of a later section. Consideration of the Arrhenius parameters, with all their faults, does however convey some useful messages.

These are most economically summarised as compensation plots. We may start by considering the many results available for the hydrogenolysis of the lower alkanes on various platinum (and palladium) catalysts:[30,33−35] the Arrhenius parameters lie within a band about a line corresponding to an isokinetic temperature of about 590 K,[33] the activation energies varying by almost a factor of six ($\sim$50 to $\sim$300 kJ mol^{-1}, Figure 13.3). Some of the values shown in this figure were included in the original compilation,[33] but most are new or newly discovered. The collections of results made by Gabor Somorjai and his colleagues[34,35] have been a great help in this respect. At first sight the data points lie randomly within the band, irrespective of alkane structure, type of platinum catalyst or experimental conditions: those in Figure 13.3 cover ethane, propane and the butanes, and there are a few points relating to C$_5$ alkanes. The sources of the information are listed in the figure legends. It is likely that the vertical width of the band is substantially due to errors in estimating metal area, although an error of ±20% on TOF converts to only ±0.2 on ln A, i.e. about the size of the points on the figures. This source of error is removed when the same catalyst (EUROPT-1, 6.3% Pt/SiO$_2$) is used to obtain results for the total reaction of C$_3$ to C$_6$ alkanes and neopentane, points for which lie almost exactly on a single line, while those for the individual hydrogenolyses and isomerisations lie about a separate lower line (Figure 13.4).[33] Points for ethane hydrogenolysis (not shown) are on a yet lower line.[33] A further reason for the variability in Figure 13.3 (but absent from the result in Figure 13.4) might be the very different reactant pressures used; hydrocarbon pressures have varied by at least 10^2, from 10^{-3} atm in UHV work to 0.15 atm in some flow-systems, and it remains to be seen whether the effect this would have on rate is accompanied by changes in the Arrhenius parameters.

We may therefore conclude that on platinum there are no *radical* differences in mechanism as chain length is increased, and that hydrogenolysis and skeletal

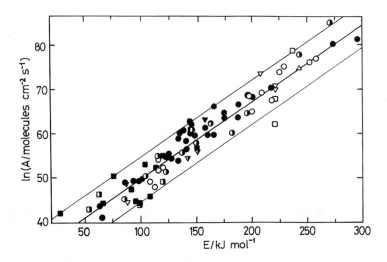

Figure 13.3. Compensation plot of Arrhenius parameters for alkane hydrogenolysis (mainly C_2 to C_4) on various platinum and palladium catalysts. 1 = Pt; 2 = Pd.

Symbols used in Figures 13.3 and 13.4–13.8

Metal	Supported	Unsupported		Alkane/Process
1	○	□		C_2H_6
2	△	▽	etc.	
1	●	■		$>C_2$/total or hydrogenolysis only
3	◇	◆	etc.	
1	◐	◧		$>C_3$/isomerisation only

The units of A are molecules cm^{-2} s^{-1}. Where rates or values of A are given in units of time^{-1}, i.e., where a number of 'sites' or surface atoms has been estimated, they have been changed by assuming 10^{15} 'sites' cm^{-2}: the mean value based on Pt(111) and Pt(100) is close to 1.5×10^{15}, but the same round figure is taken as a mean for all metals of Groups 8 to 11.

isomerisation are processes that are closely linked. Ethane does however distinguish itself from the higher alkanes.

   Closer examination of the location of points within Figure 13.3 and the tables of results on which it is based lead to the following further conclusions. (i) While the points for single crystals lie well within the band, those for metal 'blacks' are towards the bottom or even below it, perhaps because their effective areas have been over-estimated. (ii) Activation energies for the alkanes usually decrease with increasing chain length, from about 200–250 kJ mol^{-1} for ethane to a lower limit reached at about C_5 or C_6 (Table 13.1); values for *iso*butane are generally higher than those for *n*-butane, and those for isomerisation are often (but not always) *higher* than those for hydrogenolysis (see Figure 13.4). As might be expected from Figure 5.15, the few additional points for palladium lie within the platinum band. Note that in this discussion, as in much that has gone before, 'activation energy'

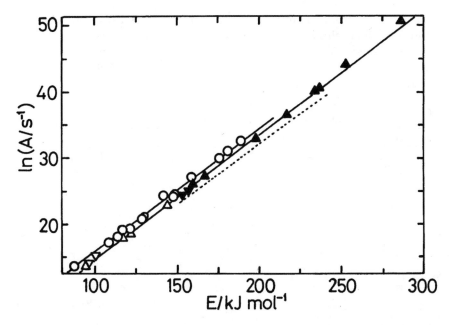

Figure 13.4. Compensation plot of Arrhenius parameters for hydrogenolysis of alkanes (C_3 to C_6) on Pt/SiO$_2$ (EUROPT-1): circles, total rates; open triangles, hydrogenolysis only; filled triangles, isomerisation only; broken line shows location of points for ethane.[33]

TABLE 13.1. Apparent Activation Energies (kJ mol^{-1}) for Hydrogenolysis (E_h) and for Isomerisation (E_i) of the Lower Alkanes

| Metal | Form | E_h | | | | E_i | | |
		C_2H_6	C_3H_8	n-C_4H_{10}	iso-C_4H_{10}	n-C_4H_{10}	iso-C_4H_{10}	References
Ni	film	243	130	142	125	—	—	210
Ni	/SiO$_2$	191	167	128	153	—	—	89
Ru	/TiO$_2$	—	137	133	118	—	—	100
Rh	/TiO$_2$	131	49	68	165	—	—	217
Rh	/SiO$_2$	—	174	174	169	—	—	12
Rh	/SiO$_2$	—	188	163	—	—	—	47
Ir	(111)	145	140	132	152	—	—	8
Ir	/SiO$_2$	174	—	174	243	—	—	9
Pt	/Al$_2$O$_3$	219	150	135	161	105	116	17
Pt	/Al$_2$O$_3$	—	205	141	163	132	—	43
Pt	/SiO$_2$	201	142	94	102	123	148	17
Pt	/SiO$_2$^{a}	199	189	129	—	—	—	41
Pt	black	221	100	96	109	100	121	32

^{a}EUROPT-1
Activation energies vary most with the least active metal (Pt).

is not prefixed by the term 'apparent', as it should be, but the distinction between apparent and true values will be drawn where it is essential.

Orders of reaction using Power Rate Law formalism are rarely reported, but a few values are given in Table 13.2; temperatures of measurement are not always quoted.

TABLE 13.2. Kinetics of Hydrogenolysis of the Lower Alkanes

Metal	Form	Alkane	n_H	n_A	T/K	E/kJ mol^{-1}	ln A^b	References
Fe	Powder	C_2H_6	−0.7	1.0	~500	—	—	207
Fe	/SiO$_2$	C_2H_6	0.5	0.6	543	107	—	24,25
Co	/SiO$_2$	C_2H_6	−0.8	1.0	492	125	58.7	24–27
Co^a	/SiO$_2$	C_3H_8	−0.72	0.79	517	93	—	37
Ni	Powder	C_2H_6	−0.7	0.9	523	163	—	60
Ni	/SiO$_2$	C_2H_6	−2.4	1.0	450	170	73.0	24–27
Ni	/SiO$_2$-Al$_2$O$_3$	C_2H_6	−1.8	0.94	493	167	—	20
Ni^a	/SiO$_2$	C_3H_8	−1.57	0.73	528	175	—	37
Ni	/SiO$_2$ model	n-C_4H_{10}	−2.2	0.8	473	113	—	107
Ni	/SiO$_2$-Al$_2$O$_3$	iso-C_4H_{10}	−2.8	1.0	493	218	—	20
Cu	/SiO$_2$	C_2H_6	−0.4	1.0	580	89.5	47.5	27
Ru	Black	C_2H_6	−0.1	0.92	456	119	59.0	38
Ru	Sponge	C_2H_6	−1.10	1.03	461	88	51.8	206^c
Ru	/Al$_2$O$_3$	C_2H_6	−2.37	0.85	479	155	68.8	206
Ru	/SiO$_2$	C_2H_6	−2.21	0.66	433	125	63.3	206
Ru	/MgO	C_2H_6	−0.73	0.98	462	88	51.1	206
Ru	Sponge	C_3H_8	−0.79	0.80	427	88	54.3	206
Ru^a	/Al$_2$O$_3$	C_3H_8	−2.07	0.54	409	120	—	204
Ru	/Al$_2$O$_3$	C_3H_8	−1.98	0.80	440	155	—	37
Ru	/Al$_2$O$_3$	C_3H_6	−0.9	0.65	464	142	68.3	206
Ru	/SiO$_2$	C_3H_8	−1.3	0.65	416	113	64.4	206
Ru	/MgO	C_3H_8	−0.82	0.92	427	75	50.1	206
Ru	Black	n-C_4H_{10}	+0.3	0.81	443	—	—	38
Ru	/Al$_2$O$_3$	iso-C_4H_{10}	−0.66	0.74	~373	151	—	97
Rh	Black	C_2H_6	−2.2	0.82	400	150	71.2	38
Rh	/SiO$_2$	C_2H_6	−2.2	0.8	487	176	73.1	24
Rh	Black	n-C_4H_{10}	−1.38	0.45	444	123	66.6	28
Pd	/SiO$_2$	C_2H_6	−2.5	0.9	627	243	77.3	24
Os	/SiO$_2$	C_2H_6	−1.2	0.6	425	146	71.0	24
Ir	Black	C_2H_6	−1.65	0.72	444	167	65.5	38
Ir	/Al$_2$O$_3$	C_2H_6	−2.9	1.0	?	174	67.4	9
Ir	/SiO$_2$	C_2H_6	−1.6	0.7	483	151	66.1	24
Ir	Black	n-C_4H_{10}	−0.3	0.42	468	100	52.6	38
Pt	Black	C_2H_6	−1.9	0.9	~690	221	62.0	38
Pt	/Al$_2$O$_3$	C_2H_6	−2.2	1.0	633	259	76.7	211
Pt	/SiO$_2$	C_2H_6	−1.5	0.95	630	199	68.3	41

^{a} Values of kinetic parameters depend on operating conditions.
^{b} A in molecules cm^{-2} s^{-1} (assuming 10^{15} sites cm^{-2}).
^{c} This paper gives results for other Ru loadings and supports, and, for C_3H_8, orders as a function of temperature (for Ru/MgO and Ru black).

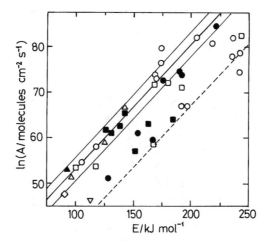

Figure 13.5. As Figure 13.3 but for cobalt, nickel and copper catalysts. See footnote to Fig. 13.3; 1 = Ni, 2 = Co, 3 = Cu.

We may consider results for other metals in the same way. Figure 13.5 shows Arrhenius parameters for cobalt, nickel and copper; most of them are for nickel.[20] Many of them lie inside a quite narrow band, but a number of those for ethane (filled points) lie well below it, as do those for higher alkanes. The reasons for these discrepancies are not immediately clear: there are few causes for a too high specific rate being recorded (*under*-estimation of the active area is one), but several for a too low rate (incomplete reduction, especially of the base metals, if areas are based on TEM, and particularly loss of active area early in use due to 'carbon' deposition). Results for ruthenium catalysts are displayed in Figure 13.6; a good compensation plot is again seen, with the values for ethane lying close to the bottom of the band. Quite a large number of results are available for rhodium catalysts (Figure 13.7); the points for ethane now lie about a distinctly separate lower line. Results for iridium catalysts are shown in Figure 13.8; parameters for ethane are now contained within the band, only a few points for metal blacks lying below it. A selection of the Power Rate Law orders with the associated Arrhenius parameters is contained within Table 13.2.

This broader survey of rates and Arrhenius parameters should harmonise with and perhaps extend the generalisation contained in Figure 5.15, which was based only on ethane, and this as we have seen tends to be somewhat less reactive than the larger alkanes. From the central lines of Figures 13.3 and 13.5–13.8 values of ln A at $E = 175$ kJ mol^{-1} are interpolated, and with values of isokinetic temperatures T_i are compared in Table 13.3. The agreement is fair (except for T_i in the case of iridium), so we may conclude that this wider comparison generally supports the earlier classification.

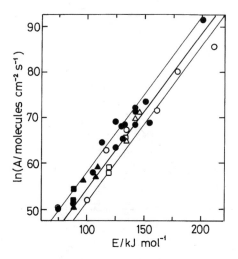

Figure 13.6. As Figure 13.3 but for ruthenium and osmium catalysts. See footnote to fig. 13.3; 1 = Ru; 2 = Os.

The Arrhenius parameters forming the substance of Figures 13.3 and 13.5–13.8 and Table 13.1 have naturally been obtained under a variety of conditions, especially of reactant pressures (which are sometimes not noted). It is surprising to say the least that, irrespective of the numbers and ranges of the variables involved, the values of the Arrhenius parameters form such a tightly grouped set. It is now necessary to introduce orders of reaction into the discussion. These are rarely measured at more than a single temperature; with Sinfelt's data set for ethane

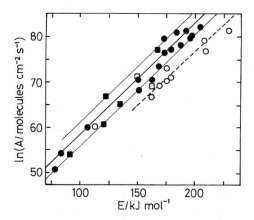

Figure 13.7. As Figure 13.3 but for rhodium catalysts. See footnote to Fig. 13.3.

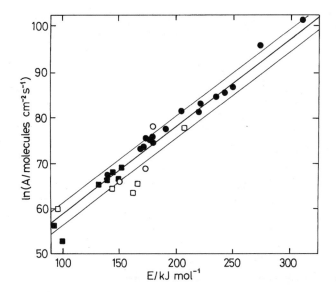

Figure 13.8. As for Figure 13.3 but for iridium catalysts. See footnote to Fig. 13.3.

hydrogenolysis we have already noted a trend of hydrogen order to become more positive as the temperature of measurement was raised, but since different metals were used the conclusion is unsafe. An extensive study[36] of a number of nickel catalysts has shown that the rates of propane hydrogenolysis to ethane + methane, and methane only, generate Arrhenius parameters that show excellent compensation, and the activation energies show a marked correlation with hydrogen orders (order, -3; $E = 270$ kJ mol^{-1}: order, zero; $E = 55$ kJ mol^{-1}). Perhaps the best confirmation of the connection in the older literature is provided by hydrogen orders for propane hydrogenolysis measured at three temperatures on four catalysts;[37] in each case the order rose smoothly, and the value at 573 K was a linear function of activation energy (see Figure 13.9). There is clearly an intimate connection between hydrogen order and activation energy. There have also been suggestions that the value of the activation energy depends on the hydrogen pressure used to measure it,[38-40] and the feeling therefore arose that, far from being divinely decided to characterise a metal, it was in fact a very moveable feast. A further trend that is sometimes apparent is for the hydrogen order to become less negative, and that for the alkane to become less positive, as alkane chain length increases, but on the whole the alkane order is relatively insensitive to variations in catalyst structure and composition, to alkane chain length, and to experimental conditions.[9,32,41,42]

On iridium[9] and platinum[43] catalysts, the activation energies for *iso*butane hydrogenolysis were always higher than that for *n*-butane, reflecting perhaps the

TABLE 13.3. Comparison of Coordinates of Compensation Lines from the Results of J.H. Sinfelt and those in Figures 13.3 and 13.5–13.8

Metal(s)	Isokinetic Temperature/K		Value of $\ln A$ at $E = 175$	
	JHS	Figures	JHS	Figures
Ni, Co	401	316	73	77
Ru	340	353	80	80
Rh	401	445	73	77
Ir	401	650	73	72.7
Pt, Pd	650	600	64	62.5

impossibility of forming more than a single C—M bond at the tertiary carbon atom. However, this distinction is not consistently shown by other metals (Table 13.1), and arguments based purely on activation energy have to be used with care,[9] especially where compensation occurs (see Figure 13.8).

13.2.3. Mechanisms and Kinetic Formulations

Interpretation of the work of Sir Hugh Taylor and his associates at Princeton was based on the following mechanism:

$$C_2H_6 \xrightleftharpoons{K_1} C_2H_x{}^* + H_2 + \left(\frac{6-x}{2}\right)H_2 \qquad (13.A)$$

$$C_2H_x{}^* + H_2 \longrightarrow CH_p{}^* + CH_q{}^* \qquad (13.B)$$

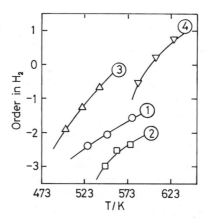

Figure 13.9. Dependence on temperature of hydrogen orders for hydrogenolysis of propane on iron (4), cobalt (3) and nickel (1 and 2) catalysts.[37]

Equilibrium was assumed to be maintained between ethane, C_2H_x and gaseous hydrogen, and the mono-carbon species were swiftly somehow converted into methane: the second process was rate-determining. This led to a rate expression of the form

$$r = k P_E^y P_H^{(1-ya)} \qquad (13.5)$$

where a is $(6 - x)/2$. John Sinfelt subsequently extended [24,25] this analysis to the following reaction scheme:

$$C_2H_6 \underset{k_{-1}}{\overset{k_1}{\rightleftarrows}} C_2H_5{}^* + H^* \overset{K_2}{\rightleftarrows} C_2H_x{}^* + \left(\frac{6-x}{2}\right)H_2 \qquad (13.C)$$

$$C_2H_x{}^* + H_2 \overset{k_3}{\longrightarrow} CH_p{}^* + CH_2{}^* \qquad (13.D)$$

which gives the rate expression

$$r = k_1 P_E/(1 + b P_H^{(a-1)}) \qquad (13.6)$$

where a as before is $(6 - x)/2$, and b is $k_1'/k_3 K_2$. C—C bond splitting was again taken to be rate-limiting. Measurement of the hydrogen orders for ethane hydrogenolysis over Co/SiO_2[26] had shown that b decreased with increasing temperature, perhaps chiefly because the value of K_2 became greater. Application of this analysis to his results for this reaction on a variety of metals led to estimates[24,25] of the values of x, which ranged from zero for rhodium, palladium and platinum, two for nickel, ruthenium, osmium and iridium, and four for cobalt: answers could not be obtained for iron or rhenium, which had shown positive orders in hydrogen. It was accepted that there was no competition between ethane and hydrogen for surface sites, so there was no provision for the independent chemisorption of hydrogen, and no detailed site counting, the single asterisk simply indicating an adsorbed state. Despite its limitations, the procedure was widely adopted by other scientists,[44] and has been extended.[45]

Subsequent developments have addressed these and other deficiencies, which give rise to the following questions.

- What is the site requirement for the initial chemisorption of the alkane, and for each of the subsequent steps? How does this vary with the size of the alkane?
- What is the mechanism by which the C_2 species disintegrates?
- Does hydrogen chemisorb in competition with the alkane?

- Are hydrogen atoms or only hydrogen molecules involved?
- Which is the rate-determining step?

In addressing these questions, the following considerations must be kept in mind. (1) The behaviour of propane and especially the butanes differs significantly from that of ethane; activation energies are lower, orders in the alkane are often fractionally positive and hydrogen orders are sometimes less negative. (2) The Power Rate Law formalism and the derivation of reaction orders as exponents of the reactant pressures overlooks the fact that the rate must become zero and not infinite when no hydrogen is present, and must therefore pass through a maximum as its pressure is changed. The unfortunate habit of showing results as log-log plots[4,16,46,47] disguises trends in the sense of dependence of rate on hydrogen pressure, although their non-linearity has often been noted, and maxima have sometimes been detected.[16,20] What is even more regrettable is the reporting of the dependence of rates upon experimental variables only in terms of some cherished theoretical framework, so that their reinterpretation by some other model becomes a virtual impossibility.

Leclercq and her colleagues[10,12] have made much use of an equation based on the mechanisms shown above, in which the alkane adsorbs by losing hydrogen atoms in pairs (not singly); they immediately combine to form a gaseous molecule (process 13.A), this needing only a single site to accommodate the dehydrogenated species. Introduction of the independent chemisorption of hydrogen and detailed consideration of the relevant form of isotherm led with some simplifications to the expression

$$r = k_3 K' P_C P_H / (K' P_C + P_H{}^{a+mx}) \tag{13.7}$$

where $K' = K_1/b_H$, m is the size of the ensemble on which reaction takes place and x is defined as

$$(1 + (b_H P_H)^{1/2}) \approx (b_H P_H)^x \tag{13.8}$$

Linearisation of this equation has allowed evaluation of k_3, K' and $(a + mx)$ for the hydrogenolysis of a number of alkanes on nickel[12,20,48] and platinum[49] catalysts (also Ir/Al_2O_3; results for alkanes with more than four carbon atoms will be mentioned in the next chapter.

Opposing views have been expressed concerning the location of the rate-determining step. Martin and his associates[50-52] have taken this to be the initial chemisorption of the alkane, which they believed to require an ensemble of m atoms, similar in size to that which their magnetic studies had suggested. Careful work on EUROPT-1[51] involving hydrogen chemisorption, transient and steady-state kinetics led them to the rate expression

$$r = kP_E(1 - \beta\theta_H)^m \tag{13.9}$$

where β ($= 1.32$) represents the maximum value of H/Pt_s: thus the slow step is the collision of an ethane molecule with a group of m platinum atoms devoid of hydrogen atoms. The methodology of their further procedures is somewhat hard to follow, as they reported values of activation energy as a function of temperature and of hydrogen orders as a function of hydrogen pressure. On a $PtFe/SiO_2$ catalyst it has however been shown[51] that the hydrogen order becomes less negative and the n-butane order less positive as the hydrogen coverage increases. Use of the Temkin equation (5.29) with a heat of hydrogen chemisorption of 84 kJ mol^{-1} gave a true activation energy of 55 kJ mol^{-1} and a pre-exponential factor that was not much less than the estimated collision number for ethane. The value of m was nine; but similar work[21] with propane hydrogenolysis on Ni/SiO_2 gave values of m of 17 for single bond breaking and 24 for the breaking of both bonds. These numbers would seem to be on the generous side.

Equally thoughtful work by Frennet and his colleagues has led to the definite conclusion that the formation of methane from C_1 fragments is rate-determining.[53-55] This conclusion was based on a comparison between rates of hydrogenolysis and of alkane exchange. Rates of ethane exchange were much the slower, but for many metals the rates of methane were comparable. Only for the base metals were they much slower, and for platinum much faster. This idea would require C_1 species to be the MASI (Section 5.3). A thorough review[56] to 1982 of proposed mechanisms wisely concludes that unequivocal identification of the slow step based on kinetic measurements alone is not possible. Clearly these theories cannot be simultaneously true; if one were to take the average, the slow step would be somewhere in the middle.

Speaking of which brings us to the excellent work of Dumesic and his associates, who have sought to apply the techniques of kinetic simulation to ethane hydrogenolysis, using measured or calculated values of H—M and C—M bond strengths.[46,57-59] This approach, directed to Sinfelt's kinetic measurements, brought them to the conclusion that on silica-supported platinum, palladium, iridium and cobalt the slow step is irreversible C—C bond rupture in C_2H_x ($x = 4$ or 3) and that hydrogenation of C_1 species is kinetically insignificant. In a further contribution,[46] new kinetic work on Pt/SiO_2, backed by DFT calculations, permitted speculation as to the nature of the C—C bond-breaking step: this was thought to be the spontaneous disruption of ethylidene (CH_3CH^{**}) by interaction with two vacant sites, and kinetic modelling based on this belief seemed to accord with experimental results obtained over a wide range of conditions (e.g. T $= 573$–673 K). Thus in a sense the wheel has come full circle; and having read these papers one wonders if there is anything left to be said on ethane hydrogenolysis.

There are however a number of other points deserving comment. It is not certain that there is only one value for x in the reactive intermediate C_2H_x, i.e. there may be several species differing in their reactivity; it was suggested some years ago that the composition of C_2H_x might depend on the reactant ratio.[40,60] A more catholic approach to deciding what mechanism best fits experimental results was adopted by Kristyan and his associates:[61,62] writing the reaction set as

$$H_2 + 2^* \overset{b_H}{\rightleftharpoons} 2H^* \tag{13.E}$$

$$C_2H_6 + (7-x)^* \overset{K_A}{\rightleftharpoons} C_2H_x{}^* + (6-x)H^* \tag{13.F}$$

$$C_2H_x{}^* + B \overset{k}{\longrightarrow} CH_p{}^* + CH_2{}^* \tag{13.G}$$

The identity of B was then taken as either * or a hydrogen atom or a hydrogen molecule. This gave the alternative rate expressions shown in Table 13.4. Their own (somewhat unsatisfactory) results for ethane hydrogenolysis on nickel and palladium powders[62] were best fitted by taking B to be a hydrogen molecule. Plus ça change. Shang and Kenney[63] pursued this broader approach and considered no less than ten rate expressions (there are unfortunately several errors in the equations in their Table 2). Their experimental work on ethane hydrogenolysis over Ru/SiO_2 showed that the position of the rate maximum as hydrogen pressure was changed increased with ethane pressure and with temperature, and that the rate versus ethane pressure also exhibited a gentle maximum at a hydrogen pressure of 0.027 atm. Their analysis concluded that B = * offered the best option, while admitting that discrimination between alternative rate expressions was not easy. This has already been demonstrated in Figure 5.9, where three alternative equations were applied to results for n-butane hydrogenolysis on a $PtRe/Al_2O_3$ catalyst: two of them give virtually indistinguishable curves.

TABLE 13.4. Rate Expressions Based on the Mechanism Given by Processes 13.E, F and G[64]

B	r	Code
*	kyG_6/D^2	ES5A
H*	kyG_7/D^2	ES5B
H₂	kyP_H/D	ES5

Definitions:

$y = K_A P_A$; $G_6 = (b_H P_H)^a$; $G_7 = (b_H P_H)^{a+1/2}$
$D = y + G_6 + G_7$; $a = (n-x)/2$ where n is the number of carbon atoms in the alkane, and x the number of H atoms in the reactive species.

TABLE 13.5. Dependence of the Values of the Constants for Ethane Hydrogenolysis on Ru/A1$_2$O$_3$ at 473 K on the Form of Rate Expression Used[63,64]

Code	R	k[+]	K_A	b_H/atm^{-1}
ES2	$kyP_H(y + P_H{}^a)$	1.44E−4	1.06E−3	—
ES3	$kP_A/(1 + b_H P_H{}^{a-1})$	8.16E−5	3.24E−2	—
ES4	$kyP_H/[y + P_H{}^a(1+(b_H P_H)^{1/2})]$	1.52E−4	1.45E+1	1.1E+7
ES5	kyP_H/D	1.53E−4	2.08E+7	7.31E+3

[+]In units of mol $g_{Ru}{}^{-1}$ s^{-1} atm^{-1}.

The values of the constants of the rate expression, however, depend critically on the form of the expression used:[64] results for the four tested by Shang and Kenney[63] are shown in Table 13.5. While values of k are reasonably constant, those for K_A vary by more than 10^{10}, presumably because their dimensions are different in each expression: but even b_H in the cases where it does appear differs by more than 10^3, although its units are the same. This sensitivity to mathematical form does not assist the process of identifying the 'right' equation. The equation coded ES5B (Table 13.4) has been selected[65] as giving values of K_A and b_H that are of similar size.[64]

The availability of microprocessor control of the reactor system greatly improves the speed and accuracy with kinetic measurements can be made, and the number of levels of the variables that can be used. The better quality permits a more rigorous comparison between alternate rate expressions, and increases the reliability of the derived constants for that selected. These advantages have been exploited in extensive studies (see 'Further Reading' list) of the hydrogenolysis of the lower alkanes on standard platinum catalysts,[66−69] as the following examples will illustrate.

Comparison of the rate dependence upon hydrogen pressure for the hydrogenolysis of ethane, propane and n-butane on Pt/Al$_2$O$_3$ (EUROPT-3) (Figure 13.10) shows that the forms of the curves differ markedly in respect of the position and sharpness of the rate maximum.[64,65,70] This explains why (if it were not already clear) the 'order' depends on the range of hydrogen pressure used, and why the 'order' measured at say 0.3 to 0.5 atm pressure becomes more positive as chain-length increases. Qualitatively it is clear that the larger the alkane, the more easily is it chemisorbed and the more effectively can it compete with hydrogen for the surface. The values of the constants of equation ES5B give quantitative support (Table 13.6): with both Pt/SiO$_2$ and Pt/Al$_2$O$_3$ the ratio K_A/b_H increases with chain-length, although with the more active Pt/SiO$_2$ at a 75 K lower temperature values of K_A are smaller, while those of b_H are similar (see below). These observations incidentally deprive the question 'which alkane is the most reactive?' of any meaning.[70]

For each alkane the form of the kinetic curve changes with temperature,[65] the rate maximum moving to higher hydrogen pressures and becoming less distinct

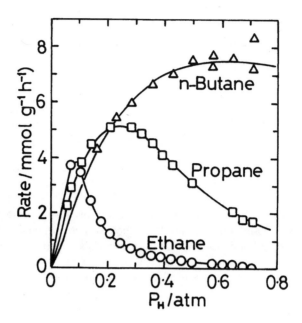

Figure 13.10. Dependence of rates of hydrogenolysis of C_2 to C_4 alkanes on hydrogen pressure over Pt/Al$_2$O$_3$ (EUROPT-3) at 608 K.[64,65]

TABLE 13.6. Values of the Constants of Equation ES5B for Hydrogenolysis of the Lower Alkanes[64]

Metal	Support	Alkane	T/K	k	K_A/atm^{a-1}	b_H/atm	a
Ru[a]	Al$_2$O$_3$	C$_2$H$_6$	418	29	0.60	18	1.83
Ru[a]	Al$_2$O$_3$	C$_3$H$_8$	418	80	17	25	2.34
Ru[a]	Al$_2$O$_3$	n-C$_4$H$_{10}$	418	116	39	34	1.34
Rh	Al$_2$O$_3$	n-C$_4$H$_{10}$	430	189	47	63	1.75
Rh	SiO$_2$	n-C$_4$H$_{10}$	430	151	30	70	0.86
Rh	TiO$_2$	n-C$_4$H$_{10}$	430	36	6.6	18	1.27
Pt[b]	Al$_2$O$_3$	C$_2$H$_6$	608	33	10	9	2.25
Pt[b]	Al$_2$O$_3$	C$_3$H$_8$	608	45	18	5	1.52
Pt[b]	Al$_2$O$_3$	n-C$_4$H$_{10}$	608	57	27	2	0.88
Pt[c]	SiO$_2$	C$_3$H$_8$	533	9	0.7	5	1.41
Pt[c]	SiO$_2$	n-C$_4$H$_{10}$	533	11	2.3	6	0.90

[a] 1% Ru/Al$_2$O$_3$ after HTR1 (see Section 13.5);
[b] EUROPT-3;
[c] EUROPT-1.

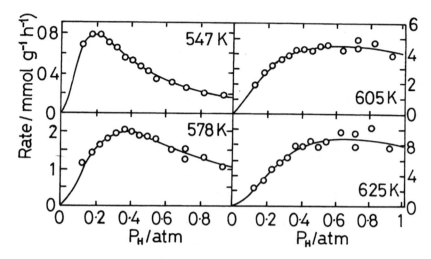

Figure 13.11. Dependence of rate of hydrogenolysis of *n*-butane on hydrogen pressure at various temperatures on Pt/Al$_2$O$_3$ (EUROPT-3).[64,65]

as temperature increases (Figure 13.11).[64,65] Hydrogen chemisorption becomes weaker and its coverage decreases, and 'order' becomes less negative. This immediately implies (if it were not already obvious) that the activation energy measured at different hydrogen pressures must *increase* with the pressure used, as is shown in Figure 13.12. *Apparent activation energy derived from rates of reaction is therefore not an absolute quantity, and varies with the reactant pressures used.*[71] Moreover activation energies obtained by altering hydrogen pressure also exhibit compensation, the parameters E_a and ln A being consistent with those found when different alkanes, and different but similar catalysts, are used (Figure 13.13). *Compensation between Arrhenius parameters can therefore occur within a single reaction system, and their dependence on operating conditions must go some way to explaining the spread of their values as revealed in Figures 13.3 and 13.5–13.8.*[71]

Application of the rate equation ES5B (Table 13.4) to results for the lower alkanes obtained with EUROPT-3 (0.3%Pt/Al$_2$O$_3$, AKZO CK303) (and to the corresponding rhenium-containing catalyst, EUROPT-4) generated [65] values of the constants k, K_3, b_H and x at various temperatures, and the use of the Arrhenius equation on k and of the van't Hoff isochore on K_3 and b_H yielded satisfactorily linear plots from which a true activation energy E_t and heats of adsorption for the alkane $(-\Delta H_A)$ and hydrogen $(-\Delta H_H)$ were derived. Values for E_t and $-\Delta H_A$ were respectively 82 and 88 kJ mol^{-1} for propane, and 76 and 79 kJ mol^{-1} for *n*-butane, but for $-\Delta H_A$ were about zero. This was excused on the basis that measurements related to high surface coverage, at which low values were to be expected. A further curious feature of the results was a marked correlation between

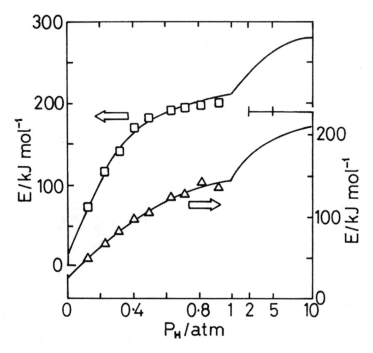

Figure 13.12. Dependence of apparent activation energies for hydrogenolysis of propane and of *n*-butane over Pt/Al$_2$O$_3$ (EUROPT-3) on hydrogen pressure: the curves are calculated using equation ES5B and the constants in Table 13.6.[64,65]

k and K_A, which may have had physical significance or may have been an artefact of the kinetic model. These procedures have been further discussed,[65] and applied to results obtained with Ru/Al$_2$O$_3$ catalysts. Three-dimensional plots of rate versus alkane and hydrogen pressures for the three linear alkanes emphasise their very different characters (Figure 13.14). The dependence of apparent activation energy on reactant pressures and hence on surface concentrations is implicit in the Temkin equation (5.29), where the contributions of the heats of adsorption to the true activation energy are moderated by the reaction orders n_i thus:

$$E_a = E_t + n_A \Delta H_A + n_B \Delta H_B \qquad (13.10)$$

A zero order reflects a lack of dependence of coverage on temperature, and so the relevant heat term disappears. E_a will then decrease as the pressure of that reactant is increased. Application of this equation to the hydrogenolysis of alkanes is not straightforward, because E_a actually *increases* with hydrogen pressure[72] (Figure 13.12). This is because the concentration of the dehydrogenated reactive intermediate *increases* with temperature because of the endothermic nature of its formation, and because the desorption of hydrogen provides more vacant sites for

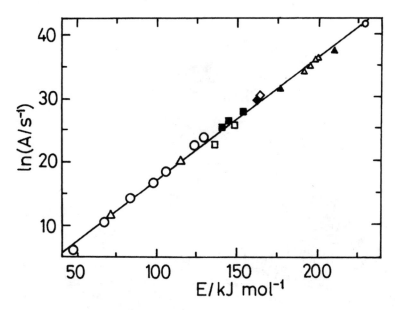

Figure 13.13. Compensation plot for the total reaction of C_2 to C_4 alkanes with hydrogen on Pt/Al_2O_3 (0.3% Pt, open points and 0.6% Pt, filled points):[33] circles, C_2H_6; triangles, C_3H_8: squares, n-C_4H_{10}; diamonds, iso-C_4H_{10}. The larger triangles and circles derive from the variation of hydrogen pressure (Figure 13.12) for C_3H_8 and iso-C_4H_{10} respectively.

it. The experimentally estimated upper and lower limits for E_a (Figure 13.12) can be compared with the values stemming from the ES5B equation, because

$$E_a(\max) = E_t + n_A \Delta H_A \qquad (13.11)$$

$$E_a(\min) = E_t + n_H \Delta H_H \qquad (13.12)$$

where A = alkane, and the n's the respective 'orders'. Agreement between the measured and calculated values of E_a is fair rather than good, but perfect agreement is not to be expected because the limiting values of the 'orders' are not easily estimated, and the rate expression used may not be entirely valid. The comparison does however supply another means by which the utility of a rate expression may be judged.

13.2.4. A Generalised Model for Alkane Hydrogenolysis

It now becomes possible to sketch a qualitative outline of a model describing kinetic aspects of hydrogenolysis of the lower alkanes.[33,71] Let us imagine three curves representing the dependence of rate on hydrogen pressure; for simplicity they are normalised to the same maximum rate (Figure 13.15). The order n_H

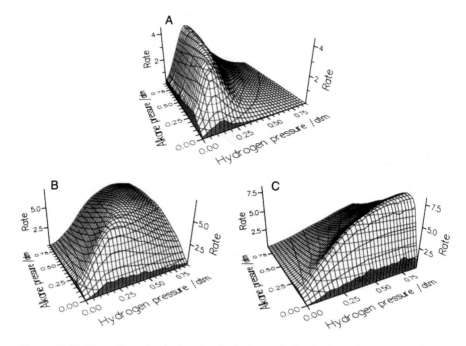

Figure 13.14. Three-dimensional plots for the hydrogenolysis of ethane (A), propane (B) and *n*-butane (C) on Pt/Al_2O_3 (EUROPT-3) at 608 K calculated using the constants for equation ES5B in Table 13.6.[65]

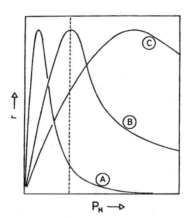

Figure 13.15. Schematic diagram showing rate dependence on hydrogen pressure in three different circumstances (see text).[30,71]

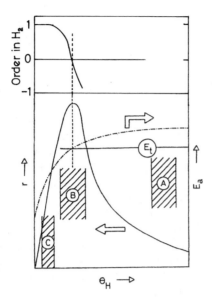

Figure 13.16. Schematic diagram showing the dependence of rate, order in hydrogen and apparent activation energy on hydrogen coverage (θ_H) (see text): the zones A, B and C are those located about the dashed line in Figure 13.15.[30,71]

measured in any fixed range of pressure will decrease in the sequence C > B > A, so that these curves could represent either (i) metals or catalysts A, B and C arranged in order of *decreasing* strength of hydrogen adsorption, or (ii) *increasing* temperature,[71] or (iii) the use of alkanes A, B and C that show *increasing* strength of adsorption or ease of activation. Now corresponding to each curve there will be an adsorption isotherm, so the three curves can be condensed onto a θ_H scale, and the ranges over which the three situations A, B and C are met can be identified (Figure 13.16). The value of E_a will be a function of θ_H. If one elects or is constrained to working at the tail of the curve (Case A), one finds low rate a high E_a and a negative 'order' in hydrogen: working near the maximum (case B) there will be faster rates, a lower E_a ($= E_t$) and an order n_H close to zero: and at low θ_H (case C) low to medium rates, a low E_a and a positive value of n_H. The kinetic form thus depends importantly on the magnitude of the hydrogen coverage, as had been anticipated in several earlier works. So the values of E_a obtained *using a fixed hydrogen pressure* under cases A, B and C will differ because the ranges of θ_H scanned by the temperature change will not be the same. Thus for example values of E_a found with different alkanes are necessarily not the same (Table 13.1, Figure 13.13) *because the θ_H ranges differ*; they decrease with increasing chain length, but are usually larger for *iso*butane than *n*-butane because it has fewer hydrogen atoms that can be lost on adsorption. The correspondence between measured activation energy and hydrogen 'orders' thereby finds a logical explanation.

The very high activation energies encountered in alkane hydrogenolysis ($50–250$ kJ mol^{-1}; even higher for *neo*pentane) are unlike the low values found in hydrogenations ($30–90$ kJ mol^{-1}) *because of the endothermic nature of the pre-equilibrium that has to be set up before the reactive state is reached*; and because their concentration is temperature-dependent, they are of necessity *apparent* and not *true*. The thermochemistry of the surface processes parallels their gas-phase counterparts, and so the ease of forming the key intermediate increases in the sequence ethane $<$ propane $<$ *n*-butane, as do their heats of dehydrogenation, which are the same as those for hydrogenation (Table 7.1) with the signs changed.

The observation that the activation energy for isomerisation E_i is greater than that for hydrogenolysis E_h is understandable[72] in terms of the Temkin equation if that process requires a more dehydrogenated intermediate (Scheme 13.3, see below): the greater summation of the heats of adsorption necessitates a higher E_a. The hydrogen pressure at which rates are maximal decreases with increase in chain branching because the adsorption of the hydrocarbon becomes progressively weaker with respect to the linear molecule.[72]

13.2.5. Alkane Hydrogenolysis on Metals Other than Platinum

The previous section has concerned reactions on platinum almost exclusively, because of the wealth of information available, but the same modelling procedure has been used for rhodium,[47] palladium,[73] and especially ruthenium[74–79] catalysts, whose curious behaviour will be considered further below. Palladium catalysts proved difficult to work with because they deactivated easily, and indeed the literature contains somewhat few references to work on this metal (Table 13.2 and Figure 13.3): rhodium presented no difficulties. Table 13.6 shows a selection of values of the constants of the ES5B equation (see Table 13.4) obtained by optimum curve fitting,[47,64] and whatever their limitations they form a handy way of comparing the effects of varying support, chainlength, and metal. (i) With Ru/Al$_2$O$_3$, k and K_A both increase with chainlength, but b_H is almost constant. (ii) With *n*-butane on Rh/SiO$_2$ and Rh/Al$_2$O$_3$ give similar results, but all constants are smaller with Rh/TiO$_2$. The operation of this equation at least succeeds in separating the effects of alkane and hydrogen in generally sensible ways.

13.3. STRUCTURE-SENSITIVITY OF RATES OF ALKANE HYDROGENOLYSIS[80,81]

It is convenient to examine structure-sensitivity as revealed only by rates, since much of the available information concerns ethane; there are however major and important effects on product selectivities in the reaction of *n*-butane especially, and these will be mentioned later. There are three manifestations of structure-sensitivity, which probably have elements in common, although it is wise to separate them. This section concerns (i) particle-size dependence of specific rates or TOFs, and (ii) their dependence on the face geometry of single crystals. The third aspect, namely, ensemble size and character in bimetallic systems, is also considered later.

Although alkane hydrogenolyses are usually reckoned to be structure-sensitive, the evidence arising from particle-size effects is sparse and somewhat contradictory.[82,83] Evaluation of the results is rendered the more difficult by the various means that have been used to alter the mean particle size, and it is hard—although perhaps not impossible[84,85]—to secure alteration without changing other features of the catalyst's structure and composition.[86] Techniques used include (i) changing metal loading (most often), (ii) changing the support, and (iii) changing the conditions of preparation (including precursor, calcination temperature[87,88] etc.). Metal-support interactions alone can cause large effects on rates,[89] so simultaneous alteration of these factors is not to be recommended. A further difficulty is that large differences in activity may necessitate use of different temperatures, so that direct comparison is not always easy.[3,14,90] A selection of the available results is shown in Table 13.7; some simplification has been used in columns 3 to 6, e.g. some other supports and alkane may have been used besides those cited, so reference to the original papers is needed for full details. The following comments can however be made.

(1) A change in rate is usually accompanied by an antipathetic change in activation energy. (2) Variable results have been obtained with ruthenium (entries 3–6), due perhaps to the use of different supports and temperatures: detailed studies (see Further Reading section) to be mentioned later (Section 13.6) resolve some of these problems. (3) With rhodium, iridium and platinum the findings are inconclusive, although those for Pt/C catalysts are particularly clear; in both cases, n-butane *isomerisation* was size-*insensitive*. (4) Only with nickel is the verdict almost unanimous, the rates decreasing with size, but only above about 4 nm (entries 1–3). (5) Orders of reaction were usually not much affected by size (entries 3, 9 and 11), but the order in hydrogen became more positive with Rh/SiO$_2$ as dispersion increased. (6) Following the argument developed in the last section, an increase in activation energy might betoken a raising of the strength of hydrogen chemisorption, and *vice versa*, but there are too few results to support the expected correlation with order of reaction.

Studies of the rates of hydrogenolysis of the butanes on single-crystal surfaces are similarly scarce and unsatisfactory. On various platinum surfaces at 573 K dehydrogenation is the predominant process, followed by isomerisation, with hydrogenolysis only a very minor component.[91] The last process was favoured by stepped or kinked surfaces (111) structure, whereas isomerisation preferred the (100) structure, but all clean single-crystal surfaces gave much higher activities

TABLE 13.7. Effect of Decreasing Particle Size or Increasing Dispersion on Rates and Activation Energies of Alkane Hydrogenolysis

Entry	Metal	Form	Variant	Range	Alkane	r_{sp}	T/K	$E/kJ\ mol^{-1}$	References
1	Ni	/SiO$_2$-Al$_2$O$_3$	Method	<4–8 nm	C$_2$H$_6$	+	540	~	209
2	Ni	/Al$_2$O$_3$	[Ni], method	2–8 nm	C$_3$H$_8$	Max	600	min	228
3	Ni	/SiO$_2$	[Ni]	—	n-C$_4$H$_{10}$	+	473		107
4	Ru	Supported	Support, [Ru]	0.09–72%	C$_3$H$_8$	~	513	+	206
5	Ru	/SiO$_2$	[Ru]	16–5 nm	n-C$_4$H$_{10}$	Max	600		220
6	Ru	/SiO$_2$	[Ru]	2–30%	C$_2$H$_6$	—	508		156
7	Ru	/Al$_2$O$_3$	[Ru]	0.07–100%	C$_3$H$_8$	—	433	+	106
8	Ru	/TiO$_2$	Method	24–84%	n-C$_4$H$_{10}$	+	433	—	99,100
9	Rh	/SiO$_2$	[Rh]	2.2–1.6 nm	n-C$_4$H$_{10}$	—	450	+	93
10	Rh	/TiO$_2$	?	30–85%	C$_2$H$_6$	+	~573	—	138
11	Rh	/SiO$_2$	[Rh], method	12.7–4 nm	C$_2$H$_6$	Max	526	~	203
12	Rh	/SiO$_2$, Al$_2$O$_3$	[Rh]	7.2–<1	n-C$_4$H$_{10}$	—	448		224
13	Ir	/SiO$_2$	[Ir]	20–1 nm	n-C$_4$H$_{10}$	~	473	?	9
14	Ir	/TiO$_2$	[Ir] etc.	10–1 nm	n-C$_4$H$_{10}$	—	473	+	105
15	Pt	/Al$_2$O$_3$	Method	15–1 nm	C$_2$H$_6$	—	633	+	211
16	Pt	/C	Method	8–50%	n-C$_4$H$_{10}$	+	603	?	95
17	Pt	/C	Support, method	6–20%	n-C$_4$H$_{10}$	+	583		96

Note: ? *indicates information not given;* + *signifies increase;* − *decrease;* ~ *little change.*

than those of supported catalysts. Isomerisation selectivity decreased with temperature, unlike the situation with supported platinum catalysts.[3,14,43] This paper[91] provides a detailed and thoughtful analysis of the role of surface structure in hydrocarbon reactions; it is therefore disconcerting to find another work in which Pt(100) and (111) surfaces were used that makes no mention at all of dehydrogenation, even although much lower pressures of hydrogen were used. Work on low-index faces of nickel,[92] rhodium[93,94] and iridium[8,92] have also been reported, but are chiefly of interest through their observations on product selectivities (see later).

It is therefore possible to harmonise some of these observations in the following way. Isomerisation selectivity *decreases* as dispersion *increases*[95,96] and as the concentration of low coordination number atoms *increases*:[91] these latter will predominate at high dispersion. Isomerisation by bond-shift also appears to be favoured at high temperature and low hydrogen pressure;[3,14,43] therefore, on platinum at least, the strength of hydrogen chemisorption should be greater at high dispersion.

In view of the possible connection between the effects of dispersion and hydrogen coverage, it is worth stressing that most of the conclusions concerning structure-sensitivity in alkane reactions stem from work performed at a single reactant ratio. Measurements or orders of reaction at different dispersions are rare, the effects where observed not being large. Activation energy values are more common (Table 13.7) and usually increase with dispersion.

13.4. SELECTIVITY OF PRODUCT FORMATION IN ALKANE HYDROGENOLYSIS

The reaction of propane with hydrogen affords methane + ethane; that of *n*-butane gives *iso*butane as well as methane, ethane and propane, while *iso*butane gives *n*-butane and the smaller alkanes from which ethane is largely absent in the initial products. The preferred method for quoting selectivities has been described in Section 13.1.4. This additional feature adds fascinating further insights into the mechanisms of alkane hydrogenolysis, and there is a rich and extensive literature describing how selectivities vary with reaction conditions, with metal, its physical form, its dispersion, and other variables. The problem of deciding the conceptual framework for this discussion has already been mentioned: instead of supposing that there are three, or even four, reactions proceeding in parallel, each with its own set of kinetic parameters, we take the line that after its initial dissociative chemisorption the alkane decides, perhaps depending on the local environment, what species to form next, and in what amounts, because this will determine the product mix.

The procedure to be adopted is that expounded by R.B. Anderson and his associates,[37,42,97] initially for *n*-butane, but later for more complex alkanes.[98] The

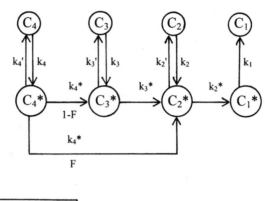

$$T_j = k_j/(k_j + k_j*)$$

Scheme 13.1. The Anderson–Kempling scheme for the hydrogenation of n-butane.

reaction network to be used is set out in Scheme 13.1, and a typical plot product selectivities as a function of conversion is shown in Figure 13.17; conversion is measured as mol hydrogen consumed per mol n-butane, and isomerisation is ignored for the moment. Steady-state analysis of this reaction scheme leads to expressions for the variation of S_2 and S_3 with conversion, depending on values as signed to the component constants. At low conversions these expressions conveniently simplify to

$$S_2/T_2 = 1 + F - S_3 \qquad (13.13)$$

$$S_3/T_3 = 1 - F \qquad (13.14)$$

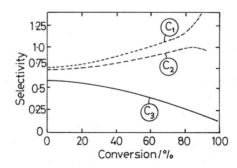

Figure 13.17. Hydrogenolysis of n-butane: manner of variation of product selectivities with conversion.[97]

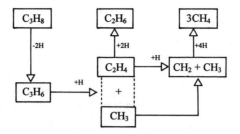

Scheme 13.2. Hydrogenolysis of propane.

where T_j ($j = 2$ or 3) is defined by

$$T_j = k'_j/(k'_j + k_j^*)$$ (13.15)

Now unfortunately with the two observables S_2 and S_3 it is impossible to estimate the three unknowns, so one of two procedures has then been used. (1) Scheme 13.2, which is a simpler version of Scheme 13.1, *starts* with propane, so that $T_2 = S_2$, and this value of T_2 can be determined and substituted for that in the above equations for n-butane, assuming the identity of the two quantities. (2) One can assume a value of unity for T_2 which, based on the reaction of propane, is often a very good approximation (see below). The success of these procedures can be judged by the ways in which the parameter values vary smoothly and independently as conditions are changed; they must lie between zero and unity, and if, following the second option above, they do not it is probably because the propane T_2 is less than unity. For *iso*butane there is only one mode of fission, i.e. $F = 0$, so $(S_2 + S_3)$ should be unity; it often is about that,[97] but not always.[91] Selectivities are usually independent of conversion over a considerable distance (i.e. about 0–30%),[97,99] so their values can be found quite precisely.[14]

The reviewer of the literature faces considerable difficulties. Quite often, those studying the reactions of propane and n-butane do not trouble to measure or report product selectivities, and when they are given it is most usually in graphical form, from which numbers of limited accuracy have to be extracted by tedious interpolation. Sometimes only ethane selectivities are quoted.[90] Measurements made in UHV systems seem to be more scattered than those made in conventional equipment, and almost all values have to be converted into the Kempling-Anderson formalism (equations 13.13 and 13.14) to make them comparable. In the accompanying Tables 13.9 and 13.10, most values of F and T_3 are obtained by method (2) above. Although for n-butane products the values of S_1 can easily be derived from S_2 and S_3, they are quoted in the tables to save the reader unnecessary labour.

Our task is now to try to summarise the ways in which product selectivities vary with temperature and reactant pressures, especially that of hydrogen.

Following earlier arguments we may expect that *rising* temperature and *falling* hydrogen pressure will produce equivalent effects, as both will tend to encourage formation of more dehydrogenated species and more vacant surface sites. As expected, changing alkane pressure has comparatively little effect, because except at the highest alkane/hydrogen ratios the surface concentration of hydrocarbon radicals remains low. The response of selectivities to changes in the levels of temperature and of hydrogen pressure depends on the strength with which hydrogen is adsorbed; if it is strong, both effects are small, but if it is weak they are much more marked. It proves impossible however to divorce these effects from those produced by changing dispersion, which as we have seen may also alter the concentration of adsorbed hydrogen. Particularly marked effects have been seen with titania-supported metals[99-103] and with Ru/Al$_2$O$_3$;[74-79] some of these are discussed in Sections 13.6 and 13.7.

The selectivity S_2 with which ethane is formed from propane is indicative of the behaviour of the butanes. With platinum S_2 was *always* close to unity (>0.985 for Pt/SiO$_2$ EUROPT-1[14] and for Pt/Al$_2$O$_3$ EUROPT-3[3]); it was also high (>0.96) for Rh/SiO$_2$ and Rh/Al$_2$O$_3$,[47] and for Ir/TiO$_2$,[102] irrespective of the mode of pretreatment. These values reflect the much greater difficulty of breaking the C—C bond in a C$_2$ species. Various values have been given for ruthenium on different supports; for Ru/TiO$_2$,[101] Ru/Al$_2$O$_3$[78] (and Os/TiO$_2$[102]), it depended markedly on pre-treatment and operating conditions (see Section 13.6). The reaction of *iso*-butane is also quite straightforward. While the breaking of a single C—C bond is the much preferred reaction at low conversions (i.e. $S_1 \approx S_3$), ethane has often been an initial product, not being formed via propane that has re-adsorbed, but through an adsorbed C$_3$ species. On Pt/Al$_2$O$_3$ at 700 K S_2 was initially zero,[98] but in another study[43] both S_2 and S_i increased with temperature (see Scheme 13.3), and it was suggested that formation of ethane might be preceded by skeletal isomerisation to a linear species. On ruthenium catalysts, S_2 was very small at 370–400 K, but

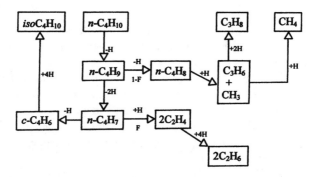

Scheme 13.3. Reactions of *n*-butane with hydrogen showing suggested intermediate species.

TABLE 13.8. Hydrogenolysis of *iso*Butane ($P_H/P_C \sim 10$): Product Selectivities

Metal	Form	T/K	S_1	S_2	S_3	S_i	References
Ru	/SiO$_2$	383	1.15	0.15	0.85	—	97
Rh	Black	420	1.65	0.51	0.45	—	38
Ir	/SiO$_2$		1.33	0.57	0.51	—	9
Pt	/Al$_2$O$_3$	460	1.15	0.12	0.87	—	12
Pt	/Al$_2$O$_3$	700	1.02	0	1.00	0.41	98
Pt	/Al$_2$O$_3$	643	0.98	0.28	0.82	—	17
Pt	(100)^a	573	0.71	0.83	0.54	0.98	91
Pt	(111)^a	573	1.23	0.25	0.76	0.90	91

^{a} Results for stepped and kinked surfaces also given.

increased with temperature.[97,100] Some values of selectivity parameters for other metals are given in Table 13.8.

Very many results for n-butane product selectivities have been published. Once again, the tabulation (Table 13.9) contains only a small selection, and the cited papers give many other product distributions, obtained by changing operating conditions and catalyst pre-treatment. On the whole, temperature coefficients tend to be quite small and dependence on reactant ratio is typically also not very significant. The range of temperature used is also small (e.g. for Rh, 440–476 K), and an excess of hydrogen (generally 10/1 or 20/1) is employed. The variations that appear for a given metal are therefore mainly due either to conversion where high values of S_1 and low values of F' suggest that product distributions are not close to initial, or to real differences in catalyst structure, for which certain significant trends are apparent.

The behaviour of single crystals and other forms of low dispersion (film, black) is quite different from that of the corresponding supported metal. For rhodium and iridium, very high values of S_2 (1.35–1.7) have been shown, irrespective of the support used, while for the macroscopic forms they were less than unity. Very high values of S_2 were for example recorded for Ir(110)(1 × 2) between 420 and 475 K at high hydrogen/n-butane ratios (100–400),[8] but not for Ir(111)[8] or iridium black;[38] small particles inside NaY zeolite gave $S_2 \approx 1.65$, but larger particles on the surface showed much lower values.[104] Central C—C bond splitting increased progressively as particle size diminished,[93,105] although it is not clear whether this decreased the proportion of the critical active centre or changed the strength of hydrogen chemisorption or altered electronic structure in a way that somehow influenced the mode of reaction. This feature, whatever it is, seemed to affect F more than T_3, which was usually greater than 0.8 and was frequently close to unity: T_3 is the analogue T_2 in propane hydrogenolysis, and it again reflects the preference for intermediate species to vacate the surface rather than suffering further bond fission. Values of F were usually low on platinum (<0.3), and also on ruthenium except when the dispersion was very high (H/Ru > 1),[106] when it

TABLE 13.9. Hydrogenolysis of *n*-Butane: Product Selectivities

Metal	Form	T/K	P_H/P_C	S_1	S_2	S_3	F	T_3	References
Ru	black	443	32	1.48	0.66	0.40	0.06	0.64	38
	/SiO$_2$	431	10	1.01	0.75	0.49	0.24	0.64	171
	/SiO$_2$	363		0.73	0.73	0.60	0.33	0.86	97
	/TiO$_2$	403	10	0.88	0.75	0.54	0.29	0.77	99
	/Al$_2$O$_3$	433	10	0.55	1.17	0.37	0.59	0.91	106
Rh	black	440	10	1.25	0.70	0.43	0.15	0.55	38
	(110)	455	20	0.89	0.71	0.56	0.27	0.77	94,217
	(111)	455	20	0.65	0.98	0.46	0.44	0.82	94,217
	/SiO$_2$	433	10	0.24	1.58	0.20	0.78	0.91	47
	/SiO$_2$	476	20	0.29	1.55	0.20	0.75	0.80	94,217
	/Al$_2$O$_3$	433	10	0.37	1.35	0.31	0.66	0.91	47
	/Al$_2$O$_3$	473	20	0.42	1.47	0.21	0.69	0.68	94,217
	/TiO$_2$	433	10	0.52	1.11	0.42	0.53	0.89	47
Pd	/SiO$_2$	627	10	0.94	0.09	0.96	0.05	1.01	73
	Film	530	10	0.91	0	1.03	0.03	~1	166
Ir	/SiO$_2$	453	20	0.26	1.48	0.26	0.74	1	9
	/SiO$_2$	483	10	0.62	0.95	0.49	0.44	0.87	171
	/Al$_2$O$_3$	~460	20	0.15	1.71	0.15	0.86	>1	9
	/TiO$_2$	473	20	0.18	1.38	0.34	0.72	0.89	105
	/TiO$_2$	510	10	0.35	1.48	0.23	0.71	0.79	102
Pt	(100)	573	10	0.47	1.18	0.39	0.58	0.93	91
	(111)	573	10	0.84	0.63	0.63	0.26	0.70	91
	/SiO$_2$	573	10	0.66	0.77	0.60	0.37	0.96	14
	/SiO$_2$	573	5	0.85	0.86	0.47	0.34	0.71	125
	/Al$_2$O$_3$	603	10	0.83	0.33	0.82	0.15	0.96	65,139
	/Al$_2$O$_3$	693	4	0.73	0.56	0.72	0.28	1	17
	/TiO$_2$	600	20	0.90	0.20	0.90	0.10	1	19
	/TiO$_2$	573	5	0.82	0.32	0.86	0.18	~1	125
	/C	603	10	0.72	0.56	0.72	0.28	1	95
	/C	583	10	1.55	0.26	0.64	—	—	96
	/NaY	625	10.7	0.80	0.51	0.72	0.23	0.93	178

Many of the cited publications give more results than those abstracted for this table, the purpose of which is to illustrate effects of metal, form, support, etc.

showed the high S_2 characteristic of rhodium and iridium. The differences between macroscopic and microscopic are however much less marked than with the other metals.

Very little has been published on hydrogenolysis of the lower alkanes using either nickel[20,107,108] or palladium[73,109] catalysts, perhaps because both are liable to suffer rapid deactivation by carbon deposits, or to show unstable behaviour. With *n*-butane, both gave predominantly terminal C—C bond fission at low conversion, but there are no detailed kinetic studies to report.

Table 13.10 tries to summarise a mass of experimental results for the dependence of *n*-butane product selectivities on temperature and hydrogen pressure,

TABLE 13.10. Hydrogenolysis of *n*-Butane: Variation of Product Selectivities and Anderson–Kempling Parameters with *Increasing* Temperature and *Decreasing* Hydrogen Pressure

Metal	Form	T/K	P_H/Torr	S_1	S_2	S_3	S_i	F	T_3	References
Rh	black	*	691	(+)	+	(−)				38
	(110)	*	200	+	+	−				93,94
		500	*	+	−	?				
	/SiO$_2$	*	623	(+)	(−)	(+)		−	(−)	47
		430	*	~	~	~		~	~	
	/Al$_2$O$_3$	*	623	(−)	(−)	~		−	−	47
		430	*	+	−	+		−	~	
Pd	/Al$_2$O$_3$	*		~	−	~	+	~	~	73
		608	*	~	+	−	~	(+)	~	
Ir	black	*	691	+	−	−				38
	(111)	*	100	+	−	−				8
		500	*	+	−	−				
	(110)-(1 × 2)	*	100	+	−	−				8
		475	*	+	max	max				
	/TiO$_2$	*	543	min	max	min				102
Pt	(100)	*	50	(+)	~	−				7
		575	*	−	~	−	+			
	(111)	*	50	(+)	min	~	~			7
		583	*	−	−	max	~			
	black	735	*	+	~	−	−			32
	/SiO$_2$	539	*	+	max	−				16
	/SiO$_2$	*	338	(+)	(−)	~	~			17
	/SiO$_2^a$	*		~	+	−	+			41
	/SiO$_2^a$	*	543	~	+	−	+	+	(−)	14
		533	*	−	+	−	+			
	/Al$_2$O$_3^b$	*	543	−	+	−		−	(−)	3,4,65
		547	*	−	+	−	+			

Columns 3 and 4: *signifies what was varied, the corresponding number gives the level. [a]EUROPT-1; [b]EUROPT-3.

with a view to seeing the extent to which similar effects have been shown, (i) by different forms of metal, and (ii) by altering the levels of these variables in the sense of *decreasing* hydrogen coverage. It is hard to discern any overall consensus: contributing factors are (i) that changes are often quite small within the range investigated, and (ii) that it is difficult to maintain low conversion while exploring a wide variation in the level of the variable. Two studies[7,91] of platinum single crystals have given somewhat different results, but those for the macroscopic forms of iridium were reasonably consistent,[8,38] and those for the reference platinum catalysts agreed well. Discrepancies between macroscopic and microscopic forms are again apparent, and the sense of the trends with iridium and platinum are quite different, suggesting that their mechanisms are not the same.[9] A more quantitative analysis of the consequences of hydrogen pressure variation is attempted in the next section. One gets the general impression that the effect of increasing temperature is somewhat greater than that of decreasing hydrogen pressure, which is perhaps not surprising where its adsorption is strong.

13.5. MECHANISMS BASED ON PRODUCT SELECTIVITIES

The term 'mechanisms' is used in the plural, because it is unsafe to assume that a single 'mechanism', as defined by the type of essential intermediates, their prevalence and ways of interaction (Section 5.3), will suffice for all metals and all conditions: indeed there is clear evidence that such is not the case.[9] This section is mainly concerned with what happens on rhodium and platinum, for which most information is available. Our task is to understand the significance of the markedly negative orders in hydrogen, and how they correlate with the variations of product selectivities with hydrogen pressure. Effects due to changing temperature are, as we have seen, most likely a consequence of change in hydrogen coverage. Orders of reaction become less negative with rise in temperature because hydrogen atom coverage becomes less and more vacant sites become available for alkane chemisorption, and in constant ranges of hydrogen pressure variation they become more positive with increasing alkane chain length as the alkane's chemisorption strengthens.[65] Changes in product selectivities will therefore vary because of the H:C ratio of the adsorbed intermediates and the coverage by hydrogen atoms, both of which decrease as hydrogen pressure falls, and the fraction of vacant cites, which naturally increases. Now the Kempling-Anderson scheme as set out in Scheme 13.1 does not specify the need for hydrogen atom acceptance or donation in each step. It has always been accepted [24,25] (and is supported by kinetic analysis [4,65] based on the ES5B equation, Section 13.23) that an alkane has to lose *at least* two, and very possibly three or even four hydrogen atoms in order to be activated for hydrogenolysis. Scheme 13.2 illustrates one possible formulation of the reaction of propane, defining needs for hydrogen atoms; it assumes adsorption by loss of only *two* hydrogen atoms, and the requirement for *one* to assist C—C bond breaking. It follows that *more* (i.e. 2) *are needed to secure formation of gaseous ethane than to break a C—C bond and form two adsorbed C_1 fragments.* The difference is exacerbated if more than two atoms are lost in the chemisorption, and if none are needed to break C—C bonds, as some claim.[46] Ethane selectivity S_2 therefore falls with decrease in hydrogen pressure and with increasing temperature. The *greater* the rates of these changes, the *weaker* is the hydrogen chemisorption. The high S_2 values typically found mean however that the extent of these changes is often small, and they can only rarely be treated quantitatively to deduce the difference in the numbers of hydrogen atoms involved in the two routes.[76,78]

The argument can be extended to the more complex case of *n*-butane, for which there have been detailed studies of the effects of varying hydrogen pressure for platinum,[4,65] palladium[73] and rhodium[47] on different supports (for ruthenium,

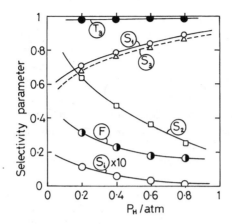

Figure 13.18. Hydrogenolysis of n-butane on 0.3% Pt/Al$_2$O$_3$ (EUROPT-3): selectivity parameters at 547 K as a function of hydrogen pressure.[65]

see Section 13.6). Inspection of the results shows that *in the higher ranges of hydrogen pressure* there were consistent changes. As the pressure was decreased, (i) the isomerisation selectivity S_i *always* increased (i.e. with Pt and Pd), (ii) S_2 *always* rose, as did F, (iii) S_1 and S_3 decreased in parallel or remained constant, the difference between them often being small, and (iv) T_3 was high and either constant, or decreased slightly (Figure 13.18). The variation in F arose because, at the point where the routes to ethane and to propane branched, the former went *via* a species that had lost *one more hydrogen atom* than that proceeding to propane, so that the ethane route was favoured by decreasing hydrogen atom concentration and increasing numbers of vacant sites to accommodate new C—M bonds. These trends were observed even when the hydrogen was very strongly chemisorbed, as it was on rhodium catalysts at about 430 K, as indicated by the occurrence of rate maxima at hydrogen pressures below 0.02 atm.[47] Only in the case of Rh/SiO$_2$ after HTR1 when F and S_2 were very large (respectively ~0.8 and ~1.5) were they independent of hydrogen pressure over most of the range. Also since on Pt/KL zeolite and on EUROPT-1[4] S_i increased with falling hydrogen pressure more quickly than did S_2, the former proceeded by the loss of *two* more hydrogen atoms than did the latter. The amplified Kempling-Anderson Scheme 13.3 shows how the relevant species might be composed.

The striking difference between platinum on the one hand and rhodium and iridium on the other lies in the former's ability to isomerise alkanes even on large particles,[95,96,110] whereas with the latter (and also with ruthenium[99]) it has only been observed at high temperatures with large particles that have been substantially deactivated. It might be thought that on these metals the high S_2 values (Table 13.9) were a consequence of a frustration of the isomerisation route, because rhodium catalysts capable of isomerising showed much lower values of S_2.[90] It must be remembered that iridium and rhodium are much more active than platinum and have therefore been studied at much lower temperatures. The 'frustration' of isomerisation may simply arise from the use of moderate temperatures, at which, because of the greater activation energy needed to form the vital intermediate, hydrogenolysis is preferred. The reason why platinum (and palladium) are always less active than metals to their left is a more difficult question. It may be that they find greater difficulty in forming *multiple* C—M bonds because their d-electron orbitals are more nearly filled.

Particle-size effects may also be addressed in terms of the model outlined in Schemes 13.2–5. In the case of the reference platinum catalysts, a decrease in size was accompanied by a marked decrease in S_2 and in S_1, and in F; this implies stronger hydrogen chemisorption on the Pt/Al$_2$O$_3$ reforming catalyst. On rhodium and iridium, however, the effect of size on S_2 and F (Table 13.9) is more probably due to a change in the surface's ability to accommodate multiple C=M bonds.

The mechanism by which skeletal isomerisation occurs has been much debated (see Further Reading section). It assumes greater importance with higher alkanes, so the following short discussion will be amplified in the next chapter. Favoured mechanisms have assumed (i) that it begins with a species that has lost either one, two or three hydrogen atoms, but (ii) that it only needs one site (or forms only one C—M bond). This last point was supported by the observations[91] that the reaction proceeded on organometallic complexes having only one metal atom, and that isomerisation selectivity increased as ensemble size decreased when diluting an active metal with an inactive one (see Section 13.7). Arguments concerning the desirable H/C ratio based on kinetic observations were not used. The essential step of the isomerisation then took one of two forms: either (1) a C—C bond was broken, and rotation of the alkene portion followed by re-assembly of the C$_4$ unit took place; or (2) a cyclopropanoid structure was formed, where the more probable breaking of the bond between the least substituted

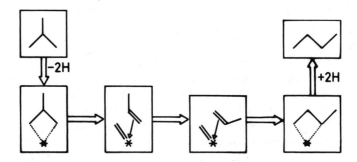

Scheme 13.4. Isomerisation of *iso*butene via a metallocyclic intermediate.

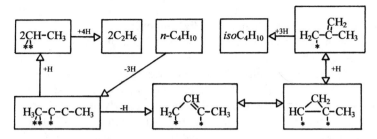

Scheme 13.5. Reactions of n-butane showing suggested structures for intermediate species.

carbon atoms produced the desired effect (Scheme 13.5). Re-formation of a C—C bond under hydrogenolysis conditions does however seem unlikely. The old simple-minded model[111] of the projection of molecular orbitals from single-crystal surfaces has been deployed[91] to explain the various isomerisation efficiencies, and indicated that the best surfaces had sites at which a metallocyclobutane structure could be formed (Scheme 13.4). Scheme 13.5 shows an example of the kinds of structure that could be responsible and that are consistent with kinetic requirements.

13.6. HYDROGENOLYSIS OF ALKANES ON RUTHENIUM CATALYSTS

Ruthenium merits a separate section because of the extreme sensitivity of Ru/TiO_2 and Ru/Al_2O_3 to conditions of pre-treatment, a sensitivity that incidentally is not shared by Ru/SiO_2.[101,103] It shares with osmium (which has been little studied[102]) the honour of being the most active metal for hydrogenolysis, and can be studied well below 373 K. The somewhat variable results shown for ruthenium in Table 13.2 and some of the later tables may find the basics for their explanation in the following paragraphs. The effects to be reported for Ru/TiO_2 involve SMSI but are also additional to it; results relating purely to SMSI with other metals are considered in Section 13.7.

The reduction by hydrogen of the precursor $RuCl_3/TiO_2$ at 758 K (HTR1) gave ruthenium particles of moderate size ($\sim$3 nm), but they were partially poisoned by chloride ion that adhered strongly to the support,[112] and so transition to the SMSI state was incomplete, but the activity for alkane hydrogenolysis was low.[99–101,103,113,114] Hydrogen chemisorption isotherms failed to show flat plateaux, and were useless for size estimation. Higher temperature (893 K) was needed to remove all the chloride ion, but a mild oxidation (623 K) (which did not give any RuO_4) also removed it, and led after mild reduction (LTR) to higher dispersions (H/Ru $\approx$ 0.5–1) and *very* much higher activities (typically by a factor of 10–70), with little change in activation energy. The effect was shown by ethane, propane and both butanes, by several types of titania, and was usually more marked at low metal loadings (0.1 and 0.5%). A second high-temperature reduction (HTR2) caused the H/Ru ratio to decrease, probably because of entry into the SMSI state (Section 13.7); activities were then very low. The O/LTR treatment (mild oxidation followed by mid reduction) clearly increased the number of active centres more than in proportion to the increase in area, since the Arrhenius parameters gave a compensation line that was well above that for the less active catalysts. Initial reduction at 893 K produced the SMSI state, and the rate was then increased by O/LTR by more than 200 times.[101]

Very significant changes in product distributions also occurred (Table 13.11). The effect of the O/LTR treatment was to decrease S_2 and F, without greatly affecting T_3; the rates at which selectivities changed with temperature also altered, becoming somewhat greater after the O/LTR.[99] Following HTR2, the selectivity parameters (except T_3) moved towards their original values, without quite achieving them. It was thought likely that the oxidation transformed ruthenium metal into the oxide, and that because of the similarity of the titania and ruthenium oxide structures (RuO_2 has the rutile structure) Ru^{4+} ions could easily migrate over the surface to form a well-dispersed layer, so that gentle reduction would create small metal particles.

Analogous but not identical effects have been seen with Ru/Al_2O_3 (see Further Reading section). Use of a chloride-free precursor (e.g. $Ru(NO)(NO_3)_3$[103] or the acetylacetonate[115]) removed any possible interference it might have caused; with only 1% metal, the first HTR gave extremely small particles (H/Ru = 0.88;[76] mean size (TEM) = 1.2 nm,[115] by EXAFS ∼12 atoms per particle[116,117]), but the TOFs obtained using 0.71 atm hydrogen were low. The reason became clear when the variation of rate with hydrogen pressure was determined.[74] It increased markedly as hydrogen pressure was lowered, giving a sharp maximum at 0.03 atm. (Figure 13.19); product selectivities and derived parameters were virtually constant (Figure 13.20) and changed only slowly with temperature; this behaviour implied very strong hydrogen chemisorption. On applying the O/LTR treatment, TOFs at 0.71 atm hydrogen increased markedly, but inspection of the effect of varying hydrogen pressure showed that this was because hydrogen inhibition had been lowered, and the rate maximum occurred at higher pressures[74] (Figure 13.19). Product selectivities then changed (Table 13.11) and became hydrogen-pressure-sensitive (Figure 13.20) and more temperature-dependent. This change in character was analysed in terms of the ES5B equation (Table 13.4), and this showed[76] the way in which the constants changed varied with the alkane (Table 13.6): the decrease in b_H became larger, and the decrease in K_A became smaller, as chain-length increased, and the number of hydrogen atoms lost in forming the reactive species fell. It was then established that the O/LTR treatment caused a *decrease* in dispersion,[75,103] and

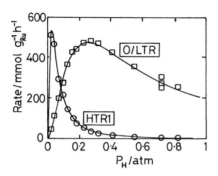

Figure 13.19. Hydrogenolysis of n-butane on 1% Ru/Al$_2$O$_3$: variation of rate at 398 K with hydrogen pressure after HTR1 and O/LTR pre-treatments.[75]

formation of aggregates of larger particles.[116] A second high-temperature reduction (HTR2) partly neutralised the effect of the O/LTR; there was no further important change in particle size and activity was much lower, but the original high values of K_A and b_H were not restored (see Table 13.12). These somewhat complex changes, involving the chemisorption of both reactants, were therefore effected by *both* a change in particle size *and* some changes in surface character. With a higher metal loading (4%), HTR1 produced larger particles (H/Ru = 0.25), and behaviour that more resembled the 1% Ru/Al$_2$O$_3$ after O/LTR, but even here O/LTR led to higher

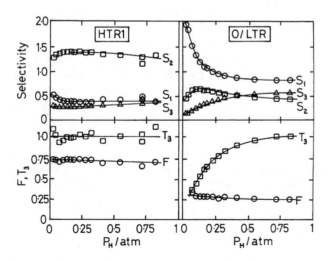

Figure 13.20. Hydrogenolysis of n-butane on 1% Ru/Al$_2$O$_3$: selectivity parameters at 398 K as a function of hydrogen pressure after HTR1 and O/LTR.[75]

TABLE 13.11. Hydrogenolysis of n-Butane on 1% Ru/TiO$_2$ and 1% Ru/Al$_2$O$_3$: Selectivity Parameters after Various Pretreatments

Support	Pretreatment	T/K	S_2	S_3	F	T_3	References
/TiO$_2$	HTR1	433	1.07	0.39	0.46	0.73	99
	O/LTR	403	0.59	0.64	0.22	0.83	99
	HTR2	433	0.85	0.43	0.29	0.60	99
/Al$_2$O$_3$	HTR1	433	1.39	0.28	0.72	1.00	77
	O/LTR	433	0.71	0.43	0.21	0.54	77
	HTR2	433	0.81	0.39	0.28	0.55	77

(1) Where in this and following tables values of F and T_3 are not those derived by assuming $T_2 = 1$ it is because the value used has been obtained from the reaction of propane under the same conditions. Where $T_3 > 1$ this must be because this procedure is invalid.

(2) The cited papers contain much further information.

activity: similar effects were also seen with ruthenium powder,[116] emphasising that the dominant feature was the type of surface given by O/LTR.

Mobility of oxidised ruthenium species on alumina therefore led to aggregates of RuO$_2$ and larger metal particles after reduction,[116] unlike the case of titania. Nevertheless in both instances there were increases *in TOF* caused by O/LTR, so similar surface reorganisations must have been responsible. Its nature however remains unclear.

Analysis of the hydrogen-pressure-dependence of product selectivities from n-butane by the Kempling-Anderson methodology on catalysts subjected to O/LTR revealed [75,76] a significant difference from the reactions over platinum (compare Figures 13.18 and 20); in the case of platinum they were caused mainly by the splitting parameter F, but with ruthenium F was almost constant, and T_3 was the principal variable (Figure 13.20). This means that the degree of dehydrogenation of the species leading to propane and to ethane was the same, the processes perhaps requiring different types of site: selectivity changes were therefore caused only by the different numbers of hydrogen atoms wanted for the desorption and C—C breaking steps. The manner of the T_3 variation led to estimates for the difference in these numbers, which was about two for n-butane,[76] so if the requirement for *one* hydrogen atom to break a C—C bond is retained (and this is a central

TABLE 13.12. Hydrogenolysis of n-Butane on 1% Ru/Al$_2$O$_3$: Constants of the ES5B Equation and Quantities Derived from Their Temperature Dependence after Various Pretreatments[76]

Pretreatment	T/K	k	K_A/atm^{a-1}	b_H/atm	a	E_t^1	ΔH_A^1
HTR1	418	111	37	27	1.4	55	79
OLTR	413	73	33	3.2	1.3	66	66
HTR2	414	7.8	3.3	1.8	1.3	39	52

[1] In kJ mol^{-1}.

assumption of the ES5B equation) the composition of the C_3 species should have been C_3H_5. Similar methodology gave the composition of the C_2 species in the reaction of propane as C_2H_2.[78] The method does not depend on the validity of the ES5B equation, but only on the Kempling-Anderson method of analysing selectivities. Fuller discussions of these results have been presented.[76] The dependence of the rates of reaction of the first three linear alkanes on hydrogen pressure have been determined at various temperatures for 1% Ru/Al_2O_3 subjected to each of the three pretreatments.[76] The form was generally similarly to that shown by Pt/Al_2O_3; apparent activation energies increased with hydrogen pressure,[76,118] and values of b_H were small and not temperature-dependent in a consistent way. The effect of particle-size on strength of hydrogen chemisorption cannot therefore be substantiated in this way, but isosteric measurements of the 1 and 4% Ru/Al_2O_3 catalysts after HTR1 do confirm the greater adsorption strength on the former.[76] A selection of values of E_t and of ΔH_A are shown in Table 13.12; they are of similar magnitude to those found for platinum catalysts at much higher temperatures. Since the concentrations of hydrogen atoms and the reactive hydrocarbon species must be equal at the point of maximum rate, whatever the temperature and their absolute values, the apparent activation energy should approximate to E_t; for 1% Ru/Al_2O_3 after HRT1 it was 62 kJ mol^{-1}.

13.7. EFFECTS OF ADDITIVES AND THE STRONG METAL-SUPPORT INTERACTION ON ALKANE HYDROGENOLYSIS

The so-called 'Strong Metal-Support Interaction' (SMSI) was first revealed by a loss of capacity for hydrogen chemisorption by the metals of Groups 8 to 10 supported on oxides of the metals of Group 5, and on titania and manganous oxide, after first heating them in hydrogen to temperatures above about 573 K (see Section 3.3.5): oxides of sp metals and those of Group 3 and 4 did not respond in this way. This early work at the Exxon laboratories led to a torrent of further publications from elsewhere, and the following main conclusions were established. (1) While all the metals of Groups 8 to 10 were affected,[119,120] there have been clear indications that all did not suffer equally,[121] and that the extent of the effect might depend on the exact conditions of catalyst preparation (e.g. in the case of Ru[99]). (2) This loss of chemisorption capacity was not primarily due to decrease in metal-particle size, but rather to a partial or almost complete coverage or encapsulation of the metal by species emanating from the partially-reduced support (e.g. TiO in the case of TiO_2). (3) The importance of the effect increased with the reduction temperature and with decrease in metal loading, i.e. small particles succumbed more easily than large ones. (4) In the case of Pt/TiO_2, the heat of hydrogen chemisorption was greatly lowered; there was a smaller effect with Pd/TiO_2. This very brief summary (and that in Section 3.3.5) does scant justice to the enormous volume of work performed to identify the cause of the effect; indeed the attention it commanded was surprising in view of its generally negative implications (except for the hydrogenation of carbon monoxide and carbonyl compounds, rates of which were enhanced by it).

The quantitative study of the SMSI was however made difficult because many of the oxides of interest did not readily lend themselves to become supports for metals. They had different surface areas and surface chemistry, so similar metal dispersions would not be obtainable, and with some their opacity to electrons would make TEM inapplicable. In a practical sense, their mechanical properties (e.g. hardness) did not encourage their large-scale use. An attractive alternative was therefore to mount the modifying oxide onto a strong conventional support (e.g. SiO_2) and then to deposit the metal on top;[122] or the modifier could be added to the silica-supported metal;[103,123,124] or both could be introduced at the same time; or the modifier could be incorporated into the support, as with TiO_2-SiO_2.[125] In this way it should be possible always to have the same metal-particle size, and to alter the modifier to metal ratio without having to change the pre-treatment temperature. These techniques were widely applied,[126–129] and in general the same effects were observed as with the equivalent metal *on* modifier.

The effect of the SMSI on catalytic activity also aroused much interest. In the context of alkane hydrogenolysis, rates fell in parallel to ability for hydrogen chemisorption, and much of these reactions' reputation for structure-sensitivity depends upon their sensitive response to the incursion of modifying species. This was qualitatively understandable in terms of a requirement for the reactive species to form several C—M bonds (Scheme 13.5), the main consequence of the SMSI being to eliminate large ensembles of metal atoms. The conversion of this qualitative concept into reliable quantitative estimates of size of active centre has not in general been possible, and it is unfortunate that most of the studies, which involved mainly ethane[130] and n-butane (see Table 13.13 for a selection of examples), reported only rates under one set of experimental conditions. A comprehensive study[120] of ethane hydrogenolysis on all the metals of Groups 8 to 10 at 478 K showed that specific rates for titania-supported metals were lower than those for silica-supported metals by factors that were generally about 10^2 to 10^3, although much larger for iron and very much smaller for ruthenium. This, and much other early work, informed more on the conditions for and extent of the modification, and revealed little on the changes wrought on the reaction mechanisms. For Rh/TiO_2, activation energies for the reactions of the four C_2 to C_4 alkanes increased progressively with reduction temperature, and exhibited compensation, but the temperature ranges had necessarily to be raised as activity was lost. In another study using Rh/TiO_2, rates for both ethane and n-butane at 623 K decreased with increasing dispersion following high-temperature reduction, the opposite of what was found at 523 K after low-temperature;[131] there was no change in the activation energy for the ethane reaction (197 kJ mol^{-1}).

TABLE 13.13. Selected References to the Modification of Groups 8 to 10 Metals for Alkane Hydrogenolysis by Oxides Causing the SMSI

Reactant	Metal	Modifier	Support[a]	References
C_2H_6	Ni	Ta_2O_5	—	126
	Ru	Nb_2O_5	SiO_2	218
	Rh	MgO	—	132
	Rh	TiO_2	—	153,217
	Rh	Nb_2O_5	(SiO_2)	127,128,219,312
	Rh	V_2O_5	—	219
	Rh	MnO_2	(SiO_2)	129,219
	Rh	La_2O_3	—	221
	Pt	TiO_2	(SiO_2)	125
	Pt	SiO_2	—	41,133
n-C_4H_{10}	Ru	many	SiO_2	123,124
	Ru	V_2O_5	SiO_2, TiO_2, Al_2O_3	122
	Ru	K	Al_2O_3	141
	Rh,Pt	V_2O_3	(SiO_2)	135,136
	Rh,Ir	TiO_2	—	90,138
	Pt	TiO_2, Al_2O_3	SiO_2	139
	Pt	cations	Clay	140
	Pt	TiO_2	Al_2O_3	19
	Pt	CeO_2	(Al_2O_3)	137
	Pt	SiO_2	—	133

[a] When the support is put in brackets, it was optionally used to support the modifier.

Apparent SMSI effects have also been reported with non-Transition Metal oxide supports.[119] Those obtained with magnesia were almost certainly due to evolution of traces of hydrogen sulfide, or of iron ions, from the bulk at the high temperatures used,[132] but real effects have been seen with silica when very high pre-treatment temperatures were employed:[41,133,134] the rate of the ethane reaction decreased more quickly than that of hydrogen chemisorption, pointing clearly to the need for multiatomic sites in hydrogenolysis.

Studies of n-butane have been somewhat more informative (Table 13.13). It has been thoroughly established that the effect of the SMSI (whatever it is) can be partially or even completely restored by oxidation, and a sequence of reduction (HTR1)...oxidation and low-temperature reduction (O/LTR)...reduction (HTR2) applied to supported and modified ruthenium catalysts has produced results that extend our understanding both of the SMSI and of those described in the last section.[103] Application of the oxides of the first row Transition Metals (Ca through Mn) to Ru/SiO_2 led[124] to the results shown in Figure 13.21. HTR1 and HTR2 produced a particularly severe depression of the n-butane hydrogenolysis rate with vanadia, with smaller but still very significant effects with chromia and manganic oxide, not fully restored by oxidation, due perhaps to the formation of stable mixed oxides. However, in each case the rate was enhanced by O/LTR and

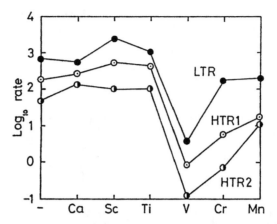

Figure 13.21. Hydrogenolysis of n-butane on 1% Ru/SiO$_2$ modified by oxides of Groups 2 to 7: rates (mmol g$_{Ru}^{-1}$ h^{-1}) at 433 K.[123]

decreased by HTR2 (the effects with Ru-Ti/SiO$_2$ were smaller than those reported before because the precursor was not calcined before HTR1). Arrhenius parameters for reactions after O/LTR fell on a separate higher line than those for the HTRs, as before. Various generally minor changes to product selectivities were seen, so that evidence for electronic or geometric alteration to the active centre was absent (except perhaps with Ru-V$_2$O$_5$/SiO$_2$ after HTR1), and it seemed possible that residual activity was due to a few particles that had escaped unscathed from the assault of the SMSI.

Pre-coating supports (SiO$_2$, Al$_2$O$_3$ and TiO$_2$) with various amounts of V$_2$O$_5$, followed by ruthenium from RuC1$_3$, gave similar but somewhat clearer results.[122] In the *n*-butane reaction, high values of F and T_3 were found after the HTRs, but were lowered by O/LTRs and partially restored by HTR2s. It appears that this method of composing the catalysts gave a closer interaction between metal and modifier than that obtained when the reverse sequence was used. Analogous but less detailed results have been found with Rh/V$_2$O$_3$ and Rh-V$_2$O$_3$/SiO$_2$, and corresponding platinum catalysts.[135,136] HTR of Pt/CeO$_2$ caused an increase in activation energy, a decrease in rate and a significant rise in methane selectivity.[137] HTR of Rh/TiO$_2$ and of Ir/TiO$_2$ virtually eliminated activity for hydrogenolysis, but allowed isomerisation and dehydrogenation to be seen at 688 K.[90,138] Deposition of titania onto Pt/SiO$_2$ (EUROPT-1) decreased rates of both hydrogenolysis and isomerisation of *n*-butane progressively,[139] the latter more than the former, and both more quickly than H/Pt, so that activity was almost killed while H/Pt had fallen only to 0.75. The formation of 'carbon' and of titania modifiers (e.g. TiO) on the metal had complementary effects, both suppressing isomerisation selectivity *and* to a smaller extent increasing S_2 and F. These effects are not explicable in terms of hydrogen atom availability (Schemes 13.3 and 5), but require the idea of a site of different structure at which the intermediates for isomerisation can be accommodated. It therefore seems that some aspects of particle-size and ensemble-size effects can be equally well or perhaps better explained by specific site requirements than by adsorbed hydrogen concentration (Section 13.5). However, reaction of *n*-butane on Pt/TiO$_2$ affords a lower S_2 than does Pt/SiO$_2$, even after reduction at only 573 K, and, with Pt/TiO$_2$-SiO$_2$ containing a large amount of titania, reduction at 773 K almost eliminates central C—C bond-breaking, due to restriction of site size.[125] There will be further news of the effects of SMSI and modifiers in Chapter 14, when reactions of larger alkanes are considered, and when in consequence a wider range of mechanistic options becomes available.

It was surprising to find that Os/TiO$_2$ did not behave in the same way as Ru/SiO$_2$:[102] O/LTR caused rates to *decrease*, and S_2 to fall quite considerably in the reactions of both propane and *n*-butane. With Ir/TiO$_2$,[102] on the other hand, the O/LTR treatment did cause the rate of the *n*-butane reaction to increase, although the effect with different metal contents was not consistent, but there was no significant change in selectivity parameters throughout the various pre-treatments, suggesting that iridium once reduced could not be oxidised as was ruthenium. To obtain the effect, it was not necessary to use titania as the support; a small amount (~1 wt.%) on silica or alumina produced equivalent changes as pre-treatments were altered.[103]

Two other forms of modification deserve to be noted. A series of platinum-containing smectite-like clays have been exchanged with the divalent cations Ni^{2+}, Co^{2+} and Mg^{2+}, and their behaviour in *n*-butane hydrogenolysis compared with that of Pt/SiO$_2$;[140] the results are summarised in Table 13.14. The first two were more active than the Pt/SiO$_2$, and the activation energies were the exact inverse of the activities. Excellent compensation was shown, but the slope of the line was greater than that of the mean slope in Figure 13.3. As reaction temperature necessarily rose, so did both S_i and S_2, suggesting that decreasing hydrogen coverage was the main factor; this would also cause activation energies to fall (see Figure 13.16). It seems likely that the platinum catalysed the reduction of the Ni^{2+} and Co^{2+} ions (but not Mg^{2+} ions), which then contributed to the activity.

TABLE 13.14. Hydrogenolysis of *n*-Butane on Pt-containing Smectic-like Clays (SM) and on Pt/SiO$_2$[140]

Support	E/kJ mol^{-1}	ln A	T_{mean}/K	S_2	S_3	S_I
SM-Ni^{2+}	178	74.3	473	0.06	0.97	0.01
SM-Co^{2+}	146	63.3	548	0.42	0.79	0.04
SM-Mg^{2+}	86	48.0	617	0.45	0.78	0.10
/SiO$_2$	63	43.6	623	0.54	0.93	0.58

The addition of potassium to Ru/Al$_2$O$_3$ has produced quite dramatic effects on propane hydrogenolysis.[141] The changes produced by O/LTR and HTR2 were abolished, and the almost constant behaviour was characterised by larger values of K_A and b_H, and by smaller values of E_t and $-\Delta H_a$, although S_2 values changed marked with hydrogen pressure. A convincing explanation of these effects is not yet available, but resistance to the effects of oxidation may be the result of the formation of KRuO$_2$. The effect of potassium on ethane hydrogenolysis catalysed by Ru/SiO$_2$ was to decrease activation energy and increase hydrogen order.[142] Inclusion of potassium into palladium[143] and platinum on LTL zeolite, and onto Pt/SiO$_2$,[144] inhibited hydrogenolysis of propane, and it also interfered with the process of hydrogen spillover.

When pyridine was added to n-butane during its hydrogenolysis over Ru/SiO$_2$, the rate was suppressed, but (as with sulfur poisoning[123]) isomerisation then became visible.[145] Pyridine also scavenged hydrocarbon radicals, which were detected as their pyridine adducts.

13.8. HYDROGENOLYSIS OF ALKANES ON BIMETALLIC CATALYSTS[24,146-148]

13.8.1. Introduction

Of the modifiers considered in the previous section, most if not all were believed to exercise their influence while remaining in a positive oxidation state. With those about to be considered there is good evidence that in the main they are in the zero oxidation state, forming real bi*metallic* catalysts. It is possible, indeed likely, that in some cases not all of the modifier is metallic;[149] some may remain on the support (e.g. in the Pt-Re system), but even with elements not easily fully reduced it appears that bimetallic particles may be formed (e.g. Pt-Mo,[150,151] Pt-Zr[152]). In principle however the effects produced are expected to be similar,[153] i.e. a lowering of the mean size of the active ensemble, with the possibility in some cases of electronic modification as well. The main additional feature is the chance that a bimetallic site will be of comparable activity to that composed only of 'active' atoms; this chance rises to near-certainty when the two metals are drawn from within Groups 8 to 10.

Its reputation for structure-sensitivity has made alkane hydrogenolysis attractive for investigation using bimetallic catalysts, and there is an extensive literature on the subject. Much of it however relates only to the ability of a pair of metals to form and retain bimetallic particles, and adds little to the understanding of reaction mechanisms. Sometimes for example the calcination of two precursor compounds gives a binary oxide that is easily reduced to a bimetal,[126,154] while at other times reduction of the precursors gives the desired product, but oxidation undoes the good work and the separated oxides cannot on reduction re-form the bimetal.[108] The combination of two metals into a bimetallic particle should be possible where mutual solubility in the bulk is extensive or complete.[109] In such cases the main concern is whether preferential segregation of the component of lower surface energy to the surface occurs, and how it is distributed over sites that differ in coordination number.[45,155,158] Such considerations initiate attempts[159] to extract the size of the active ensemble from the dependence of rate upon composition based on simple statistics. We may however expect to see variations in product selectivity with composition where each component reaction has its specific site requirement.[159,160]

The discovery[161,162] that bulk mutual solubility is not an essential pre-requisite for the formation of a stable *surface* bimetallic phase opened the way for extensive research, particularly on the ruthenium-copper system. The platinum-rhenium system has also proved of interest because of its use in petroleum reforming, but the platinum-iridium system, which is of comparable importance, has not been so widely examined with the smaller alkanes.[163] It is convenient to classify the material to be considered in the following way.

(1) (A) Metals of Groups 10 and 11;
    (B) metals of Groups 9 and 11;
    (C) metals of Groups 8 and 11.
(2) Metals of Groups 8 to 10 and Groups 13 to 14.
(3) Metals of Groups 8 to 10 and Groups 4 to 7.
(4) Metal pairs formed within Groups 8 to 10.

Emphasis will be placed on those studies that illuminate reaction mechanisms.

13.8.2. Metals of Groups 8 to 10 plus Group 11

The early and systematic work by John Sinfelt and his associates on ethane hydrogenolysis using nickel-copper powders showed[24,164] that specific rates at 589 K fell catastrophically as copper content was increased, that at 74% copper being only 10^{-5} that of nickel. Together with the lack of dependence of the rate of cyclohexane dehydrogenation on composition, these results—perhaps the most often cited if not the best understood in bimetallic catalysis—were influential in distinguishing structure-sensitive reactions from the structure-insensitive. Over most of the composition range, excluding pure nickel, activation energies were constant, and orders of reaction changed slightly in the sense expected because of the necessary increase in temperature needed to determine them, because of the activity decrease. Rates therefore depended on the pre-exponential term (Figure 13.22) and hence on the *number* of active centres, the composition of which was uniform,

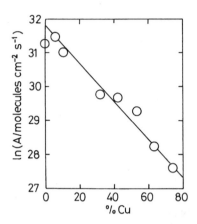

Figure 13.22. Hydrogenolysis of ethane on nickel-copper powders: ln (A/molecules cm^{-2} s^{-1}) versus composition.[164]

there being no obvious 'electronic factor'. The amount of hydrogen chemisorbed under ambient conditions was almost independent of composition, suggesting that Sachtler's 'cherry' model was applicable; this predicted a constant surface composition between about 3 and 80% copper. However phase separation is only thermodynamically possible below about 470 K, so that at reaction temperatures (>600 K) homogeneous bimetallic particles were probably formed. In that case the rate would have depended on the chance of finding ensembles of the required size,[45] and the observed effect would be qualitatively understood. Similar but less orderly results were obtained by John Clarke.[165] There have been no reports of the use of nickel-silver or nickel-gold for reactions of the smaller alkanes.

With NiCu/SiO$_2$ catalysts, rates of ethane hydrogenolysis declined with increasing copper content less fast than with the powders because the particles were smaller, but the kinetic parameters were similar.[159] Experiments with propane and n-butane[159,160] produced only one surprise: with the former, the ethane selectivity S_2 rose appreciably as (0.45 to 0.85) the copper content was increased from zero to 35%. This favouring of reactive desorption, which as we have seen needs more hydrogen atoms for its completion than the further bond-breaking, cannot be due to restricted site size, but must be a consequence of a weakening of the C–M bond strength. This is the only indication of any electronic interaction between the components.

In the reaction of n-butane over (111)-oriented palladium-copper and -silver films,[166] the isomerisation selectivity jumped from about 10% to much higher values (25–75%) for the bimetallics, while the ethane selectivity was lowered, but rose again for copper contents above 50%. The use of various temperatures to accommodate the activity changes complicates interpretation. Deposition of gold onto platinum foil caused the rate of *iso*butane isomerisation at 573 K to decrease less quickly than its hydrogenolysis,[167] while with graphite-supported platinum-gold particles derived from colloidal suspensions[168,169] the rate of *n*-butane dehydrogenation at 633 K increased with gold content, but rates of isomerisation and (*a fortiori*) hydrogenolysis decreased. Temperature-independent product selectivities were only slightly changed by the inclusion of gold. Addition of silver to Rh/TiO$_2$[138] caused the rate of ethane hydrogenolysis at 673 K to decrease by 10^4, but a PtAu/SiO$_2$ catalyst had about the same activity for *iso*butane hydrogenolysis as Pt/SiO$_2$.[170]

The behaviour of the ruthenium-copper system has been widely studied (the osmium-copper system less so, but it was similar in all respects[161]). The effects are due to the extent to which copper atoms cover the surface of the ruthenium particles; thus with ruthenium powder of low dispersion a trace of copper (1.5%) had a dramatic effect,[162] more than halving hydrogen uptake and lowering the specific rate of ethane hydrogenolysis by 10^3. With small silica-supported ruthenium particles,[108,156,157,161] the same effect has only been found at a ruthenium/copper ratio of about unity. Mild oxidation (623 K) destroyed the bimetallic particles, and reduction at 433 K gave rates (for *n*-butane hydrogenolysis) characteristic of ruthenium, but a second reduction at 623 K gave larger metal particles, the activity of which fell faster with increasing copper content than did that of the original smaller ($\sim$1 nm) particles.[171] The choice of support is also important: most studies have used silica (see Further Reading section), and hydrogenolysis rates have usually fallen logarithmically with increasing copper content, but with alumina and magnesia minima have been seen,[172] possibly due to the effect of support impurities. Hydrogenolysis of ethane took place without loss of activity,[156] although with *n*-butane some deactivation occurred.[155]

Although temperature effects have been studied, results have often been presented as compensation plots[171] or graphically[155,156,161] or as plots of TOF versus copper content at several temperatures,[156] so that precise information on Arrhenius parameters is hard to come by. The activation energy for ethane hydrogenolysis decreased from 134 to 105 kJ mol^{-1} with increasing copper content,[162] and then stayed the same, in consequence of which TOFs became larger with bimetallic catalysts than with pure ruthenium: this may be a result of lesser 'carbon' deposition. However, in another work, activation energy *increased* from 138 to 184 kJ mol^{-1} with increasing copper content.[173] The mechanism of the reaction on RuCu/SiO$_2$ has been derived using isotopic transient methodology.[174] A compensation plot containing a number of Arrhenius parameters for *n*-butane hydrogenolysis exhibited two distinct bands, the upper containing those for ruthenium alone and the lower those for bimetallic catalysts.[156] This suggests that copper has some effect that is additional to lowering the number of ruthenium ensembles of the necessary size, such as a mild electronic modification of the active centre. This view is supported by the finding that with Ru$_{82}$Cu$_{18}$/SiO$_2$ (and with Ru$_{60}$Ag$_{40}$/SiO$_2$) the order

TABLE 13.15. Hydrogenolysis of *n*-Butane on Ru/Al$_2$O$_3$ and RuGe/Al$_2$O$_3$ Catalysts Variously Pretreated[79]

%Ru	%Ge	Pretreatment	r^a	S_2	S_3	F	T_3
1	0	HTR1	307	1.39	0.28	0.72	1.00
1	0.7	HTR1	42	1.05	0.37	0.50	0.74
4	0	HTR1	895	0.87	0.85	0.36	0.56
4	0.4	HTR1	70	0.76	0.39	0.26	0.53
1	0	O/LTR	1216	0.71	0.43	0.14	0.50
1	0.7	O/LTR	3454	0.71	0.38	0.15	0.45
1	0	HTR2	313	0.81	0.39	0.20	0.49
1	0.7	HTR2	28	0.83	0.31	0.16	0.37

^{a}In mmol g_{Ru}^{-1} h^{-1}.

in hydrogen at 508 K became distinctly more negative (-2.3 compared to -1.4), which signifies a *stronger* hydrogen chemisorption. Sensitivity to copper content appeared to vary somewhat with the alkane, being particularly large for *iso*butane,[108] the Arrhenius parameters for which lay on a separate compensation line.

In the reaction of *n*-butane,[155,171] product selectivities were not much changed by the introduction of copper, high values of S_2 (1.05–1.20) being found at 453–473 K: values of F were 0.5 to 0.6 and of T_3 0.75 to 1.0. In one case, values of S_3 were reported to exceed those of S_1; this, while impossible, was not commented on. The same occurrence with platinum catalysts under some conditions has been ascribed to a preferential incorporation of C_1 units into the 'carbon' deposit.

Similar effects have been obtained by adding either silver or gold to ruthenium,[156,158] but these systems have not been widely investigated.

13.8.3. Metals of Groups 8 to 10 plus Groups 13 or 14

A thorough investigation has been reported of the effects of Group 14 (Ge, Sn, Pb) additives to Ru/Al$_2$O$_3$ catalysts made in various ways.[79] Prior indications were that germanium might be randomly dispersed over the ruthenium, thus lowering ensemble size, while tin and lead might decorate edge sites selectively and hence simulate large particle behaviour. However, for this concept to succeed, it is necessary to start with ruthenium particles that are sufficiently large for these possibilities to develop. With very highly dispersed Ru/Al$_2$O$_3$ (Ru1; ∼12 atoms per particle, see Section 13.6), germanium had little effect except to lower activity until a Ge/Ru ratio of 0.31 was reached (Table 13.15). EXAFS showed that the germanium was in a positive oxidation state,[116] so the effect resembled an SMSI more than bimetal formation. Perhaps the smallest particles became encapsulated leaving a few larger particles free, as the values of the selectivity parameters moved towards those of Ru3 (H/Ru = 0.25) (Table 13.15). Adding germanium to Ru3 caused further movement in the same direction; putting tin on Ru1 decreased activity markedly, but lead had little effect and neither altered selectivities noticeable. Oxidation (O/LTR) disassembled the components, and in the larger particles thus produced[116] (Section 13.6) germanium was beneficial; a second reduction (HTR2) partially restored the connection. Determination of hydrogen orders and their interpretation by the ES5B equation (Table 13.4) showed that the effects produced by germanium on both catalysts were due to a decrease in b_H and to a greater decrease in K_A.

The beneficial effects of admixing tin with platinum for alkane dehydrogenation were noted in Section 12.23; the selectively-lowered activity for hydrogenolysis and for 'carbon' formation have been examined.[175–177] Using single-crystal surfaces of platinum modified by deposition of tin so as to give p(2×2) Sn/Pt(111) (i.e. Pt_3Sn) and ($\sqrt{3} \times \sqrt{3}$)R30°Sn/Pt(111) (i.e. Pt_2Sn) (see Sections 3.2 and 12.2.2), it was found[176] that rates of n-butane hydrogenolysis at all temperatures were Pt_2Sn < Pt < Pt_3Sn. Product selectivities were changed by tin to give much higher values of S_2 and lower values of both S_1 and S_3, suggesting that the preferred route became that involving more dehydrogenated species (Scheme 13.3), but the results were only shown graphically, and as is usual for UHV systems were somewhat scattered. The activity of the corresponding Ni_3Sn surface was also less than that of Ni(111) but the extent of deactivation by carbon was much reduced.[177]

The inclusion of indium with Pt/Al$_2$O$_3$ had similar effects to those of tin.[175] Heats of adsorption of hydrogen were lowered and rates of n-butane hydrogenolysis were smaller by about 10^2, due mainly to a decrease in the pre-exponential factor. Isomerisation selectivities were somewhat increased, possible at the expense of S_2; the weaker hydrogen adsorption encouraged routes for species more fully dehydrogenated (Scheme 13.3). Addition of indium to Pt/NaY zeolite on the hand *increased* hydrogen adsorption strength at 625 K, as the order in hydrogen became progressively more negative and that of n-butane more positive.[178] Both indium and tin therefore act to lower the mean size of the reactive ensemble and to lessen excessive dehydrogenation, without any significant electronic effect.

13.8.4. Platinum and Iridium plus Zirconium, Molybdenum, and Rhenium

The advent of the bimetallic PtRe/Al$_2$O$_3$ catalyst revolutionised the practice of petroleum reforming, by reason chiefly of its greater stability and longevity. There have been several studies directed towards understanding what difference in the roles of the *metallic* function might contribute to the benefits in industrial practice (see Further Reading section). In the hydrogenolysis of ethane on Pt(111) modified by deposition of rhenium, orders in hydrogen became less negative at 623 K, signalling a weakening of hydrogen chemisorption,[179] but interpretation of the kinetics obtained with Pt/Al$_2$O$_3$ and PtRe/Al$_2$O$_3$ led to the opposite conclusion.[65] There were significant changes in product selectivities[3,43,65] (Table 13.16). With propane, S_2 decreased when rhenium was present, to an extent that depended on

TABLE 13.16. Hydrogenolysis of Propane ($T_2 = S_2$) and of n-Butane (Kinetic and Selectivity Parameters) on Pt/A1$_2$O$_3$ and PtRe/A1$_2$O$_3$ Catalysts at 603 K[3]

%Pt	%Re	T_3	r^1	E^2	S_2	S_3	F	T_3	Note
0.3	0	0.992	0.4	134	0.32	0.83	0.16	0.99	a
0.6	0	0.986	8.4	150	0.34	0.82	0.17	0.98	b
0.3	0.3	0.844	5.9	145	0.79	0.45	0.39	0.74	c
0.45	0.45	0.834	3.4	140	0.68	0.63	0.44	>1	d

^{1}In mmol g_{cat}^{-1} h^{-1}. ^{2}In kJ mol^{-1}.
Catalysts have been designated as follows: ^{a}CK303 (EUROPT-3); ^{b}CK306; ^{c}CK433 (EUROPT-4); ^{d}CK455.

the conditions of pre-treatment, which presumably affected surface composition; rates increased as S_2 fell.[3] With n-butane the methane selectivity increased with the rhenium/platinum ratio.[180,181] There were observed numerous subtle differences with this reaction when Pt/Al$_2$O$_3$ and PtRe/Al$_2$O$_3$ were compared[3,43,65] (Table 13.6), but two effects stood out: (i) the effect of thermal cycling on selectivities was much smaller with PtRe/Al$_2$O$_3$, showing that rhenium minimised 'carbon' deposition; and (ii) the splitting factor F was raised, and T_3 lowered, when rhenium was present. These effects can all be attributed to an increase in the C—M bond strength, for which the interpretation of rate dependences on hydrogen pressure by the ES5B equation provides some support.[65]

The relative values of the surface energies of platinum and of rhenium would suggest the former should segregate to the surface, but the particles are likely to be so small (1 to 1.5 nm) that there is effectively little or no interior, so the surface must contain atoms of both kinds, with the platinum atoms preferentially occupying low coordination-number sites.[3,182] Now rhenium is a catalyst that is vigorous but non-selective for hydrogenolysis,[6,149,180,181] but Pt$_1$Re$_1$/Al$_2$O$_3$ showed neither of these characteristics, appearing rather to look like a modified platinum catalyst. This view is supported by the common compensation line that the two types of catalyst share[3,65] (Figure 13.23). Any surface rhenium atoms must therefore be so well dispersed that they cannot form ensembles large enough to display their true nature (Figure 13.24). Studies of PtRe/Al$_2$O$_3$ using IR spectroscopy of chemisorbed carbon monoxide[183,184] and XANES[184] suggest however that it is the rhenium atoms that segregate preferentially to the surface, probably occupying low CN sites. It is unclear why considerations of surface energy[182] do not apply in this case; perhaps they are irrelevant for single atoms, and the character of bonding to adjacent atoms of the same or different kinds may be more important. XANES results[184] do however confirm some mutual electronic influence.

Somewhat different conclusions have been reached in a study[180,185] of n-butane hydrogenolysis on a series of PtRe/α-Al$_2$O$_3$ catalyst covering the whole composition range. There was an astronomic factor ($\times$ 10^3) between the rates shown by platinum and the Pt$_{25}$Re$_{75}$ catalyst (Figure 13.25), and as little as 12.5% rhenium was sufficient to increase methane selectivity and to move the Arrhenius

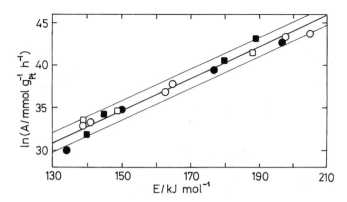

Figure 13.23. Compensation plot of Arrhenius parameters for hydrogenolysis of propane (O) and of n-butane (□) on Pt/Al$_2$O$_3$ (EUROPT-3) and PtRe/Al$_2$O$_3$ (EUROPT-4). The two sets of results (ref. 43, filled points; ref. 3, open points) are from independent studies on different batches of catalyst, published respectively in 1989 and 1996.

parameters away from the characteristic platinum line (Figure 13.26). Conversion to rhenium-type behaviour was complete at 37.5% rhenium. This difference may have been a consequence of the use of various forms of pre-treatment (see also Section 14.5.3). Under petroleum reforming conditions, PtRe/Al$_2$O$_3$ catalysts are sulfided, and the ReS species thus formed act only as inert diluents.

Remarkably large promotional effects have been observed when molybdenum was added to Pt/SiO$_2$[150,151,186,187] or Pt/NaY zeolite.[188] The rate of ethane hydrogenolysis was raised by 10^2 and activation energy lowered from 226 to 138 kJ mol^{-1}; the order in hydrogen became less negative, suggesting a weakening of its chemisorption.[150] The opposite effect has however been found with n-butane,[151,187] where a similar increase in rate achieved its maximum at 40%

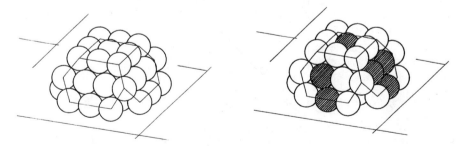

Figure 13.24. Models of metal particles containing 46 platinum atoms (size ~1.2 nm, 72% dispersion) or 23 atoms each of platinum and rhenium: of the 13 interior atoms, all are rhenium, and of the remaining 10 there are six visible (shaded).[3]

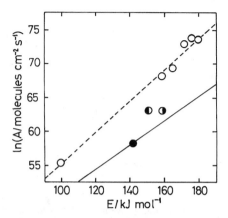

Figure 13.25. Compensation plot for *n*-butane hydrogenolysis on PtRe/α-Al$_2$O$_3$ catalysts of various compositions:[180,185] filled point, Pt; open points, 50–100% Re; half-filled points, intermediate values.

molybdenum. Examination of Arrhenius parameters suggests that PtMo/SiO$_2$ catalysts containing 12 to 50% molybdenum belonged to a separate more active class, while at higher concentrations there was another less active class. Rates expressed per hydrogen atom adsorbed did not however decrease.[151] The cause of the effect has not been established for certain, but it seems possible that Mo^{5+} ions decorate the platinum surface, and that a 'bimetallic' site is more active than a purely platinum site. Skeletal isomerisation was suppressed and S_2 increased. Analogous effects have also been seen[187] with PdMo/SiO$_2$ and with PtW/SiO$_2$. IrMo/Al$_2$O$_3$

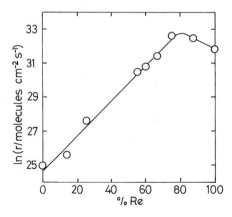

Figure 13.26. Activity of PtRe/α-Al$_2$O$_3$ catalysts for *n*-butane hydrogenolysis as a function of rhenium content at 513 K (ln (rate/molecules cm^{-2} s^{-1})).[180,185]

(Ir$_3$Mo) prepared from the bimetallic carbonyl complex was somewhat more active for the n-butane reaction at 488 K than a pure iridium catalyst,[189] but high ethane selectivity was retained (1.43). The Ir$_2$Mo$_2$ complex gave a less active catalyst, with a lower value of S_2 (1.08). The addition of zirconium to platinum (Pt$_{75}$Zr$_{25}$/C) lowered the rate of ethane hydrogenolysis about 20-fold, and raised the activation energy.[152]

13.8.5. Bimetallic Catalysts of Metals of Groups 8 to 10

As we have noticed previously, results of work done with bimetallic catalysts having components drawn from within Groups 8 to 10 rarely cause much excitement; they are usually intermediate between those shown by each metal separately (although synergism of rates is quite commonly met[48]), and what has been done with alkane hydrogenolysis is no exception. There is little evidence for bimetallic sites giving effects radically different from those shown by monometallic sites. Systems examined include iron-platinum,[40,190] rhodium-platinum[191,192] and iridium-platinum,[48,163,193] but studies of the latter have done little to illuminate its particular virtues for petroleum reforming. The rhodium-iridium[194–195] and ruthenium-iridium[171] systems have also been looked at, using n-butane as the reactant: in the nickel-rhodium system,[10,197] the high ethane selectivity characteristic of rhodium grows with the rhodium content, reaching a value of 1.4 at Ni$_{60}$Rh$_{40}$. Other combinations (e.g. Ir-Os[185]) have shown marked synergism.

13.9. APOLOGIA

The extensive attention that has been paid to the reactions of the lower alkanes on metal catalysts reflects the wide range of phenomena encountered with structure-sensitive reactions, as opposed to those reactions met with earlier, the insensitivity of which limited the importance of variables such as particle size, crystal face and composition of bimetallic systems. Far more attention has also been paid to the careful measurement of reaction kinetics, and their interpretation by various models. This, one hopes, explains even if it does not excuse the length of this chapter.

REFERENCES

1. G.C. Bond, *Appl. Catal. A: Gen.* **148** (1997) 3.
2. Z. Paál and P. Tétényi in: *Specialist Periodic Reports: Catalysis,* (G.C. Bond and G. Webb, eds.), *Roy. Soc. Chem.* Vol.5 (1982), p. 80.
3. G.C. Bond and R.H. Cunningham, *J. Catal.* **163** (1996) 328.

4. G.C. Bond and Xu Lin, *J. Catal.* **168** (1997) 207; **169** (1997) 76.

5. G.C. Bond and Xu Lin, *J. Catal.* **169** (1997) 76.

6. G.C. Bond and M.R. Gelsthorpe, *Catal. Lett.* **2** (1989) 257; *J. Chem. Soc. Faraday Trans. 1* **87** (1991) 2479.

7. S.L. Anderson, J. Szanyi, M.T. Paffett and A.K. Datye, *J. Catal.* **159** (1996) 23.

8. J.R. Engstrom, D.W. Goodman and W.H. Weinberg, *J. Am. Chem. Soc.* **110** (1988) 8305; **108** (1986) 4653.

9. K. Foger and J.R. Anderson, *J. Catal.* **59** (1979) 325.

10. G. Leclercq, S. Pietrzyk, J.F. Lamonier, L. Leclerq, L.M. Bouleau and R. Maurel, *Appl. Catal. A: Gen.* **123** (1995) 161.

11. Z. Paál and P. Tétényi, *Nature* **267** (1977) 234.

12. G. Leclercq, L. Leclercq and R. Maurel, *J. Catal.* **44** (1976) 68; **50** (1977) 87.

13. G.C. Bond, *J. Catal.* **115** (1989) 286.

14. G.C. Bond and Lou Hui, *J. Catal.* **137** (1992) 462.

15. S. Gao and L.D. Schmidt, *J. Catal.* **111** (1988) 210; **115** (1989) 356.

16. L. Guczi, K. Matusek, A. Sárkány and P. Tétényi, *Bull. Soc. Chim. Belg.* **88** (1979) 497.

17. S.D. Jackson, G.J. Kelly and G. Webb, *J. Catal.* **176** (1998) 225.

18. Z. Paál, *Adv. Catal.* **29** (1980) 273.

19. L. Bonneviot and G.L. Haller, *J. Catal.* **130** (1991) 359.

20. G. Leclercq, L. Leclercq, L.M. Bouleau, S. Pietrzyk and R. Maurel, *J. Catal.* **88** (1984) 8.

21. M.F. Guilleux, J.A. Dalmon and G.A. Martin, *J. Catal.* **62** (1980) 235.

22. E.I. Ko and P.A. Burke, *J. Catal.* **127** (1991) 453.

23. J.H. Sinfelt, *Catal. Today,* **53** (1999) 303.

24. J.H. Sinfelt, *Adv. Catal.* **23** (1973) 91.

25. J.H. Sinfelt *Catal. Rev.* **3** (1969) 175.

26. J.H. Sinfelt and W.F. Taylor, *Trans. Faraday Soc.* **64** (1968) 3086.

27. J.H. Sinfelt, W.F. Taylor and D.J.C. Yates, *J. Phys. Chem.* **69** (1965) 95.

28. J.H. Sinfelt and D.J.C. Yates, *J. Catal.* **8** (1967) 82.

29. J.H. Sinfelt and D.J.C. Yates, *J. Catal.* **10** (1968) 362.

30. G.C. Bond, *Catal. Today* **49** (1999) 41.

31. P. Tétényi, L. Guczi and A. Sárkány, *Acta Chim. Acad. Sci. Hung.* **97** (1978) 221.

32. L. Guczi, A. Sárkány and P. Tétényi *J. Chem. Soc. Faraday Trans. 1* **70** (1974) 1971.

33. G.C. Bond, *Appl. Catal. A: Gen.* **191** (2000) 23.

34. F.H. Ribeiro, A.E. Schach von Wittenau, C.H. Bartholomew and G.A. Somorjai, *Catal. Rev.-Sci. Eng.* **39** (1997) 49.

35. G.A. Somorjai, *Introduction to Surface Chemistry and Catalysis,* Wiley: New York (1994).

36. P.A. Burke and E.I. Ko, *J. Catal.* **116** (1989) 230.

37. C.J. Machiels and R.B. Anderson, *J. Catal.* **58** (1979) 253, 260, 268; **60** (1979) 339.

38. A. Sárkány, K. Matusek and P. Tétényi, *J. Chem. Soc. Faraday Trans. 1* **73** (1977) 1699.

39. J.R. Anderson and N.R. Avery, *J. Catal.* **5** (1966) 446.

40. B.S. Gudkov, L. Guczi and P. Tétényi, *J. Catal.* **74** (1982) 207.

41. G.C. Bond and Xu Yide, *J. Chem. Soc. Faraday Trans. 1* **80** (1984) 969.

42. J.C. Kempling and R.B. Anderson, *Proc. 5th Internat. Congr. Catal.*, (J.W. Hightower, ed.), North Holland: Amsterdam **2** (1972) 1099.

43. G.C. Bond and M.R. Gelsthorpe, *J. Chem. Soc. Faraday Trans. 1,* **85** (1989) 3767; *Catal. Lett.* **3** (1989) 359.

44. D.F. Hollis and H. Taheri, *AIChE Journal* **22** (1976) 1112.

45. J.J. Burton and E. Hyman, *J. Catal.* **37** (1975) 114; J.J.Burton, E. Hyman and B.G. Fedak, *J. Catal.* **37** (1975) 106.

46. R.D. Cortright, R.M. Watwe, B.E. Spiewak and J.A. Dumesic, *Catal. Today* **53** (1999) 395.

47. G.C. Bond, J. Calhoun and A.D. Hooper, *J. Chem. Soc. Faraday Trans.*, **92** (1996) 5117.
48. G. Leclercq, L. Leclercq and R. Maurel, *Bull. Soc. Chim. Belg.* **88** (1979) 599.
49. J. Barbier and P. Marécot, *Nouv. J. Chim.* **5** (1981) 393.
50. G.A. Martin, *Bull. Soc. Chim. Belg.* **105** (1996) 131.
51. G.A. Martin, R. Dutartre, Shibin Yuan, C. Márquez-Alvarez and C. Miradatos, *J. Catal.* **177** (1998) 105.
52. G.A. Martin, *J. Catal.* **60** (1979) 345, 452.
53. A. Frennet, A. Crucq, L. Degols and G. Lienard, *Acta Chim. Acad. Sci. Hung.* **111** (1982) 499.
54. A. Frennet in: *Hydrogen Effects in Catalysis* (Z. Paál and P.G. Menon, eds.), Marcel Dekker: New York (1988), p. 399.
55. A. Frennet, L. Degols, G. Lienard and A. Crucq, *J. Catal.* **35** (1974) 18.
56. L. Guczi, A. Frennet and V. Ponec, *Acta Chim. Hung.* **112** (1983) 127.
57. S.A. Goddard, M.D. Amiridis, J.E. Rekoske, N. Cardona-Martinez and J.A. Dumesic, *J. Catal.* **117** (1989) 155.
58. D.F. Rudd and J.A. Dumesic, *Catal. Today* **10** (1991) 147.
59. R.D. Cortright, R.H. Watwe and J.A. Dumesic, *J. Molec. Catal. A: Chem.* **163** (2000) 91.
60. L. Guczi, B.S. Gudkov and P. Tétényi, *J. Catal.* **24** (1972) 187.
61. S. Kristyan and R.B. Timmins, *J. Chem. Soc. Faraday Trans. I* **83** (1987) 2825.
62. S. Kristyan and J. Szamosi, *J. Chem. Soc. Faraday Trans. I* **80** (1984) 1645; **84** (1988) 917.
63. S.B. Shang and C.N. Kenney, *J. Catal.* **134** (1992) 134.
64. G.C. Bond, *Ind. Eng. Chem. Res.* **36** (1997) 3173; G.C. Bond, A.D. Hooper, J.C. Slaa and A.D. Taylor, *J. Catal.* **163** (1996) 319.
65. G.C. Bond and R.H. Cunningham, *J. Catal.* **166** (1997) 172.
66. G.C. Bond and Z. Paál, *Appl. Catal. A: Gen.* **86** (1992) 1.
67. G.C. Bond, *J. Molec. Catal.* **81** (1993) 99.
68. G.C. Bond, F. Garin and G. Maire, *Appl. Catal.* **41** (1988) 313.
69. G.C. Bond and P.B. Wells, *Appl. Catal.* **18** (1985) 221, 225.
70. G.C. Bond, R.H. Cunningham and J.C. Slaa, *Topics in Catal.* **1** (1994) 19.
71. G.C. Bond, M.A. Keane, H. Kral and J.A. Lercher, *Catal. Rev.- Sci. Eng.* **42** (2000) 323.
72. F.G. Gault, *Adv. Catal.* **30** (1981) 1.
73. G.C. Bond and A. Donato, *J. Chem. Soc. Faraday Trans.* **89** (1993) 3129.
74. G.C. Bond and J.C. Slaa, *Catal. Lett.* **23** (1994) 293.
75. G.C. Bond and J.C. Slaa, *J. Molec. Catal.* **89** (1994) 221.
76. G.C. Bond and J.C. Slaa, *J. Molec. Catal. A: Chem.* **98** (1995) 81.
77. G.C. Bond and J.C. Slaa, *J. Molec. Catal. A: Chem.* **101** (1995) 243.
78. G.C. Bond and J.C. Slaa, *J. Chem. Tech. Biotechnol.* **65** (1996) 15.
79. G.C. Bond and J.C. Slaa, *J. Molec. Catal. A: Chem.* **106** (1996) 135.
80. A. Masson in: *NATO ASI Series* (J. Davenas and P.A. Rabette, eds.) Martinus Nijhoff Publishers, (1986), p. 295.
81. M. Che and C.O. Bennett, *Adv. Catal.* **36** (1989) 55.
82. D. Nazimek and J. Ryczkowski, *React. Kinet. Catal. Lett.* **40** (1989) 137, 145.
83. M.J. Yacaman and A. Gómez, *Appl. Surf. Sci.* **19** (1984) 348.
84. Y. Hadj Romdhane, B. Bellamy, V. de Gouveia, A. Masson and M. Che in: *Proc. 8th Internat. Congr. Catal.* Verlag Chemie: Weinheim **IV** (1984) 333.
85. I. Zuburtikudis and H. Saltsburg, *Science* **258** (1992) 1337.
86. G.C. Bond and R. Burch in: *Specialist Periodical Reports: Catalysis* Vol. 6 (G.C. Bond and G. Webb, eds.), *Roy. Soc. Chem.* (1983) 27.
87. C. Lee, L.D. Schmidt, J.F. Mounder and T.W. Rusch, *J. Catal.* **99** (1986) 472.
88. L.D. Schmidt and K.R. Krause, *Catal. Today,* **12** (1992) 269.
89. S.D. Jackson, G.J. Kelly and G. Webb, *Phys. Chem. Chem. Phys.* **1** (1999) 2581.

90. D.E. Resasco and G.L. Haller, *J. Phys. Chem.* **88** (1984) 4552.

91. S.M. Davis, F. Zaera and G.A. Somorjai, *J. Am. Chem. Soc.* **104** (1982) 7453; *J. Catal.* **77** (1982) 439; **85** (1984) 206.

92. D.W. Goodman, *Catal. Today,* **12** (1992) 189.

93. D. Kalakkad, S.L. Anderson, A.D. Logan, J. Peña, E.J. Braunschweig, C.H.F. Peden and A.K. Datye, *J. Phys. Chem.* **97** (1993) 1437.

94. A.K. Datye, B.F. Hegarty and D.W. Goodman, *Faraday Disc. Chem. Soc.* **87** (1989) 337.

95. F. Rodríguez-Reinoso, I. Rodríguez-Ramos, C. Moreno-Castilla, A. Guerrero-Ruiz and J.D. López-González, *J. Catal.* **107** (1987) 1.

96. C. Moreno-Castilla, A. Porcel-Jimenez, F. Carrasco-Marin and E. Utera-Hidalgo, *J. Molec. Catal.* **66** (1991) 329.

97. J.C. Kempling and R.B. Anderson, *Ind. Eng. Chem. Proc. Des. Dev.* **11** (1972) 146; **9** (1970) 116;

98. J. Monnier and R.B. Anderson, *J. Catal.* **78** (1982) 419.

99. G.C. Bond, R.R. Rajaram and R. Yahya, *J. Molec. Catal.* **69** (1991) 359.

100. G.C. Bond, R.R. Rajaram and R. Burch, *J. Phys. Chem.* **90** (1986) 4877.

101. G.C. Bond and Xu Yide, *J. Chem. Soc. Faraday Trans. I* **80** (1984) 3103; *J. Chem. Soc. Chem. Comm.* (1983) 1248.

102. G.C. Bond and R. Yahya, *J. Chem. Soc. Faraday Trans. I* **87** (1991) 775.

103. G.C. Bond. R.R. Rajaram and R. Burch, *Proc. 9th Internat. Congr. Catal.*, (M.J. Phillips and M. Ternan, eds.), Chem. Inst. Canada: Ottawa **3** (1988) 1130.

104. D.M. Somerville, M.S. Nashner, R.G. Nuzzo and J.R. Shapley, *Catal. Lett.* **46** (1997) 17.

105. K. Foger, *J. Catal.* **78** (1982) 406.

106. G.C. Bond, R. Yahya and B. Coq, *J. Chem. Soc. Faraday Trans.* **86** (1990) 2297.

107. Y. Hadj Romdhane, A. Masson, B. Bellamy, V. de Gouveia and M. Che, *Compt. Rend. Acad. Sci. Paris,* **303** (Ser.II) (1986) 129.

108. G.C. Bond and Yide Xu, *J. Molec. Catal.* **25** (1984) 141; *Proc. 8th Internat. Congr. Catal.*, Verlag Chemie: Weinheim **IV** (1984) 577.

109. Z. Karpiński, *J. Catal.* **77** (1982) 118.

110. Tran Manh Tri, J. Massardier, P. Gallezot and B. Imelik, *Proc. 7th Internat. Congr. Catal.* (T. Seiyama and K. Tanabe, eds.), Elsevier: Amsterdam (1980) 266.

111. G.C. Bond, *Discuss. Faraday Soc.* **41** (1966) 200.

112. G.C. Bond, R.R. Rajaram and R. Burch, *Appl. Catal.* **27** (1986) 379.

113. R. Burch, G.C. Bond and R.R. Rajaram, *J. Chem. Soc. Faraday Trans. I* **82** (1986) 1985.

114. G.C. Bond and R. Yahya, *J. Molec. Catal.* **68** (1991) 243.

115. B. Coq, E. Crabb, M. Warawdeker, G.C. Bond, J.C. Slaa, S. Galvagno, L. Mercadante, J. García Ruiz and M.C. Sanchez Sierra, *J. Molec. Catal.* **99** (1994) 1.

116. G.C. Bond, B. Coq, R. Dutartre, J. Garcia Ruiz, A.D. Hooper, M. Grazia Proietti, M.C. Sanchez Sierra and J.C. Slaa, *J. Catal.* **161** (1996) 480.

117. M.C. Sanchez Sierra, J. García Ruiz, M.G. Proietti and J. Blasco, *J. Molec. Catal. A: Chem.* **96** (1995) 65; *Physica B* **208-209** (1995) 705.

118. G.C. Bond, A.D. Hooper, J.C. Slaa and A.O. Taylor, *J. Catal.* **163** (1996) 319.

119. *Metal-Support Interactions, Sintering and Redispersion* (S.A. Stevenson, J.A. Dumesic, R.T.K. Baker and E. Ruckenstein, eds.) Van Nostrand Reinhold: New York (1987).

120. E.I. Ko and R.L. Garten, *J. Catal* **68** (1981) 233.

121. J.B.A. Anderson, R. Burch and J.A. Cairns, *J. Catal.* **107** (1987) 351, 364.

122. G.C. Bond and S. Flamerz, *J. Chem. Soc. Faraday Trans. I* **87** (1991) 767.

123. G.C. Bond, M.R. Gelsthorpe, R.R. Rajaram and R. Yahya in: *Structure and Reactivity of Surfaces* (C. Morterra, A. Zecchina and G. Costa, eds.), Studies in Surface Science and Catalysis, Elsevier: Amsterdam **48** (1989) 167.

124. G.C. Bond and R. Yahya, *J. Molec. Catal.* **69** (1991) 75.

125. K. Ebitani, T.M. Salama and H. Hattori, *J. Catal.* **134** (1992) 751.
126. D. Nishio, H. Shindo and K. Kunimori, *J. Catal.* **151** (1995) 460.
127. K. Kunimori, Y. Doi, K. Ito and T. Ochijima, *J. Chem. Soc. Chem. Comm* (1986) 965.
128. K. Kunimori, H. Nakamura, Z. Hu and T. Uchijima, *Appl. Catal.* **53** (1989) L11.
129. K. Kunimori, T. Watasugi, Z. Hu, H. Oyanagi, M. Imai, H. Asano and T. Uchijima, *Catal. Lett,* **7** (1990) 337.
130. E.I. Ko, J.M. Hupp and N.J. Wagner, *J. Catal.* **86** (1984) 315.
131. D.E. Resasco and G.L. Haller, *SSSC* **11** (1982) 105.
132. Jialiang Wang, J.A. Lercher and G.L. Haller, *J. Catal.* **88** (1984) 18.
133. G.A. Martin, R. Dutartre and J.A. Dalmon, *React. Kinet. Catal. Lett.***16** (1981) 329.
134. G.-A. Martin and J.A. Dalmon, *React. Kinet. Catal. Lett.* **16** (1981) 35.
135. You-Jyh Lin, D.E. Resasco and G.L. Haller, *J. Chem. Soc. Faraday Trans. I* **83** (1987) 2091.
136. G.C. Bond and M.A. Duarte, *J. Catal.* **111** (1988) 189.
137. D. Kalakkad and A.K. Datye, *Appl. Catal. B.* **1** (1992) 191.
138. G.L. Haller, D.E. Resasco and A.J. Rouco, *Faraday Disc. Chem. Soc.* **72** (1981) 109.
139. G.C. Bond and Lou Hui, *J. Catal.* **142** (1993) 512.
140. M. Arai, S.-L. Guo, M. Shirai, Y. Nishiyama and K. Torü, *J. Catal.* **161** (1996) 704.
141. G.C. Bond and A.D. Hooper, *React. Kinet. Catal. Lett.* **68** (1999) 5.
142. T.E. Hoost and J.G. Goodwin Jr. *J. Catal.* **130** (1991) 283.
143. B.L. Mojet, J.T. Miller, D.E. Ramaker and D.C. Koningsberger, *J. Catal.* **186** (1999) 373.
144. J.T. Miller, B.L. Meyers, F.S. Modica, G.S. Lane, M. Vaarkamp and D.C. Koningsberger, *J. Catal.* **143** (1993) 395.
145. K.-W. Huang and J.G. Ekerdt, *J. Catal.* **92** (1983) 232.
146. J.K.A. Clarke, *Chem. Rev.* **75** (1975) 291.
147. V. Ponec, *Appl. Catal. A: Gen.* **222** (2001) 31.
148. C.T. Campbell, *Ann. Rev. Phys. Chem.* **41** (1990) 775.
149. R.M. Edreva-Kardjieva and A.A. Andreev, *J. Catal.* **94** (1985) 97.
150. Yu. I. Yermakov, B.N. Kuznetsov and Yu. A. Ryndin, *J. Catal.* **42** (1976) 73.
151. G. Leclercq, A. El Gharbi and S. Pietrzyk, *J. Catal.* **144** (1993) 118.
152. R. Szymanski and H. Charcosset, *J. Molec. Catal.* **25** (1984) 337.
153. D.E. Resasco and G.L. Haller, *J. Catal.* **82** (1983) 279.
154. T. Uchijima, *Catal. Today,* **28** (1996) 105.
155. M. Sprock, X. Wu and T.S. King, *J. Catal.* **138** (1992) 617.
156. M.W. Smale and T.S. King, *J. Catal.* **120** (1990) 335.
157. M.W. Smale and T.S. King, *J. Catal.* **119** (1989) 441.
158. J.K. Strohl and T.S. King, *J. Catal.* **116** (1989) 540.
159. J.A. Dalmon and G.-A. Martin, *J. Catal.* **66** (1980) 214.
160. I. Alstrap, V.E. Petersen and J.R. Rostrop-Nielsen, *J. Catal.* **191** (2000) 401.
161. J.H. Sinfelt, *J. Catal.* **29** (1973) 308.
162. J.H. Sinfelt, Y.L. Lam, J.A. Cusumano and A.E. Barnett, *J. Catal.* **42** (1976) 227.
163. D. Garden, C. Kemball and D.A. Whan, *J. Chem. Soc. Faraday Trans. I* **82** (1986) 3113.
164. J.H. Sinfelt, J.L. Carter and D.J.C. Yates, *J. Catal.* **24** (1973) 283.
165. T.J. Plunkett and J.K.A. Clarke, *J. Chem. Soc. Faraday Trans. I* **68** (1972) 600.
166. Z. Karpiński, W. Juczczyk and J. Stachurski, *J. Chem. Soc. Faraday Trans. I* **81** (1985) 1447.
167. D.I. Hagen and G.A. Somorjai, *J. Catal.* **41** (1976) 466.
168. P.A. Sermon, K. Keryou, J.M. Thomas and G.R. Millward in: *Mat. Res. Soc. Symp. Proc.* **111** (1988) 13.
169. P.A. Sermon, J.M. Thomas, K. Keryou and G.R. Millward, *Angew. Chem. Ind. Edn. Eng.* **26** (1987) 918.
170. J.-Y. Chen, J.M. Hill, R.M. Watwe, S.G. Podkolzin and J.A. Dumesic, *Catal. Lett.* **60** (1991) 1.

171. H. Hamada, *Appl. Catal.* **27** (1986) 265.
172. C. Crisafulli, R. Maggiore, G. Schembari, S. Sciré and S. Galvagno, *J. Molec. Catal.* **50** (1989) 67.
173. Bin Chen and J.G. Goodwin Jr., *J. Catal.* **154** (1995) 1.
174. Bin Chen and J.G. Goodwin Jr. *J. Catal.* **158** (1996) 228.
175. F.B. Passos, M. Schmal and M.A. Vannice, *J. Catal.* **160** (1996) 106.
176. J. Szanyi, S. Anderson and M.T. Paffett, *J. Catal.* **149** (1994) 438.
177. A.D. Logan and M.T. Paffett, *Proc. 10th Internat. Congr. Catal.*, (L. Guczi, F. Solymosi and P. Tétényi, eds.), Akadémiai Kiadó: Budapest **B** (1992) 1595.
178. P. Mériaudeau, A. Thangaraj, D.F. Dutel, P. Gelin and C. Naccache, *J. Catal.* **163** (1996) 338; P. Mériaudeau, A. Thangaraj, J.F. Dutel and C. Naccache, *J. Catal.* **167** (1997) 180.
179. D.J. Godbey, F. Garin and G.A. Somorjai, *J. Catal.* **117** (1989) 144.
180. C. Betizeau, G. Leclercq, R. Maurel, C. Bolivar, H. Charcosset, R. Frety and L. Tournayan, *J. Catal.* **45** (1976) 179.
181. I.H.B. Haining, C. Kemball and D.A. Whan, *J. Chem. Res.* **5** (1977) 170.
182. R.W. Joyner and E.S. Shpiro, *Catal. Lett.* **9** (1991) 233.
183. J.A. Anderson, F.K. Chong and C.H. Rochester, *J. Molec Catal. A: Chem.* **140** (1999) 65.
184. F.K. Chong, J.A. Anderson and C.H. Rochester, *Phys. Chem. Chem. Phys.* **2** (2000) 5730.
185. G. Leclercq, H. Charcosset, R. Maurel, C. Betizeau, C. Bolivar, R. Frety, D. Jaunez, H. Mendes and L. Tournayan, *Bull. Soc. Chim. Belg.* **88** (1979) 577.
186. G. Leclercq, A. El Gharbi, L. Gengembre, T. Romero, L. Leclercq and S. Pietrzyk, *J. Catal.* **148** (1994) 550.
187. G. Leclercq, S. Pietrzyk, T. Romero, A. El Gharbi, L. Gengembre, J. Grimblot, F. Aïssi, M. Guelton, A. Latef and L. Leclercq, *Ind. Eng. Chem. Res.* **36** (1997) 4015.
188. Tran Manh Tri, J. Massardier, P. Gallezot and B. Imelik *J. Molec. Catal.* **25** (1984) 151; *J. Catal.* **85** (1984) 244.
189. J.R. Shapley, W.S. Uchiyama and R.A. Scott, *J. Phys. Chem.* **94** (1990) 1190.
190. L. Guzci, K. Matusek and M. Eszterle, *J. Catal.* **60** (1979) 121.
191. J.A. Oliver and C. Kemball, *Proc. Roy. Soc. A* **429** (1990) 17.
192. Teik Chen Wong, L.C. Chang, G.L. Haller, J.A. Oliver, N.R. Scaife and C Kemball, *J. Catal.* **87** (1984) 389.
193. Teik Chen Wong, L.F. Brown, G.L. Haller and C. Kemball, *J. Chem. Soc. Faraday Trans. I* **77** (1981) 519.
194. M. Ichikawa, L.-F. Rao, T. Kimura and A. Fukuoka, *J. Molec. Catal.* **62** (1990) 15.
195. D. Garden, C. Kemball and D.A. Whan, *J. Chem. Soc. Faraday Trans. I* **82** (1986) 3113.
196. M. Ichikawa, Lingfen Rao, T. Ito and A. Fukuoka, *Faraday Discuss. Chem. Soc.* **87** (1989) 321.
197. G. Leclercq, S. Pietrzyk, L. Gengembre and L. Leclercq, *Appl. Catal.* **27** (1986) 299.
198. Z. Karpiński, *Adv. Catal.* **37** (1990) 45.
199. J.R. Anderson and B.G. Baker in: *Chemisorption and Reactions on Metal Films* (J.R. Anderson, ed.) Academic Press: London, Vol. 2 (1971), p. 64.
200. G.C. Bond, *Acc. Chem. Res.* **26** (1993) 490.
201. J.R. Anderson, *Adv. Catal.* **23** (1973) 1.
202. J.K.A. Clarke and J.J. Rooney, *Adv. Catal.* **25** (1976) 125.
203. A.J. Hong, A.J. Rouco, D.E. Resasco and G.L. Haller, *J. Phys. Chem.* **91** (1987) 2665.
204. D.J.C. Yates and J.H. Sinfelt, *J. Catal.* **8** (1967) 348.
205. P.K. Tsjeng and R.B. Anderson, *Canad. J. Chem. Eng.* **54** (1976) 101.
206. D.G. Tajbl, *Ind. Eng. Chem. Proc. Des. Dev.* **8** (1969) 364.
207. S. Galvagno, J. Schwank, G. Gubitosa and G.R. Tauszik, *J. Chem. Soc. Faraday Trans. I,* **78** (1982) 2509.
208. A. Cimino, M. Boudart and H.S. Taylor, *J. Am. Chem. Soc.* **58** (1954) 796.
209. G.J. Haddad and J.G. Goodwin Jr. *J. Catal.* **157** (1995) 25.

210. J.L. Carter, J.A. Cusumano and J.H. Sinfelt, *J. Phys. Chem.* **70** (1966) 2257.
211. J.R. Anderson and B.G. Baker, *Proc. Roy. Soc. A* **271** (1963) 402.
212. J. Barbier, A. Morales and R. Maurel, *Bull. Soc. Chim. Fr.* **1–2** (1978) I-31.
213. Ihl Hyun Cho, Seung Bin Park, Sung June Cho and Ryong Ryoo, *J. Catal.* **173** (1998) 295.
214. K.J. Blankenberg and A.K. Datye, *J. Catal.* **128** (1991) 186.
215. R.A. Campbell, Jie Guan and T.E. Madey, *Catal. Lett.* **27** (1994) 273.
216. P.A. Sermon, K.M. Keryou and F. Ahmed, *Phys. Chem. Chem. Phys.* **2** (2000) 5723.
217. A. Sárkány, L. Guczi and P. Tétényi, *Acta Chim. Acad. Sci. Hung.* **96** (1978) 27.
218. M.J. Holgado and V. Rives, *React. Kinet. Catal. Lett.* **32** (1986) 215, 221; *Appl. Catal.* **41** (1988) L1.
219. K. Kunimori, H. Shindo, D. Nishio, T. Sugiyama and T. Uchijima, *Bull. Chem. Soc. Japan* **67** (1994) 2567.
220. T. Ichijima, *Catal. Today* **28** (1996) 105.
221. D. Nazimek, *React. Kinet. Catal. Lett.* **27** (1985) 273.
222. G.R. Gallaher, J.G. Goodwin Jr. and L. Guczi, *Appl. Catal.* **73** (1991) 1.
223. R. Brown and C. Kemball, *J. Chem. Soc. Faraday Trans.* **89** (1993) 585.
224. P. Tétényi, *React. Kinet. Catal. Lett.* **53** (1994) 369.
225. D. Kalakkad, S.L. Anderson and A.K. Datye, *Proc. 10th Internat. Congr. Catal.* (L. Guczi, F. Solymosi and P. Tétényi, eds.), Akadémiai Kiadó: Budapest C (1992) 2411.
226. D.R. Rainer and D.W. Goodman, *J. Molec. Catal. A: Chem.* **131** (1998) 259.
227. P. Mahaffey and R.S. Hansen, *J. Chem. Phys.* **71** (1979) 1853.
228. K.-I. Tanaka, T. Miyazaki and K. Aomura, *J. Catal.* **81** (1983) 328.
229. D. Nazimek and J. Ryczkowski, *Appl. Catal.* **26** (1986) 47.

FURTHER READING

Hydrogenolysis of lower alkanes on single metals; 34, 56, 66, 68, 198–200
Hydrogenolysis of lower alkanes on standard platinum catalysts; 3, 4, 14, 64, 65, 70, 73
Hydrogenolysis on ruthenium catalysts; 70, 74–79, 100, 101, 103, 106, 115, 117, 121, 122
Mechanism of skeletal isomerisation; 9, 21, 91, 201, 202
Hydrogenolysis on RuCu/SiO$_2$ catalysts; 108, 156, 157, 162, 173, 203
PtRe/Al$_2$O$_3$ catalysts: industrial use; 3, 43, 65, 178–180

For further information on **Arrhenius parameters and orders of reaction** (mainly), see the following references:
Iron 24, 25, 208
Cobalt 24–27, 31, 209
Nickel 15, 20, 24, 25, 27, 31, 60, 89, 141, 210, 211, 224, 226, 228
Copper 27
Ruthenium 15, 24, 31, 38, 97, 99, 100, 106, 117, 156, 157, 163, 172, 173, 192, 205–207, 219, 221, 223
Rhodium 12, 15, 24, 47, 90, 93, 136, 139, 191, 204, 218, 220, 222, 225
Palladium 24, 31, 39, 109, 217, 226
Osmium 102, 181
Iridium 8, 9, 24, 31, 34, 38, 57, 102, 104, 105, 108, 181, 227
Platinum 7, 12, 15, 17, 19, 31, 32, 41, 91, 95, 110, 138, 141, 151, 152, 176, 179, 212, 216

REACTIONS OF HIGHER ALKANES WITH HYDROGEN

PREFACE

This final major chapter brings us to a large and complex area of metal-catalysed reactions of hydrocarbons, which has been stimulated by the greatest industrial application of catalysis by metals, namely, the reforming of petroleum to produce fuels of higher quality and feedstocks for the petrochemical industry. It had long been known that treatment of petroleum fractions by acidic solids induced some of the desired changes, but it was not until the introduction in the 1950s of bifunctional catalysts having both metallic and acidic functions overcame the disastrously rapid deactivation caused by 'carbon' deposition, and even then the further development of bimetallic bifunctional catalysts was needed to increase catalyst life from months to years.

As a result of intensive academic studies, it slowly came to be appreciated that the metallic function alone was capable of effecting many of the transformations that bifunctional catalysts achieved, and for this the position of platinum has remained unchallenged—for reasons that remain somewhat obscure. This chapter endeavours to analyse and classify the enormous literature that has been generated, but apart from a short section that describes simply the way in which bifunctional catalysts operate, our concern will be purely with those reactions that the metal function can accomplish by itself. To extend this to cover those processes that solid acids can bring about with the aid of a metal would require another book.

14.1. INTRODUCTION: PETROLEUM REFORMING AND REACTIONS OF HIGHER ALKANES WITH HYDROGEN

14.1.1. The Scope of This Chapter

The subject matter of this chapter has been kept to last because it calls to a greater or lesser extent on the material presented in all the earlier chapters, and also because the complexity of the reactions involved requires abbreviated treatment of concepts already introduced (especially in Chapters 5 and 13); without these prior discussions, the contents of this chapter would make even less sense. In the last chapter we were concerned only with hydrogenolysis, and with isomerisation that had very limited scope; extension to alkanes to those containing five to seven or eight carbon atoms increases dramatically the range of possible processes. In hydrogenolysis we shall have to consider *selectivity* in a much broader context; the relative reactions of various types of C—C bond vary quite widely, and simple treatments such as the Kempling-Anderson method found so useful in Chapter 13 are hard to develop and apply. The range of possible skeletal isomerisations, and the mechanisms that may underlie them, are also greatly extended, but even more importantly it is possible for *cyclisation* to occur by the formation of new C—C bonds; and this provides a route to the making of aromatic compounds by further dehydrogenation.

It will be interesting to see to what extent the observations and concepts relating to the lower alkanes also apply to larger alkanes. Points of direct comparison will include the relative activities of metals, and the kinetics (i.e. orders of reaction and Arrhenius parameters) of hydrogenolysis and skeletal isomerisation. The additional types of reaction made possible by the greater size vastly complicate any simple-minded contrast, and new methods for defining product selectivities will be needed. Finding an orderly and systematic method for presenting the available information within a limited number of pages has proved a daunting task: that adopted, after much thought and several false starts, is far from perfect, but better than some alternatives. It has been necessary to try to identify a number of *themes* that the various publications address; the obstacles to doing this, and the nature of those revealed, are the subject of Section 14.1.4.

14.1.2. Bifunctional Catalysis: Principles of Petroleum Reforming[1-7]

We must first differentiate *metal-support interactions* that undoubtedly occur when a metal particle is placed on a support that has distinctly either acidic or basic character from true *bifunctional catalysis*. Such interactions have been shown to introduce delicate but significant alteration to the organisation of the metal's valence electrons, and these have important catalytic consequences: but reactions that are affected are still metal-catalysed, and should not be regarded as instances of bifunctional catalysis. Examples of these effects will be considered in Section 14.5.4. They may be contrasted to the enhancement of catalytic activity shown[8] when a metal particle resides on a strongly acidic ('super acid') support such as sulfated zirconia; hydrocarbon intermediates are then thought to have ionic character.

The Earth provides abundant but not infinite supplies of hydrocarbons, both as 'natural gas' and as crude oil; unfortunately not all the major sources are located in the areas where demand is greatest, and this can create problems. The crude oil is first separated into a fraction that is volatile below about 670 K and a non-volatile 'residual oil' or 'resid'; the volatile part is then further divided by fractional distillation into a number of fractions ranging from C_1–C_4 hydrocarbons to kerosene and light gas oil. These various fractions find a variety of applications and form the mainstay of present-day civilisation. 'Natural gas' (chiefly methane) and the light alkanes (light petroleum gas, LPG) are used domestically: other fractions are needed as fuels for vehicles (internal-combustion or diesel engine powered) and aircraft, and also for domestic heating. The part having highest molar mass range, arising from further treatment of 'resid', provides lubricating oils and greases. Less volatile fractions can be subjected to *cracking*, which lowers the mean molar mass, originally performed thermally, but later catalytically using clays, then synthetic aluminosilicates, and most recently zeolites.

All fractions except the lightest contain molecules having either oxygen or nitrogen or sulfur atoms, the concentration of which increases with the boiling point, and for many uses it is necessary to eliminate them or at least reduce their amounts by processes collectively known as *hydrotreating*.[4,9] Of these perhaps the most important is *hydrolesulfurisation* (HDS), for which a Co-Mo/Al_2O_3 catalyst has long been used. The technology of petroleum refining and hydrotreating has been well described in a number of publications.[3–5] All in all, little is wasted—although the disposal of the sulfur arising from HDS can be a worry.

While each application has its own set of criteria for acceptability, more or less severe, depending on the intended use, we will focus on what is desired for the efficient operation of the internal-combustion engine (ICE). The energy released in the combustion of the fuel in the engine cylinder depends upon its *octane rating* (OR), which is based on a comparison with mixtures of *n*-heptane (OR = 0) and *iso*octane (OR = 100). In brief, high octane ratings are given by aromatics, by branched alkanes and by alkenes, although the last are not favoured because they produce unwanted gum in the engine. Thus the processes originally desired were (i) dehydrogenation of cyclohexane and its derivatives ('naphthenes') (see Chapter 12), (ii) dehydrocyclisation of alkanes ('paraffins'), and (iii) isomerisation of linear alkanes into branched alkanes. Hydrogenolysis ('hydrocracking') is not wanted. Now reactions of types (ii) and (iii) in particular proceed via carbocationic intermediates, which are formed on acidic solids, most easily by adding a proton to an alkene. The great advance that was made in the 1950s was the realisation that combining metallic and acidic functions into one catalyst speeded up alkane dehydrogenation, creating more alkene and hence faster isomerisation, as well as helping the formation of aromatics. So the *bifunctional* (or *dual-function*) catalyst was born. Further benefits were that 'carbon' deposit was slight, lifetime was considerable, and *in situ* regeneration could be performed. After early attempts to use nickel as the metal, platinum came to be adopted, and has remained a vital component, either alone or combined with rhenium or iridium (see Section 14.5), of petroleum reforming catalysts. The acidic component has been provided by amorphous silica-alumina, by chlorided alumina, and by zeolites.

The way in which the two components interact is shown schematically in Scheme 14.1A.[7] It was thought at first that they ought to be close together,[10] but it was later found that alkenes could move quite effectively through the gas phase from metal to acid and back, even at low concentration, because a physical mixture of, for example, $Pt/SiO_2 + SiO_2-Al_2O_3$, performed as well as $Pt/SiO_2-Al_2O_3$.[7] Scheme 14.1B shows the principal routes by which *n*-hexane is transformed into isohexanes and benzene, with methylcyclopentane as the key intermediate, on a bifunctional catalyst. As we shall see, most of these transformations can also be brought about by the metallic function alone, most notably by platinum. Early work using film[11,12] and black,[13,14] and particles on neutral supports, showed that the acidic function was not essential: however it continues to be used in industrial practice, and the contribution of the metal in bifunctional catalysis under industrial conditions (except for hydrogenation-dehydrogenation) is hard to assess.

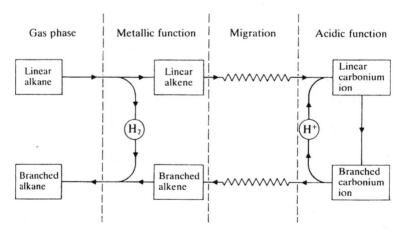

Scheme 14.1A. Schematic representation of the mechanism of skeletal isomerisation on a bifunctional catalyst.

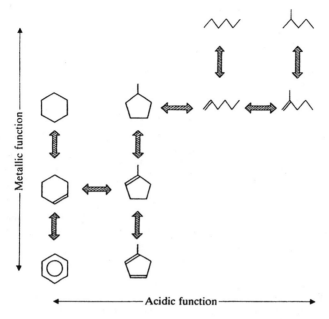

Scheme 14.1B. Pathways for the reaction of *n*-hexane on a bifunctional reforming catalyst.

Aromatic molecules are now not so acceptable as components of ICE fuel, due to the carcinogenic behaviour of benzene, although they are still wanted by the petrochemical industry. Other additives including tetraethyl-lead and oxygenated molecules (alcohols, ethers) have been approved from time-to-time as anti-knock or octane-rating enhancers, but each has its disadvantages or environmental hazards as well, and none is totally free from objection. Palladium has sometimes been used with platinum to control 'carbon' formation.

Emphasis is now being placed on improving the quality of mid-distillate diesel fuel by lowering sulfur content and the concentration of multi-ring aromatic compounds; in this way the density is lessened, and the cetane number (which measures the concentration of alkanes) is increased.[15] This is achieved by hydrogenating compounds such as naphthalene to the fully saturated analogue (i.e. decalin, see Chapter 10), and further converting it by hydrogenolysis to an alkylcyclohexane and thence to a branched C_{10} alkane (Section 14.2.5). The ring opening is desirably achieved selectively, i.e. without forming light alkanes, and this is easier if the C_6 ring is first isomerised to an alkylcyclopentane. These are all metal-catalysed reactions, but some concomitant hydrocracking of larger alkanes to bring them within the necessary boiling range is also needed, so that the use of acidic supports for the metal is recommended.

14.1.3. Reactions of the Higher Alkanes with Hydrogen

Three alkanes in particular have been chosen for fundamental studies: (i) *n*-hexane, (ii) methylcyclopentane, and (iii) *neo*pentane (2,2-dimethylpropane). The first is a significant component of industrial feedstocks and the second a likely important intermediate (Scheme 14.3); both can undergo many different reactions. Metal-catalysed reactions of *n*-hexanes are depicted in Scheme 14.3; not all possible processes are shown, e.g. there are other routes to 'carbon'. Methylcyclopentane, shown there as being formed by dehydrocyclisation (DHC) of *n*-hexane, is attractive as a reactant because of the variety of possible reactions, namely, (i) ring-opening, (ii) multiple hydrogenolysis (to lighter alkanes), (iii) demethanation, (iv) ring-expansion, (v) aromatisation, and (vi) dehydrogenation. The attraction of *neo*pentane lies in its having only one type of C—C bond, and its difficult hydrogenolysis must give methane and *iso*butane as initial products; also it cannot undergo other reactions without first isomerising to 2-methylbutane. Relative rates in this rich panoply of reactions may be expected to depend on the structure and composition of the catalyst used, as well as on operating conditions, and so it has turned out. The greater the size of the alkane, the greater is the variety of possible reactions, and the greater becomes the difficulty of quantitative study and kinetic modelling. So for example *n*-heptane can give five dimethylcyclopentanes (including two pairs of *Z-E* isomers), while *n*-octane can lead to all three xylene isomers. Fortunately no new principles or kind of mechanism emerge from work on alkanes higher than C_6, so we shall need to touch only lightly on these further complications.

Other alkanes have not however been neglected. *n*-Pentane has been less studied, perhaps because the more limited scope of its reactions, and, although the C_5 ring is strain-free, *n*-pentane is less easily cyclised than *n*-hexane, and indeed *n*-heptane cyclises even more effectively[16] (see Table 14.1). neoHexane (2,2-dimethylbutane)[17] has three types of C—C bond, the reactivities of which differ, while 2,2,3,3-tetramethylbutane[18] has only two types of C—C bond, the breaking of which gives either two molecules of *iso*butane or methane + trimethylbutane. These have been usefully employed to characterise catalysts, and the different reactivities of C—C bond in other branched alkanes have also been examined (Section 14.2.4).

To the study of this great family of reactions a wide variety of techniques has been brought. In addition to the use of UHV equipment for examining single-crystal surfaces (which as noted before usually provides results that are somewhat scattered), simple flow or recirculatory reactors have generally been used, although the reactor type does sometimes affect the results, especially in the early stages of an experiment. The mechanisms of skeletal isomerisation have been illuminated by experiments designed with high intelligence and performed with consummate skill, using alkanes labelled with either ^{13}C or ^{14}C, the products being analysed by mass-spectrometry,[19,20] radiochemical methods[21] and magic-angle-spinning

TABLE 14.1. Product Selectivities for the Reactions of n-Alkanes with Hydrogen over Platinum Catalysts[16,52] (T = 603 K, $P_H \approx 16$ kPa)

Form	Alkane	TOF/s^{-1} × 10^3	$S_{<n}$	S_i	S_{C_5}	S_{C_6}	S_{arom}
Pt/SiO$_2$	n-C$_5$H$_{12}$	3.6	30	51	19	—	—
Pt black	//	—	60	18	22	—	—
Pt/SiO$_2$	n-C$_6$H$_{14}$	8.2	13	25	41.5	8	12.5
Pt black	//	—	35	11	9	—	45
Pt/SiO$_2$	n-C$_7$H$_{16}$	0.8	1	6	79.5	—	13.5
Pt black	//	—	59	8	18	—	14

(1) TOF is for reactant removal.
(2) $S_{<n}$, selectivity to lower alkanes; S_i, to skeletal isomers; S_{C5}, to C$_5$ cyclic molecules; S_{C6} to C$_6$ cyclic molecules; S_{arom}, to benzene or toluene.
(3) The Pt/SiO$_2$ is EUROPT-1.

NMR (MASNMR)[22] (Section 14.3). Of course, without gas-chromatography almost nothing could have been achieved.

14.1.4. The Scope and Limitations of the Literature

The difficulty of providing a short but informative account of the extensive literature has already been noted (Section 14.1.1); we must now see wherein the complications lie. (1) Almost every paper deals with two or more different reactants or catalysts, and for each it often explores the effects of either conversion, temperature, hydrogen pressure or some other variable: collecting information on any particular facet therefore requires a large number of papers to be scanned. (2) Graphical presentation of results is most common, but often at such a density and on such a scale that their significance is obscured. Tabular presentation is rare.[23] (3) Except in a few cases,[24-26] quantitative modelling is not attempted: each product is assumed to derive from an independent reaction with its own site demand and kinetic parameters, and the way in which its formation depends on the experimental variable is expressed either by TOFs or rates or by selectivities. (4) Experimental procedures are sometimes not well described, and it may be unclear whether the results pertain to a catalyst in its initial or stable (i.e. partially deactivated) state. It is useful to know, if the level of a variable has been changed, whether this has led to an irreversible change in the extent of 'carbon' deposit, or whether the catalyst has been reactivated before the next experiment.[25] (5) Finally, a point that has been frequently made before, comparison of the 'activities' of a family of catalysts in some defined state is most usually based on *rates* obtained under a single set of operating conditions, and one is left to wonder whether using some other set would lead to the same conclusion. Nevertheless, in spite of all these difficulties, we have plenty of material to work with, indeed, one might say an *embarras de richesse:* or perhaps 'enough is enough, and plenty is too much'.

A recurring self-imposed task of many of the publications has been to identify the 'sites' responsible for each type of reaction, so that structure-sensitivity has been a dominant theme. What has received much less attention is the possibility that, for example, particle size might determine the strength of hydrogen chemisorption, so that the use of constant operating conditions on a series of catalysts might produce results mainly decided by the surface concentration of hydrogen atoms. The dependence of kinetic parameters on particle size or other catalyst feature has been rarely examined.

14.1.5. The Principal Themes

The following classification of the information available will be used. (1) *The activities of metals* for the relevant reactions, expressed quantitatively as far as possible: the identity of the metal is the focus, and the exact state of the catalyst (e.g. metal dispersion) may not be known. (Section 14.2.1). (2) *The effect of conversion* on rate and on product selectivities in representative cases: this points to the extent of 'carbon' deposition and its effect on selectivities, and to the difficulty of ascertaining the behaviour of the catalyst in its initial clean state (Section 14.2.2). (3) The following three sections will concentrate on *the effects of operating conditions* (chiefly temperature and hydrogen pressure) on rates and selectivities for linear (Section 14.2.5) alkanes: effects of catalyst structure, and of chain-length and molecular complexity of the alkane, will be noted. (4) Studies directed mainly to the relevance of *the state of the catalyst surface* either as induced by the reaction itself or by the method of preparation (precursor, pre-treatment, calcination etc.) will be the subject of Section 14.2.6. We may note here that the great preponderance of publications describe work on platinum-containing catalysts. Although important work has been done with ruthenium and palladium, platinum is unique in its ability to catalyse transformations other than hydrogenolysis. So if the metal used is not specified every time, it is safe to assume it is platinum; if it is not, its identity will be specified.

In an effort to achieve a clear analysis of the literature, Section 14.3 addresses work directed to the understanding of the *mechanisms* of skeletal isomerisation (Section 14.3.2) and of dehydrocyclisation (Section 14.3.3), covering superlative work with the use of isotopic labels. Structure-sensitivities of the component reactions have also received much careful attention, and papers addressing this matter specifically are considered in Section 14.4. Finally there has been much work on the *modification* of platinum catalysts to minimise 'carbon' formation and hydrogenolysis: studies using rhenium and similar additives (Sections 14.5.3 and 14.5.4), bimetallic systems (Section 14.5.5), sulfur (Section 14.5.6), and supports liable to give the Strong Metal-Support Interaction (Section 14.5.7) are the subject of the final section. While a modicum of repetition, and anticipation of what is to come, will be inevitable, every effort will be made to make each section as self-contained as possible.

One further general comment may be in order. With a few notable exceptions, each of the main laboratories has concentrated on the use of one particular technique or operating variable to access the heart of the problem, which is to understand what determines how a catalyst attains its unique properties. Thus Hungarian scientists, ably led by Zoltán Paál, have been emphasised, quite properly, the great importance of the amount of hydrogen on the surface, while the late François Gault and his colleagues at Strasbourg studied intently the mechanisms of reactions without employing extensive variations of operating parameters (except particle size). Some limitation of scope is of course inevitable, because life is short (tragically so in Gault's case) and so are resources, but the conjunction of separate but related studies to construct a unified picture is thereby made less easy.

14.2. REACTIONS OF HIGHER ALKANES WITH HYDROGEN: RATES AND PRODUCT SELECTIVITIES

14.2.1. Activities of Pure Metals

There are few sets of results available for comparing the activities of metals for reactions of alkane greater than C_4 by means of Arrhenius parameters based on specific rates or TOFs. Non-specific rates for hydrogenolysis of cyclopentane led[27,28] to Arrhenius parameters, the compensation plot for which divides the metals examined into three groups, as follows:

$$Ru, Rh, Os, Ir > Co, Ni, Pt > Pd$$

Metals in each group showed similar activity at 455 K, but the difference between each group was about 10^4. Activation energies ranged from 54 (Ru) to 192 (Pd) kJ mol^{-1}, and various supports and metal concentrations were used. Japanese workers[29] have examined the reactions of n-pentane on silica- and carbon-supported of the metals of Groups 8 to 10 (except Os), and reported orders in alkane that were mainly close to unity, and orders in hydrogen between -1.3 and -1.6 at 533 to 673 K, using pressures notably higher than those normally used (up to 40 atm). Arrhenius parameters were also given, but unfortunately, while the text says they were based on rate *constants*, the tabulated values were stated in units appropriate to *rate*. Values of ln A were substantially higher than expected by comparisons with lower (and higher) alkanes, perhaps due to the higher pressures of n-pentane that were used. Nevertheless, iron and palladium emerged as the least active, while the activities of the remainder were generally similar.

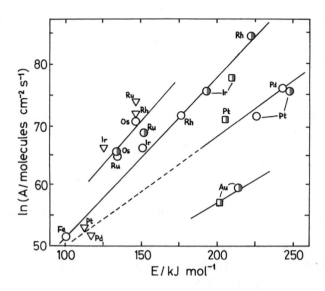

Figure 14.1. Compensation plot of Arrhenius parameters for the reactions of (i) *neo*pentane[30] and (ii) *n*-heptane[32] with hydrogen on various supported metals. They are compared with selected values for ethane hydrogenolysis, the lines being those used to classify the activities of metals for that reaction in Chapter 13 (see Figures 13.3–13.8). Ethane hydrogenolysis[31] ☉; *n*-heptane hydrogenolysis[32] ▽; *neo*pentane hydrogenolysis[30] ◑; *neo*pentane isomerisation[30] ◨.

There are two sets of results expressed in specific units that may be used to compare with Sinfelt's extensive set for ethane hydrogenolysis.[3] In Figure 14.1 selected points for this reaction are shown as a compensation plot, and they and the lines are those depicted in Figures 13.3 to 13.8. These are then used as a framework against which the Arrhenius parameters for *neo*pentane[30] and *n*-heptane[31,32] hydrogenolysis can be compared. Those for *neo*pentane, both for hydrogenolysis and isomerisation, agree well with those for ethane, although gold appears as having very low activity. In the *n*-heptane reaction,[31,32] only palladium and platinum seem to conform; ruthenium, rhodium and iridium were all more active than expected, due in the two last cases to a lower activation energy.

Product distributions were recorded, at necessarily very different temperatures (Pd, 573 K; Ru, 361 K): the five metals (Os was not studied) all gave some isomerisation, but it was only significant with platinum. With palladium, bond breaking was almost exclusively terminal, and mainly so with rhodium, but with the other three metals it was largely statistical. This behaviour conforms to that found with the butanes (Chapter 13) and with other alkanes, as we shall see.

Reflection on some of these results leads to the conclusion that iridium and rhodium sometimes ally themselves with the most active group of metals (e.g. in reactions of cyclopentane and of n-heptane) and sometimes with the group having moderate activities (e. g. for *neo*pentane). There is no obvious reason for this oscillation.

14.2.2. Effect of Varying Conversion

The extraction of meaningful information on the products formed in complex reactions is fraught with difficulties. It is of interest to know what are the initially formed products, i.e. before they are transformed by sequential reaction and before the catalyst has had time to acquire its equilibrium amount of 'carbon'. There are several ways in which conversion can be systematically varied: (1) by changing the flow-rate and hence the contact time, (2) by altering the concentration of reactants, (3) by altering the temperature, and (4) by allowing the catalyst to deactivate spontaneously, noting the changes in rate and products with time-on-stream. Method (1) is undoubtedly the most satisfactory, because the other methods necessarily cause changes in the amount of 'carbon' deposited, in the concentrations of adsorbed reactants, and in the H/C ratio of adsorbed species, but even with method (1) the composition of the surface may not stay constant, so it is necessary to know whether the catalyst has been cleaned and reactivated between measurements. With a catalyst or under conditions where hydrogenolysis is the only reaction, it is sometimes possible to monitor how product selectivities change with conversion, and to deduce not only their initial values but also their reactivities as they move towards their final value, which is inevitably 100% methane if there is enough hydrogen. In such cases, changes of selectivities with conversion are often slow, so that initial values can be obtained very accurately;[29,33-38] but when multiple types of product are formed, as for example in the reaction of n-hexane with hydrogen over a platinum catalyst,[23,39] the changes occur very quickly, and even recording product compositions at conversions of only a few percent[40] or less[39] is not adequate to define initial selectivities.

We may illustrate these two extreme situations by reference to selected examples from the literature. A simple case is the hydrogenolysis of 2,2-dimethylbutane over a cobalt catalyst at 518 K;[34] products were mainly methane and *neo*pentane, selectivities for other molecules being less than 0.1 (Figure 14.2A); but even in the more complex case of n-hexane on Ru/Al$_2$O$_3$ at 422 K,[34] rates of change were low and initial selectivities easily determined (Figure 14.2B). Publications by Paál and his associates provide numerous examples of the complexity of the early stages of the reactions of n-hexane[23,39] and other alkanes[13,16] on platinum catalysts. The extent of the changes depended on the hydrogen/alkane ratio used: when this was high (12 to 48) it increased rapidly with conversion but 'carbon' formation was limited, and each adsorbed species was at all times largely surrounded by hydrogen atoms. When it was low (e.g. 3), however, 'carbon' formation was important, and its change materially altered the composition of each reacting

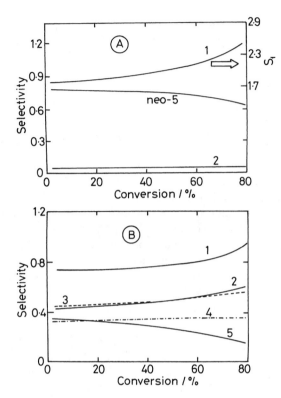

Figure 14.2. Hydrogenolysis of (A) 2,2-dimethylbutane on Co/MgO-SiC at 518 K and (B) n-hexane on Ru/Al$_2$O$_3$ at 422 K : product selectivities as a function of conversion.[34] 1 = CH$_4$; 2 = C$_2$H$_6$; 3 = C$_3$H$_8$; 4 = n-C$_4$H$_{10}$; 5 = n-C$_5$H$_{12}$.

centre. In such cases, smaller alkanes were the main or perhaps the exclusive initial product, but at high ratios isomers and cyclic products also appeared early on. With this system it is next to impossible to disentangle the effects of formation of surface 'carbon' and changing reactant ratio from the natural progression of the reaction as determined by the reactivity of the products. Rhodium catalysts on the other hand, causing only hydrogenolysis with a little cyclisation, showed (as with Ru and Co) comparatively slow changes of selectivities up to high conversion ($\sim$60%).[35,36]

14.2.3. Reactions of Linear Alkanes with Hydrogen

Of the reactions of the various types of alkane with hydrogen, those of the n-alkanes, especially n-hexane have been the focus of attention because of their prime importance in petroleum reforming: they have been ably and extensively reviewed.[11,13,14,19,41−45] In this section, attention is concentrated on the effects of varying hydrogen pressure, and the related effects of temperature, on rates of product formation and on selectivities. The alternative modes of presenting the results have already been noted (Section 14.1.4): expressing them as rates or TOFs emphasises the importance of the variable (e.g. hydrogen pressure), but so does the total rate, and the depiction of a rate versus hydrogen pressure at several temperatures can sometimes lead to negative apparent activation energies,[46] which are hard to explain. On the whole, the use of selectivities is to be preferred,[47] although this is also not without its problems, Treating each product as being formed by an independent reaction implies that the reactant at its initial chemisorption is destined to give ultimately a single defined product, and the options available for the interconversion of intermediates are thereby neglected. There is much convincing evidence to show that the structure of the site on which the alkane first chemisorbs (or to which it later moves), together with the ambient hydrogen atom concentration, determines its subsequent fate, but this does not mean that each product necessarily stems from a site of unique and specific geometry.

The reactions of linear alkanes catalysed by EUROPT-1 (6.3% Pt/SiO$_2$) have been intensively studied by Paál and his associates(see Further Reading sections 1 and 2 at the end of the chapter). Most of these studies have used n-hexane, but n-heptane[46,49] and n-nonane[50] have also been employed. Detailed measurements have been made on the rates and selectivities of product formation over a range of hydrogen/n-hexane pressures and temperatures, and more recently the results have been subjected to kinetic analysis.[25,26] In looking at these results, we must try to imagine that the surface coverage by the over-dehydrogenated species we refer to as 'carbon' will be variable,[51] but may play a deciding role in determining what products are formed (Section 14.2.6) especially at low hydrogen: alkane ratios. It is evident, but not surprising, that at low ratios *alkenes* (i.e. hexenes) are the main products. Their formation is a complication not usually encountered with the lower alkanes, because of the less favourable thermochemistry, and especially at higher temperatures some further dehydrogenation to *alkadienes* may occur: these may either cyclise into benzene by a C_6 *dehydrocyclisation*[42,48] or may form unreactive 'carbon' (see Scheme 14.2). A typical form of dependence of selectivities on hydrogen pressure is shown in Figure 14.3.[26] Results such as these, together with those found by varying the conversion (which will also affect the reactant ratio), and temperature, enable us to envisage the H/C ratio in the key intermediate leading to each product, at least in a qualitative way, and so to deduce the kind of reaction scheme shown in Scheme 14.2. The methylcyclopentane (MCP) selectivity passes through a maximum, showing that its formation requires hydrogen atoms when they are scarce, but is inhibited by higher hydrogen pressures due to suppression of the alkene from which it is formed. Skeletal isomerisation is also dependent on hydrogen availability, which is understandable if isomers are formed

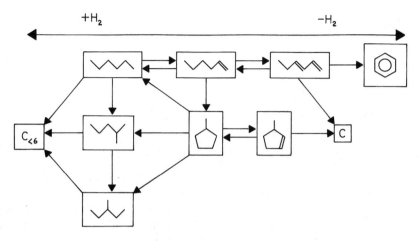

Scheme 14.2. Simplified scheme for metal-catalysed reactions of *n*-hexane.

mainly through MCP. The yield of fragments ($<C_6$) by hydrogenolysis decreased as temperature increased, unlike the situation with *n*-butane, but with *n*-hexane there is a greater variety of routes by which they may be formed. A comprehensive and quantitative model for the dependence of products of the *n*-hexane reaction on process variables seems to have eluded the efforts of the best minds to have attempted it.

A substantial amount of work has also appeared on the reactions of *n*-hexane catalysed by *platinum black* (see Further Reading section 2). Use of the unsupported metal avoids any possible complications due to support effects, but the particle size was large; 'carbon' formation was more noticeable,[53] but it was more active for hydrogenolysis than the highly-dispersed EUROPT-1. The manner of variation of the other products with hydrogen pressure was however generally similar,[52] but in the reaction of *n*-heptane[49] the formation of toluene and of 3-methylhexane was more suppressed by high hydrogen pressure. In this reaction Pt/Al$_2$O$_3$ also gave more alkane fragments than EUROPT-1.[49] Product distributions in the *n*-hexane reaction were generally similar on Pt/Al$_2$O$_3$ and Pt/SiO$_2$, although activities (and therefore temperatures) differed; Pt/C gave mainly terminal fission.[54,55] Other work on Pt/Al$_2$O$_3$ catalysts has been directed more towards the formation and removal of 'carbon'.[57] Platinum in KL zeolite is renowned for its efficient aromatisation of *n*-alkanes, and it has been extensively studied:[50,58] it will receive further mention in Section 14.5. Results for various single-crystal surfaces have been compared with those for Pt/SiO$_2$.[49,52,59] They will be reviewed in the context of particle-size and surface geometry effects (Section 14.4). A Pt/silicalite catalyst also effected DHC of *n*-hexane.[60]

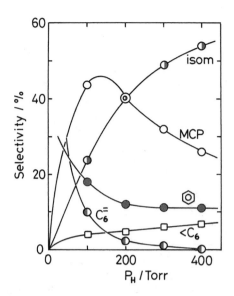

Figure 14.3. Hydrogenolysis of n-hexane over EUROPT-1 (Pt/SiO$_2$): product selectivities as a function of hydrogen pressure at 603 K.[26]

Product selectivities observed with platinum catalysts vary markedly with the chain length of the alkane (Table 14.1).[16,52] Products of hydrogenolysis, always greater on platinum black than on Pt/SiO$_2$ (markedly so in the case of n-heptane),[61] were a minimum with n-hexane, and skeletal isomerisation decreased with chain length as other options such as C$_5$ cyclisation became available. We should note that the existence of five carbon atoms in a chain is not enough to ensure efficient cyclisation, and the progressively greater flexibility provided us chain-length is increased allows more opportunities for cyclisation to occur, the more so on the smaller platinum particles of EUROPT-1.[62] All C—C bonds in n-alkanes have comparable probabilities of breaking, but there is a tendency for the chance to decrease on moving towards the centre of the molecule.[19,51]

Arrhenius parameters calculated[26] for n-hexane removal at various hydrogen pressures for EUROPT-1 two types of platinum black give an excellent compensation plot (Figure 14.4); this is not unexpected, as the same behaviour had been found with the lower alkanes (Section 13.2.2; Figure 13.12 and 13.13) and is explicable by the Temkin equation (Section 5.2.5) and the general model presented in the last chapter. The points obtained at or above the rate maximum agree well with the line for the total reaction of lower linear alkanes on EUROPT-1 (Figure 13.4); those obtained below the rate maximum, although not inconsistent with those for higher pressures, lie somewhat below the standard line. This may reflect the difficulty of getting perfectly clean surfaces at low hydrogen pressures. Parameter

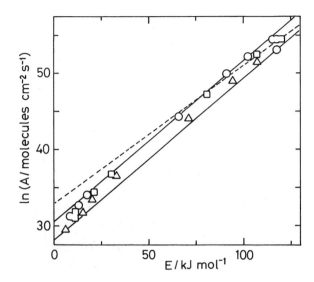

Figure 14.4. Compensation plot of Arrhenius parameters for the total reaction of *n*-hexane on two kinds of platinum black (Pt-N, □; Pt-HCO, △) and EUROPT-1 (Pt/SiO₂ O) measured at various hydrogen pressures.[26] The parallel lines delineate a band within which the points fall; the broken line is that for EUROPT-1 taken from Figure 13.4. Note that Pt-HCHO is somewhat less active than the other catalysts.

values estimated[25] for the major contributing processes on EUROPT-1 also show compensation (Figure 14.5), and with care and a little imagination separate lines for each product can be identified. The propriety of applying the Arrhenius equation to reactions that are not rate-limiting has already been questioned, and it is doubtful whether this exercise greatly advances our understanding of the system. Interpretation of the isokinetic parameters presents many pitfalls for the unwary.[26]

Much less work has been done on palladium catalysts, which show comparable activity to platinum for the *n*-alkanes, but with very different behaviour.[63] Tremendous differences were shown between unsupported palladium (foil and (111) surface)[64] and Pd/Al₂O₃ treated in various ways[65,66] (Table 14.2). The former gave dehydrogenation as one of the major routes, hydrogenolysis being the other, with small amounts of cyclic alkanes and benzene. Of the lower alkanes, methane was the major component, its production being increased with the hydrogen/*n*-hexane ratio. On Pd/Al₂O₃ reduced at 573 K (LTR) hydrogenolysis was the chief route,[66] but isomerisation and cyclisation to MCP also occurred, in amounts that tended to rise with metal loading. Reduction at 873 K (HTR) produced dramatic changes: activity at low metal loading (0.3%) was increased some 300 times and activation energy lowered (Figure 14.6), and isomerisation then became the major route (~90%). The rise in rate was less at higher metal loadings, although the

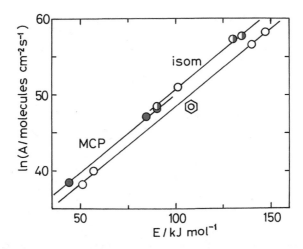

Figure 14.5. Compensation plot of Arrhenius parameters for the production of branched isomers (◖), methylcyclopentane (●) and benzene (○) from n-hexane on EUPOPT-1.[25]

change in products was retained. Lewis acid sites generated at 873 K, adjacent to palladium particles, may have been responsible. Regeneration after HTR by oxidation and LTR gave catalysts that showed intermediate behaviour. Arrhenius parameters showed compensation,[66] the general level of activities being close to the upper side of the band that encompassed platinum (and palladium) catalysts in their activity towards lower alkanes (Figure 13.3). Dehydrogenation, which was not reported, may have been suppressed by the use of high hydrogen pressure. The main difference between platinum and palladium lies in the inability of the latter to bring about cyclisation, although, once accomplished, dehydrogenation to benzene is easy, especially at higher temperatures.[64] Cyclic products did however amount to 42% of the total on regenerated 2.8% Pd/Al$_2$O$_3$.[66]

We turn now to the metals that are more active than palladium and platinum for the reactions of alkanes with hydrogen. Rhodium catalysts have been the subject of a number of investigations:[35,36,67-71] they were active for n-hexane hydrogenolysis between about 420 and 500 K, and were characterised by giving

TABLE 14.2. Hydrogenolysis of n-Hexane on Various Palladium Catalysts

Form	T/K	P_H/kPa	$TOF/s^{-1} \times 10^3$	$S_{<6}$	S_i	S_{C_5}	S_{C_6}	S_{arom}	References
Foil	573	8.5	52	52	—	5	4	39	64
(111)	573	8.5	44	27	—	4	5	64	64
/Al$_2$O$_3$LTR	563	127	4	79	10	10	1	—	65,66
/Al$_2$O$_3$HTR	563	127	1370	5	90	4	1	—	65,66

See footnote to Table 14.1.

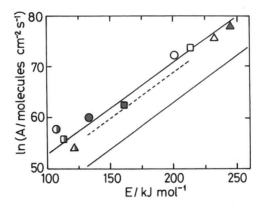

Figure 14.6. Arrhenius parameters for hydrogenolysis of n-hexane on Pd/Al_2O_3 shown as a compensation plot: effect of various metal loadings and pre-treatments (see text). The parallel lines delineate the zone that contains the points for lower alkanes on catalysts in Figure 13.3; the broken line is that for EUROPT-1.[66] O, 0.3% Pd; □, 0.6% Pd; △, 2.8% Pd. Open points, LTR; half-filled points, HTR; filled points, regenerated.

mainly hydrogenolysis (Table 14.3) The small amount of cyclisation that occurred decreased with increase in hydrogen pressure;[36] it did not happen at all on rhodium film. With Rh/Al_2O_3 and Rh/SiO_2 reduced at either 603 or 1253 K, high hydrogen coverages (i.e. high pressure, low temperature) gave random single C—C bond fission, while the opposite conditions encouraged multiple breaking and methane formation.[71] There was a trend from terminal to internal C—C fission as hydrogen pressure was increased. Rh/SiO_2 resembled Rh(111) more than Rh(100) in showing a high activation energy (197 kJ mol^{-1}) and a preference for internal C–C bond breaking with n-pentane hydrogenolysis:[70] pre-oxidation gave higher rates with lower activation energy, and no products other than lower alkanes were noted.

n-Pentane reacted with hydrogen at 423 K on Ir/Al_2O_3 to give mainly ethane and propane, with traces of cyclisation and homologation.[51] On iridium film,[51] n-hexane gave all the lower alkanes but no benzene below 544 K, its yield increasing above 598 K with methane as the other chief product. Hydrogenolysis was also the principal route on iridium single crystal faces[72] and on iridium foil.[73]

TABLE 14.3. Product Selectivities for the Reaction of n-Hexane with Hydrogen over Rhodium Catalysts at 498 K.[36]

[Rh]/%	Support	H_2:n-C_6/Torr	$S_{<6}$	S_i	S_{C_5}
10	Al_2O_3	60 : 10	87	8	6
//	//	480 : 10	92	7	0.5
0.3	//	480 : 10	97.5	1.5	0.3
5	SiO_2	480 : 10	98	2	0.4

On Ir/Al$_2$O$_3$ containing various chlorine contents, n-hexane at 513 K afforded mainly the lower alkanes, the selectivity for which increased with temperature and hydrogen pressure;[74,75] small amounts of other products varied with operating conditions much as with platinum catalysts. Depth of hydrogenolysis signalled by the ζ factor increased rapidly with rising temperature.[76] DHC of n-heptane was more effective on Ir/Al$_2$O$_3$ than on Pt/Al$_2$O$_3$, but there was also more hydrogenolysis, which could not be suppressed by sulfiding.[77]

Ruthenium catalysts are also noted for their high activity for hydrogenolysis, but are capable of giving some surprises. An early study[34] of the reaction of n-hexane with hydrogen on Ru/Al$_2$O$_3$ showed that at low conversion (422 K) the breaking of C—C bonds was more or less statistical; conversions were followed to 80%. Ru/TiO$_2$ catalysts showed[78] the same marked variations in activity for n-hexane hydrogenolysis as the pre-treatment was changed as were found with the lower alkanes (Section 13.6); exceptionally fast rates were observed with 5% Ru/TiO$_2$ after LTR following reduction at 893 K. Catalysts having 0.1 or 0.5% ruthenium were however able to show skeletal isomerisation when in the SMSI state,[79] after reduction at 758 or 893 K, a value of S_i of 94% being found at 633 K with 0.1% Ru/TiO$_2$ made by ion exchange using the [Ru(NH$_3$)$_6$]$^{3+}$ion. Residual hydrogenolysis activity could be further lowered by treatment of 0.5% Ru/TiO$_2$ with thiophene.[78] These results nicely illustrate the priority given to reactions demanding large ensembles of atoms when such are available. When they are not, isomerisation becomes possible.

Ru/ZSM-5 catalysts modified by inclusion of either rhenium or rhodium or nickel have been examined[80] for n-hexane hydrogenolysis at 403 and 433 K: methane was the main product in most cases, but Ru-Ni/ZSM-5 gave chiefly isoheptane from the reaction of n-heptane at 423 K.

Nickel resembles palladium in giving principally demethylation of alkanes,[81−83] although C$_2$ to C$_4$ alkanes were also initial products of n-hexane with Ni/MgO-SiC at 528 K:[34] less methane was formed from n-pentane on Ni/SiO$_2$ as hydrogen pressure was increased.[82] The possibility of reaction proceeding *via* a π-allylic intermediate again needs to be considered. Cobalt catalysts show greater tendency to multiple fragmentation to methane.[34]

Rhenium film gave minor amounts of benzene and isomers in the products of the reaction of n-hexane.[84]

14.2.4. Reactions of Branched Alkanes with Hydrogen

The introduction of single or double branches (i.e. of tertiary or quaternary carbon atoms) into alkane molecule immediately further differentiates the C—C bonds; thus for example 2-methylpentane has four. The presence of branches also allows a greater variety of modes of attachment to the surface by dissociation of C—H bonds: it is generally assumed that reactive species must be σ-diadsorbed,

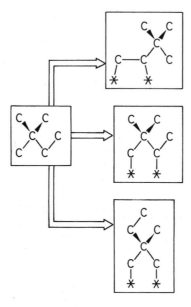

Scheme 14.3. Reactions of *neo*hexane.

although for skeletal isomerisation a single point bonding is sometimes preferred.[85] So *neo*hexane (2,2-dimethylbutane) can be diadsorbed in either the $\alpha\beta$-, the $\alpha\gamma$ or the $\alpha\gamma'$ modes; those and the products to which they may give rise are shown in Scheme 14.3. This molecule has been the subject of intensive study.[17,86] While all branched alkanes can undergo hydrogenolysis, possible alternative products are circumscribed by the molecule's structure. *Neo*pentane (2,2-dimethylpropane) is initially limited to isomerisation to 2-methylbutane, but molecules containing five or more carbon atoms in a straight line can also cyclise.[2] The range of possible hydrogenolysis products is also much increased by the presence of branches, and there has been great interest shown in different reactivities of various types of C—C bond, although the results have not always been explained in terms of the preferred forms of chemisorption. Quite complex molecules can however show very simple reaction paths: thus both 2,2,3,3-tetramethylbutane and 2,2,4,4-tetramethylpentane[87] contain only two distinguishable sorts of C—C bond. The former has been particularly widely used[18,67,88−90] in the expectation that formation of the $\alpha\delta$-diadsorbed state, presumed to be needed for breaking the central C—C bond, might be difficult in very small particles, so that the product distribution would be sensitive to surface structure, i.e. particle size and composition.

   The principal themes of work in branched alkanes have therefore been (1) selectivities for isomerisation and cyclisation as opposed to hydrogenolysis, (2) the nature of the products of hydrogenolysis, and (3) the structures of intermediate species and their reaction mechanisms. Most emphasis has been placed on the effect of surface structure (crystal orientation, particle size) and composition (surface state, presence of modifiers) on rates and the above three themes. These are developed in Sections 14.4, 14.2.6 and 14.5. There has been comparatively little straightforward *kinetic* work to explore the effects of experimental variables, and what there has been is not especially helpful. Positions of equilibria between the C_6 isomers, benzene and methylcyclopentane have been calculated and compared with experimental values.[2,70]

The multiplicity of bond-breaking for 3-methylpentane at low conversion for various metal blacks was much as expected: rhodium, palladium, iridium and platinum give predominantly two fragments,[91] the first formed new species thus desorbing quickly, while osmium in particular gave mainly methane, with other metals showing intermediate behaviour. The environment of the metal does matter, however; with 2-methylpentane, Co/Al$_2$O$_3$ gave chiefly deep hydrogenolysis, but Co/NaY zeolite was said to be an excellent isomerisation catalyst.[92] As usual the most plentiful (and interesting) results have been obtained with platinum catalysts.

The derivation of reaction networks for deriving selectivity equations relevant to hydrogenolysis of branch alkanes has been extended from *iso*butane to 2-methylbutane[33,37] and the dimethylbutanes.[34,38] Arrhenius parameters for skeletal isomerisation and hydrogenolysis of 2-methylbutane and of *n*-pentane on 10% Pt/Al$_2$O$_3$ are shown[93] as a compensation plot in Figure 14.7. Values for the movement of a ^{13}C label within the structure but without change to its chemical identity are included; activation energy for this 'self-isomerisation' of *n*-pentane is notably high (300 kJ mol^{-1}), and only the data for its demethylation are seriously wide of the mark. As we now expect, there was a distinct trend of the order of reaction in

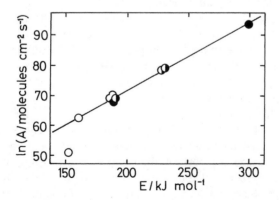

Figure 14.7. Reactions of 2-methylbutane and of *n*-hexane on 10% Pt/Al$_2$O$_3$: Arrhenius parameters as a compensation plot.[93] Hydrogenolysis (open); isomerisation (half-filled); self-isomerisation (filled).

hydrogen becoming more negative with increase in activation energy. The core of this important paper is the mechanism of skeletal isomerisation of labelled pentanes, and it will be revisited on Section 14.3. The order in hydrogen for isomerisation of 2-methylbutane was much more negative than that for hydrogenolysis on 0.2% Pt/Al_2O_3 heated by microwave radiation,[94] so that isomerisation selectivity S_i was greatest at low hydrogen pressure, as found for the reactions of n-butane (Section 13.4) and in qualitative agreement with what was found with 10% Pt/Al_2O_3.[93]

The manner of variation of rate with hydrogen pressure/concentration depends somewhat in the mode of reaction used. In the cyclisation of 3-methylpentane on platinum black, sharper maxima were seen when a recirculation reactor was used than when a pulse mode was employed, and inhibition at low hydrogen pressure due to 'carbon' deposition, which was marked 633 K in the former case, was absent in the latter.[95] In both cases, the hydrogen pressure giving rate maxima increased with temperature.

While with linear and singly-branched alkanes there is clear but not extensive evidence that on platinum catalysts the intermediates for isomerisation and for hydrogenolysis differ in their extents of dehydrogenation, with doubly-branched alkanes as exemplified by *neo*pentane (2,2-dimethylpropane) the situation appears not the same. In an extensive review of Arrhenius parameters for its reactions,[96] activation energies for the two reactions were found to be of the same order,[85,97] as were orders of reaction (for Pt/KL and Pt/KY zeolites[98,99]). On EUROPT-1 and on 'oriented' model platinum catalysts, activation energies for total reaction increased markedly with hydrogen pressure, as indeed they should.[85] The two reaction paths thus seem to go *via* the same intermediate, which might be the $\alpha\gamma$-diadsorbed species.[100]

The palladium-catalysed reactions present a different picture, however:[63,95,98,99,101-104] activation energies for hydrogenolysis were uniformly higher ($\sim$300–370 kJ mol^{-1}) than for isomerisation ($\sim$200–250 kJ mol^{-1}), and orders in hydrogen much more negative (-4 compared to -1.9 for Pd/KL;[98] -3.6 compared to -0.6 for Pd/SiO$_2$[99]). On palladium, therefore, the hydrogenolysis intermediate must be the more dehydrogenated, being perhaps an $\alpha\alpha\gamma$-species. This difference does not however prevent the data points for the two metals from sharing a common compensation line.

The possible reaction paths available to *neo*hexane (2,2-dimethylbutane) were shown in Scheme 14.3. On various supported platinum catalysts, values of S_i (to all isomers) were between 36 and 74%, depending no doubt on factors such as dispersion and surface composition, but only 9% on platinum black.[17,106] Detailed product analysis revealed that most of the products (50–70%) were formed through the $\alpha\gamma'$ route and most of the remainder through the $\alpha\gamma$ route. 2.2-Dimethylbutane reacted similarly on Pt/SiO$_2$ at 568 K.[107] This is striking evidence of the ease with which such species can be found on platinum surfaces, to

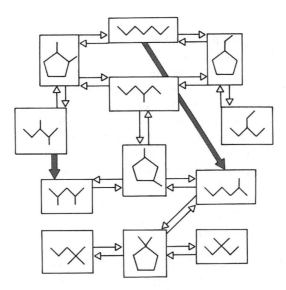

Scheme 14.4. Cyclic intermediates in the skeletal isomerisation of C_7 alkanes. Routes only feasible by bond-shift are shown by dark arrows.

the almost total exclusion of the $\alpha\beta$ route. Other doubly branched alkanes have been examined,[17,55,107,108] and rates on platinum black as a function of hydrogen pressure sometimes passed through extraordinarily sharp maxima, usually at very low hydrogen pressure.[52] Poorly-dispersed Pt/Al$_2$O$_3$ (H/Pt = 0.09) hydrogenolysed 2,2,3,3-tetramethylbutane almost exclusively by the $\alpha\delta$ mode, as expected,[109] but this continued the major way up to H/Pt = 0.99; the predicted correlation of reaction mode with particle size is clearly not straightforwardly obeyed.

Reactions of the C_7 isomeric alkanes (Scheme 14.4) have been studied on platinum black and on EUROPT-1,[16,52,110] and rates of formation of each product followed as a function of hydrogen pressure. C_5 cyclisation products peaked at moderate pressures (20–30 kPa), while rates for isomerisation, aromatisation and hydrogenolysis usually rose continuously. From such results it is not easy to divine any general principle concerning the optimum degree of dehydrogenation for each process, although C_5 cyclisation clearly requires a moderately dehydrogenated species.

The importance of the metal's identity and degree of dispersion is dramatically illustrated by the reactions of neohexane over iridium catalysts. Ir/SiO$_2$ (unlike Pt/SiO$_2$) gave only hydrogenolysis, 94% of which occurred in the $\alpha\beta$ mode (see Scheme 14.4), but iridium black (like platinum black) also gave much hydrogenolysis, but chiefly by the $\alpha\gamma'$ route. 2-Methyl- and 2,2-dimethylbutane reacted with hydrogen over Ir/Al$_2$O$_3$ in the range 423–495 K giving mainly methane and the corresponding branched alkane;[51] variation of hydrogen pressures gave rate maxima that decreased with the number of branches. The small amounts of isomers decreased as hydrogen pressure was raised. It was noted that mechanisms for isomerisation starting with a mono-σ-bonded species are not consistent with the observed dependence of rate on hydrogen pressure. *Neo*pentane was not isomerised on iridium film,[42] and 2-methylpentane isomerised to 3-methylpentane on Ir/Al$_2$O$_3$ and iridium sponge by the cyclic mechanism.[111]

The effects of hydrogen pressure variation on rates and product selectivities in the reaction of 2-methylpentane have been reported[36] for Rh/Al$_2$O$_3$ catalysts having 10 and 0.3% metal, and for the less active 5% Rh/SiO$_2$. At 483 K hydrogenolysis predominated, although up to about 20% isomerisation also occurred, mostly by the bond-shift mechanism (Section 14.4). S_i was greatest on 10% Rh/Al$_2$O$_3$, where it was independent of hydrogen pressure, but on 0.3% Rh/Al$_2$O$_3$ it decreased, and on the Rh/SiO$_2$ it increased, with hydrogen pressure. Small amounts of cyclisation also happened, their selectivity decreasing with rising hydrogen pressure in all cases. Unlike the situation with *n*-butane, no maximal TOFs were observed.

Highly dispersed rhodium on various supports favoured hydrogenolysis of 2,2,3,3-tetramethylbutane by the $\alpha\gamma$ mode, giving methane and trimethylbutane; only the poorly active Rh/MgO gave isobutane as the chief product.[67] A kinetic study[88] of Rh/Al$_2$O$_3$ in states of high and low dispersion (H/Rh respectively 1.17 and 0.08) showed (i) a more negative order in hydrogen on the latter, i.e. stronger adsorption of hydrogen, (ii) mainly fission by the $\alpha\delta$ mode on the latter, its selectivity decreasing as hydrogen pressure rose, (iii) mainly fission by the $\alpha\beta$ mode on the former, its selectivity decreasing at low hydrogen pressures, and (iv) activation energies that were higher for $\alpha\delta$ mode, although for both modes they increased (as expected) with hydrogen pressure. These observations suggest that the $\alpha\delta$ intermediate is more highly dissociated than that for the $\alpha\beta$ mode. The contrast in the hydrogen orders between this reactant and 2-methylpentane was very marked. Similar results have been obtained[112] with Rh/Al$_2$O$_3$ catalysts at 453 K using 2,2,3-trimethylbutane as reactant.

Ruthenium is noted for its tendency to give multiple hydrogenolysis at low conversions,[91] although this was less at high dispersion,[89,90] and for its inability to show isomerisation except under special circumstances: small amounts of the latter, and some cyclisation,[8,90] has however been reported with 2-methylpentane on Ru/Al$_2$O$_3$ of moderate to high dispersion (H/Ru > 0.35). On 0.5% Ru/Al$_2$O$_3$ 2-methylbutane reacted with the same activation energy as *neo*pentane (182 kJ mol^{-1}), but some 200 times faster,[33] suggesting that somewhat special sites are needed to allow $\alpha\gamma$ diadsorption to take place. Demethanation of 2,2-dimethylbutane to *neo*pentane was the almost exclusive at 473 K, very similar results being obtained with supported nickel and cobalt catalysts:[34] with 2,3-dimethylbutane, demethanation again predominated, but *iso*butane (from double-demethantion) as well as ethane and propane were also initial products.[34]

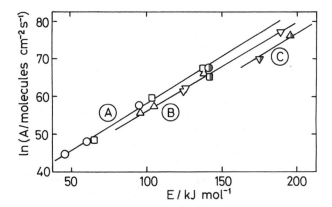

Figure 14.8. Hydrogenolysis of alkanes (*n*-hexane, 2-methylpentane and 2,2,3,3-tetramethylbutane, the last shown as half filled points) on Ru/Al$_2$O$_3$ of various dispersions.[89] Line A covers low dispersions (Sponge, O; H/Ru = 0.07, □); line B covers high dispersions (H/Ru = 0.37, △; H/Ru = 1.1, ▽); line C covers the tetramethylbutane on the high-dispersion catalysts where the $\alpha\beta$ mode predominates. Note that TOF for this reactant is more sensitive to changes in dispersion than the others.

2,2,3,3-Tetramethylbutane reacted much more slowly than either *n*-hexane or 2-methylpentane, due to notably higher activation energies; a selection of the Arrhenius parameters for these reactants on Ru/Al$_2$O$_3$ of various dispersions[89] is shown in Figure 14.8. Activation energies did not however change smoothly with dispersion. Only the data points for the tetramethylbutane on poorly-dispersed catalysts lie well below the lines embracing the others; this is because the expected but slower $\alpha\delta$ mode is more significant in these cases. Variation of hydrogen pressure revealed[113,114] that on small (1 nm) particles the rate maximum for the $\alpha\gamma$ process occurred at a *higher* pressure than for the $\alpha\delta$ process, the $\alpha\gamma/\alpha\delta$ ratio increasing continuously with hydrogen pressures above 40 Torr. On large (4 nm) particles the $\alpha\delta$ rate was maximal at a very low hydrogen pressure and showed a large negative order (-2.9). Activation energies were independent of hydrogen pressure above 25 kPa ($\alpha\gamma$ mode, 139; $\alpha\delta$ mode, 159 kJ mol^{-1}) for 1 nm particles, but were variable between 110 and 210 kJ mol^{-1} for 4 nm particles. These results imply that (i) the $\alpha\delta$ intermediate was the more dehydrogenated, and (ii) hydrogen chemisorption was strong, so that its coverage was high over most of the pressure range on both types of catalyst. Comparison between them is not easy because of the variable contributions of the two modes; it appears that it was the TOF for the $\alpha\delta$ mode that was chiefly sensitive to dispersion. A parallel study with propane and *n*-butane[115] concluded that hydrogen chemisorption was stronger on the smaller particles.

The base metals of Groups 8 to 10 resemble ruthenium quite closely in their preference for demethanation and easy multiple hydrogenolysis. Results are available for nickel,[34,38,81,89,91,107,108] cobalt[34,38] and iron[34,38,116]. With

2,2-dimethylbutane, splitting by the $\alpha\beta$ mode between C3 and C4 was the principal route on Ni/MgO-SiC, but with 2-methylbutane[37] it was the $\alpha\beta$ C1—C2 fission that was preferred, and with 2,3-dimethylbutane[38] double demethanation also occurred at low conversion. A detailed and systematic study[81] of a number of branched alkanes on 20% Ni/SiO$_2$ confirmed the preference for demethanation, the reactivity of the C—C bond decreasing as the multiplicity of the atom attached to the terminal atom increased (i.e. C_I—C_{II} > C_I—C_{III} > C_I—C_{IV}, where C_I is primary, C_{II} secondary etc.). It was concluded that $\alpha\gamma$- and $\alpha\delta$-species adequately explained the preference shown by nickel for demethanation.

Mechanisms of aromatisation of several branched alkanes have been investigated on nickel catalysts, with the aid of ^{13}C labelling using 3-methyl*-pentane.[117] In addition to a C$_5$ cyclisation plus ring enlargement route, two other routes participated: (i) an ethenyl-shift of 3-methyl*-pentane to n-hexane, followed by 1,6-cyclisation, and (ii) an addition-abstraction route involving mono-carbon species (i.e. 3-methylpentane + C$_1$ $\rightarrow$ toluene $-$ C$_1$ $\rightarrow$ benzene).

14.2.5. Reactions of Cyclic Alkanes with Hydrogen[118]

The reactions of cyclic alkanes with hydrogen have attracted enormous interest. Cyclic C$_5$ molecules have been strongly implicated as intermediates in the skeletal isomerisation of alkanes (see Section 14.3), either as visible transitory products or as inferred adsorbed (virtual) species not vacating the surface:[19,100] their role in aromatisation of alkanes through ring-enlargement on catalysts having a purely metallic function is however more debatable (see later). For these reasons, cyclic C$_5$ molecules, especially methylcyclopentane, have been extensively studied; the variety of reactions that they undergo[119] (Section 14.1.3), and in particular selectivities towards the various isomers that ring-opening generates, have proved irresistible magnets for those who have hoped to use catalytic reactions as means of characterising metal surfaces.[120] Although some useful generalisations have emerged, the conclusions reached have not perhaps fully repaid the effort invested. Cyclopentane itself has not been much studied, except to compare the activities of various metals[28,121] and to assess the particle-size dependence of its hydrogenolysis rate;[122–124] its tendency to deactivate rhodium and palladium catalysts has also been noted.[125] Cyclohexane reacts mainly be dehydrogenation and aromatisation,[13,126] and its hydrogenolysis has been neglected.[127] Larger ring systems undergo reactions of considerable interest,[13,19] although these too have not received detailed attention.

Although the C—C—C bond angles in the C$_5$ ring are close to that for tetrahedral carbon, the fission of a C—C bond by hydrogenolysis occurs more readily than that of linear alkanes.[2] Reaction takes place at lower temperatures and with lower activation energies, and the order of reaction in hydrogen (for methylcyclopentane) is positive, where for an acyclic alkane it would be negative.[128] This is clearly shown at least for platinum catalysts, where the corresponding acyclic alkanes constitute the major if not the sole initial products.[24][128-131] These are rapidly desorbed, although with some other metals more active for hydrogenolysis further fragmentation occurs.[113] The molecule is therefore more strongly chemisorbed than an ordinary alkane, possibly by reaction with a hydrogen atom to give a cyclopentyl radical in a rate-determining step. It certainly appears that the essential intermediate is more hydrogen-rich than those for all other transformations. Subsequent steps have not been clearly defined[19,113] (see Section 14.3.1), but the suggestion that dehydrogenation to an $\alpha_2\beta_2$-tetra-adsorbed species must precede C—C bond breaking is not in accord with the kinetics. The process has been aptly described as 'a peculiar sort of C—C bond rupture".[13]

The introduction of a methyl group differentiates the C—C bonds of the C_5 ring into three types according to whether the product is n-hexane, 2- or 3-methylpentane. Although discussion of the effects of particle size and surface geometry on *rates* is to be deferred to Section 14.4, it is impossible to consider the reactions of methylcyclopentane (MCP) without reference to particle-size and related effects. The literature[13,19] recognises two extremes of mechanisms according to the extent that the substituent shields the adjacent C—C bonds. By the so-called *selective mechanism* this shielding is complete and no n-hexane is made; this situation has been found with 10% Pt/Al$_2$O$_3$,[132] but only at a single temperature (506 K); at higher temperatures, n-hexane became a significant product.[132,133] It has also been said to be absent from the products on Pt(100) and Pt (111) at 540 to 650 K,[134] but this observation was not confirmed on Pt(111) or Pt(557) (or Pt foil) at 623 K.[133] When it is certainly observed, however, it is with *large* particles[109,132] or extended surfaces.[134] With the *non-selective mechanism* this shielding effect is absent, and all three types of bond are reactive: if they were equally so, the products n-hexane, 2-and 3-methylpentane would be as 2:2:1. This type of distribution has been most often closely approached by catalysts containing *small* platinum particles[21,109,129,135] (Table 14.3), but extremely small platinum particles have inexplicably given only 10% n-hexane.[130] With Pt/Al$_2$O$_3$, the mechanism moved towards the non-selective form as dispersion and hydrogen pressure were decreased and as temperature increased.[19,113] While in the great majority of cases the observed selectivities lay between the expected limits, the ratio of 2-methylpentane 3-methylpentane has been not infrequently less than two (e.g. with Pt/KL zeolite[130]) and sometimes a little greater.[129,135] Fact other than the purely statistical can clearly affect the case of breaking of the C2—C3 and C3—C4 bonds. A third *partially selective mechanism* has been proposed, but it is hard to see what this might entail. It is probably safer to believe that the condition of the surface and the nature of the sites available determine the extent to which the methyl group interferes with the process of chemisorption or the stability the adsorbed state, so that a gradual transition between the two limits might be envisaged. It is not however clear what circumstance could prevent the chemisorption of MCP by loss of the hydrogen atoms at C1 and C2 on the *unobstructed* side. However this may be, we may assign a *degree of selectivity S* to the ring-opening by linear interpolation of the observed amount of *n*-hexane between the theoretical limits of zero and 40%.[19] The non-selective route may proceed by an *adlineation mechanism* involving adjacent sites on the metal and the support.[136,137] The evidence suggested the MCP was dissociatively chemisorbed on Pt/SiO$_2$ at 623 K, since some -OD groups on the support were changed to -OH.[138]

Table 14.4 contains a small selection of the reported product distributions, but a fuller recapitulation is not warranted because in much of the earlier work (and some of the later) there is a lack of awareness of the importance of experimental variables in determining product selectivities. Relevant factors include (i) surface cleanliness (especially 'carbon') and type of pre-treatment,[139] (ii) temperature,[95] and (iii) the hydrogen/MCP ratio used.[24,130,135] There is however some disagreement as to the importance of the latter.

TABLE 14.4. Hydrogenolysis of Methylcyclopentane on Platinum Catalysts: Selectivity Parameters

Form	D/d	T/K	S$_{<6}$	S$_2$	S$_3$	S$_n$	S	References
Black	—	573	—	61	22	16	60	95
Foil	—	623	—	55	25	20	50	133
/SiO$_2$^{a}	1.8 nm	723	0.5	52	16	31	23	130
/SiO$_2$^{a}	1.8 nm	530	<5	38	18	44	~0	136,180
//	//	548	2	46	14	40	0	303
/SiO$_2$	0.12	503	—	66	20	14	65	129
/SiO$_2$	small	758	—	37	24	39	2	135
/SiO$_2$^{b}	1.7nm	520	—	38	18	44	~0	136
/SiO$_2$^{b}	0.15	520	—	57	26	17	58	180
//	10.4nm	520	<5	61	27	12	70	136
/Al$_2$O$_3$^{c}	0.12	506	—	78	22	0	100	19,132
/Al$_2$O$_3$^{c}	0.12	589	—	61	28	11	63	132
/Al$_2$O$_3$	1.2 nm	513	—	39	19	42	0	109
/Al$_2$O$_3$	12.3 nm	513	—	80	20	0	100	109
/Al$_2$O$_3$^{d}	small	523	—	42	21	37	5	132
/Al$_2$O$_3$^{e}	—	483	—	66	23	9	78	132
/Al$_2$O$_3$^{e}	small	573	—	37	24	39	2	302
/TiO$_2$	0.59	483	2	54	22	22	45	129
/MgO	0.45	520	<5	58	21	21	48	136,180
/K-LTL	small	723	1.7	66	23	9	78	130
/K-L	small	623	~5	39	27	34	15	128

Column headings: D = fractional dispersion, *d* = size (nm): *S* subscripts; < 6 = smaller alkanes; 2 = 2-methylpentane; 3 = 3-methylpentane; *n* = *n*-hexane: S = mechanistic selectivity (%).
^{a}EUROPT-1 (6% Pt/SiO$_2$) :
^{b}'model' catalysts made by vacuum deposition :
^{c}[Pt] = 10%;
^{d}[Pt] = 0.2%,
^{e}[Pt] = 1%.

Other products formed in the reaction of MCP with hydrogen are usually minor (Table 14.4). They include benzene, the formation of which increased with decreasing hydrogen pressure[140] and with increasing temperature[128] (which has the same effect on the concentration of adsorbed hydrogen). It is now thought to arise mainly by 1,6-dehydrocyclisation of n-hexane via linear unsaturated C_6 species (Scheme 14.3), and is therefore sometimes included with it.[130] Alkanes were very major products of the reactions of cyclopentane and MCP over platinum black and EUROPT-1 at 603 K,[141,142] in amounts that understandably decreased with increasing hydrogen pressure. The same trend was observed in $<C_6$ fragments, which must therefore have arisen from thoroughly dehydrogenated intermediates; their amounts have been analysed in detail.[139]

In the further consideration of the platinum-catalysed reaction, we focus on a few quite recent papers that contain kinetic information, and from which references other than those already cited can be gleaned. There have been two studies relating partly or wholly to EUROPT-1 (6.3% Pt/SiO$_2$), but unfortunately they are in substantial disagreement, and different models were used to explain the results. The first[24] gave the dependence of rate on hydrogen pressure at three temperatures; the pressure giving maximum rate increased with temperature, and the activation energy increased from 90 to 220 kJ mol^{-1} as the hydrogen: MCP ratio was increased from 6.7 to 83. The second[130] gave the intermediate value of 138 kJ mol^{-1} at a ratio of 40. A basis for explaining this variation was suggested in Section 13.2.4. The first paper[24] also showed that the order in MCP increased as the fixed hydrogen pressure and temperature were raised. This showed that MCP was competitively and exothermically adsorbed. This paper also reported that selectivities depended on MCP and hydrogen pressures, in the sense that the purely non-selective reaction only took place at high hydrogen and low MCP pressures. In the second paper[130], selectivities were independent of hydrogen/MCP ratio over much of the range covered, but they were not expressed in terms of separate reactant pressures. This paper also reported minor benzene formation; the other did not see it. The implied participation of two 'mechanisms', the contributions of which vary with experimental conditions, makes the modelling exercise difficult. The analyses proposed in both these papers can be criticised on a number of grounds. One model suggested different modes of adsorption of MCP as being responsible for the two 'mechanisms', and a rate-controlling step involving molecular hydrogen.[24] The slow step in the other model[130] apparently involved several hydrogen atoms. Comparison of these two papers illustrates how the diversity of experimental and theoretical procedures so often used in this field renders arrival at agreed conclusions a hazardous business. *For if the trumpet give an uncertain voice, who shall prepare himself for war?*

There appears to be some disagreement as to whether ring-enlargement of MCP to cyclohexane and thence rapidly to benzene can occur on catalysts having only a metallic function.[143] It does not take place on single-crystal platinum surfaces,[134,135] nor on supported catalysts that are strictly neutral;[128,130] it does however appear to go readily on platinum black,[13] although it may result from 1,6-cyclisation of n-hexane. Some benzene was also observed on EUROPT-1.[139]

Detailed studies have also been made on the dimethylcyclopentane isomers,[141] using EUROPT-1 and platinum black at 603 K; these help to illuminate factors governing the adsorption of the C_5 ring. In 1,1-dimethylcyclopentane, the C1—C2 bond was strongly deactivated, presumably by steric interference, but the other bonds were very reactive, giving mainly fragments at low hydrogen pressure, large amounts of aromatic products and only small amounts of alkene. The behaviours of the Z- and E-1,2-dimethyl isomers differed significantly: both gave large amounts of the cyclic 1-alkene even at high hydrogen pressures as the consequential flattening of the ring reduced strain in the adsorbed state. The Z-isomer was the more reactive, and only this gave n-heptane as one of the products, by breaking of the C1—C2 bond. Z-E isomerisation was observed, equilibrium being attained at high hydrogen pressure; the mechanism must involve breaking and reforming the C1—C2 bond. Demethanation also took place, but is importance decreased with increasing hydrogen pressure. Reaction mechanisms were exhaustively discussed. On rhodium, palladium and platinum films, 1,1,2- and 1,1,3-trimethlcyclopentanes gave either aromatisation or demethylation to the 1,1-dimethyl compound without ring-opening.[19,144] The intervention of C_7 cyclic species in the interconversion of the heptane isomers is illustrated in Scheme 14.4; certain transformations do however necessitate a bond-shift mechanism.

Methylcyclopentane reacted with hydrogen on Rh/SiO$_2$ and Rh/Al$_2$O$_3$ at $\sim$500 K to give quantities of smaller alkanes that fell as the hydrogen pressure was raised;[35,36,145–148] C_5 product selectivities ($\geq$60 %[36]) showed little dependence on this variable or on conversion up to $\sim$70%.[35] In another study with Rh/Al$_2$O$_3$, the selectivity S decreased (from 95 to 65%) as dispersion $increased$.[18,67] The activation energy for fragmentation exceeded that for ring-opening, because the intermediates were more hydrogen-deficient; Arrhenius parameters for both reactions showed compensation.[146] Over Ru/Al$_2$O$_3$ at 458 K, the reaction gave much fragmentation at low dispersion (H/Ru = 0.07)[89], but the C_6 selectivities were unaffected, and approached those expected for the selective mode ($S \geq$ 83%);[149,150] on Ru/SiO$_2$ a lower value (42%) was reported[15] at 548 K. Ir/Al$_2$O$_3$, Ir/SiO$_2$ and iridium sponge all gave extremely small amounts of n-hexane ($S \geq$ 99%),[15,111,151] and ratios of 2-methylpentane/3-methylpentane greater than two, as is often the case. On 10% Pd/SiO$_2$ at 496 K, ring opening was essentially non-selective.[105]

With various alkyl-substituted cyclopentanes, ring-opening selectivity with Ir/Al$_2$O$_3$ depended[15] on the number of CH$_2$—CH$_2$ bonds, which argues for the dicarbene mechanism, in which the intermediate is a 1,1,2,2-σ_4 species. The chemisorbed state of cyclopentene was reported[152] to be more or less perpendicular to the Ir(111) surface by a NEXAFS study, in support of this mechanism.

Complex product distributions were observed with 1,2,4-trimethylcyclohexane using Ir/Al$_2$O$_3$. Those obtained with Pd/Al$_2$O$_3$ at 573 K depended neither on temperature or particle size, S being about 40%.[153] Thus it appears that only with platinum and rhodium is there clear evidence for a particle-size effect on the direction of ring-opening; however, with platinum the effect only appeared below about 2 nm and with rhodium below 1.2 nm. On cobalt films, where complete degradation was minimised by use of low temperatures, reaction mechanisms have been studied with the help of deuterium labelling.[19] Extension of the alkyl side-chain to C$_5$ lowered ring-opening selectivity,[15] but it remained above 70% for Pt/Al$_2$O$_3$, Rh/Al$_2$O$_3$ and even Ru/Al$_2$O$_3$, while for Ir/Al$_2$O$_3$ it was still 92%.

Ring-opening of alkyl-substituted cyclohexanes is slower and much less selective; selectivity was only about 5% with platinum, although Ir/Al$_2$O$_3$ gave 87% C$_7$ alkanes. Lower values were found with n-butylcyclohexane.[15]

Larger and more complex ring systems undergo other types of transformation, especially on platinum catalysts, but they have not been subjected to quantitative treatment. With spiro(4,4)nonane, one of the two C$_5$ rings was preferentially opened in all possible ways, but surprisingly one of the bonds adjacent to the quaternary carbon atom was the most reactive, so that n-butylcyclopentane was the main product. Isomerisation of the reactant also led *via* indane to *o*-ethyltoluene. Spiro[4,5]decane gave naphthalene and n-butylbenzene, spiro[5,5]undecane mainly n-pentylbenzene, and spiro[5,6]dodecane gave biphenyl.[13] Cycloheptane reacted to give toluene, and benzene by demethylation. Rings containing eight or more carbon atoms undergo intra-annular dehydrocyclisation: cyclooctane gave Z-pentalene (bicyclo[3.3.0]octane), and cyclononane gave bicyclo[3.4.0]nonane. In these and other similar reactions the hydrogen atoms removed were those closest to each other in the stablest conformation of the ring, which is presumably maintained in the adsorbed state. On Ir/Al$_2$O$_3$, perhydroindan (bicyclo(4.2.0)nonane) was reduced to alkylcyclohexanes much faster than decalin, and bicyclo(3.3.0)octane gave 72% ring-opening selectivity at high conversion.

14.2.6. The Environment of the Active Site: Effect of 'Carbon'[17,42,154]

It is desirable at this point to try to draw together a few of the threads that have permeated the previous discussion, in order to give them the prominence they deserve. We have seen that the product distributions and rates of reaction of the higher alkanes with hydrogen are dependent upon operational variables, especially temperature, reactant pressures, time-on-stream and the state of the surface, as well as on the nature of the metal, its support (if any) and its dispersion. Unfortunately the variables that are controllable, namely temperature and reactant pressures, do not give results that are immediately suitable for modelling, because these variables, and others, also affect the coverage of the surface by unreactive *carbonaceous residues* ('carbon' for short), which are hydrogen-deficient species derived from the hydrocarbon reactant.[155-157] This of course is not a new problem; we have mentioned it before (e.g. Section 14.2.2), but it is more prominent with the higher alkanes and has received more explicit attention in this context. The use of short reaction pulses, beneficially used with smaller alkanes (Section 13.1.3), has not been much used here, nor is it always clear that random alteration of the variables and adequate back-checking has always been employed. It therefore seems to be accepted that in most cases one has to be prepared to live with this situation, and to make the best of it. Antal Sárkány[56,158] has proposed a qualitative classification of various states of the surface as (i) Pt-H, where carbonaceous species are absent but the surface is hydrogen-covered, (ii) Pt-HC, where adsorbed species are not too much dehydrogenated, and (iii) Pt-C, where the H:C ratio in the adsorbed species is low. The character of reaction in each of these states can then be considered. Four factors may be at work: (1) lowering of the number of available sites results in loss of activity; (2) reduction in the mean size of the remaining free sites may render certain modes of reaction inoperable, and may facilitate others not formerly possible (this may be a consequence of a reduced availability of hydrogen atoms[89]), (3) the electronic character of the free sites may be influenced by the adjacent adsorbed species;[159,160] and (4) hydrogen associated with the 'unreactive' species may participate in the continuing reaction. One is therefore left wondering whether it will ever be possible to obtain a measure of the true catalytic character of a metal uninfluenced by these factors, and whether important parameters such as particle size can ever be truly evaluated because their importance is itself dependent on the nature of the metal.

These difficulties may be exemplified by reference to some of the publications already cited, which will also illustrate additional refinements. The extent of 'carbon' deposition (this term covers both Pt-HC and Pt-C states) increases with molar mass of the alkane, but not smoothly;[161] we saw before that ethane and propane cause few problems, but with higher alkanes they are unavoidable. There is no direct link between hydrogenolysis to smaller alkanes and 'carbon' deposition. On relatively clean surfaces, the former is favoured by *high* hydrogen pressures, but the *depth* of the process goes oppositely, methane usually being a major product at *low* hydrogen pressures.[36] Deposition of 'carbon' occurs also most readily at low hydrogen pressures,[43] the H/C ratio decreasing as the hydrogen/alkane pressure ratio falls. This suggests that (i) isomerisation needs less fully dehydrogenated species than hydrogenolysis, (ii) C—C bond breaking needs fewer hydrogen atoms than reactive desorption of intermediates (conclusions already reached in Chapter 13), and (iii) 'carbon' comes mainly from the reactant alkane rather than fragments, although polymerisation of C_1 species may occur.[162] The state of the surface affects the types of product formed: thus on surfaces largely 'carbon'-covered, selectivity towards hydrogenolysis is lowered, and is less affected by varying hydrogen pressure.[43] The view has been advanced that the way hydrogen pressure affects selectivities is through change in the 'carbon' coverage and not simply to variation in coverage by hydrogen, although as noted above the two are not easily separated.[51,128]

The occurrence of 'carbon' formation is most directly sensed by changes in reaction parameters with time-on-stream (TOS):[163] so for example with MCP and Rh/Al$_2$O$_3$ the selectivity to fragments decreased with TOS and increased with temperature, while at and below 468 K the proportions of the C$_6$ products also changed with TOS.[36,146] Effects due to 'carbon' were also responsible for the different results obtained with the form of reactor used (pulse vs. continuous recirculation[95]). It also has to be remembered that large particles may be deactivated faster than small ones, so that effects of TOS etc. may be due to gradual elimination of certain classes of active site. The sense of variation of hydrogenolysis selectivity also appears to depend on particle size; with platinum black preheated to 633 K, selectivity decreased with increasing hydrogen pressure for several C$_7$ alkanes, but if preheating was only to 433 K, and the particles therefore being smaller, it increased.[164]

Variation in conditions of pre-treatment can give major alterations in product selectivities: variables applied include reduction temperature[119,146] (or temperature of hydrogen treatment[39]), oxidation,[147,165] and manner of storage.[86] Heating by microwave radiation during reduction or use also has had major effects.[166,167] The precise effects of these changes are not always clear, but they probably affect surface contamination, particle size or roughness.

It would be nice to find a simple explanation for the outstanding activity of platinum for skeletal isomerisation and other desirable reactions. It is due in part to its inactivity for hydrogenolysis, which in turn follows from its inability to break C—M bonds: thus 'carbon' layers are relatively stable under reaction conditions, and carbon contamination is not easily removed by hydrogen.[62,147] Thus only small ensembles of free atoms remain in the steady state, and these mercifully are capable of doing what is wanted. *In my end is my beginning.* With rhodium, however, its much greater activity for hydrogenolysis limits its utility for other reactions;[147] the same goes for iridium. Iron was converted into a mixture of carbide phases at high *iso*pentane/hydrogen ratios at ∼600 K, and there were changes in selectivities favouring intermediate products.[116]

This section has focused on self-generated effects that can be limited but not eliminated by appropriate choice of conditions. There are of course many other ways in which the environment of the active site can be influenced by deliberate alteration to the design of the catalyst. These include the use of zeolitic supports, and modification by other elements (Re, Sn, S) and inactive metals, adventitious poisons, and the Strong Metal-Support Interaction. These items will be discussed in Section 14.5.

14.3. MECHANISMS OF ALKANE TRANSFORMATIONS

14.3.1. A General Overview[168,169]

Much has been written about the mechanisms of the reactions undergone by alkanes having five or more carbon atoms (see Further Reading section 3), but little of note has been added to our understanding for the past 25 years or so. It is helpful to start by listing the types of input that have led to these mechanistic statements, and to assess the weight that should be given to them. (1) First and most importantly, there is simply the nature of the products formed, and their relative amounts. (2) Of equal importance is the way in which the amounts vary with experimental conditions, especially hydrogen pressure and temperature, and with the nature and form of the catalyst: the significance of these last variables means that 'mechanism' needs to be related to a particular catalyst, there being few if any general statements that can be valid. (3) Routes by which skeletal isomerisations proceed have been elucidated in depth by using isotopically labelled reactants, in a way that would otherwise have been impossible.

Concerning the interpretation, there have been two approaches: (i) input from the general body of organic chemistry and in particular reactions of ligands attached to organometallic complexes, and (ii) quantitative modelling of the reaction kinetics. While undoubtedly some heterogeneously-catalysed reactions have their counterparts in metal-complex-mediated processes, these analogies have led to the thesis[170] that single metal atoms are sufficient to bear the key intermediates of the former; this of course is not impossible, and there is some evidence for an important role for atoms at steps and edges, but some of the proposed adsorbed states cannot be reconciled with the observed kinetics.[19] Indeed it has generally been the case that mechanisms have been advanced in ignorance of the kinetics, and only in a few cases have orders of reaction been used as criteria.[19,93,113] There is in fact remarkably little firm ground on which to build: even the interpretation of orders of reaction is debatable, largely because of the unknown relevance of 'carbon' and its possible dependence on reactant pressures, and as we saw in Chapter 13 activation energies are very frail reeds on which to lean. It is however unfortunate that the extensive results[13,41] on effects of hydrogen pressure on rates and selectivities collected by the Budapest group have not proved susceptible to quantitative modelling.

What if anything can we then be certain of? The routes whereby skeletal isomerisations proceed are very clearly indicated by isotopic labelling experiments (Section 14.3.2), but even here the structures devised to explain them have relied heavily on organic chemical intuition, and have often involved multiple carbon-metal bonds (i.e. carbenes, C=M, and carbynes, C≡M) that stretch the imagination to near-breaking point. Formal multiple C—M bonds of this type, i.e. having a π-component, are now thought unlikely, and representation of, for example, C=M as a di-σ CM$_2$ is more probably correct. This reformulation places great strain on the availability of the necessary metal orbitals, so that the manner of bonding of $\alpha_3\varepsilon_3$-hexa-adsorbed n-pentane were it to occur, would need careful thought. Isomerisation mechanisms will be considered further in the next section.

There is little more that can be usefully said about the mechanism of hydrogenolysis than was set down in Chapter 13. What is new concerns the particular form of hydrogenolysis that is responsible for the ring-opening of cyclopentane and its derivatives. On platinum, and to a lesser extent on other metals this process is easier than that of breaking C—C bonds in acyclic species, and requires a *less* dehydrogenated species[13] (Section 14.25). It may be that the inflexibility of the C$_5$ ring imposes additional strain in the C—C bond in the $\alpha\beta$-diadsorbed state, causing it to break easily. It is this that justifies the supposition that cyclic C$_5$ species can be intermediates in skeletal isomerisation (see below). The two modelling studies[24,130] already mentioned neither confirm nor deny this possibility, and there are no kinetic studies of C$_6$ ring-opening to help us further.

14.3.2. Mechanisms of Skeletal Isomerisation[2,13,19,42,113,135]

There is now no doubt that the thermochemical advantages of increasing the degree of branching of alkanes can be realised in practice by metallic catalysts, and that an acidic function is not needed. The elegant and demanding work performed, with the assistance of his colleagues, by the late François Gault before his untimely decease in 1979, and prosecuted subsequently by Maire, Garin and others, has led to a deep understanding of the extent and subtlety of hydrocarbon transformations. The following short account does scant justice to this outstanding work; the several major reviews and the original papers will have to be studied to appreciate their major contribution to the science of catalysis.

There are two separate and distinct mechanisms by which skeletal isomerisation can occur: (i) *the bond shift mechanism*, and (ii) *the C$_5$ cyclic mechanism*. The first is clearly the only possibility when there are less than five carbon atoms in the chain: so the way of isomerisation of n-to *iso*butane has to be by bond-shift. Two somewhat different mechanisms with a number of minor variations have been proposed.[19,42] The first involves an actual or virtual cyclopropanoid species formed by loss of four (or thereabouts) hydrogen atoms, with the bond-shift occurring by the subsequent breaking of a C—C bond other than that just made (Scheme 14.5). Depending on how the ring is formed, the reaction may lead to either a methyl or an ethyl shift. The second involves forming an $\alpha\gamma$-diadsorbed species attached to a single metal atom (i.e. a metallocyclobutane), which then dissociates, and following rotation of the alkene part re-assembles and is released as the isomer (Scheme 14.6). The intermediate may be formed directly on platinum, as there is evidence from deuterium exchange (Chapter 6) that this is a favoured mode of

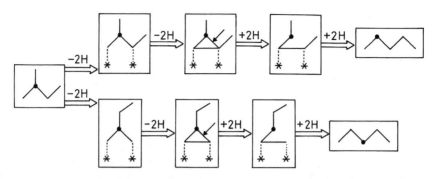

Scheme 14.5. Skeletal isomerisation of 2-methylbutane-2-[13]C: the bond-shift mechanism *via* a cyclopropanoid intermediate.

alkane adsorption on this metal, but indirectly *via* a π-alkenylic species on palladium. If it fails to come back together, the two parts may acquire hydrogen atoms and form smaller fragments, and preferred formation of such a species at the end of the chain could account for the prevalence of demethylation on palladium (and nickel). However, formation of this intermediate requires the loss of only two hydrogen atoms, and if this is the slow step it does not provide a way of distinguishing isomerisation from hydrogenolysis on kinetic grounds. Reactions proceed in this way on single metal atoms in organometallic complexes,[19,171] but other opportunities (e.g. hydrogenolysis) are then absent, and the analogies therefore not quite sound. There is however evidence that single atoms, or at least small ensembles, are quite sufficient as an active site for isomerisation (see Section 14.5). Analysis of the products of isomerisation of labelled 2- and 3-methylpentanes on Ir/SiO$_2$ has shown[151] that some must have arisen from the latter by a *1,3-ethyl shift*; this third mechanism must have involved a C$_4$ cyclic intermediate.

The differences between platinum and palladium appear in a number of guises, but nowhere more clearly than in the reactions of *neo*pentane.[63,98] Although caution has constantly to be recommended in the use of values of activation energy for mechanistic discrimination, large differences may well betoken the intervention of alternative intermediates, differing in their degrees of dehydrogenation. Values

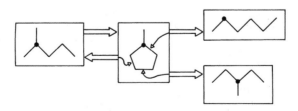

Scheme 14.6. Skeletal isomerisation of 2-methylpentane-2-[13]C: cyclic mechanism.

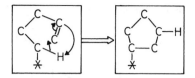

Scheme 14.7. A mechanism for 1,5-cyclisation.

for *neo*pentane hydrogenolysis on palladium are uniformly higher than those for isomerisation, whereas for platinum they are similar, and orders in hydrogen support the view that on palladium hydrogenolysis (but not isomerisation) requires a substantially dehydrogenated species[98] (Section 4.2.4). It is possible that because *neo*pentane cannot immediately form a π-alkenyl species because of its quaternary carbon atom, it is driven *faute de mieux* to find an alternative route in which the two processes differ, and differ from those occurring on platinum. Activation energy values have also been used to argue for *two* bond-shift mechanisms,[19] one applying to reactions in which the degree of branching is unchanged, and another in which it is increased or decreased.

The C_5 cyclic mechanism (Scheme 14.7) is clearly established as a strong possibility where the necessary chain exists: it is preferred by palladium and *small* platinum particles, and is the *sole* means of isomerisation on iridium catalysts. However metals are further distinguished, on the basis of their behaviour in the hydrogenolysis of methylcyclopropane, as giving products by either selective or non-selective breaking of the C_5 cycle: with platinum (small particles) and palladium, the non-selective route predominated, while on iridium it was non-existent.[113] This startling difference between adjacent metals is one of the minor mysteries of catalysis, but there does not appear to have been any *kinetic* study of reactions on iridium to help explain what is happening. Where the selective mechanism of ring *opening* operates, its reverse is ruled out as a route in isomerisation; *n*-hexane cannot then isomerise to methylpentanes. Recent work has however shown[74,75] that Ir/Al$_2$O$_3$ catalysts are able to effect the dehydrocyclisation of n-hexane to a limited extent in competition with the predominant hydrogenolysis; such catalysts ought then to show a degree of non-selectivity in MCP hydrogenolysis. Disagreements of this kind may be resolvable in terms of variables such as surface cleanliness or operating conditions; it has for example been recently shown[172] that with Ir/SiO$_2$ the routes followed in the reaction of 1,4-dimethylcyclohexane changes dramatically with time-on-stream, from mainly hydrogenolysis to mainly dehydrogenation, with small amounts of other products in between. The species involved in the cyclisation step has been considered;[113] since for the selective mechanism two adjacent methylene groups are needed, it has been argued that an $\alpha_3\varepsilon_3$-hexa-adsorbed species is required. Such a deeply dehydrogenated species is however unlikely to possess the necessary reactivity, and other alternatives are more likely.

It was explained in Section 13.5 that the increase in n-butane isomerisation selectivity with *decreasing* hydrogen pressure was due to the loss of a further hydrogen atom from the first-formed C_4H_7 species. In the case of n-heptane isomers, however, molecules formed by the bond-shift route increased progressively with hydrogen pressure, while those arising from a C_5 cyclisation route passed through maxima. It is hardly to be expected that these alkanes would isomerise by the bond-shift route through different intermediates; a possible explanation might be that with the C_7 isomers the final step becomes rate-limiting, this requiring several hydrogen atoms (see Scheme 13.5) and hence proceeding best at high hydrogen pressures. Precise structures, specifying H/C ratios, for intermediates in C_5 cyclic and bond-shift isomerisation are rarely set down.

In the case of the isomerisation of 2- to-3-methylpentane, the switch from the non-selective to the selective C_5 mechanism starts (in the case of platinum) as the particle size is decreased below about 2.5 nm, and progresses continuously: however, the bond-shift mechanism (where possible) is the major mechanism above 1 nm size. The extents of these changes depend on the structure of the reactant, and at low dispersions it contributes much more to the 2- to 3-methylpentane reaction than to the 2-methylpentane to n-hexane process. The above statements concerning critical sizes for apparent changes in mechanism rest on TEM studies of the catalysts rather than on mean dispersions obtained by hydrogen chemisorption.[19,113] Single-crystal studies (see Section 4.4) show that stepped surfaces are the origin of both types of isomerisation,[173] and it has been thought therefore that low-coordination number atoms in small supported particles are responsible.[19] Attempts to find a geometric explanation for the totality of the results have however failed, and opinion now favours the change in electronic structure that accompanies decrease in size (Section 2.5) as being the cause of the changes noted above.[19,113] A detailed explanation of how this effect might operate is however still awaited. Activation energies for small platinum particles (in 0.2% Pt/Al_2O_3) are uniformly higher than those for the larger particles in 10% Pt/Al_2O_3; a possible explanation for this,[19] following the argument developed in Section 13.2.4, is that hydrogen is more strongly chemisorbed on the smaller particles; but there are no kinetic measurements to bear this out.

14.3.3. Dehydrocyclisation[2,57,135,174]

The process of converting alkanes into cyclic compounds by loss of hydrogen and formation of new C—C bonds, i.e. *dehydrocyclisation* (DHC), is an important contributor to the matrix of reactions that comprise petroleum reforming (Section 14.1.2); the products thus made, i.e. alicyclic molecules and aromatics, give added value to the output as a fuel, although aromatics are now disliked because of their carcinogenicity. We have already met DHC in the context of the reactions of linear alkanes (Section 14.2.3) and in the reversible formation of C_5 rings as a route to skeletal isomerisation (Sections 14.2.3 and 14.3.2). It simply remains to comment briefly on the actual process of C—C bond formation.

The literature recognises the occurrence of both 1,5- and 1,6-cyclisation.[2] The latter takes place through unsaturated linear species (dienes, trienes, Scheme 14.3); it is naturally most prevalent when the concentration of adsorbed hydrogen atoms is low. The process accounting for the former has been much debated;[2,175] it differs in that a much less dehydrogenated species is adequate, but simple elimination of two hydrogen atoms at C1 and C5 seems unlikely. Alkanes and alkenes reacted at similar rates, suggesting that the common alkyl radical was involved.[2] A mechanism based on an alkenyl radical appears more probable, however (Scheme 14.8): this may be regarded as *alkyl-alkene insertion*. Carbene-alkyl insertion has also been considered, but the carbon atom is an $\alpha\alpha$-bonded species is probably sp^3 rather than sp^2.

14.4. STRUCTURE–SENSITIVITY[2,42]

14.4.1. Reactions on Single-Crystal Surfaces

Frequent mention has already been made of the effects of surface structure as thought to be affected by changing particle size on product selectivities and reaction mechanisms. In this section, emphasis falls chiefly on effects on rates or TOFs, with cross-reference to mechanistic information where necessary.

Most of the work that has been reported on reactions of the higher alkanes on single-crystal surfaces concerns platinum.[176,177] There have been detailed studies[59,133,173] of the reaction of *n*-hexane and other alkanes on Pt(111), Pt(100) and several stepped and kinked surfaces: the effects of changing temperature and hydrogen pressure were observed.[59] Arrhenius plots were non-linear due to self-poisoning at higher temperatures, and activation energies tended to rise with hydrogen pressure. Product selectivities depended on surface structure in interesting ways: (100) terraces favoured internal C—C bond rupture, while (111) terraces promoted terminal breaking, and (111) microfacets also favoured formation of benzene. The bond-shift mechanism was more important on stepped surfaces than on low-index planes.[133,178] The dependence of rates and selectivities on operational parameters was much as found with supported platinum. It did not however prove possible to identify the facets exposed on supported particles by comparing products formed on them with those appearing on single crystals because amounts of MCP on the latter often exceeded those on the former.[59] MCP reacted faster on Pt(100) than on Pt(111), but the activity of stepped surfaces was like that of flat surfaces because the steps became inactivated by 'carbon'.[134] This is unexpected, because it is generally thought that 'carbon' formation occurs preferentially on terraces. n-Hexane yields at 623 K were significant on Pt(111), Pt(557) (and platinum foil), in disagreement with a later observation that the ring-opening was wholly selective.[133] Reactions of C_6 hydrocarbons on Pt(111) have been compared with those on palladium foil, but not with other single-crystal surfaces,[64] and n-pentane hydrogenolysis on Rh(100) and Rh(111) has been compared with the reaction on Rh/SiO$_2$ treated in various ways.[179] Ir(111) and (755) were less active than the same platinum faces for DHC of n-heptane at 423 and 523 K.[72]

Preparation of supported micro-crystals by vacuum-evaporation of platinum onto crystalline supports gives 'model' catalysts exposing specific planes.[85,97,136,180,181] Deposition of the metal onto the (100) or (111) face of NaCl followed by evaporation of the support (Al$_2$O$_3$, SiO$_2$ etc.) and dissolution of the NaCl in water has given metal particles exposing only the face originally in contact with the NaCl. Detailed studies have confirmed particle-size effects on the selectivity of MCP ring-opening, but effects on rates were small.[136] The technique did however permit study of the mobility of the support at moderate temperatures, in harmony with the encapsulation seen with ordinary supported catalysts at much higher temperatures.

14.4.2. Particle-Size Effects with Supported Metals

There have been numerous studies of the reactions of the higher alkanes on supported metal catalysts, the particle size of which has been controlled by metal loading,[36,92,182] calcination,[181,183] sintering or some other method. Before attempting a brief summary, we should remind ourselves that *mean* size (as determined by hydrogen chemisorption or X-ray diffraction) is only a poor guide to the true state of a catalyst, unless the distribution happens to be narrow. Within a broad distribution there may be significant differences in activity, especially as the tendency to lose activity by 'carbon' deposition is itself variable. It has even been speculated that apparent particle-size dependence of reaction parameters may be *induced* by deposited 'carbon'[160] (see also Section 12.3). TEM is a much better indicator of what is in a catalyst.[67] Furthermore, the simple measurement of a TOF under a single set of conditions, without determination of kinetics, is only of limited value. A short survey will therefore suffice, especially since some of the trends seen merely confirm those found with the smaller alkanes (Chapter 13).

The structure-sensitivity of alkane hydrogenolysis catalysed by platinum was first established by Oles Poltorak many years ago;[184] since then, comparatively little work has been done on this system. The TOF for cyclopentane decreased about five-fold on Pt/Al$_2$O$_3$ catalysts of 7 to 65% dispersion at 573 K, this dependence being similar to that of the stepwise exchange of methane, but less than that for multiple exchange.[122] The trends of isomerisation and hydrogenolysis selectivities with particle size observed with n-butane (Section 13.3) have been confirmed with n-hexane.[165] Perhaps the most interesting results have been obtained[185,186] with 'cluster-derived' catalysts made either from (A) Chini complexes $[Pt_3(\mu_2CO)_3(CO)_3]_n^{-2}$ ($n = 2 - 5$) or (B) from $Pt_3(\mu_2CO)_3L_4$ (L = PPh$_3$ or PEt$_3$); they have been compared with those made by normal methods and have been characterised by TEM. Type A catalysts were less prone to deactivation than classical materials, and showed some differences in 2-methylpentane isomerisation; Type B catalysts showed very selective demethylation of MCP, attributed to the presence of residual phosphorus.

Very careful characterisation of catalysts having a variety of metal loadings is necessary if misleading conclusions are to be avoided. We have seen (Section 2.4) that two types of metal are possible; one very highly dispersed, formed first as the loading is increased, and reaching a limit of typically 1%, and a second, less well dispersed, the concentration of which increases with loading, and therefore predominating at high loadings. In the reaction of neopentane on Pt/γ-Al$_2$O$_3$, there was no change in TOF or activation energy at the changeover point,[187] but stepwise hydrogenolysis was associated with the first type and isomerisation with the second. S_i therefore increased with metal loading or sintering. These important differences are often overlooked, as they are not revealed by hydrogen chemisorption measurements and may even be missed by TEM (where attention is naturally focused on what is most easily seen). Decreasing particle size in Pt/Al$_2$O$_3$ favoured hydrogenolysis at the expense of aromatisation.[188] The difficulty of obtaining an unequivocal connection between activity and particle size is underscored by the observation[189] that mode of pre-treatment is in fact more important than size. HRTEM measurements on Pt/SiO$_2$ catalysts suggested that particle shape *per se* did not affect product selectivity[190] (see however results[86] for Pt/Al$_2$O$_3$).

Values of TOF for hydrogenolysis of cyclopentane[124,191] and MCP[67,113,123] on Rh/Al$_2$O$_3$ at 353 K passed through maxima at 20–30% dispersion (H/Rh $\sim$0.4), lowest values occurring at high dispersion (Figure 14.9): MCP was much the less

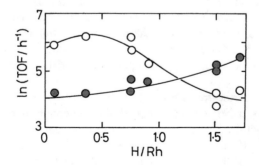

Figure 14.9. Hydrogenolysis of n-hexane ($\bullet$) and MCP (O) on Rh/Al$_2$O$_3$ at 493 K: effect of dispersion (H/Rh) on TOF.[67]

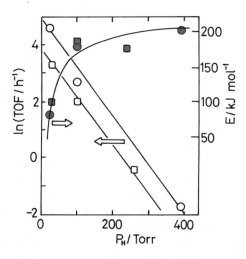

Figure 14.10. Hydrogenolysis of 2,2,3,3-tetramethylbutane on Rh/Al_2O_3: effect of hydrogen pressure on TOF and on activation energy for $\alpha\beta$ bond-splitting.[88] O, H/Rh = 0.08; □, H/Rh = 1.7.

reactive. [Note: the use of a logarithmic scale minimises the scale of the effect; for MCP the change in TOF was about × 7]. With Rh/SiO_2 the effect was less regular, although the trend was similar.[67,123] n-Pentane hydrogenolysis on rhodium on various supports indicated[192] that activity correlated with surface roughness and the concentration of B_5 sites. With Rh/Al_2O_3 n-hexane showed somewhat irregular treads *in the opposite sense*, (Figure 14.9) as did 2,2,3,3-tetramethylbutane: values of TOF for the total reaction of this latter reactant are shown as a function of hydrogen pressure for Rh/Al_2O_3 of low and high dispersion (H/Rh = 1.7 and 0.08) in Figure 14.10. The order in hydrogen is slightly more negative on the larger particles, consistent with the greater integral heat of adsorption determined separately; activation energies (shown only for the $\alpha\beta$-fission process, Section 14.2.4) increase with hydrogen pressure in the expected way. The effect of particle size on the rates of the two bond-breaking modes in this molecule has already been considered (Section 14.2.4). These results confirm the unusual nature of the ring-opening reaction of MCP, compared to C—C bond breaking in acyclic alkanes.

At 453 K, chloride-free Ru/Al_2O_3 (unlike Rh/Al_2O_3) gave TOFs that increased 40-fold as dispersion *decreased*, as did the depth of hydrogenolysis measured by the ζ factor[89,193] (Figure 14.11); a few percent of isomerised products were seen at high H/Ru. High dispersions and low TOFs were also obtained with other precursors.[90] TOFs for MCP (and other branched alkanes) also decreased as dispersion rose.[89] Arrhenius parameters for the reactions of several alkanes on variously dispersed Ru/Al_2O_3 were shown in Figure 14.8, and the effects of

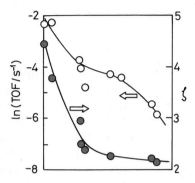

Figure 14.11. Hydrogenolysis of *n*-hexane on Ru/Al$_2$O$_3$ at 458 K: TOF and fragmentation factor ζ as a function of dispersion (H/Ru).[89]

dispersion of the ways of breaking 2,2,3,3-tetramethylbutane were also discussed in Section 14.2.4. For the reaction of MCP on Ru/Al$_2$O$_3$, see Section 14.2.5.

Palladium differs from the other metals in that there are only negligible or irregular effects of dispersion on TOF for cyclopentane,[123] MCP[123,194] or branched alkanes[102,153] with alumina or silica as support. Pd/Al$_2$O$_3$ made using Pd(NO$_3$)$_2$ did not give lower alkanes from MCP,[194] and selectivities for ring-opening were 50–60%. Reduction of chloride-free Pd/Al$_2$O$_3$ at 873 –1073 K resulted in a PdAl$_x$ bimetallic phase ($x_{\text{max}} = 0.1$), which prevented the formation of carbide and was reversed by oxidation. It caused S_i in the reaction of *neo*pentane at 527 K to increase from 20 to 80%, and the activation energy to decrease from 243 to only 92 kJ mol^{-1}. Chloride ion inhibited formation of the bimetallic phase.

The above results emphasis the need to appreciate that structure-sensitivity is a property of the whole system, not simply of the reaction. We have had examples of where the effect of dispersion varies with hydrocarbon (compare MCP and *n*-hexane on Rh/Al$_2$O$_3$[89]), with the support (compare Rh/Al$_2$O$_3$ with Rh/SiO$_2$ etc[67,123]) and of course with the metal. TOFs of hydrogenolysis usually decrease as dispersion *rises*,[71] (platinum, ruthenium), but not with Rh/Al$_2$O$_3$ (Figure 14.9) or palladium catalysts. MCP behaves as an ordinary alkane on platinum, ruthenium and rhodium (at dispersions above 30%, Figure 14.9); it is only the behaviour of *n*-hexane on Rh/Al$_2$O$_3$ that provides a contrast with that of MCP.[67] The best efforts—and there have been very many—to find a geometric basis for structure-sensitivity have been at best only partially successful. Variable extents of 'carbon' deposits and of strengths of hydrogen chemisorption have also been thought responsible,[147] but positive identification of the culprit has not yet been possible. While the experimental observations are not in doubt, it remains uncertain whether geometric or electronic factors are the main cause, or whether some other factors that derive from the basic ones are not even more significant.

14.5. MODIFICATION OF THE ACTIVE CENTRE

14.5.1. Introduction

We have already considered how 'carbon' formed by excessive dehydrogenation of a reactant alkane can modify the characteristics of the active centre, by restricting its size, altering its electronic structure and possibly by providing a source of hydrogen atoms (Section 14.2.6). This chapter concludes with a review of deliberately introduced modifications. These may be classified as follows: (1) alteration of the environment of the metal particle by placing it within zeolite framework, which controls the movement of molecules around it (Section 14.5.2), (2) modification by addition of rhenium (chiefly to platinum, Section 14.5.3) or elements of Group 14 (Section 14.5.4), (3) effects produced by admixing platinum with other inactive metals (mainly those of Group 11) or metals of different activity (Ru, Ir etc.) (Section 14.5.5, (4) effects due to other toxins either deliberately or accidentally added (S, Cl, H, etc.), and finally (5) effects attributable to metal-support interactions, especially of the SMSI kind.

The motivation for these studies is fairly obvious: it is to limit parasitic reactions such as hydrogenolysis and 'carbon' deposition when the target reactions are skeletal isomerisation, cyclisation or aromatisation. What is to be described will amplify the information contained in Sections 12.5.3, 13.7, and 13.8.

14.5.2. Metal Particles in Zeolites

Attention is confined to neutral or basic zeolites: it is unnecessary to discuss the structures of those that have been used, so we simply note the designations of those used most often. These include L, Y (see Further Reading section 4), β,[195] mordenite,[195,202] and ZSM-5;[203] the nature of the balancing cation is denoted by the prefix. The complexity of zeolite structures has unfortunately obstructed any attempt to devise a suitable way of naming them so as to reveal what the structure is.

The Pt/KL system has attracted much attention, as it is very effective for the aromatisation of n-hexane,[206] but its disadvantage is a high sensitivity to sulfur, which is inevitably present in at least trace amounts in hydrocarbon feedstocks. The presence of the potassium appears essential, since it increases aromatisation and suppresses isomerisation,[195,197] but at the same time decreases thiotolerance. Aromatisation selectivity has been correlated with terminal C—C bond breaking,[198] and sulfur has been thought to poison the reaction of n-hexene.[199] Pt/KL has been shown to be superior to Pt/KY for aromatisation,[201] although activation energies for this and for MCP formation are the same on both: benzene and MCP were made without any intermediate molecules being detected. High electron concentration on the platinum particles (which are necessarily very small) seemed to favour aromatisation,[195] and the heat of hydrogen chemisorption was greater on Pt/KL than on Pt/SiO$_2$ or the acidic Pt/H-LTL.[130] This agrees with the rate maxima in the reactions of n-hexane and n-butane (Section 13.23) which occur at small hydrogen pressures: in the latter case, the rate was strongly inhibited by higher pressures. These properties are unique to platinum; Ir/KL did not give an unusual amount of aromatisation.[200] *neo*Pentane reacted with hydrogen only by hydrogenolysis on Pd/L containing either lithium, potassium or calcium; Arrhenius parameters were reported.[103]

The mechanism of aromatisation has been investigated in detail by ^{13}C MASNMR using n-hexane-1-^{13}C as reactant;[22] basic catalysts (Pt/MgO-Al$_2$O$_3$, Pd/MgO-Al$_2$O$_3$, Pt/KL) were shown to function both by C$_6$ cyclisation (via hexadienes and hexatriene, the chief route on platinum catalysts) and *via* C$_5$ cyclisation on palladium. Mechanisms of MCP formation, isomerisation and hydrogenolysis were also examined: platinum catalysts were much more active than the palladium catalyst, this being explained by the need through π-alkenic species on the latter but not the former. Part of the benefit of using the KL zeolite arose through steric restriction on forming the C$_5$ cyclic intermediate. It was concluded that small metal particles were electron-deficient on acidic supports but electron-rich on basic supports. Calcination of Pt/KL is however liable to move metal particles from within to the surface.[196]

14.5.3. Platinum-Rhenium Catalysts[114,135,207]

The combination of rhenium with platinum has provided one of the two outstandingly successful catalysts for petroleum reforming, the other being platinum-iridium (see Section 14.5.5). PtRe/Al$_2$O$_3$ catalysts have therefore been subjected to intensive academic study, and a vast patent literature also exists. Commercial catalysts normally contain equal amounts of the two components (either 0.3 or 0.6% of each), although fundamental work has explored the whole composition range. For some period through the 1970s and 1980s there was much debate about the *modus operandi* of the bimetallic catalyst: the main question concerned the state of reduction of the rhenium in the working catalyst, i.e. whether it was all in the zero-valent state and associated with the platinum, or whether some if not all was on the alumina support in a positive oxidation state. Work on the recovery of values from spent catalyst at a very early stage revealed that the second alternative was at least partly true. Rhenium is, as we have seen (Chapter 13), very active for hydrogenolysis, and the performance of the binary catalyst in the unsulfided state shows that separate rhenium particles, or even large ensembles of rhenium atoms in a bimetallic particle, cannot be present, because it is not notably more active for hydrogenolysis than Pt/Al$_2$O$_3$. We noted in Section 13.8.4 that surface energy considerations ought to lead to segregation of platinum to the surface, with some possible modification in their properties due to electronic interaction with neighbouring rhenium atoms; although this model has been proposed and used there is clear evidence[208,209] that the reverse is true, with rhenium atoms preferentially occupying edge or step sites; mutual electronic interaction is plainly indicated by XANES measurements.[208] It is also evident that no *large* assemblies of rhenium atoms are normally present,[210] because activity for hydrogenolysis is limited. Although a number of significant studies have been made using unsulfided catalysts,[70,211] industrial practice demands as complete elimination as possible of parasitic reactions, and sulfided catalyst are therefore used.[212,213] It is now clear that sulfur atoms latch onto surface rhenium atoms, so that the working surface comprises small platinum ensembles separated by ReS species.[214,215] In industrial use, the catalyst performs in a bifunctional manner (see Scheme 14.1) because the support is rendered acidic by chloride ion; a contribution to its role by some rhenium cations is not however impossible.

The early work on this system has been reviewed by Charcosset,[216] and theoretical ideas to account for its success have been succinctly explained.[217] As is often the case with catalysis, it is likely that a number of effects work in concert to produce the final result.[218] Many of the early studies focussed on the dependence of rates of contributing processes on composition. With n-hexane[217] and n-heptane[219] and unsulfided catalyst, hydrogenolysis selectivity rose with rhenium content, but the TOF for total rate fell linearly.[219] Cyclopentane hydrogenolysis at 773 K showed a marked maximum rate at 75% rhenium; the rate of this reaction acted as a measure of the number of binary sites, the PtRe unit being some 40 times more active than platinum itself.[131,220,221] Arrhenius parameters for this reaction on catalysts of different composition showed a convincing compensation effect.[28] However, in general, activation energies have not been measured, and kinetic measurements have been confined to more "realistic" conditions. Deposition of rhenium onto Pt(111) naturally accelerated hydrogenolysis of n-hexane, while addition of sulfur suppressed it and allowed cyclisation to remain:[222] the maximum effect was produced by a monolayer of rhenium atoms. However on PtRe/Al$_2$O$_3$, sulfidation encouraged isomerisation rather than cyclisation.[217] A small increase in methanation of 2-methylpentane on performing the reaction under microwave radiation has been reported.[166]

A particular feature of the PtRe/Al$_2$O$_3$ system is the extreme sensitivity of the surface composition to the conditions of pre-treatment applied. The need to re-activate used catalyst by removal of 'carbon' and to re-constitute it in its active configuration has generated extensive studies of the effects of oxidation, reduction and oxychlorination: when properly performed, these procedures are eminently successful. Recent publications[180,209−211] by Anderson, Rochester and their associates treat these matters in detail, and cite many relevant references. The presence of chlorine, needed to provide acidity, modifies the surface composition, decreasing the surface segregation of the rhenium.[180] Excess rhenium, not employed in forming bimetallic particles, is located on the alumina support.[205] Structural changes my also occur during use,[180] as well as 'carbon' deposition.

The palladium-rhenium system has been little studied,[223] and rhenium's influence on other metals hardly at all.[224] It rapidly negated the isomerisation activity of palladium, and the reaction of *neo*pentane with hydrogen had a maximum rate (mainly hydrogenolysis) at 50% rhenium.[225]

The improved selectivity of sulfided $PtRe/Al_2O_3$ for non-parasitic reactions requires a short discussion. Although the PtRe unit has been implicated in greater rates of *hydrogenolysis* (perhaps because the rhenium atom can more easily accommodate a multiple C—M bond), the higher selectivity for isomerisation and cyclisation must be a feature of a *small platinum ensemble*. This conforms to the effect of dispersion, where small particles also show high isomerisation selectivity.

14.5.4. Modification by Elements of Groups 14 and 15 and Some Others

The platinum-tin system, the use of which was found to be most beneficial for dehydrogenations, has also been widely investigated for the characteristic reactions of the higher *n*-alkanes. Previous sections (e.g. 5.5 and 12.3.3) have described the bimetallic structures that can be formed, and likely configurations in alumina-supported catalysts; alumina is the support of choice for practical purposes. A brief résumé is therefore all that is needed. There was much discussion in the earlier publications concerning the extent of reduction of the tin, and the form of its interaction with the platinum. It is now clear that at low Sn/Pt ratios (i.e. ≤ 1) most if not all the tin is reduced, and bimetallic particles are formed:[226,227] at higher ratios, much of the tin remains as Sn^{II} on the support.[228] The homogeneity of the metallic phase and other features depend very much on the method of preparation;[188] procedures used include impregnation with H_2PtCl_6 and $SnCl_2$,[229,230] prior application of the tin component,[226] coprecipitation,[231] and the use of bimetallic complexes.[229,230] Reaction of SnR_4 ($R = n\text{-}C_4H_9$) with hydrogen covered platinum particles is also effective.[109] Structural studies have been greatly helped by the use of Mössbauer spectroscopy on the ^{119}Sn nucleus.[229,230]

The extensive literature necessitates a short summary that catches only the main highlights; fortunately, most of the reported studies show similar or related conclusions, so the task is simplified. An early study[232] using films revealed a trend that has frequently observed since,[230,233−235] namely, a decrease in hydrogenolysis selectivity with increasing tin content. An additional effect was seen,[232] not often specifically noted more recently: increasing tin content raised the C_6 cyclic yield and decreased the C_5 cyclic yield. A related effect was reported[235] with $PtSn/Al_2O_3$ and with $PtPb/Al_2O_3$, i.e. increased yields of hexadienes from *n*-hexane; these are the necessary precursors to C_6 cycles. As always, the beneficial effects are bought at the expense of activity, although in the reaction of MCP a rate maximum was seen at a low Sn/Pt ratio.[235] This is accounted for by the tin blocking hyperactive sites at which disruption of the alkane and formation of 'carbon' occurs; sulfur also acted in the same way as tin when deposited onto platinum foil.[234] The presence of tin also raised the ratio of 2-methylpentane to 3-methylpentane in the reaction of n-hexane;[234] the isomer composition is not however always reported.[229] Catalysts prepared from bimetallic complexes have been used for the n-hexane reaction;[229] they showed much higher isomer selectivities relative to those for benzene and MCP than those made by simultaneous impregnation of the salts. The latter behaved much as Pt/Al$_2$O$_3$, suggesting that bimetallic particles were not created by this method. The DHC of n-octane has also been followed on PtSn/Al$_2$O$_3$ catalysts;[218,236] sodium and other basic additives increased aromatisation selectivity with n-heptane.[237] PtSn/Al$_2$O$_3$ subjected to oxychlorination lost much of its activity for aromatisation.[188]

Divergent views have been expressed on the way in which the tin acts. Its role in limiting the size of platinum ensembles is not in question; what is at issue is whether there is any electronic modification of the active centre. It was claimed that, in the reaction of MCP with hydrogen, tin produced positive effects on dehydrogenation and aromatisation that were not shown by either 'carbon' or sulfur; they were attributed to an electronic action,[226,227] for which Mössbauer spectroscopy provided some evidence. In view of the proposal interpretation of the effect of sulfur on butadiene hydrogenation (Section 8.3) it would not be surprising if tin also influenced the platinum ensembles to some degree.

Platinum-germanium catalysts have also been used for the reactions of higher alkanes.[161,238,239] Those prepared by co-impregnation of salts onto alumina gave higher aromatisation selectivities than those made by sequential impregnation,[240] and bimetallic particles (PtGe, Pt$_3$Ge$_2$ and Pt$_3$Ge) were detected in them; catalysts made by latter method were not however much superior to straight Pt/Al$_2$O$_3$. Oxidation of Pt$_{97}$Si$_3$ alloy produced an active catalyst for reactions of 2-methylpentane.[241]

The modifying effects of molybdenum[242−244] and chromium[245−247] have been looked at. The latter, introduced to Pt/Al$_2$O$_3$ as chromyl chloride or potassium dichromate, changed the direction of the reaction of n-pentane with hydrogen towards cyclisation rather than isomerisation or hydrogenolysis, due it was thought to electron transfer from reduced chromium ions to platinum.

Coq, Figuéras and their associates have conducted wide-ranging investigations of surface modification of supported platinum,[109] rhodium,[170,248] and especially ruthenium[89,113,114,248−250] catalysts by treating them when hydrided with alkyl compounds of aluminium, zinc, antimony, germanium, tin or lead. The purpose of this work was to explore the locations of the modifying atoms on the surface of the active metal particles, and to see whether in any case there was evidence for the selective blocking of sites on either low co-ordination number

(CN) sites on planar parts or high (CN) sites at edges and corners. The effect being sought was termed *topological* segregation; the influences at work would be (i) the relative surface energies, which in most cases would be expected to affect edge and corners sites, and (ii) the size of the modifier. To obtain selective effects, it was necessary to use particles of medium size (3 to 4 nm) displaying comparable amounts of the two classes of site; on small or large particles, one class only would predominate.

With the aid of a suitably sensitive reactant, namely, 2,2,3,3-tetramethylbutane, topological segregation was observed with a number of systems. On Rh/Al$_2$O$_3$[170] and Ru/Al$_2$O$_3$,[89,250] tin and lead selectively decorated (or occupied) low CN sites, and the reaction mode then changed to give more isobutane, characteristic of *large* particles; the same effect was seen with aluminium, zinc and tin on Pt/Al$_2$O$_3$.[109] Germanium on the other hand showed no topological effects, decorating both classes of site indifferently. Addition of germanium to Rh/Al$_2$O$_3$ favoured demethylation over central C—C bond breaking;[251] this simulated the behaviour of small rhodium particles, and FTIR of chemisorbed carbon monoxide showed that at sub-monolayer loading the atoms preferred to occupy low Miller index microfacets. Other molecules such as *n*-hexane and MCP however were not much affected by modifiers, suggesting that site environment is not greatly important in these cases: addition of tin to Pt/Al$_2$O$_3$ (H/Pt = 0.99) did however *decrease* the selectivity of MCP ring-opening to the point where it was perfectly non-selective.[109] In the case of Ru/Al$_2$O$_3$, neither tin nor germanium had much effect on hydrogen orders or activation energies,[114] although naturally TOFs were much smaller.[248]

14.5.5. Other Bimetallic Catalysts[252–256]

It is hardly surprising that the reactions of the higher alkanes with hydrogen on bimetallic catalysts have attracted interest (see Further Reading section 5), because the scope for variation in product selectivities is so great, and understanding of the factors affecting their structure is now well advanced. It might be thought, in view of what has been said about bimetallic systems in earlier chapters, and about other means of modification already in this chapter, that it should be possible to predict with fair accuracy what effects to expect. To a certain extent this is so, but there are still some surprises and puzzles. It appears to be difficult to make generalisations: especially with the Groups 8 to 10 and Group 11 systems, what happens seems to depend on the nature of the alkane reactant, and on the manner in which the catalyst has been constructed. The various methods used affect behaviour in ways not yet well understood.

We may start with the platinum-Group 11 systems. Deposition of gold onto Pt(111) gave epitaxial islands of the former, and activity for the reaction of *n*-hexane decreased linearly with significant changes to selectivities to reach a minimum value when a gold monolayer had been formed.[257] Heating a multilayer of gold gave surface *alloys* and with these the selectivities at 573 K for hydrogenolysis and aromatisation fell, and that for isomerisation rose, as gold concentration increased. Contrary results were however obtained with a 'bimetallic molecular cluster precursor'-derived PtAu/SiO$_2$ catalyst,[258] which gave with *n*-hexane at the higher temperature of 675 K more hydrogenolysis, at the expense of isomerisation and 1,6-cyclisation than Pt/SiO$_2$, but it deactivated less quickly. These results are not easily harmonised without knowing the temperature coefficients of the component processes. Single platinum atoms in a very dilute PtAu/SiO$_2$ catalyst are reported to catalyse skeletal isomerisation of alkanes.[259] With PtCu/SiO$_2$, made either conventionally[260] or *via* a cluster compound,[258] hydrogenolysis selectivity was increased, due it was thought to the PtCu sites. Little use has been made of bimetallic single crystal surfaces for these reactions, but the DHC of 4-phenyl-1-butene to naphthalene has been followed on PtCu$_3$(111) below 500 K.[261] This surface also catalysed the cyclisation of 1-hexene to benzene at 405 K,[262] the rate being limited by C—C bond formation rather than loss of hydrogen atoms. With platinum-gold films, the rate of *neo*pentane decreased with increasing gold concentration;[263] both activation energy and pre-exponential factor rose, the one partially compensates the other (Figure 14.12A). The mechanism of *n*-hexane isomerisation on nickel-platinum crystals has been followed using ^{13}C-labeled molecules.[264]

Palladium-Group 11 systems provide some interesting contrasts to the platinum-Group 11 systems. With PdAu/SiO$_2$ catalysts made by Barbier's direct redox method (Section 2.32), activation energies for the *neo*pentane reaction *decreased* with rising gold content,[265] and were again partially compensated by changes in pre-exponential factor (Figure 14.12B): isomerisation selectivities (S_i) were not much altered, their values being between 20 and 40%. The precise trend is not easily seen, because of the need to use different temperatures in order to keep conversions low. The PdAu/SiO$_2$ system needs careful characterisation, because the size distribution may be binodal, each fraction having a different composition.[266] The astronomically high activation energy for palladium (324 kJ mol^{-1}) compared to the more modest value for platinum (116 kJ mol^{-1}) has been noted before (Section 14.24) and must reflect the much greater difficulty of forming the $\alpha\gamma$ species on palladium in cases where the π-alkenyl route is note available. Exactly why this process becomes easier with the bimetallics is not at all clear; similar observations have also been reported for palladium-gold powders[267] and films.[268] Significant differences between the effects of copper and silver in palladium-based films have also been demonstrated.[269] In the *neo*pentane reaction, values of S_i were maximal at about Pd$_{85}$Cu$_{15}$, rates at about 573 K falling to low values at a copper content about 40%, but with silver as the inert component, rates fell precipitately at silver contents of only a few per cent.[270] This may also suggest that binary sites (e.g. PdCu) can show activity.

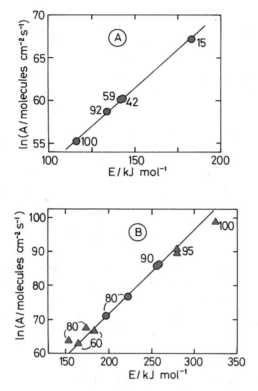

Figure 14.12. Compensation effects for the Arrhenius parameters of *neo*pentane hydrogenolysis on (A) platinum-gold films[263] and (B) PdAu/SiO$_2$ catalysts[265] (series A, ●; series B, ▲) concentrations of the active metal (%) are shown.

With rhodium-copper films (10 and 50% copper), *n*-pentane isomerisation (but little cyclisation) occurred above 510 K,[68] while with *n*-hexane DHC selectivity was greater than with pure rhodium.[271] Formation of benzene occurred more readily when MCP was used at 540 K, the ring-enlargement mechanism being thought to differ from that for bond-shift isomerisation.[271] With RhCu/SiO$_2$ catalysts, 2,2,3,3-tetramethylbutane reacted at 493 K chiefly by the $\alpha\gamma$ mode until the Rh$_{25}$Cu$_{75}$ composition was reached, and in this range MCP reacted by a mainly selective mechanism.[272] It was inferred that copper preferred low CN sites (as indeed it ought), and that no ligand effect was operative.

There is little to report on other systems involving a Group 11 metal. Films comprising iridium and gold gave higher aromatisation selectivities than iridium alone in the reaction of *n*-hexane above 600 K.[76,273] Early studies of the iridium-copper[224] and nickel-copper[274,275] systems have not been followed through.

It remains to consider bimetallic systems formed between metals of Group 8 to 10. While studies of such systems have often been uninformative, two have attracted particular attention: the platinum-ruthenium system, and the platinum-iridium system, the latter because of its utility in petroleum reforming, where undesirable 'carbon' deposition is moderated by the iridium,[276] which is more active for hydrogenolysis than platinum. This binary system is therefore less given to deactivation,[218] but sulfidation is necessary to control iridium's activity; this has been shown to reduce hydrogenolysis of n-hexane and to increase S_i.[277] Both PtIr/SiO$_2$ and PtIr/Al$_2$O$_3$ were used, but it appeared that the alumina contributed to isomerisation by an acid-catalysed process. The reactivity of other alkanes on unsulfided PtIr/Al$_2$O$_3$ decreased progressively with iridium content,[278] but the effect was much less with cyclopentane than with linear alkanes. The different activities of the two metals have been explained in the following way.[276] Their bulk physical properties are quite similar, although iridium has much the higher melting temperature and is therefore less inclined to sinter:[218] but in the very highly dispersed state, such considerations are irrelevant and metal atoms will behave more as free atoms. The separation of the $5d$ and $6p$ levels increases from iridium to platinum,[279] so promotion of a $5d$ electron to higher levels is easier with the former, which can thereby more easily accept electrons from the adsorbate into its d-shell. Multiply-bonded (e.g. $\alpha\alpha$-) species of the kind thought to be essential for hydrogenolysis are therefore the more easily formed. It must be remembered however that the ability of metal atoms to form a C=M, that is, to bind an $\alpha\alpha$-diadsorbed species to a *single* metal atom has been questioned, although such bonds are frequently met in organometallic complexes.

The platinum-ruthenium combination has been of interest in fuel-cell technology, because ruthenium imparts some resistance to poisoning by carbon monoxide; as with platinum-iridium, it is important to see whether joining metals of very different activities in hydrogenolysis creates binary centres having new properties. The few available studies[149,150,280,281] all confirmed ruthenium's superior activity for hydrogenolysis (except for cyclopentane[121]) and its inability to do much else under normal circumstances. It does however induce other reactions characteristic of platinum at a temperature (493 K) well below that at which that metal would be active by itself.[280] With Pt/Al$_2$O$_3$, X-ray absorption spectroscopy showed PtRu interaction,[150] but in other work[281] the surface composition depended on the support used. The clearest evidence for the operation of a binary centre was provided by the reaction of 2-methylpentane, where intermediate compositions gave maximal amounts of ethane and isobutane.[149]

Work on the remaining pairs of metals in Groups 8 to 10 can be dealt with swiftly. With PtCo/NaY, the importance of the cyclic mechanism for isomerisation of 2-methylpentane was enhanced at cobalt concentrations above 22%;[92,282,283] PdCo/NaY was more active for isopentane isomerisation than Pd/NaY,[284] but as we have already seen Co/NaY was itself very selective for skeletal isomerisation,[92] unlike Co/SiO_2.[284] $PtPd/SiO_2$ catalysts showed higher selectivity for n-pentane cyclisation than either metal alone and better values of S_i with n-hexane; extensive results on effects of temperature and hydrogen pressure were reported.[140] The platinum-rhodium system is another in which the second metal is much the more active,[157,232,285] although rates and product selectivities for $PtRh/SiO_2$ were intermediate and binary PtRh sites may have been active:[285] consideration of the Arrhenius parameters for a range of alkanes and compositions suggest however that the mixtures partake more of the character of platinum than of rhodium.

14.5.6. The Role of Sulfur

Reference has already been made at several points to the important role that sulfur has to play in petroleum reforming operations; it is desirable to offer a brief summary of these effects, together with some further information (see also Further Reading section 6).

There have been a number of fundamental studies of the chemisorption of sulfur atoms on platinum singe-crystal surfaces;[286] they are usually deposited from hydrogen sulfide or thiophene. The Pt—S bond is essentially covalent, but the sulfur atom (0.184 nm) is larger than the platinum atom (0.138 nm), and so its influence will extend over several adjacent metal atoms (see the discussion on butadiene hydrogenation, Section 8.3). Its interaction with small metal particles is more complex: covalent bond formation is weakened if the particle is small and thus somewhat electron-deficient through a metal-support interaction, although the precise electronic consequences may be more subtle (see Section 2.6 and the following section). Difference in electron density between plane and edge atoms, together with steric considerations, determine that sulfur will preferentially reside in trigonal or octahedral sites where it can form bonds to several metal atoms, leaving edge and corner sites clear unless the sulfur concentration is high. These sites will be the seat of activity in industrial operations using Pt/Al_2O_3 with feedstocks containing up to 20 ppm of sulfur. The Ir—S bond is stronger than the Pt—S, which may be explained along the lines set out above (Section 14.55). The Pd—S bond appears to be weaker,[287] as $PdPt/Al_2O_3$ is more thiotolerant than Pt/Al_2O_3; palladium atoms should preferentially occupy *low* CN sites (the latent heat of sublimation is lower), and the integrity of the active area should be better preserved.

The modification of the Pt/Al_2O_3 by sulfur is essential (and in a sense inevitable, as feedstocks cannot be made absolutely sulfur-free) if its admittedly limited tendency to hydrogenolysis is to be suppressed, and if therefore it can be made to perform its intended tasks. It has other consequences as well; for example the preferred product in the reaction of n-hexane becomes MCP rather than benzene,[214,215] but sulfur present on platinum black as sulfate allows its dehydrogenation to hexenes.[215] The importance of sulfur in removing the activity for hydrogenolysis that would otherwise be seen in PtRe/A1$_2$O$_3$[222,288,289] and PtIr/A1$_2$O$_3$[290] has already been noted; in the former case, in the sulfided state, the rhenium also encourages acid-catalysed reactions occurring on the support.[289]

Both sulfate and sulfide have been detected by XPS on sulfided Pt/Al$_2$O$_3$ and platinum black;[291] sulfide ions selectively eliminated aromatisation and hydrogenolysis of n-hexane, while sulfate ions stopped all processes except dehydrogenation. At high sulfur loadings, where both forms were present, activity was reduced by 95%.

Sulfiding Pt/A1$_2$O$_3$ altered the *kinetics* of hydrogenolysis of cyclopentane, increasing the order in the hydrocarbon from 0.1 to 0.6 and decreasing the hydrocarbon pressure at which the rate was maximal:[290] thus paradoxically it strengthened hydrogen adsorption but weakened the hydrocarbon adsorption. It affected different reactions in various ways; thus covering Pt/A1$_2$O$_3$ with 0.39 of a sulfur monolayer decreased the rate for cyclopentane by a factor of 25, while for ethane hydrogenolysis it was 280. Similar effects were obtained with PtRe/A1$_2$O$_3$ and PtIr/A1$_2$O$_3$.

Sulfur can also moderate the behaviour of ruthenium. Ru/TiO$_2$ treated with thiophene gave a value of S_i in the reaction of n-hexane of 66% at 753 K;[78] untreated it was active for hydrogenolysis at 413 K, so only a small fraction of the original surface remained free. With Ru/SiO$_2$-A1$_2$O$_3$, sulfur emerging from the support by reduction of residual sulfate on produced similar effects.

14.5.7. Metal-Support Interactions

These are of two kinds: (i) those of moderate character, experienced by very small metal particles on ceramic (i.e. irreducible) oxides having various acid/base characters, and (ii) strong interactions (SMSI) where the support is partially reduced, or a reducible component has been added, and the metal is decorated by species of indeterminate type, stemming from the support. These effects have been outlined in Section 2.6; their consequences for the reactions of higher alkanes will now be considered. The last shall be first.

One might expect that the presence of reduced entities such as TiO$_x$ partially covering a metal particle would have the effect of decreasing the mean size of the active ensemble, and thus suppressing hydrogenolysis and facilitating other reactions. Results presented in Section 13.7 gave some support to this view, which has been confirmed by a study of the n-hexane reaction on Pt/A1$_2$O$_3$ containing titania:[292] cyclisation benefited from the loss of activity for hydrogenolysis, while isomerisation (which had a notably high activation energy) was little affected. A comparison of Pt/TiO$_2$ and Pt/CeO$_2$ with Pt/SiO$_2$ for the n-hexane reaction at 613 K showed that after reduction at 773 K all three showed quite similar selectivities.[293] The effects of raising the reduction temperature from 623 to 773 K were however different; for Pt /TiO$_2$, S_i was decreased and C$_5$ and C$_6$ cyclisation products increased, while for Pt/CeO$_2$ hydrogenolysis by suppressed and only MCP selectivity rose. A very high activity for Pt/CeO$_2$ was recorded after 623 K reduction, but the activity loss on increasing this to 773 K was greatest for Pt/TiO$_2$.

Unfortunately in this particular field one dare not hope to see any simple picture emerging. Reduction of Pt/TiO$_2$ at the more moderate temperatures of 473 to 573 K gave[294] catalysts having very small particles (0.8–1 nm) but H/Pt ratios of only 0.3 to 0.16. They were in a partial SMSI state, and showing exceptionally high *hydrogenolysis* selectivities for *n*-pentane and *neo*hexane between these temperatures. MCP suffered some demethylation, the ring-opening reaction being about 50% selective; 2,2,3,3-tetramethylbutane also gave selective demethylation at 468 K, but the $\alpha\delta$ mode increased with temperature. In a similar study,[129] reduction at 573 K gave particles of 1.4 nm and H/Pt = 0.59, and the MCP reaction showed product selectivities and TOFs, similar to those given by Pt/SiO$_2$ (~65% selective), with no demethylation; this catalyst was only partially afflicted by the reduction. After 773 K reduction, however, all activity for MCP was removed, although some for *n*-hexane was kept. Partial restoration by oxidation gave catalysts of similar activity to the initial one, and the Arrhenius parameters for the component processes showed compensation (Figure 14.13). They fall on three lines, and suggest (perhaps for the first time) that the route to 2-methylpentane differs from that to the other isomers. Corresponding parameters for Pt/SiO$_2$ were much more closely spaced, and may well have been within experimental error of each other for each process; their variation with Pt/TiO$_2$ implies that variation of pre-treatment affected the reaction environment, but for Pt/SiO$_2$ it did not. The mechanisms of isomerisation of the methylpentanes on Pt/TiO$_2$ reduced at only 473 K have been analysed using ^{13}C-labelled molecules:[92] all showed

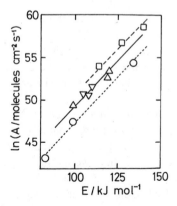

Figure 14.13. Compensation effects for the Arrhenius parameters of MCP hydrogenolysis on Pt/TiO$_2$ variously pretreated (see text):[129] □, 2-methylpentane; △, 3-methylpentane; ▽, *n*-hexane; O, C$_1$-C$_5$ alkanes.

predominance of the bond-shift mechanism, the importance of which decreased for 87 to 58% as metal loading increased from 0.2 to 10%. This was quite unlike the behaviour of Pt/Al_2O_3 (Section 14.3) where the *cyclic* mechanism was preferred on small particles.

All these results have been discussed very fully, without firm conclusions being reached. One possible line of thinking is to suppose that the lower temperature reduction (473 K) only affected the support, and that small particle size allowed an electronic influence that negated non-destructive reactions, and led to a preference for bond-shift over cyclic mechanisms for isomerisation. This may be the electron-transfer process

$$Pt^0 + Ti^{+3} \rightarrow Pt^{-1} + Ti^{+4} \qquad (14.A)$$

originally advocated by Horsley, although significant charge-transfer is unlikely and the effect may be more subtle (see the final paragraph). The consequences of this partial or incipient SMSI have been likened to those produced by basic additives and by tin, where similar influences may be at work. At higher reduction temperatures, it may be that small particles are entirely encapsulated, while residual activity is due to larger particles that are not much affected.

Reduction of 0.1% Ru/TiO_2 at 758 K led to high values of S_i in the *n*-hexane reaction at 633 K, reacting 94% when the precursor had been made by ion exchange; lower values were found with 0.5% Ru/TiO_2, and very low values after reduction at only 433 K. Thus as with the effect of sulfur, even the character of the most active hydrogenolysis catalyst is susceptible to modification.

The behaviour of platinum (and other metals) has also been altered by incorporating vanadium[295] and niobium[296] additives, and using niobia[297] as support. Pt/UO_2 catalysts containing either 0.2 or 8% metal reduced to 473 K both had small particles (3 nm), notwithstanding the low area of the support (5 m^{2}g^{-1}), but activities for the reactions of MCP and of 2-methylpentane were lower than for Pt/TiO_2 and *much* lower than for Pt/Al_2O_3.[298] There have also been strong indications that SMSI-like effects can be produced by magnesia,[299] although it has been suggested that traces of sulfur from the support may be responsible.

Some years ago there were a number of reports that high-temperature reduction of metals supported not only on titania, but also on silica[300] and alumina,[301] resulted in a drastic loss of activity for alkane hydrogenolysis, as well as loss of hydrogen chemisorption capacity: there were other physical changes as well, and the same effects were shown by platinum black.[302] Numerous explanations were offered, including SMSI, hydrogen spillover, and superficial alloy formation, but none of these apply to the unsupported metal. The effect does not seem to be related to that of reduction temperature on hydrogen at the interface or on particle shape (see the next paragraph). The most likely explanation is that atomic hydrogen penetrates into sub-surface sites where it is quite stable and only removable by oxidation. New vibrational bands assigned to such species have been detected by inelastic neutron scattering spectroscopy.[303] The probable role of this kind of hydrogen in hydrogenation of unsaturated hydrocarbons was considered in Section 14.2.6. These observations have not been incorporated into mainstream thinking on hydrocarbon transformations, and the lack of kinetic information and isotopic tracing are again keenly felt.

Milder but still significant metal-support interactions are observed with ceramic oxides, especially alumina; raising the reduction temperature of Pt/Al_2O_3 from 573 to 723 K was shown to remove a layer of hydrogen atoms from the interface, bringing metal and support into direct contact, and changing the particles' morphology from hemisphere to pancake.[304] This however made very little difference to the TOFs for *neo*pentane or MCP hydrogenolysis; product selectivities for the latter were probably the same within error, but S_i for *neo*pentane increased from 45 to 68%. The use of AXAFS spectroscopy (Section 2.6) led[305] to an explanation of the toxic effects of base (K^+) on the activities of Pt/LTL for hydrogenolysis of *neo*pentane and propane, and for Pt/LTL and Pt/SiO_2 for the former, in terms of alteration to a 'change in the energy position of the metal valence-orbitals'. It is however nothing short of tragic that such highly refined characterisation, which led to an intimate understanding of how metal-support interactions work, was applied to such a limited number of catalytic reactions and to such a small range of operational conditions. The results were confined to one table and ten lines of text. Specifically, it is impossible to correlate changes in the electronic arrangements in the metal atom to any fundamental parameter of the reaction, in the absence of kinetic information. Although this mode of operation of supports and modifiers has been hinted at in earlier sections, much remains to be done to exploit these novel techniques and concepts.

REFERENCES

1. J.H. Sinfelt in: *Catalysis—Science and Technology*, Vol. 1 (J.R. Anderson and M. Boudart, eds.) Springer-Verlag: Berlin (1981), p. 257.
2. P.G. Menon and Z. Paál, *Ind. Eng. Chem.* **36** (1997) 3282.
3. M.D. Edgar in: *Applied Industrial Catalysis*, Vol. 1 (B.E. Leach, ed.), Academic Press: New York (1983).
4. R.J. Farrauto and C.H. Bartholomew, *Fundamentals of Industrial Catalytic Processes,* Chapman and Hall: London (1997).
5. C.N. Satterfield, *Heterogeneous Catalysis in Practice,* 2nd Edn., McGraw-Hill: New York (1991).
6. J.H. Sinfelt, *Adv Catal.* **23** (1973) 91.
7. G.C. Bond, *Heterogeneous Catalysis–Principles and Applications,* 2nd Edn., Oxford U.P.: Oxford (1987).
8. G. Fitzsimmonds, J.K.A. Clarke, M.R. Smith and J.J. Rooney, *Catal. Lett.* **52** (1998) 69; *Appl. Catal. A: Gen.* **165** (1997) 357.

9. H. Topsøe, B.S. Clausen and F.E. Massoth, *Hydrotreating Catalysts,* Springer-Verlag: Berlin (1996).
10. J.H. Sinfelt, H. Hurwitz and J.C. Rohrer, *J. Phys. Chem.* **64** (1960) 892.
11. J.R. Anderson, *Adv. Catal.* **23** (1973) 1.
12. J.R. Anderson and B.G. Baker in: *Chemisorption and Reactions on Metallic Films,* Vol. 2 (J.R. Anderson, ed.) Academic Press: London (1971), p. 64.
13. Z. Paál, *Adv. Catal.* **29** (1980) 273.
14. Z. Paál and P. Tétényi in: *Specialist Periodical Reports: Catalysis,* (G.C. Bond and G. Webb, eds.), *Roy. Soc. Chem.* Vol. 5 (1982), p. 80.
15. G.B. McVicker, M. Daage, M.S. Touvelle, C.W. Hudson, D.P. Klein, W.C. Baird Jr., B.R. Cook, J.G. Chen, S. Hantzer, D.E.W. Vaughan, E.S. Ellis and O.C. Feeley, *J. Catal.* **210** (2002) 137.
16. Z. Paál, K. Matusek and H. Zimmer, *J. Catal.* **141** (1993) 648.
17. R. Burch and Z. Paál, *Appl. Catal. A: Gen.* **114** (1994) 9.
18. B. Coq and F. Figuéras, *Coord. Chem. Rev.* **178–180** (1998) 1753.
19. F.G. Gault, *Adv. Catal.* **30** (1981) 1.
20. M. Hajek, S. Corolleur, C. Corolleur, G. Maire, A. O Cinneide and F.G. Gault, *J. Chim. Phys.* **71** (1974) 1329.
21. H. Zimmer and Z. Paál, *React. Kinet. Catal. Lett.* **39** (1989) 227.
22. I.I. Ivanova, A. Pasau-Claerbout, M. Seirvert, N. Blom and E.G. Derouane, *J. Catal.* **158** (1996) 521; **164** (1996) 347.
23. Z. Paál, H. Groeneweg and H. Zimmer, *Catal. Today* **5** (1989) 199; *J. Chem. Soc. Faraday Trans.* **86** (1990) 3159.
24. Y.-P. Zhuang and A. Frennet, *Chin. J. Catal.* **17** (1996) 178; *Chin. J. Catal.* **18** (1997) 271; *Appl. Catal. A: Gen.* **177** (1997) 205.
25. A. Wootsch and Z. Paál, *J. Catal.* **185** (1999) 199.
26. A. Wootsch and Z. Paál, *J. Catal.* **205** (2002) 86.
27. G. Leclercq, H. Charcosset, R. Maurel, C. Betizeau, C. Bolivar, R. Frety, D. Jaunez, H. Mendes and L. Tournayan, *Bull. Soc. Chim. Belg.* **88** (1979) 577.
28. R. Maurel and G. Leclercq, *Bull. Soc. Chim. France* (1971) 1234.
29. E. Kikuchi, M. Tsurunu and Y. Morita, *J. Catal.* **22** (1971) 226.
30. M. Boudart and L.D. Ptak, *J. Catal.* **16** (1970) 90.
31. J.H. Sinfelt, *Catal. Rev.* **3** (1969) 175; *Adv. Catal.* **23** (1973) 91.
32. J.L. Carter, J.A. Cusumano and J.H. Sinfelt, *J. Catal.* **20** (1971) 223.
33. J.C. Kempling and R.B. Anderson in: *Proc. 5ᵗʰ Internat. Congr. Catal.* (J.W. Hightower, ed.), North Holland: Amsterdam **2** (1972) 1099.
34. C.J. Machiels and R.B. Anderson, *J. Catal.* **58** (1974) 268.
35. K. de Oliveira, D. Teschner, L. Oliveiro and Z. Paál, *React. Kinet. Catal. Lett.* **75** (2002) 185.
36. D. Teschner, D. Duprez and Z. Paál, *J. Molec. Catal. A: Chem.* **179** (2002) 201.
37. C.J. Machiels and R.B. Anderson, *J. Catal.* **60** (1979) 339.
38. C.J. Machiels and R.B. Anderson, *J. Catal.* **58** (1974) 260.
39. Z. Paál, Zhan Zhaoqi, I. Manninger and M. Muhler, *Appl. Catal.* **66** (1990) 301.
40. Z. Paál, X.L. Xu, J. Paál-Lukacs, W. Vogel, M. Muhler and R. Schlögl, *J. Catal.* **152** (1995) 252.
41. Z. Paál, *Catal. Today* (1992) 297.
42. J.K.A. Clarke and J.J. Rooney, *Adv. Catal.* **25** (1976) 125.
43. G.C. Bond and Z. Paál, *Appl. Catal. A* **86** (1992) 1.
44. G.C. Bond, *J. Molec. Catal.* **81** (1993) 99.
45. G.C. Bond, F. Garin and G. Maire, *Appl. Catal.* **41** (1988) 313.
46. Z. Paál, *J. Catal.* **91** (1985) 181.
47. Z. Paál and Xu Xian Lun, *Appl. Catal.* **43** (1988) L1.
48. Z. Paál, B. Brose, M. Räth and W. Gombler, *J. Molec. Catal.* **75** (1992) L13.

49. Z. Paál, H. Zimmer and P. Tétényi, *J. Molec. Catal.* **25** (1984) 99.
50. K. Matusek and Z. Paál, *React. Kinet. Catal. Lett.* **67** (1999) 241.
51. A. Sárkány, *J. Chem. Soc. Faraday Trans. I* **85** (1989) 1511, 1523.
52. H. Zimmer, M. Dobrovolszky, P. Tétényi and Z. Paál, *J. Phys. Chem.* **90** (1986) 4758; H. Zimmer, Z. Paál and P. Tétényi, *Acta. Chim. Hung* **124** (1987) 13.
53. Z. Paál and D. Marton, *Appl. Surf. Sci.* **26** (1986) 161.
54. Z. Paál, A. Gyóry, I. Uszkurat, S. Olivier, M. Guérin and C. Kappenstein, *J. Catal.* **168** (1997) 164.
55. I.I. Levitskii, A.M. Gyal'maliev and E.A. Udal'tzova, *J. Catal.* **58** (1979) 144.
56. A. Sárkány in: *Structure and Reactivity of Surfaces,* (C. Morterra, A. Zecchina and G. Costa, eds.), Studies in Surface Science and Catalysis', Elsevier: Amsterdam, **48** (1989) 835.
57. Buchang Shi and B.H. Davis, *J. Catal.* **162** (1996) 134; **157** (1995) 626; **168** (1997) 129; **147** (1994) 38.
58. I. Manninger, Z. Zhan, X.L. Xu and Z. Paál, *J. Molec. Catal.* **66** (1991) 223.
59. S.M. Davis, F. Zaera and G.A. Somorjai, *J. Catal.* **85** (1984) 206.
60. Mériaudeau, A. Thangaraj, C. Naccache amd S. Narayanan, *J. Catal.* **146** (1994) 579.
61. Z. Paál, *J. Catal.* **91** (1985) 181.
62. K. Matusek, A. Wootsch, H. Zimmer and Z. Paál, *Appl. Catal. A: Gen.* **191** (2000) 141.
63. Z. Karpiński, *Adv. Catal.* **37** (1990) 45.
64. A.L.D. Ramos, Seong Han Kim, Peilin Chen, Jae Hee Song and G.A. Somorjai, *Catal. Lett.* **66** (2000) 5.
65. W. Juszczyk, Z. Karpiński, I. Ratajczokowa, J. Stanasiuk, J. Zieliński, L.-L. Shen and W.M.H. Sachtler, *J. Catal.* **120** (1989) 68.
66. M. Skotak, D. Łomot and Z. Karpiński, *Appl. Catal. A: Gen.* **229** (2002) 103; M. Skotak and Z. Karpiński, *Chem. Eng. J.* **90** (2002) 89.
67. B. Coq, R. Dutartre, F. Figuéras and T. Tazi, *J. Catal.* **122** (1990) 438.
68. J.K.A. Clarke, K.M.G. Rooney and T. Baird, *J. Catal.* **111** (1988) 374.
69. B. Coq, F. Figuéras and T. Tazi, *Z. Phys. D—Atoms, Molecules and Clusters,* **12** (1989) 579.
70. Z. Paál, G. Székely and P. Tétényi, *J. Catal.* **58** (1979) 108.
71. L. Oliveiro and Z. Paál, *React. Kinet. Catal. Lett.* **74** (2001) 233.
72. B.E. Nieuwenhuys and G.A. Somorjai, *J. Catal.* **46** (1977) 259.
73. A.L. Bonivardi, F.H. Ribeiro and G.A. Somorjai, *J. Catal.* **160** (1996) 269.
74. A. Majesté, S. Balcon, M. Guérin, C. Kappenstein and Z. Paál, *J. Catal.* **187** (1999) 486.
75. A. Charron, C. Kappenstein, M. Guérin and Z. Paál, *Phys. Chem. Chem. Phys.* **1** (1999) 3817.
76. T.J. Plunkett and J.K.A. Clarke, *J. Catal.* **35** (1974) 330.
77. R.W. Rice and K. Lu, *J. Catal.* **77** (1982) 104.
78. R. Burch, G.C. Bond and R.R. Rajaram, *J. Chem. Soc. Faraday Trans. I* **82** (1986) 1985.
79. G.C. Bond, M.R. Gelsthorpe, R.R. Rajaram and R. Yahya in: *Structure and Reactivity of Surfaces* (C. Morterra, A. Zecchina and G. Costa, eds.), Studies in Surface Science and Catalysis', Elsevier: Amsterdam, **48** (1989) 167.
80. V.M. Akhmedov and S.H. Al-Khowaiter, *Appl. Catal. A: Gen.* **197** (2000) 201.
81. G. Leclercq, S. Pietrzyk, M. Peyrovi and M. Karroua, *J. Catal.* **99** (1986) 1.
82. E. Kikuchi and Y. Morita, *J. Catal.* **15** (1969) 217.
83. K. Kochloefl and U. Bažant, *J. Catal.* **10** (1968) 140.
84. J.K.A. Clarke and J.F. Taylor, *J. Chem. Soc. Faraday Trans. I* **71** (1975) 2063.
85. E. Gehrer and K. Hayek, *J. Molec. Catal.* **39** (1987) 293.
86. R. Burch and L.C. Garla, *React. Kinet. Catal. Lett.* **16** (1981) 315.
87. O.E. Finlayson, J.K.A. Clarke and J.J. Rooney, *J. Chem. Soc. Faraday Trans. I* **80** (1984) 191.
88. B. Coq, T. Tazi, R. Dutartre and F. Figuéras in: *Proc. 10ᵗʰ Internat. Congr. Catal.* (L. Guczi, F. Solymosi and P. Tétényi, eds.), Akadémiai Kiadó: Budapest **C** (1993) 2367.

89. B. Coq, A. Bittar and F. Figuéras, *Appl. Catal.* **59** (1990) 103.
90. B. Coq, A. Bittar, R. Dutartre and F. Figuéras, *Appl. Catal.* **60** (1990) 33.
91. Z. Paál, P. Tétényi and M. Dobrovolszky, *React. Kinet. Catal. Lett.* **37** (1988) 163.
92. F. Garin, P. Girard, G. Maire, G. Lu and L. Guzci, *Appl. Catal. A: Gen.* **152** (1997) 237.
93. F. Garin and F.G. Gault, *J. Am. Chem. Soc.* **97** (1975) 4466.
94. F. Garin, personal communication; L. Seyfried, F. Garin and G. Maire, *J. Catal.* **148** (1994) 281.
95. Z. Paál, K. Matusek and P. Tétényi, *Acta Chim. Acad. Sci. Hung,* **94** (1977) 119.
96. G.C. Bond, *Appl. Catal. A: Gen.* **191** (2000) 23.
97. R. Brown, A.S. Dolan, C. Kemball and G.S. McDougall, *J. Chem. Soc. Faraday Trans.* **88** (1992) 2405.
98. P.V. Menacherry and G.L. Haller, *J. Catal.* **167** (1997) 425.
99. J. Juszczyk and Z. Karpiński, *React. Kinet. Catal. Lett.* **37** (1988) 367; Z. Karpiński, W. Juszczyk and J. Pielaszek, *J. Chem. Soc. Faraday Trans.* **83** (1987) 1293.
100. Y. Barron, G. Maire, J.M. Cornet, J.M. Muller and F. Gault, *J. Catal.* **2** (1963) 152; **5** (1966) 428.
101. A. Sárkány, L. Guczi and P. Tétényi, *Acta Chim. Acad. Sci. Hung.* **96** (1978) 27.
102. Z. Karpiński, J.B. Butt and W.M.H. Sachtler, *J. Catal.* **119** (1989) 521.
103. Z. Karpiński, S.N. Gandhi and W.M.H. Sachtler, *J. Catal.* **141** (1993) 337.
104. S.T. Hoymeyer, Z. Karpiński and W.M.H. Sachtler, *Rec. Trav. Chim.* **109** (1990) 81.
105. M. Hayek, S. Corolleur, C. Corolleur, G. Maire, A. O Cinneide and F.G. Gault, *J. Chim. Phys.* **71** (1974) 1329.
106. R. Burch and V. Pitchon, *Catal. Today* **10** (1991) 315.
107. H. Matsumoto, Y. Saito and Y. Yoneda, *J. Catal.* **19** (1970) 107; **22** (1971) 182.
108. H. Zimmer, P. Tétényi, and Z. Paál *J. Chem. Soc. Faraday Trans.* **78** (1982) 3573.
109. B. Coq, *Appl. Catal. A: Gen.* **82** (1992) 231.
110. H. Zimmer, Z. Paál and P. Tétényi, *Acta Chim. Acad. Sci. Hung,* **111** (1982) 513.
111. F. Weisang and F.G. Gault, *J. Chem. Soc. Chem. Comm.* (1979) 519.
112. L. Pirault-Roy, D. Teschner, Z. Paál and M. Guerin, *Appl. Catal. A: Gen.* **245** (2003) 15.
113. F.G. Gault, V. Amir-Ebrahimi, F. Garin, P. Parayre and F. Weisang, *Bull. Soc. Chem. Belg.* **88** (1979) 475.
114. B. Coq, E. Crabb and F. Figuéras, *J. Molec. Catal. A: Chem.* **96** (1995) 35.
115. G.C. Bond and J.C. Slaa, *J. Molec. Catal. A: Chem.* **98** (1995) 81.
116. J. Monnier and R.B. Anderson, *J. Catal.* **82** (1983) 479; J.M.G. Dénès and R.B. Anderson, *Canad. J. Chem. Eng.* **62** (1984) 419.
117. A. Sárkány, *J. Catal.* **105** (1987) 65.
118. Z. Paál in: *Encyclopaedia of Catalysis*, Vol. 6 (I.T. Horvath, ed.), Wiley: New York (2002), p. 116.
119. A. da Costa Faro Jr. and C. Kemball, *J. Chem. Soc. Faraday Trans.* **91** (1995) 741.
120. J.P. Buchet, J. Buttet, J.J. van der Klink, M. Graetzel, E. Newson and T.B. Truong, *J. Molec. Catal.* **43** (1987) 213.
121. R. Gomez, G. Corro, G. Diaz, A. Maubert and F. Figuéras, *Nouv. J. Chim.* **4** (1980) 677.
122. J. Barbier, A. Morales, P. Marécot and R. Maurel, *Bull. Soc. Chim. Belg.* **88** (1979) 569.
123. G.A. del Angel, B. Coq, G. Ferrat and F. Figuéras, *Surf. Sci.* **156** (1985) 943.
124. S. Fuentes and F. Figuéras, *J. Catal.* **61** (1980) 443.
125. S. Fuentes, F. Figuéras and R. Gomez, *J. Catal.* **68** (1981) 419.
126. M. Guenin, M. Breysse and R. Frety, *J. Molec. Catal.* **25** (1984) 119.
127. Z.-C. Hu, A. Maeda, K. Kunimori and T. Uchijima, *Chem. Lett.* (1986) 2079.
128. C. Dossi, R. Psaro, A. Bartsch, A. Fusi, L. Sordelli, R. Ugo, M. Bellaltreccia, Z. Zanoni and G. Vlaic, *J. Catal.* **145** (1994) 377.
129. J.B.F. Anderson, R. Burch and J.A. Cairns, *J. Catal.* **107** (1987) 351, 364.
130. M. Vaarkamp, D. Dijkstra, J. van Grondelle, J.T. Miller, F.S. Modica, D.C. Koningsberger and R.A. van Santen, *J. Catal.* **151** (1995) 330.

131. S.M. Augustine and W.M.H. Sachtler, *J. Catal.* **106** (1987) 417.
132. G. Maire, G. Plouidy, J.C. Prudhomme and F.G. Gault, *J. Catal.* **4** (1965) 556.
133. F. Garin, S. Aeiyach, P. Légaré and G. Maire, *J. Catal.* **77** (1982) 323.
134. F. Zaera, D. Godbey and G.A. Somorjai, *J. Catal.* **107** (1986) 73.
135. B.H. Davis, *Catal. Today* **53** (1999) 443.
136. R. Kramer and H. Zuegg, *J. Catal.* **80** (1983) 446; **85** (1984) 530.
137. H. Glassl, K. Hayek and R. Kramer, *J. Catal.* **68** (1981) 397.
138. B. Török, J.T. Kiss, A. Molnár and M. Bartók, *Abstracts EUROPACAT-II* P.636.
139. E. Fülöp, V. Gnutzmann, Z. Paál and W. Vogel, *Appl. Catal.* **66** (1990) 319.
140. T. Koscielski, Z. Karpiński, and Z. Paál, *J. Catal.* **77** (1982) 539.
141. H. Zimmer and Z. Paál, *J. Molec. Catal.* **51** (1989) 261.
142. Z. Paál, *Catal. Today* **2** (1988) 595.
143. Y. Barron, G. Maire, J.M. Muller and F. Gault, *J. Catal.* **5** (1966) 428.
144. G. Maire, G. Plouidy, J.C. Prudhomme and F.G. Gault, *J. Catal.* **4** (1965) 556.
145. D. Teschner and Z. Paál, *React. Kinet. Catal. Lett.* **68** (1999) 25.
146. D. Teschner, K. Matusek and Z. Paál, *J. Catal.* **192** (2000) 335.
147. U. Wild, D. Teschner, R. Schlögl and Z. Paál, *Catal. Lett.* **67** (2000) 93.
148. D. Teschner, Z. Paál and D. Duprez, *Catal. Today* **65** (2001) 185.
149. G. Diaz, P. Esteban, L. Guczi, F. Garin, P. Bernhardt, J.-L. Schmitt and G. Maire in: *Structure and Reactivity of Surfaces* (C. Morterra, A. Zecchina and G. Costa, eds.), Studies in Surface Science and Catalysis', Elsevier: Amsterdam, **48** (1989) 363.
150. G. Diaz, P. Esteban, L. Guczi, F. Garin, P. Bernhardt, J.L. Schmitt and G. Maire, *J. Chim. Phys.* **86** (1989) 1741.
151. F. Garin, P. Girard, F. Weisang and G. Maire, *J. Catal.* **70** (1981) 205.
152. Mentioned in reference 15; J.G. Chen (University of Delaware), private communication.
153. F. Le Normand, K. Kili and J.L. Schmitt, *J. Catal.* **139** (1993) 234.
154. G.C. Bond, *Appl. Catal. A Gen.* **148** (1997) 3.
155. N.M. Rodriguez, P.E. Anderson, A. Wootsch, V. Wild, R. Schlögl and Z. Paál, *J. Catal.* **197** (2001) 365.
156. G.A. del Angel, B. Coq, G. Ferrat and F. Figuéras, *Surf. Sci.* **156** (1985) 943.
157. G. del Angel, B. Coq and F. Figuéras, *J. Catal.* **95** (1985) 167.
158. A. Sárkány, *J. Chem. Soc. Faraday Trans.* **84** (1988) 2267; *Catal. Today* **5** (1989) 173.
159. Z. Paál, *J. Molec. Catal.* **94** (1994) 225.
160. P.P. Lankhorst, H.C. de Jongste and V. Ponec in: *Catalyst Deactivation* (B. Delmon and G.F. Froment, eds.) Elsevier: Amsterdam (1980), p. 43.
161. J.M. Parera, C.A. Querini, J.N. Beltramini and N.S. Figoli, *Appl. Catal.* **32** (1987) 117.
162. A. Sárkány, H. Lieske, T. Szilágyi and L. Tóth, *Proc. 8th Internat. Congr. Catal.*, Verlag Chemie: Weinheim **2** (1984) 613.
163. F. Luck, S. Aeiyach and G. Maire, *Proc. 8th Internat. Congr. Catal.*, Verlag Chemie: Weinheim **2** (1984) 613.
164. Z. Paál, H. Zimmer, J.R. Günter, R. Schlögl and M. Muhler, *J. Catal.* **119** (1989) 146.
165. Z. Paál, R. Schlögl and G. Ertl, *Catal. Lett.* **12** (1992) 331; *J. Chem. Soc. Faraday Trans.* **88** (1992) 1179.
166. G. Roussy, S. Hilaire, J.M. Thiébaut, G. Maire, F. Garin and S. Ringler, *Appl. Catal. A: Gen.* **156** (1997) 167.
167. L. Seyfried, F. Garin, G. Maire, J.-M. Thiébaut and G. Roussy, *J. Catal.* **148** (1994) 281.
168. W.M.H. Sachtler, *Faraday Discuss. Chem. Soc.* **72** (1981) 7.
169. E.H. Broekhoven and V. Ponec, *Prog. Surf. Sci.* **19** (1985) 351.
170. B. Coq, A. Goursot, T. Tazi, F. Figuéras and D.R. Salahub, *J. Am. Chem. Soc.* **113** (1991) 1485.
171. W.R. Patterson and J.J. Rooney, *Catal. Today,* **12** 1992) 113.
172. F. Locatelli, D. Uzio, G. Nicolái, J.M. Basset and J.P. Candy, *Catal. Comm.* **4** (2003) 189.

173. A. Dauscher, F. Garin and G. Maire, *J. Catal.* **105** (1987) 233.
174. Z. Paál in: *Encyclopaedia of Catalysis*, Vol. 3 (I.T. Horvath, ed.) Wiley: New York (2002) p. 9.
175. O.E. Finlayson, J.K.A. Clarke and J.J. Rooney, *J. Chem. Soc. Faraday Trans. I* **80** (1984) 191.
176. F. Zaera, *Appl. Catal. A: Gen.* **229** (2002) 75.
177. F. Zaera, S. Tjandra and T.V.W. Janssens, *Langmuir* **14** (1998) 1320.
178. F. Zaera and G.A. Somorjai, *Langmuir* **2** (1986) 686.
179. A.D. Logan, K. Sharoudi and A.K. Datye, *J. Phys. Chem.* **95** (1991) 5568.
180. R. Kramer and H. Zuegg in: *Proc. 8th. Internat. Congr. Catal.*, Verlag Chemie: Weinheim **5** (1984) 275.
181. K. Hayek, *J. Molec. Catal.* **51** (1989) 347.
182. J. Barbier, P. Marécot and R. Maurel, *Nouv. J. Chim.* **4** (1980) 385.
183. J.A. Anderson, M.G.V. Mordente and C.H. Rochester, *J. Chem. Soc. Faraday Trans. I* **85** (1989) 2991.
184. O.M. Poltorak and V.S. Boromin, *Russ. J. Phys. Chem.* **39** (1965) 781, 1329; **40** (1966) 1436.
185. O. Zahraa, F. Garin and G. Maire, *Faraday Disc. Chem. Soc.* **72** (1981) 45.
186. F. Garin, O. Zahraa, C. Crouzet, J.L. Schmitt and G. Maire, *Surf. Sci.* **106** (1981) 466.
187. H.C. Yao and M. Shelef, *J. Catal.* **73** (1982) 76.
188. G.J. Arteaga, J.A. Anderson and C.H. Rochester, *J. Catal.* **187** (1999) 219.
189. J.B. Butt, *Appl. Catal.* **15** (1985) 161.
190. A.S. Kamachandran, S.L. Anderson and A.K. Datye, *Ultramicroscopy* **51** (1993) 282.
191. J. Barbier, P. Marécot, A. Morales and R. Maurel, *Bull. Soc. Chim. Fr.* **7–8** (1978) I-309.
192. S. Fuentes, F. Madera and M.J. Yacaman, *J. Chim. Phys. Phys.-Chim. Biol.* **80** (1983) 379.
193. Z. Paál and P. Tétényi, *Nature* **267** (1977) 234.
194. A.B. Gaspar and L.C. Dieguez, *Appl. Catal. A: Gen.* **201** (2000) 241.
195. L.-X. Dai, Y. Hashimoto, H. Tominaga and T. Tatsumi, *Catal. Lett.* **45** (1997) 107.
196. W.E. Alvarez and D.E. Resasco, *J. Catal.* **164** (1996) 467.
197. T. Fukunaga and V. Ponec, *Appl. Catal. A: Gen.* **154** (1997) 207.
198. P.V. Menacherry and G.L. Haller, *J. Catal.* **177** (1998) 175.
199. 199. G. Jacobs, C.L. Padro and D.E. Resasco, *J. Catal.* **178** (1998) 43.
200. N.D. Triantafillou, J.T. Miller and B.C. Gates, *J. Catal.* **155** (1995) 131.
201. G.S. Lane, F.S. Modica and J.T. Miller, *J. Catal.* **129** (1991) 145.
202. A. van de Runstraat, J. van Grondelle and R.A. van Santen, *J. Catal.* **167** (1997) 460.
203. K.A. Altynbekova and N.A. Zakarina in: *Abstracts, 9th Internat. Symp. on Relations between Homogeneous and Heterogeneous Catalysis,* Roy. Soc. Chem: Southampton (1998), p. 141.
204. R.A. Dalla Betta and M. Boudart in: *Proc. 5th Internat. Congr. Catal.* (J.W. Hightower, ed.), North Holland: Amsterdam (1972) 1329.
205. Z. Paál, Z.Q. Zhan, I. Manninger and W.M.H. Sachtler, *J. Catal.* **155** (1995) 43.
206. P.W. Tamm, D.H. Mohr and C.R. Wilson in: *Catalysis 1987* (J.W. Ward, ed.), Studies in Surface Science and Catalysis', Elsevier: Amsterdam, **38** (1987) 335.
207. J. Biswas, G.M. Bickle, P.G. Gray, D.D. Do and J. Barbier, *Catal. Rev.- Sci. Eng.* **30** (1988) 161.
208. Fai Kait Chong, J.A. Anderson and C.H. Rochester, *Phys. Chem. Chem. Phys.* **2** (2000) 5730.
209. F.K. Chong, J.A. Anderson and C.H. Rochester, *J. Catal.* **190** (2000) 327.
210. J.R. Anderson, F.K. Chong and C.H. Rochester, *J. Molec. Catal. A: Chem.* **140** (1999) 65.
211. M. Fernández-García, F.K. Chong, J.A. Anderson, C.H. Rochester and G.L. Haller, *J. Catal.* **182** (1999) 199.
212. P.A. van Trimpont, G.B. Marin and G.F. Froment, *Appl. Catal.* **24** (1986) 53.
213. P.A. van Trimpont, G.B. Marin and G.F. Froment, *Appl. Catal.* **17** (1985) 161.
214. M.J. Sterba and V. Haensel, *Ind. Eng. Chem. Prod. Res. Dev.* **15** (1976) 2.
215. Z. Paál, K. Matusek and M. Muhler, *Appl. Catal. A: Gen.* **149** (1997) 113.
216. H. Charcosset, *Rev. Inst. Fr. Petr.* **34** (1979) 238; *Internat. Chem. Eng.* **23** (1983) 187, 411.

217. V. Ponec, *Catal. Today* **10** (1991) 251.
218. N. Macleod, J.R. Fryer, D. Stirling and G. Webb, *Catal. Today* **46** (1998) 37.
219. L. Tournayan, R. Bacaud, H. Charcosset and G. Leclercq, *J. Chem. Research (S)* (1978) 290.
220. S.M. Augustine and W.M.H. Sachtler, *J. Catal.* **106** (1987) 417.
221. S.M. Augustine and W.M.H. Sachtler, *J. Phys. Chem.* **91** (1987) 5953.
222. Changmin Kim and G.A. Somorjai, *J. Catal.* **134** (1992) 179.
223. W. Juszczyk and Z. Karpiński, *Appl. Catal. A: Gen.* **206** (2001) 67.
224. J.P. Brunelle, R.E. Montarnal and A.A. Sugier in: *Proc. 6th Internat. Congr. Catal.*, (G.C. Bond, P.B. Wells and F.C. Tompkins, eds.), *Roy. Soc. Chem.*: London **2** (1976) 844.
225. M. Bonarowska, A. Malinowski and Z. Karpiński, *Appl. Catal. A: Gen.* **138** (1999) 145.
226. B. Coq and F. Figuéras, *J. Molec. Catal.* **25** (1984) 87.
227. B. Coq and F. Figuéras, *J. Catal.* **85** (1984) 197.
228. G.J. Arteaga, J.A. Anderson, S.M. Becker and C.H. Rochester, *J. Molec. Catal A: Chem.* **145** (1999) 183.
229. X. Li, Y. Wei, J. Cheng and R. Li, *Proc. 10th Internat. Congr. Catal.* (L. Guczi, F. Solymosi and P. Tétényi, eds.), Akadémiai Kiadó: Budapest **C** (1993) 2407.
230. K. Matusek, C. Kappenstein, M. Guérin and Z. Paál, *Catal. Lett.* **64** (2000) 33.
231. P. Kirszensztejn and L. Wachowski, *React. Kinet. Catal. Lett.* **60** (1997) 93.
232. Z. Karpiński and J.K.A. Clarke, *J. Chem. Soc. Faraday Trans. I* **71** (1975) 893.
233. F.B. Passos, D.A.G. Aranda and M. Schmal, *J. Catal.* **178** (1998) 478.
234. T. Fujikawa, F.H. Ribeiro and G.A. Somorjai, *J. Catal.* **178** (1998) 58.
235. Z. Paál, M. Dobrovolszky, J. Völter and G. Liesz, *Appl. Catal.* **14** (1985) 33.
236. Yuguo Wang, Buchang Shi, R.D. Guthrie and B.H. Davis, *J. Catal.* **170** (1997) 89.
237. G.L. Szabo in: *Metal Support and Metal-Additive Effects in Catalysis*, (B. Imelik, C. Naccache, G. Coudurier, H. Praliaud, P. Meriaudeau, P. Gallezot, G.A. Martin and J.C. Védrine, eds.), Studies in Surface Science and Catalysis, Elsevier: Amsterdam, **11** (1982) 349.
238. C.A. Querini, N.S. Figoli and J.M. Parera, *Appl. Catal.* **53** (1989) 53.
239. A. Wootsch, L. Pirault-Roy, J. Leverd, M. Guérin and Z. Paál, *J. Catal.* **208** (2002) 490.
240. Z. Huang, J.R. Fryer, C. Park, D. Stirling and G. Webb, *J. Catal.* **175** (1998) 226.
241. G. Maire, *Catal. Today* **12** (1992) 201.
242. G. Leclercq, S. Pietrzyk, T. Romero, A. El Gharbi, L. Gengembre, J. Grimblot, F. Aïssi, M. Guelton, A. Latef and L. Leclercq, *Ind. Eng. Chem. Res.* **36** (1997) 4015.
243. X.-X. Guo, Y.-H. Yang, M.-C. Den, H.-M. Li and Z.Y. Lin, *J. Catal.* **99** (1986) 218.
244. G. Leclercq, A. El Gharbi, L. Gengembre, T. Romero, L. Leclercq and S. Pietrzyk, *J. Catal.* **148** (1994) 550.
245. L. Vlaev, D. Damyanov and M.M. Mohamed, *Appl. Catal.* **65** (1990) 11.
246. L.T. Vlaev, M.M. Mohamed and D.P. Damyanov, *Appl. Catal.* **63** (1990) 293.
247. R.W. Joyner, K.M. Minachev, P.A.D. Pudney, E.S. Shpiro and G. Tuleouva, *Catal. Lett.* **5** (1990) 257.
248. B. Coq, A. Bittar, T. Tazi and F. Figuéras, *J. Molec. Catal.* **55** (1989) 34.
249. B. Coq, A. Bittar, R. Dutartre and F. Figuéras, *J. Catal.* **128** (1991) 275.
250. B. Coq, A. Bittar and F. Figuéras in: *Structure and Reactivity of Surfaces* (C. Morterra, A. Zecchina and G. Costa, eds.), Studies in Surface Science and Catalysis', Elsevier: Amsterdam, **48** (1989), 327.
251. L. Pirault-Roy, D. Teschner, Z. Paál and M. Guérin, *Appl. Catal. A: Gen.* **245** (2003) 15.
252. R.D. Gonzalez, *Appl. Surf. Sci.* **19** (1984) 181.
253. R.L. Moss in: *Specialist Periodical Reports: Catalysis*, (C. Kemball and D.A. Dowden, eds.), *Roy. Soc. Chem.* Vol. 1 (1977), p. 37; Vol. 4 (1981), p. 31.
254. D.A. Dowden in: *Specialist Periodical Reports: Catalysis*, (C. Kemball and D.A. Dowden, eds.), *Roy. Soc. Chem.* Vol. 2 (1978), p. 1.

255. V. Ponec, *Appl. Catal. A: Gen.* **222** (2001) 31.
256. M.W. Vogelzang, M.J.P. Botman and V. Ponec, *Faraday Discuss. Chem. Soc.* **72** (1981) 33.
257. J.W.A. Sachtler and G.A. Somorjai, *J. Catal.* **81** (1983) 77.
258. B.D. Chandler, A.B. Schnabel and L.H. Pignolet, *J. Catal.* **193** (2000) 186.
259. J.R.H. van Schaik, R.P. Dessing and V. Ponec, *J. Catal.* **38** (1975) 273.
260. A.J. den Hartog, P.J.M. Rek and V. Ponec, *J. Chem. Soc. Chem. Comm.* (1988) 1470.
261. A.T. Mathauser and A.V. Teplyakov, *Catal. Lett.* **73** (2001) 207.
262. A.V. Teplyakov and B.E. Bent, *Catal. Lett.* **42** (1996) 1; *J. Phys. Chem.* **101** (1997) 9052.
263. K. Foger and J.R. Anderson, *J. Catal.* **59** (1978) 325, **61** (1980) 140.
264. S. Aeiyach, F. Garin, L. Hilaire, P. Légaré and G. Maire, *J. Molec. Catal.* **25** (1984) 183.
265. M. Bonorowska, J. Pielaszek, W. Juszczyk and Z. Karpiński, *J. Catal.* **195** (2000) 304.
266. W. Juszczyk, Z. Karpiński, D. Łomot, J. Pielaszek and J.W. Sobczak, *J. Catal.* **151** (1995) 67.
267. C. Visser, J.G.P. Zuidwijk and V. Ponec, *J. Catal.* **35** (1974) 407.
268. Z. Karpiński, *J. Catal.* **77** (1982) 118.
269. Z. Karpiński, W. Juszczyk and J. Stachurski, *J. Chem. Soc. Faraday Trans. I* **81** (1985) 1447.
270. J.K.A. Clarke, I. Manninger and T. Baird, *J. Catal.* **54** (1978) 230.
271. A. Péter and J.K.A. Clarke, *J. Chem. Soc. Faraday Trans. I* **72** (1976) 1201.
272. B. Coq, R. Dutartre, F. Figuéras and A. Rouco, *J. Phys. Chem.* **93** (1989) 4094.
273. Z. Karpiński and J.K.A. Clarke, *J. Chem. Soc. Faraday Trans. I* **71** (1975) 2310.
274. W.G. Reman, A.H. Ali and G.C.A. Schuit, *J. Catal.* **20** (1971) 374.
275. D.F. Hollis and H. Taheri, *AIChE Journal* **22** (1976) 1112.
276. A.V. Ramaswamy, P. Ratnasamy, S. Sivasanker and A.J. Leonard in: *Proc. 6th. Internat. Congr. Catal.*, (G.C. Bond, P.B. Wells and F.C. Tompkins, eds.), Chem. Soc.: London **2** (1976) 855.
277. M.J. Dees and V. Ponec, *J. Catal.* **115** (1988) 347.
278. A.C. Faro Jr. and C. Kemball, *J. Chem. Soc. Faraday Trans. I* **82** (1986) 3125.
279. L. Pauling, *Nature of the Chemical Bond,* 3rd Edn., Cornell U.P.: Ithaca (1960).
280. G. Diaz, F. Garin and G. Maire, *J. Catal.* **82** (1983) 13.
281. G. del Angel, C. Medina, R. Gomez, B. Rejai and R.D. Gonzalez, *Catal. Today* **5** (1989) 395.
282. P. Tétényi and V. Galsán, *Appl. Catal. A: Gen.* **229** (2002) 181.
283. Z. Karpiński, Z. Zhang and W.M.H. Sachtler, *Catal. Lett.* **13** (1992) 123.
284. J.A. Oliver and C. Kemball, *Proc. Roy. Soc. A* **429** (1990) 17.
285. J. Oudar in: *Metal-Support and Metal-Additive Effects in Catalysis,* (B. Imelik, C. Naccache, G. Coudurier, H. Praliaud, P. Meriaudeau, P. Gallezot, G.A. Martin and J.C. Védrine, eds.), Studies in Surface Science and Catalysis, Elsevier: Amsterdam, **11** (1982) 255.
286. Jeong-Kyu Lee and Hyun-Ku Rhee, *J. Catal.* **177** (1998) 208.
287. S.M. Augustine, G.N. Alameddin and W.M.H. Sachtler, *J. Catal.* **115** (1989) 217.
288. A.J. den Hartog, P.J.M. Rek, M.J.P. Botman, C. de Vrengd and V. Ponec, *Langmuir* **4** (1988) 1100.
289. J. Barbier in: *Metal-Support and Metal-Additive Effects in Catalysis,* (B. Imelik, C. Naccache, G. Coudurier, H. Praliaud, P. Meriaudeau, P. Gallezot, G.A. Martin and J.C. Védrine, eds.), Studies in Surface Science and Catalysis, Elsevier: Amsterdam, **11** (1982) 293.
290. Z. Paál, M. Muhler and K. Matusek, *J. Catal.* **175** (1998) 245.
291. Z.-K. Ruan, Z.-Y. Pei Y.-Z. Shi and X.-X,. Guo, *Cuihua Xuebao* **11** (1990) 358.
292. P. Meriaudeau, J.F. Dutel, M. Dufaux and C. Naccache in: *Metal-Support and Metal-Additive Effects in Catalysis,* (B. Imelik, C. Naccache, G. Coudurier, H. Praliaud, P. Meriaudeau, P. Gallezot, G.A. Martin and J.C. Védrine, eds.), Studies in Surface Science and Catalysis, Elsevier: Amsterdam, **11** (1982) 95.
293. J.K.A.Clarke, R.J. Dempsey and T. Baird, *J. Chem. Soc. Faraday Trans.* **86** (1990) 2789.
294. A. Dauscher, F. Garin, F. Luck and G. Maire in: *Metal-Support and Metal-Additive Effects in Catalysis,* (B. Imelik, C. Naccache, G. Coudurier, H. Praliaud, P. Meriaudeau, P. Gallezot, G.A.

Martin and J.C. Védrine, eds.), Studies in Surface Science and Catalysis, Elsevier: Amsterdam, **11** (1982) 113.

295. A.J. den Hartog, A.G.T.M. Bastein and V. Ponec, *J. Molec. Catal.* **52** (1989) 129.

296. F.B. Passos, D.A.G. Aranda, R.R. Soares and M. Schmal, *Catal. Today* **43** (1998) 3.

297. R. Brown and C. Kemball, *J. Chem. Soc. Faraday Trans.* **92** (1996) 281;

298. M. Romeo, A. Dauscher, L. Hilaire, W. Muller and G. Maire, *SSSC* **48** (1989) 799.

299. J.K.A. Clarke, M.J. Bradley, L.A.J. Garvie, A.J. Craven and T. Baird, *J. Catal.* **143** (1993) 122.

300. G.A. Martin, D. Dutartre and J.A. Dalmon, *React. Kinet. Catal. Lett* **16** (1981) 329.

301. L. Kępiński, M. Wołcyrz and J.M. Jabłoński, *Appl. Catal.* **54** (1989) 267; W. Juszczyk, D. Lomot, Z. Karpiński and J. Pielaszek, *Catal. Lett.* **31** (1995) 37.

302. P.G. Menon and G.F. Froment in: *Metal-Support and Metal-Additive Effects in Catalysis*, (B. Imelik, C. Naccache, G. Coudurier, H. Praliaud, P. Meriaudeau, P. Gallezot, G.A. Martin and J.C. Védrine, eds.), Studies in Surface Science and Catalysis, Elsevier: Amsterdam, **11** (1982) 171.

303. A.J. Renouprez, T.M. Tejero and J.P. Candy in: *Proc. 8th Internat. Congr. Catal.* Verlag Chemie: Weinheim **3** (1985) 47.

304. R. Kramer and M. Fischbacher, *J. Molec. Catal.* **51** (1989) 247.

305. B.L. Mojet, J.T. Miller, D.E. Ramaker and D.C. Koningsberger, *J. Catal.* **186** (1999) 373.

306. M. Vaarkamp, J.T. Miller, F.S. Modica and D.C. Koningsberger, *J. Catal.* **163** (1996) 294.

307. J.K.A. Clarke, *Chem. Rev.* **75** (1979) 291.

308. J. Barbier, E. Lamy-Pitara, P. Marécot, J.P. Boitiaux, J. Cosyns and F. Verma, *Adv. Catal.* **37** (1990) 279.

309. K. Thomas, C. Binet, T. Chevreau, D. Cornet and J.-P. Gilson, *J. Catal.* **212** (2002) 63.

310. U.S.P. 6063 724 to Univ. Oklahoma State.

311. F. Garin, *Catal. Today* **89** (2004) 255.

312. M. Boutahala, B. Djellouli, N. Zouaoui and F. Garin, *Catal. Today* **89** (2004) 379.

FURTHER READING

1 **Reactions catalysed by EUROPT-1:** 14, 16, 23, 25, 39, 41, 43, 44, 47–50

2 **Reactions catalysed by platinum black:** 16, 23, 25 ,26, 40, 50–52

3 **Mechanisms of hydrocarbon transformation:** 2, 12–14, 17, 19, 42, 113, 168, 169, 309

4 **Metal particles in zeolites - L:** 103, 130, 141, 195–201
                             - Y: 98, 104, 198, 201, 204, 205

5 **Bimetallic catalysts:** 42, 168, 169, 207, 252–256, 206, 310

6 **Poisoning by sulfur:** 307–309

> ... *but the continuing to the end, until it be thoroughly finished, yields the true glory.*
>
> Sir Francis Drake

INDEX

657